Methods in Enzymology

Volume 311
Sphingolipid Metabolism and Cell Signaling
Part A

METHODS IN ENZYMOLOGY

EDITORS-IN-CHIEF

John N. Abelson   Melvin I. Simon

DIVISION OF BIOLOGY
CALIFORNIA INSTITUTE OF TECHNOLOGY
PASADENA, CALIFORNIA

FOUNDING EDITORS

Sidney P. Colowick and Nathan O. Kaplan

Methods in Enzymology

Volume 311

Sphingolipid Metabolism and Cell Signaling
Part A

EDITED BY

Alfred H. Merrill, Jr.

EMORY UNIVERSITY SCHOOL OF MEDICINE
ATLANTA, GEORGIA

Yusuf A. Hannun

MEDICAL UNIVERSITY OF SOUTH CAROLINA
CHARLESTON, SOUTH CAROLINA

ACADEMIC PRESS

San Diego   London   Boston   New York   Sydney   Tokyo   Toronto

Academic Press
A Harcourt Science and Technology Company
525 B Street, Suite 1900, San Diego, California 92101-4495, USA
http://www.academicpress.com

Academic Press Limited
24-28 Oval Road, London NW1 7DX, UK
http://www.hbuk.co.uk/ap/

International Standard Book Number: 0-12-182212-5

PRINTED IN THE UNITED STATES OF AMERICA
99 00 01 02 03 04 MM 9 8 7 6 5 4 3 2 1

Table of Contents

Section I. Sphingolipid Metabolism

A. Biosynthesis

C. Genetic Approaches

Section II. Inhibitors of Sphingolipid Biosynthesis

Section III. Chemical and Enzymatic Synthesis

Contributors to Volume 311

Article numbers are in parentheses following the names of contributors.
Affiliations listed are current.

AKIRA ABE (6, 12, 38), *Division of Nephrology, Department of Internal Medicine, University of Michigan Medical School, Ann Arbor, Michigan 48109-0676*

DENNIS C. ARGENTIERI (19), *R. W. Johnson Pharmaceutical Research Institute, Raritan, New Jersey 08869*

SANDRA BAJJALIEH (24), *Department of Pharmacology, University of Washington, Seattle, Washington 98195*

ARMINDA G. BARBONE (19), *R. W. Johnson Pharmaceutical Research Institute, Raritan, New Jersey 08869*

KEITH D. BARLOW (25), *Department of Biochemistry and Molecular Biology, Georgetown University Medical Center, Washington, DC 20007*

OLIVER BARTELSEN (16), *Kekulé-Institut für Organische Chemie und Biochemie, Universität Bonn, D-53121 Bonn, Germany*

MANJU BASU (31), *Department of Chemistry and Biochemistry, University of Notre Dame, Notre Dame, Indiana 46556-5601*

SUBHASH C. BASU (31), *Department of Chemistry and Biochemistry, University of Notre Dame, Notre Dame, Indiana 46556-5601*

ROBERT BATCHELOR (24), *Department of Pharmacology, University of Washington, Seattle, Washington 98195*

ALICJA BIELAWSKA (42–44), *Department of Biochemistry and Molecular Biology, Medical University of South Carolina, Charleston, South Carolina 29425*

UWE BIERFREUND (29), *Kekulé-Institut für Organische Chemie und Biochemie, Universität Bonn, D-53121 Bonn, Germany*

DAVID N. BRINDLEY (27), *Signal Transduction Laboratories, Lipid and Lipoprotein Research Group, Department of Biochemistry, University of Alberta, Edmonton, Alberta T6G 2S2, Canada*

ANGELEAH BROWDY (14), *Department of Food Science and Nutrition, University of Rhode Island, West Kingston, Rhode Island 02892*

ANATOLIY S. BUSHNEV (45), *Department of Chemistry, Emory University, Atlanta, Georgia 30322*

SUBROTO CHATTERJEE (9), *Sphingolipid Signaling and Vascular Biology Laboratory, Division of Lipid Research Atherosclerosis, Department of Pediatrics, School of Medicine, Johns Hopkins University, Baltimore, Maryland 21287-3654*

PRASSON CHATTURVEDI (11), *Department of Biomedical Sciences, E. K. Shriver Center, Waltham, Massachusetts 02452*

VANNA CHIGORNO (50), *Study Center for the Functional Biochemistry of Brain Lipids, Department of Medical Chemistry and Biochemistry, L.I.T.A.–Segrate, University of Milan, 20090 Segrate, Milan, Italy*

NAMJIN CHUNG (34), *Duke University Medical Center, Durham, North Carolina 27710*

AIDA E. CREMESTI (14), *Department of Food Science and Nutrition, University of Rhode Island, West Kingston, Rhode Island 02892*

CHRISTOPHER CURFMAN (39), *Department of Chemistry, Emory University, Atlanta, Georgia 30322*

ROBERT C. DICKSON (1), *Department of Biochemistry, University of Kentucky College of Medicine, Lexington, Kentucky 40536-0096*

RICK T. DOBROWSKY (21), *Department of Pharmacology and Toxicology, University of Kansas, Lawrence, Kansas 66045*

RUI-DONG DUAN (30), *Department of Cell Biology 1, EB-Blocket, University Hospital of Lund, S-22185 Lund, Sweden*

JACQUES FANTINI (49), *Laboratoire de Biochimie et Biologie de la Nutrition, Faculté des Sciences de St. Jérôme, 13397 Marseille Cedex 20, France*

ANTHONY S. FISCHL (14), *Department of Food Science and Nutrition, University of Rhode Island, West Kingston, Rhode Island 02892*

VALESWARA-RAO GAZULA (21), *Department of Pharmacology and Toxicology, University of Kansas, Lawrence, Kansas 66045*

MARK GIRZADAS (31), *Department of Chemistry and Biochemistry, University of Notre Dame, Notre Dame, Indiana 46556-5601*

MICHELLE M. GRILLEY (2), *Department of Biology, Utah State University, Logan, Utah 84322-5305*

YUSUF A. HANNUN (17, 18, 42–44), *Department of Biochemistry and Molecular Biology, Medical University of South Carolina, Charleston, South Carolina 29425*

GUY H. HARRIS (35), *Department of Natural Products Drug Discovery, Merck Research Laboratories, Rahway, New Jersey 07065-4607*

DANIEL F. HASSLER (20), *GlaxoWellcome, Research Triangle Park, North Carolina 27709*

YOSHIO HIRABAYASHI (33), *Laboratory for Cellular Glycobiology, Frontier Research Program, The Institute of Physical and Chemical Research (RIKEN), Wako-shi, Saitama 351-0198, Japan*

SHINICHI ICHIKAWA (33), *Laboratory for Cellular Glycobiology, Frontier Research Program, The Institute of Physical and Chemical Research (RIKEN), Wako-shi, Saitama 351-0198, Japan*

TAKAO IKAMI (46), *Drug Discovery Research Department, Sanwa Kagaku Kenkyusho Company, Limited, Hokusei-cho, Mie 511-0406, Japan*

HIDEHARU ISHIDA (46), *Department of Applied Bioorganic Chemistry, Gifu University, Gifu 501-1193, Japan*

MAKOTO ITO (32, 52), *Laboratory of Marine Biochemistry, Faculty of Agriculture, Kyushu University, Higashi-ku, Fukuoka 812-8581, Japan*

HIROYUKI IZU (32), *Biotechnology Research Laboratories, Takara Shuzo Company Limited, Fukuoka 812-8581, Japan*

ALISA C. JACKSON (19), *R. W. Johnson Pharmaceutical Research Institute, Raritan, New Jersey 08869*

RENATA JASINSKA (27), *Signal Transduction Laboratories, Lipid and Lipoprotein Research Group, Department of Biochemistry, University of Alberta, Edmonton, Alberta T6G 2S2, Canada*

KARL-HEINZ JUNG (40), *Fakultät Chemie, Universität Konstanz, D-78457 Konstanz, Germany*

FIROZE B. JUNGALWALA (11), *Department of Biomedical Sciences, E. K. Shriver Center, Waltham, Massachusetts 02452*

CHRISTOPH KAES (10), *Institute of Physiological Chemistry, University of Bonn, D-53115 Bonn, Germany*

YASUSHI KAMISAKA (7), *National Institute of Bioscience and Human Technology, Tsukuba, Ibaraki 305, Japan*

PATRICK KELLY (31), *Department of Chemistry and Biochemistry, University of Notre Dame, Notre Dame, Indiana 46556-5601*

MAKOTO KISO (46), *Department of Applied Bioorganic Chemistry, Gifu University, Gifu 501-1193, Japan*

KATSUHIRO KITA (32, 52), *Laboratory of Marine Biochemistry, Faculty of Agriculture, Kyushu University, Higashi-ku, Fukuoka 812-8581, Japan*

FRIEDERIKE KNOLL (47), *Kekulé-Institut für Organische Chemie und Biochemie, Universität Bonn, D-53121 Bonn, Germany*

THOMAS KOLTER (29, 47), *Kekulé-Institut für Organische Chemie und Biochemie, Universität Bonn, D-53121 Bonn, Germany*

ARI M. P. KOSKINEN (41), *Department of Chemistry, University of Oulu, FIN-90570 Oulu, Finland*

PÄIVI M. KOSKINEN (41), *Department of Chemistry, University of Oulu, FIN-90570 Oulu, Finland*

TOYOHISA KURITA (32), *Biotechnology Research Laboratories, Takara Shuzo Company Limited, Fukuoka 812-8581, Japan*

RONALD M. LAETHEM (20), *GlaxoWellcome, Research Triangle Park, North Carolina 27709*

STEFANIE LANSMANN (16, 23), *Kekulé-Institut für Organische Chemie und Biochemie, Universität Bonn, D-53121 Bonn, Germany*

LIHSUEH LEE (38), *Division of Nephrology, Department of Internal Medicine, University of Michigan Medical School, Ann Arbor, Michigan 48109-0676*

TEN-CHING LEE (13), *Basic and Applied Research Unit, Oak Ridge Associated Universities, Oak Ridge, Tennessee 37831-0117*

ROBERT L. LESTER (1), *Department of Biochemistry, University of Kentucky College of Medicine, Lexington, Kentucky 40536-0096*

ZHIXIONG LI (31), *Department of Chemistry and Biochemistry, University of Notre Dame, Notre Dame, Indiana 46556-5601*

THOMES LINKE (23), *Kekulé-Institut für Organische Chemie und Biochemie, Universität Bonn, D-53121 Bonn, Germany*

DENNIS C. LIOTTA (39, 45), *Department of Chemistry, Emory University, Atlanta, Georgia 30322*

BIN LIU (17, 18), *Neuropharmacology Section, Laboratory of Pharmacology and Chemistry, National Institute of Environmental Health Sciences, Research Triangle Park, North Carolina 27709*

YONGSHENG LIU (14), *Department of Food Science and Nutrition, University of Rhode Island, West Kingston, Rhode Island 02892*

DANIEL V. LYNCH (15), *Department of Biology, Williams College, Williamstown, Massachusetts 01267*

SUZANNE M. MANDALA (35), *Department of Infectious Disease, Merck Research Laboratories, Rahway, New Jersey 07065-4607*

CUNGUI MAO (26), *Departments of Medicine, General Internal Medicine, and Geriatrics, Medical University of South Carolina, Charleston, South Carolina 29425*

DAVID L. MARKS (7), *Mayo Clinic and Foundation, Rochester, Minnesota 55905*

FILMORE I. MEREDITH (37), *Toxicology and Mycotoxin Research Unit, Russell Agricultural Research Center, United States Department of Agriculture/ARS, Athens, Georgia 30604-5677*

ALFRED H. MERRILL, JR. (3, 22), *Department of Biochemistry, Emory University School of Medicine, Atlanta, Georgia 30322-3050*

CHRISTOPH MICHEL (4), *Kekulé-Institut für Organische Chemie und Biochemie, Universität Bonn, D-53121 Bonn, Germany*

SUSUMU MITSUTAKE (52), *Laboratory of Marine Biochemistry, Faculty of Agriculture, Kyushu University, Higashi-ku, Fukuoka 812-8581, Japan*

M. MAREK NAGIEC (1), *Pharmacia and Upjohn, Kalamazoo, Michigan 49007*

MARVIN R. NATOWICZ (11), *Division of Medical Genetics, E. K. Shriver Center, Waltham, Massachusetts 02452*

DAVID S. NEWBURG (11), *Department of Biomedical Sciences, E. K. Shriver Center, Waltham, Massachusetts 02452*

MARIANA NIKOLOVA-KARAKASHIAN (5, 22), *Department of Physiology and Sanders Brown Center on Aging, University of Kentucky, Lexington, Kentucky 40536-0084*

ÅKE NILSSON (30), *Department of Medicine, University Hospital of Lund, S-22185 Lund, Sweden*

LINA M. OBEID (26, 34), *Department of Medicine, General Internal Medicine, and Geriatrics, Medical University of South Carolina, Charleston, South Carolina 29425*

ANA OLIVERA (25), *Department of Biochemistry and Molecular Biology, Georgetown University Medical Center, Washington, DC 20007*

RICHARD E. PAGANO (7), *Mayo Clinic and Foundation, Rochester, Minnesota 55905*

PASCAL PAUL (7), *Synthelabo Research, 92500 Rueil-Malmaison, France*

RONALD D. PLATTNER (36), *Mycotoxin Research Unit, National Center for Agricultural Research, United States Department of Agriculture/ARS, Peoria, Illinois 61604*

GOTTFRIED POHLENTZ (10), *Institute of Physiological Chemistry, University of Bonn, D-53115 Bonn, Germany*

LAURA RIBONI (51), *Study Center for the Functional Biochemistry of Brain Lipids, Department of Medical Chemistry and Biochemistry, L.I.T.A.–Segrate, University of Milan, 20090 Segrate, Milan, Italy*

RONALD T. RILEY (36), *Toxicology and Mycotoxin Research Unit, Russell Agricultural Research Center, United States Department of Agriculture/ARS, Athens, Georgia 30604-5677*

DAVID M. RITCHIE (19), *R. W. Johnson Pharmaceutical Research Institute, Raritan, New Jersey 08869*

KONRAD SANDHOFF (10, 16, 23, 29, 47), *Kekulé-Institut für Organische Chemie und Biochemie, Universität Bonn, D-53121 Bonn, Germany*

RICHARD R. SCHMIDT (40), *Fakultät Chemie, Universität Konstanz, D-78457 Konstanz, Germany*

HEIKE SCHULZE (4), *Kekulé-Institut für Organische Chemie und Biochemie, Universität Bonn, D-53121 Bonn, Germany*

GÜNTER SCHWARZMANN (48), *Kekulé-Institut für Organische Chemie und Biochemie, Universität Bonn, D-53121 Bonn, Germany*

JAMES A. SHAYMAN (6, 12, 38), *Division of Nephrology, Department of Internal Medicine, University of Michigan Medical School, Ann Arbor, Michigan 48109-0676*

LIMING SHU (38), *Division of Nephrology, Department of Internal Medicine, University of Michigan Medical School, Ann Arbor, Michigan 48109-0676*

GARY K. SMITH (20), *Department of Molecular Biochemistry, GlaxoWellcome, Research Triangle Park, North Carolina 27709*

SANDRO SONNINO (50), *Study Center for the Functional Biochemistry of Brain Lipids, Department of Medical Chemistry and Biochemistry, L.I.T.A.–Segrate, University of Milan, 20090 Segrate, Milan, Italy*

SARAH SPIEGEL (25), *Department of Biochemistry and Molecular Biology, Georgetown University Medical Center, Washington, DC 20007*

HEIN SPRONG (8), *Department of Cell Biology and Histology, Academic Medical Center, University of Amsterdam, 1105 AZ Amsterdam, The Netherlands*

NORIYUKI SUEYOSHI (32), *Laboratory of Marine Biochemistry, Faculty of Agriculture, Kyushu University, Higashi-ku, Fukuoka 812-8581, Japan*

ZDZISLAW SZULC (42, 44), *Department of Biochemistry and Molecular Biology, Medical University of South Carolina, Charleston, South Carolina 29425*

JON Y. TAKEMOTO (2), *Department of Biology, Utah State University, Logan, Utah 84322-5305*

MOTOHIRO TANI (52), *Laboratory of Marine Biochemistry, Faculty of Agriculture, Kyushu University, Higashi-ku, Fukuoka 812-8581, Japan*

GUIDO TETTAMANTI (50, 51), *Study Center for the Functional Biochemistry of Brain Lipids, Department of Medical Chemistry and Biochemistry, L.I.T.A.–Segrate, University of Milan, 20090 Segrate, Milan, Italy*

PETER VAN DER SLUIJS (8), *Department of Cell Biology, Utrecht University School of Medicine, 3584 CX Utrecht, The Netherlands*

GERHILD VAN ECHTEN-DECKERT (4), *Kekulé-Institut für Organische Chemie und Biochemie, Universität Bonn, D-53121 Bonn, Germany*

GERRIT VAN MEER (8), *Department of Cell Biology and Histology, Academic Medical Center, University of Amsterdam, 1105 AZ Amsterdam, The Netherlands*

PAUL P. VAN VELDHOVEN (28), *Dep. Moleculaire Celbiologie, Afdeling Farmakologie, Katholieke Universiteit Leuven–Campus Gasthuisberg, B-3000 Leuven, Belgium*

PAOLA VIANI (51), *Study Center for the Functional Biochemistry of Brain Lipids, Department of Medical Chemistry and Biochemistry, L.I.T.A.–Segrate, University of Milan, 20090 Segrate, Milan, Italy*

DAVID W. WAGGONER (27), *Cell Therapeutics Inc., Seattle, Washington 98119*

ELAINE WANG (3), *Department of Biochemistry, Emory University School of Medicine, Atlanta, Georgia 30322-3050*

JIM XU (27), *Signal Transduction Laboratories, Lipid and Lipoprotein Research Group, Department of Biochemistry, University of Alberta, Edmonton, Alberta T6G 2S2, Canada*

Preface

Sphingolipids are a highly diverse family of compounds constructed from over 60 different sphingoid base backbones, numerous long-chain fatty acids, and hundreds of different headgroups, which include highly complex carbohydrates and even some covalently linked proteins. They are found in all eukaryotic organisms, as well as in some prokaryotes and viruses, primarily as components of membranes, but, also, of other lipid-rich structures such as lipoproteins and skin.

Research over the past decades has provided a conceptual framework and methodologies for elucidation of the structures, biophysical properties, biosynthesis and turnover, trafficking, and biological functions of these complex molecules as determinants of specialized membrane structures, modulators of growth factor receptors and extracellular matrix proteins, and intracellular mediators for a growing list of agonists and toxins. As this knowledge base has grown, investigators from a wide range of scientific disciplines have become "sphingolipidologists," or at least have begun to explore the role of sphingolipids in their experimental system.

This volume of *Methods in Enzymology* and its companion Volume 312 present techniques that are useful for such investigations, including assays of enzymes of sphingolipid biosynthesis and turnover, plus their purification, cloning, expression, and characterization; approaches for the selection of genetic mutations; preparation of structurally defined sphingolipids, radio-labeled compounds, analogs, and inhibitors by chemical, enzymatic, and microbial syntheses; analysis of sphingolipid structures and biophysical properties; quantitation of sphingolipids and bioactive sphingolipid metabolites by a variety of techniques; and characterization of protein–sphingolipid interactions, especially with respect to cell regulation and signal transduction.

These methods, plus the increasing availability of sphingolipids, sphingolipid analogs, antibodies to sphingolipids, and other tools from commercial sources, should improve the ease and sophistication of research on the structures, metabolism, and functions of sphingolipids, as well as on their roles in the etiology, prevention, and treatment of disease.

ALFRED H. MERRILL, JR.
YUSUF A. HANNUN

METHODS IN ENZYMOLOGY

VOLUME 73. Immunochemical Techniques (Part B)
Edited by JOHN J. LANGONE AND HELEN VAN VUNAKIS

VOLUME 74. Immunochemical Techniques (Part C)
Edited by JOHN J. LANGONE AND HELEN VAN VUNAKIS

VOLUME 75. Cumulative Subject Index Volumes XXXI, XXXII, XXXIV–LX
Edited by EDWARD A. DENNIS AND MARTHA G. DENNIS

VOLUME 76. Hemoglobins
Edited by ERALDO ANTONINI, LUIGI ROSSI-BERNARDI, AND EMILIA CHIANCONE

VOLUME 77. Detoxication and Drug Metabolism
Edited by WILLIAM B. JAKOBY

VOLUME 78. Interferons (Part A)
Edited by SIDNEY PESTKA

VOLUME 79. Interferons (Part B)
Edited by SIDNEY PESTKA

VOLUME 80. Proteolytic Enzymes (Part C)
Edited by LASZLO LORAND

VOLUME 81. Biomembranes (Part H: Visual Pigments and Purple Membranes, I)
Edited by LESTER PACKER

VOLUME 82. Structural and Contractile Proteins (Part A: Extracellular Matrix)
Edited by LEON W. CUNNINGHAM AND DIXIE W. FREDERIKSEN

VOLUME 83. Complex Carbohydrates (Part D)
Edited by VICTOR GINSBURG

VOLUME 84. Immunochemical Techniques (Part D: Selected Immunoassays)
Edited by JOHN J. LANGONE AND HELEN VAN VUNAKIS

VOLUME 85. Structural and Contractile Proteins (Part B: The Contractile Apparatus and the Cytoskeleton)
Edited by DIXIE W. FREDERIKSEN AND LEON W. CUNNINGHAM

VOLUME 86. Prostaglandins and Arachidonate Metabolites
Edited by WILLIAM E. M. LANDS AND WILLIAM L. SMITH

VOLUME 87. Enzyme Kinetics and Mechanism (Part C: Intermediates, Stereochemistry, and Rate Studies)
Edited by DANIEL L. PURICH

VOLUME 88. Biomembranes (Part I: Visual Pigments and Purple Membranes, II)
Edited by LESTER PACKER

VOLUME 89. Carbohydrate Metabolism (Part D)
Edited by WILLIS A. WOOD

VOLUME 90. Carbohydrate Metabolism (Part E)
Edited by WILLIS A. WOOD

Section I

Sphingolipid Metabolism

A. Biosynthesis
Articles 1 through 15

B. Turnover
Articles 16 through 32

C. Genetic Approaches
Articles 33 and 34

[1] Serine Palmitoyltransferase

By ROBERT C. DICKSON, ROBERT L. LESTER, and M. MAREK NAGIEC

Introduction

The committed step in *de novo* sphingolipid synthesis begins with the condensation of L-serine and palmitoyl-CoA to produce a C_{18} carbon unit, D-3-ketosphinganine or 3-ketodihydrosphingosine (D-2-amino-1-hydroxy-octadecan-3-one). This reaction, catalyzed by serine palmitoyltransferase (SPT)(EC 2.3.1.50, also called 3-ketosphinganine synthase), was first demonstrated in cell-free extracts made from the yeast *Hansenula ciferrii*.[1-4] Shortly thereafter its existence was shown in extracts prepared from rat liver[5] and mouse brain.[6] Snell and co-workers recognized that the enzyme requires pyridoxal phosphate for activity.[1-3] Since these initial reports the enzyme has been assayed in many cell types which are summarized in Table I.

Serine palmitoyltransferase activity has not been purified from any organism, despite many attempts, and only about a 100-fold enrichment in activity has been achieved.[7,8] In all cells that have been examined, SPT activity has been found in the membrane fraction, specifically the endoplasmic reticulum.[9,10] Membrane association may be one reason the enzyme has not yet been purified. There may be other reasons why the enzyme has been difficult to purify, including loss of subunits during purification. In *Saccharomyces cerevisiae*, at least two genes, *LCB1*[11] and *LCB2*,[12,13] are necessary for SPT activity. The Lcb2 protein is implicated in catalysis

[1] P. E. Braun and E. E. Snell, *Proc. Natl. Acad. Sci. U.S.A.* **58**, 298 (1967).
[2] P. E. Braun and E. E. Snell, *J. Biol. Chem.* **243**, 3775 (1968).
[3] R. N. Brady, S. J. Di Mari, and E. E. Snell, *J. Biol. Chem.* **244**, 491 (1969).
[4] W. Stoffel, D. LeKim, and G. Sticht, *Hoppe-Seyler's Z. Physiol. Chem.* **348**, 1570 (1967).
[5] W. Stoffel, D. LeKim, and G. Sticht, *Hoppe-Seyler's Z. Physiol. Chem.* **349**, 664 (1968).
[6] P. E. Braun, P. Morell, and N. S. Radin, *J. Biol. Chem.* **245**, 335 (1970).
[7] M. Lev and A. F. Milford, *Arch. Biochem. Biophys.* **212**, 424 (1981).
[8] A. H. Merrill, Jr., *Biochim. Biophys. Acta* **754**, 284 (1983).
[9] W. M. Holleran, M. L. Williams, W. N. Gao, and P. M. Elias, *J. Lipid Res.* **31**, 1655 (1990).
[10] E. C. Mandon, I. Ehses, J. Rother, G. van Echten, and K. Sandhoff, *J. Biol. Chem.* **267**, 11144 (1992).
[11] R. Buede, C. Rinker-Schaffer, W. J. Pinto, R. L. Lester, and R. C. Dickson, *J. Bacteriol.* **173**, 4325 (1991).
[12] M. M. Nagiec, J. A. Baltisberger, G. B. Wells, R. L. Lester, and R. C. Dickson, *Proc. Natl. Acad. Sci. U.S.A.* **91**, 7899 (1994).
[13] C. Zhao, T. Beeler, and T. Dunn, *J. Biol. Chem.* **269**, 21480 (1994).

TABLE I

CELL TYPES IN WHICH ENZYME HAS BEEN ASSAYED

Source of enzyme activity	Reference[a]
Hansenula ciferri	a
Rat liver	b–d
Rat brain	e
Bacteroides melaninogenicus	f
Chinese hamster ovary	g, h
Morris hepatoma 7777 (rat origin)	d
Rat lung, heart, spleen, muscle, stomach, kidney, pancreas, intestine, testes, ovary	i
Rabbit aorta	j
Human keratinocytes	k
Mouse cerebellar cells grown in culture	l
Mouse liver	m
Saccharomyces cerevisiae	n
Summer squash fruit (*Cucurbita pepo*)	o
Mouse cytotoxic T-cell line CTLL-2	p
Human embryonic kidney cells (HEK 293)	q
Human HeLa cells	r

[a] Key to references: (a) P. E. Braun and E. E. Snell, *Proc. Natl. Acad. Sci. U.S.A.* **58,** 298 (1967). (b) W. Stoffel, D. LeKim, and G. Sticht, *Hoppe-Seyler's Z. Physiol. Chem.* **348,** 1570 (1967). (c) R. D. Williams, E. Wang, and A. H. Merrill, Jr., *Arch. Biochem. Biophys.* **228,** 282 (1984). (d) R. D. Williams, D. W. Nixon, and A. H. Merrill, Jr., *Cancer Res.* **44,** 1918 (1984). (e) P. E. Braun, P. Morell, and N. S. Radin, *J. Biol. Chem.* **245,** 335 (1970). (f) M. Lev and A. F. Milford, *Arch. Biochem. Biophys.* **212,** 424 (1981). (g) A. H. Merrill, Jr., *Biochim. Biophys. Acta* **754,** 284 (1983). (h) K. Hanada, M. Nishijima, and Y. Akamatsu, *J. Biol. Chem.* **265,** 22137 (1990). (i) A. H. Merrill, Jr., D. W. Nixon, and R. D. Williams, *J. Lipid Res.* **26,** 617 (1985). (j) R. D. Williams, D. S. Sgoutas, G. S. Zaatari, and R. A. Santoianni, *J. Lipid Res.* **28,** 1478 (1987). (k) W. M. Holleran, M. L. Williams, W. N. Gao, and P. M. Elias, *J. Lipid Res.* **31,** 1655 (1990). (l) G. van Echten, R. Birk, G. Brenner-Weiss, R. R. Schmidt, and K. Sandhoff, *J. Biol. Chem.* **265,** 9333 (1990). (m) E. C. Mandon, I. Ehses, J. Rother, G. van Echten, and K. Sandhoff, *J. Biol. Chem.* **267,** 11144 (1992). (n) W. J. Pinto, G. W. Wells, and R. L. Lester, *J. Bacteriol.* **174,** 2575 (1992). (o) D. V. Lynch and S. R. Fairfield, *Plant. Physiol.* **103,** 1421 (1993). (p) Y. Miyake, Y. Kozutsumi, S. Nakamura, T. Fujita, and T. Kawasaki, *Biochem. Biophys. Res. Commun.* **211,** 396 (1995). (q) B. Weiss and W. Stoffel, *Eur. J. Biochem.* **249,** 239 (1997). (r) S. M. Mandala, B. R. Frommer, R. A. Thornton, M. B. Kurtz, M. B. Young, M. A. Cabello, O. Genilloud, J. M. Liesch, J. L. Smith, and W. S. Horn, *J. Antibiot.* **47,** 376 (1994).

because of amino acid sequence similarity to aminolevulinate synthases and because it carries a domain that is believed to form a Schiff base with pyridoxal phosphate.[12]

Mammalian cDNA homologs of both *LCB1*[14,15] and *LCB2*[15,16] have been identified. The Lcb1 protein is a component of the SPT enzyme, as a His$_6$-tagged version of the Chinese hamster ovary (CHO) Lcb1 protein binds SPT activity to a nickel affinity column.[14] Whether the Lcb1 protein plays a catalytic or a regulatory role, or both, awaits determination. Finally, there may be other unidentified subunits based on our observation that a 10-fold overproduction of the Lcb1 and Lcb2 proteins in *S. cerevisiae* yields only a 2- to 3-fold increase in SPT enzyme activity (unpublished data).

Assay

The assay of SPT activity is based on the conversion of water-soluble [^{3}H]serine to the chloroform-soluble product, 3-ketosphinganine. The most widely used format of this assay, Protocol I, is based on the original work of Williams *et al.*[17] and on recent modifications.[18] Protocol II is a modified format that employs an alternative extraction procedure developed by Wells and Lester (unpublished results) and is used routinely for assay of SPT activity in *S. cerevisiae*.

Protocol I

The final reaction volume of 0.1 ml contains 100 mM HEPES (pH 8.3), 5 mM dithiothreitol (DTT), 2.5 mM EDTA (pH 7), 50 μM pyridoxal phosphate, 200 μM palmitoyl-CoA, 1 mM ^{3}H- or ^{14}C-labeled L-serine (specific activity: 10–30 mCi/mmol), and 50–150 μg of microsomal protein. The reaction is initiated by the addition of palmitoyl-CoA as a precaution to avoid depletion of this substrate by fatty acyl-CoA hydrolases, which is more of a problem in mammalian than in yeast cells. The reaction is incubated at 37° with shaking for 10 min and is generally linear for up to 20 min.

The reaction is terminated by the addition of 1.5 ml of chloroform–methanol (1:2, v/v). The radiolabeled enzymatic product, 3-ketosphinganine, is separated from radiolabeled serine by phase partitioning. Sphinganine (25 μl of a 1-mg/ml ethanolic solution) is added as a carrier, fol-

[14] K. Hanada, T. Hara, M. Nishijima, O. Kuge, R. C. Dickson, and M. M. Nagiec, *J. Biol. Chem.* **272,** 32108 (1997).

[15] B. Weiss and W. Stoffel, *Eur. J. Biochem.* **249,** 239 (1997).

[16] M. M. Nagiec, R. L. Lester, and R. C. Dickson, *Gene* **177,** 237 (1996).

[17] R. D. Williams, E. Wang, and A. H. Merrill, *Arch. Biochem. Biophys.* **228,** 282 (1984).

[18] A. H. Merrill, Jr., and E. Wang, *Methods Enzymol.* **209,** 427 (1992).

lowed by the addition of 1 ml of $CHCl_3$ and 2 ml of 0.5 N NH_4OH. The sample is vortexed vigorously and centrifuged briefly in a bench-top centrifuge to separate the two phases. The upper aqueous phase is removed and discarded. The lower chloroform layer is washed twice with 2 ml of water to remove any remaining unincorporated radiolabeled serine and then 0.8 ml, about 50% of the total, is transferred to a scintillation vial. The sample is evaporated to dryness in a stream of hot air, resuspended in 4 ml of scintillation fluid, and counted in a liquid scintillation spectrometer. Enzyme activity is expressed as picomoles of 3-ketosphinganine formed per minute per milligram of protein.

A control reaction to determine background radioactivity is run in the same manner except that palmitoyl-CoA is omitted. This control corrects for contaminants found in most preparations of radiolabeled serine. The impurity can be removed by chromatography on AG50W-X8 resin (Bio-Rad, Richmond, CA). Radiolabeled serine is applied in 0.01 N HCl to a 1-ml column of the hydrogen form of the resin, and after washing the column with 10 volumes of deionized water, the serine is eluted with 1 M HEPES, pH 8.3.

Protocol II

The reactions are set up in a similar way as in Protocol I except that 5 mM serine (30 kcpm/nmol) is used and the final volume of the reaction is 200 μl. More serine is used because the $S.$ $cerevisiae$ SPT enzyme has a K_m for serine of 4 mM.[19] The reaction is stopped by the addition of 2 ml ethanol:diethyl ether (3:1, v/v). Carrier sphinganine (50 μg in 50 μl of 95% ethanol) is added followed by 1.2 ml of petroleum ether. The sample is vortexed, mixed with 2.4 ml of 1 M KCl, and then centrifugated in a tabletop centrifuge for 5 min. The upper organic phase containing radioactive 3-ketosphinganine is removed and mixed with a sham lower phase made by extracting 200 μl of water in exactly the same way as the enzyme reaction. The washed upper phase is transferred to a new tube, evaporated to dryness, and dissolved in 0.5 ml of $CHCl_3:CH_3OH:H_2O$ (16:16:5, v/v/v). Radioactivity is determined in a 50-μl sample in the same way as described in Protocol I.

Substrates

SPT has an absolute requirement for serine because no other natural amino acid serves as a substrate. The apparent K_m for L-serine varies from

[19] W. J. Pinto, G. W. Wells, and R. L. Lester, $J.$ $Bacteriol.$ **174,** 2575 (1992).

0.1 to 4 mM, depending on the source of the enzyme.[7,19,20,21] The preferred fatty acyl-CoA is C_{16} (palmitoyl-CoA) the K_m value of which varies slightly, depending on the source, but which is generally less than 1 mM.[7,9,19,20,21] The enzyme will, however, use other fatty acyl-CoAs, ranging in chain length from C_{12} to C_{24}.[7,9,19,20,21] Some organisms and cell types do use different chain length CoAs and the biological reason for this remains to be determined. For example, *S. cerevisiae* cells make more C_{18} than C_{16} CoA when the incubation temperature rises (37°, e.g.), resulting in an increased ratio of C_{20}- to C_{18}-containing sphingoid long-chain bases.[22,23]

Pyridoxal 5'-phosphate will dissociate from the enzyme on extended dialysis against cysteine. Catalytic activity can be reconstituted by the addition of pyridoxal phosphate, which has an apparent K_m of 1 μM.[21]

Other Considerations

It is important to ensure that product formation is linear with time and protein concentration, as some types of cells may contain fatty acyl-CoA hydrolases that can deplete the cosubstrate. When assaying a new cell type, it is important to optimize the palmitoyl-CoA concentration. SPT forms a Schiff base with pyridoxal phosphate; therefore, Tris and other compounds with a free amino group should be avoided because they may prevent binding of the cofactor. SPT has a relatively high K_m for serine, which contributes to a relatively low percentage (typically 0.05%) conversion of substrate to product. EDTA is included in the reaction to inhibit phosphatidylserine synthase (serine exchange enzyme) activity and the consequent formation of phosphatidylserine, a possible chloroform-soluble contaminant in the assay. Although uniformly labeled serine is the least expensive type of radiolabel to use, it gives less accurate specific activities than L-[3-^{14}C]- or L-[2-^{3}H]serine because the α hydrogen is lost during the reaction.[24] Reactions are performed conveniently in 13 × 100-mm screw-capped glass disposable tubes (e.g., VWR Cat. No. 60826-188) or reusable tubes (e.g., Kimax, VWR Cat. No. 60828-208) that have been acid washed after use.

The assay has been adapted to microtiter plates for high throughput screening of inhibitors.[25] The assay has also been adapted for use with

[20] S. J. Di Mari, R. N. Brady, and E. E. Snell, *Arch. Biochem. Biophys.* **143**, 553 (1971).
[21] R. D. Williams, D. W. Nixon, and A. H. Merrill, Jr., *Cancer Res.* **44**, 1918 (1984).
[22] R. C. Dickson, E. E. Nagiec, M. Skrzypek, P. Tillman, G. B. Wells, and R. L. Lester, *J. Biol. Chem.* **272**, 30196 (1997).
[23] G. M. Jenkins, A. Richards, T. Wahl, C. G. Mao, L. Obeid, and Y. Hannun, *J. Biol. Chem.* **272**, 32566 (1997).
[24] K. Krisnangkura and C. C. Sweeley, *J. Biol. Chem.* **251**, 1597 (1976).
[25] S. M. Mandala, R. A. Thornton, B. R. Frommer, S. Dreikorn, and M. B. Kurtz, *J. Antibiot.* **50**, 339 (1997).

cultures of suspended cells[8,26] and for assay of colonies of cultured cells.[27] Alternative ways of detecting the reaction product 3-ketosphinganine[18] include thin-layer chromatography and trapping of $^{14}CO_2$ liberated from [1-^{14}C]serine or uniformly labeled [^{14}C]serine.

Inhibitors

Several inhibitors of SPT activity have now been described. The first were mechanism-based or "suicide" inhibitors, e.g., cycloserine[28] and β-haloalanines such as β-chloro-L-alanine and β-fluoro-L-alanine.[29] These are not necessarily specific for SPT and likely affect other enzymes that use pyridoxal phosphate as a cofactor. Several inhibitors with greater specificity have now been identified from screens of natural products, including the sphingofungins,[30] the immunosuppressant ISP-1/Myriocin,[31] lipoxamycin,[32] and the viridiofungins.[25] The natural product inhibitors are not available commercially but can be isolated by the method described in Article [36].

Regulation

SPT activity is likely to be regulated because it catalyzes the initial and committed step in sphingolipid synthesis. However, there are relatively few findings demonstrating regulation and even less is known about the regulatory mechanisms. Data up to about 1990 that support regulation have been summarized thoroughly[33] so we will concentrate our comments on more recent experiments.

There is a transient increase in sphinganine of up to 10-fold following the transfer of many types of cultured cells to fresh medium.[34,35] Because the increase is inhibited by β-fluoro-L-alanine, a known SPT inhibitor,[29] it is possible that increased SPT activity is responsible for the changes in

[26] T. O. Messmer, E. Wang, V. L. Stevens, and A. H. Merrill, Jr., *J. Nutr.* **119**, 534 (1989).

[27] K. Hanada, M. Nishijima, and Y. Akamatsu, *J. Biol. Chem.* **265**, 22137 (1990).

[28] K. S. Sundaram and M. Lev, *Biochem. Biophys. Res. Commun.* **119**, 814 (1984).

[29] K. A. Medlock and A. H. Merrill, Jr., *Biochemistry* **27**, 7079 (1988).

[30] M. M. Zweerink, A. M. Edison, G. B. Wells, W. Pinto, and R. L. Lester, *J. Biol. Chem.* **267**, 25032 (1992).

[31] Y. Miyake, Y. Kozutsumi, S. Nakamura, T. Fujita, and T. Kawasaki, *Biochem. Biophys. Res. Commun.* **211**, 396 (1995).

[32] S. M. Mandala, B. R. Frommer, R. A. Thornton, M. B. Kurtz, M. B. Young, M. A. Cabello, O. Genilloud, J. M. Liesch, J. L. Smith, and W. S. Horn, *J. Antibiot.* **47**, 376 (1994).

[33] A. H. Merrill, Jr., and D. D. Jones, *Biochim. Biophys. Acta* **1044**, 1 (1990).

[34] Y. Lavie, J. K. Blusztajn, and M. Liscovitch, *Biochim. Biophys. Acta* **1220**, 323 (1994).

[35] E. R. Smith and A. H. Merrill, Jr., *J. Biol. Chem.* **270**, 18749 (1995).

sphinganine, although other explanations, such as increased substrate availability, cannot be excluded at this time.

Several types of data suggest that SPT activity is regulated in mammalian epidermis where sphingolipids are known to be necessary for preventing water loss. When the mouse epidermal barrier is depleted, there is a rapid increase in sphingolipid synthesis, which is accompanied by a transient 50% increase in SPT activity.[36] These findings demonstrate that SPT activity is regulated, but the mechanism remains to be established. Experiments with cultured human keratinocytes have shown that exposure to ultraviolet light, UVB, stimulates sphingolipid synthesis primarily through increases in both LCB2 mRNA and protein levels.[37]

Treatment of cultured primary rat cerebellar neurons with 50 μM sphingosine and its homologs reduced SPT activity by 80% after 24 hr.[38] In the same neuronal cell model, SPT activity decreased 50% following treatment with cis-4-methylsphinosine.[39] In both cases it appears that reduced SPT activity is not the result of a direct interaction between SPT and the sphingosine homologs; rather there appears to be a more complex, but uncharacterized, mechanism at work.

[36] W. M. Holleran, K. R. Feingold, M. Q. Man, W. N. Gao, J. M. Lee, and P. M. Elias, *J. Lipid Res.* **32,** 1151 (1991).
[37] A. M. Farrell, Y. Uchida, M. M. Nagiec, I. R. Harris, R. Dickson, P. Elias, and W. Holleran, *J. Lipid Res.* **39,** 2031 (1998).
[38] E. C. Mandon, G. van Echten, R. Birk, R. R. Schmidt, and K. Sandhoff, *Eur. J. Biochem.* **198,** 667 (1991).
[39] G. van Echten-Deckert, A. Zschoche, T. Baer, R. R. Schmidt, A. Raths, T. Heinemann, and K. Sandhoff, *J. Biol. Chem.* **272,** 15825 (1997).

[2] Assay of the *Saccharomyces cerevisiae* Dihydrosphingosine C-4 Hydroxylase

By MICHELLE M. GRILLEY and JON Y. TAKEMOTO

Introduction

Most of the sphingolipids of fungi and plants contain phytosphingosine (or 4-hydroxysphinganine) as the sphingoid long chain base backbone.[1] Although sphingosine predominates in the sphingolipids of animal cells, a significant portion with the 4-hydroxylated long chain base occurs in epithe-

[1] R. L. Lester and R. C. Dickson, *Adv. Lipid Res.* **26,** 253 (1993).

0076-6879/99 $30.00

lial tissues[2] and in adenocarcinomas.[3] Despite its widespread occurrence, the biological significance of sphingoid base 4-hydroxylation is not known. Recently, however, it was observed in the yeast *Saccharomyces cerevisiae* that defects in this step correlate with resistance to the growth inhibitory effects of the antifungal metabolite syringomycin E.[4]

In yeast, the dihydrosphingosine (or sphinganine) 4-hydroxylase is encoded by the gene *SYR2* (also called *SUR2*) of chromosome IV.[4,5] The enzyme is located in the endoplasmic reticulum[6] and contains an eight histidine motif characteristic of diiron-containing enzymes that catalyze O_2-dependent modifications of hydrocarbon substrates.[7] Consistent with these findings, a reduced pyridine nucleotide-dependent dihydrosphingosine and dihydroceramide hydroxylation activity was measured in microsomal extracts from a *SYR2*-overexpressing yeast strain.[4] The enzyme has not been purified, and the equivalent enzymes or genes from plants and animals have not been described.

This article describes procedures for the assay of sphingoid base 4-hydroxylation by measuring the conversion of dihydrosphingosine to phytosphingosine. Although the methods were developed using yeast preparations, they should be generally applicable to other organisms. The assays facilitate purification and characterization of these 4-hydroxylases, which, in turn, will help reveal their as yet undefined physiological roles.

Preparation of Yeast Microsomes

Dihydrosphingosine 4-hydroxylase activity is assayed conveniently using a microsomal membrane preparation from either wild-type *S. cervisiae* or a *SYR2*-overexpressing strain such as W303C(pYSYR2a).[4] The following preparation is modified from published procedures to maintain maximal hydroxylase activity.[4,8]

A 500-ml culture grown to a density of approximately 1×10^8 cells/ml is harvested by centrifugation, washed with water, and resuspended in 25 ml of 100 m*M* Tris–sulfate, pH 9.4, 10 m*M* dithiothreitol (DTT). Following a 15-min incubation at room temperature and a wash with 10 m*M* Tris–HCl,

[2] K.-A. Karlsson, *Lipids* **5,** 878 (1970).
[3] H.-J. Yang and S.-I. Hakomori, *J. Biol. Chem.* **246,** 1192 (1971).
[4] M. M. Grilley, S. D. Stock, R. C. Dickson, R. L. Lester, and J. Y. Takemoto, *J. Biol. Chem.* **273,** 11062 (1998).
[5] D. Haak, K. Gable, T. Beeler, and T. Dunn, *J. Biol. Chem.* **272,** 29704 (1997).
[6] P. Cliften, Y. Wang, D. Mochizuki, T. Miyakawa, R. Wangspa, J. Hughes, and J. Y. Takemoto, *Microbiology* **147,** 477 (1996).
[7] J. Shanklin, E. Whittle, and B. G. Fox, *Biochemistry* **33,** 12787 (1994).
[8] R. Kato, T. Yasumori, and Y. Yamazoe, *Methods Enzymol.* **206,** 183 (1991).

pH 7.5, 0.6 M sorbitol, 0.1 mM DTT, 0.1 mM EDTA, cells are incubated for 1 hr at 30° in 7.5 ml 10 mM Tris–HCl, pH 7.5, 2 M sorbitol, 0.1 mM DTT, 0.1 mM EDTA, 0.1 mg/ml Zymolyase 100T (Seikagaku Corp., Tokyo). Cells are then washed gently with 10 mM Tris–HCl, pH 7.5, 2 M sorbitol and are resuspended in 5 ml cold 10 mM Tris–HCl, pH 7.5, 0.65 M sorbitol, 0.1 mM phenylmethylsulfonyl fluoride, 1 μg/ml leupeptin, 1 μg/ml pepstatin. All subsequent procedures are conducted at 0–4°. Disruption is by sonication (5 × 1-min bursts, Fisher Scientific Sonic Dismembrator 550 with microtip, power level 3). Cell debris and mitochondria are removed by centrifugation (5 min at 1600g, then 20 min at 12,000g). Microsomal membranes are collected by centrifugation for 90 min at 100,000g. Pellets are homogenized in 0.6–1 ml cold 10 mM Tris–HCl, pH 7.5, 20% glycerol. If preparations are not to be used the same day, they may be frozen in liquid nitrogen and stored at −70°. Samples may be thawed and refrozen once with little loss of activity. Protein concentrations of the following microsomal preparations were determined using the Pierce Coomassie protein assay reagent with bovine serum albumin as standard.

Assay of Dihydrosphingosine Hydroxylase Activity

Reagents

Tris buffer: 1 M adjusted to pH 7.5 with HCl
NADPH: 10 mM in 10 mM Tris–HCl, pH 7.5, 1 mM EDTA
CHAPS (3-[(cholamidopropyl)dimethylammonio]-1-propane sulfonate): 1.5% (w/v) in water
DHS (DL-erythro-dihydrosphingosine) (Sigma or Matreya, Inc.): 50 mM in chloroform–methanol (1 : 1, v/v)
Buffered saline solution: 135 mM NaCl, 10 mM Tris–HCl, pH 7.5, 15 mM EDTA
90% methanol: HPLC grade methanol–water (90 : 10, v/v)
Biphenylcarbonyl chloride (Sigma): 5% (w/v) in tetrahydrofuran
Triethylamine solution: 0.5% (v/v) in tetrahydrofuran
Ethanolamine solution: 20% (v/v) in tetrahydrofuran
Ninhydrin solution: 100 g in 500 ml butanol

Enzyme Reaction

Substrate is prepared by combining 50 nmol DHS with 4 μl CHAPS in a 1.5- to 1.8-ml microcentrifuge tube, drying under nitrogen to remove organic solvents, and resuspending in 0.09 ml water and 0.01 ml Tris buffer. Sonication for 1 min in a bath sonicator aids resuspension. Water and yeast microsomes, prepared as described earlier, are added to give a final volume of 192 μl and a final membrane protein concentration of 1.5–4 mg/ml. The reaction is initiated by the addition of 8 μl NADPH. Incubation is 30 min at 25°. To stop the reaction, 0.38 ml methanol and 0.19 ml chloroform are added. The tubes are mixed and spun for 30 sec in a microcentrifuge to separate the phases. The lower chloroform phase containing the long chain bases is transferred to a fresh tube, and the remaining aqueous phase is washed twice with 0.2 ml chloroform. The combined chloroform phases are washed with 0.2 ml buffered saline solution and dried under nitrogen.

Chromatography

Two chromatographic techniques for the resolution and detection of substrate and product long chain bases are detailed here. Both provide comparable results when using *S. cerevisiae* microsomes. The choice of technique may be based on facilities available, number of samples to be assayed, and the presence of interfering contaminants in specific applications.

For thin-layer chromatography (TLC) separation, dried long chain bases are resuspended in a small volume of chloroform–methanol (v/v, 1 : 1) and applied to a 10-cm silica gel G TLC plate (Whatman 4420221). Plates are developed in chloroform–methanol–concentrated NH_4OH (v/v/v, 15 : 3.8 : 0.8). Long chain bases are visualized by spraying with ninhydrin solution and baking at 80° for 10 min (Fig. 1). Approximate R_f values are 0.55 for DHS and 0.35 for PHS. Results can be quantitated using programs such NIH Image (available by anonymous FTP form zippy.nimh. nih.gov[128.231.98.32]) or commercially available scanning and analysis software.

Alternatively, for reversed-phase high-performance liquid chromatography (HPLC) separation, long chain bases are first derivatized with the amine-reactive detection reagent 4-biphenylcarbonyl chloride.[9,10] Dried samples are resuspended in 0.1 ml triethylamine solution, followed by the addition of 0.01 ml biphenylcarbonyl chloride and incubation at room temperature for 90 min. The reaction is quenched with 0.01 ml ethanolamine solution for 30 min at room temperature. A portion of the quenched reaction, e.g., 0.02 ml, is injected onto an analytical C_{18} HPLC column (Alltech Econosil C_{18}, 250 × 4.6 mm with a 10-mm guard column). Elution is isocratic with 90% methanol. Absorbance of the effluent is monitored at 280 nm (Fig. 2). Exact elution times will vary with the HPLC apparatus used, but an absorbance peak representing the derivatized enzyme product, biphe-

[9] F. B. Jungalwala, J. E. Evans, E. Bremer, and R. H. McCluer, *J. Lipid Res.* **24**, 1380 (1983).
[10] R. C. Dickson, G. B. Wells, A. Schmidt, and R. L. Lester, *Mol. Cell. Biol.* **10**, 2176 (1990).

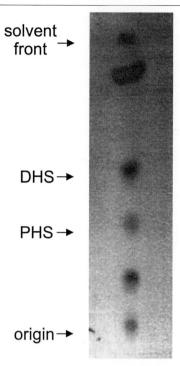

FIG. 1. Typical TLC analysis of the dihydrosphingosine hydroxylase assay. Chromatography and detection with ninhydrin are as described in the text. Origin, solvent front, and positions of dihydrosphingosine (DHS) and phytosphingosine (PHS), as determined by cochromatography with standards, are indicated.

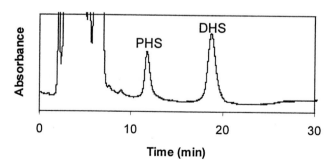

FIG. 2. Typical HPLC analysis of dihydrosphingosine hydroxylase. Chromatography of biphenylcarbonyl-derivatized reaction products are as described in the text. Elution positions of derivatized dihydrosphingosine (DHS) and phytosphingosine (PHS), as determined by cochromatography with standards, are indicated.

nylcarbonyl phytosphingosine, will appear at approximately 12 min, whereas the remaining derivatized substrate, biphenylcarbonyl-DHS, will elute about 8–10 min later.

If samples cause an unacceptable increase in back pressure of the column, problematic contaminants can be removed from the derivatized long chain bases by passage through a 1-ml Amberlite CG-50 (Sigma) column, equilibrated in 90% methanol. After application of the sample, the column is washed with 90% methanol and the void volume containing the derivatized long chain bases is collected. Long chain bases are dried under nitrogen and resuspended in 0.12 ml 90% methanol prior to HPLC as described earlier. The Amberlite column can be reused at least 20 times with no noticeable loss in performance if it is washed between samples with 100% methanol.

Use of Dihydroceramide as Substrate

Saccharomyces cerevisiae microsomes can catalyze 4-hydroxylation of both dihydrosphingosine and dihydroceramide.[4] The substrate specificities of the putative sphingoid base 4-hydroxylases from other organisms are unknown. The 4-hydroxylation of dihydroceramide can be assayed as described earlier with the substitution of dihydroceramide for dihydrosphingosine and the addition of a hydrolytic step before chromatography to release the long chain bases for analysis. Hydrolysis is affected by reflux of the dried chloroform extract of the assay mixture in 4 ml of 1 M HCl in methanol–water (82:18, v/v) at 80° for approximately 20 hr. After the addition of 12 ml concentrated NH_4OH; freed long chain bases are extracted with 3 × 3 ml chloroform.[10] Combined chloroform phases are back extracted with 3 ml concentrated NH_4OH, dried under nitrogen, and subjected to either 4-biphenylcarbonyl chloride derivatization and HPLC analysis or TLC separation, as described earlier.

Acknowledgments

This work was supported by Eli Lilly and Co. and the Utah State University Biotechnology Center.

[3] Ceramide Synthase

By Elaine Wang and Alfred H. Merrill, Jr.

Ceramide synthase catalyzes the acylation of sphinganine, sphingosine, and other long chain sphingoid bases to form their *N*-acyl derivatives, which are usually referred to as dihydroceramides (*N*-acylsphinganines) and ceramides (*N*-acylsphingosines).[1] Ceramide synthase is, therefore, involved in both the early steps of *de novo* sphingolipid biosynthesis (the acylation of sphinganine) and the recycling of free sphingoid bases generated in cells (e.g., from ceramide hydrolysis) or taken up from exogenous sources.[2] Ceramide synthase activity is conventionally regarded as the acylation of sphingoid bases using fatty acyl-CoA of varying fatty acid chain length,[1,2] although there have been occasional reports of ceramide synthesis via the condensation of sphingosine and a free fatty acid, perhaps by the reversal of ceramidase.[3,4] The significance of this latter pathway for sphingolipid metabolism is unclear, therefore, this article deals only with the fatty acyl-CoA-dependent reaction.

Ceramide synthase is of interest not only because it forms the lipid backbone for complex sphingolipid biosynthesis, but also as the target of a number of microbial toxins[1] (and see Articles [35] and [37]) and as a possible participant in the induction of apoptosis by various stimuli and toxins.[5–7]

Assay of Ceramide Synthase

Ceramide synthase is commonly assayed by the acylation of [³H]sphingosine with unlabeled fatty acyl-CoA, or acylation of an unlabeled sphingoid

[1] A. H. Merrill, Jr., and C. C. Sweeley, *in* "Biochemistry of Lipids, Lipoproteins, and Membranes" (D. E. Vance and J. Vance, eds.), Chapter 12. Elsevier Science, Amsterdam, 1996.

[2] M. Nikolova-Karakashian, T. R. Vales, E. Wang, D. S. Menaldino, C. Alexander, J. Goh, D. C. Liotta, and A. H. Merrill, Jr., *in* "Sphingolipid-Mediated Signal Transduction, Molecular Biology Intelligence Unit" (Y. A. Hannun, ed.), p. 159. R. G. Landes Co., 1997.

[3] S. J. Gatt, *J. Biol. Chem.* **238,** PC3131 (1963).

[4] W. Stoffel, E. Kruger, and I. Melzner, *Hoppe-Seyler's Z. Physiol. Chem.* **361,** 773 (1980).

[5] R. M. Bose, M. Verheij, A. Haimovitz-Friedman, K. Scotto, Z. Fuks, and R. Kolesnick, *Cell* **82,** 405 (1995).

[6] A. Haimovitz-Friedman, *Radiat. Res.* **150,** S102 (1998).

[7] J. Xu, C. H. Yeh, S. Chen, L. He, S. L. Sensi, L. M. Canzoniero, D. W. Choi, and C. Y. Hsu, *J. Biol. Chem.* **273,** 16521 (1998).

base (or analog) using a radiolabeled fatty acyl-CoA.[8] Many of these compounds are available commercially, and there are numerous methods for the synthesis of radiolabeled sphingoid bases (see Articles [40] and [42]) and fatty acyl-CoA.[9] Synthesis of ceramides (as standards) is described in Article [43].

The assay mixture (totalling 100 μl) contains 25 mM potassium phosphate buffer (pH 7.4), 0.5 mM dithiothreitol, varying concentrations of [³H]sphingosine (1–10 μM), 25–150 μg of protein (e.g., 40 μg of liver microsomes), and 50–100 μM palmitoyl-CoA (or another fatty acyl-CoA). The fatty acyl-CoA is added last to initiate the reaction. Initial assays should have several controls: incubation of the fatty acyl-CoA and sphingoid base without enzyme, the fatty acyl-CoA (if radiolabeled) plus enzyme without sphingoid base, and the sphingoid base (if radiolabeled) plus enzyme without fatty acyl-CoA.

The sphingosine can be added in a number of ways: (1) by direct addition of a small volume from a concentrated stock in ethanol (in some cases, the ethanol can be added to the test tube used to prepare the assay stock, dried under nitrogen, and sonicated after addition of the other ingredients); (2) use of another vehicle for delivering the sphingoid base, such as bovine serum albumin[10]; or (3) as part of phosphatidylcholine liposomes. The latter has the advantage that the physical state of such liposomes is reasonably well known, and sphingosine can undergo rapid equilibration between liposomes and the membranous enzyme preparations (A. Kau *et al.,* manuscript in preparation). The liposomes are formed by mixing the phosphatidylcholine (e.g., egg phosphatidylcholine) and long chain base in an organic solvent at a molar ratio of 2:1, removal of the solvent under a stream of nitrogen, and addition of 25 mM potassium phosphate buffer (pH 7.4) for a stock solution that is five times the concentration of the sphingoid base in the final assay. The milky suspension is incubated for ca. 10 min at 37°, vortexed, and sonicated for several minutes (until clear). Whichever method is used, its success in solubilizing the sphingoid base should be confirmed by counting (if radiolabeled) or high-performance liquid chromatography (HPLC).[11]

The assay mixture is usually incubated in a shaking water bath for 15 min at 37° (longer times are not recommended because the fatty acyl-CoA may be depleted by thioesterases) and is then stopped by the addition of 1 ml of methanol and 0.5 ml of CHCl₃. To improve recovery of the products,

[8] A. H. Merrill, Jr., and E. Wang, *Methods Enzymol.* **209,** 427 (1992).
[9] A. H. Merrill, Jr., S. Gidwitz, and R. M. Bell, *J. Lipid Res.* **23,** 1368 (1982).
[10] Y. A. Hannun, A. H. Merrill, Jr., and R. M. Bell, *Methods Enzymol.* **201,** 316 (1991).
[11] A. H. Merrill, Jr., E. Wang, R. E. Mullins, W. C. L. Jamison, S. Nimkar, and D. E. Liotta, *Anal. Biochem.* **171,** 373 (1988).

25 μg of unlabeled ceramide is added as carrier, 1 ml of $CHCl_3$ is added, the mixture is vortexed, approximately 3 ml of slightly basic water is added, and the mixture is vortexed again. The aqueous phase is removed (usually after centrifugation) and the $CHCl_3$ layer is washed with an additional 3 ml of water, passed through anhydrous (granular) sodium sulfate (usually packed in a small column prepared from a Pasteur pipette), and evaporated to dryness under a stream of nitrogen or *in vacuo* (e.g., using a Speed-Vac concentrator, Savant Instruments, Farmingdale, NY, without heating).

If a radiolabeled fatty acyl-CoA is used, glycerolipids are cleaved by incubation for 1 hr at 37° in 0.1 M KOH in methanol (followed by reextraction as described earlier). For analysis by TLC, the samples are dissolved in 20 μl of $CH_3OH : CHCl_3$ (1 : 2) and spotted onto Brinkman type 60 silica TLC plates. The plates are developed in diethyl ether : CH_3OH (99 : 1, v/v), the positions of ceramide versus other labeled compounds (i.e., fatty acid or sphingoid base, depending on which is used) are determined by comparison with standards (unlabeled compounds can be visualized with I_2; however, saturated substrates such as palmitic acid are difficult to visualize with I_2 and use of a radiolabeled standard may be preferable), and scintillation counting. For analysis by HPLC, the samples are dissolved in 200 μl of mobile phase ($C_2H_5OH : CH_3CN$, 60 : 40 or 50 : 50) and injected onto a C_{18} reversed-phase column eluted with mobile phase (typically at a flow rate of 2 ml/min). The amount of radiolabeled ceramide is determined with an on-line radioactivity detector or by collecting the appropriate eluate for scintillation counting with the appropriate corrections for quenching. Product formation is calculated from the dpm ceramide formed minus the dpm in the appropriate control (e.g., the assays minus the unlabeled co-substrate).

Comments

This description gives a range of concentrations for some of the substrates because the optimal conditions can vary from preparation to preparation (due, mainly, to the amounts of other lipids that are present, see discussion later). Therefore, it is advisable to optimize these parameters for each enzyme source.

Under usual conditions, the dpm in the controls will be ≤10% of that in the complete assay; when more is found, it is usually due to decomposition of the radiolabeled substrate (especially if [³H]sphingosine is used) to products that comigrate with ceramide. Minor modifications of the TLC or HPLC conditions to resolve these compounds may solve the problem, but in some cases, purification of the substrate[8] (or purchase of a different preparation) may be necessary.

Some enzyme preparations have high amounts of endogenous sphingolipids that may alter the specific activity of the radiolabeled substrate. For example, we have found that rat brain microsomes contain 427 ± 74 pmol of endogenous sphinganine and 17 ± 1 pmol of sphingosine/mg protein. Significant amounts of endogenous ceramide and ceramidase can also be present, resulting in increases in endogenous sphingosine during the assay (with rat brain microsomes, 77 ± 44 pmol of sphingosine was "liberated"/mg protein during a 10-min assay) as well as hydrolysis of the product. In most cases, the contribution from these sources is insignificant because only small amounts (μg) of protein are added to the assay. Many microsomal preparations have thioesterases that cleave fatty acyl-CoA, necessitating short assay times. As for any enzymatic assay, initial experiments should verify that expected product is formed, that product formation is linear over the time chosen for the assay, that the rate is proportional to the enzyme amount, and so on.

Kinetic Properties of Ceramide Synthase

Studies of this enzyme have been conducted with cell extracts (mostly microsomes), with the exception of one study with a preparation of ceramide synthase purified 99-fold from a bovine liver mitochondrial-enriched fraction.[12] The apparent K_m values for sphingosine and sphinganine with this preparation were 171 and 144, μM, respectively, and for different fatty acyl-CoA (with sphingosine as the cosubstrate) were 141 μM for palmitoyl-CoA (C16:0), 146 μM for stearoyl-CoA (C18:0), and 299 μM for behenoyl-CoA (C22:0). The V_{max} with these substrates ranged from 3 to 11 nmol ceramide formed/min/mg of protein, with a somewhat higher V_{max} with sphingosine (11 nmol/min/mg protein) than sphinganine (9 nmol/min/mg protein).

These estimates are higher than those seen in brain microsomes, where the apparent K_m values for sphinganine and stearoyl-CoA were 35 and 3 μM.[13] This is probably due to the high amounts of lipids in their assay, which contains liposomes prepared from 170 to 200 μg of mitochondrial lipids,[12] which probably raises the K_m by surface dilution. The relative order of activities with various fatty acyl-CoA also differs somewhat for the purified enzyme[12] versus microsomes; for the purified enzyme the order is stearoyl-CoA > palmitoyl-CoA = oleoyl-CoA > behenoyl-CoA in a ratio of 4:3:3:1, whereas two studies with microsomes found the ranking

[12] H. Shimeno, S. Soeda, M. Sakamoto, T. Kouchi, T. Kowakame, and T. Kihara, *Lipids* **33**, 601 (1998).

[13] A. H. Merrill, Jr., G. van Echten, E. Wang, and K. Sandhoff, *J. Biol. Chem.* **268**, 27299 (1993).

stearoyl-CoA > lignoceroyl-CoA > palmitoyl-CoA > oleoyl-CoA $(60:12:3:1)^{14}$ and stearoyl-CoA = lignoceroyl-CoA ≫ palmitoyl-CoA $(7:7:1)$.[13] This suggests that there may be a mitochondrial isoenzyme that prefers fatty acyl-CoA of 16 and 18 carbon atoms and a microsomal enzyme that is most active with fatty acyl-CoA with ≥18 carbon atoms. It has been reported that the properties of the activities that utilize stearoyl-CoA and lignoceroyl-CoA differ considerably, which further indicates that there may be more than one enzyme.[15]

D-*erythro*-Sphinganine and sphingosine are acylated by both the purified mitochondrial enzyme[12] and microsomes[13] with similar maximal activities. However, on the basis of V_{max}/K_m, the purified enzyme has sphingosine slightly favored over sphinganine (0.066 versus 0.059),[12] whereas the microsomal preparations show a higher V_{max}/K_m for sphinganine (0.248) than sphingosine (0.090).[13] D-*erythro-cis*-Sphingosine is also acylated (with a V_{max}/K_m of 0.057),[2] which suggests that the microsomal ceramide synthase has little selectivity for the conformation at positions 4 and 5 of the sphingoid base.

Some studies[16] failed to show stereoselectivity toward the long chain base; however, others clearly show a preference for the naturally occurring *erythro* forms of sphingosine and sphinganine.[2,17,18] For example, the V_{max}/K_m for D-*erythro*-sphinganine is 0.248 versus 0.013 for L-*threo*-sphinganine.[18] Nonetheless, all four stereoisomers are acylated, with the following activities relative to D-*erythro* sphingosine (100%): D-*threo*-sphingosine (34%), L-*erythro*-sphingosine (13%), and L-*threo*-sphingosine (14%) (the sphinganine stereoisomers are also acylated).[2] Noting that DL-*threo*-sphinganine is sometimes used as an inhibitor of sphingosine kinase,[19] it should be kept in mind that this results in synthesis of *N*-acyl-DL-*threo*-sphinganine (and presumably some inhibition of the acylation of endogenous sphingoid bases). Other structure–function studies have shown that removal of the 1-hydroxyl group has a relatively minor effect on activity.[18]

Subcellular Localization

As already stated, activity is found in both microsomes and mitochondria, although, until recently,[12] mostly the former has been studied. The

[14] P. Morell and N. Radin, *J. Biol. Chem.* **245,** 342 (1970).

[15] M. Ullman and N. Radin, *Arch. Biochem. Biophy.* **152,** 767 (1972).

[16] M. Sribney, *Biochim. Biophys. Acta* **125,** 542 (1966).

[17] T. Sagisaka, M. Nakano and Y. Fujino, *Agric. Biol. Chem.* **36,** 1983 (1972).

[18] H. U. Humpf, E. M. Schmelz, F. I. Meredith, H. Vesper, T. R. Vales, E. Wang, D. S. Menaldino, D. C. Liotta, and A. H. Merrill, Jr., *J. Biol. Chem.* **273,** 19060 (1998).

[19] B. M. Buehrer and R. M. Bell, *Adv. Lipid Res.* **26,** 59 (1993).

active site(s) of the microsomal ceramide synthase is localized to the cytosolic aspect of the endoplasmic reticulum, which is the same as for the earlier enzymes of this pathway as well as with the side where fatty acyl-CoA is acquired.[20,21] The intracellular trafficking of ceramide has mostly been studied with fluorescent sphingolipids[22] and appears to be sensitive to the stereochemistry of the sphingoid base.[23]

Inhibitors of Ceramide Synthase

Ceramide synthase is inhibited by a variety of microbial secondary metabolites, including fumonisins (the most prevalent of which is fumonisin B_1, FB_1),[24] *Alternaria* toxins,[25] and australifungins.[26] FB_1 has been studied the most extensively because it is responsible for a number of diseases of veterinary animals and humans[27] and is commercially available for use to manipulate sphingolipid metabolism by cells in culture.[28]

FB_1 inhibits ceramide synthase *in vitro* with an apparent K_i in the range of 0.05 to 0.1 μM, depending on the concentrations of sphinganine (or sphingosine) and the fatty acyl-CoA.[28] The inhibition of ceramide synthase by FB_1 is competitive versus the long chain (sphingoid) base and mixed with the fatty acyl-CoA,[13] and it appears that fumonisins interact with the binding sites for both substrates.[38] Consistent with this model, removal of the tricarballylic acid side chain (which mimics the CoA moiety in charge), converts FB_1 from merely an inhibitor of ceramide synthase to a substrate.[18]

Sphingolipid metabolism is inhibited by FB_1 in, as far as we are aware, every eukaryotic cell type that has been tested,[28] although the concentrations (usually 25 to 50 μM FB_1 is effective) and conditions that are needed for inhibition vary among different cell types. Incubation with FB_1 blocks the incorporation of radiolabeled serine into the sphingoid base backbone of (dihydro)ceramides and complex sphingolipids and decreases total cellular

[20] E. C. Mandon, I. Ehses, J. Rother, G. van Echten, and K. Sandhoff, *J. Biol. Chem.* **267,** 11144 (1992).

[21] K. Hirschberg, J. Rodger, and A. H. Futerman, *Biochem. J.* **290,** 751 (1993).

[22] A. G. Rosenwald and R. E. Pagano, *Adv. Lipid Res.* **26,** 101 (1993).

[23] J. W. Kok, M. Nikolova-Karakashian, K. Klappe, C. Alexander, and A. H. Merrill, Jr., *J. Biol. Chem.* **272,** 21128 (1997).

[24] E. Wang, W. P. Norred, C. W. Bacon, R. T. Riley, and A. H. Merrill, Jr., *J. Biol. Chem.* **266,** 14486 (1991).

[25] A. H. Merrill, Jr., E. Wang, D. G. Gilchrist, and R. T. Riley, *Adv. Lipid Res.* **26,** 215 (1993).

[26] S. M. Mandala, R. A. Thornton, B. R. Frommer, J. E. Curotto, W. Rozdilsky, M. B. Kurtz, R. A. Giacobbe, G. F. Bills, M. A. Cabello, I. Martin, F. Pelaez, and G. H. Harris, *J. Antibiot.* **48,** 349 (1995).

[27] W. F. O. Marasas, *Adv. Exp. Med. Biol.* **392,** 1 (1996).

[28] A. H. Merrill, Jr., D. C. Liotta, and R. T. Riley, *Trends Cell Biol.* **6,** 218 (1996).

sphingolipid mass. In general, FB_1 causes a more rapid depletion of sphingolipid mass in cells undergoing rapid membrane synthesis and turnover; for example, the amount was reduced by half in growing LLC-PK1 cells,[29] whereas there was no significant change for mouse cerebellar neurons in culture over several days.[13] There is also an accumulation of sphinganine, and sphingosine is sometimes also elevated when cells are undergoing sphingolipid turnover with reacylation of sphingosine. The increased formation of sphingoid bases additionally increases the formation of sphinganine 1-phosphate and the production of substantial amounts of ethanolamine 1-phosphate (plus fatty aldehyde) via sphingoid bases. For example, in FB_1-treated J774 cells, it has been estimated that one-third of the ethanolamine in phosphatidylethanolamine arose from sphingoid base catabolism.[30]

FB_1 has been used extensively for studies of the functions of endogenous sphingolipids; however, results with fumonisins must be interpreted with caution because it also elevates sphingoid bases and the 1-phosphates, which are highly bioactive compounds that might account for observed cell changes. Furthermore, as with most inhibitors, additional targets for FB_1 are being found (protein threonine phosphatase[31] and MAP kinase[32]), which may influence the response of a given system. In fact, the cellular responses to FB_1 that can be most definitively interpreted are those that arise from the accumulation of sphinganine, as these can be differentiated by the upstream inhibition of serine palmitoyltransferase by ISP1.[33] This application even extends to animals, where fumonisins elevate sphingoid bases[34,35] and this can be blocked by ISP1.[36]

[29] H. Yoo, W. P. Norred, E. Wang, A. H. Merrill, Jr., and R. T. Riley, *Toxicol. Appl. Pharmacol.* **114,** 9 (1992).

[30] E. R. Smith and A. H. Merrill, Jr., *J. Biol. Chem.* **270,** 18749 (1995).

[31] H. Fukuda, H. Shima, R. F. Vesonder, H. Tokuda, H. Nishino, S. Katoh, S. Tamura, T. Sugimura, and M. Nagao, *Biochem. Biophys. Res. Commun.* **220,** 1160 (1996).

[32] E. V. Wattenberg, F. A. Badria, and W. T. Shier, *Biochem. Biophys. Res. Commun.* **227,** 622 (1996).

[33] E. M. Schmelz, M. A. Dombrink-Kurtzman, Y. Kozutsumi, and A. H. Merrill, Jr., *Toxicol. Appl. Pharmacol.* **148,** 252 (1998).

[34] E. Wang, P. F. Ross, T. M. Wilson, R. T. Riley, and A. H. Merrill, Jr., *J. Nutr.* **122,** 1706 (1992).

[35] R. T. Riley, E. Wang, and A. H. Merrill, Jr., *J AOAC Int.* **77,** 533 (1993).

[36] R. T. Riley, K. A. Voss, W. P. Norred, C. W. Bacon, F. I. Meredith, and R. P. Sharma, *Environ. Toxicol. Pharmacol.* **7,** 109 (1999).

[4] Dihydroceramide Desaturase

By Heike Schulze, Christoph Michel, and
Gerhild van Echten-Deckert

Introduction

Ceramide is the biosynthetic precursor of sphingomyelin and all glyco-sphingolipids. It anchors them in the outer leaflet of the plasma membrane so that their hydrophilic residues face the extracellular space. Because of its biosynthetic topology, glucosylceramide is, however, anchored in the inner leaflet of the plasma membrane, with its glucose residue facing the cytosol.

The biosynthesis of ceramide requires four sequential steps. The first and rate-limiting step is the condensation of serine and palmitoyl-CoA catalyzed by the pyridoxal phosphate-dependent serine palmitoyltransferase. The 3-dehydrosphinganine formed is immediately reduced by the NADPH-dependent 3-dehydrosphinganine reductase yielding D-erythro-sphinganine. For a long time there has been no conclusive information on the sequence of the next two steps in ceramide biosynthesis. Whether sphinganine is first desaturated to form sphingosine and then acylated to yield ceramide or first acylated to dihydroceramide and then desaturated remained unclear for over a decade. Initial studies suggested that sphingosine is made by the dehydrogenation of sphinganine.[1] Later the dehydrogenation of 3-ketosphinganine was proposed,[2] but in vivo studies using labeled sphinganine and double-labeled dihydroceramide strongly suggested that the desaturation of sphinganine occurs at the level of N-acylsphinganine.[3] [14C]Serine incorporation studies also argued in favor of an introduction of the double bond after addition of the amide-linked fatty acid.[4] Further support[5,6] and a direct demonstration[7] for this pathway came from studies

[1] R. O. Brady and G. Koval, J. Biol. Chem. 233, 26 (1958).
[2] Y. Fujino and M. Nakano, Biochim. Biophys. Acta 239, 273 (1971).
[3] D. E. Ong and R. O. Brady, J. Biol. Chem. 248, 3884 (1973).
[4] A. H. Merrill, Jr., and E. Wang, J. Biol. Chem. 261, 3764 (1986).
[5] E. Wang, W. P. Norred, C. W. Bacon, R. T. Riley, and A. H. Merrill, Jr., J. Biol. Chem. 266, 14486 (1991).
[6] A. H., Merrill, Jr., G. van Echten, E. Wang, and K. Sandhoff, J. Biol. Chem. 268, 27299 (1993).
[7] J. Rother, G. van Echten, G. Schwarzmann, and K. Sandhoff, Biochem. Biophys. Res. Commun. 189, 14 (1992).

with fumonisin B1, a potent inhibitor of sphinganine N-acyltransferase. Thus, in the presence of fumonisin B1, labeled dihydroceramide (but not labeled sphinganine) was desaturated in cultured cells.[7] Although in some biochemistry books, the introduction of the double bond is still sited at the level of sphinganine involving a flavoprotein-dependent dehydrogenation reaction, recent *in vitro* measurements confirm the introduction of the 4,5-*trans*-double bond at the level of dihydroceramide.[8–10] Furthermore, it became clear from the *in vitro* measurements that, in contrast to previous belief, the conversion of dihydroceramide to ceramide is catalyzed by a desaturase, not by an oxidase or a dehydrogenase.[8,9] The reaction catalyzed by dihydroceramide desaturase appears to be particularly important for cells, as ceramide, the product of this reaction, has been shown to alter growth and trigger apoptosis, whereas its saturated analog dihydroceramide, the substrate of this reaction, shows less, if any, of these effects.[11,12] Thus, it seems likely that the enzyme that introduces the double bond plays an important role in the tight regulation of the turnover of ceramide used as a lipid second messenger.

This review gives detailed instruction for the determination of dihydroceramide desaturase activity *in vitro* by measuring the formed ceramide. An alternative method that employs the measurement of formed tritiated water during the desaturation reaction is summarized briefly.

Determination of Dihydroceramide Desaturase Activity *in Vitro* by Measuring the Formed Ceramide[8]

In this assay, first described by Michel *et al.*,[8] protein from the microsomal fraction of rat liver is used as the enzyme source. Note that similar results were obtained with homogenates of rat lung, rat brain, mouse liver, and mouse brain, as well as with homogenates from cultured cells (primary cultured neurons and neuroblastoma cell lines). In principle the enzyme source is incubated with a radioactively labeled dihydroceramide analog as substrate and either NADH or NADPH as cosubstrate. The molecular structure of the substrate used, as well as that of the reaction product, is

[8] C. Michel, G. van Echten-Deckert, J. Rother, K. Sandhoff, E. Wang, and A. H. Merrill, Jr., *J. Biol. Chem.* **272,** 22432 (1997).

[9] L. Geeraert, G. P. Mannaerts, and P. P. van Veldhoven, *Biochem. J.* **327,** 125 (1997).

[10] T. Mikami, M. Kashiwagi, K. Tsuchihashi, T. Akino, and S. Gasa, *J. Biochem.* **123,** 906 (1998).

[11] A. Bielawska, H. M., Crane, D. Liotta, L. M. Obeid, and Y. A. Hannun, *J. Biol. Chem.* **268,** 26226 (1993).

[12] D. A. Wiesner and G. Dawson, *J. Neurochem.* **66,** 1418 (1996).

shown in Fig. 1. Ceramide formed by desaturation of this substrate is detected by autoradiography (Fig. 1, left) after extraction with organic solvents and separation by thin-layer chromatography (TLC).

Preparation of Rat Liver Microsomes

All solutions are prepared a day before use and stored at 0–4°. Male Wistar rats (50–55 days old, 200–250 g from Charles River Wiga GmbH, Sulzfeld, Germany) are starved for 12 hr and killed by decapitation. The livers are excised and rinsed twice with 250 mM sucrose in ice-cold buffer A (100 mM NaH$_2$PO$_4$/Na$_2$HPO$_4$, pH 7.4; sucrose and salts were purchased from Merck, Darmstadt, Germany). To prevent enzyme inactivation, all subsequent procedures are carried out at 0–4°. The livers are weighed, minced with a scapel, added to buffer A (0.5 g of liver/ml), and homogenized with five up-and-down strokes at 600 rpm in a Braun glass homogenizer with a loose-fitting Teflon pestle. The microsomal fraction is obtained by

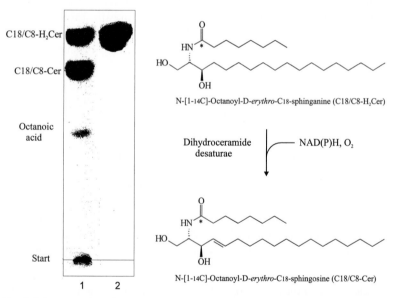

Fig. 1. Substrate and product of the *in vitro* dihydroceramide desaturase assay. Dihydroceramide desaturase was assayed as described in the text using C18/C8-H$_2$Cer as the lipid substrate and intact rat liver microsomes (lane 1) or heat-inactivated microsomes (10 min, 95°, lane 2) as the enzyme source. The lipids were extacted and separated as described. Radioactive spots were visualized by autoradiography. The structures of C18/C8-H$_2$Cer (substrate) and C18/C8-Cer (product) are shown on the right side. The ^{14}C radiolabel is indicated by an asterisk.

successive centrifugation steps. The homogenate is centrifuged at 680g for 10 min. The supernatant is saved, and the pellet is resuspended and centrifuged as described earlier. The combined supernatants are centrifuged at 10,000g for 10 min. The resulting supernatant is saved and recentrifuged at 105,000g for 1 hr. The resulting pellet is resuspended in buffer B (100 mM NaH$_2$PO$_4$/Na$_2$HPO$_4$, pH 8.0) (pellet deriving from 1 g of liver/ml). After recentrifugation at 105,000g for 1 hr, the pellet is finally resuspended in buffer A (pellet deriving from 2.5 g of liver/ml). After freezing in liquid nitrogen, aliquots are stored at −80°. The activity of dihydroceramide desaturase is quite stable at low temperature. Half-maximal enzyme activity could be measured after 1 year, 7 days, or 3 hr when kept at −80, 4, or 37°, respectively.

Preparation of Silica Gel/Borate-Impregnated Thin-Layer Chromatography Sheets

Borate-impregnated TLC sheets are generally used to separate compounds that differ in their grade of saturation.[13] To separate dihydroceramide from ceramide, which differ only in the 4,5-double bond in the sphingosine moiety, we obtained optimal separation with borate-impregnated aluminum TLC sheets 20 × 20-cm silica gel 60 from Merck (Darmstadt, Germany). Prior to use the sheets are dipped into a 70 mM methanolic solution of sodium borate for 30 min and dried at room temperature for 2 hr.

Preparation of ^{14}C-Labeled Substrate (N-[1-^{14}C]Octanoyl-D-erythro-sphinganine, C18/C8-H$_2$Cer)

The labeled substrate is synthesized by acylation of sphinganine (Calbiochem, Bad Soden, Germany) with 1-[^{14}C]octanoic acid with a specific radioactivity of 5.8 Ci/mmol (Amersham, Braunschweig, Germany). Prior to the acylation reaction the acid is activated with N-hydroxysuccinimide and dicyclohexylcarbodiimide, as described previously in this series.[14] The crude product mixture is dried under a stream of nitrogen. Purification involves consecutive TLC separation steps. For all these steps, silica gel 60 glass-backed plates from Merck are used. The product mixture is dissolved in 30 μl of chloroform/methanol (1:1, v/v) and applied to the TLC plate. Chloroform/methanol/water (80:10:1, v/v) is used as the developing system. Detection of radioactive spots is achieved by autoradiography on the bases of R_f values. The product spot is scraped from the TLC plate and

[13] W. R. Morrison, *Biochim. Biophys. Acta* **176,** 537 (1969).
[14] G. Schwarzmann and K. Sandhoff, *Methods Enzymol.* **138,** 319 (1987).

reextracted by stirring with chloroform/methanol (2:1, v/v; approximately 2 ml). Separation from 1-[^{14}C]octanoic acid is achieved by TLC with diethyl either/chloroform (99:2, v/v) as the developing system. The product spot is again scraped from the TLC plate and reextracted with chloroform/ methanol (2:1, v/v).

Solubilization of Lipid Substrate

Method A: BSA–Substrate Complexes. BSA–H$_2$Cer complexes are pre-pared as already described for other lipids.[15] A stock solution containing 15 nmol of substrate in chloroform/methanol (2:1, v/v) is dried under a stream of nitrogen and dissolved in 10 μl of ethanol. Thirty nanomoles of bovine serum albumin (BSA; free fatty acid, from Sigma, Munich, Ger-many) is dissolved in 90 μl of buffer A. The ethanolic solution is added to the dissolved BSA under gentle stirring. The solution is then vortexed vigorously for 30 sec. After 1 min this step is repeated twice. Due to the low ethanol concentration used (10%, v/v), its removal by dialysis can be omitted.

Method B: Solubilization Using Zwitterionic Detergent CHAPS. A stock solution containing 15 nmol of substrate in chloroform/methanol (2:1, v/v) is evaporated under a stream of nitrogen in a 1.5-ml Eppendorf tube. CHAPS (1.1 mg in 10 μl of buffer A) is added, mixed thoroughly, and sonicated for 3 min (Sonifier250, Branson, Danbury, CT).

Method C: Solubilization Using Ethanol/Dodecane (98:2, v/v).[16] A stock solution containing 15 nmol of substrate in chloroform/methanol (2:1, v/v) is evaporated under a stream of nitrogen in a 1.5-ml Eppendorf tube. After dissolving in 1 μl of ethanol/dodecane (98:2, v/v), 99 μl of buffer A is added and the solution is sonicated for 3 min.

Dihydroceramide Desaturase Reaction

The microsomal suspension (stored at −80°) is defrosted at 4°. Assays are conducted in a final volume of 300 μl that contains microsomes (con-taining 600 μg of protein), lipid substrate (15 nmol, solubilized in either way described earlier), and buffer A added to an Eppendorf tube and mixed thoroughly. After preincubation at 37° for 5 min, the reaction is started by adding 1 μmol of NADH in 30 μl of buffer A. After shaking gently at 37° for 60 min, the reaction is terminated by the addition of 200 μl chloroform/methanol (83:17, v/v) on ice.

[15] R. E. Pagano, *Methods Cell Biol.* **29,** 75 (1989).
[16] L. Ji, G. Zhang, S. Uematsu, Y. Akahori, and Y. Hirabayashi, *FEBS Lett.* **358,** 211 (1995).

Lipid Extraction and Quantification

Lipids are extracted by adding 343 μl of methanol and 22 μl of chloroform and mixing vigorously for 20 min. Phases are separated by centrifugation and the lower organic phase is collected. The extraction procedure is repeated twice with 200 μl of chloroform/methanol (83:17, v/v) each. More than 95% of the radioactivity in the reaction mixture could be recovered by this extraction procedure. The combined organic phases are dried under a stream of nitrogen. The dried lipid extract is dissolved in 30 μl of solvent C (chloroform/methanol, 1:1, v/v) and applied to the borate-impregnated TLC sheets (prepared as described earlier) using glass capillaries (length 50 mm, diameter 0.5 mm) from Hilgenberg (Malsfeld, Germany). To avoid edge effects, 2-cm margins are left on each side of the sheet. The empty sample tubes are rinsed twice with 15 and 10 μl of solvent C, respectively, and also applied to the TLC sheet. In the middle of each TLC plate, substrate and N-[1-^{14}C]octanoyl-D-*erythro*-sphingosine are applied as references. TLC sheets are developed with freshly mixed solvent D (chloroform/methanol, 9:1) in a rectangular glass-separating chamber with a polished glass lid from Desaga (Heidelberg, Germany). The solvent is added to the tank to a depth of 1 cm (the margin of the TLC sheet is at least 2 cm). Filter paper is placed on the inside glass of the chamber to help equilibrate the vapor phase before placing the TLC sheet in the tank (see Article [6], Part B, second volume). To improve closure, weights are placed on the lid during development of the TLC. Lipids are visualized and quantified by autoradiography with the bioimaging analyzer Fujix Bas 1000 (from Raytest, Straubenhardt, Germany), using software TINA 2.08. Alternative techniques for the determination of lipids by autoradiography can also be applied (see Article [4], this volume). R_f values for N-[1-^{14}C]octanoyl-D-*erythro*-sphinganine and N-[1-^{14}C]octanoyl-D-*erythro*-sphingosine are 0.7 $\pm$ 0.1 and 0.6 $\pm$ 0.1, respectively. A ΔR_f of at least 0.1 was always observed.

Determination of Dihydroceramide Desaturase Activity *in Vitro* by
    Detecting Formed Water[9]

An alternative way to determine dihydroceramide desaturase activity is described by Geeraert *et al.*[9] In this method, desaturase activity is determined by following the formation of tritiated water that accompanies the 4,5-double bond formation if the substrate is labeled appropriately. The truncated dihydroceramide N-hexanoyl-[4,5-^{3}H]-D-*erythro*-sphinganine is used as substrate and NADPH or NADH as cosubstrate. The conversion of dihydroceramide into ceramide was studied in both intact and permeabilzed cultured rat hepathocytes as well as in rat liver homogenate.

Enzyme Preparation

Male Wistar rats (200 g) are killed by decapitation, and the livers are excised and homogenized in 0.25 M sucrose/5 mM MOPS (pH 7.2) buffer containing 0.1% (v/v) ethanol and fractionated as described in van Veldhoven and Mannaerts.[17]

Preparation of N-Hexanoyl-[4,5-³H]-D-erythro-sphinganine (C18/C6-H₂Cer)

This truncated dihydroceramide is obtained by N-acylation of the respective labeled sphinganine with hexanoic acid as described earlier for N-[1-¹⁴C]octanoyl-D-*erythro*-sphinganine. The 4,5-tritiated sphinganine can be obtained either by hydrolysis of tritiated dihydrosphingomyelin[18] or, more elegantly, by tritiating D-*erythro*-sphinganine as described previously in this series.[14] Tritium-labeled sphinganine is diluted with unlabeled sphinganine to a specific radioactivity of 86.7 μCi/μmol.

Solubilization of Substrate

The substrate is added to the assay as lipid–BSA complex (1:1) (see earlier discussion).

Dihydroceramide Desaturase Reaction

Two hundred microliters of homogenate or subcellular fraction (derived from 200 μg of tissue and diluted appropriately in homogenization medium) is added to 800 μl of reaction mixture. The reaction mixture contains 40 μM N-hexanoyl-[4,5-³H]-D-*erythro*-sphinganine (1:1 complex with BSA), 2 mM NADPH, 20 mM bicine, pH 8.5, 50 mM NaCl, and 50 mM sucrose. The mixture is incubated for 20 min at 37°. The reaction is terminated by the addition of 100 μl of 8% BSA (w/v), immediately followed by 100 ml of 72% trichloroacetic acid (w/v). To remove denatured protein, the reaction mixture is centrifuged at 1100g for 20 min at 4°. To 800 μl of the supernatant is added 300 μl of 1 M Na₂HPO₄ to bring the pH to 5.5, and then the supernatant is passed over a Varian Bond Elut C₁₈ column (500 mg). The flow-through fraction and a wash fraction of 2 ml of water are collected and radioactivity is determined.

---

[17] P. P. van Veldhoven and G. P. Mannaerts, *J. Biol. Chem.* **266,** 12502 (1991).
[18] P. De Ceuster, G. P. Mannaerts, and P. P. van Veldhoven, *Biochem J.* **311,** 139 (1995).

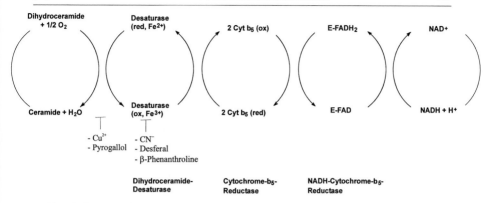

FIG. 2. Postulated model for the dihydroceramide desaturase complex. The sites and effectors known to interfere with conversion of dihydroceramide to ceramide are indicated.

Conclusions

The assay described here for the determination of dihydroceramide desaturase activity *in vitro* enabled us[8,19] and others[9,10] to characterize this enzyme further. The requirement for an electron donor (rather than an acceptor) suggests that conversion of dihydroceramide to ceramide is catalyzed by a desaturase rather than by a dehydrogenase or an oxygenase. Moreover, molecular oxygen was shown to be the electron acceptor. Therefore, it has been proposed that dihydroceramide desaturase, like other well-known desaturase systems, might involve a series of coupled reactions that transport electrons from NAD(P)H to a terminal desaturase that reduces oxygen (Fig. 2). Different effectors known to interfere with the activity of Δ^9-stearoyl-coenzyme A desaturase[20] and Δ^1-alkyl desaturase[21] were shown to also decrease dihydroceramide desaturase activity, as illustrated in Fig. 2. Note that bathophenantrolin sulfonate, an iron(II) chelator, known to inhibit Δ^9-desaturase by removing nonheme bound and catalytically important iron(II), had no effect on dihydroceramide desaturase of rat liver. This finding suggests that dihydroceramide desaturase lacks catalytically important nonheme bound iron(II) or, more likely, that the iron is not accessible to this chelator. Because all known desaturases receive the required electrons from the cytochrome b_5 electron transport system localized

[19] C. Michel and G. van Echten-Deckert, *FEBS Lett.* **416,** 153 (1997).
[20] F. Paltauf, *in* "Ether Lipids: Biochemical and Biochemical Aspects" (H. K. Mangold and F. Paltrauf, eds.), p. 107. Academic Press, New York, 1983.
[21] N. Oshino, Y. Imai, and R. Sato, *Biochim. Biophys. Acta* **128,** 13 (1966).

on the cytosolic face of the ER membrane, it is not suprising that dihydroceramide desaturase shares this localization and topology.[19] Furthermore, all
three enzymes leading to the formation of dihydroceramide are also active
at the cytosolic face of the ER.[22] Thus, the substrate dihydroceramide is
directly accessible to the desaturation enzyme. A certain disadvantage of
both assays described in this article is the employment of unnatural semi-
truncated dihydroceramide analogs to improve solubility in the aqueous *in
vitro* system. In our hands, the *in vitro* activity in rat liver microsomes
decreased as the chain length of the amide-linked fatty acid of dihydroceramide was increased (i.e., C18/C8 > C18/C12 > C18/C18). In fetal rat skin
and liver homogenates, however, C18/C14-H_2Cer was a better substrate
for desaturation than the dihydroceramide analogs containing fatty acids
with 18, 10, 6, or 2 carbon atoms[10] (the latter two were not desaturated
at all). Unfortunately, C18/C8 H_2Cer was not tested in this system. An
explanation for this difference could be either the slightly changed assay
conditions used or differences in the substrate specificity of fetal and adult
tissues. The stereochemistry of the sphinganine moiety of the substrate
appears to have a large effect on enzyme activity. Desaturation of the D-
erythro isomere was much higher than that of the L[8] or D[10]-*threo* isomeres.
Whereas sphinganine and dihydroglucosylceramide were not desaturated
in vitro, dihydrosphingomyelin (D-*erythro*-C18/C8-H_2SM) was a relatively
good substrate, yielding 20% of the activity of the D-*erythro*-C18/C8-H_2Cer,
suggesting that cells could minimize the amount of free ceramide that is
produced during membrane biogenesis by utilizing the hypothetical pathway H_2Cer → H_2SM → SM. The existence of this pathway *in vivo* seems
unlikely, as in contrast to the cytosolic topology established for desaturation,
sphingomyelin formation has been assigned to the luminal side of Golgi
membranes.[23] However, mechanisms triggering translocation of (H_2)SM
across the cellular membranes can not be excluded at the present time. The
in vitro measurement of dihydroceramide desaturase not only delineates the
pathway for the introduction of the 4,5-*trans* double bond of sphinganine,
but also enables more sophisticated studies concerning the regulation of
de novo biosynthesis of ceramide and sphingomyelin.

[22] E. C. Mandon, I. Ehses, J. Rother, G. van Echten, and K. Sandhoff, *J. Biol. Chem.* **267**,
11144 (1992).
[23] A. H. Futerman, B. Stieger, A. L. Hubbart, and R. E. Pagano, *J. Biol. Chem.* **265**, 8650 (1990).

[5] Assays for the Biosynthesis of Sphingomyelin and Ceramide Phosphoethanolamine

By MARIANA NIKOLOVA-KARAKASHIAN

Introduction

Sphingomyelin (*N*-acyl sphingosine phosphorylcholine, SM) is the primary sphingophospholipid in mammalian cells. As much as two-thirds of the total cellular sphingomyelin resides at the plasma membrane, where it plays important structural and functional roles. Membrane structural order, fluidity and asymmetry, cholesterol transport, esterification and *de novo* synthesis, and membrane vesiculation and trafficking are related to the level of SM in plasma membranes. SM is also the major source of bioactive sphingolipid "second messengers," ceramide and sphingosine, which are involved in cellular responses to stress and inflammation. Although sphingomyelin turnover has been studied intensively since the mid-1980s, little is known about the enzymes that carry out its synthesis.

Overview of Enzymes of Sphingomyelin Synthesis

Synthesis of Sphingomyelin

The major pathway for the synthesis of sphingomyelin is by transfer of the phosphorylcholine group from phosphatidylcholine to ceramide, yielding diacylglycerol and sphingomyelin (Fig. 1). This reaction is catalyzed by phosphatidylcholine : ceramide cholinephosphotransferase (PC; ceramide-PCh transferase), also known as sphingomyelin synthase.[1,2] CDP-choline can act as a phosphorylcholine donor *in vitro;* however, the reaction prefers an unnatural analog of ceramide in which the sphingoid base is in a *threo* configuration and is acylated by a short chain fatty acid.

Characterization of SM synthase has been difficult because this enzyme is an integral membrane protein and even mild detergent treatment causes loss of activity. Treatment of membranes with phospholipase C (PlaseC) has a similar effect; however, the loss is reversible on restoration of the membrane bilayer.[3] The enzyme mechanism is intramembranous because

[1] M. D. Ullman and N. Radin, *J. Biol. Chem.* **249,** 1506 (1974).

[2] W. D. Marggraf, H. Diringer, M. D. Koch, and F. A. Anderer, *Hoppe Zeyler's Z. Physiol. Chem.* **353,** 1761 (1972).

[3] W. D. Marggraf and J. N. Kanfer, *Biochim. Biophys. Acta* **897,** 57 (1987).

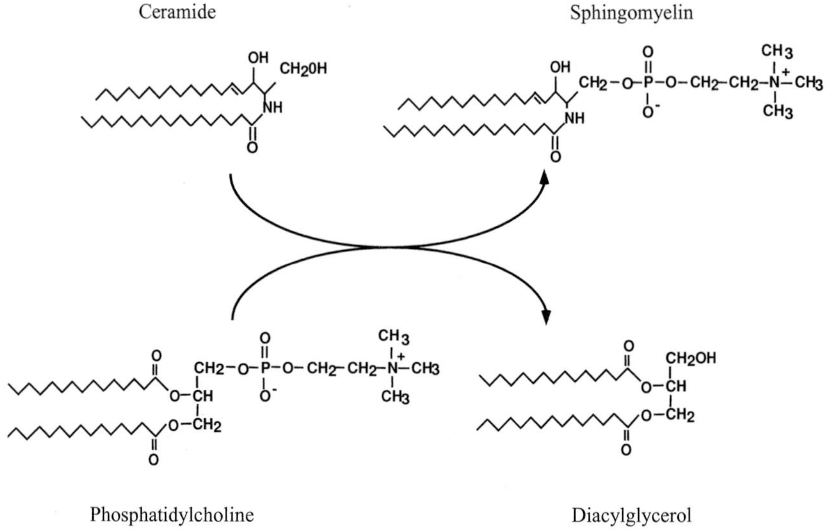

Ceramide   Sphingomyelin

Phosphatidylcholine   Diacylglycerol

FIG. 1. Substrates and products of PC:ceramide-PCh transferase.

incorporation of the substrates into the surrounding bilayer is required for the reaction to occur.[1] Exogenous PC and ceramide are poor donors or acceptors of the phosphocholine group. Kinetic studies on PC:ceramide-PCh transferase suggest a ping-pong mechanism and existence of phospho-choline–enzyme intermediate.[1–5]

When measured *in vitro,* PC:ceramide-PCh transferase has a sharp pH maximum. In Tris buffer, it is between pH 7.0 and 7.4, whereas in imidazole buffer it is shifted to pH 6.5.[1] The addition of EDTA, Ca^{2+}, Mg^{2+}, or Zn^{2+} does not affect the rate of the reaction. There are conflicting reports on the effects of Mn^{2+}, which has been found either to be ineffective or to induce twofold activation.

PC:ceramide-PCh transferase activity is present in isolated plasma membranes[4] and microsomal preparations.[1] There is controversy, however, as to which organelle, plasma membrane, or Golgi is the major site of SM synthesis *in vivo. In vitro,* the activity in plasma membranes exceeds that in Golgi; however, the latter is stimulated significantly by adding ceramide,[5] demonstrating that the availability of ceramide in Golgi is a rate-limiting factor in the reaction. Accordingly, treatment of cultured cells with 1-phenyl-2-decamoylamino-3-morpholino-1-propanol · HCl (PDMP), an in-hibitor of glucosyltransferases, stimulates the synthesis of SM, which sug-

[4] D. R. Voelker and E. P. Kennedy, *Biochemistry* **21,** 2753 (1982).
[5] W. D. Marggraf, R. Zertani, F. A. Anderer, and J. N. Kanfer, *Biochim. Biophys. Acta* **710,** 314 (1982).

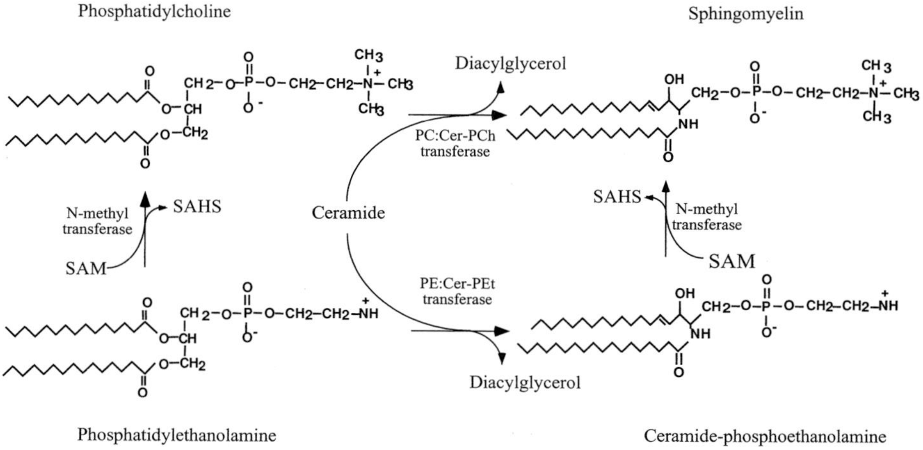

FIG. 2. A scheme relating synthesis of ceramide phosphoethanolamine and sphingomyelin.

gests that SM synthase and complex sphingolipid synthases compete for the pool of ceramide.

Active synthesis of sphingomyelin has been reported in kidney, lung, liver, spleen, and heart. Surprisingly, the brain, an organ that is a rich source of SM, has little PC : ceramide-PCh transferase activity, which raises an intriguing question about the pathway of SM synthesis in brain.

Synthesis of Ceramide Phosphoethanolamine

An alternative pathway for sphingomyelin synthesis is via transfer of a phosphoethanolamine group from phosphatidylethanolamine to ceramide with the formation of ceramide phosphoethanolamine and diacylglycerol (PE : ceramide-PEth transferase).[1,6,7] The former is further converted to sphingomyelin by methylation in a reaction analogous to the methylation of phosphatidylethanolamine (PE) to PC by S-adenosylmethionine (Fig. 2). This pathway has been shown in isolated membrane fractions from rat brain and liver using ethanolamine-labeled dioleoylphosphatidylethanolamine (DOPE) as a substrate. The activity in brain synaptosomes or microsomes is higher when compared to that in liver; however, the activity has not been tested in other organs. The PE : ceramide-PEth transferase activity has properties similar to that of PC : ceramide-PCh transferase: it is an integral membrane protein, requires lipid bilayer for activity, and appears to be cation independent.

[6] M. Malgat, A. Morice, and J. Baraud, *J. Lipid Res.* **27**, 251 (1986).
[7] B. A. Muehlenberg, M. Sribney, and M. K. Duffe, *Can. J. Biochem.* **50**, 166 (1972).

In Vitro Assay of SM Synthase

Isolated plasma membranes or microsomal vesicles are used as the enzyme source in most assays of SM synthase. Methods for the isolation and purification of these fractions are published elsewhere,[1,8] The intramembranous nature of the enzyme and its sensitivity to detergent pose two major obstacles for the *in vitro* assay of SM synthase: (i) delivery of the exogenous hydrophobic substrate (either ceramide or PC) into an aqueous solution without detergent and (ii) incorporation of the substrate into the membrane.

Use of Radiolabeled Substrate

Either radiolabeled PC or radiolabeled ceramide can be used as the exogenous substrate. Only dipalmitoyl-[^{3}H-methyl]phosphatidylcholine, however, is available commercially (Amersham) and is not the optimal substrate for SM synthase, which prefers unsaturated over saturated species of PC. Radioactive PC containing predominantly unsaturated fatty acids (or radioactive ceramides) can be prepared. Preparation of radioactive ceramide is significantly easier than that of PC. However, when radiolabeled ceramide is used as a substrate, separation of the final product of the reaction, SM, requires thin-layer chromatography (TLC). In contrast, using labeled PC as a substrate allows measurement of the formation of SM by phase partitioning.

Preparation of Radiolabeled Substrates. Many cell lines (BHK21, fibroblasts, and hepatomas) have been used to prepare radiolabeled PC.[4] Cells are incubated with [^{3}H]choline (1.5 mM, specific activity of 80 Ci/mM) for 5 hr in choline-free medium. The cells are harvested and the lipids are extracted by the standard Bligh and Dyer procedure. The total lipid extract is treated with 0.5 units of SMase (*Staphylococcus aureus*) to degrade SM and, after reextraction, the remaining lipids are subjected to preparative TLC in chloroform : methanol : acetic acid : water (50 : 25 : 8 : 2, by volume). Spots corresponding to PC are scraped and eluted from the silica with chloroform : methanol (2 : 1, by volume), and the clear supernatant is concentrated.

On a larger scale, radiolabeled PC is prepared by injecting rats intravenously with 200 μCi of methyl-[^{14}C]choline chloride (specific activity of 56 mCi/mmol., Amersham) 1 hr before decapitation. Specific organs, such as liver or lung, are harvested and homogenized, and the lipids are extracted as described earlier. Radiolabeled PC is isolated further by column chromatography. The specific activity of the synthesized PC is calculated after

[8] A. L. Hubbar, D. A. Wall, and A. Ma, *J. Cell Biol.* **96,** 217 (1983).

measuring the amount of radioactivity and the mass of phosphate in an aliquot of the PC stock. Phosphate measurements are made by the method of Kohavkova and Odavich[9] as follows.

A small aliquot of PC is applied to a TLC plate, sprayed with 50% H_2SO_4, and baked in an oven for at least 1 hr at 180°, and the dark brown spot is scraped and put in a glass tube. One milliliter of Hanh's solution and 4 ml of water are added and the sample is boiled for 30 min. After cooling to room temperature, the absorbance is measured at 700 nm. For proper calibration of the assay, blank and phosphate standards must be run simultaneously with the sample. Hanh's solution is prepared by mixing 6.84 g $NaMoO_4$ with 400 mg of hydrazine sulfate and 100 ml of water. One hundred milliliters of concentrated H_2SO_4 is added, and the volume is brought to 1 liter. The sensitivity of this method is approximately 2–3 nmol of lipid-bound phosphate. If greater sensitivity is needed, the Ames method[10] can be used.

The alternative substrate for PC:ceramide-PCh transferase, [^{14}C]hexanoic ceramide, is prepared from the N-hydroxysuccinimide ester of 1-[^{14}C]hexanoic acid (Sigma).[11] Radiolabeled C2-ceramide can also be used as substrate. This compound is synthesized easily by acylation of sphingosine with [^{3}H] $(CH_3CO)_2O$.[12]

Delivery of Substrate to Membrane and Incubation Procedures. As mentioned earlier, the successful assay of PC:ceramide-PCh transferase requires incorporation of the exogenous substrate into the membrane bilayer, which can be achieved by any of the following methods.

PC-SPECIFIC LIPID TRANSFER PROTEIN (LTP). The transfer of PC from PC-containing vesicles to biological membranes is facilitated in the presence of PC-specific LPT. A method for partial purification of PC-specific LTP and its application for PC:ceramide-PCh transferase activity assay has been described by Voelker and Kennedy.[13]

BOVINE SERUM ALBUMIN (BSA). The most widely used method to deliver exogenous substrate for PC:ceramide-PCh transferase is by supplying the substrate as a complex with fatty acid-free BSA (1:1 by molar). To prepare the complex, 0.5 ml of 1 mM BSA stock solution in PBS is mixed with 0.01 ml of a 50 mM stock solution of radilabeled ceramide in ethanol. The solution is mixed vigorously, incubated at 37° for 30 min, and used immediately. The specific activity of the complex may vary depending

[9] J. Kohavkova and R. Odavic, *J. Chromatogr.* **40,** 90 (1969).
[10] B. N. Ames, *Methods Enzymol.* **8,** 115 (1966).
[11] G. Scharzman and K. Sandhoff, *Methods Enzymol.* **138,** 319 (1987).
[12] P. Herold, *Helv. Chim. Acta* **71,** 354 (1988).
[13] D. R. Voelker and E. P. Kennedy, *Methods Enzymol.* **98,** 596 (1983).

on the specific activity of ceramide; however, values around 10,000 dpm/ nmol have been shown to give reproducible results.[14]

The incubation buffer contains 50 mM Tris–HCl, pH 7.4, 25 mM KCl, 0.5 M EDTA, and 20 μl of BSA : ceramide mixture (final ceramide concentration is 20 nmol per incubation), and 0.150–0.300 mg of membrane protein for a final volume of 0.5 ml. After incubation for 1 hr at 37° in a shaking water bath, the reaction is stopped by adding 3 ml of chloroform : methanol (1 : 2 by volume), and the lipids are extracted by the standard Bligh and Dyer procedure. The amount of radioactive SM generated during the reaction is determined after its separation from nonreacted ceramide by TLC in chloroform : methanol : CaCl$_2$ (60 : 35 : 8, by volume). The SM is identified by autoradiography, and the respective spots are scraped and counted with correction for quenching by the silica. The same protocol can be used to assay the activity toward exogenous PC, in which case 50–100 nmol of PC (0.05–0.1 ml from the lipid : BSA complex) is used per sample.

SMALL UNILAMELLAR VESICLES (SUV). Even without LPT or BSA, lipid molecules are incorporated into a membrane bilayer through spontaneous transfer from donor vesicles or by fusion between the acceptor membrane and the vesicle. The efficiency of this process may be increased by extending the incubation time (up to 5 hr) and by increasing the concentration of exogenous substrate in the donor vesicles. This approach has been applied successfully in microsomal and plasma membranes from rat liver and brain.[6] The protocol is as follows.

For the preparation of small unilamellar PC-containing vesicles for 10 assays, dry 1000 nmol of ^{14}C-labeled DPPC (specific activity of 27 mCi/ mmol) or "home-made" radiolabeled PC (specific activity of approximately 1500 cpm/nmol) under nitrogen and resuspend in 1 ml of 50 mM Tris–HCl, pH 7.4, containing 5.7 μg/ml Triton X-100. Detergent is required to facilitate formation of the SUV; however, it should be kept minimal to avoid denaturing the transferase. During the assay, the final detergent concentration is 0.002%. The SUV are dispersed in a probe-type ultrasonicator and the efficiency of the dispersion is verified by counting the amount of radioactivity in a small aliquot of the suspension.

Membrane fractions (0.25–0.35 mg protein) and 0.1 ml of the substrate suspension are incubated in a final volume of 0.35 ml of 50 mM Tris–HCl, pH 7.4, 0.25 M sucrose, and 0.15 mM KCl for 3 hr. The reaction is stopped by adding 1 ml of 0.3 N NaOH in methanol to convert the nonreacted PC into water-soluble compounds, whereas the product of the reaction, SM, remains chloroform soluble. After incubation in a water bath at 37°, the

---

[14] C. Luberto and Y. A. Hannun, *J. Biol. Chem.* **273,** 14550 (1998).

solution is neutralized with molar equivalents of HCl for 1 hr, and 2 ml of chloroform : methanol (2 : 1 by volume) is added. The chloroform-soluble lipids are washed twice with 1 ml of 50 mM KCl. After each addition, the samples are vortexed, the phases are separated by centrifugation, and the upper phase is discarded. The lower phase is transferred to scintillation vials and, after evaporation of the chloroform, the radioactivity in SM is determined using a scintillation counter. When initially establishing this procedure, it is important to analyze the chloroform phase for the presence of nondegraded PC by TLC using chloroform : methanol : ammonia : water (72 : 48 : 2 : 9 by volume) as a developing system. The addition of carrier SM (10 nmol/sample) increases recovery of the product.

Using NBD-Ceramide as Substrate

A relatively easy way to measure SM synthase activity is to use NBD-ceramide (Molecular Probes, Matreya) as a substrate. The following protocol uses isolated plasma membranes from rat liver as an enzyme source. The amount of reagent given is for 10 assays.

Fifty-six nanomoles of NBD-ceramide (from a stock solution in ethanol) and 392 μl of 1% fatty acid free BSA in phosphate-buffered saline (PBS) are dried under nitrogen and resuspended in 600 μl of 50 mM Tris–HCl, pH 7.4, 0.25 M sucrose, and 0.15 mM KCl. The mixture is incubated at 37° for 30 min and placed on ice. After the substrate mixture cools, isolated plasma membranes (0.5 mg) are added and the volume is adjusted to 1 ml with the same Tris, sucrose, and KCl-containing buffer. The mixture is incubated on ice for 10 min and 0.1 ml is transferred to each assay in glass tissue culture tubes containing 0.2 ml of the same buffer. The reaction proceeds for 30 min at 37° and is stopped by adding 2 ml of chloroform : methanol (1 : 2). Lipids are extracted by the standard Bligh–Dyer procedure and the results are analyzed by TLC (glass HPTLC plates, Merck) in a developing system containing chloroform : methanol : acetic acid : water (25 : 15 : 4 : 2 by volume). The NBD-SM is detected under UV light, the corresponding spots are scraped, and the lipid material is extracted from the silica with chloroform : methanol (1 : 2 by volume). To ensure uniformity in the final volume before reading, the samples are dried and resuspended in 1 ml of heptane. The fluorescence is measured with excitation at 455 nm and emission at 530 nm.

Alternatively, the reaction is stopped by the addition of 1 ml of mobile phase (methanol : water : 85% phosphoric acid, 850 : 150 : 1.5 by volume), and the samples are incubated at 37° for 1 hr. The insoluble material is removed by centrifugation. Small aliquots of the clear supernatant are injected onto a reversed-phase column (Nova-Pack, C18, Waters Corp.).

NBD lipids are eluted with the mobile phase of methanol:water:85% phosphoric acid (850:150:1.5 by volume) at a flow rate of 2 ml/min. In this system, the elution time for NBD-ceramide is 10.3 min, whereas NBD-SM elutes at 13.5 min. The NBD fluorescence is analyzed with excitation at 455 nm and emission at 530 nm and the mass of NBD-SM produced is measured by comparison with standard NBD-SM.[15]

The following factors should be considered in optimizing the assay: (i) linearity with the time of incubation (this is particularly important for the longer incubation times) and (ii) dilution of the labeled by unlabeled endogenous substrate. This latter factor is critical for PC:ceramide-PCh transferase because of the limited capacity of the membranes to incorporate exogenous lipids. Control experiments that measure the "coefficient of dilution" are useful. Such experiments can be done by measuring the mass of phosphatidylcholine or ceramide present in the membranes. Alternatively, the effect of dilution by the endogenous substrate can be monitored by the (lack of) linearity between the increase in the amount of membranes used in the assay and the increase in SM production.

In Vitro Assay of Ceramide Phosphoethanolamine Transferase

Analysis of the alternative pathway for SM synthesis carries the same difficulties as the PC:ceramide-PCh transferase assay due to the hydrophobic nature of the substrate, the limitation on detergent use, and the intramembranous nature of the reaction. Furthermore, CPE, the product of the reaction, is not available as a standard, which hampers proper identification of the reaction product.

PE:ceramide-PEth transferase activity is measured using [14C]ethanolamine-labeled dioleoylphosphatidylethanolamine (available from Amersham) supplied as SUV.[6] Because it is in an aqueous phase, PE spontaneously forms inverted hexagonal structures rather than liposomes, and 10% phosphatidic acid is added to help facilitate vesicle formation. Furthermore, an inhibitor of methylation, β-hydroxyethylhydrazine (5 mM), is also added to the incubation to prevent methylation of the substrate by endogenous N-methyltransferases.

To prepare the substrate for 10 assays of PE:ceramide-PEth transferase, 1320 nmol of [14C]ethanolamine-labeled dioleoylphosphatidylethanolamine (specific activity of 0.92 mCi/mmol) and 132 nmol of phosphatidic acid are mixed and dried under nitrogen, resuspended in 1 ml of assay

---

[15] M. N. Nikolova-Karakashian, E. T. Morgan, C. Alexander, D. C. Liotta, and A. H. Merrill, Jr., *J. Biol. Chem.* **272,** 18718 (1997).

buffer containing 5.7 μg/ml Triton X-100, and sonicated using a probe sonicator until a homogeneous suspension is formed. The incubation mixture (final volume of 0.35 ml) contains 0.1 ml of lipid mixture, 0.25–0.35 mg of membrane protein, and 0.035 ml of 50 mM β-hydroxyethylhydrazine in buffer of 50 mM Tris, pH 7.4, 0.25 M sucrose, and 0.15 mM KCl. The incubations are performed for 3 hr in a water shaking bath and the unreacted substrate is cleaved by mild alkaline methanolysis in 1 ml of 0.3 N NaOH in methanol for 1 hr at 37°. The samples are neutralized with an equimolar amount of HCl, and 2 ml of chloroform : methanol (2 : 1 by volume) is added to each tube. The chloroform soluble lipids (newly generated CPE and other endogenous sphingo or ether lipids) are washed with 50 mM KCl and the chloroform-soluble material is either counted with a scintillation counter or analyzed by silica TLC in chloroform : methanol : acidic acid : water (25 : 15 : 4 : 2 by volume). In this system, the R_f values for SPE and SM are 0.36 and 0.18–0.20, respectively.

In Situ Assay of SM Synthesis

Pulse-chase experiments allow measurement of the rate of SM synthesis in intact cells. SM can be labeled by numerous metabolic precursors, such as ceramide, sphingosine, serine, (phosphoryl)choline, (phosphoryl)ethanolamine, S-adenosylmethionine, and acetate, and each method shows different aspects of SM synthesis. Protocols for the three most often used methods to label SM are as follows.

Labeling with NBD-Ceramide

This fluorescent analog of ceramide is taken up by cells readily, transported to the Golgi apparatus, and converted to SM or glycosphingolipids. Hepatocytes, HT-29 cells, and fibroblasts are labeled efficiently with NBD-ceramide (final concentration of 4 μM) as follows: Fresh tissue culture medium (e.g., 25 ml) is transferred to sterile 50-ml tissue culture tubes and covered with parafilm. Ten microliters of NBD-ceramide (from a 10 mM stock solution in ethanol) is injected into the medium under constant vortexing for 1 min and is then added fresh to the cells, replacing the old medium. The cells are cultured for 6 to 12 hr, harvested by a standard procedure, pelleted, resuspended in 0.3 ml of PBS, and transferred to borosilicate glass tubes. After taking an aliquot for protein measurements, 1 ml of the mobile phase (methanol : water : 85% phosphoric acid, 850 : 150 : 1.5 by volume) is added and the tubes are incubated for 1 hr at 37°. The insoluble material is pelleted and the clear supernatant is analyzed by HPLC as described earlier.

Because of the specific transport of NBD-ceramide to the Golgi, the major limitation of this assay is that the method may preferentially represent SM synthesis at this organelle and, to a lesser extent, SM synthesis at the plasma membrane.

Radioactive Choline

In an alternative approach, the rate of sphingomyelin synthesis can be monitored by the rate of incorporation of [³H]choline chloride into SM. This method represents the SM synthesis via PC : ceramide-PCh transferase only. The following protocol is adapted from Luberto and Hannun.[14]

Human fibroblasts (2×10^5 cells per 100-mm petri dish) are cultured for 2 days and the medium is changed to a medium containing choline methyl-³H-labeled choline chloride (specific activity of 85 mCi/mmol, final concentration 0.5 mCi/ml medium). The cells are cultured for an additional 60 hr, scraped, and the lipids extracted by the standard Bligh and Dyer procedure. Incorporation of the label in SM is determined after separation of SM by TLC.

SM synthesis can be stimulated by the addition of bacterial SMase, a treatment that generates ceramide and induces resynthesis of SM. This approach is particularly appropriate for experiments in which the rate of resynthesis of SM in different cell types or treatments is compared. A major disadvantage of this method, however, is that it does not account for the pool of SM generated from PC formed by the "salvage" pathway rather than the classical Kennedy pathway, e.g., PC synthesized via methylation of PE. Similarly, the pool of SM produced as a result of direct methylation of CPE remains unlabeled. The contribution of the methylation-mediated synthesis can be assessed by labeling the cells with *S*-adenosylmethionine. In liver and brain these pathways are very effective.[16]

Radioactive Serine

An alternative approach is to use [¹⁴C]serine labeling.[17] Serine is a common precursor for *de novo* synthesis of all sphingolipids. Typically, the cells are incubated for up to 48 hr in a serine-free medium supplemented with 5 μCi/ml [¹⁴C]serine (specific activity of 25 mCi/mmol). The cells are washed with PBS, harvested, and, after extraction by a standard Bligh–Dyer procedure, the lipids are analyzed by TLC. Incorporation of the label into SM is determined after scraping the spots corresponding to SM and counting

[16] Z. Kiss, *Biochem. J.* **168,** 387 (1997).
[17] E. Wang, W. Norred, C. Bacon, R. Riley, and A. H. Merrill, Jr., *J. Biol. Chem.* **266,** 14486 (1991).

the radioactivity. However, the rate of SM labeling by serine is not very high. In addition, some studies suggest the existence of an "inert" pool of ceramide that is difficult to label *de novo;* rather, it is actively involved in resynthesis of SM.

Regulation of Sphingomyelin Synthesis

Substrate Specificity

The activity of PC:ceramide-PCh transferase for unsaturated PC (such as oleoyl-containing PC) is more than 10-fold higher than for saturated species, such as DPPC. The physiological significance of this specificity is difficult to establish because it has been shown only toward exogenous substrate and may represent a lower rate of incorporation of DPPC into the membrane rather than enzyme selectivity. Nevertheless, studies with lung lamellar bodies, where DPPC accounts for 80% of the total phosphatidylcholine, have shown a high activity of PC:ceramide-PCh transferase,[18] suggesting that the selectivity of SM synthase may contribute to the enrichment of DPPC in lung by clearing unsaturated PC via conversion to SM.

Lipid and Cation Dependence

PC:ceramide-PCh and PE:ceramide-PEth transferase activities are affected by the presence of a lipid bilayer. PC is required for the successful reconstitution of SM synthase in Chinese hamster ovary cellular membranes[19]; in contrast, the enrichment of membranes with phosphatidylserine, phosphatidylinositol, DAG, and lysophosphatidylcholine inhibits PC:ceramide-PCh transferase. Phosphatidylserine has also been shown to inhibit PE:ceramide-PEth transferase.[20] The existence of such inhibitory lipids has led some investigators to suggest that the increase of SM synthase activity in the presence of BSA during an *in vitro* assay is due not only to the more efficient delivery of the substrate, but also to the removal of particular inhibitory lipid factors from the membrane.

The two products of the reaction, SM and DAG, also inhibit enzyme activity. In contrast, an increase in membrane cholesterol (by 30 to 40%) stimulates both PC:ceramide-PCh and PE:ceramide-PEth transferase activity.[21] Depletion of membrane cholesterol appears to have the opposite

[18] J. Lecerf, L. Fouilland, and J. Gaduiarre, *Biochim. Biophys. Acta* **918,** 48 (1987).

[19] K. Hanada, M. Horii, and Yakamazu, *Biochim. Biophys. Acta* **1086,** 151 (1991).

[20] D. Petkova, M. Nikolova, S. Koshlukova, and K. Koumanov, *Int. J. Biochem.* **23,** 689 (1991).

[21] M. Nikolova-Karakashian, N. J. Gavrilova, D. H. Petkova, and M. S. Setchenska, *Biochem. Cell Biol.* **70,** 613 (1992).

effect. The spontaneous transfer of cholesterol between lipid bilayers is relatively fast; therefore, longer incubations of biological membrane with lipid vesicles that do not contain cholesterol may result in partial depletion of membrane cholesterol and inhibition of the activity.

Inhibitors

A phospholipase C inhibitor, D609, was found to inhibit SM synthase.[14] Addition of this inhibitor to both membrane extracts (at a concentration of 50 to 200 μg/ml) or intact fibroblasts (concentration of 25 μg/ml) caused significant decreases in PC:ceramide-PCh transferase activity and in the mass of SM. This finding reopens an intriguing question about the identity of SM synthase and that of PC-specific PLase C. Both enzyme activities have the same substrate and product, PC and DAG, respectively. They differ only in the acceptor of the phosphorylcholine group, which, in the case of SM synthase, is ceramide, whereas for PLase C it is water. Therefore, unless simultaneous measurements of PC, phosphocholine, DAG, and SM mass are performed, it is difficult to distinguish the two activities biochemically. Moreover, an early study has shown that commercial preparations of *Clostridium perfingers* PLase C are capable *in vitro* of synthesizing SM from phosphatidylcholine and *N*-oleoylceramide.[22] Therefore, it is unclear whether some of the cellular functions associated with PLase C activation may, in fact, be manifestations of SM synthase activity.[14]

[22] J. Kanfer and C. Spielvogel, *Lipids* **10,** 391 (1975).

[6] Glucosylceramide Synthase: Assay and Properties

By James A. Shayman and Akira Abe

Introduction

Glucosylceramide is the primary monohexosylceramide (or cerebroside) resulting from the first glycosylation step in the formation of over 200 different glycosphingolipids. Glucosylceramide is synthesized from UDP-glucose and ceramide by a ceramide glycosyltransferase (EC 2.4.1.80). This enzyme, recently sequenced, is distinct from the synthase responsible for the formation of the other major cellular cerebroside, galactosylceramide synthase.[1] The enzyme has an estimated molecular weight of 44,900 and lacks homology with other known glycosyltransferases.

Glucosylceramide synthase was expression cloned from a melanoma cell line, GM-95, which lacked the glycosyltransferase. Although the formation of glucosylceramide and glucosylceramide-based glycosphingolipids was not essential for the growth and survival of this cell line and may not be critically involved in embryonic development, numerous reports have implicated glucosylceramide in critical cellular functions. These include renal growth,[2] neuronal differentiation,[3] the establishment of the water permeability barrier in keratinocytes,[4] and multidrug resistance in cancer cells.[5]

Because glucosylceramide formation is a necessary step in the formation of glycosphingolipids, glucosylceramide synthase is a potential therapeutic target for inherited spingolipidoses such as Gaucher, Fabry and Tay-Sachs disease.[6] Specific inhibitors of glucosylceramide synthase have been developed that block the glycosyltransferase activity and depelete cell and tissue levels of glucosylceramide- and glucosylceramide-based glycosphingolipids.[7] N-Butyldeoxynojirimycin, a nonspecific inhibitor of glucosylceramide synthase, has been reported to block the expression of ganglioside accumulation in a knockout model of Tay-Sachs disease and to reverse the phenotype associated with glycosphingolipid deposition.[8]

Enzyme Assay

Principle

Glucosylceramide synthase catalyzes the transfer of glucose from UDP-glucose to the hydroxyl group at C1 of ceramide. Prior work has demonstrated that octanoylsphingosine is a better glucose acceptor than ceramides of longer acyl chain lengths.[9] In the following assay system, ^{3}H-labeled

[1] S. Ichikawa, H. Sakiyama, G. Suzuki, K. I. Hidari, and Y. Hirabayashi, *Proc. Natl. Acad. Sci. U.S.A.* **93,** 4638 (1996).

[2] J. A. Shayman, *J. Am. Soc. Nephrol.* **7,** 171 (1996).

[3] S. Boldin and A. H. Futerman, *J. Neurochem.* **68,** 882 (1997).

[4] C. S. Chujor, K. R. Feingold, P. M. Elias, and W. M. Holleran, *J. Lipid Res.* **39,** 277 (1998).

[5] Y. Lavie, H. T. Cao, A. Volner, A. Lucci, T. Y. Han, V. Geffen, A. E. Giuliano, and M. C. Cabot, *J. Biol. Chem.* **272,** 1682 (1997).

[6] N. S. Radin, *Glycoconj. J.* **13,** 153 (1996).

[7] N. S. Radin, J. A. Shayman, and J. Inokuchi, *Adv. Lipid Res.* **26,** 183 (1993).

[8] F. M. Platt, G. R. Neises, G. Reinkensmeier, M. J. Townsend, V. H. Perry, R. L. Proia, B. Winchester, R. A. Dwek, and T. D. Butters, *Science* **276,** 428 (1997).

[9] R. R. Vunnam and N. S. Radin, *Biochim. Biophys. Acta* **573,** 73 (1979).

UDP-glucose and octanoylsphingosine, a nonradioactive ceramide, are used as the sugar donor and acceptor, respectively. A single-step extraction procedure is employed to simplify the recovery of radiolabeled glucosylceramide. Unlabeled glucosyloctanoylsphingosine is added during the extraction step to increase the recovery of the radiolabeled product:

Octanoylsphingosine + UDP-[1-^{3}H]glucose

$$\rightarrow 1\text{-}[^3\text{H}]\text{glucosylceramide} + \text{UDP}$$

Reagents

2 mM uridine diphosphate (UDP)-glucose
70 mM NAD$^+$
100 mM dithiothreitol
250 mM Tris–HCl (pH 7.4)
500 mM MgCl$_2$
100 mM EGTA (sodium salt, pH 7)
0.1 mCi/ml (15.3 Ci/mmol) UDP-[1-^{3}H]glucose
5% (w/v) Na$_2$SO$_4$
25 mg/ml dioleoylphosphatidylcholine in chloroform
10 mg/ml octanoylsphingosine in chloroform
2.5 mg/ml sulfatide-sodium in chloroform/methanol (2/1)

Procedures

Preparation of Octanoylsphingosine and Glucosyloctanoylsphingosine. The acylation of sphingosine to form octanoyl sphingosine is carried out by adding 1 mmol octanoyl chloride dropwise to a stirred mixture of 0.8 mmol of sphingosine (acetate salt) in 8 ml tetrahydrofuran and 8 ml 50% sodium acetate (w/v) at 5°. The mixture is stirred for 30 min on ice and then 30 min at room temperature. The product is extracted with ethyl acetate and the organic phase is washed sequentially with water, 2 N HCl, water, sodium bicarbonate, and water. The amide is purified with a silica gel column. Octanoylsphingosine is eluted with either hexane/isopropanol (90 : 10, v/v) or chloroform/methanol (98.5 : 1.5, v/v). Octanoylsphingosine has a melting point of 99–100°, a molecular weight of 427.7, and an R_f of 0.26 when separated on thin layer plates with chloroform/methanol (95 : 5, v/v).

Glucosyloctanoylsphingosine is made from the acylation of glucosylsphingosine.[10] Glucosylsphingosine, the precursor, is made from a modifi-

---

[10] N. S. Radin, *Lipids* **9,** 358 (1974).

cation of the procedure of Taketomi and Yamakawa[11] as described previously.

Preparation of Glucosylceramide Synthase from Tissues and Cultured Cells. Cultured cells are washed twice with phosphate-buffered saline, suspended in ice-cold 0.25 *M* sucrose, 1 m*M* EDTA (pH 7.4), and disrupted with a probe sonicator three times with 10-sec pulses while on ice. The protein concentration is adjusted to 5 mg/ml. For assays of tissue samples, the fresh tissue of interest is homogenized in 9 volumes of water. The homogenate is pipetted into the incubation vessel preloaded with the lipoidal substrate and the assay cocktail.

Preparation of Liposomes. Liposomes consist of dioleoylphosphatidylcholine (70 mol%), octanoylsphingosine (20 mol%), and sulfatide (10 mol%). The constituent lipids are dissolved in chloroform:methanol (1:1, v/v), vortexed, and dried under nitrogen. Water is added to the dried lipids (1 ml per 6 μmol lipid phosphorus). The lipids are dispersed into the water with a probe sonicator for 8 min at 0°.

Glucosylceramide Synthase Assay and Product Extraction. The reaction mixture is placed in 16 × 125-mm screw-capped test tubes and consists of 100 μ*M* UDP-[1-^{3}H]glucose (5000 cpm/nmol), 100 m*M* Tris–HCl (pH 7.4), 10 m*M* MgCl$_2$, 1 m*M* dithiothreitol, 1 m*M* EGTA, 2 mM NAD, liposomes (600 nmol lipid phosphorus), and enzyme source (50–100 μg cell protein) in a total volume of 200 μl. The reaction is initiated by the addition of the enzyme to the reaction cocktail. The reaction is allowed to proceed for 30–60 min at 37° in an ultrasonic bath. The reaction is terminated by the immersion of the reaction tubes in ice and the addition of 1 ml of cold isopropyl alcohol. Use of the ultrasonic water bath has been reported to be advantageous in facilitating the dispersion of the reaction mixture in the test tube. The use of an ultrasonic water bath is probably more important for assays of tissue homogenates where tissue particles tend to adhere to the sides of the test tubes. When cell homogenates are used, however, there is probably little advantage over traditional reciprocating-action water baths where comparably low variability in assay results is observed.

[^{3}H]Glucosyloctanoylsphingosine is extracted by the addition of 0.8 ml Na$_2$SO$_4$ (5%, w/v) and 5 ml of *t*-butyl methyl ether. The mixture is vortexed and then centrifuged for 5 min at 800*g*. Five milliliter of the upper phase is transferred to a scintillation vial using a transfer pipette and dried under a stream of air in a 40° water bath. Five milliliters of scintillation cocktail is added to the dried extract and counted.

The use of *t*-butyl methyl ether allows for a single partitioning step without backwashes. The appearance of the enzyme product in the upper

[11] T. Taketomi and T. Yamakawa, *Jpn. J. Exp. Med.* **37,** 505 (1967).

phase permits transfer with significantly less contamination by the aqueous phase, which is highly radioactive. This permits a high degree of reproducibility.

Properties of Glucosylceramide Synthase

Substrate Specificity and Liposomal Composition

Lipoidal properties of the substrate for glucosylceramide synthase led Vunnam and Radin[9] to consider whether short chain ceramides may serve as better substrates for the cerebroside synthase, perhaps because of an improved ability to be dispersed in an aqueous solution. A series of ceramides differing in acyl chain length were compared. Octanoylsphingosine was found to be a significantly better substrate than decanoyl sphingosine and stearoyl sphingosine. Additionally, octanoyl dihydrosphingosine and decanoyl dihydrosphingosine were comparatively poor substrates. Truncated long chain bases, octanoyl decasphinganine and decanoyl sphinganine, were not significantly better in forming their respective glucosylceramides than those ceramides containing sphingosine as the long chain base.

The lipoidal substrate can be incorporated into a liposome or, alternatively, included in the assay mixture as coated Celite (typically 20 mg Celite Analytical Filter Aid coated by evaporation with 50 μg ceramide). In general, liposomes yield 20% more product than Celite and are easier to pipette. Different compositions of liposomes have not been studied extensively. Phosphatidylcholine is included because it was observed to stimulate glucosylceramide synthase in both brain and kidney.[12] Sulfatide has been reported to stimulate ceramide galactosyltransferase.[13]

Nucleotide Effects

Early comparisons of glucosylceramide synthase activity among tissues revealed a surprisingly low activity in kidney. Shukla and Radin[14] perceptively noted that increasing the amount of kidney homogenate in the incubation mixture did not result in a proportionately greater activity. It was demonstrated that the kidney homogenates contained a nucleotide pyrophosphatase, known to interfere with several glycosyl transferase assays.

[12] P. Morell, E. Constantino-Ceccarini, and N. S. Radin, *Arch. Biochem. Biophys.* **141,** 738 (1970).
[13] A. Brenkert and N. S. Radin, *Brain Res.* **36,** 183 (1972).
[14] G. S. Shukla and N. S. Radin, *Arch. Biochem. Biophys.* **283,** 372 (1990).

Various nucleotides were studied for their ability to compete with the pyrophosphatase. Only nicotinamide nucleotides produced enhancement of the glucosylceramide synthase activity (Table I). NAD, NADH$^+$, NADP, and NADPH$^+$ are all effective at stimulating renal cerebroside synthase. NAD is the best nucleotide, having an optimal concentration of 2 mM.

Effects of Divalent Cations and pH Optimum

Mg^{2+}, Mn^{2+}, and Ca^{2+} are all stimulatory under the basic assay conditions with peak effects at 10 mM. Mg^{2+} has the greatest stimulatory effect. In contrast, ferrous ions, Zn^{2+}, and Cu^{2+} are inhibitory. Under these assay conditions a pH optimum of 7.4 is observed. Doubling the Tris–Cl concentration from 0.1 to 0.2 M is associated with a 58% loss of activity. Conduritol B epoxide, a known inhibitor of the glucosylceramide cerebrosidase, has no effect on measured enzyme activity. Thus under normal assay conditions, the [^{3}H]glucosylceramide formed is not accessible to the glucocerebrosidase.

TABLE I

EFFECT OF ADDED NUCLEOTIDES ON
GLUCOSYLCERAMIDE SYNTHESIS BY
KIDNEY HOMOGENATE[a]

Nucleotide	Percentage of control
ATP	105
UMP	106
UDP-glucuronic acid	105
UDP-galactose	108
ADP-glucose	101
CDP-glucose	87
GDP-glucose	111
CDP-choline	93
NAD	196
NADH$^+$	187
NADP	175
NADPH$^+$	168
NADP + NAD	250

[a] Standard incubations were used without NAD and with 4 mg of kidney. The added nucleotides were at 0.5 mM except for ATP, which was 2 mM. The yield of [^{3}H]glucosylceramide in the control incubations was 259 pmol formed in 30 min.

Enzyme Induction with Glucosylceramide Synthase Inhibitors and Ceramide

A well-studied inhibitor of glucosylceramide synthase is D-*threo*-PDMP [(*R,R*)-1-phenyl-2-decanoylamino-3-morpholino-1-propanol].[15] This compound structurally resembles ceramide with adjacent fatty acylamide and alcohol groups. A phenyl group replaces the long alkenyl chain of sphingosine and a cyclic tertiary amine replaces the primary alcohol group. The addition of PDMP to cultured MDCK cells causes the time-dependent depletion of glucosylceramide and accumulation of ceramide.[16] When studied in MDCK cells, 20 μM PDMP led to the rapid induction of glucosylceramide synthase activity, detectable within 1 hr of incubation.[17]

The induction was inhibited by the prior addition of either actinomycin D or cycloheximide to the culture medium, consistent with the interpretation that this was a transcriptionally regulated change. Because PDMP treatment is associated with both glucosylceramide depletion and ceramide accumulation, further studies were conducted to determine whether the change in glycosyltransferase activity was secondary to substrate accumulation or product depletion.

In order to dissociate these metabolic effects, endogenous ceramide levels were altered in MDCK cells by three additional means. The DL-*erythro*-diastereomer of PDMP may raise ceramide levels through inhibition of an alternative pathway, the formation of 1-*O*-acylceramide.[18] This change occurs in the absence of glucosylceramide depletion. The cell-permeant ceramide *N*-acetylsphingosine raises ceramide through deacetylation and conversion of free sphingosine to long chain ceramides.[19] There is an increase in glucosylceramide levels under these conditions. Bacterial sphingomyelinase raises ceramide levels through the hydrolysis of sphingomyelin. This results in a 170% increase in glucosylceramide content. In each case where ceramide levels increased, there was a significant induction in glucosylceramide synthase activity (Table II).

Not all of the effects of glucosylceramide synthase inhibitors on induction of the glycosyltransferase can be explained based on changes in ceramide content. For example, the addition of cycloserine, an inhibitor of long chain base acylation and ceramide formation in the presence of PDMP,

[15] J. A. Shayman, G. D. Deshmukh, S. Mahdiyoun, T. P. Thomas, D. Wu, F. S. Barcelon, and N. S. Radin, *J. Biol. Chem.* **266,** 22968 (1991).
[16] C. S. Rani, A. Abe, Y. Chang, N. Rosenzweig, A. R. Salteil, N. S. Radin, and J. A. Shayman, *J. Biol. Chem.* **270,** 2859 (1995).
[17] A. Abe, N. S. Radin, and J. A. Shayman, *Biochim. Biophys. Acta* **1299,** 333 (1996).
[18] A. Abe and J. A. Shayman, *J. Biol. Chem.* **273,** 8467 (1998).
[19] A. Abe, J. A. Shayman, and N. S. Radin, *J. Biol. Chem.* **271,** 14383 (1996).

TABLE II
INDUCTIVE EFFECT OF CERAMIDE ON GLUCOSYLCERAMIDE SYNTHASE

Treatment	Change in glucosylceramide synthase specific activity	Change in ceramide concentration	Change in glucosylceramide concentration
DL-*threo*-PDMP	+139%	+56%	−54%
DL-*erythro*-PDMP	+154%	+98%	−14%
N-Acetylsphingosine	+140%	High	High
Sphingomyelinase	+143%	+380%	+170%

leads to an induction of the glucosylceramide synthase. Similar results are obtained for N-butyldeoxynojirimycin treatment, an agent that inhibits the synthase in the absence of significant changes in ceramide. One potential explanation is that inhibitors that bind reversibly to the synthase prevent its degradation. When diluted from cell homogenates in assaying the synthase in the 200-μl reaction volume, a higher specific activity for the glucosylceramide synthase is detected. In support of this interpretation is the observation that no measured induction is observed for cells treated with the palmitoyl and pyrrolidino-substituted homolog of PDMP, 1-phenyl-2-palmitoylamino-3-pyrrolidino-1-propanol. This homolog has a significantly higher affinity for glucosylceramide synthase and does not dissociate from the enzyme during normal assay conditions.

Summary

Glucosylceramide synthesis is a key step in the formation of most mammalian glycosphingolipids. The expanding number of cellular functions that may be glycosphinolipid dependent and the identification of this glucosylceramide synthase as a potential therapeutic target for several sphingolipid storage disorders necessitate the availability of a reliable assay for glucosylceramide synthase. Coupled with the recent sequencing of this enzyme, the liposome-based assay utilizing a single extraction step should aid in the understanding of this critical early pathway in glycosphingolipid formation.

[7] Methods for Studying Glucosylceramide Synthase

By David L. Marks, Pascal Paul, Yasushi Kamisaka,
and Richard E. Pagano

Introduction

Glucosylceramide synthase (GCS), also referred to as giucocerebroside synthase and ceramide glucosyltransferase,[1,2] glucosylates ceramide using UDP-glucose (UDP-Glc) as a hexose donor. Glucosylceramide (GlcCer) is the precursor of most plasma membrane glycosphingolipids. Thus, the regulation of the expression and activity of GCS has relevance to numerous biological and pathological processes (e.g., development, differentiation, tumorigenesis, and host/pathogen interactions) in which glycosphingolipids appear to play a role.[3–5] Further, inhibitors of GCS are being investigated as candidate drugs for the treatment of cancer.[2] Finally, identification of GCS as an integral membrane protein of the Golgi complex[6–8] poses questions concerning the mechanisms of targeting and localization of this enzyme. Thus, it is important to develop techniques for the study of GCS expression, activity, catalytic mechanism, and localization.

Although GCS activity was first identified in 1968,[9] progress in studying this protein has been slow. This difficulty is understandable now that it is known that GCS is an extremely hydrophobic, integral membrane protein that is apparently expressed at very low levels.[10,11] Studies concerning the solubilization and partial purification of active GCS,[11] the cloning of the human GCS sequence,[10] and the development of antibodies against GCS[12]

---

[1] S. Ichikawa and Y. Hirabayashi, *Methods Enzymol.* **311** [33] 1999 (this volume).
[2] J. Shayman and A. Abe, *Methods Enzymol.* **31** [6] 1999 (this volume).
[3] J. Shayman and N. Radin, *Am. J. Physiol.* **260,** F292 (1991).
[4] R. W. Ledeen, G. Wu, Z. H. Lu, D. Kozireski-Chuback, and Y. Fang, *Ann. N.Y. Acad. Sci.* **845,** 161 (1998).
[5] S. Hakomori and Y. Igarashi, *J. Biochem.* **118,** 1091 (1995).
[6] H. Coste, M.-B. Martel, G. Azzar, and R. Got, *Biochim. Biophys. Acta* **814,** 1 (1985).
[7] A. H. Futerman and R. E. Pagano, *Biochem. J.* **280,** 295 (1991).
[8] D. Jeckel, A. Karrenbauer, K. N. J. Burger, G. van Meer, and F. Wieland, *J. Cell Biol.* **117,** 259 (1992).
[9] S. Basu, B. Kaufman and S. Roseman, *J. Biol. Chem.* **243,** 5802 (1968).
[10] S. Ichikawa, H. Sakiyama, G. Suzuki, K. I.-P. J. Hidari, and Y. Hirabayashi, *Proc. Natl. Acad. Sci. U.S.A.* **93,** 4638 (1996).
[11] P. Paul, Y. Kamisaka, D. L. Marks, and R. E. Pagano, *J. Biol. Chem.* **271,** 2287 (1996).
[12] D. Marks, K. Wu, P. Paul, Y. Kamisaka, R. Watanabe, and R. Pagano, *J. Biol. Chem.* **274,** 451 (1999).

have led to significant methodological advances in our ability to study this enzyme. This article deals with these advances in the characterization of GCS.

Analytical Methods

Until recently, GCS could be quantified only in terms of its activity. With the development of GCS antibodies, the GCS polypeptide has been recognized on Western blots and it has been possible to estimate GCS relative enzyme mass. These dual measurements have been used to study the upregulation of GCS protein in differentiating keratinocytes[13] and to allow the potential for studying the relative specific activity of GCS in different fractions. Without the purification of measurable quantities of GCS (see later) it has not been possible to calculate the true specific activity (i.e., nanomole GlcCer formed/mg protein/hr) of GCS.

Enzymatic Activity Assays

Several different enzymatic assays have been described differing mainly in the form of ceramide (or UDP-Glc) substrate presented and the detection of GlcCer formed. These include the incubation of protein samples with UDP-Glc radiolabeled on the glucose moiety,[9,14] radiolabeled short chain ceramide,[7] or fluorescent derivatives of ceramide (also see Article [6] by Shayman and Abe).[11,15] In each of these examples, lipids are extracted and the GlcCer formed is separated from ceramide by thin-layer chromatography. An alternative method uses ceramide immobilized on silica gel beads and UDP-[^{14}C]Glc with the immobilized product ([^{14}C]GlcCer) being separated from UDP-[^{14}C]Glc by centrifugation or filtration of the beads.[16] This article presents an *in vitro* assay for GCS activity using the fluorescent short chain ceramide, N-[7-(4-nitrobenzo-2-oxa-1,2-diazole)]-6-aminocaproyl-D-*erythro*-sphingosine (C_6-NBD-Cer), as a substrate. The assay has been used successfully with nonsolubilized sources of GCS (suspensions of tissue homogenates, lysates of bacterial cells expressing recombinant GCS) and detergent solubilized samples (see later).

Assay for GCS Activity with C_6-NBD-Cer

Preparation of C_6-NBD-Cer Bovine Serum Albumin (BSA) Complexes. Dissolve 1 mg of C_6-NBD-Cer in 1 ml of chloroform:MeOH (1:2, v:v)

---

[13] R. Watanabe, K. Wu, P. Paul, D. L. Marks, T. Kobayashi, M. R. Pittelkow, and R. E. Pagano, *J. Biol. Chem.* **273,** 9651 (1998).

[14] H. Coste, M.-B. Martel, and R. Got, *Biochim. Biophys. Acta* **858,** 6 (1986).

[15] R. E. Pagano and O. C. Martin, *Biochemistry* **27,** 4439 (1988).

[16] N. Matsuo, T. Nomura, and G. Imokawa, *Biochim. Biophys. Acta* **1116,** 97 (1992).

and transfer to a 1.5-ml microfuge tube. Place the open tube under a stream of nitrogen until the C_6-NBD-Cer is visibly dry and then continue to dry under vacuum for $\geq$30 min. Redissolve the C_6-NBD-Cer in 50 μl 100% EtOH. Add C_6-NBD-Cer drop by drop, while vortexing, to 3.42 ml of 0.5 mM defatted BSA in serum-free balanced salt solution [e.g., 10 mM 4-(2-hydroxyethyl)-1-piperazineethanesulfonic acid-buffered minimal essential medium, pH 7.4, without indicator]. The final concentration is 0.5 mM C_6-NBD-Cer. Prepare aliquots (0.5–1.0 ml) and store at $-20°$ until use.

Enzyme Assay. The assay is performed in screw-top test tubes (13 $\times$ 100 mm). If protein samples lack endogenous phosphatidylcholine (e.g., some detergent-solubilized samples), the tubes may be precoated with dioleoylphosphatidylcholine (50 nmol added in chloroform and dried down under nitrogen). To begin the assay, add the following to each tube: 10 μl of C_6-NBD-Cer/BSA complex, protein samples (usually 50 μl or less), and then assay buffer (50 mM HEPES, pH 7.4, 25 mM KCl, 5 mM MnCl$_2$, 2.5 mM UDP-glucose prepared fresh) for a final volume of 500 μl. A mini stir bar is then added to each tube and the samples are incubated at 37° for 20–30 min with stirring.

After removal from heat, 3 ml of chloroform : methanol (1 : 2, v : v), 1 ml chloroform, and 1.3 ml acid saline (9 g NaCl, 1.24 ml concentrated HCl in 1 liter distilled water) are added. The tubes are capped, vortexed, and centrifuged at 2000–3000 rpm for 15 min. The top (aqueous) layer of each tube is then removed by aspiration and discarded. The bottom layer is transferred to 12 $\times$ 75-mm glass tubes and dried under nitrogen. Samples are then redissolved in 25 μl of spotting solvent (19 : 1 chloroform : MeOH) and run on Silica 60 thin-layer chromatography plates in chloroform : methanol : 15 mM CaCl$_2$ (60 : 35 : 8) along with appropriate C_6-NBD-lipid standards.

The fluorescent lipids on the plates are visualized on a UV light box and can be quantified by scraping each spot followed by use of a fluorometer[17] or, more conveniently, by computer analysis of an image of the plate acquired via a video camera.[18] Results are calibrated using similarly run C_6-NBD-GlcCer standards (0.05–1 nmol).

Modifications. The possible addition of phosphatidylcholine to the reaction has been noted earlier. Other phospholipids have been added similarly to determine the specificity and concentration dependence of GCS phospholipid requirements.[11] The inclusion of Mn^{2+}, Mg^{2+}, or Ca^{2+} in the assay buffer stimulates GCS activity by 50–100%, but inclusion of divalent cations is not necessary for activity.[7,16] Various pH optima between 6.5 and 7.8

[17] N. G. Lipsky and R. E. Pagano, *Proc. Natl. Acad. Sci. U.S.A.* **80** (1983).
[18] M. Koval and R. E. Pagano, *J. Cell Biol.* **108**, 2169 (1989).

have been reported for GCS using differing sources of enzyme and assay conditions.[6,7,9,11,16] Assays are generally performed at pH 7.2–7.4 to minimize the hydrolysis of newly synthesized GlcCer by acid glucosylceramidase (β-glucocerebrosidase).[7] Inhibitors of acid glucosylceramidase (e.g., conduritol β-epoxide[19]) or UDP-glucose hydrolysis (e.g., NAD[20]) may also be added to the assay components to minimize competing reactions.

SDS–PAGE and Western Blotting of GCS

We have developed polyclonal antibodies against GCS[12,13] by immunizing rabbits with peptides based on the sequence of human GCS.[10] Antibodies against three different peptides recognized rat and human GCS as a polypeptide migrating with an apparent molecular weight of ~38,000 on SDS–PAGE gels.[10] Although each antibody may have unique characteristics, several general features of our Western blotting technique may be of general use. First, GCS expression is apparently very low in mammalian tissues and cells; we were only able to detect GCS on Western blots in enriched fractions (e.g., Golgi membranes or immunoprecipitates) and not in crude lysates or homogenates.[12] Second, thiol-reducing agents (e.g., dithiothreitol) have no effect on the migration of GCS, but their exclusion from SDS–PAGE sample buffers decreases the background on blots. Finally, GCS is best visualized when the SDS–PAGE sample buffer contains 8 M urea and GCS-containing samples are heated only to low temperatures (e.g, 37°) prior to electrophoresis.

Sources of GCS

Initial studies of GCS activity were performed using crude brain homogenates.[9,21] GCS activity and substrate specificity have also been studied to a limited extent in cultured cells.[15,22] However, most investigations of GCS have utilized various preparations from mammalian tissues.[7,8,11,14,20,23] Although liver probably has been the most widely used source of GCS activity, a survey of male rat tissues shows that GCS activity is most enriched in brain, followed by spleen, testes, and lung, with much lower levels in liver

[19] J. A. Shayman, G. D. Deshmukh, S. Mahdiyoun, T. P. Thomas, D. Wu, F. S. Barcelon, and N. S. Radin, *J. Biol. Chem.* **266,** 22968 (1991).
[20] G. Shukla and N. Radin, *Arch. Biochem. Biophys.* **283,** 372 (1990).
[21] S. Basu, B. Kaufman, and S. Roseman, *J. Biol. Chem.* **248,** 1388 (1973).
[22] N. D. Ridgway and D. L. Merriam, *Biochim. Biophys. Acta* **1256,** 57 (1995).
[23] M. Trinchera, M. Fabbri, and R. Ghidoni, *J. Biol. Chem.* **266,** 20907 (1991).

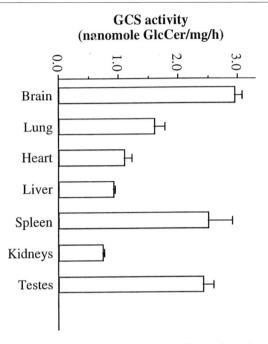

GCS activity
(nanomole GlcCer/mg/h)

FIG. 1. GCS specific activity of rat tissue homogenates. Tissues from six rats were homogenized and assayed for GCS activity with C_6-NBD-Cer for 30 min at 37 ° as described in the text. Equal amounts of protein (200 μg) were used for each sample. Lipids were extracted, separated by thin-layer chromatography, and GlcCer quantified by image analysis as described in the text. Data are means of four measurements.

(Fig. 1). No significant differences in the relative activities of these tissues were found when these assays were performed in the presence of the acid glucosylceramidase inhibitor, conduritol β-epoxide (data not shown), suggesting that the results reflect differences in GCS activity rather than in acid glucosylceramidase activities. We have also detected high levels of GCS in differentiated human keratinocytes.[12,13]

Preparation of Golgi Fractions

Several groups have demonstrated that GCS is localized predominantly in the Golgi apparatus and has a cytosolic orientation.[7,8,14,23] In addition, procedures for subfractionation of Golgi compartments have been used to determine GCS distribution within the Golgi.[7,8] In these studies, Golgi fractions were prepared using established procedures based on the buoyant density of Golgi cisternae or vesicles in sucrose gradients.[24–26] This section describes a simplified variation of the method of Bergeron et al.[26] for preparing Golgi fractions from rat liver. The method has been optimized for maximal yield and enrichment (~60-fold compared to homogenate) of GCS in a minimum of time rather than for purity of Golgi fractions. The technique has only been used for rat liver, but may also be suitable for Golgi preparations from rat brain, spleen, testes, or lung that have higher apparent GCS specific activities than liver (see Fig. 1).

Procedure. Sprague–Dawley rats (5 weeks old) are fasted overnight. The rats are then anesthetized with diethyl ether, decapitated with a guillotine, and their livers perfused *in situ* with ice-cold 0.9 M NaCl. Livers are minced with scissors to pieces of <1 cm, the saline is poured off, and the tissue is homogenized (25%, w/v) in ice-cold 0.25 M sucrose in buffer A (50 mM Tris–HCl, pH 7.4, 25 mM KCl) with protease inhibitors [10 μg/ml each of leupeptin, tosylarginylmethyl ester, and aprotinin; 1 μg/ml each of pepstatin and antipain; and 25 mM 4-amidophenylmethanesulfonyl fluoride (all from Sigma Chemical Co., St. Louis, MO)] using a polytron (Brinkmann Instruments Inc., Westbury, NY) at setting 1 for ~30 sec (or until most large pieces of tissue are gone). All remaining procedures are performed at 4°. The homogenate is then filtered through cheesecloth and then adjusted [150 parts homogenate: 95 parts 2 M sucrose in buffer A (v/v)] to a final concentration of 1.07 M sucrose. The adjusted homogenate (19 ml per tube) is loaded into SW28 tubes (Beckman); 9 ml each of 0.9 and 0.2 M sucrose in buffer A are then sequentially overlaid above the homogenate. The tubes are then ultracentrifuged in an SW28 rotor (Beckman) for 2 hr at 83,000g with no braking during deceleration. Crude Golgi fractions are collected at the 0.2/0.9 M sucrose interface using a 10-ml syringe and a wide-bore needle. Fractions are flash frozen in liquid nitrogen and stored at −80° until needed.

For studies of GCS topology, we prepare right-side out, "intact" Golgi membranes from fresh or frozen Golgi fractions isolated as described earlier.[12] Crude Golgi fractions are mixed at 1:1 (v/v) with buffer B (50 mM HEPES, pH 7.4, 100 mM KCl, 20% glycerol plus protease inhibitors as described earlier), and centrifuged for 90 min at 200,000g. Golgi membrane pellets are resuspended with a pipette tip at one-twentieth of their original volume in 0.25 M sucrose in buffer B, frozen, and stored as described

[24] D. Morré, R. Hamilton, H. Mollenhauer, R. Mahley, W. Cunningham, R. D. Cheetham, and V. Lequire, *J. Cell Biol.* **44,** 484 (1970).

[25] D. Leelavathi, L. Estes, D. S. Feingold, and B. Lombardi, *Biochim. Biophys. Acta* **211,** 124 (1970).

[26] J. J. M. Bergeron, R. A. Rachubinski, R. A. Sikstrom, B. I. Posner, and J. Paiement, *J. Cell Biol.* **92,** 139 (1982).

earlier. Golgi membranes prepared in this way are oriented right-side out and are not "leaky" as demonstrated by the inaccessibility of sphingomyelin synthase (a Golgi lumenal enzyme) to proteases in controlled proteolysis studies.[12]

For further enrichment of GCS in Golgi membranes, we mix crude Golgi fractions at 1 : 1 (v/v) with 0.1% N-lauroylsarcosine in buffer B, stir for 30 min at 4°, and centrifuge as described earlier. The pellets are resuspended at one-twentieth of their original volume in buffer B and are homogenized by five passes each through 18-, 22-, and 26-gauge needles attached to a 5-ml syringe. Fractions are frozen and stored as described earlier. N-Lauroylsarcosine-washed Golgi membranes are typically enriched ~200-fold in GCS.[11]

Expression of Recombinant Mammalian GCS

The recently cloned human GCS was reported to be expressed in active form in bacteria.[10] We cloned rat GCS, expressed the sequence in the pET-3d vector (Novagen, Madison, WI) in *Escherichia coli,* and found it to be highly active (K. Wu, D. L. Marks, and R. E. Pagano, unpublished results). Recombinant GCS was active both in disrupted bacterial cells (e.g, by sonication) or when solubilized in an appropriate detergent (see later). No exogenous phospholipids[11] were required for activity, even though bacteria do not synthesize phosphotidylcholine. Presumably, endogenous phospholipids (e.g., phosphatidylethanolamine) satisfied the phospholipid requirement of GCS.

In these preliminary studies, the level of expression of GCS in bacteria has been extremely low (comparable to the levels present in rat liver Golgi membranes) as assessed by Western blotting. Expression levels of the human GCS sequence in bacteria also appear to be low,[10] judging by activity assays. It is uncertain if these low levels of expression are due to the presence of bacterially rare codons in the GCS sequences,[27] protein instability or toxicity in the bacterial cells, or to other factors. Modifications of the GCS sequence (e.g., elimination of rare codons) may increase the yield of bacterially expressed GCS. Alternatively, recombinant GCS may be expressed in a eukaryotic system (e.g., yeast, insect or mammalian) with higher yields. In either case, the availability of recombinant GCS is sure to provide a new source of the enzyme.

[27] S. Macrides, *Microbiol. Rev.* **60,** 512 (1996).

Detergent Solubilization of GCS

GCS is soluble in a number of detergents with retention of activity (Table I). Low concentrations (0.1–0.2%) of nonionic detergents [Triton X-100, Nonidet P-40 (NP-40)] are compatible with GCS activity, as are concentrations of up to 2% for most cholesterol-like detergents (CHAPS, CHAPSO, sodium cholate, and digitonin). Interestingly, diheptanoylphosphatidylcholine (DHPC) also solubilizes GCS efficiently. This short chain PC thus acts as a detergent, as well as fulfilling the requirement of GCS for a phospholipid cofactor.[11] DHPC, however, does partially inhibit GCS activity at concentrations above 1 mM, as do other phospholipids.[11] In general, we have found CHAPS and CHAPSO to be most useful for the purification of active GCS[11] because of their dialyzability and small micelle size.

Procedure. Add detergent (see earlier discussion) to GCS-containing fraction (e.g., Golgi membranes) at 1 g protein/10 mg detergent in buffer B. Disperse membranes in detergent solution by passage through 25-gauge needles (see earlier discussion). (This is much more efficient than using a Dounce homogenizer.) After dispersion, mix the membranes slowly with a magnetic stir bar for 30 min at 4°. Ultracentrifuge the extract at ~400,000g for 20 min and separately collect the supernatant and pellet. GCS is usually

TABLE I
COMPATIBILITY OF GCS ACTIVITY WITH DETERGENTS[a]

Compatible	Incompatible
Triton X-100 (≤0.2%)	Triton X-100 (1%)
Nonidet P-40 (≤0.2%)	Nonidet P-40 (1%)
CHAPSO (≤2%)	Zwittergent 3-14 (0.2%)
CHAPS (≤2%)	Deoxycholate (1%)
Sodium cholate (≤1%)	N-Dodecylmaltoside (1%)
Digitonin (≤1%)	SDS (0.2%)
Diheptanoylphosphatidylcholine (14–20 mM)	

[a] GCS in Golgi membranes was solubilized at the detergent concentrations listed in parentheses, centrifuged, and then supernatants and pellets were assayed for activity after a 10-fold dilution of detergent. Compatible detergents were found to solubilize 50–80% of total GCS activity detected in supernatant and pellet combined. Those detergents listed as incompatible had little or no activity in either pellet or supernatant but solubilized GCS efficiently as assessed by Western blotting. CHAPSO, 3-[(3-cholamidopropyl)methylammonio]-2-hydroxy-1-propanesulfonate; CHAPS, 3-[(3-cholamidopropyl)methylammonio]-1-propanesulfonate.

60–80% extractable under these conditions. The stability of solubilized GCS is improved markedly by the inclusion of 10–20% glycerol (or ethylene glycol) in solubilizing buffers. GCS is unaffected by the inclusion of DTT or EDTA in the buffer.

Reconstitution

For studies of GCS topology, catalytic activity, and exploring the mechanism of GlcCer transmembrane transport,[7,8,28] it may be useful to reconstitute solubilized GCS into detergent-free lipid vesicles. The following method has been used to reconstitute CHAPSO- or CHAPS-solubilized GCS.

Procedure. Solubilized GCS in 0.5% CHAPSO (or CHAPS), 50 mM HEPES, pH 7.4, 0.15 M KCl, 20% glycerol, and 1 mM dioleoylphosphatidylcholine plus protease inhibitors (see earlier discussion) is dialyzed against 3 × 400 volumes of 50 mM HEPES, pH 7.4, 25 mM KCl, 10% glycerol, and 1 mM UDP-Glc for 24 hr at 4°. In preliminary studies, the protein/lipid ratio was ~1 : 1 (w : w); however, the protein/lipid ratio has not been optimized. The resulting dialyzate retains ~60% of initial GCS activity. The orientation of GCS in these lipid vesicles has not been established.

Purification Techniques

No large-scale purification of GCS has been performed yet because of the limited quantities thus far available in Golgi membranes. We have previously utilized several techniques for the partial purification of GCS. A two-step dye agarose column method based on the passage of GCS through dye agarose in the presence of UDP-Glc and the binding to dye agarose when UDP-Glc is removed gave a >25-fold enrichment of GCS compared to N-lauroylsarcosine Golgi membranes (already enriched ~200-fold compared to homogenate; see earlier) for a total enrichment of >5000-fold relative to homogenate.[11] Recovery of GCS, however, was very low (<2%) in this method, and GCS was not visible on silver-stained SDS–PAGE gels. Partial immunopurification of CHAPS-solubilized, active GCS has also been demonstrated.[12] Although GCS was bound to and eluted efficiently from immunocolumns, the specific activity of GCS isolated in this manner was reduced ~80% compared to the starting material (as judged by GCS activity vs protein detected on Western blots). This loss of specific activity is most likely due to the harsh elution reagent used (3 M MgCl$_2$/25% ethylene glycol). Finally, GCS in solubilized Golgi membranes has

---

[28] O. C. Martin and R. E. Pagano, *J. Cell Biol.* **125,** 769 (1994).

been partially separated from other proteins in 6–25% glycerol density gradients using either CHAPSO or NP-40.[11,12]

Conclusions

The purification of measurable quantities of GCS for studies of secondary structure, catalytic mechanism, inhibitor design, and Golgi localization will probably only be possible by the overexpression of GCS in a bacterial or eukaryotic cell system. The design of fusion proteins in which GCS is fused to tags (e.g., polyhistidine) for purification or to large hydrophilic carrier proteins for better solubility and decreased aggregation will likely facilitate GCS purification. The methods presented in this article on the stabilization, solubilization, and detection of GCS should continue to prove useful as these developments unfold.

[8] Analysis of Galactolipids and UDP-Galactose: Ceramide Galactosyltransferase

By HEIN SPRONG, GERRIT VAN MEER, and PETER VAN DER SLUIJS

Galactolipids and Galactosyltransferase

Glycosphingolipids form a highly polymorphic class of lipids, and several hundreds of the more than 2000 possible molecular species[1] have been characterized.[2] There are at least 20 different ceramide (Cer) backbones due to differences in sphingoid base, mostly sphingosine (4-sphingenine) and phytosphingosine (4-hydroxysphinganine), and acyl chain. The head groups can vary from 1 to 60 sugars. Glycosphingolipids in mammals can be subdivided into two major classes—galacto- and glucosphingolipids—based on the presence of Gal or Glc as the first sugar moiety. Most complex glycolipids are based on Gal β1-4 Glc β1-1 Cer, lactosylceramide (LacCer). Galactosylceramide (Gal β1-1 Cer or GalCer) serves as a precursor for a few simple glycolipids: the sulfatide SGalCer (SO$_3$-3 GalCer), galabiosylceramide (Gal α1-4 GalCer or Ga$_2$Cer), and the ganglioside sialo-GalCer (I^{3}NeuAc-GalCer or "G$_{M4}$"). Gal and SGal are also found on diglycerides: Gal β1-3 diacylglycerol (GalDAG), Gal β1-3 alkyl-acyl-glycerol (GalAAG), digalactosyldiglyceride, and seminolipid (SO$_3$-3 GalAAG).[3]

[1] S.-I. Hakomori and Y. Igarashi, *J. Biochem.* **118,** 1091 (1995).
[2] J. A. Shayman and N. S. Radin, *Am. J. Physiol.* **260,** F291 (1991).
[3] J. P. Vos, M. Lopes-Cardozo, and B. M. Gadella, *Biochim. Biophys. Acta* **1211,** 125 (1994).

Glycosphingolipids are enriched in the outer leaflet of the plasma membrane of most eukaryotic cells where they are thought to be involved in cell recognition and signaling.[1] Although glycosphingolipids constitute only a few mol% of the lipids in most membranes, they are major components of the myelin sheath,[4] where GalCer and SGalCer are involved in axonal insulation, myelin function, and stability.[5,6] The apical plasma membrane of epithelial cells in the gastrointestinal and urinary tracts is enriched in glycosphingolipids. In rodents these are typically glucolipids,[2,7] whereas in humans most are galactolipids.[8–10] Glycosphingolipids play a structural role in rigidifying and protecting the apical cell surface. Their role in sorting lipids and proteins to various membranes along the exocytotic and endocytotic transport routes is not fully understood.[7,11]

The foremost enzyme involved in the biosynthesis of galactosphingolipids is the UDP-galactose : ceramide galactosyltransferase, CGalT or GalT-1.[12] CGalT catalyzes the transfer of galactose from UDP-galactose to Cer yielding GalCer[13] and has a relatively promiscuous substrate specificity. Whether there are one or more CGalT enzymes with distinct specificity and cellular localization has been a controversial issue.[14–18] Importantly, knockout mice do not make GalCer,[5,6] showing that there is only one CGalT. *In vitro* studies demonstrated that partially purified CGalT from brain has a >15-fold preference for hydroxy fatty acid (HFA) over nonhydroxy fatty acid (NFA) containing Cer.[13,19] This has been confirmed for CGalT after transfection into CGalT-negative cells.[18,20] *In vivo,* however,

[4] P. Morell, R. H. Quarles, and W. T. Norton, *in* "Basic Neurochemistry" (G. J. Siegel, B. W. Agranoff, R. W. Albers, and P. B. Molinoff, eds.), p. 117. Raven Press, New York, 1994.

[5] A. Bosio, E. Binczek, and W. Stoffel, *Proc. Natl. Acad. Sci. U.S.A.* **93,** 13280 (1996).

[6] T. Coetzee, N. Fujita, J. Dupree, R. Shi, A. Blight, K. Suzuki, K. Suzuki, and B. Popko, *Cell* **86,** 209 (1996).

[7] G. van Meer, *Annu. Rev. Cell Biol.* **5,** 247 (1989).

[8] N. Yahi, J.-M. Sabatier, P. Nickel, K. Mabrouk, F. Gonzalez-Scarano, and J. Fantini, *J. Biol. Chem.* **269,** 24349 (1994).

[9] K.-E. Falk, K.-A. Karlsson, H. Leffler, and B. E. Samuelsson, *FEBS Lett.* **101,** 273 (1979).

[10] B. Siddiqui, J. S. Whitehead, and Y. S. Kim, *J. Biol. Chem.* **253,** 2168 (1978).

[11] K. Simons and E. Ikonen, *Nature* **387,** 569 (1997).

[12] M. Basu, T. De, K. K. Das, J. W. Kyle, H.-C. Chon, R. J. Schaeper, and S. Basu, *Methods Enzymol.* **138,** 575 (1987).

[13] P. Morell and N. S. Radin, *Biochemistry* **8,** 506 (1969).

[14] E. Costantino-Ceccarini, A. Cestelli, and G. H. DeVries, *J. Neurochem.* **32,** 1175 (1979).

[15] C. Sato, J. A. Black, and R. K. Yu, *J. Neurochem.* **50,** 1887 (1988).

[16] C. Sato and R. K. Yu, *Dev. Neurosci.* **12,** 153 (1990).

[17] K. N. J. Burger, P. van der Bijl, and G. van Meer, *J. Cell Biol.* **133,** 15–28 (1996).

[18] P. van der Bijl, M. Lopes-Cardozo, and G. van Meer, *J. Cell Biol.* **132,** 813 (1996).

[19] S. Basu, A. M. Schultz, M. Basu, and S. Roseman, *J. Biol. Chem.* **246,** 4272 (1971).

[20] N. Schaeren-Wiemers, P. van der Bijl, and M. E. Schwab, *J. Neurochem.* **65,** 2267 (1995).

CGalT is responsible for the galactosylation of HFA- as well as NFA-containing Cer. NFA–CGalT activity found previously in the Golgi[15,17,18] has now been demonstrated to be an *in vitro* activity of the Golgi ceramide glucosyltransferase (CGlcT).[21] CGalT is also responsible for the galactosylation of diglycerides.[18]

The localization of CGalT has long been enigmatic.[14,22–25] We have shown that the enzyme is exclusively localized to the endoplasmic reticulum (ER) by immunogold electron microscopy on ultrathin cryosections.[21] CGalT is a high mannose-type glycoprotein that is N-glycosylated at Asn-78 and Asn-333[26] and contains a putative carboxy terminal Lys–Lys–Val–Lys ER-retrieval signal.[20,27,28] Surprisingly, the conceptual translation product exhibits no amino acid sequence similarity with other glycosyltransferases. Instead, CGalT is related to the superfamily of UDP-glucuronosyltransferases.

Thus, while most glycosylation steps of sphingolipids occur in the Golgi complex, CGalT enzyme activity resides in the lumen of the ER (Fig. 1).[21] Cer is synthesized at the cytosolic surface and is sufficiently hydrophobic to diffuse freely across cellular membranes. How the other substrate, UDP-Gal, reaches the active center of CGalT is unclear. CHOlec8 cells, which are deficient in UDP-Gal import into the Golgi apparatus,[29] are also impaired in UDP-Gal import into the ER.[21] Whether UDP-Gal import in the ER and in the Golgi complex is mediated by the same or distinct UDP-Gal importers remains to be resolved. GalCer is converted to Ga_2Cer[30] and sulfatide[31] in the lumen of the Golgi, from where these products cannot reach the cytosolic surface.[17] In contrast, GalCer can translocate from the lumenal to the cytosolic leaflet of the ER membrane,[17] where it may interact with cytosolic galactose-binding lectins[32] or, in contrast to present dogma, may oligomerize and form microdomains in the cytosolic leaflet.

[21] H. Sprong, B. Kruithof, R. Leijendekker, J. W. Slot, G. van Meer, and P. van der Sluijs, *J. Biol. Chem.* **273**, 25880 (1998).

[22] H. P. Siegrist, T. Burkart, U. N. Wiesmann, N. N. Herschkowitz, and M. A. Spycher, *J. Neurochem.* **33**, 497 (1979).

[23] O. Koul, K. H. Chou, and F. B. Jungalwala, *Biochem. J.* **186**, 959 (1980).

[24] A. Carruthers and E. M. Carey, *J. Neurochem.* **41**, 22 (1983).

[25] O. Koul and F. B. Jungalwala, *Neurochem. Res.* **11**, 231 (1986).

[26] S. Schulte and W. Stoffel, *Eur. J. Biochem.* **233**, 947 (1995).

[27] S. Schulte and W. Stoffel, *Proc. Natl. Acad. Sci. U.S.A.* **90**, 10265 (1993).

[28] N. Stahl, H. Jurevics, P. Morell, K. Suzuki, and B. Popko, *J. Neurosci. Res.* **38**, 234 (1994).

[29] S. L. Deutscher and C. B. Hirschberg, *J. Biol. Chem.* **261**, 96 (1986).

[30] D. A. Wenger, K. Subba Rao, and R. A. Pieringer, *J. Biol. Chem.* **245**, 2513 (1970).

[31] K. Honke, M. Tsuda, Y. Hirahara, A. Ishii, A. Makita, and Y. Wada, *J. Biol. Chem.* **272**, 4864 (1997).

[32] S. H. Barondes, D. N. W. Cooper, M. A. Gitt, and H. Leffler, *J. Biol. Chem.* **269**, 20807 (1994).

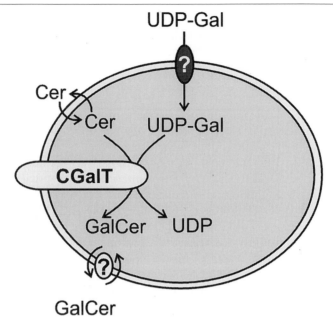

FIG. 1. Schematic organization of GalCer synthesis in the ER membrane.

Detection of CGalt by Its Products

Until recently, the presence of the CGalT could only be assessed via the presence of its products or by enzyme assay. GalCer and S-GalCer were discovered originally as major lipids in human brain by Thudichum in 1884,[33] whereas glycerol-based galactolipids were discovered by Carter et al.[34]

Chemical Detection of Galactolipids

Tissue can be analyzed for galactolipids chemically. Routinely, lipids are first extracted in chloroform/methanol (one-phase) at elevated temperatures for maximal yield. For sphingolipid analysis, glycerolipids are removed by alkaline hydrolysis, and acidic and neutral sphingolipids are separated by a DEAE column. Nonpolar lipids and sphingomyelin can then be removed by acetylation, column chromatography, and deacetylation. Next, the glycosphingolipids are subfractionated by thin-layer chromatography (TLC). Including dialysis steps and additional columns, this procedure may

[33] A. Makita and N. Taniguchi, in "Glycolipids" (H. Wiegandt, ed.). Elsevier, Amsterdam, 1985.
[34] H. E. Carter, P. Johnson, and E. J. Weber, Annu. Rev. Biochem. 34, 109 (1965).

take two weeks.[35] A simplified analysis starts with a two-phase extraction,[36] after which the more polar lipids like sulfatides, which partition to some extent into the aqueous phase, can be recovered by adsorption to a reversed-phase cartridge. Lipids can be separated by two-dimensional TLC.[18,37] Separation of GalCer from GlcCer requires the use of borate-impregnated Whatman paper or TLC plates.[12,18,37–40] Spots are visualized classically by charring or staining by a variety of reagents.[12,35,39]

Galactolipids can be radiolabeled conveniently using galactose, acetate, fatty acid, and sulfate, whereas sphingolipids will be efficiently labeled also by serine, palmitate, sphingosine, sphinganine, or a ceramide containing a $C_6(2\text{-OH})$ chain (Figs. 2A and 2B). Fluorescent galactolipids can be produced from C_6-NBD-Cer, but more efficiently from $C_6(2\text{-OH})$-NBD-Cer,[37,41] whereas C_6-NBD-DAG can be used to obtain C_6-NBD-GalDAG (Fig. 2C). Radiolabels and fluorescence are detected and quantitated by phosphorimaging or fluorography and scintillation counting[13,18,37] and by fluorimaging or fluorometry.[37]

Originally, galactolipids on TLC plates were identified by chemical determination of the sphingoid base or glycerol, fatty acid, galactose, or sulfate.[34] Often, sufficient information is obtained from comigration with standards, sensitivity of the lipid to enzymes such as α- or β-galactosidase, and in cell lines, after radiolabeling with specific precursors or treatment of the cells with inhibitors of glycolipid synthesis or sulfation.[18,37] The precise structure of a galactolipid can be obtained with mass spectrometry in combination with nuclear magnetic resonance (NMR) spectroscopy.[42] Whereas even one two-dimensional TLC separation of total lipids may yield galactolipid spots of sufficient purity to allow identification by mass spectrometry,[37] high-performance liquid chromatography (HPLC) remains the method of choice for this purpose.[43] Amounts in the picomole range can now be quantified with nano-electrospray tandem mass spectrometry.[44]

[35] M. E. Breimer, G. C. Hansson, K.-A. Karlsson, and H. Leffler, *Exp. Cell Res.* **135,** 1 (1981).

[36] E. G. Bligh and W. J. Dyer, *Can. J. Biochem. Physiol.* **37,** 911 (1959).

[37] P. van der Bijl, G. J. Strous, M. Lopes-Cardozo, J. Thomas-Oates, and G. van Meer, *Biochem. J.* **317,** 589 (1996).

[38] O. M. Young, and J. N. Kanfer, *J. Chromatogr.* **19,** 611 (1965).

[39] E. L. Kean, *J. Lipid Res.* **7,** 449 (1966).

[40] I. L. van Genderen, G. van Meer, J. W. Slot, H. J. Geuze, and W. F. Voorhout, *J. Cell Biol.* **115,** 1009 (1991).

[41] R. E. Pagano and O. C. Martin, *Biochemistry* **27,** 4439 (1988).

[42] Y. Yachida, M. Kashiwagi, T. Mikami, K. Tsuchihashi, T. Daino, T. Akino, and S. Gasa, *J. Lipid Res.* **39,** 1039 (1998).

[43] R. H. McCluer, M. D. Ullman, and F. B. Jungalwala, *Adv. Chromatogr.* **25,** 309 (1986).

[44] B. Brügger, G. Erben, R. Sandhoff, F. T. Wieland, and W. D. Lehmann, *Proc. Natl. Acad. Sci. U.S.A.* **94,** 2339 (1997).

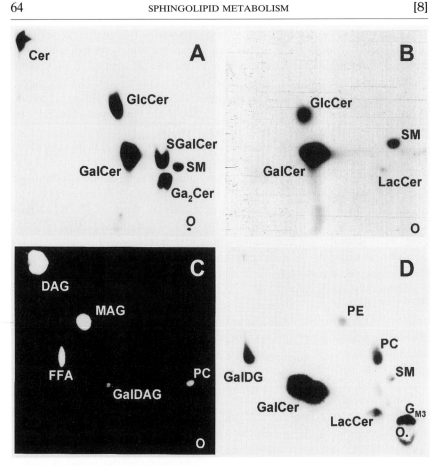

Fig. 2. Lipid synthesis in cell lines expressing CGalT. TLC analysis of the lipid products synthesized during 1 hr at 37° from C$_6$OH-[^{3}H]Cer in dog kidney MDCK II cells (A), in Chinese hamster ovary cells transfected with CGalT (CGalT-CHO) (B), and during 2 hr from NBD-DAG in CGalT-CHO cells (C).[1] (D) The fluorograph of CGalT-CHO lipids after an overnight incubation with [^{3}H]galactose [P. van der Bijl, G. J. Strous, M. Lopes-Cardozo, J. Thomas-Oates, and G. van Meer, *Biochem. J.* **317,** 589 (1996)]. FFA, C$_6$-NBD-hexanoic acid, free fatty acid; GalDG, sum of GalDAG and GalAAG; G$_{M3}$, sialo-LacCer; MAG, monoacylglycerol; PC, phosphatidylcholine; PE, phosphatidylethanolamine; SM, sphingomyelin. For solvents and further details, see text. Panel A was reproduced from the *J. Cell Biol.* **132,** 813 (1996) by copyright permission of The Rockefeller University Press. Panels C and D were reproduced with permission from the *Biochem. J.* **317,** 589 (1996). © Biochemical Society.

Often, a combination of the methods described here is required to define the precise galactolipid content of a sample.[18,37,45]

Immunological Detection of Galactolipids

Some lipids can be identified by antibody-overlay techniques.[46] Antibodies are available that recognize GalCer, GalDAG, GalAAG, Ga_2Cer, and their sulfated forms[47–59] with a degree of specificity.[60,61] A variation on this theme is the use of bacterial toxins recognizing GalCer,[62] the ectodomain of human immunodeficiency virus gp120 that recognizes GalCer and sulfatides,[63–70] or mammalian proteins that recognize sulfatides.[71–75] A common problem of these assays is their lack of specificity.

[45] A. E. Backer, M. E. Breimer, B. E. Samuelsson, and J. Holgersson, *Glycobiology* **7,** 943 (1997).

[46] J. L. Magnani, S. L. Spitalnik, and V. Ginsburg, *Methods Enzymol.* **138,** 195 (1987).

[47] T. Uchida and Y. Nagai, *J. Biochem.* **87,** 1829 (1980).

[48] C. Lingwood and H. Schachter, *J. Cell Biol.* **89,** 621 (1981).

[49] K. Sakakibara, T. Momoi, T. Uchida, and Y. Nagai, *Nature* **293,** 76 (1981).

[50] I. Sommer and M. Schachner, *Dev. Biol.* **83,** 311 (1981).

[51] B. Ranscht, P. A. Clapshaw, J. Price, M. Noble, and W. Seifert, *Proc. Natl. Acad. Sci. U.S.A.* **79,** 2709 (1982).

[52] C. Goujet-Zalc, A. Guerci, G. Dubois, and B. Zalc, *J. Neurochem.* **46,** 435 (1986).

[53] J. A. Benjamins, R. E. Callahan, I. N. Montgomery, D. M. Studzinski, and C. A. Dyer, *J. Neuroimmunol.* **14,** 325 (1987).

[54] S. J. Crook, R. Stewart, J. M. Boggs, A. I. Vistnes, and B. Zalc, *Mol. Immunol.* **24,** 1135 (1987).

[55] P. Fredman, L. Mattsson, K. Andersson, P. Davidsson, I. Ishizuka, S. Jeansson, J. E. Mansson, and L. Svennerholm, *Biochem. J.* **251,** 17 (1988).

[56] H. Nakakuma, M. Arai, T. Kawaguchi, K. Horikawa, M. Hidaka, K. Sakamoto, M. Iwamori, Y. Nagai, and K. Takatsuki, *FEBS Lett.* **258,** 230 (1989).

[57] H. Murakami, Z. Lam, B. C. Furie, V. N. Reinhold, T. Asano, and B. Furie, *J. Biol. Chem.* **266,** 15414 (1991).

[58] D. Arvanitis, M. Dumas, and S. Szuchet, *Dev. Neurosci.* **14,** 328 (1992).

[59] S. Bhat, *J. Neuroimmunol.* **41,** 105 (1992).

[60] R. Bansal, A. E. Warrington, A. L. Gard, B. Ranscht, and S. E. Pfeiffer, *J. Neurosci. Res.* **24,** 548 (1989).

[61] B. M. Gadella, T. W. J. Gadella, Jr., B. Colenbrander, L. M. G. van Golde, and M. Lopes-Cardozo, *J. Cell Sci.* **107,** 2151 (1994).

[62] S. Kozaki, J. Ogasawara, Y. Shimote, Y. Kamata, and G. Sakaguchi, *Infect. Immun.* **55,** 3051 (1987).

[63] S. Bhat, S. L. Spitalnik, F. Gonzalez-Scarano, and D. H. Silberberg, *Proc. Natl. Acad. Sci. U.S.A.* **88,** 7131 (1991).

[64] S. Bhat, T. Otsuka, and A. Srinivasan, *DNA Cell Biol.* **13,** 211 (1994).

[65] L. H. van den Berg, S. A. Sadiq, S. Lederman, and N. Latov, *J. Neurosci. Res.* **33,** 513 (1992).

[66] D. G. Cook, J. Fantini, S. L. Spitalnik, and F. Gonzalez-Scarano, *Virology* **201,** 206 (1994).

[67] D. Long, J. F. Berson, D. G. Cook, and R. W. Doms, *J. Virol.* **68,** 5890 (1994).

[68] T. McAlarney, S. Apostolski, S. Lederman, and N. Latov, *J. Neurosci. Res.* **37,** 453 (1994).

[69] N. Yahi, J.-M. Sabatier, S. Baghdiguian, F. Gonzalez-Scarano, and J. Fantini, *J. Virol.* **69,** 320 (1995).

Expression patterns of galactolipids may be established by immuno-labeling methods. For light microscopy, a primary galactolipid-binding protein is visualized with fluorescently or otherwise labeled antibodies. For electron microscopy, protein A conjugated with colloidal gold is the detection method of choice. Because of the potential cross-reactivity of the galactolipid-binding protein, morphological techniques must always be confirmed by lipid analysis. Immunolabeling of (glyco)lipids is hampered by artifacts that include relocation and solubilization of the antigen during fixation with organic solvents and permeabilization with detergents. Immunolabeling of thawed cryosections may also result in the redistribution of lipid molecules. The best method so far is freeze substitution.[76,77] Glycolipids are thought to be enriched in patches in the membrane[7,11,77] However, antibody labeling may cluster glycolipids artificially, even after fixation. This can only be prevented by a second round of fixation after binding of the first antibody.[78]

Assays for CGalT Enzyme Activity

The enzyme activity producing GalCer was first demonstrated by Morell and Radin[13] and, since then, it has been characterized under numerous conditions. A technical problem is the difficulty of controlling the Cer concentration in the membrane containing CGalT as Cer is regulated tightly in the ER membrane *in vivo*.[79] Moreover, natural ceramides do not exchange efficiently between membranes *in vitro,* limiting the possibilities to manipulate Cer levels of isolated ER membranes. Cer has been supplied efficiently in detergent.[12,19,80] Detergent assays test enzyme activity under standard but nonphysiological conditions, as the ER membrane has been dissolved. Moreover, enzyme activity is reduced manyfold. Cer has also

[70] D. Hammache, G. Pieroni, N. Yahi, O. Delezay, N. Koch, H. Lafont, C. Tamalet, and J. Fantini, *J. Biol. Chem.* **273,** 7967 (1998).
[71] D. D. Roberts, D. M. Haverstick, V. M. Dixit, W. A. Frazier, S. A. Santoro, and V. Ginsburg, *J. Biol. Chem.* **260,** 9405 (1985).
[72] A. Aruffo, W. Kolanus, G. Walz, P. Fredman, and B. Seed, *Cell* **67,** 35 (1991).
[73] Y. Suzuki, Y. Toda, T. Tamatani, T. Watanabe, T. Suzuki, T. Nakao, K. Murase, M. Kiso, A. Hasegawa, K. Tadano-Aritomi *et al., Biochem. Biophys. Res. Commun.* **190,** 426 (1993).
[74] K. L. Crossin and G. M. Edelman, *J. Neurosci. Res.* **33,** 631 (1992).
[75] T. Kobayashi, K. Honke, K. Miyazaki, K. Matsumoto, T. Nakamura, I. Ishizuka, and A. Makita, *J. Biol. Chem.* **269,** 9817 (1994).
[76] I. van Genderen and G. van Meer, *J. Cell Biol.* **131,** 645 (1995).
[77] R. G. Parton, *J. Histochem. Cytochem.* **42,** 155 (1994).
[78] C. Butor, E. H. K. Stelzer, A. Sonnenberg, and J. Davoust, *Eur. J. Cell Biol.* **56,** 269 (1991).
[79] A. van Helvoort, M. L. Giudici, M. Thielemans, and G. van Meer, *J. Cell Sci.* **110,** 75 (1997).
[80] N. M. Neskovic, P. Mandel, and S. Gatt, *Methods Enzymol.* **71,** 521 (1981).

been presented from Celite[13] or phosphatidylethanolamine "membranes."[81] Disadvantages are the low efficiency, undefined local ceramide concentrations, and, in some cases, uncontrolled effects on the CGalT-containing membrane (e.g. by fusion). As an alternative, short chain ceramides provide a very efficient assay for enzyme activity in the ER membrane.[20,28,37,82,83] However, they yield indirect data on kinetics and substrate specificity.

Assay for CGalT Activity in Cells Using Short Chain Ceramides. The method used to detect CGalT enzyme activity is based on measuring the incorporation of fluorescent or radioactive, short chain Cer into GalCer. Because of the short fatty acyl chain, these ceramides and their products will display a higher off rate from membranes than the natural membrane lipids. For that reason, short chain lipid analogs can be efficiently presented to or depleted from membranes by a back-exchange against liposomes or bovine serum albumin (BSA) in the absence of detergent.[18,84] The reaction requires UDP-Gal, which, for *in vitro* studies, must be added exogenously. Lipids are extracted, separated by two-dimensional TLC, and quantitated by fluorescence of radioactivity.

Reagents

Phosphate-buffered saline (PBS; ice cold) containing 0.9 mM Ca^{2+} and 0.5 mM Mg^{2+}

Cell incubation mixture: Hanks' balanced salt solution, 20 mM HEPES–NaOH, pH 7.2, 1% (w/v) BSA (fraction V from Sigma, St. Louis, MO), and 35 nM of C$_6$OH-[^{3}H]Cer

Homogenization buffer (HB): 250 mM sucrose, 10 mM HEPES–NaOH, pH 7.2, and 1 mM EDTA

Reaction mixture: HB containing 2% (w/v) BSA, 4 mM UDP-glucose, 4 mM UDP-galactose, 4 mM MgCl$_2$, 4 mM MnCl$_2$, 1 μg/ml protease inhibitors, and 50 μM of NBD-Cer or NBD-DAG or 35 nM of C$_6$OH-[^{3}H]Cer

Ceramides: Fluorescent *N*-6(7-nitro-2,1,3-benzoxadiazol-4-yl)-aminohexanoylceramide (NBD-Cer) is obtained commercially (Molecular probes, Eugene, OR). The radiolabeled short chain ceramides hexanoyl-[^{3}H]Cer (C$_6$-[^{3}H]Cer) and 2-hydroxyhexanoyl-[^{3}H]Cer (C$_6$OH-[^{3}H]Cer) (800 MBq/μmol) are synthesized according to Ong and Brady.[18,85] Ceramides are dried from stock solutions in chloroform/

[81] E. Costantino-Ceccarini and A. Cestelli, *Methods Enzymol.* **72**, 384 (1981).
[82] K. R. Warren, R. S. Misra, R. C. Arora, and N. S. Radin, *J. Neurochem.* **26**, 1063 (1976).
[83] N. M. Neskovic, G. Roussel, and J. L. Nussbaum, *J. Neurochem.* **47**, 1412 (1986).
[84] N. G. Lipsky and R. E. Pagano, *J. Cell Biol.* **100**, 27 (1985).
[85] D. E. Ong and R. N. Brady, *J. Lipid Res.* **13**, 819 (1972).

methanol (2 : 1, v/v) under nitrogen, dissolved in ethanol (final concentration less than 0.2% v/v), and injected into BSA buffer under vortexing to yield the reaction mixture. This is incubated 30 min on ice, allowing BSA complexes of the ceramides to be formed prior to the addition of the enzyme source.

Fluorescent 1-palmitoyl-2,6(7-nitro-2,1,3-benzoxadiazol-4-yl)-amino-hexanoyldiacylglycerol (NBD-DAG) is prepared from NBD-phosphatidylcholine (Avanti Polar Lipids, Alabaster, AL) using phospholipase C[37]

TLC plates (Si60, Merck, Darmstadt, FRG) are dipped in 2.5% (w/v) boric acid in methanol[39] and dried prior to usage. Borate treatment is required to separate GlcCer and GalCer analogs.

All reactions and lipid extractions are performed in Corex or Pyrex glassware. Chromatography solvents are of Pro Analyse quality. All lipid stocks are stored in chloroform/methanol (2 : 1, v/v) at −20°. Solutions are stored under nitrogen and should be checked routinely for concentration and purity.

CGalT Source. Chinese hamster ovary (CHO) cells transfected with CGalT (CGalT-CHO cells[37]) are cultured in Eagle's minimum essential medium (MEM)-α (with nucleotides) with 10% fetal calf serum (FCS), 10 mM HEPES, and 500 μg/ml G418. To prepare a postnuclear supernatant (PNS), a 10-cm-diameter dish of CGalT-CHO cells is washed twice with ice-cold PBS and scraped gently in 1 ml ice-cold HB. Cells are pelleted and resuspended in 400 μl HB. The cells are homogenized by 12 to 14 passages through a 25-gauge needle and centrifuged for 15 min at 375g at 4° to remove nuclei and unbroken cells. Protein in the PNS is measured using the BCA assay (Pierce, Rockford, IL) and adjusted to 2 mg/ml with HB. In some cases, 0.4% (w/v) saponin is added to the PNS to permeabilize membranes during a 30-min incubation on ice prior to the experiment. MDCK II cells are grown as monolayers in MEM with 10 mM HEPES and 5% FCS.

Incubation. A 3-cm dish of CGalT-CHO cells or a 24-mm filter with MDCK cells is incubated with 1 ml cell incubation mixture. When PNS is used, 1 volume of reaction mixture is added to the PNS and the samples are incubated for 1 or 2 hr at 37°. The reaction is stopped by transferring the samples to an ice bath and by starting the lipid extraction.

Lipid Analysis. Lipids from cells, media, or PNS are extracted by a two-phase extraction.[36] The aqueous solution used for the phase separation contains 20 mM acetic acid and (for radiolabeled lipids) 120 mM KCl. An additional chloroform wash of the upper (aqueous) phase is performed. The organic (lower) phase is dried under N$_2$ at 37°, and the lipids are applied to borate-treated TLC plates using chloroform/methanol (2 : 1, v/v). Thin-layer chromatography plates are developed in the first dimension using chloroform/methanol/25% (v/v) NH_4OH/water (65:35:4:4, v/v) and in chloroform/acetone/methanol/acetic acid/water (50:20:10:10:5, v/v) in the second dimension. Fluorescent spots are quantitated using a STORM imager (Molecular Dynamics, Sunnyvale, CA) using ImageQuant software. Alternatively, spots are detected under UV, scraped, and extracted from the silica in 2 ml chloroform/methanol/20 mM acetic acid (1:2.2:1, v/v) for 30 min. After pelleting the silica for 10 min at 2000g, NBD fluorescence in the supernatant is quantified in a fluorometer at 470/535 nm using the appropriate controls and after calibration of the fluorometer using the Raman band of water at 350/397 nm. Radiolabeled spots are detected by fluorography after dipping the TLC plates in 0.4% (v/v) 2,5-diphenyl oxazole in 2-methylnaphthalene with 10% (v/v) xylene.[86] Preflashed film (Kodak X-Omat S, France) is exposed to the TLC plates for several days at −80°. The radioactive spots are scraped from the plates and the radioactivity is quantified by liquid scintillation counting in 0.3 ml Solulyte (J.T. Baker Chemicals, Deventer, The Netherlands) and 3 ml of Ultima Gold (Packard Instrument Company, Downers Grove, IL).

Results. The results of this assay are highly reproducible. In dog kidney MDCK cells, C_6OH-[^3H]Cer is converted to GalCer, Ga_2Cer, and SGalCer, whereas GlcCer and sphingomyelin are also formed (Fig. 2A). In contrast, transfection of CHO cells with rat CGalT results in a shift from incorporation into GlcCer and sphingomyelin to the production of C_6OH-[^3H]GalCer (Fig. 2B). In homogenates from both cell types, CGalT has a great preference for ceramides containing a 2-OH fatty acid.[13,19,37] Interestingly, tissues expressing high CGalT activity also contain high levels of 2-OH fatty acids. GalCer produced in CGalT-CHO cells contained exclusively nonhydroxy fatty acids,[37] which suggests that in the genome CGalT and the enzymes responsible for the synthesis of 2-OH fatty acids are coordinately controlled. This is apparently also the case for the α1,4-galactosyltransferase responsible for the synthesis of Ga_2Cer and the sulfotransferase synthesizing SGalCer. In contrast to the parental CHO cells, CGalT-CHO cells synthesized GalDAG from C_6-NBD-DAG (Fig. 2C) and a mixture of GalDAG and GalAAG from [^3H]galactose (Fig. 2D).

It should be noted that cellular factors may influence the CGalT activity measured. For example, the synthesis of GalCer is dependent on UDP-Gal import into the lumen of the ER. Some cell lines, such as CHOlec8 cells, have an impaired UDP-Gal import. A PNS of CGalT-CHOlec8 cells displayed low CGalT activity. This activity could be restored by permeabilizing membranes prepared from CGalT-CHOlec8 cells with saponin, sug-

---

[86] W. M. Bonner and J. D. Stedman, *Anal. Biochem.* **89,** 247 (1978).

gesting that the ER in CHOlec8 cells does not import UDP-Gal. These cells are known to lack the Golgi UDP-Gal transporter,[29] suggesting that the two transporter activities may reside within the same protein.

Enzyme assays have suggested the existence of two CGalTs with different intracellular locations (see earlier discussion). In our own studies, this finding was caused by an artifact of the CGalT assay. After the observation that the second CGalT activity had many properties in common with CGlcT in the Golgi,[17,37] a comparison between CGalT-negative cells that did or did not express CGlcT demonstrated that CGlcT can synthesize GalCer when assayed in the absence of UDP-Glc.[21] Similar observations were made using a purified CGlcT.[87] In the presence of UDP-Glc (as in living cells), UDP-Gal was essentially competed out. Alternatively, GalCer synthesis by CGlcT can be inhibited by a specific CGlcT inhibitor, such as D-threo-PDMP.[88]

Detection of CGalT Protein in Cells

Until recently, the characteristics of the CGalT could only be addressed by measuring its activity in isolated subcellular fractions (Ref. 17 and references therein). Although antibodies have been available for some time,[83] only the antibodies raised against recombinant CGalT have facilitated analysis of the protein. Histidine-tagged fusion proteins representing different regions of rat CGalT were used to generate rabbit polyclonal antisera that specifically recognize different lumenal regions of rat CGalT.[21] CGalT antisera work well for Western blotting, immunoprecipitation, and immunofluorescence microscopy. Cross-reactivity in other species has not been tested yet.

To study the properties of CGalT in cultured cells, newly synthesized proteins are labeled metabolically with radioactive amino acids and are chased with unlabeled amino acids for various time periods. Now different aspects of CGalT can be studied in more detail, such as its biosynthetic maturation and its membrane topology. Assays for analysis of its co- and posttranslational modifications can also be found elsewhere.[27] Radiolabeled CGalT is isolated by immunoprecipitation, followed by separation on sodium dodecyl sulfate–polyacrylamide (SDS–PAA) gels and analysis by phosphorimaging.

Reagents

Depletion medium: Cysteine- and methionine-free minimum essential medium (MEM α, Sigma, M3786), 20 mM HEPES, pH 7.3, at 37°

[87] P. Paul, Y. Kamisaka, D. L. Marks, and R. E. Pagano, *J. Biol. Chem.* **271**, 2287 (1996).
[88] J.-i. Inokuchi, K. Momosaki, H. Shimeno, A. Nagamatsu, and N. S. Radin, *J. Cell. Physiol.* **141**, 573 (1989).

Pulse medium: Depletion medium containing 250 μCi/ml Tran[^{35}S]-label (>1000 Ci/mmol; ICN, Costa Mesa, CA) at 37°

Chase medium: MEM supplemented with 5 mM methionine, 5 mM cysteine, and 20 mM HEPES, pH 7.4, at 37°

Stop buffer: PBS, 20 mM N-ethylmaleimide (NEM), ice cold. An alkylating agent, such as NEM or iodoacetamide, should be included in the stop and lysis buffer to prevent artificial formation of disulfide bonds.

Lysis buffer: PBS, 0.5% (v/v) Triton X-100 (TX-100), 1 mM EDTA, 20 mM NEM, 1 mM phenylmethylsulfonyl fluoride (PMSF), and 1 μg/ml of aprotinin, chymostatin, leupeptin, pepstatin A (ice cold). Because alkylating agents and protease inhibitors have short half-lives in aquous solutions, they should be added to buffers immediately prior to use.

Wash buffer: 150 mM NaCl, 2 mM EDTA, 100 mM Tris–HCl, pH 8.3, 0.1% (w/v) SDS, 0.5% (w/v) Nonidet P-40, 0.5% (w/v) sodium deoxycholate

Homogenization buffer

TE: 20 mM Tris–HCl, pH 6.8, 1 mM EDTA

4× reducing sample buffer: 800 mM Tris–HCl, pH 6.8, 12% (w/v) SDS, 40% (v/v) glycerol, 4 mM EDTA, 0.01% (w/v) bromophenol blue, 300 mM dithiothreitol

Biosynthetic Processing of CGalT. CGalT-CHO cells grown in 6-cm tissue culture dishes are rinsed with PBS and once with depletion medium. To deplete cellular cysteine and methionine levels, cells are starved for 30 min in depletion medium. Cells are labeled in pulse medium for 5 min at 37°. Cells are rinsed with chase medium once and are incubated at 37° in chase medium. To follow biosynthetic processing of CGalT, the cells are put on ice after different periods of time, washed with stop buffer, and incubated for 20 min with stop buffer on ice. Cells are lysed in PBS, 1% (v/v) TX-100 and are centrifuged at 14,000g for 10 min at 4°. Cleared lysates are subjected to immunoprecipitation.

Protease Protection Assay. For a protease protection assay, CGalT-CHO cells are labeled metabolically for 15 min as described, followed by a chase period of 10 min, and PNS as described earlier. Fifty microliters 1 (±50 μg) PNS is incubated with 0.1 mg/ml of proteinase K or trypsin (pretreated with L-1-tosylamide-2-phenylethyl chloromethyl ketone) for 60 min at 10° in the presence or absence of 0.5% (w/v) saponin. The digestion should be performed in a small volume in order to keep the total amount of protease as low as possible. Samples are transferred to ice and the reaction is stopped by adding PMSF (2.5 mg/ml), leupeptin (0.25 mg/ml), aprotinin (0.25 mg/ml), pepstatin A (0.25 mg/ml), and trypsin inhibitor (1.0 mg/ml) to the indicated concentrations. Membranes are solubilized in 0.5%

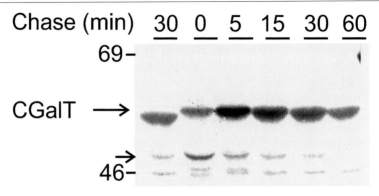

FIG. 3. Maturation of CGalT. CGalT-CHO cells were pulse labeled for 5 min with Tran[35]S-labeled amino acids and were then, $t = 0$, chased for different time intervals. After cell lysis, CGalT was imunoprecipitated with antiserum 635. Proteins were resolved under reducing conditions by SDS–polyacrylamide gel electrophoresis on a 10% gel. Please note the small shift in mobility of the mature CGalT (large arrow) and the disappearance of the immature CGalT of 50 kDa (small arrow) in time. The $t = 30$ sample was run twice to facilitate comparison with the $t = 0$ sample.

(v/v) TX-100, and CGalT is immunoprecipitated from the detergent lysates in the presence of protease inhibitors.

Immunoprecipitation. Protein A–Sephacryl CL4B beads are washed five times with ice-cold PBS, 0.5% (w/v) BSA and are incubated with anti-CGalT rabbit serum 635[21] for at least 1 hr at 4°. Beads are pelleted by centrifugation at 14,000g for 1 min at 4°. The supernatant is removed and the pellet is resuspended in ice-cold PBS, 0.5% (w/v) BSA. Cell lysates are incubated with the 60 μl of 10% beads for at least 1 hr at 4°. Beads are pelleted and washed three times with wash buffer. Eventually, the beads are resuspended in 30 μl TE, and 10 μl 4× reducing sample buffer is added. Samples are incubated for 5 min at 95° and centrifuged briefly at 14,000g. Samples are separated by SDS–PAA gel electrophoresis.[89] Gels are dried onto Whatman 3MM filter paper and exposed to a phosphor-imaging screen.

Results. Immature CGalT appears as a 50-kDa precursor protein that is N-glycosylated rapidly, resulting in a band of 54 kDa. A small but significant shift to a higher mobility form of CGalT occurred in the first hour after the pulse. This shift represents processing of N-linked oligosaccharides in the ER (Fig. 3).

The predicted molecular mass of the CGalT is approximately 64 kDa, and several studies describe an apparent molecular mass of 50–70 kDa.[6,26,27] We, however, consistently detected mature CGalT in different assay sys-

[89] U. K. Laemmli, *Nature* **227**, 680 (1970).

tems and in distinct cell lines as a band with an apparent molecular mass of 54 kDa.[21] In order to obtain sufficient resolution to separate mature and newly synthesized CGalT, we used 7.5 or 10% SDS polyacrylamide gels. Possibly the high content of hydrophobic amino acids in the lumenal portion of CGalT is responsible for the anomalous behavior of the protein on SDS–PAA gels.

Important questions to be solved in the near future are the coordinate transcriptional regulation of the enzymes involved in galactosphingolipid synthesis and the unraveling of the cellular functions of each of the various products.

[9] Assay of Lactosylceramide Synthase and Comments on Its Potential Role in Signal Transduction

By Subroto Chatterjee

Introduction

Lactosylceramide synthase (GalT-2) catalyzes the transfer of galactose from UDP-galactose to glucosylceramide (GlcCer) to form lactosylceramide via the following reaction:

$$\text{UDP-[}^{14}\text{C or }^3\text{H]Gal} + \text{GlcCer} \xrightarrow[\text{Mg}^{2+} \text{ and/or m}^{2+}]{\text{GalT-2}} \text{Gal}\beta\text{1-4GlcCer} + \text{UDP}$$

An essential feature of this galactosyltransferase is the requirement for manganese ions and detergent for optimal activity. Most, if not all, GalT-2 activity has been associated with the Golgi apparatus. However, evidence indicates that some enzymatic activity may also be associated with the cell membrane.[1,2]

Increases in GalT-2 activity have been associated with cell fusion in small Fu-1 cells, atherosclerotic tissue, and polycystic kidney diseased cells, as well as in renal cancer. This is accompanied by the accumulation of large amounts of LacCer in these tissues.[3–10] Interestingly, atherosclerotic tissue

[1] S. Chatterjee, *Arterio. Thromb. Vasc. Biol.* **18**, 1523 (1998).
[2] S. Basu and M. Basu, *in* "The Glycoconjugates" (M. Horowitz, ed.), Vol. 3, p. 265. Academic Press. New York, 1982.
[3] L. D. Cambrous and K. C. Leskawa, *Mol. Cell. Biochem.* **130**, 173 (1994).
[4] S. Chatterjee, S. Dey, W.-Y. Shi, K. Thomas, and G. M. Hutchins, *Glycobiology* **7**, 57 (1997).
[5] S. Chatterjee and E. Castiglione, *Biochim. Biophys. Acta* **923**, 136 (1987).
[6] S. Chatterjee, W.-Y. Shi, P. Wilson, and A. Mazumdar, *J. Lipid Res.* **37**, 1334 (1996).

is also enriched with oxidized low-density lipoproteins (Ox-LDL) and pro-inflammatory cytokines, such as tumor necrosis factor-α (TNF-α).[11] A possible role of GalT-2 in cell proliferation was suggested by the finding that Ox-LDL could activate GalT-2 at an early step in the signaling pathway.[12] In turn, LacCer served as a lipid second messenger that orchestrated an "oxidant-sensitive" signal transduction pathway, ultimately leading to cell proliferation in human aortic smooth muscle cells, as well as the proliferation of human kidney proximal tubular cells.[6,13–17] In human endothelial cells, TNF-α was found to activate GalT-2 and produce lactosylceramide. LacCer stimulates the expression of nuclear factor (NF)-kB and the expression of cell adhesion molecules, such as intracellular cell adhesion molecule (ICAM-1).[18] Interestingly, both of these signal transduction pathways leading to cell proliferation and/or cell adhesion include LacCer-mediated activation of an NADPH oxidase that produces superoxide. Superoxide, in turn, mediates the oxidant-sensitive transcription pathway that stimulates cell proliferation and/or cell adhesion in these two kinds of vascular cells. In human aortic smooth muscle cells, the LacCer-mediated generation of superoxide stimulates the GTP loading of p21^{ras}. Subsequently, the kinase cascade (Raf-1, Mek-2, and p^{44} MAPK) is activated. The phosphorylated form of p^{44} MAPK translocates from the cytoplasm to the nucleus and engages c-*fos* expression,[16] proliferating cell neutral antigen (PCNA) (cyclin) activation and cell proliferation.[19] Interestingly, D-*threo*-1-phenyl-2-decanoylamino-3-morpholino-1-propanol (D-PDMP), an inhibitor of GalT-2, can abrogate the Ox-LDL-mediated activation of GalT-2, the signal kinase cascade noted earlier, as well as cell proliferation in human aortic smooth muscle cells and/or cell adhesion in endothelial cells. Additional

---

[7] S. Chatterjee, *Biochem. Biophys. Acta* **1167**, 339 (1993).

[8] S. Chatterjee, C. S. Serkerke, and P. O. Kwiterovich, Jr., *J. Lipid Res.* **23**, 513 (1982).

[9] S. Chatterjee, P. O. Kwiterovich, Jr., P. Gupta, Y. Erozan, C. A. Alving, and R. Richard, *Proc. Natl. Acad. Sci. U.S.A.* **80**, 1313 (1983).

[10] D. Mukhin, F. F. Chao, and H. S. Kruth, *Arterio. Thromb. Vasc. Biol.* **15**, 1607 (1995).

[11] R. Ross, *Nature* **362**, 801 (1993).

[12] S. Chatterjee and N. Ghosh, *Glycobiology* **6**, 303 (1996).

[13] S. Chatterjee, A. K. Bhunia, H. Han, and A. Snowden, *Glycobiology* **7**, 703 (1997).

[14] S. Chatterjee, *Mol. Cell. Biochem.* **111**, 143 (1992).

[15] S. Chatterjee, *Biochem. Biophys. Res. Commun.* **181**, 554 (1991).

[16] A. K. Bhunia, H. Han, A. Snowden, and S. Chatterjee, *J. Biol. Chem.* **271**, 10660 (1996).

[17] A. K. Bhunia, H. Han, A. Snowden, and S. Chatterjee, *J. Biol. Chem.* **272**, 15642 (1997).

[18] A. K. Bhunia, T. Arai, G. Bulkley, and S. Chatterjee, *J. Biol. Chem.* **273**, 34349 (1998).

[19] S. Chatterjee, *Ind. J. Biochem. Biophys.* **34**, 56 (1997).

studies showed that the L-isoform of PDMP that activates GalT-2 can stimulate the proliferation of aortic smooth muscle cells.[13] The expression of ICAM-1 on the surface of endothelial cells may contribute to the adhesion of neutrophils and/or monocyte and thus initiate the process of atherosclerosis. However, the LacCer-mediated proliferation of aortic smooth muscle cells may contribute to the progression of atherosclerosis and restenosis following balloon angioplasty. This article presents a method to measure the activity of GalT-2 and focuses on the biological role of GalT-2 regarding signal transduction pathways leading to cell proliferation and cell adhesion.

GalT-2 Assay Method

The activity of GalT-2 is measured using $[^{14}C]$UDP-galactose as the galactose donor and GlcCer as the acceptor, essentially according to the procedure of Basu et al.[20] Either cultured mammalian cells or mammalian tissue homogenates can be used as a source of GalT-2. The product of this enzymatic reaction, LacCer, is partitioned using organic solvents and the radioactivity is measured by scintillation spectrophotometry. When first used with a new enzyme source, the product is further characterized to confirm its identity.

Reagents: Glucosylceramide (stearoyl glucocerebroside, Sigma, St. Louis, MO) [10 mM in chloroform–methanol, 2:1 (v/v)], Triton X-100–Cutscum [mixed at a ratio of 1:2 (v/v), at 21 mg/ml in chloroform–methanol, 1:2 (v/v)], cacodylate buffer (pH 6.8, 1.0 M), MnCl$_2$ and MgCl$_2$ (0.1 M each), 0.5 M KCl, and UDP-$[^{14}C]$galactose (10 mM at 1.5 × 10^5 dpm/10 μl)

High-performance liquid chromatography (HPLC) plates

Enzyme source: Transformed cultured human embryonic kidney cells (HEK-293 ATCC CRL-1573). The enzyme preparation is obtained as follows: Approximately 1 ml of packed HEK-293 cells is harvested in phosphate-buffered saline (PBS), pelleted, and 200 μl of cacodylate buffer (same as above) is added (this can be stored at −20° until use). Approximately 0.8 ml of fresh buffer containing 15 μg/ml phenylmethylsulfonyl fluoride and 10 μg/ml of leupeptin is added, and the cells are homogenized using a tight-fitting glass pestle. The homogenate is centrifuged at 500g for 15 min at 4°, the supernatant is collected for use as a source

[20] M. Basu, T. De, K. K. Das, J. W. Kyle, H. C. Chen, R. J. Schaefer, and S. S. Basu, *Methods Enzymol.* **138,** 575 (1987).

of GalT-2, and the nuclear debris is discarded. The protein content of the supernatant is measured.
Liquiscint (New England Nuclear, Boston, MA)

Procedure

The assay mixture for the measurement of GalT-2 activity consists of 20 mM of cacodylate buffer (pH 6.8), 1 mM Mg^{2+} and Mn^{2+}, 0.2 mg/ml Triton X-100 and Cutscum [ratio 1:2 (v/v)], 30 nmol of GlcCer, and 0.05 mole of UDP-galactose (1.5 × 10^5 dpm) in a total volume of 100 μl. After the addition of the enzyme, the mixture is incubated for 2 hr at 37°.

Assays without exogenous GlcCer serve as blanks (they should yield approximately 15–20 dpm, which are subtracted from the dpm of the full assay tubes). The assay is terminated by the addition of 10 μl of 0.25 M EDTA (pH 7.0), 10 μl of 0.5 M KCl, and 500 μl of chloroform–methanol [2:1 (v/v)], 5 μg of LacCer is added as carrier, the contents of the tubes are mixed for 30 sec, and the products are separated by centrifugation. The lower layer (approximately 300 μl) is saved, washed with 200 μl of chloroform–methanol–0.1 M/KCl [3:47:48 (v/v)], and centrifuged. The lower phase is collected and washed once again with 200 μl of chloroform–methanol–0.1 M/KCl [3:47:48 (v/v)], collected and dried under nitrogen, resuspended in 100 μl of chloroform methanol [2:1 (v/v)], and applied onto a high-performance thin-layer chromatography (HPTLC) plate. Standard glycosphingolipids of known structures such as glucosylceramide, LacCer, GbOseCer$_3$, and GbOseCer$_4$ are applied simultaneously and the plate is developed with chloroform–methanol–water [100:42:6 (v/v)]. The area corresponding to LacCer is transferred to a glass scintillation vial, 10 ml of Liquiscint' is added and mixed, and the radioactivity is measured with a scintillation counter. Alternatively, the lower phase is dried in nitrogen atmosphere and radioactivity is measured directly without further separation by HPTLC.

Characterization of Products

To determine the product formed due to GalT-2 activity, this assay is done employing at least 5 to 10 times the amount of enzyme and a proportional amount of nucleotide sugar. The product is purified by silicic acid column chromatography as follows: The lower phase mixture is dried under nitrogen, redissolved in chloroform (500 μl), and applied to a Sep-Pak silicic acid column. The column is washed with 5 ml of chloroform and the wash is discarded. The column is then washed with acetone–methanol [9:1

(v/v)] and the effluent is collected, dried under nitrogen, and separated by HPTLC as described earlier or separated by HPLC following perbenzoylation.[21]

General Properties of GalT-2

GalT-2 activity is linear with time (30–240 min), temperature (24–37°), enzyme (50–150 μg), and substrate (0.5–2 nmol/100 μl) concentration.[5] Under these conditions, GalT-2 activity (derived from cultured human kidney proximal tubular cells) is ~2–10 nmol/hr/mg/protein. Similar results have been obtained using the HEK-293 cell line.

pH Optima, pI Value and Size. Employing [^{14}C]UDP galactose, optimal enzyme activity is observed at pH 6.8. The enzyme is composed of a double band having apparent molecular weights of 58,000 and 60,000, and a p*I* of 4.55.[22]

Metal Ion and Detergent Requirements

GalT-2 requires Mg^{2+} and Mn^{2+}, a detergent mixture of Cutscum and Triton X-100, and a temperature of 37° for optimal activity.

Substrates and Inhibitors

The preferred substrate is glucosylceramide. Moreover, UDP-Gal is a preferred nucleotide sugar donor, and other nucleotide sugar donors are ineffective.

In normal human kidney proximal tubular cells, D-PDMP (10 μM) inhibits the activity of GalT-2. The activity of GalT-2 in purified preparation is also inhibited by 2.5 μM D-PDMP.[23]

Stability and Storage

Purified GalT-2 from human kidney is highly unstable even when stored in the presence of protease inhibitors. The purified GalT-2 is free of contaminating enzymes, such as β-galactosidase, and β-glucosidase.[22]

[21] R. H. McClure and M. D. Ullman, "Cell Surface Glycolipids." Am. Chem. Soc. Washington, DC, 1980.

[22] S. Chatterjee, N. Ghosh, and S. Khurana, *J. Biol. Chem.* **267,** 7148 (1991).

[23] S. Chatterjee, T. Cleveland, W. Y. Shi, J. C. Inokuchi, and N. J. Radin, *Glycoconjugates* **13,** 481 (1996).

However, HEK-293 cell enzyme preparations have stable enzyme activity up to 2 months when stored at $-20°$. In our laboratory, we divide up fresh GalT-2 preparations into small aliquots at $-20°$. Whenever necessary, one of the vials is thawed rapidly with warm water $(30°)$ and utilized for the enzyme assay.

Role of GalT-2 in Signal Transduction Pathways

An important feature regarding the regulation of GalT-2 activity was a finding that lipoproteins, such as human plasma low-density lipoproteins (LDL), exert a time- and concentration-dependent inhibition of the activity of this enzyme and, consequently, the biosynthesis of LacCer[24,25] in several mammalian cells. The binding, internalization, and degradation of LDL were essential in the LDL-mediated downregulation of GalT-2 activity, thus implicating the potential role of LDL receptors in this process.[24] However, various lipid components in LDL failed to suppress the activity of GalT-2 in cultured human kidney proximal tubular cells.[26] Collectively, these studies suggested that the apoprotein-B moiety through its interaction with the LDL receptors was an essential step for the suppression effect of LDL on the activity of GalT-2. This was supported by a series of observations employing mutant cells and normal cells. For example, fibroblasts from LDL receptor negative familial hypercholesterolemic homozygotes have an elevated activity of GalT-2. This phenomenon was also recapitulated in urinary epithelial cells derived from these patients.[5,24] An additional line of support arrived from work with human kidney tumor cells that lack LDL receptors. These studies revealed that LDL markedly stimulated the activity of GalT-2, as well as LacCer biosynthesis in tumor cells.[7] Interestingly, when the lysine residues in apolipoprotein-B in LDL were blocked by reductive methylation, such modified LDL was unable to enter the cells through the LDL receptor, which was associated with a marked stimulation in the activity of GalT-2. These findings suggest that entry of modified LDL into the cells through the scavenger pathway leads to upregulation, rather than a downregulation of GalT-2 activity.[24]

Role of GalT-2 in Cell Proliferation

The interaction of LDL with vascular cells such as endothelial cells, aortic smooth muscle cells, and macrophages results in the modification of

[24] S. Chatterjee, N. Ghosh, E. Castiglione, and P. O. Kwiterovich, Jr., *J. Biol. Chem.* **263,** 13017 (1988).
[25] S. Chatterjee, *Ind. J. Biochem. Biophys.* **25,** 85 (1988).
[26] S. Chatterjee and N. Ghosh, *Ind. J. Biochem. Biophys.* **27,** 375 (1990).

LDL, and LDL can also be oxidized in the laboratory with $CuSO_4$ (5 mM).[14] Oxidized LDL exerts a time- and concentration-dependent stimulation of the activity of GalT-2 and LacCer biosynthesis in cultured human aortic smooth muscle cells.[12] Subsequent studies revealed that LacCer could markedly stimulate the proliferation of human aortic smooth muscle cells.[16] D-PDMP inhibited Ox-LDL-induced GalT-2 activity and, consequently, cell proliferation (inhibition of this pathway was bypassed by the addition of LacCer, but not GlcCer or Cer); conversely, L-PDMP activated GalT-2 and stimulated the proliferation of the aortic smooth muscle cells.[13] Collectively, such observations revealed that GalT-2 is a target for Ox-LDL action. Furthermore, although D-PDMP was suggested previously to be a specific target for GlcT-1,[27] our collaborative studies revealed that it is a nonspecific inhibitor of a number of glycosphingolipid glycosyltransferases.[23]

The *in vivo* relevance of these findings was addressed by employing studies of the oxidative modification of LDL with red blood cells[28] (because hemoglobin is released during bleeding episodes and after myocardial infarction, there is an increase in myoglobin in the circulating blood from ruptured cardiomyocytes).[28,29] Employing oxygen levels of 60–70%, as are found in the venous circulation, LDL was modified more rapidly by hemoglobin under hypoxic conditions as compared to Cu^{2+}-mediated LDL oxidation. When such minimally modified LDL was added to cultured human aortic smooth muscle cells, it markedly increased cell proliferation via GalT-2 activation and p44 MAPK activation.

Although the studies just described indicated a potential role of LacCer as a lipid second messenger in mediating Ox-LDL-induced aortic smooth muscle cell proliferation via GalT-2 activation, whether LacCer itself or some other molecule was responsible for this phenomena was not understood. Subsequent studies revealed that LacCer stimulated the production of reactive oxygen species (ROS), particularly the superoxide radical in aortic smooth muscle cells.[17] Reactive oxygen species serve as a signaling intermediate for several pro-inflammatory cytokines, including TNF-α and interleukin-1.[30] Reactive oxygen species include the $O_2^{-\cdot}$ H_2O_2, nitric oxide (NO), and OH^- radical.[31] The production of ROS in oxidative stress has been shown to play an important role in different pathological conditions,

[27] N. S. Radin, J. A. Shayman, and J. I. Inokuchi, *Adv. Lipid Res.* **26**, 183 (1993).
[28] C. Balagopalakrishna, A. K. Bhunia, A. Snowden, J. M. Rifkind, and S. Chatterjee, *Mol. Cell. Biochem.* **170**, 85 (1997).
[29] C. Balagopalakrishna, R. Nirmala, J. M. Rifkond, and S. Ebetterice, *Adv. Exp. Med. Biol.* **18**, 337 (1997).
[30] J. P. Kehrer, *Crit. Rev. Toxicol.* **23**, 21 (1993).
[31] Y. C. Lo and T. F. Cruz, *J. Biol. Chem.* **270**, 11727 (1995).

such as atherosclerosis and cancer.[32,33] In addition, these molecules have been implicated in the induction of growth, in the regulation of kinase activity, and in the activation of endothelial-derived relaxation factor (NO).[34,35] By virtue of being very small, rapidly diffusible, and highly reactive, ROS are now thought to activate several transcription factors and the induction of gene expression.[36,37]

The incubation of aortic smooth muscle cells with LacCer stimulated the activity of NADPH oxidase, the production of $O_2^{-\cdot}$, and the activation of a kinase cascade involving Raf, Mek-2, and p^{44} mitogen-activated protein kinase (p^{44} MAPK activity).[17] MAPK is a cytosolic enzyme and its activation/phosphorylation by lactosylceramide results in its translocation to the nucleus where it stimulates the expression of c-*fos*, a protooncogene involved intimately in cell proliferation.[13,16,17] Additional studies revealed that oxidized LDL and/or LacCer could stimulate the expression of proliferating cell nuclear antigen (PCNA), also known as cyclin. Interestingly, in cultured aortic smooth muscle cells, D-PDMP inhibited the Ox-LDL-mediated stimulation in the expression of PCNA in these cells.[19] The function of LacCer as a lipid second messenger in the oxidant-sensitive transcription pathway described earlier was substantiated by the use of a variety of inhibitors known to abrogate these reactions. For example, the LacCer-induced NADPH oxidase activity could be abrogated by the pre incubation of cells with diphenylene iodonium (DPI), an inhibitor of NADPH oxidase. In contrast, KCN, an inhibitor of NADH oxidase, had no effect on this reaction. Allopurinol, an inhibitor of xanthine/oxidase, and *N*-methyl arginine, an inhibitor of nitric oxide synthase, did not abrogate the oxidized LDL-induced activation of GalT-2 or LacCer-induced $O_2^{-\cdot}$ generation and cell proliferation. Glutathione decreased, whereas buthionine sulfoximine (an inducer of $O_2^{-\cdot}$ generation) augmented LacCer-induced phenomena. The LacCer-induced $O_2^{-\cdot}$ generation and cell proliferation was independent of PKC,[16] as staurosporine and high doses of phorbol myristic acetate did not prevent the proliferative effects of LacCer. Interestingly, none of these inhibitors, as well as Cu^{2+}, altered the activity of GalT-2 in human aortic smooth muscle cells. Additional studies showed that platelet-derived growth and epidermal growth factors activate GalT-2 and recruit LacCer in inducing cell proliferation. These findings suggest that GalT-2 activation via

[32] B. Halliwell and J. M. C. Gutteridge, *Methods Enzymol.* **186**, 1 (1990).

[33] J. P. Kehrer, *Crit. Rev. Toxicol.* **23**, 21 (1993).

[34] R. Schreck, P. Rieber, and P. A. Baeuerle, *Eur. Mol. Biol. Organ J.* **10**, 2247 (1991).

[35] R. Larsson and P. Cerutti, *J. Biol. Chem.* **263**, 17452 (1988).

[36] P. L. Puri, M. L. Avantaggiati, V. L. Burgio, P. Chirillo, D. Collepardo, G. Natoli, C. Balsano, and M. Levero, *J. Biol. Chem.* **270**, 22129 (1995).

[37] R. Pinkus, L. M. Weiner, and V. Daniel, *J. Biol. Chem.* **270**, 13422 (1996).

diverse agonists and the subsequent production of LacCer are key events in signaling phenomena leading to cell proliferation in human aortic smooth muscle cells and human kidney proximal tubular cells.[15,6]

Role of GalT-2 in Cell Adhesion

The intracellular adhesion molecule-1 (ICAM-1) is inducible on the endothelial cell surface following stimulation with TNF-α.[38,39] ICAM-1 is a specific ligand for lymphocyte function-associated antigen-1 (LFA)[40] and Mac-1 (CD-11b, CD-18) expressed on the surface of neutrophils and monocytes[41] and in human atherosclerotic lesions.[42] Incubation of human umbilical vein-derived endothelial cells (HUVEC) with TNF-α results in the activation of GalT-2 in a time- and concentration-dependent fashion. Tumor necrosis factor-α-induced expression of ICAM-1 was abrogated by D-PDMP, and the addition of LacCer reversed the D-PDMP effect on TNF-α-induced ICAM-1 expression in HUVEC.[18] Overexpression of endogenous CuZn-superoxide dismutase inhibited LacCer-induced ICAM-1 expression, suggesting that LacCer induced an oxidant-sensitive transcriptional pathway in endothelial cells that led to ICAM-1 expression. Interestingly, LacCer did not stimulate the expression of E-selectin or VCAM in HUVECs. Because TNF-α can induce the expression of all of these adhesion molecules in endothelial cells, this suggests that the TNF-α-mediated GalT-2 activation and LacCer production is a specific phenomena involving ICAM-1 expression. The expression of ICAM-1 in HUVEC by lactosylceramide was accompanied by the adhesion of neutrophils, which was visualized by confocal microscopy as well as quantitative cell adhesion assays.[18]

Acknowledgments

This work was supported by NIH Grant RO-1-DK31722 and RO-1-HL47212. I gratefully acknowledge the collaboration of Drs. Peter Kwiterovich, Nupur Ghosh, Anil Bhunia, Srabani Dey, Dimitry Mukhin, Grover Hutchins, and Avi Mazumdar and Ms. Hui Han, Ms. Tavia Cleveland, Ms. Ann Snowden, and Ms. Irina Dobromilskaya.

[38] M. S. Diamond, D. E. Staunton, S. D. Marlin, and T. A. Springer, *Cell* **65**, 961 (1991).
[39] M. P. Bevilaqua, S. Stengelin, M. A. Gimbrone, and B. Seed, *Science* **243**, 1160 (1989).
[40] S. D. Marlin and T. A. Springer, *Cell* **51**, 813 (1987).
[41] T. A. Springer, *Nature* **346**, 425 (1990).
[42] M. Cybulsky and M. A. Gimbrone, Jr., *Science* **251**, 788 (1991).

[10] *In Vitro* Assays for Enzymes of Ganglioside Synthesis

By GOTTFRIED POHLENTZ, CHRISTOPH KAES, and KONRAD SANDHOFF

Introduction

Glycosyltransferases of ganglioside biosynthesis transfer monosaccharide from an activated sugar (sugar nucleotide) to either ceramide or the growing oligosaccharides of the glycolipids. A number of ganglioside glycosyltransferases from different sources have been cloned and expressed in a variety of cell lines,[1–5] and even knockout mice with disrupted *N*-acetylgalactosaminyltransferase (GalNAc-T) genes have been created.[6] Despite the great advances in molecular biology and their important contribution to the investigation of ganglioside biosynthesis, *in vitro* assays for these enzymes still remain necessary for the determination of transferase properties such as substrate specificities, kinetic data, ion requirements, and identification and structure analysis of the products. For example, the basis for the generally accepted model of ganglioside biosynthesis shown in Fig. 1 has been evaluated from the results of *in vitro* competition experiments.[7,8]

Ganglioside glycosyltransferases are type II membrane proteins with active sites that face the lumen of the Golgi apparatus and utilize a membrane-resident lipid acceptor and a water-soluble activated sugar donor (probably following a two-dimensional Michaelis–Menten kinetics, as exemplified by Scheel *et al.*[9]). In general, glycosyltransferases involved in ganglioside biosynthesis (Fig. 1) catalyze the following reactions:

[1] G. Zeng, L. Gao, T. Ariga, and R. K. Yu, *Biochem. Biophys. Res. Commun.* **226,** 319 (1996).

[2] N. Kurosawa, T. Hamamoto, M. Inoue, and S. Tsuji, *Biochim. Biophys. Acta* **1244,** 216 (1995).

[3] S. Yamashiro, M. Haraguchi, K. Furukawa, K. Takamiya, A. Yamamoto, Y. Nagata, K. O. Lloyd, H. Shiku, and K. Furukawa, *J. Biol. Chem.* **270,** 6149 (1995).

[4] K. Nara, Y. Watanabe, I. Kawashima, T. Tai, Y. Nagai, and Y. Sanai, *Eur. J. Biochem.* **238,** 647 (1996).

[5] M. S. Lutz, E. Jaskiewicz, D. S. Darling, K. Furukawa, and W. W. Young, *J. Biol. Chem.* **269,** 29227 (1994).

[6] K. Takamiya, A. Yamamoto, K. Furukawa, S. Yamashiro, M. Shin, M. Okada, S. Fukumoto, M. Haraguchi, N. Takeda, K. Fujimura, M. Sakae, M. Kishikawa, K. Furukawa, and S. Aizawa, *Proc. Natl. Acad. Sci. U.S.A.* **93,** 10662 (1996).

[7] G. Pohlentz, D. Klein, G. Schwarzmann, D. Schmitz, and K. Sandhoff, *Proc. Natl. Acad. Sci. U.S.A.* **85,** 7044 (1988).

[8] H. Iber, R. Kaufmann, G. Pohlentz, G. Schwarzmann, and K. Sandhoff, *FEBS Lett.* **248,** 18 (1989).

[9] G. Scheel, E. Acevedo, E. Conzelmann, H. Nehrkorn, and K. Sandhoff, *Eur. J. Biochem.* **127,** 245 (1982).

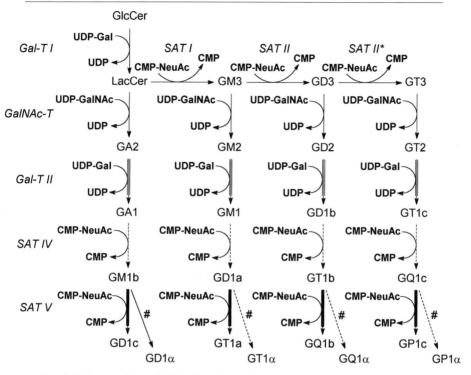

FIG. 1. Pathways of ganglioside biosynthesis (for structures of gangliosides, see Fig. 2). The identity of GA2/GM2/GD2 synthase (GalNAc-T), GA1/GM1/GD1b synthase (Gal-T II), and GM1b/GD1a/GT1b synthase (SAT IV), respectively, from rat liver has been demonstrated by *in vitro* competition experiments [G. Pohlentz, D. Klein, G. Schwarzmann, D. Schmitz, and K. Sandhoff, *Proc. Natl. Acad. Sci. U.S.A.* **85,** 7044 (1988); H. Iber, R. Kaufmann, G. Pohlentz, G. Schwarzmann, and K. Sandhoff, *FEBS Lett.* **248,** 18 (1989)]. Pound sign: In rat liver, sialylation of GM1b leads to GD1a, as shown by the large-scale preparation of the product *in vitro* and subsequent structure analysis [K. I.-P. J. Hidari, I. Kawashima, T. Tai, F. Inagaki, Y. Nagai, and Y. Sanai, *Eur. J. Biochem.* **221,** 603 (1994)]. Asterisk: Expression cloning of human GT3 synthase demonstrated that GD3 and GT3 synthases are identical [J. Nakayama, M. N. Fukuda, Y. Hirabayashi, A. Kanamori, K. Sasaki, T. Nishi, and M. Fukuda, *J. Biol. Chem.* **271,** 3684 (1995)].

Galactosyltransferases (Gal-T):

$$\text{UDP-Gal} + \text{HO-acceptor} \xrightarrow{\text{Gal-T}} \text{Gal-O-acceptor} + \text{UDP}$$

N-Acetylgalactosaminyltransferases:

$$\text{UDP-GalNAc} + \text{HO-acceptor} \xrightarrow{\text{GalNAc-T}} \text{GalNAc-O-acceptor} + \text{UDP}$$

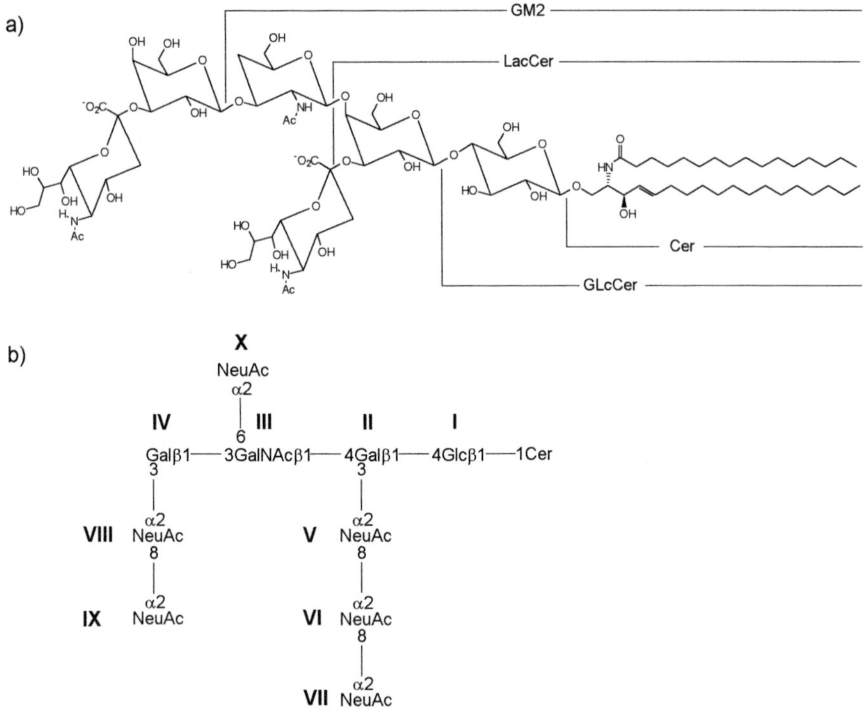

a)

b)

The following glycosphingolipid structures are part of the hexasialoganglioside structure shown above and contain in addition to the ceramide (Cer) moiety the monosaccharides indicated by Roman numbers:

GlcCer - glucosylceramide, **I**	GD1b - **I** to **VI**	GT1α - **I** to **V, VIII, X**
LacCer - lactosylceramide, **I, II**	GT1c - **I** to **VII**	GQ1α - **I** to **VI, VIII, X**
GM3 - **I, II, V**	GM1b - **I** to **IV, VIII**	GP1α - **I** to **VII, VIII, X**
GD3 - **I, II, V, VI**	GD1a - **I** to **V, VIII**	
GT3 - **I, II, V** to **VII**	GT1b - **I** to **VI, VIII**	
GA2 - **I** to **III**	GQ1c - **I** to **VII, VIII**	
GM2 - **I** to **III, V**	GD1c - **I** to **IV, VIII, IX**	
GD2 - **I** to **III, V, VI**	GT1a - **I** to **V, VIII, IX**	
GT2 - **I** to **III, V** to **VII**	GQ1b - **I** to **VI, VIII, IX**	
GA1 - **I** to **IV**	GP1c - **I** to **IX**	
GM1 - **I** to **V**	GD1α - **I** to **IV, VIII, X**	

FIG. 2. Structures of (a) ganglioside GD1a and (b) a hexasialoganglioside and related glycosphingolipids. The nomenclature of gangliosides is according to Svennerholm [*J. Neurochem.* **10,** 613 (1963)].

Sialyltransferases (SAT):

$$\text{CMP-NeuAc} + \text{HO-acceptor} \xrightarrow{\text{SAT}} \text{NeuAc-O-acceptor} + \text{CMP}$$

Methods for assaying galactosyltransferases, N-acetylgalactosaminyltransferases, and sialyltransferases described in this article are used routinely in our laboratory. These procedures will be presented in detail and compared to two very common alternative methods.

Materials

Thin-layer chromatography (TLC) plates and silica gel 60 can be purchased from Merck (Darmstadt, FRG). Uridine 5'-diphospho-[U-^{14}C]galactose (UDP-[^{14}C]Gal), uridine 5'-diphospho-N-acetyl[1-^{14}C]galactosamine (UDP-[^{14}C]GalNAc), and cytidine 5'-monophospho-N-acetyl[4,5,6,7,8,9-^{14}C]neuraminic acid (CMP-[^{14}C]NeuAc) are currently available from Amersham-Buchler (Braunschweig, FRG) and are used after dilution with the respective unlabeled sugar nucleotide from Sigma (Deisenhofen, FRG). Sephadex G-25 superfine is purchased from Pharmacia (Uppsala, Sweden). X-ray film XAR-5 is obtained from Eastman Kodak Company (Rochester, NY). All reagents and solvents used are of analytical grade quality.

Solvents

A: Chloroform/methanol 2:1 (by volume)
B: Chloroform/methanol 1:1 (by volume)
C: Chloroform/methanol/water 24:12:1:6 (by volume)
D: Chloroform/methanol/water 80:20:2 (by volume)
E: Chloroform/methanol/water 65:25:4 (by volume)
F: Chloroform/methanol/0.2% CaCl$_2$ (w/v) 60:35:8 (by volume)
G: Chloroform/methanol/0.2% CaCl$_2$ (w/v) 55:45:10 (by volume)
H: Chloroform/methanol/water 60:35:8 (by volume)
I: Chloroform/methanol/water 55:45:10 (by volume)

Methods

Glycosyltransferase Assays

All determinations should be done at least in duplicate and an appropriate blank without exogenous acceptor should be performed for each experiment.

General methods are presented for Gal-T and GalNAc-T and for SAT, respectively, because the assays for galactosyltransferases and N-acetylga-

TABLE I

COMPOSITION OF BUFFER MIXTURES FOR GALACTOSYL- AND
N-ACETYLGALACTOSAMINYLTRANSFERASE STANDARD ASSAYS

	Gal-T assay		GalNAc-T assay	
Stock solution	μl per assay	Final concentration[a]	μl per assay	Final concentration[a]
Sodium cacodylate/HCl	8	64 mM	8	64 mM
MnCl$_2$	8	20 mM	8	20 mM
CDP-choline	4	10 mM	4	10 mM
UDP-[^{14}C]Gal	10	0.5 mM	—	—
UDP-[^{14}C]GalNAc	—	—	10	0.5 mM
H$_2$O or additive	10	—	10	—
Total volume of buffer mixture per assay (μl)	40	—	40	—

[a] Final concentrations are calculated for the total assay volume of 50 μl.

lactosaminyltransferases differ only by the sugar nucleotide and the acceptor and those for all sialyltransferases are the same except for the acceptors used in individual SAT assays.

Galctosyl- and N-Acetylgalactosaminyltransferases

For Gal-T and GalNAc-T assays the following stock solutions are needed:

Sodium cacodylate/HCl, pH 7.35, 400 mM
MnCl$_2$, 125 mM
CDP-choline, 125 mM
UDP-[^{14}C]Gal, 2.5 mM, 90,000 dpm/nmol
UDP-[^{14}C]GalNAc, 2.5 mM, 90,000 dpm/nmol
Triton X-100, 0.5% (w/v) in solvent A
Acceptor gangliosides, 0.1–1 mM in solvent A or B

To achieve a fast and convenient performance of the assays, the preparation of so-called "buffer mixtures" is recommended. These mixtures contain the aqueous stock solutions in appropriate amounts resulting in the desired final concentrations of buffer, divalent cations, and nucleotides in the transferase assays, as shown in Table I. The total volume for glycosyltransferase assays is 50 μl and consists of a 40-μl buffer mixture and 10 μl of transferase preparation. (Buffer mixtures should be freshly prepared for each experiment.)

Procedure. Ten microliters of the Triton X-100 solution [resulting in a final concentration of 0.1% (w/v)] and the desired volume of the acceptor solution in solvents A and/or B, respectively, are applied to the test tubes. The organic solvents are removed under a stream of nitrogen. Forty microliters of the buffer mixture is added and the mixture is shaken vigorously and each sample is sonicated for 30 sec in a cup horn sonicator at 100 W. Ten microliters of transferase preparation is applied and, after shaking the samples again, they are incubated for 15 to 120 min (depending on the specific activity of the enzyme source). Finally, the reaction is stopped by adding 1 ml of solvent A. Further treatment of the assays (which is the same for SAT, Gal-T, and GalNAc-T) is described in the following paragraphs.

Sialyltransferases

For SAT assays, the following stock solutions are needed:
Sodium cacodylate/HCl, pH 6.6, 750 mM
MgCl$_2$, 125 mM
Mercaptoethanol, 125 mM
CMP-[^{14}C]NeuAc, 5 mM, 55,000 dpm/nmol
Triton CF-54, 0.75% (w/v) in solvent A
Acceptor gangliosides, 0.1–1 mM in solvent A or B
The procedure for SAT assays is essentially the same as described for Gal-T and GalNAc-T, with buffer mixtures according to Table II. After application of 10 μl Triton CF-54 solution (final concentration 0.15%) and the appropriate amount of acceptor in solvent A and/or B and evaporation of the solvent, 40 μl of the buffer mixture is added and the samples are shaken and sonicated as described earlier. Ten microliters of enzyme preparation is applied and, after shaking the samples again, they are incubated for 15 to 120 min (depending on the specific activity of the enzyme source). Finally, the reaction is stopped by adding 1 ml of solvent A. Further treat-

TABLE II
Composition of Buffer Mixtures for Sialyltransferase Standard Assays

Stock solution	μl per assay	Final concentration[a]
Sodium cacodylate/HCl	10	150 mM
MgCl$_2$	4	10 mM
Mercaptoethanol	4	10 mM
CMP-[^{14}C]NeuAc	10	1mM
H$_2$O or additive	12	—
Total volume of buffer mixture per assay (μl)	40	—

[a] Final concentrations are calculated for the total assay volume of 50 μl.

ment of the assays (which is the same for SAT, Gal-T, and GalNAc-T) is described in the following paragraph.

Processing of Glycosyltransferase Assays

For the separation of lipids and polar constituents of glycosyltransferase assays, a combination of gel chromatography and affinity chromatography is utilized. The following procedure was introduced by Wells and Dittmer[10] and is very efficient in retaining polar nonlipid contaminants. The recovery of lipids is >99%, as has been shown with radiolabeled gangliosides.[11]

For this procedure, a suspension of Sephadex G-25 (superfine) preswollen in water is prepared. Twenty-five grams of gel is suspended in 300 ml of water by shaking the mixture vigorously. Swelling should last 3 to 4 hr at room temperature or overnight at 4°. Before use, approximately 100 ml of the supernatant water should be decanted.

Pasteur pipettes (150 mm) are stuffed with glass wool for use as columns. Pasteur pipettes without a neck should be used because they are filled more easily than those with a neck. Before filling the Pasteur pipettes, the Sephadex G-25 is brought to room temperature and shaken vigorously to obtain a homogeneous suspension. After filling five columns (to the top), the gel suspension should be shaken again to achieve a uniform gel bed in all columns (Sephadex G-25 settles rather quickly). After the water has drained, the gel is overlayed with methanol to the top of the column, and the solvent is allowed to drain through the gel (during this procedure the gel bed shrinks to approximately one-third of the starting volume). The gel is then washed with 1.5 ml of methanol followed by 3 ml of solvent C (the solvent mixture that corresponds to the "stopped" samples, i.e., the assay mixture plus 1 ml of solvent A).

Either scintillation vials (for the direct determination of glycosyltransferase activities) or glass tubes (for further processing, e.g., product indentification, see later) are placed under the columns. The "stopped" samples are then applied to the columns and the sample tubes are washed with 1 ml of solvent A, which is also added to the columns. Elution of the products is completed by adding two 1-ml aliquots of solvent A.

Determination of Transferase Activities

The solvent in the scintillation vials is evaporated under a stream of nitrogen, the residue is dissolved in 10 ml of scintillation cocktail, and the

[10] M. A. Wells and J. C. Dittmer, *Biochemistry* **2,** 1259 (1963).
[11] H. K. M. Yusuf, G. Pohlentz, G. Schwarzmann, and K. Sandhoff, *Eur. J. Biochem.* **134,** 47 (1983).

radioactivity is determined in a liquid scintillation counter. The specific activity of the glycosyltransferase may be calculated from the measured radioactivities as follows:

$$\text{specific activity} \left[\frac{\text{nmol}}{\text{mg} \cdot \text{hr}}\right] = \frac{(\text{cpm}_{\text{sample}} - \text{cpm}_{\text{blank}}) \, [\text{cpm}]}{\text{SA}_{\text{sn}} \left[\dfrac{\text{cpm}}{\text{nmol}}\right]}$$

$$\cdot \frac{1}{\text{amount}_{\text{prot}} \, [\text{mg}] \cdot \text{time} \, [\text{hr}]}$$

where SA_{sn} is the specific radioactivity of the sugar nucleotide.

Product Identification

For this procedure, the Sephadex eluates are collected in glass tubes and the solvent is evaporated under a stream of nitrogen. The residue is first moistened with 10 μl water and then dissolved in 150 μl of solvent B. The radioactivity of an aliquot (10 μl) is determined as described earlier and the remaining solution is applied to a TLC plate. The chromatogram is developed in solvent F or G. After drying the TLC plate, radioactive products are visualized by exposure to an X-ray film (autoradiography). Figure 3 shows an example of a TLC separation of sialyltransferase products obtained from different acceptors.

The radioactive products can also be localized with a TLC linear analyzer (e.g., Multy Tracemaster, Berthold, Wildbach, Germany) or a Phosphoimager (Fuji Bas 1000, Raytest, Straubenhard, Germany). Both methods allow quantitative determination of the product distribution.

Structure Analyses of Products

Because of the relatively low amounts of product formed in glycosyltransferase assays, nuclear magnetic resonance experiments are only feasible in a few cases (e.g., when an enzyme preparation with high specific activity can be used to obtain sufficient material). The method of choice, therefore, is mass spectrometry, preferably fast atom bombardment mass spectrometry (FAB MS). This method allows analyses of gangliosides with rather high molecular weights (up to 5000) with high sensitivity; moreover, FAB spectra show characteristic fragmentation patterns that reveal information about the structure.

For product analysis, 20–100 assays (depending on the yield of product of a single assay) with the desired acceptor and unlabeled sugar nucleotide are performed as described earlier. (Note that simply increasing the scale

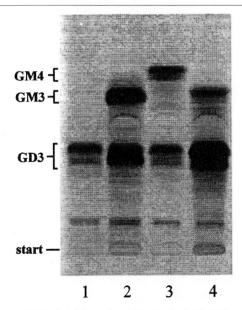

FIG. 3. Separation by TLC of sialyltransferase products obtained from different acceptors. Golgi vesicles were incubated with the acceptors as indicated and CMP-[^{14}C]NeuAc as described in the text. The radioactive products were separated with solvent A and visualized by autoradiography. Exogenously added acceptors are as follows: Lane 1, none; lane 2, LacCer; lane 3, GalCer; and lane 4, GM3. The products are lane 1, GD3* from endogenous GM3; lane 2, GM3* and GD3*; lane 3, GM4* (and GD3* from endogenous GM3); and lane 4, GD3* (and some ^{14}C-labeled GM3*, most probably from the subsequent action of endogenous sialidase and sialyltransferase). Asterisk indicates that multiple bands are due to heterogeneity in the ceramide moiety.

of the standard assay often results in a decreased product yield.) Additionally, one to five standard assays with radioactive donor are prepared to obtain a tracer. The Sephadex eluates are combined and the solvent is evaporated either in a stream of nitrogen or under reduced pressure. The products are purified either by TLC or by column chromatography on silica gel 60. (Needless to say, for these procedures, extra clean glassware and redistilled solvents should be used to avoid contaminations, e.g., with softeners or detergents.)

For TLC purification, the residue is moistened with a few microliters of water and dissolved in the smallest possible volume of solvent B. The solution is applied to a TLC plate and the chromatogram is developed in solvent F or G. The radioactive products are localized by autoradiography, radioscanning, or a phosphoimager. The corresponding areas are scraped from the plate and the silica gel is extracted at least twice (at room temperature for 30 min) with either solvent B or methanol, depending on the polarity of the product. A third extract should be prepared, but should only be combined with the first two if it contains considerable amounts of radioactive material. The combined extracts are dried under a stream of nitrogen. Remaining silicic acid oligomers (from the extraction procedure), which sometimes disturb the FAB MS analysis, can be removed by repeated dissolving of the residue in solvent B, transfering to a fresh tube, and evaporating the solvent. The final residue is subjected to FAB MS analysis. Figures 4 and 5 show the FAB spectrum and the corresponding fragmentation scheme, respectively, of GM1b obtained as a sialyltransferase product from GA1 using the just-described procedure, as an example.[12]

For purification by column chromatography, a silica gel 60 column (bed volume 10 ml) equilibrated in solvent D is prepared. The residue from the combined Sephadex eluates is dissolved in a small volume of the same solvent (preferably 0.5–1 ml) and the solution is applied to the column. To remove the detergent, the column is first eluted with 10 bed volumes of solvent D. The products are subsequently eluted by a stepwise gradient of solvents E, H, and I (20 bed volumes each). The eluate is collected in 2-ml fractions and the fractions are monitored for radioactivity. The corresponding fractions are combined and the solvent is evaporated. The residue is analyzed by FAB MS.

Alternative Methods

Because the requirements for most of the mammalian glycosyltransferases concerning pH, divalent cations, detergents, etc. are almost the same, even when the enzymes derive from different species, the assay conditions used by other scientists are very similar to those described in this article. In contrast, a number of different procedures have been described for the subsequent processing of the assays and the determination of the transferase activities.

For example, enzyme-linked immunosorbent assay (ELISA) methods have been described[13,14] where the incubations and the quantifications of the products are performed on microtiter plates. This method has the advantages of simplicity, speed, no use of radioactivity, separation of the product is not needed, and the simultanous analysis of many samples is possible. However, this procedure can only be employed when a highly specific antibody directed against the product is available. In addition,

[12] G. Pohlentz, D. Klein, D. Schmitz, G. Schwarzmann, J. Peter-Katalinic, and K. Sandhoff, *Biol. Chem. Hoppe-Seyler* **369,** 55 (1988).
[13] T. Taki, S. Nishiwaki, K. Ishii, and S. Handa, *J. Biochem.* **107,** 493 (1990).
[14] M. Nakamura, A. Tsunoda, and M. Saito, *Anal. Biochem.* **198,** 154 (1991).

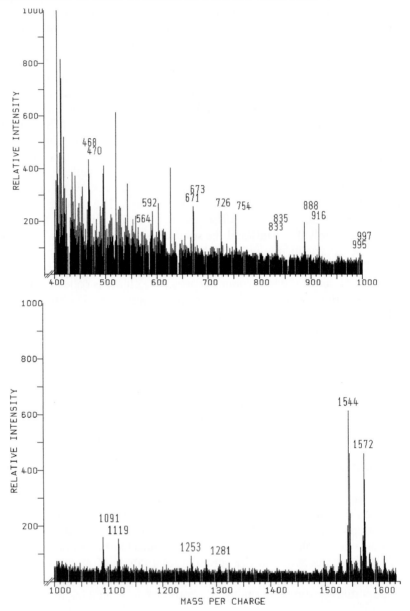

Fig. 4. Negative ion FAB mass spectrum of GM1b synthesized by rat liver Golgi sialyltransferase IV.

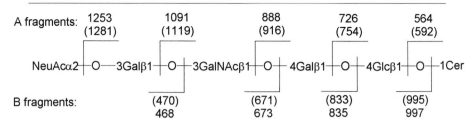

A fragments: 1253   1091   888   726   564
              (1281)   (1119)   (916)   (754)   (592)

NeuAcα2─O──3Galβ1─O─3GalNAcβ1─O─4Galβ1─O─4Glcβ1─O─1Cer

B fragments: (470)   (671)   (833)   (995)
                           468   673   835   997

FIG. 5. FAB(-) fragmentation scheme of GM1b synthesized by rat liver Golgi sialyltransferase IV. The A fragments and the molecular ion (at m/z 1544, 1572) appear as double peaks
for C_{18} and C_{20} sphingosine. The B fragment peaks are accompanied by B-H_2 peaks.

sufficient amounts of an authentic product standard must be available for
calibration of the assay.

Other authors use ion-exchange chromatography to isolate ganglioside
glycosyltransferase products.[15,16] In addition to the disadvantage that these
methods are restricted to charged products, the product solutions contain
high amounts of salts (i.e., potassium or ammonium acetate) that must
be removed in an extra step for product identification by TLC or for
structure analysis.

The most widely used methods are descending paper chromatography[17–20] and reversed-phase chromatography using C_{18} Sep-Pak cartridges,[21–24] which are described briefly.

Paper Chromatography. Enzyme reactions are stopped by adding
EDTA solution, and the resulting mixtures are applied to chromatography
paper (e.g., Whatman 3MM paper). The chromatogram is developed with
a descending solvent tray using 1% aqueous sodium tetraborate solution.
The glycosyltransferase products remain at the origin, whereas sugar nucleotides and their degradation products move with the front. Areas of the
transferase products are excised and the radioactivities are determined with
a liquid scintillation counter. The major disadvantages of this procedure
are the high quenching of the paper strips (especially when ³H-labeled

[15] K. M. Walton and R. L. Schnaar, *Anal. Biochem.* **152,** 154 (1986).
[16] M. Kawano, K. Honke, M. Tachi, S. Gasa, and A. Makita, *Anal. Biochem.* **182,** 9 (1989).
[17] M. Basu, T. De, K. K. Das, J. W. Kyle, H.-C. Chon, R. J. Schaeper, and S. Basu, *Methods Enzymol.* **138,** 575 (1987).
[18] M. Trinchera, M. Fabbri, and R. Ghidoni, *J. Biol. Chem.* **266,** 20907 (1991).
[19] A. L. Sherwood and E. H. Holmes, *J. Biol. Chem.* **267,** 25328 (1992).
[20] S. Ghosh, J. W. Kyle, S. Dastgheib, F. Daussin, Z. Li, and S. Basu, *Glycoconj. J.* **12,** 838 (1995).
[21] M. A. Williams and R. H. McCluer, *J. Neurochem.* **35,** 266 (1980).
[22] L. Melkerson-Watson, K. Kanemitsu, and C. C. Sweeley, *Glycoconj. J.* **4,** 7 (1987).
[23] Y. Hashimoto, S. Sekine, K. Iwasaki, and A. Suzuki, *J. Biol. Chem.* **268,** 25857 (1993).
[24] F. Irie, S. Kurono, Y.-T. Li, Y. Seyama, and Y. Hirabayashi, *Glycoconj. J.* **13,** 177 (1996).

sugar nucleotides are used) and the necessity to extract the products from chromatography paper prior to product identification.

Reversed-Phase Chromatography. The assays are mixed with 0.1 M KCl solution and applied to C_{18} Sep-Pak cartridges. The cartridges are washed with (0.1 M KCl solution and) water and the ganglioside products are eluted with methanol and chloroform/methanol mixtures (1:1 or 2:1 by volume). The combined eluates are evaporated to dryness and the residues are monitored for radioactivity in a liquid scintillation counter. Compared to the method presented in this article, this procedure has a number of drawbacks. First, there is an additional washing step(s) before the lipids can be eluted. For every washing and elution step, the cartridge (for each sample) has to be detached and attached to a luer lock syringe, which makes the procedure very time-consuming, especially when a larger number of samples are processed. (Using the method described in this article, up to 50 samples may be handled—including incubation, preparation of the Sephadex columns, and elution of the products—within 2 hr.) Because the mantle of the cartridges is made of plastic, elution with organic solvents leads to extraction of softeners (e.g., dioctylphthalate) that disturb the product analysis by mass spectrometry.

[11] Analysis of Sulfatide and Enzymes of Sulfatide Metabolism

By Firoze B. Jungalwala, Marvin R. Natowicz, Prasoon Chaturvedi, and David S. Newburg

Sulfatide or ceramide galactosyl-3'-sulfate is a major sulfoglycolipid of the mammalian nervous system, kidney, and spleen and is present in low concentrations in other tissues. In the nervous system, sulfatide is an integral constituent of myelin membranes and is biosynthesized by oligodendroglial cells in the central nervous system and by Schwann cells in the peripheral nervous system. Extensive reviews on the chemistry, metabolism, and functional distribution of sulfoglycolipids, including sulfatide, have appeared.[1,2] This article discusses methods of analysis of sulfatide by high-performance liquid chromatography (HPLC) and of the enzymes sulfatidase (EC 3.1.6.8) and sulfotransferase (EC 2.8.2.11) that are involved in its metabolism.

[1] I. Ishizuka, *Prog. Lipid Res.* **4,** 245 (1997).
[2] J. P. Vos, M. Lopes-Cardozo, and B. M. Gadella, *Biochim. Biophys. Acta* **1211,** 25 (1994).

The analysis of sulfatide and the characterization of sulfatidase have been studied intensively since the early 1970s due to the recognition that sulfatide accumulates in tissues, particularly brain, of patients with a lysosomal storage disorder, metachromatic leukodystrophy (MLD).[3] Different clinical forms of MLD have been described and nearly all cases are due to the deficient enzyme activity of sulfatidase, more commonly known as arylsulfatase A (ASA).[3] Mutations in the ASA gene, in turn, are responsible for this autosomal recessive disorder. Rare cases of ASA deficiency are due to a deficiency of an activator protein, saposin B, that works in concert with ASA to hydrolyze the glycolipid substrate. This latter disorder, also inherited as an autosomal recessive condition, is due to mutations in the prosaposin gene. All forms of MLD are associated with a failure to adequately catabolize sulfatide with a consequent toxic accumulation of this glycolipid.

As noted earlier, there are multiple clinical forms of ASA deficiency. The classic or late-infantile form of MLD presents at around 1 year of age with ataxia and evolves with development of spasticity, dementia, seizures, and death within several years. Forms that present later in childhood or even in the adult years are also described.[3] Greater residual ASA activity in the cells of individuals with late-onset forms of MLD is the major reason for the less severe phenotypes of the juvenile and adult-onset forms of MLD compared to the late-infantile form.

Although the clinical recognition and diagnostic evaluation of the classic late-infantile form of MLD are usually straightforward, the diagnostic evaluation of the late-onset forms of MLD is occasionally difficult. Several explanations account for this difficulty. First, some clinically affected individuals with MLD have low but not deficient ASA activity, i.e., values of enzyme activity that are greater than those observed in the majority of cases of MLD. This scenario is encountered most commonly in the evaluation of individuals with juvenile and adult-onset MLD. Conversely, some individuals come to clinical attention due to a variety of neurological complaints and are noted to have very low ASA activity but lack sulfatiduria. These individuals have deficient *in vitro* but adequate *in vivo* ASA activity due to benign or pseudodeficiency mutations in the ASA gene.[3-5] These individuals do not have MLD.

[3] E. H. Kolodny and A. L. Fluharty, *in* "The Metabolic and Molecular Bases of Inherited Diseases" (C. R. Scriver, A. L. Beaudet, W. S. Sly, and D. Valle, eds.), Vol. 2, Chapter 88, p. 2693. McGraw-Hill, New York, 1995.
[4] M. Ameen and P. L. Chang, *FEBS Lett.* **219,** 130 (1987).
[5] V. Gieselmann, A. Polten, J. Kreysing, and K. Von Figura, *Proc. Natl. Acad. Sci. U.S.A.* **86,** 9436 (1989).

Some of the difficulties in diagnosing MLD can be resolved by analysis of the ability of the cultured fibroblasts to degrade exogenously added radioactive sulfatide.[6] MLD fibroblasts have a reduced ability to degrade exogenously added sulfatide compared to normal, with the level of reduction corresponding to the amount of residual ASA activity. However, this assay is technically challenging, expensive, and only available at a few diagnostic centers. In addition, it not always possible to differentiate between individuals with adult-onset MLD and persons who are homozygous for a pseudodeficiency mutation or who are compound heterozygotes for a pseudodeficiency mutation and a disease-causing ASA mutation. Determination of the urinary sulfatide concentration has been employed to evaluate complex or atypical cases, based on the fact that normal individuals excrete little urinary sulfatide, but individuals with a clinically significant deficiency of ASA or of the activator protein excrete large amounts of sulfatide. The following procedure for the analysis of urinary sulfatide has been developed at the Shriver Center for the clinical diagnosis of MLD.[7]

Analysis of Urinary Sulfatide by HPLC

Reagents. Purified bovine brain sulfatide is obtained from Sigma Chemical Co. (St. Louis, MO). The solvents are HPLC grade and are obtained from Fisher Scientific (Pittsburgh, PA). Other chemicals are reagent grade and are also obtained from Fisher Scientific, except for the dry methanolic HCl (from Supelco, Bellefonte, PA), C_{18} silica (Sepralyte, Analytichem International, Natick, MA), and DEAE-52 (Whatman, Clifton, NJ).

Sulfatide Analysis. The complete method is outlined in Fig. 1.

Extraction of Lipids. Aliquots of 10–20 ml of urine of known creatinine concentration are frozen at $-70°$ and then lyophilized in 50-ml conical plastic tubes. The residue is resuspended in 3 ml of 0.1 mM KCl and is transferred to 30-ml glass screw-cap tubes. After the addition of methanol (4 ml) and chloroform (8 ml), the solution is mixed vigorously at ambient temperature for at least 2 hr. The resulting two phases are separated by centrifugation (400g, 10 min) and the upper phase is discarded. The lower phase is filtered through glass wool and is evaporated to dryness under nitrogen.

Alkaline Hydrolysis. The samples are subjected to mild alkaline hydrolysis to eliminate glycerophospholipids. One milliliter of 0.6 M methanolic NaOH is added to each sample. After 1 hr at ambient temperature, the reaction mixture is neutralized with 1.5 ml of 0.4 M HCl. The salts are

[6] R. L. Stevens, A. L. Fluharty, H. Kihara, M. M. Kaback, L. J. Shapiro, B. Marsh, K. Sandhoff, and G. Fischer, *Am. J. Hum. Genet.* **33,** 900 (1981).
[7] M. R. Natowicz, E. M. Prence, P. Chaturvedi, and D. S. Newburg, *Clin. Chem.* **42,** 232 (1996).

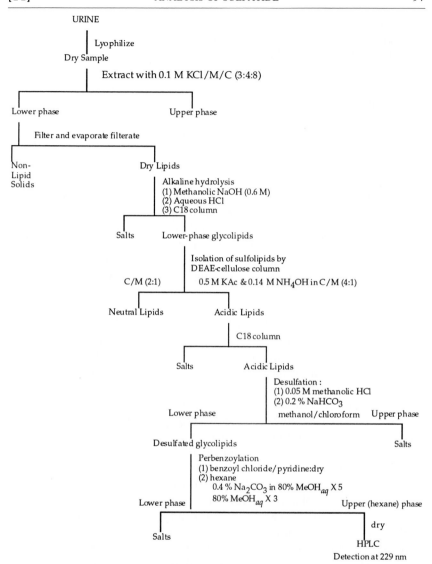

FIG. 1. Isolation and analysis of urinary sulfatides. C, chloroform; M, methanol; W, water. Adapted from M. R. Natowicz *et al., Clin. Chem.* **42,** 232 (1996).

removed by chromatography on a reversed-phase silica column as follows. The sample is mixed with 1.0 ml of methanol and 0.5 ml of water, passed twice through a 0.4-ml column of C_{18} silica (Sepralyte) that has been pre-washed with 2.5 ml of chloroform/methanol [2:1 by volume (C/M 2:1)], followed by C/M (1:1), 65% ethanol, and water. The column is then washed with 3.5 ml of 0.1 M KCl in 50% methanol/water, followed by 3.5 ml of water, and the lipids are eluted from the column with 5 ml of methanol. The sample is dried under nitrogen and redissolved in 1 ml of C/M (2:1).

Isolation of Sulfatide. Chromatography on a DEAE-cellulose column is performed to isolate acidic glycolipids as follows. The sample is applied to a 2-ml column of DEAE-52 that has been equilibrated previously with 0.8 M methanolic sodium acetate and washed with 8 ml of methanol, followed by 8 ml of C/M (2:1). Residual sample that might have been left in the tube is applied to the column with two 1-ml washes of C/M (2:1). Neutral (nonacidic) lipids are eluted from the column with 15 ml of C/M (2:1), after which the sulfatide fraction is eluted with 20 ml of 0.5 M potassium acetate and 0.14 M ammonium hydroxide in C/M (4:1). The sulfatide fraction is dried under a stream of N_2 and redissolved in 20 ml of methanol/water (1:1). The salts are removed as follows. The redissolved fraction is applied twice (i.e., the first eluate is added again) to a 0.4-ml C_{18} silica column that has been equilibrated by prior washing with 2.5 ml of 0.1 M KCl in 50% methanol/water. The column is washed successively with 2.5 ml of methanol/water (1:1) and 2.5 ml of water, and the sulfatides are eluted with 5 ml of methanol. This eluate is evaporated to dryness under N_2 and traces of moisture are removed by lyophilization.

Desulfation. Desulfation is accomplished by mild acid methanolysis and Folch partitioning as follows.[8] The dried sample is treated with 1 ml of dry 0.05 M methanolic HCl at ambient temperature for 16 hr, after which 3 ml of chloroform, 0.5 ml of methanol, and 1.1 ml of 0.2% (w/v) aqueous sodium bicarbonate are added. The two phases are separated by centrifugation and the upper phase is discarded. The lower phase is washed with 2 ml of C/M/0.1 M aqueous KCl (3:48:47) and then with 2 ml of C/M/water (3:48:47). The upper phases are discarded, and the lower phase (containing sulfatide-derived galactosylceramide) is dried and lyophilized.

Perbenzoylation. Desulfated lipids are derivatized to their benzoyl derivatives for analysis by HPLC as follows[9]: 0.5 ml of 10% (v/v) benzoyl chloride in dry pyridine is added to each sample; after 16 hr at 37°, the perbenzoylated samples are brought to dryness under N_2 and redissolved in 3 ml of hexane. The hexane solution is washed by prolonged vigorous mixing five times

[8] J. Folch, M. Lees, and G. H. Sloane Stanley, *J. Biol. Chem.* **226,** 497 (1957).
[9] M. D. Ullman and R. H. McCluer, *Methods Enzymol.* **138,** 117 (1987).

with 2 ml of 0.4% (w/v) Na_2CO_3 in methanol/water (4:1), followed by three washings with methanol/water (4:1); and, the upper hexane layer is evaporated to dryness at ambient temperature under N_2 and redissolved in 0.5 ml CCl_4.

HPLC. Aliquots of the perbenzoylated glycolipids in CCl_4 are analyzed by normal-phase HPLC on a 500 × 2.1-mm column of Zipax silica-coated glass beads (Rockland Technologies, Gilbertsville, PA) using a 16-min linear gradient from 1.0 to 20.0% dioxane in hexane at a flow rate of 2 ml/min.[9] The absorbance of the eluant is monitored at 229 nm, and data are collected on a Macintosh computer with the use of Dynamax software (Rainin, Emeryville, CA).

Analysis of Data. The amount of sulfatide in each sample is calculated by converting the combined area under the normal and α-hydroxy fatty acid containing galactosylceramide peaks into nanomoles of glycosphingolipid per sample, with correction for losses, based on the yield (typically 30%) of a concurrently analyzed external sulfatide standard (5.6 μM).[7] The concentration of urinary sulfatide is calculated as

$$\underset{\text{(nmol/mg creatinine)}}{\text{Sulfatide}} = \frac{\text{sulfatide measured (nmol)}}{\text{volume of urine (ml)}} \times \frac{1}{\text{creatinine (mg/ml)}}$$
$$\times \frac{100}{\text{\% yield}}$$

Figure 2 illustrates the HPLC-based resolution of urinary sulfatide from a healthy control, a typical MLD patient, and glycolipid standards. The sulfatide separates into two peaks: one with hydroxy fatty acid containing sulfatide and the other nonhydroxy (normal chain) fatty acid containing sulfatide. The limit of detection of this assay is 100 pmol of sulfatide in 20 ml urine. The method gives linear results from 100 to 6400 pmol/20 ml. The intraassay imprecision of urinary sulfatide measured is 27–30%. The recovery of the standard sulfatide (100 μg in 20 ml of water) is 29 ± 11% ($N = 17$). In urine samples, 64–67% of sulfatide is distributed in the urinary sediment so it is important to mix each urine sample thoroughly before taking an aliquot for analysis. In our laboratory, urine of normal controls contains 0.16 ± 0.07 nmol of sulfatide/mg creatinine (range 0.07–0.34), whereas individuals with MLD have 7.6 ± 6.1 nmol/mg creatinine (range 1.2–24.2).[7]

Levels of sulfatide in urine are very low relative to other urinary constituents. For the accurate analysis of sulfatide, it must be separated from other interfering materials as described in this method. For analysis of sulfatide in other tissues where the levels are relatively high (e.g., brain), the steps involved in the removal of salts by C_{18} reversed-phase chromatog-

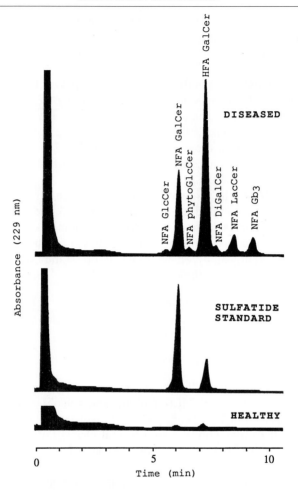

Fig. 2. Representative HPLC chromatograms of urinary sulfatides. (Top) Analysis of sulfatides from a typical MLD patient's urine, (middle) sulfatide standard, and (bottom) sulfatides from urine of a normal healthy person. Copied with permission from M. R. Natowicz *et al., Clin. Chem.* **42,** 232 (1996).

raphy, after alkaline hydrolysis and after DEAE-cellulose chromatography, may be replaced by a simpler Folch partitioning procedure.[8] Sulfatide partitions into the lower phase, with minimal losses in the upper phase, provided that the salt concentration is 0.2 *M* in the upper phase.

The method is used to evaluate clinically atypical cases in which a diagnosis of MLD is a clinical consideration. It can also be used to evaluate enzymatically atypical cases where ASA activity determinations are ambiguous, in the diagnosis of the activator deficiency form of MLD, and in the diagnosis of multiple sulfatase deficiency disease.[3]

Assay of Sulfatidase

Arylsulfatase activity can be measured using an artificial substrate, *p*-nitrocatechol sulfate (NCS). *p*-Nitrocatechol sulfate, however, is also hydrolyzed by arylsulfatase B (ASB), which, unlike ASA, has no activity toward sulfatide. In our clinical laboratory, we routinely use the method of Lee-Vaupel and Conzelmann[10] for the determination of ASA. The method is based on the fact that ASA hydrolyzes NCS at 0° at a rate of 24% of that at 37°. ASB, however, is almost inactive at 0°. For the specific determination of sulfatidase (cerebroside-3-sulfate-3-sulfohydrolase, EC 3.1.6.8), the physiological substrate sulfatide is used. The following procedure was developed by Raghavan *et al.*[11] for the assay of sulfatidase in leukocyte or fibroblast extracts for the diagnosis of MLD.

Reagents. Sulfatide (bovine) is obtained from Matreya Inc. (Pleasant Gap, PA) and is purified by preparative thin-layer chromatography (TLC). [³H]Sodium borohydride (250 mCi/nmol) is from New England Nuclear (Boston, MA). Reagent grade organic solvents, salts, precoated TLC plates, and Scintiverse are obtained from Fisher Scientific Co. (Medford, MA). DEAE-cellulose (DE-52) is obtained from Whatman, Inc. (Clifton, NJ) and sodium taurodeoxycholate from Calbiochem (La Jolla, CA).

Preparation and Purification of [³H]Sulfatide. Details of these procedures are given by Raghavan *et al.*[11] Briefly, ³H-labeled sulfatide is prepared by catalytic hydrogenation of double bonds in the sulfatide molecule with [³H]sodium borohydride.[12] The radiolabeled sulfatide is then purified by chromatography on a DEAE-cellulose column.[11] Leukocytes are isolated from freshly drawn heparanized blood by dextran sedimentation as described by Kolodny and Mumford[13] and stored at −20° until assayed. Leukocyte cell pellets from 10 ml of whole blood are suspended in 0.5 ml of distilled water and are subjected to 15-sec bursts of pulsed ultrasound (Sonifier Model W185, Heat Systems-Sonication, Inc., Plainview, Long Island, NY) in an ice bath. The sample is dialyzed overnight against 5 liters of distilled water at 4°. Protein is determined according to the method of Lowry *et al.*[14]

[10] M. Lee-Vaupel and E. Conzelmann, *Clin. Chim. Acta* **164,** 171 (1987).
[11] S. S. Raghavan, A. Gajewski, and E. H. Kolodny, *J. Neurochem.* **36,** 724 (1981).
[12] G. Schwarzmann, *Biochim. Biophys. Acta* **529,** 106 (1978).
[13] E. H. Kolodny and R. A. Mumford, *Clin. Chim. Acta* **70,** 247 (1976).
[14] O. H. Lowry, N. J. Rosebrough, H. L. Farr, and R. J. Randall, *J. Biol. Chem.* **193,** 265 (1951).

Enzyme Assay of Sulfatidase. The incubation mixture contains 70 nmol of [³H]sulfatide (1500 cpm/nmol), 25 mM sodium acetate buffer, pH 5.0, 5 mM MnCl$_2$, 0.3% sodium taurodeoxycholate, and 90–120 μg protein from dialyzed leukocyte sonicate in a final volume of 200 μl. After incubation for 2 hr, the reaction is stopped with 1 ml methanol. The mixture is passed through a DEAE-cellulose (DE-52) column (prepared as described previously in the section on "Isolation of Sulfatide") and packed in methanol to a height of 6 cm in a short Pasteur pipette. The column is washed further with 2 ml methanol that first was used to rinse the reaction tubes. The combined eluates, which contain the cerebroside released from sulfatide (unreacted sulfatide remains bound to the DEAE-cellulose column), are collected in a scintillation vial, taken to dryness, redissolved in Scintiverse, and counted. Boiled enzyme controls or reaction tubes without enzyme are included routinely. Blank values from the unhydrolyzed substrate are less than 0.3% of the added radioactivity. Blanks containing unlabeled substrate and labeled cerebroside in amounts equivalent to the product formed in the assay can be used to verify the quantitative recovery of the cerebroside in the methanol eluant from the DE-52 column. The reaction can also be stopped with 2 ml C/M (1:1), instead of methanol, and the cerebroside separated as described earlier on a DE-52 column using C/M (1:1). The results obtained are identical with either of these methods for product separation.

Properties of Leukocyte Sulfatidase. Leukocyte sulfatidase exhibits a sharp pH optimum of 5.0 in 25 mM sodium acetate buffer, with only 10 and 3% of the maximal activity present at pH 4.0 and pH 6.0, respectively. The reaction requires the presence of the anionic detergent sodium taurodeoxycholate, which at 0.3% final concentration yields maximal hydrolysis. Higher concentrations of detergent inhibit the reaction. A four- to fivefold stimulation results with 5 mM MnCl$_2$, but higher concentrations are inhibitory. Dialysis of the leukocyte extract against water at 4° enhances sulfatidase activity, suggesting the presence of dialyzable inhibitors in the leukocyte extract. The reaction is linear with time up until 3 hr. The enzyme follows Michaelis–Menten kinetics with an apparent K_m of 0.17 mM. The mean sulfatidase activity in leukocytes is 12.9 ± 3.2 nmol/mg protein/2 hr and the activity is low to undetectable in most MLD patients. In the obligate heterozygotes, the mean specific activity is 5.1 ± 3.0 units, amounting to 39% of the mean control activity.[11]

The gene for ASA is located near the end of the long arm of chromosome 22.[3] The protein sequence is encoded on eight exons and is translated from a 2.1-kb mRNA transcript. The gene extends for a considerable distance in the 3′ direction and gives rise to two longer mRNA species of unknown significance. ASA contains three potential glycosylation sites.[3]

Assay of Galactosylceramide : Sulfotransferase

Galactosylceramide : sulfotransferase (EC 2.8.2.11) catalyzes the formation of sulfatide from galactosylceramide by adenosine-3′-phosphate 5′-phosphosulfate (PAPS).[15,16]

Reagents. Galactocerebroside from bovine brain is purchased from Matreya Inc. (Pleasant Gap, PA) and is a mixture of α-hydroxy and nonhydroxy fatty acid containing species. [^{35}S]PAPS (specific activity about 1.0 Ci/mmol) is purchased from New England Nuclear (Boston, MA). Prepacked C18 (10 cc) Bond Elute reversed-phase cartridges are obtained from Analytichem International (Harbor City, CA).

Incubation. Galactocerebroside (80 μg) and Triton X-100 (1 mg) are dissolved in chloroform/methanol (2:1, v/v), then evaporated to dryness in an incubation tube. To the residue, 100 mM Tris–HCl, pH 7.1 (at 37°), 2.5 mM ATP, 20 mM MgCl$_2$, 0.23 μCi of [^{35}S]PAPS, and the enzyme source are added to make a final volume of 0.5 ml. The mixture is incubated in a shaker bath for 1 hr at 37°. The radioactive sulfatide that is synthesized is separated from the [^{35}S]PAPS by one of the two following procedures.[8,16]

Folch Partitioning.[8] After the incubation, 10 ml of chloroform/methanol (2:1) is added. The chloroform/methanol extract is partitioned after adding 0.2 volume of 0.9% NaCl and the upper phase is removed. The lower phase is then washed three times with 0.2 volume of the theoretical upper phase (TUP) containing methanol/saline/chloroform (48/47/3) and 10 mM sodium sulfate as a carrier, followed by three washes with TUP without the carrier. Alternatively, the radioactive sulfatide is separated from the radioactive nucleotide by chromatography on a C$_{18}$ Bond Elut reversed-phase cartridge as described next.

Separation Using Reversed-Phase C$_{18}$ Columns.[16] For samples processed with C$_{18}$ reversed-phase columns, the enzyme reaction is terminated by the addition of 5.0 ml of TUP prepared with 0.1 M KCl instead of saline. Up to 10 samples can be processed simultaneously when Bond Elut C$_{18}$ columns are placed in a Vac Elute column holder. All column procedures are carried out using constant gentle vacuum. Prior to loading of the incubation mixture on the columns, the columns are washed successively with 5.0 ml of chloroform/methanol (2:1, v/v), 5.0 ml of distilled water, and again with 5.0 ml of chloroform/methanol (2:1). The C$_{18}$ columns are then equilibrated with 5.0 ml of TUP containing 10 mM sodium sulfate as a carrier. The incubation mixture is then passed through the column twice, followed by washes with 5.0 ml of TUP and 10.0 ml of distilled water. For each wash,

[15] F. B. Jungalwala, *J. Lipid Res.* **15,** 114 (1974).
[16] D. A. Figlewicz, C. E. Nolan, I. N. Singh, and F. B. Jungalwala, *J. Lipid Res.* **26,** 140 (1985).

the original reaction mixture tube is rinsed with the solvent before loading onto the column to assure complete transfer. The tubes are then rinsed with 7.0 ml of chloroform/methanol (2:1) and the solvent is passed through the column to elute the sulfatide. Lipids eluted from the column with the last solvent are collected and the solvent evaporated. A portion of the radioactive products is redissolved or suspended in Scinti-Verse I scintillant and counted in a scintillation counter. The specific activity of the enzyme source is then calculated from the amount of incorporated radiolabel. Triplicate determinations are made along with appropriate controls for each assay. Both extraction procedures provide low blank values and the recoveries are comparable.

Properties of Cerebroside Sulfotransferase. This enzyme has been purified to apparent homogeneity from rat kidney,[17] testis,[18] mouse brain,[19] and human renal cancer cells.[20] The gene has also been cloned from human renal cancer cells. The deduced amino acid sequence showed it to consist of 423 amino acids with a calculated molecular mass of 48,763 Da and two N-glycosylation sites, which agrees with the 54-kDa mass of the purified human enzyme.[21] However, the size of the enzyme from rat kidney[17] is reported to be 64 kDa, and testis[18] and mouse brain[19] enzymes have molecular masses of 54 and 31 kDa, respectively. The reasons for the variability are not clear. The K_m values for PAPS and cerebroside for the mouse brain enzyme are $1.2 \times 10^{-6} M$ and $2.6 \times 10^{-5} M$, respectively.[19] The pH optimum is 7.0 and cerebroside concentrations >80 pmol/ml are inhibitory. The activity of the mouse enzyme is activated by vitamin K plus phosphate by a mechanism involving phosphorylation,[22] whereas pyridoxal 5'-phosphate (PLP) is a strong inhibitor.[22] The testis enzyme, however, is not activated by vitamin K, but phosphorylation is required for its activation.[18] The sulfotransferase activity of human renal carcinoma cells is enhanced by epidermal growth factor,[23] transforming growth factor-α[24] and hepatocyte

[17] G. Tennekoon, S. Aitchison, and M. Zaruba, *Arch. Biochem. Biophys.* **240,** 932 (1985).
[18] D. Sakac, M. Zachos, and C. A. Lingwood, *J. Biol. Chem.* **267,** 1655 (1992).
[19] K. S. Sundaram and M. Lev, *J. Biol. Chem.* **267,** 24041 (1992).
[20] K. Honke, M. Yamane, A. Ishii, T. Kobayashi, and A. Makita, *J. Biol. Chem.* **119,** 421 (1996).
[21] K. Honke, M. Tsuda, Y. Hirahara, A. Ishii, A. Makita, and Y. Wada, *J. Biol. Chem.* **272,** 4864 (1997).
[22] K. S. Sundaram and M. Lev, *Biochem. Biophys. Res. Commun.* **169,** 927 (1990).
[23] T. Kobayashi, K. Honke, S. Gasa, N. Kato, T. Miyazaki, and A. Makita, *Int. J. Cancer* **55,** 448 (1993).
[24] T. Kobayashi, K. Honke, S. Gasa, S. Imai, J. Tanaka, T. Miyazaki, and A. Makita, *Cancer Res.* **53,** 5638 (1993).

growth factor.[25] Tyrosine kinases have been shown to be involved in the expression of the sulfotransferase in cancer cells.[26]

[25] T. Kobayashi, K. Honke, S. Gasa, T. Miyazaki, H. Tajima, K. Matsumoto, T. Nakamura, and A. Makita, *Eur. J. Biochem.* **219,** 407 (1994).
[26] M. Balbaa, K. Honke, and A. Makita, *Biochim. Biophys. Acta* **1299,** 141 (1996).

[12] 1-O-Acylceramide Synthase

By James A. Shayman and Akira Abe

Introduction

The reaction covered by this article is ceramide + phosphatidyletha-nolamine → 1-O-acylceramide + lysophosphatidylethanolamine.

The emergence of ceramide as a potentially important cellular messenger has led to a renewal of attention in the metabolism of this lipid. This interest has included studies on the pharmacology of compounds, some structurally analogous to ceramide, which result in changes in endogenous cellular ceramide concentrations. 1-Phenyl-2-decanoylamino-3-morpho-lino-1-propanol (PDMP) is one such inhibitor. The D-*threo* enantiomer of PDMP specifically blocks glucosylceramide synthase, resulting in the depletion of cell glucosylceramide and more complex glucosylceramide-based glycosphingolipids.[1] Early work with this inhibitor revealed that, in the presence of PDMP, cell ceramide levels would increase in parallel with glucosylceramide depletion. This observation led to the interpretation that ceramide, a substrate for the cerebroside synthase, increased as a result of inhibition of the cerebroside synthase.

However, the development of more active homologs of PDMP and studies on the enantiomers of these compounds that were inactive against glucosylceramide synthase demonstrated that *erythro* diastereomers could raise cell ceramide levels at concentrations equivalent to those for *threo* diastereomers.[2] In addition, aliphatic homologs of PDMP, 1-morpholino-1-deoxyceramides, demonstrated activity against the glucosylceramide synthase with little effect on ceramide concentrations.[3] These observations led

[1] J. A. Shayman, S. Mahdiyoun, G. Deshmukh, F. Barcelon, J. Inokuchi, and N. S. Radin, *J. Biol. Chem.* **265,** 12135 (1990).
[2] A. Abe, N. S. Radin, J. A. Shayman *et al., J. Lipid Res.* **36,** 611 (1995).
[3] K. G. Carson, B. Ganem, N. S. Radin, A. Abe, and J. A. Shayman, *Tetrahed. Lett.* **35,** 2659 (1994).

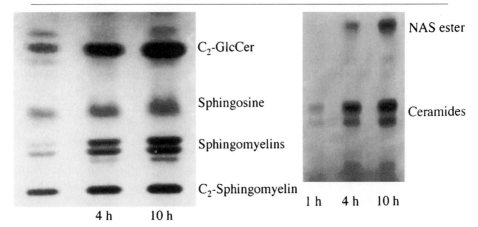

FIG. 1. Products of *N*-acetylsphingosine metabolism in MDCK cells. MDCK cells (1.35×10^6) were seeded into a 15-cm dish containing 21 ml of defined Dulbecco's modified Eagle's medium and incubated for 24 hr in a CO_2 incubator. The medium was replaced with fresh medium containing 10 μM *N*-acetyl-[3-^{3}H]sphingosine (6.9×10^5 cpm/dish). The cells were then incubated for 1, 4, or 10 hr, washed twice with 20 ml cold phosphate-buffered saline, and transferred with methanol. The total lipids were extracted with 3 ml chloroform : methanol (2 : 1), washed by partitioning against aqueous NaCl, and the resultant lipid extract dried down under a stream of nitrogen. Plates in were developed with chloroform : methanol : water (65 : 30 : 8) (left) and chloroform : methanol : acetic acid (90 : 1 : 9) (right). Plates were subjected to fluorography following spraying with En^{3}Hance (DuPont) at $-80°$.

to the conclusion that PDMP-mediated changes in cell ceramide concentrations were the result of an activity at a second, uncharacterized site. Attempts to identify this second site of action by screening for inhibitory or stimulatory effects of *erythro* and *threo* diastereomers of PDMP homologs on known enzymes of ceramide metabolism were unsuccessful. Specifically, no effects were observed on ceramidase, sphingomyelinase, sphingomyelin synthase, or ceramide synthase activities.

An alternative strategy was then employed. Cultured MDCK cells were incubated with *N*-acetyl-[3-^{3}H]sphingosine. Under these conditions, a highly nonpolar product was observed, subsequently demonstrated to be 1-*O*-acylceramide[4] (Fig. 1). This metabolite was originally discovered in 1979 by Okabe and Kishimoto.[5] However, the acyl donor and enzyme responsible for its formation remained uncharacterized. This article describes the characterization of this product, its anabolic pathway, and the purification and characterization of 1-*O*-acylceramide synthase.

[4] A. Abe, J. A. Shayman, and N. S. Radin, *J. Biol. Chem.* **271,** 14383 (1996).
[5] H. Okabe and Y. Kishimoto, *J. Biol. Chem.* **252,** 7068 (1977).

Characterization of the Pathway for 1-O-Acylceramide Formation

Conversion of N-Acetylsphingosine to a Fatty Acyl Ester in Cell Homogenates

Initial studies were performed using a homogenate of MDCK cells to study the conversion of N-acetyl-[3-^{3}H]sphingosine to the unknown lipid. The precursor and product were separated in two solvent systems. In a neutral system, chloroform : methanol : water (60 : 35 : 8), the R_f values for N-acetyl-[3-^{3}H]sphingosine and the unknown were ~0.74 and ~0.92, respectively. In an acidic system, chloroform : methanol : acetic acid (90 : 2 : 8), the R_f values were ~0.21 and ~0.72, respectively. Authentic 1-O-palmitoyl-N-acetylspingosine was synthesized and observed to migrate in the second solvent system with a R_f value of ~0.72.

It was observed that the conversion was favored at pH 4.5 over neutral pH. The N-acetyl-[3-^{3}H]sphingosine was converted to the more lipoidal unknown product, reaching a peak level at 1 hr and then being converted to the free N-acetyl-[3-^{3}H]sphingosine. The similarity of the R_f values between the palmitate ester and the unknown and the rapid conversion of the unknown to its precursor suggested that the unknown lipid was a fatty acyl ester.

Identification of the Unknown Lipid

When the tritiated unknown product was extracted from the thin-layer chromatography (TLC) plate and subjected to alkaline methanolysis with chloroform : methanolic 0.21 N NaOH for 1 hr, the unknown product disappeared and there was a corresponding increase in N-acetyl-[3-^{3}H]sphingosine. Almost all of the radioactivity was recovered as N-acetyl-[3-^{3}H]sphingosine. This finding was consistent with the presence of a carboxylic acid in ester linkage in the unknown.

The position of the O-acyl group was determined by treating the tritiated unknown product with 2,3-dichloro-5,6-dicyanobenzoquinone. This reagent oxidizes α,β-unsaturated alcohols to the corresponding ketones. This agent would therefore convert an unacylated C-3 hydroxyl group to a ketone, forming a new product. If acylated, the C-3 hydroxyl should be inert, as should be the free C-1 hydroxyl group. After treatment with 2,3-dichloro-5,6-dicyanobenzoquinone, the unknown product was converted to a new product with a higher R_f value. The location of this new product corresponded to that of 3-keto-O-palmitoyl-N-acetyl-sphingosine, prepared in the same manner. The unknown product was therefore 1-O-acyl-N-acetyl-sphingosine (Fig. 2).

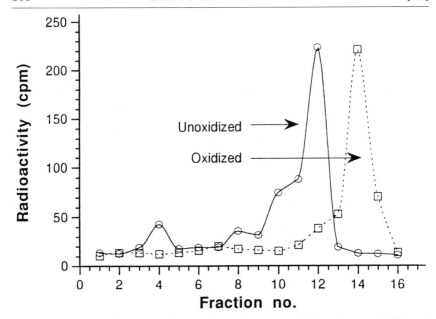

Fɪɢ. 2. Identification of the product. The putative N-acetylsphingosine ester (300 cpm) was isolated from by thin-layer chromatography (TLC) and incubated for 48 hr at 37° with or without 3% 2,3-dichloro-5,6-dicyanobenzoquinone in 40 μl dioxane and then dried down under a nitrogen stream. The dried sample was washed once with 3 ml chloroform : methanol (2:1) plus 0.6 ml of 0.1 N NaOH. The lower layer obtained after a brief centrifugation was washed twice with 2 ml MeOH : 0.1 N NaOH (1 : 1) and twice more with 2 ml of methanol : water (1 : 1). The lipid in the lower layer was chromatographed with chloroform : methanol : acetic acid (90 : 1 : 9). The TLC plate was divided into 0.5-cm fractions, starting at the origin, and examined by liquid scintillation counting.

Identification of Acyl Donor

A reaction mixture consisting of 46.5 mM sodium citrate (pH 4.5), 10 μM N-acetysphingosine (either labeled or unlabeled), and 202 μg/ml of cell homogenate was used to screen a variety of lipids for their ability to serve as acyl donors. When 1, 10, and 100 μM [^{3}H]palmitic acid were included in the incubation mixture, no conversion to the 1-O-[^{3}H]palmitoyl-N-acetylsphingosine was observed. Stearoyl-CoA (0.7 and 7 μM) slightly enhanced the formation of 1-O-acyl-N-acetylsphingosine. However, 200 μM stearoyl-CoA strongly inhibited the formation of the product. These studies ruled out free fatty acid or acyl-CoA as the source of the acyl ester.

The cellular homogenate was then tested in the presence of liposomes. Liposomes consisted of dioleoylphosphatidylcholine (128 μM) or dioleoylphosphatidylcholine and a second lipid in a 70 : 30 molar ratio. The formation of the *N*-acetylsphingosine ester was moderately stimulated with a dioleoylphosphatidylcholine/phosphatidylethanolamine liposome. Mixtures containing diolein, phosphatidylserine, or phosphatidylinositol were inhibitory.

In order to enhance the activity of the 1-*O*-acyl-*N*-acetylsphingosine assay, endogenous lipids were removed from the cellular homogenate. The MDCK cell homogenate was centrifuged at 100,000*g* for 1 hr at 4°. Most of the enzyme activity was observed in the soluble fraction. Filtration of the 100,000*g* supernatant through cellulose acetate membranes (0.45 μm pore size) lowered the basal activity of the enzyme even further. This filtrate was used for further characterizations.

Using the more sensitive preparation, a variety of liposomes were evaluated as acyl donors using *N*-acetyl-[3-^{3}H]sphingosine as the substrate. Both phosphatidylcholine and phosphatidylethanolamine were observed to serve as acyl donors, with the latter having greater activity. Inclusion of sulfatide in the liposomal preparation enhanced activity further (Table I).

To characterize the role of phosphatidylethanolamine as a potential acyl donor further, 1-palmitoyl-2-[^{14}C]arachidonoylphosphatidylethanolamine was incorporated into dioleoylphosphatidylcholine:sulfatide liposomes. In the absence of *N*-acetylsphingosine as an acceptor, only one radioactive product was formed following extraction and separation by TLC in the acidic solvent system: [^{14}C]arachidonic acid ($R_f \sim 87$). When *N*-acetylsphingosine was added to the incubation mixture, two products were formed: [^{14}C]arachidonic acid and another ($R_f \sim 69$) that migrated close to that of *O*-palmitoyl-*N*-acetylsphingosine. This latter product

TABLE I
COMPARISON OF DIFFERENT ACYL DONORS FOR TRANSACYLATION[a]

Phospholipid concentration (μM)	Transacylation activity (pmol/min/mg protein)			
	DOPC	DOPC:PE	DOPC:sulfatide	DOPC:PE:sulfatide
12.8	15, 11			
12.8	40, 62	41, 43	93, 100	265, 261
128	18, 16	97, 59	288, 298	474, 486
640	0, 0	272, 268	95, 109	150, 144

[a] The reaction mixture contained 47 m*M* sodium citrate (pH 4.5), 10 μM *N*-acetyl-[^{3}H]sphingosine, 30–60 μg/ml of the MDCK cell supernatant, and liposomal phospholipid. Molar ratios of the mixed liposomes were 70:30 for DOPC:PE, 86:14 for DOPC:sulfatide, and 61:25:14 for DOPC:PE:sulfatide. Incubations were for 20 min at 37°. The transacylase activity was calculated from the ratio, observed cpm (blank activity)/mg protein × minutes × *N*-acetyl-[^{3}H]sphingosine specific activity (cpm/pmol). The two values are from duplicate incubations.

yielded [^{14}C]methyl arachidonate when extracted from the TLC plate and subjected to alkaline methanolysis. The ratio of free arachidonic acid to 1-O-arachidonoyl-N-acetylsphingosine formed decreased when increasing concentration of N-acetylsphingosine were included in the transacylase assay (Fig. 3). This observation raised the possibility that a single enzyme had both deacylase and transacylase activity.

Specificity of Transacylase for Other Sphingolipid Acceptors

N-Acetylsphingosine was the best acceptor for the transacylase. N-Acetyl-dihydrosphingosine, N-octanoylsphingosine, and N-oleoylsphingosine had ~50, 10, and 5% of the maximal activity, respectively, as acyl acceptors. Sphingosine, N-octanoyl glucosylsphingosine, and lysophosphatidylcholine were inactive as acyl acceptors.

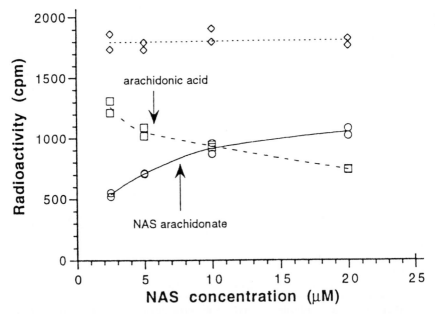

FIG. 3. Deacylase and transacylase activities as a function of N-acetylsphingosine concentration. The reaction was carried out with 47 mM sodium citrate (pH 4.5), 40 μg/ml of cytosolic enzyme, and 128 μM of liposomal DOPC:1-palmitoyl-2-[^{14}C]arachidonoylphosphatidyletha-nolamine (160,000 cpm):sulfatide (70:0.2:30). The upper horizontal line shows the sum of the two activities, which is independent of N-acetylsphingosine concentration.

Purification of 1-*O*-Acylceramide Synthase

Solutions

    25 mg/ml dioleoylphosphatidylcholine in chloroform
    10 mg/ml phosphatidylethanolamine from bovine brain in chloroform
    10 mg/ml dicetylphosphate in chloroform/methanol (2/1)
    500 μM *N*-acetylsphingosine in dimethyl sulfoxide
    N-Acetyl-[3-^{3}H]sphingosine in chloroform/methanol (1/1)
    0.1 *M* phenylmethylsulfonyl fluoride (PMSF) in isopropyl alcohol
    10% (w/v) Triton X-100
    50 m*M* sodium citrate (pH 4.5)
    Buffer A: 10 m*M* Tris–HCl (pH 7.4), 1 m*M* EDTA
    Buffer B: 25 m*M* Tris–HCl (pH 7.4), 1 m*M* EDTA
    Buffer C: 0.15 *M* NaCl, 25 m*M* Tris–HCl (pH 7.4), 1 m*M* EDTA
    Buffer D: 10 m*M* sodium phosphate (pH 6.0), 1 m*M* EDTA
    Buffer E: 0.5 *M* NaCl, 1 m*M* CaCl$_2$, 1 m*M* MnCl$_2$, 25 m*M* Tris–HCl
      (pH 7.4)

Assay of Transacylase Activity

    The general assay for the purification of transacylase uses a tritiated short chain ceramide, *N*-acetylsphingosine, as this is a better acceptor than long chain ceramides:

N-acetyl-[3-^{3}H]sphingosine + phosphatidylethanolamine →
    1-*O*-acyl-*N*-acetyl-[3-^{3}H]sphingosine + lysophosphatidylethanolamine

The enzyme activity is determined by measuring the activity of 1-*O*-acyl-*N*-acetyl-[3-^{3}H]sphingosine. Liposomes are used as an acyl group donor and consist of phosphatidylcholine, phosphatidylethanolamine, and dicetylphosphate (molar ratio of 7/3/1). The reaction mixture consists of 40 m*M* sodium citrate (pH 4.5), liposomes (128 μM phospholipid), 10 μg/ml bovine serum albumin (BSA), 10 μM *N*-acetyl-[3-^{3}H]sphingosine (2000 cpm/nmol), and enzyme in a total volume of 500 μl in a 16 × 125-mm glass tube with a screw cap.
    When *N*-acetylsphingosine is dispersed into an aqueous solution, the lipid is immediately and tightly bound to the glass tube wall. The addition of *N*-acetylsphingosine into the assay solution in the presence of liposomes and BSA prevents absorption on the wall. The reaction is initiated by the addition of the enzyme and is maintained at 37°. The reaction is terminated by the addition and mixing of 3 ml of chloroform/methanol (2/1) plus 0.3 ml of 0.9% NaCl. The mixture is centrifuged for 10 min at 800*g* at room

temperature. After removing the upper phase, the lower phase is transferred into a 13 × 100-mm glass tube and dried under a stream of nitrogen in a water bath at 45°. The dried lipids are dissolved in chloroform/methanol (95/5) and applied to a high-performance (HP) TLC plate. The plate is developed in a solvent system consisting of chloroform/acetic acid (9/1) and sprayed with primulin. Under ultraviolet light, spots corresponding to 1-O-acyl-N-acetylsphingosine are identified, scraped, and counted.

Enzyme activity may also be determined using nonradioactive N-acetylsphingosine as an acceptor. The reaction product is visualized by charring after thin layer chromatography and is analyzed quantitatively using an image scanner.

Synthesis of N-Acetyl-[3-³H]Sphingosine

N-Acetyl-[3-³H]sphingosine is synthesized by the modified method of Garver and Sweeley.[6] Fifty microcuries of D-*erythro*-[3-³H]sphingosine (22 Ci/mmol, New England Nuclear) in ethanol is dried down under a stream of nitrogen and dissolved in 50 μl of dry pyridine. One microliter of acetic anhydride is added to D-*erythro*-[3-³H]sphingosine and incubated for 12 hr at room temperature. The mixture is then dried down under nitrogen. To remove the O-acetyl group, alkaline methanolysis is performed. The products are extracted by the method of Bligh and Dyer[7] and isolated by thin-layer chromatography using a solvent system consisting of chloroform/acetic acid (9/1). The spot corresponding to N-acetylsphingosine is scraped and transferred into a screw cap glass tube and extracted twice with 4 ml of chloroform/methanol (1/1). N-Acetyl-[3-³H]sphingosine is stored at −20°.

Purification Procedure

The transacylase has been purified 193,000-fold in an eight-step procedure yielding a protein with a molecular weight of 40,000 as estimated by sodium dodecyl sulfate–polyacrylamide gel electrophoresis (SDS–PAGE)(Table II). The yield from 10 calf brains (2400 g) is 4.5 μg with a specific activity of 25 μmol/min/mg protein.

Step 1: Postmitochondrial Supernatant. Fresh calf brains are stored at −80° as thin blocks after removal of the pia matter. Frozen brains (600 g) are allowed to thaw partially at 4° overnight and are cut in small pieces, after which all manipulations are carried out at 4°. A 30% homogenate is prepared with 2.5 volumes of buffer A plus 1/200 volume of 0.1 M PMSF by use of a Polytron for 1 min at 20,000 rpm. The mixture is centrifuged

[6] R. C. Garver and C. C. Sweeley, *J. Am. Chem. Soc.* **88**, 3643 (1966).
[7] E. G. Bligh and J. J. Dyer, *Can. J. Biochem. Physiol.* **37**, 911 (1959).

TABLE II
PURIFICATION OF TRANSACYLASE FROM CALF BRAIN

Step	Volume (ml)	Protein (mg)	Specific activity[a] (nmol/min/mg)	Recovery (%)	Purification factor
12,000g supernatant	10,850	54,009	0.130[b]	100	1
20–70% saturated ammonium sulfate	2,326	23,413	0.159	52.4	1.22
DEAE-Sephacel (unbound)	7,920	3,036	2.5	103	19.2
Phenyl-Sepharose (50% ethylene glycol)	780	250	18.9	67.2	145
S-Sepharose (fast)	95	81.1	15.8	20.6	122
Sephadex G-75[c]	31	7.41	50.8	5.36	391
S-Sepharose (slow)	86	26.4	77.8	29.2	598
Sephadex G-75[d]	37	9.21	96.7	12.7	744
Con A agarose	17.6	0.0372	5.325	2.82	40,962
Heparin Sepharose	3.7	0.0045	25,149	1.62	193,450

[a] Transacylase activity was measured as described.

[b] In order to obtain the enzyme activity, the postmitochondrial supernatant was centrifuged for 60 min at 100,000g and the specific activity was determined using the membrane-free supernatant.

[c] Fraction "S-Sepharose fast" was used.

[d] Fraction "S-Sepharose slow" was used. Active fractions from this Sephadex G-75 chromatography step were pursued for further purification.

for 30 min at 14,300g. The pellet is resuspended as a 30% homogenate and the procedure is repeated. Both supernatants are then combined.

Step 2: Ammonium Sulfate Precipitation. Solid ammonium sulfate is added to the postmitochondrial supernatant to attain 20% saturation (10.6 g/100 ml). After stirring for 1 hr, the precipitate is removed by centrifugation for 30 min at 14,300g. To precipitate the enzyme, the resultant supernatant is adjusted to 70% saturation with solid ammonium sulfate (31.2 g/100 ml) and stirred for 30 min. After standing overnight, the clear supernatant above the precipitate is siphoned off. The precipitate is collected by centrifugation for 30 min at 14,300g, suspended in 120 ml of buffer A, and dialyzed five times over 2 days against 4 liters of buffer B.

Step 3: DEAE-Sephacel Chromatography. The protein concentration of the dialysate is adjusted to less than 20 mg/ml with buffer B. Insoluble material is removed from the dialysate by centrifugation for 1 hr at 100,000g. The supernatant is applied to a column (5 × 22 cm) of DEAE-Sephacel preequilibrated with buffer B at a flow rate of 100 ml/hr. The enzyme is eluted with 2 liters of buffer B. The material bound to the column is eluted with 2 liters of buffer B containing 0.3 M NaCl. The enzyme activity is mostly recovered in the initial eluant.

Step 4: Phenyl-Sepharose CL-4B Chromatography. The unbound fraction is adjusted to 3 M NaCl and applied to a column (2.5 × 20 cm) of Phenyl-Sepharose CL-4B preequilibrated with buffer B containing 3 M NaCl at a flow rate of 80 ml/hr. The column is rinsed with 1000 ml of buffer B containing 3 M NaCl, 450 ml of buffer B containing 1.5 M NaCl, and 400 ml of buffer B containing 0.15 M NaCl (buffer C). The transacylase activity is eluted with 400 ml of buffer C containing 50% (w/v) ethylene glycol and stored at −80°. The same procedure, steps 1 to 4, is repeated separately four times. The active fractions obtained after the step 4 are combined, adjusted to 0.5 mM PMSF, and dialyzed against buffer D.

Step 5: S-Sepharose Chromatography. After removal of insoluble materials using filter paper, the dialysate is applied to a column (1.4 × 20 cm) of S-Sepharose preequilibrated with buffer D at a flow rate of 40 ml/hr. The column is rinsed with 1000 ml of buffer D. The bound protein is eluted with a 480-ml linear gradient of NaCl from 0 to 0.3 M in buffer D. The enzyme activity is recovered as two peaks in fractions eluting between 60 and 130 mM NaCl. Each peak is collected separately and named S-S fast fraction and S-S slow fraction by following the order of elution. Each fraction is concentrated using Centriprep-10 (Amicon).

Step 6: Sephadex G-75 Chromatography. Each fraction is applied separately to a column (2.6 × 90 cm) of Sephadex G-75 preequilibrated with buffer C at a flow rate of 40 ml/hr. The protein is eluted with buffer C. For S-S fast fraction, the enzyme activity is eluted as two peaks. The first and the second peaks correspond to a molecular weight of about 80,000 and about 40,000, respectively. For the S-S slow fraction, the enzyme activity is eluted as a single peak that corresponds to a molecular weight of about 40,000. The active fractions obtained from the slow fraction possess higher specific enzyme activity and are used for further purification (Table II).

Step 7: Concanavalin A Agarose Chromatography. The active fraction is adjusted to 1 mM CaCl$_2$ and 1 mM MnCl$_2$ and applied to a column (0.85 × 1.8 cm) of Con A agarose (Seikagaku Corporation) preequilibrated with buffer E at a flow rate 1.5 ml/hr. The column is rinsed with 20 ml of buffer E and 10 ml of buffer E containing 10 mM α-methyl-D-mannopyranoside. The enzyme activity is mostly eluted with 10 ml of buffer E containing 500 mM α-methyl-D-mannopyranoside. To stabilize the enzyme, the active fractions are combined, adjusted to 0.005% (w/v) of Triton X-100, and dialyzed against buffer A containing 0.005% (w/v) of Triton X-100. At this stage, two-dimensional gel electrophoresis is performed. In the first dimension the protein is separated in two separate runs on 10% acrylamide gels at pH 4.3 in the absence of SDS. One gel is is sliced into 2-mm segments for assay of the transacylase activity. Each segment is minced with a glass homogenizer with 650 μl of 50 mM sodium citrate (pH 4.5) and vortexed overnight at 4°. The segments are centrifuged for 10 min at 2000g to remove the gel particles, and the transacylase activity is assayed using 400 μl of the supernatant. The second gel is equilibrated for 2 hr against 2.3% SDS, 5% 2-mercaptoethanol, 10% glycerol, and 62.5 mM Tris–HCl (pH 6.8) and

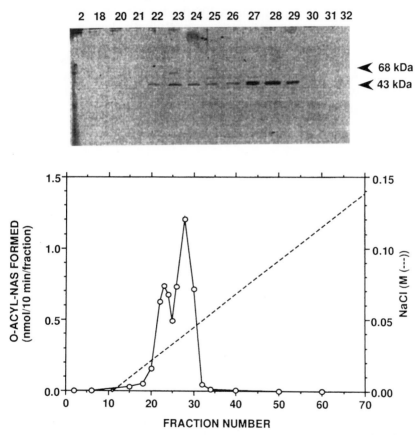

FIG. 4. Heparin Sepharose column chromatography of purified transacylase. (Bottom). Active fractions obtained from the concanavalin A column were pooled, adjusted to 77.3 μM Triton X-100, and dialyzed against 10 mM Tris–HCl (pH 7.4), 1 mM EDTA, and 77.3 μM Triton X-100. The column was eluted as described, and 3.42- and 0.57-ml fractions were collected for tubes 1 to 10 and 11 to 70, respectively. Ten-microliter aliquots were assayed for transacylase activity (○). The dashed line denotes the NaCl concentration. (Top) SDS–PAGE of heparin Sepharose chromatography. The bands were visualized by silver staining.

then electrophoresed on a 12% slab gel in the presence of SDS. This separation identifies a protein having transacylase activity with an R_f of 0.19 to 0.35 in the first dimension and a molecular weight of about 40,000 observed in the SDS–PAGE separation.

Step 8: Heparin Sepharose Chromatography. The dialysate is applied to a column (HiTrap Heparin, 1 ml, Pharmacia) of heparin Sepharose preequilibrated with buffer A containing 77.3 μM of Triton X-100 at a flow rate of 5 ml/hr. Unbound protein is eluted with 15 ml of buffer A containing 77.3 μM of Triton X-100. The column is then rinsed with 15 ml of buffer B containing 77.3 μM of Triton X-100 (buffer B′). The enzyme activity appears as two peaks with a 60-ml linear gradient of sodium chloride from 0 to 0.2 M in buffer B′. The elution profile of the enzyme activity corresponds exactly to that of a protein with a molecular weight of 40,000 as estimated by SDS–PAGE. The protein in the second peak is homogeneous. The specific activity of the enzyme in the second peak is 25,149 nmol/min/mg of protein (Fig. 4).

Properties

Storage and Stability

The highly purified enzyme, even low protein concentration, is considerably stable in the presence of 77.3 μM Triton X-100 and is able to retain

TABLE III
COMPARISON OF 1-O-ACYLCERAMIDE SYNTHASE TO MAJOR PHOSPHOLIPASE A₂S

Type	I, II, III	IV	Canine myocardial and macrophage	1-O-Acylceramide synthase
Location	Secreted	Cytosolic	Cytosolic	Lysosomal
Molecular mass (kDa)	14	85	40/80	40
Cysteines	10–14	9	Unknown	Present
Arachidonate preference	No	Yes	Yes	Unknown
Calcium requirement	mM	μM	None	None
pH optimum	Neutral	Neutral	Neutral	Acid
ATP cofactor	No	No	Yes	No
Regulatory phosphorylation	No	Yes	No	Unknown
Phospholipase A₁ activity	No	Yes	No	No
Transacylase activity	No	Yes	No	Yes

activity when stored at 4° over 3 months. However, the enzyme activity is abolished completely in the presence of greater than 773 μM of Triton X-100.

Physicochemical Properties

The enzyme is a glycoprotein of a molecular weight of about 40,000 with a single polypeptide chain as estimated from its behaviors of SDS–PAGE, Sephadex G-75 gel filtration, and concanavalin A agarose chromatography.

Biochemical Properties

The purified enzyme has a pH optimum at 4.5. The divalent cations, Ca^{2+} and Mg^{2+}, enhance but are not essential for transacylase activity. Neither activation nor inactivation of the enzyme activity is observed in the presence of 2 mM ATP or 2 mM dithiothreitol. Preincubation of the enzyme with 1 mM N-ethylmaleimide, 1 mM PMSF, or 3.1 μM bromoenol lactone, a potent inhibitor of cytosolic calcium-independent phospholipase A_2, has no significant effect on enzyme activity. Activity is inhibited slightly in the presence of 28 μM AACOCF$_3$, a potent inhibitor of cytosolic phospholipase A_2. Partial inhibition of the enzyme activity is observed in the presence of 10–100 μg/ml of heparin. In the absence of N-acetylsphingosine, the enzyme acts as a phospholipase A_2. The enzyme purified here is distinct from other known phospholipase A_2s and is thought to be a novel calcium-independent phospholipase A_2 as well as transacylase (Table III).

[13] N-Acetylation of Sphingosine by Platelet-Activating Factor: Sphingosine Transacetylase

By TEN-CHING LEE

Introduction

Most mammalian cells appear to have a signaling system through the sphingomyelin pathway. Sphingomyelin is hydrolyzed by sphingomyelinase (SMase) to generate ceramides.[1,2] Ceramides are important lipid second messengers involved in mediating a variety of cell functions, including cell cycle arrest, cell differentiation, and apoptosis,[1,2] C$_2$-ceramide (N-acetyl-

[1] Y. A. Hunnun, *Science* **274**, 1855 (1996).
[2] S. Spiegel, D. Foster, and R. N. Kolesnick, *Curr. Opin. Cell Biol.* **8**, 159 (1996).

sphingosine) has been used extensively by many investigators as an unnatural, cell-permeable analog of ceramides. However, cumulative evidence indicates that C_2-ceramide can mimic many, but not all, of the effects of ceramides that were produced during signaling through the sphingomyelin pathway.[3-5] For instance, SMase treatment antagonized the mitogenic effect of the tumor promoter 12-O-tetradecanoylphorbol-13-acetate, whereas C_2-ceramide had no effect.[3] Likewise, tumor necrosis factor (TNF) has been reported to increase ceramides in cells, stimulating both $I\kappa B\alpha$ degradation and p105 processing.[4] However, C_2-ceramide had only a marginal effect on $I\kappa B\alpha$ degradation but promoted the processing of p105 to its p50 product strongly.[4]

We have identified[6,7] a CoA-independent platelet-activating factor (PAF): sphingosine transacetylase (TA_s) in the membrane fraction of HL-60 cells that can transfer the acetate group from PAF to sphingosine, forming N-acetylsphingosine (C_2-ceramide) by the following reaction:

l-alkyl-2-acetyl-sn-glycero-3-phosphocholine (alkylacetyl-GPC)
$\quad\quad\quad$ + sphingosine $\rightarrow$ alkyllyso-GPC + C_2-ceramide

In addition, TA_s activity can be induced by nerve growth factor and lysophosphatidic acid.[8] Also, C_2-ceramide in micromolar concentrations was detected in HL-60 cells, which is the concentration range where C_2-ceramide exerts most of its biological effects.[6] Taken together, these data suggest the possibility that some of the observed physiological effects by several factors, such as cytokines and environmental stresses through sphingomyelin signaling, may be accounted for by the action of C_2-ceramide and that C_2-ceramide should be considered a naturally occurring lipid mediator. This article describes assay procedures and properties of PAF: sphingosine transacetylase, the only enzyme known so far to be responsible for the synthesis of C_2-ceramide in mammalian systems.

[3] A. Olivera, A. Romanowski, C. S. Sheela Rani, and S. Spiegel, *Biochim. Biophys. Acta* **1348**, 311 (1997).
[4] M. P. Boland and L. A. J. O'Neill, *J. Biol. Chem.* **273**, 15494 (1998).
[5] M. Verheij, R. Bose, X. H. Lin, W. D. Jarvis, S. Grant, M. J. Birrer, E. Szabo, L. I. Zon, J. M. Kyriakis, A. Halmovitz-Friedman, Z. Fuks, and R. N. Kolesnick, *Nature* **380**, 75 (1996).
[6] T-c. Lee, M-c. Ou, K. Shinozaki, B. Malone, and F. Snyder, *J. Biol. Chem.* **271**, 209 (1996).
[7] T-c. Lee, *in* "Platelet-Activating Factor and Related Lipid Mediators in Health and Disease" (S. Nigam, G. Kumkel, and S. M. Prescott, eds.), p. 113. Plenum, New York, 1996.
[8] X. Qiu, A. Sebok, G. Tigyi, and T-c. Lee, *FASEB J.* **12**, A1379 (1998).

Assays

Reagents

    Tris–HCl buffer, 1 mM (pH 8.0)
    EDTA, 100 mM
    Sodium acetate, 20 mM
    Hexadecyl-[^{3}H]acetyl-GPC, 0.15 mM [after removing the organic solvent present in the vials from vendor with N_2, suspending 3.75 nmol (1 μCi) in 25 μl 0.1% bovine serum albumin (BSA)-saline]
    Sphingosine (or other substrate analogs), 0.5 mM (removing organic solvent from the substrate as just described and then suspending 12.5 nmol in 25 μl 3.3% BSA saline)

Preparation of Enzyme Source

    Total homogenates, membrane fractions from cells in culture, or various tissues of adult male (Sprague–Dawley) rats can be used as the enzyme source. Total homogenates are prepared by either sonicating the cells suspensions in homogenizing media [0.25 M sucrose, 10 mM Tris–HCl, pH 7.3, 1 mM dithiothreitol (DTT), and 1 μg/ml leupeptin] for 15 sec three to five times at a setting of 4 (Model W-375, Ultrasonic, Inc.) or homogenizing the tissue (10%, w/v) with four strokes of a motor-driven Potter–Elvehjem homogenizer.

    Membrane fractions are prepared by centrifuging the homogenates at 500g for 10 min, and the supernatants are centrifuged further at 100,000g for 60 min to obtain the membrane pellets. The membrane fraction is then suspended in 0.25 M sucrose, 10 mM HEPES, pH 7.0, and 1 mM DTT (3 ml/g of tissue or 1 ml/10^7–10^8 cells).

Enzymatic Assays and Identification of Lipid Products

    Assay mixtures consist of 25 μl of sphingosine, 25 μl of Tris–HCl, 12.5 μl of EDTA, 12.5 μl of sodium acetate, 25 μl of hexadecyl[^{3}H]acetyl-GPC, and 50 μl of cell homogenates or up to 100 μg of membrane proteins in a final volume of 250 μl and are incubated at 37° for up to 60 min. Reactions are terminated by lipid extraction,[9] except that methanol containing 2% acetic acid is used in the lipid extraction mixture.

    [^{3}H]PAF and N-[^{3}H]acetylsphingosine (C_2-ceramide), along with authentic standards (PAF from Avanti Polar Lipids, Inc. and acetylsphingosine, a product of Matreya, Inc.), are separated on silica gel H plates

---

[9] E. G. Bligh and W. J. Dyer, *Can. J. Biochem. Physiol.* **37**, 911 (1959).

developed in chloroform : methanol (90 : 10). The resolved lipids are visualized by exposing the silica gel layers to iodine vapor, and the radioactivity associated with each lipid class is quantitated by either zonal scanning or area scraping in conjunction with liquid scintillation spectroscopy.[10] The radioactivities of the lipid products present in the incubation mixture in the absence of an enzyme serve as a control blank.

Properties of Enzyme

Reaction Kinetics

During the initial phase of investigation on C_2-ceramide formation by the transacetylase in the membrane fraction of HL-60 cells, we noticed that the reaction rate was not linear with time, it exhibited a lag period, and a small but significant amount of C_2-ceramide was formed in the absence of an enzyme source.[6] However, these problems could be circumvented if sphingosine was prepared in the presence of BSA with a molar ratio of BSA/sphingosine = 1, albeit the enzyme activity was decreased by including BSA in the assay media.[6] Under these conditions, the apparent K_m for PAF was determined to be 5.4 μM. A Lineweaver–Burk plot gave a curved reciprocal line with an upward divergence for sphingosine. This type of kinetic behavior makes the determination of the Michaelis constant for sphingosine difficult.[6]

Substrate Specificity

Four stereoisomers for sphingosine, namely D-erthyro-, L-erthyro-, D-threo-, and L-threosphingosine exist, but only the D-erthyro isomer of sphingosine found in biological systems can serve as an acceptor for [³H]acetate from [³H]PAF.[6] In addition, among various sphingosine analogs tested, including sphinganine (dihydrosphingosine), stearylamine, ceramide, sphingomyelin, sphingosine 1-phosphate, and sphingosylphosphorylcholine, only sphinganine exhibited some capability to accept the acetate from PAF.[6]

In addition, when choline phospholipids containing long chain acyl groups such as [³H]dipalmitoyl-GPC or hexadecyl-[³H]arachidonoyl-GPC instead of [³H]PAF were incubated with sphingosine using the same conditions under which PAF : sphingosine transacetylase was assayed, no measurable amount of long chain acyl groupings could be transferred to sphingosine to form long chain acyl ceramide.[6] Similarly, sphingosine could not be acetylated by the lyso-PAF : acetyl-CoA acetyltransferase.[6]

---

[10] F. Snyder and D. Smith, *Sep. Sci.* **1,** 709 (1966).

We have previously identified the presence of PAF:lysophospholipids transacetylase in the membrane fractions of HL-60 cells.[11] This enzyme transfers the acetate group from PAF to a variety of lysophospholipids to form different kinds of analogs of PAF in a CoA-independent manner. Mixed substrate experiments indicated that sphingosine at an equal molar concentration with that of lysoplasmalogen inhibited the acetylation of lysoplasmalogens. However, lysoplasmalogens at 50 μM had no effect on the generation of C_2-ceramide by 50 μM sphingosine.[6] These results suggested that either the CoA-independent transacetylase has a higher substrate affinity for sphingosine than any of these other substrate analogs or that two isoforms of the transacetylase might be involved in the transfer of the acetate from PAF to sphingosine and lysophospholipids. Data discussed in the "Purification Procedures" section implicate the likelihood for the existence of both possibilities.

pH Optimum and Subcellular and Tissue Distribution

A broad optimal pH between 8.0 and 9.0 was observed for PAF:sphingosine transacetylase.[7] However, a different pH optimum (pH 7.0–8.0) was reported for PAF:lysophospholipid transacetylase. The amount of PAF:sphingosine transacetylase activity appeared to be distributed equally between mitochondrial and microsomal fractions, but the specific activity of the enzyme was higher in mitochondrial fractions than in microsomal fractions. Nuclear fractions possessed <10% of the total activity in the homogenates.[7] We also noticed in preliminary experiments that the PAF:lysophospholipid transacetylase activity in certain tissue, such as lung tissue, was highest in cytosolic fractions (data not shown). It is currently uncertain whether the same observations apply to the PAF:sphingosine transacetylase activity.

Kidney had the highest PAF:sphingosine transacetylase activity, whereas both kidney and lung had the highest PAF:lysoplasmalogen transacetylase activities. In general, the patterns of tissue distribution for both transacetylases were similar with the exception of rat lung and the undifferentiated HL-60 cells. PAF:sphingosine transacetylase activity was much higher than that of PAF:lysoplasmalogen transacetylase in undifferentiated HL-60 cells, and the reverse was true for the two transactylase activities in the lung.[6]

Effects of Inhibitors

Both PAF:sphingosine transacetylase and PAF:lysophospholipid transacetylase were inhibited in the range of 0.1–1.0 mM by serine esterase

[11] T-c. Lee, Y. Uemura, and F. Snyder, *J. Biol. Chem.* **267,** 19992 (1992).

inhibitors (diisopropylfluorophosphate and p-aminoethylbenzenesulfonyl fluoride[12]), reagents reacting with the -SH group of cysteine [5,5,-dithio-bis(nitrobenzoic acid) and N-ethylmaleimide], and the histidine modifier (diethyl pyrocarbonate),[6,13] but to a different degree. PAF: sphingosine transacetylase was more sensitive to the inhibition by sulfhydryl regents than that of PAF: lysophospholipid transacetylase.

Purification Procedures

We have achieved the purification of PAF: lysophospholipid trans-acetylase from the membrane fractions of rat kidneys to a single band with an apparent molecular mass of 40 KDa[13,14] and have identified five peptides of the purified enzyme (GTLDPYEGQEVMVR, AMLAFLQK, LLFSSGTR, IKEGEKEFHVR, and LPVSWNGPFK) showing amino acid sequence homologs with that of intracellular acetylhydrolase II.[15]

Furthermore, the purified PAF: lysophospholipid transacetylase contains two additional enzyme activities. PAF: sphingosine transacetylase and acetylhydrolase.[13] As with the crude enzyme preparations from the membrane fractions, the purified enzyme has also a higher affinity for sphingosine than lysophoslipids. At equal molar concentrations of sphinogine and lysoplasmalogens (50 μM), lysoplasmalogens had no effect on the acetylation of sphingosine. In constrast, sphingosine inhibited the formation of acetylated lysoplasmalogens. At present, we cannot rule out the possibilities that other PAF: sphingosine transacetylases may exist and/or that there is a unique mechanism(s) involved in the regulation of these three catalytic activities within the same protein molecule.

Acknowledgments

This work was supported by the National Heart, Lung, and Blood Institute (Grant HL52492). I thank Ms. Joanna Wilkins for editing the manuscript.

[12] C. Dentan, A. D. Tselepis, M. J. Chapman, and E. Ninio, *Biochim. Biophys. Acta* **1299,** 353 (1996).

[13] K. Karasawa, X. Qiu, and T-c. Lee, *J. Biol. Chem.* **274,** 8655 (1999).

[14] K. Karasawa and T-c. Lee, *FASEB J.* **12,** A1380 (1998).

[15] K. Hattori, H. Adachi, A. Matsuzawa, K. Yamamoto, J. Aoki, M. Hattori, H. Aria, and K. Inoue, *J. Biol. Chem.* **271,** 33032 (1996).

[14] Inositolphosphoryl Ceramide Synthase from Yeast

By ANTHONY S. FISCHL, YONGSHENG LIU, ANGELEAH BROWDY,
and AIDA E. CREMESTI

Introduction

Inositolphosphoryl ceramide (IPC) synthase catalyzes the transfer of inositol phosphate from phosphatidylinositol (PI) to ceramide forming diacylglycerol and IPC, a major yeast sphingolipid and precursor of the yeast sphingolipids, mannosylinositolphosphoryl ceramide and mannosyldiinositolphosphoryl ceramide.[1] The enzyme, encoded by the yeast *AUR1* gene, and its product, IPC, are essential to the growth and viability of *Saccharomyces cerevisiae*.[2,3] IPC synthase activity is associated with the microsomal membrane fraction of yeast,[1,4] and the expression of IPC synthase activity is regulated by inositol and the growth phase of *S. cerevisiae*.[5] Methods to measure IPC synthase activity from yeast microsomal membranes[1,3,5] and Triton X-100-solubilized yeast microsomes[6,7] have been reported. This article describes the purification and properties of the enzyme.

Growth of Yeast Cells

Wild-type *S. cerevisiae* strain S288C (*MATa gal 2*) is used for the purification of IPC synthase. The cells are grown in 1% yeast extract, 2% peptone, and 2% glucose at 28° until mid- to late exponential phase where IPC synthase activity is maximally expressed.[5] Cells are harvested by centrifugation and stored at −80°.[8,9]

Preparation of Enzyme Substrates

CDP-diacylglycerol (CDP-DAG) is prepared from phosphatidic acid and CMP morpholidate by the method of Agranoff and Suomi[10] with the

[1] G. W. Becker and R. L. Lester, *J. Bacteriol.* **142,** 747 (1980).
[2] G. B. Wells and R. L. Lester, *J. Biol. Chem.* **258,** 10200 (1983).
[3] M. M. Nagiec *et al., J. Biol. Chem.* **272,** 9809 (1997).
[4] A. Puoti, C. Desponds, and A. Conzelmann, *J. Cell Bio.* **113,** 515 (1991).
[5] J. Ko, S. Cheah, and A. S. Fischl, *J. Bacteriol.* **176,** 5181 (1994).
[6] S. M. Mandala *et al., J. Biol. Chem.* **272,** 32709 (1997).
[7] J. S. Ko, S. Cheah, and A. S. Fischl, *J. Food Biochem.* **19,** 253 (1995).
[8] G. M. Carman and A. S. Fischl, *Methods Enzymol.* **209,** 305 (1992).
[9] A. S. Fischl and G. M. Carman, *J. Bacteriol.* **154,** 304 (1983).
[10] B. W. Agranoff and W. D. Suomi, *Biochem. Prep.* **10,** 47 (1963).

modifications of Carman and Fischl.[11] Radioactive phosphatidyl-[³H]inositol ([³H]PI) is prepared enzymatically using myo-[2-³H]inositol and Triton X-100-solubilized yeast PI synthase as described by Fischl and Carman.[9] Solubilized PI synthase (3.5–4 U/mg) is incubated in 50 mM Tris–HCl buffer (pH 8.0) containing 2.4 mM Triton X-100, 2.5 mM MnCl$_2$, 0.2 mM CDP-DAG, 0.8–1 mg/ml protein, and 15 μCi myo-[2-³H]inositol in a total volume of 0.1 ml at 30° for 2.5 hr. The reaction is terminated by the addition of 0.5 ml of 0.1 N HCl in methanol, and the product [³H]PI is extracted using 2 ml of chloroform and 3 ml of 1 M MgCl$_2$. To separate radioactive [³H]PI from unreacted CDP-DAG, the radioactive extract is applied onto a column (1 ml) of CM-52 (Whatman), sodium form, that is equilibrated in chloroform. A small amount of glass beads (0.5 mm) is added to the top of the packed resin in order to prevent disturbances of the column bed during subsequent solvent changes. The column is washed with 2 × 2.5 ml of chloroform/methanol (9:1) and 2 × 2.5 ml of chloroform/methanol (9:3). Radioactive [³H]PI is eluted with chloroform/methanol (4:6). Fractions containing [³H]PI are pooled and dried under a stream of nitrogen gas. The [³H]PI, in chloroform/methanol (9:1), is stored at −20° and is used within a week.

Inositolphosphoryl Ceramide Synthase Assay

Radioactive Assay Using Phosphatidyl [³H]Inositol

IPC synthase activity is measured by following the incorporation of phospho-[³H]inositol from [³H]PI [20,000 disintegrations per minute (dpm)/min/nmol] into an alkaline stable product (IPC) as described by Ko et al.[5] Reactions are conducted in 50 mM bis–Tris–HCl buffer (pH 6.5) containing 0.5 mM [³H]PI, 5 mM Triton X-100, 1 mM MnCl$_2$, 5 mM MgCl$_2$, 0.2 mM ceramide (N-acylsphingosine containing hydroxy fatty acids), and purified IPC synthase (0.005–0.1 mg/ml) in a final assay volume of 0.1 ml. The reaction is terminated by the addition of 0.64 ml chloroform/methanol (1:1). Methanolic NaOH (0.6 N NaOH in methanol), 0.1 ml, is added, and samples are vortexed and incubated at 37° for 60 min in order to deacylate unreacted [³H]PI. Following incubation, the assay mixture is placed in an ice bath to stop the reaction. The sphingolipid product, [³H]IPC, is extracted by the addition of 2 N HCl in methanol (2 ml), chloroform (2 ml), 1 M NaCl (3 ml), and carrier lipids (phosphatidic acid and PI, 100 μg each). The system is mixed, and the phases are separated by a 2-min centrifugation at 100g. The upper aqueous phase is removed by aspiration, and a 1-ml aliquot of the chloroform phase is taken and transferred to a 7-ml scintilla-

[11] G. M. Carman and A. S. Fischl, J. Food Biochem. 4, 53 (1980).

tion vial and evaporated to dryness in a 50° water bath. Liquid scintillation fluid (Ecoscint H, National Diagnostics, 3 ml) is added to each vial and radioactivity is determined by scintillation counting. Because IPC is the only radioactive product detectable following silica gel-impregnated paper thin-layer chromatography analysis of microsomal and Triton X-100-solubi- lized enzyme preparations,[5] the final organic phase is analyzed by direct scintillation counting. Each sample is assayed in triplicate, along with a blank consisting of all assay reagents inactivated with 0.64 ml of chloroform/ methanol (1:1) prior to the start of the reaction.

Fluorescent Assay Using NBD-Ceramide

IPC synthase activity can be monitored by following the incorporation of NBD-ceramide (Avanti Polar Lipids), a fluorescent analog of ceramide containing a six carbon N-acyl side chain, into NBD-IPC. Reactions are conducted in 50 mM bis–Tris–HCl buffer (pH 6.5) containing 0.5 mM PI, 5 mM Triton X-100, 1 mM MnCl$_2$, 5 mM MgCl$_2$, 0.1 mM NBD-ceramide, and purified IPC synthase (0.005–0.1 mg/ml) in a final assay volume of 0.1 ml. The reaction is terminated by the addition of 0.5 ml of 0.1 N HCl in methanol. Chloroform (1 ml) and 1 M MgCl$_2$ (1.5 ml) are added, the system is mixed, and the phases are separated by a 2-min centrifugation at 100g. The chloroform soluble product, NBD-IPC, is analyzed by analytical thin- layer chromatography on silica gel 60 plates (EM Science) using the solvent system chloroform/methanol/water (65:25:4). NBD-IPC is identified and quantified by direct fluorescence using a Molecular Dynamics 840 STORM unit. The R_f values of NBD-ceramide and NBD-IPC using this solvent system are 0.77 and 0.13, respectively.

One unit of enzymatic activity is defined as the amount of enzyme that catalyzes the formation of 1 nmol of product per minute. Specific activity is defined as units per milligram of protein. Protein is determined by the method of Bradford.[12] Buffers identical to those containing the protein samples are used as blanks. The presence of Triton X-100 does not interfere with the protein determination, provided the blank contains a final concen- tration of detergent identical to that of the sample.[9]

Purification of IPC Synthase

All purification steps are performed at 4°.

Step 1: Preparation of Cell Extract

The cell extract is prepared from 200 g (wet weight) of yeast cells by disruption with glass beads using a Bead-Beater (Biospec Products) as

---

[12] M. M. Bradford, *Anal. Biochem.* **72,** 248 (1976).

described by Fischl and Carman.[9] Yeast cells (50 g, wet weight) are suspended in 200 ml of 50 mM Tris–HCl buffer (pH 7.0) containing 1 mM sodium EDTA, 0.3 M sucrose, 10 mM 2-mercaptoethanol, 0.2 mM phenylmethylsulfonyl fluoride (PMSF), 0.1 mM benzamidine, and 0.5 μg/ml leupeptin. The cell suspension is added to a Biospec Bead Beater chamber containing 300 g of prechilled glass beads, 0.5 mm in diameter. Cells are disrupted by homogenization for five 1-min bursts, with a 5-min pause between bursts. Glass beads, unbroken cells, and cell debris are removed by centrifugation at 1500g for 10 min to obtain the cell extract (supernatant).

Step 2: Preparation of Microsomes

The microsomal membrane fraction is isolated from the cell extract by differential centrifugation.[9] Mitochondria are removed by centrifugation of the cell extract at 27,000g for 25 min. The microsomal membrane fraction is obtained by centrifugation of the 27,000g supernatant at 100,000g for 90 min. The microsomal membrane pellets are washed with 50 mM Tris–HCl buffer (pH 7.0) containing 5 mM 2-mercaptoethanol, 10% glycerol, 1 mM PMSF, 0.1 mM benzamidine, and 0.5 μg/ml leupeptin, combined and resuspended in the same buffer to a final protein concentration of 30–40 mg/ml. Microsomes are frozen at $-80°$ until used for purification.

Step 3: Preparation of Triton X-100 Extract

The microsomal membrane fraction is suspended in 50 mM bis–Tris–HCl buffer (pH 6.0) containing 10% glycerol, 5 mM 2-mercaptoethanol, 1% Triton X-100, 1 mM MnCl$_2$, 1 mM PMSF, 0.1 mM benzamidine, and 0.5 μg/ml leupeptin at a final protein concentration of 10 mg/ml.[7] After incubation for 1 hr at 4°, the suspension is centrifuged at 100,000g for 90 min to obtain the Triton X-100 extract (supernatant). The solubilized enzyme is frozen immediately and stored at $-80°$.

Step 4: DEAE-Cellulose Column Chromatography

IPC synthase does not bind to DEAE-cellulose (DE-52, Whatman) at pH 7.0. This property is exploited in the following purification. The Triton X-100 extract (724 units) is titrated to pH 7.0 using 0.1 M Tris–HCl (pH 9.5) and loaded onto a DE-52 column (2.5 × 17 cm) preequilibrated with 50 mM Tris–HCl buffer (pH 7.0) containing 5 mM MgCl$_2$, 1 mM MnCl$_2$, 5 mM 2-mercaptoethanol, 0.05% Triton X-100, 10% glycerol, 1 mM PMSF, 0.1 mM benzamidine, and 0.5 μg/ml leupeptin. The column is washed with 1.5 column volumes of the same buffer. Because the enzyme is not retained by DE-52, a few large fractions are collected and assayed for IPC synthase activity. The DE-52 run through and wash fractions are collected and used for the next step in the purification scheme.

Step 5: Q-Sepharose Chromatography

A column (2.5 × 9 cm) of Q-Sepharose fast flow (Pharmacia Biotech) is equilibrated in 50 mM bis–Tris–HCl buffer (pH 6.0) containing 5 mM MgCl$_2$, 1 mM MnCl$_2$, 5 mM 2-mercaptoethanol, 0.05% Triton X-100, 10% glycerol, 1 mM PMSF, 0.1 mM benzamidine, and 0.5 μg/ml leupeptin. The DE-52-purified enzyme (650 units) is applied to the column at a flow rate of 60 ml/hr. The column is washed with 4 column volumes of equilibration buffer followed by elution of IPC synthase activity in 10-ml fractions with 11 column volumes of a linear NaCl gradient (0–0.5 M NaCl) in equilibration buffer. The peak of IPC synthase activity elutes from the column at a NaCl concentration between 0.15 and 0.25 M. The most active fractions are pooled and used for the next step of the purification.

Step 6: Octyl-Sepharose Chromatography

A column (2.5 × 9.5 cm) of Octyl-Sepharose 4 fast flow (Pharmacia Biotech) is equilibrated with 50 mM Tris–HCl buffer (pH 7.0) containing 5 mM MgCl$_2$, 1 mM MnCl$_2$, 5 mM 2-mercaptoethanol, 10% glycerol, 0.2 M NaCl, 1 mM PMSF, 0.1 mM benzamidine, and 0.5 μg/ml leupeptin. It is necessary to omit Triton X-100 from the equilibration buffer for the binding of IPC synthase to Octyl-Sepharose. The Q-Sepharose-purified enzyme (709 units) is applied to the Octyl-Sepharose column at a flow rate of 60 ml/hr. The column is washed with 2 column volumes of equilibration buffer followed by 4 column volumes of equilibration buffer without NaCl. IPC synthase activity is eluted from the column with 10 column volumes of a linear Triton X-100 gradient (0–1%) in 50 mM Tris–HCl buffer (pH 7.0) containing 5 mM MgCl$_2$, 1 mM MnCl$_2$, 5 mM 2-mercaptoethanol, 10% glycerol, 1 mM PMSF, 0.1 mM benzamidine, and 0.5 μg/ml leupeptin. The peak of IPC synthase activity elutes from the column at a Triton X-100 concentration of about 0.5%. The most active fractions are pooled and used for the next step in the purification scheme.

Step 7: Mono Q I Chromatography

A Mono Q column (0.5 × 5 cm; Pharmacia Biotech) is equilibrated in 50 mM bis–Tris–HCl buffer (pH 6.0) containing 5 mM MgCl$_2$, 1 mM MnCl$_2$, 5 mM 2-mercaptoethanol, 0.05% Triton X-100, 10% glycerol, 1 mM PMSF, 0.1 mM benzamidine, and 0.5 μg/ml leupeptin. Purification of the Octyl-Sepharose-purified IPC synthase is performed using a 0.8- to 1.0-mg aliquot of the enzyme preparation. The Octyl-Sepharose-purified enzyme (60 units) is applied to the column at a flow rate of 60 ml/hr. The column is washed with 4 column volumes of equilibration buffer followed by enzyme elution with 25 column volumes of a linear NaCl gradient (0–0.5 M) in equilibration buffer. IPC synthase elutes in a sharp peak at a NaCl concentration of about 0.2 M. Fractions containing activity are pooled and used for the next step in the purification scheme.

Step 8: Phenyl-Resource Chromatography

A Phenyl-Resource column (1 × 2 cm; Pharmacia Biotech) is equilibrated in 50 mM Tris–HCl buffer (pH 7.0) containing 1.5 M NaCl, 5 mM MgCl$_2$, 1 mM MnCl$_2$, 5 mM 2-mercaptoethanol, 10% glycerol (w/v), 1 mM PMSF, 0.1 mM benzamidine, and 0.5 μg/ml leupeptin. It is necessary to add NaCl and omit Triton X-100 from the equilibration buffer for binding of IPC synthase to Phenyl-Resource. Just prior to the chromatography step, NaCl is added to the Mono Q-purified enzyme to a final concentration of 1.5 M and then one-half of the enzyme extract (103 units) is applied onto the Phenyl-Resource column at a flow rate of 60 ml/hr. The column is washed with 2 column volumes of equilibration buffer followed by 6 column volumes of equilibration buffer without NaCl. IPC synthase activity is eluted from Phenyl-Resource with 25 column volumes of a linear Triton X-100 gradient (0–1%) in 50 mM Tris–HCl buffer (pH 7.0) containing 5 mM MgCl$_2$, 1 mM MnCl$_2$, 5 mM 2-mercaptoethanol, 10% glycerol, 1 mM PMSF, 0.1 mM benzamidine, and 0.5 μg/ml leupeptin. Prior to purification of the remainder of the enzyme extract, the Phenyl-Resource column is washed with 5 column volumes of deionized water and 20 column volumes of 20% ethanol. This removes bound Triton X-100, which interferes with binding of IPC synthase by the Phenyl-Resource resin. The remainder of the enzyme extract (103 units) is purified as described earlier. The most active fractions are pooled and used for the next step in the purification scheme.

Step 9: Mono Q II Chromatography

A second Mono Q column (0.5 × 5 cm; Pharmacia Biotech) is equilibrated with 50 mM bis–Tris–HCl buffer (pH 6.0) containing 5 mM MgCl$_2$, 1 mM MnCl$_2$, 5 mM 2-mercaptoethanol, 0.05% Triton X-100, 10% glycerol, 1 mM PMSF, 0.1 mM benzamidine, and 0.5 μg/ml leupeptin. The Phenyl-Resource-purified enzyme (207 units) is applied to the column at a flow rate of 60 ml/hr. The column is washed with 4 column volumes of equilibration buffer followed by enzyme elution with 40 column volumes of a linear NaCl gradient (0–0.5 M) in equilibration buffer. Fractions containing activ-

TABLE I
PURIFICATION OF IPC SYNTHASE FROM *Saccharomyces cerevisiae*[a]

Purification step	Total units (nmol/ min)	Protein (mg)	Specific activity (units/mg)	Purification (fold)	Yield
Cell-free extract	752	4700	0.16	1	100
Microsomes	739	2250	0.33	2.1	98
Triton X-100 extract	724	452	1.6	10	96
DEAE-cellulose	650	214	3.0	19	86
Q-Sepharose	709	17.7	40	250	94
Octyl-Sepharose	480	6.65	72	450	64
Mono QI	488	4.76	102	668	65
Phenyl Resource	207	1.26	163	1025	28
Mono Q II	170	0.31	548	3425	23

[a] Starting with 200 g of cell paste from *S. cerevisiae* strain S288C.

ity are pooled and stored at $-80°$. The purified enzyme is completely stable for at least 3 months of storage at $-80°$.

A summary of the purification of IPC synthase is presented in Table I. The overall purification of IPC synthase over the cell extract is 3425-fold with an activity yield of 23%. Examination of the purified enzyme by sodium dodecyl sulfate gel electrophoresis reveals four major protein bands, suggesting that the enzyme is highly purified but not to homogeneity. The four major protein bands have minimum subunit molecular weights of 97,000, 66,000, 50,000 and 46,000. Attempts to purify the enzyme further results in total loss of enzyme activity.

Properties of IPC Synthase

Maximal IPC synthase activity is measured at $30°$ and pH 6.5 in the presence of 5 mM Triton X-100, 1 mM manganese, and 5 mM magnesium ions.[7] IPC synthase activity is dependent on the surface concentration of PI and ceramide in the Triton X-100 mixed micellar assay system,[7] and PI activates IPC synthase activity in a cooperative manner with a Hill constant of 3. The apparent K_m values are 5 mole % for the surface concentration of PI and 1.35 mole % for the surface concentration of ceramide in the Triton X-100 mixed micellar assay system. Thioreactive reagents,[7] sphingoid bases,[13] Aureobasidin A,[3] Rustmicin,[14] and Khafrefungin[6] inhibit IPC synthase activity.

[13] W.-I. Wu et al., J. Biol. Chem. **270**, 13171 (1995).
[14] S. M. Mandala et al., J. Biol. Chem. **273**, 14942 (1998).

Synthetic and Analytical Uses

The Mono Q I-purified enzyme, which is easier to prepare, can be used for synthetic and analytical uses. Purified IPC synthase can be used to synthesize fluorescent NBD-IPC from NBD-ceramide and PI and radiolabled [³H]IPC from ceramide and [³H]PI. Purified IPC synthase can be used to determine quantitatively the intracellular ceramide concentration of yeast by following the formation of [³H]IPC from [³H]PI. Ceramides are extracted from trichloroacetic acid (TCA)-treated yeast cells with chloroform/methanol (1 : 1, v/v) at 50° for 30 min.[15] Sphingoid bases, which inhibit IPC synthase activity, are removed from lipid extracts by passage over a 0.5-ml BioRex-70 (H⁺) (200–400 mesh) column.[15] Standard curves are generated using ceramides containing hydroxy fatty acids (200–2000 pmol) and Mono Q I-purified IPC synthase (7–15 μg/assay). All incubations are conducted at 30° for 70 min as described earlier. The assay is sensitive to 200 pmol of ceramide, and conversion of ceramide to [³H]IPC is near 100% between 200 and 2000 pmol of ceramide.

[15] G. B. Wells, R. C. Dickson, and R. L. Lester, *J. Biol. Chem.* **273**, 7235 (1998).

[15] Enzymes of Sphingolipid Metabolism in Plants

By DANIEL V. LYNCH

Introduction

The predominant sphingolipids in plant tissues are glucosylceramides[1] and complex glycophosphosphingolipids, the inositolphosphorylceramides (InsPCers).[2] The latter have been demonstrated in only a few plant tissues but comprise a structurally diverse group of molecules, exhibiting variability in the number and arrangement of monosaccharides linked to the inositol head group. Ceramides and free long chain bases are also present at low concentrations in plant tissues.[1]

Considerable structural diversity exists among the glucosylceramides from various plant tissues with respect to long chain base and fatty acid composition: these results have been summarized previously.[1] The most

[1] D. V. Lynch, *in* "Lipid Metabolism in Plants" (T. S. Moore, ed.), p. 285. CRC Press, Boca Raton, FL, 1993.
[2] R. L. Lester and R. C. Dickson, *Adv. Lipid Res.* **26**, 253 (1993).

abundant long chain bases in plant glucosylceramides (and ceramides) are isomers of 4,8-sphingadienine, 4-hydroxy-8-sphingenine, and 8-sphingenine. α-Hydroxy fatty acids account for >90% of the total fatty acids of glucosylceramide. Saturated C_{16} to C_{24} acyl chains are most abundant, although monounsaturated hydroxy fatty acids are common in some cereals. 4-Hydroxysphinganine and 4-hydroxysphingenine, together with saturated C_{24} hydroxy fatty acid, are prevalent in InsPCers.[2] Note that the prevalent long chain base of many mammalian sphingolipids, sphingosine (4-sphingenine), is virtually absent from plant sphingolipids. Sphinganine and 4-hydroxysphinganine are present at low concentrations as free bases in plant tissues.

Glucosylceramides are minor components of plant lipid extracts, typically accounting for <5% of the total lipid. Determination of the InsPCer content in plant tissues has been hindered by the propensity for these polar lipids to partition in the aqueous phase during conventional lipid extraction procedures. The low abundance of sphingolipids in plant extracts may have led some plant lipidologists to overlook their possible significance. However, glucosylceramides are quantitatively important components of the plasma membrane, comprising 7 to 30 mol% of the membrane lipid, depending on the plant tissue.[1,3] Glucosylceramides are also quantitatively important components of the tonoplast, constituting 12 to 17 mol% of the membrane lipid.[1,4] Glucosylceramide is present almost exclusively in the outer (apoplastic) monolayer of squash plasma membrane,[5] but its transbilayer distribution in the tonoplast has not been determined. The intracellular location and transbilayer distribution of InsPCers in plants have not been investigated.

In contrast to the tremendous volume of published research on bioactive sphingolipids in mammalian systems, there is a paucity of published studies using plant systems. It has been found that mycotoxins such as fumonisin and AAL toxin, which disrupt sphingolipid metabolism by inhibiting sphinganine N-acyltransferase, promote apoptosis in tomato[6] and can induce a lethal accumulation of long chain bases in a variety of plant tissues.[7] These results point to the potential importance of sphingolipids in regulating cellular processes in plant cells, although mechanistic evidence of a sphin-

[3] M. Uemura and P. L. Steponkus, *Plant Physiol.* **104,** 479 (1994).

[4] E. Tavernier, D. LeQuoc, and K. LeQuoc, *Biochim. Biophys. Acta* **167,** 242 (1993).

[5] D. V. Lynch and A. J. Phinney, *in* "Plant Lipid Metabolism" (J. C. Kader and P. Mazliak, eds.), p. 239. Kluwar Academic, Dordrecht, The Netherlands, 1995.

[6] H. Wang, J. Li, R. M. Bostock, and D. G. Gilchrist, *Plant Cell.* **8,** 375 (1996).

[7] H. K. Abbas, T. Tanaka, S. O. Duke. J. K. Porter, E. M. Wray, L. Hodges, A. E. Sessions, E. Want, A. H. Merrill, Jr., and R. T. Riley, *Plant Physiol.* **106,** 1085 (1994).

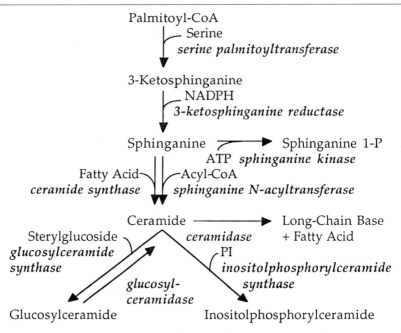

FIG. 1. Enzymatic steps of sphingolipid metabolism demonstrated in plants. Only cosubstrates, not by-products, of each reaction are shown. Note that the term ceramide synthase is used to identify ceramide synthesis by a poorly understood mechanism that utilizes free fatty acids; the term is not redundant with sphinganine N-acyltransferase. The term glucosylceramide synthase is used to distinguish the proposed enzymatic reaction in plants using steryl glucoside as the glucose donor from the reaction catalyzed by ceramide glucosyltransferase that uses UDP-glucose as substrate.

golipid signal transduction pathway is lacking. Evidence shows that InsPCers may serve as anchors for GPI-linked proteins.[8]

Sphingolipid metabolism in plants has been investigated only recently and has focused on demonstrating and characterizing the *in vitro* activities of the major enzymatic steps (Fig. 1). Assay procedures for these reactions, optimized for plant preparations, are given in this article. Methods of preparing membrane fractions appropriate for these assays are described at the end of the article.

[8] N. Morita, H. Nakazato, H. Okuyama, Y. Kim, and G. A. Thompson, Jr., *Biochim. Biophys. Acta* **21,** 53 (1996).

Synthesis of Sphinganine: Characterization of Serine
  Palmitoyltransferase and Ketosphinganine Reductase Activities

The activity of serine palmitoyltransferase (palmitoyl-CoA:L-serine C-palmitoyltransferase) was first characterized in a plant system using a microsomal membrane preparation from summer squash fruit, and, subsequently, activity was demonstrated in microsomes from a number of other plant tissues, including bean hypocotyl.[9] Activity is assayed by monitoring the incorporation of L-[³H]serine into the chloroform-soluble product, 3-ketosphinganine. The assay procedure used is similar to that described previously,[10,11] but is optimized for the plant enzyme.

Assay

The assay contains 100 mM HEPES–KOH (pH 8.0), 2.5 mM EDTA, 4 mM dithiothreitol (DTT), 50 μM pyridoxal 5'-phosphate, 200 μM palmitoyl-CoA, 2 mM L-[³H(G)]serine (from NEN; adjusted to a specific activity of 5 μCi/μmol, 1 μCi added to each tube), and 50–100 μg of membrane protein in a total volume of 100 μl. A paired tube, prepared along with each assay tube and containing all of the assay components except palmitoyl-CoA, is necessary to account for the recovery of radioactivity in the chloroform extract from a contaminant in the labeled serine (see later).

To perform the assay, all components except membrane are added to each 13 × 100-mm test tube on ice. The tubes are equilibrated at 30° in a shaking water bath, membrane is added, and each tube is mixed briefly to initiate the reaction. Alternatively, the reaction can be initiated by the addition of palmitoyl-CoA to assay tubes containing membrane. After incubating for 10 min at 30° with mild shaking, the reaction is stopped by adding 200 μl of 0.1 N NH$_4$OH. Lipids are extracted by adding 1.5 ml of chloroform/methanol (1:2, v/v) to each tube with mixing, then adding 20 μg of sphinganine carrier (in 20 μl ethanol), 1 ml of chloroform, and 2 ml of 0.1 N NH$_4$OH. After mixing, the tubes are centrifuged briefly at medium speed in a tabletop centrifuge to facilitate phase separation. The upper phase is removed and the bottom phase is washed twice with 0.8 ml of water, mixing and centrifuging each time. After removing the upper phase from the second wash, 0.8 ml of the lower chloroform phase from each tube is transferred to a 7-ml scintillation vial. The solvent is evaporated

[9] D. V. Lynch and S. R. Fairfield, *Plant Physiol.* **103**, 1421 (1993).
[10] R. D. Williams, E. Wang, and A. H. Merrill, Jr., *Arch. Biochem. Biophys.* **228**, 282 (1984).
[11] A. H. Merrill, Jr. and E. Wang, *Methods Enzymol.* **209**, 427 (1992).

under a stream of nitrogen, 5 ml of scintillation cocktail is added, and the radioactivity is measured by liquid scintillation counting.

To calculate enzyme activity, the background radioactivity (from a paired tube containing all components but palmitoyl-CoA) is subtracted, and the resulting value is converted to nanomoles of 3-ketosphinganine formed per minute per milligram protein, taking into account the liquid scintillation counting efficiency, the proportional volume of chloroform counted (53%), the specific radioactivity of the [³H]serine used, the incubation time, and the amount of protein added.

The amount of serine contaminant recovered with product in the chloroform extract is dependent on the assay conditions, particularly the amount of protein added. The amount of labeled contaminant present in the serine stock varies with its lot and age and can be removed from the stock using an ion-exchange column[10,11] or by washing with ethyl ether prior to using it in the assay.

To optimize assay conditions when using different plant tissues as the source of the enzyme, several variables should be examined. These include (1) the temperature and duration of the assay (whereas 37° and 5 min were optimal for squash, 30° and 10 min were optimal for bean and are used routinely); (2) the optimal palmitoyl-CoA concentration (typically between 50 and 200 μM), as higher concentrations may be inhibitory; and (3) the amount of membrane protein added, as well as the procedure used for membrane isolation. Two other points merit mention. First, some plant tissues (e.g., rye leaves) may have heat-labile inhibitors of the reaction, most likely acyl-CoA hydrolases. This can be determined by adding suspect membranes (native or heat-inactivated) to an assay using a membrane preparation exhibiting reliable enzyme activity and determining whether the reaction is inhibited.[9] Second, evidence showed that free fatty acids may inhibit enzyme activity,[9,12] but this can be ameliorated by the addition of as much as 40 μg of fatty acid-free bovine serum albumin to the assay.

Substrate Specificity and Properties of the Reaction

The enzyme exhibits a strong preference for palmitoyl-CoA, consistent with the preponderance of C_{18} sphingoid long chain bases in plant tissues. The apparent K_m for serine is approximately 1.8 mM. Two known mechanism-based inhibitors of the mammalian enzyme, L-cycloserine and β-chloro-L-alanine, are effective inhibitors of plant serine palmitoyltransferase activity. Results to date indicate that the properties and catalytic mecha-

---

[12] D. V. Lynch, R. A. Spence, K. M. Theiling, K. W. Thomas, and M. T. Lee, *in* "Biochemical and Molecular-Biological Aspects of Membrane and Storage Lipids of Plants" (C. R. Somerville and N. Murata, eds.), p. 183. A.S.P.P., Rockville, MD, 1993.

nism of plant serine palmitoyltransferase are similar to those of the animal, fungal, and bacterial enzyme in most respects. However, the specific activities of the enzyme in plant microsomes isolated from actively expanding tissues are 2- to 20-fold higher than those reported previously for preparations from animal tissues.[10]

In this same assay system, the conversion of 3-ketosphinganine to sphinganine via an NADPH-dependent ketosphinganine reductase (dihydrosphingosine : NADP$^+$ 3-oxidoreductase) can also be demonstrated. The enzymatic reduction is carried out by adding 1.5 mM NADPH or a NADPH-regenerating system consisting of four units of glucose 6-phosphate dehydrogenase, 1 mM glucose 6-phosphate, and 1 mM NADP$^+$.[9,13] To separate labeled 3-ketosphinganine and sphinganine, the chloroform extract is applied to a silica gel thin-layer chromatography (TLC) plate and developed in chloroform/methanol/2 N NH$_4$OH (80:30:4, v/v). The former has an R_f of 0.7 whereas the latter has an R_f of 0.25. Radioactivity in the two bands may be monitored by liquid scintillation counting of the appropriate regions scraped from the TLC plate or by radiometric scanning of the TLC plate (using a Berthold, Bioscan, or comparable instrument). More specific procedures for assaying 3-ketosphinganine reductase activity in plant membrane preparations have not been developed. However, assay conditions using [^{3}H]ketosphinganine to investigate the reductase from other sources have been presented previously[11] and should be applicable to the plant enzyme.

Synthesis of Ceramide: Characterization of Acyl-CoA-Dependent and Acyl-CoA-Independent Mechanisms

Two different enzymatic mechanisms of ceramide formation have been demonstrated in plant membrane preparations, including (1) acyl-CoA : sphinganine N-acyltransferase activity utilizing acyl-CoA as the acyl donor and (2) ceramide synthesis by an undetermined mechanism apparently utilizing free fatty acid as the acyl donor.[12]

Acyl-CoA-dependent sphinganine N-acyltransferase activity has been demonstrated in microsomal membranes of squash fruit, bean hypocotyls, and corn shoots.[12] Using [^{3}H]sphinganine or [^{14}C]palmitoyl-CoA, the formation of labeled ceramide is monitored following the separation of product and substrate by TLC. Both assay procedures, optimized for bean hypocotyl microsomes, are described.

---

[13] P. E. Braun, P. Morrell, and N. S. Radin, *J. Biol. Chem.* **245**, 335 (1970).

Assay

The assay using [³H]sphinganine as a labeled substrate contains 25 m*M* potassium phosphate (pH 7.5), 1 m*M* ATP, 40 μ*M* palmitoyl-CoA, and 50 μ*M* D-*erythro*-[³H]sphinganine (adjusted to a specific activity of 10–20 μCi/μmol, 0.05–0.1 μCi added to each tube), and 40–60 μg of membrane protein in a total volume of 100 μl. Labeled sphingosine and sphinganine are available commercially, and a procedure for reducing the double bond of sphingosine with tritium to yield D-*erythro*-[³H]sphinganine is given at the end of the article.

To perform the assay, [³H]sphinganine (in methanol) is first added to each 13 × 100-mm test tube, and the solvent is evaporated under a nitrogen stream. Immediately upon drying, the remaining assay components (except membrane) are added. The tubes are sonicated to disperse the [³H]sphinganine and are then equilibrated in a 37° water bath. The reaction is initiated by the addition of membrane with brief mixing. After incubation for 20 min in a 37° water bath with mild agitation, the reaction is terminated by adding 200 μl of 0.1 *N* NH₄OH. Lipids are extracted by adding 1.5 ml of chloroform/methanol (1:2, v/v) to each tube with mixing and then adding 20 μl each of sphinganine and ceramide carriers (1.0 mg/ml in ethanol), followed by 500 μl of chloroform and 500 μl of water. The tubes are mixed thoroughly and centrifuged in a tabletop centrifuge for 10 min at medium speed. The upper aqueous phase of the mixture is removed and discarded and the lower phase is transferred into a 12 × 75-mm test tube. After the solvent is evaporated under a nitrogen stream the lipids are resuspended in 25–50 μl of chloroform/methanol (6:1, v/v) and applied quantitatively to a silica gel TLC plate (from EM Science). The TLC plate is developed in chloroform/methanol/acetic acid (90:2:8, v/v). Following development and drying, the distribution of radioactivity in substrate and product is monitored by radiometric scanning of the TLC plate.

To calculate enzyme activity, the proportion of radioactivity comigrating with ceramide (having an R_f of approximately 0.4) is converted to nanomoles of ceramide formed per minute per milligram protein, taking into account the total radioactivity added to the assay, the specific radioactivity of the [³H]sphinganine used, the assay duration, and the amount of protein added. To account for potential nonenzymatic ceramide formation or degradation of radiolabeled sphinganine, a control assay containing heat-inactivated membrane is also performed, and the radioactivity present in the region corresponding to ceramide is subtracted as background.

Alternatively, the assay may be performed as described earlier, but containing 40 μ*M* [1-¹⁴C]palmitoyl-CoA (from NEN; adjusted to a specific activity of 10 μCi/μmol, 0.04 μCi added to each tube) as the labeled substrate and including 50 μM unlabeled long chain base (sphingosine or D-*erythro*-sphinganine). A long chain base (dissolved in organic solvent) is added to each 13 × 100-mm test tube and the solvent is evaporated under nitrogen. This is followed by the addition of potassium phosphate buffer, ATP, and [^{14}C]palmitoyl-CoA. The tubes are sonicated to disperse the long chain base and are then equilibrated in a 37° water bath. The reaction is initiated by the addition of membrane with brief mixing. After incubation for 20 min in a 37° water bath with mild agitation, the reaction is terminated by the addition of 1 ml of 0.6 N methanolic NaOH and is allowed to sit at room temperature for 1 hr. This mild alkaline hydrolysis releases labeled palmitate from palmitoyl-CoA, allowing recovery of all radioactivity in the chloroform phase during subsequent extraction. To terminate the hydrolysis and extract the lipids, 950 μl of 1 N HCl and 400 μl of water are added with mixing. This is immediately followed by the addition of 1.0 ml of chloroform, 10 μl of palmitic acid carrier (2.5 mg/ml in ethanol), and 25 μl of ceramide carrier (1 mg/ml in ethanol), mixing with each addition. The tubes are centrifuged in a tabletop centrifuge for 10 min at medium speed, followed by the removal of the top aqueous layer. The lower phase is subsequently processed and analyzed by TLC as described earlier for the assay using [^{3}H]sphinganine as the substrate.

To optimize assay conditions when using different plant tissues as the source of enzyme, variables such as the temperature and duration of the assay, amount of protein added, and the concentrations of both substrates should be manipulated. In particular, the specificity of the enzyme for differing chain lengths of acyl-CoA may vary, depending on the source of the enzyme (and appears to parallel the acyl chain profile of the glucosylceramides isolated from the source; see later). Therefore, it is wise to test the activity in separate assays using 20–40 μM palmitoyl-CoA and lignoceroyl-CoA, especially if the glucosylceramide composition of the source tissue is not known. When performing this assay using either labeled sphinganine or labeled palmitoyl-CoA, it is prudent to regularly confirm the recovery of labeled lipid following extraction by removing a specific aliquot of the chloroform phase for scintillation counting. From this, the total radioactivity in the entire chloroform extract can be calculated and compared to the radioactivity added initially to the assay.

Substrate Specificity and Properties of the Reaction

D-*erythro*-Sphinganine (the naturally occurring isomer) as well as sphingosine (D-*erythro*-sphingenine) serve as substrate, but DL-*threo*-sphinganine and phytosphingosine (4-hydroxysphinganine) do not. The relative activities of squash, bean, and corn sphinganine N-acyltransferase using acyl-

CoA molecules varying in chain length from 16 to 24 carbons parallel the distribution of hydroxy fatty acyl chains in glucosylceramides from the respective sources.[12] These results suggest that this enzyme plays a key role in determining the respective acyl chain compositions of glucosylceramides found in different plant tissues. Hydroxy acyl chains apparently do not serve as substrate, evidence that acyl chain hydroxylation occurs following ceramide formation. Fumonisin B_1, a mycotoxin demonstrated to inhibit ceramide synthesis in animal cells,[14] is a potent inhibitor of sphinganine N-acyltransferase activity in bean ($IC_{50} = 30$ nM) and corn ($IC_{50} = 4$ nM).[12,15] The specific activity of the enzyme in bean microsomes is comparable to that of bean serine palmitoyltransferase and is significantly greater than those reported for sphinganine N-acyltransferase in animal preparations.[16–19]

A second mechanism for ceramide formation, which utilizes free fatty acid rather than acyl-CoA (and referred to as ceramide synthase in Fig. 1), has been characterized in these same tissues, but under different assay conditions and using a postnuclear membrane fraction.[12] Incubation of membrane with [^{14}C]palmitic acid in the absence of CoASH, ATP, or other activating molecule results in the formation of labeled ceramide. Although its role *in vivo* remains unknown, ceramide synthesis utilizing free fatty acid can be demonstrated *in vitro* using the following assay procedure.

Assay

The assay contains 20 mM HEPES–NaOH (pH 6.8), 300 μM DL-*erythro*-sphinganine, 10 μM [1-^{14}C]palmitic acid (from NEN; adjusted to a specific activity of 10 μCi/μmol, 0.05 μCi added to each tube), and 100–200 μg of membrane protein in a total volume of 500 μl. Note that sphinganine and labeled palmitic acid are each prepared as 50× stock solutions in ethanol and that 10 μl of each stock is added directly with the aqueous components. The preparation of the membrane used in this assay is described at the end of the article.

To perform the assay, all components except [^{14}C]palmitic acid are added to each 13 × 100-mm test tube on ice. The tubes are equilibrated

[14] E. Wang, W. P. Norred, C. W. Bacon, R. T. Riley, and A. H. Merrill, Jr., *J. Biol. Chem.* **266,** 14486 (1991).

[15] R. Tandon and D. V. Lynch, unpublished results (1994).

[16] P. Morrell and N. S. Radin, *J. Biol. Chem.* **245,** 342 (1970).

[17] M. D. Ullman and N. S. Radin, *Arch. Biochem. Biophys.* **152,** 767 (1972).

[18] H. Akanuma and Y. Kishimoto, *J. Biol. Chem.* **254,** 1050 (1979).

[19] E. C. Mandon, J. Ehses, J. Rother, G. van Echten, and K. Sandhoff, *J. Biol. Chem.* **267,** 11144 (1992).

at 37° in a water bath, and the labeled substrate is added to each tube with mixing to initiate the reaction. After incubation for 60 min at 37° with mild shaking, the reaction is terminated by the addition of 1.0 ml of 0.6 N methanolic NaOH. After 1 hr at room temperature, the hydrolysis is stopped by adding 950 μl of 1 N HCl. Lipids are extracted by adding 1.0 ml of chloroform, 10 μl each of palmitic acid carrier (2.5 mg/ml in ethanol) and ceramide carrier (1 mg/ml in ethanol), and mixing vigorously. The tubes are centrifuged in a tabletop centrifuge at medium speed for 10 min. The aqueous phase is discarded, and the chloroform phase is removed and placed in screw-cap vials. The solvent is evaporated under nitrogen and the lipid is dissolved in 100 μl of chloroform. For each sample, 20 μl of the final chloroform solution is transferred to a scintillation vial for liquid scintillation counting. This step is performed in order to calculate the total radioactivity recovered in the lipid extract, as the preferential loss of a portion of labeled palmitic acid is not uncommon. Another 20 μl of the final chloroform solution is applied to a TLC plate that is developed in chloroform/methanol/acetic acid (90:2:8, v/v) and analyzed by radiometric scanning of the TLC plate.

To calculate enzyme activity, the proportion of radioactivity comigrating with ceramide (having an R_f of approximately 0.4) is converted to nanomoles of ceramide formed per minute per milligram protein, taking into account the total radioactivity recovered in the lipid extract, the specific radioactivity of the [^{14}C]palmitic acid, the assay duration, and the amount of protein added. To account for potential nonenzymatic ceramide formation, a control assay containing heat-inactivated membrane is also performed, and the radioactivity present in the region corresponding to ceramide is subtracted as background. Note that endogenous fatty acids are not accounted for in the calculations given previously so the value obtained for activity may be an underestimate.

Substrate Specificity and Properties of the Reaction

The greatest activity is observed with palmitic acid, whereas lower activity is observed with stearic, oleic, or lignoceric acid. Enzyme activity is stimulated by the addition of the *erythro-* but not the *threo-* isomer of sphinganine. Sphingosine is a suitable substrate, but phytosphingosine inhibits ceramide synthesis. Note that fumonisin B_1 does not inhibit this reaction. The pH optimum is 6.6–6.8, with activity declining sharply below pH 6. Our studies suggest that the reaction does not require acyl-CoA (i.e., free fatty acid is indeed the true substrate) and does not reflect a reversal of acidic ceramidase activity (see later). The results are consistent with amide bond formation by a mechanism similar to that reported for the acylation of the free amino group of phosphatidylethanolamine to form N-acyl PE.[20,21]

Biological Significance of the Two Reactions

That two distinct mechanisms for ceramide synthesis, one utilizing acyl-CoA and the other free fatty acid, exist in plant membrane preparations *in vitro* raises interesting questions concerning the respective contributions of the two reactions to ceramide formation *in vivo*. Comparison of the specific activities of the two enzymes (sphinganine N-acyltransferase activity is approximately 80-fold greater than that of the acyl-CoA independent reaction in bean), the respective acyl chain specificities of the two reactions, and the effects of the sphinganine N-acyltransferase inhibitor fumonisin B_1 on long chain base accumulation in intact tissues and whole plants[7] indicate that ceramide formation *in vivo* occurs predominantly, if not exclusively, by sphinganine N-acyltransferase. Although its role is not yet defined, the acyl-CoA-independent mechanism may act as a retailoring mechanism or as a salvage pathway for fatty acids and long chain bases.

Synthesis of Glucosylceramide: Ceramide Glucosylation by a Novel Mechanism

Following incubation of plant microsomal membranes with UDP-[^{14}C]glucose in an assay mixture, radioactivity is incorporated into glucosylceramide, as well as steryl glucoside and acylated steryl glucoside. Although this is consistent with UDP-glucose serving as the glucose donor for the reaction catalyzed by ceramide glucosyltransferase, as demonstrated in mammalian systems,[22–24] several lines of experimental evidence suggested that UDP-glucose does not donate glucose directly to ceramide, but that steryl glucoside (or possibly acylated steryl glucoside) serves as the immediate glucose donor for glucosylceramide formation in plants.[25] Nevertheless, the activity of the enzyme catalyzing ceramide glucosylation (termed glucosylceramide synthase in Fig. 1) can be monitored indirectly using labeled UDP-glucose as the penultimate glucose donor in an assay procedure optimized for bean hypocotyl microsomes.[25]

[20] K. D. Chapman and T. S. Moore, *Arch. Biochem. Biophys.* **301,** 21 (1993).
[21] K. D. Chapman and T. S. Moore, *Plant Physiol.* **102,** 761 (1993).
[22] A. Brenkert and N. S. Radin, *Brain Res.* **36,** 183 (1972).
[23] G. S. Shukla and N. S. Radin, *Arch. Biochem. Biophys.* **283,** 372 (1990).
[24] A. H. Futerman and R. E. Pagano, *Biochem. J.* **280,** 295 (1991).
[25] D. V. Lynch, A. K. Criss, J. L. Lehoczky, and V. T. Bui, *Arch. Biochem. Biophys.* **340,** 311 (1997).

Assay

The assay contains 50 mM Tris–HCl (pH 7.0), 25 mM KCl, 10 mM MnCl$_2$, 500 μM ATP, 180 μM UDP-[^{14}C(U)]glucose (from NEN; adjusted to a specific radioactivity of 2.22 μCi/μmol, 0.08 μCi added to each tube), and 100–150 μg of membrane protein in a total volume of 200 μl.

To perform the assay, all components except membrane are added to each 13 × 100-mm test tube on ice. The tubes are equilibrated at 37° prior to adding membrane, with mixing, to initiate the reaction. After incubation for 90 min at 37° with mild shaking, the reaction is terminated by the addition of 1.2 ml of chloroform/methanol (1 : 2, v/v), followed by vigorous mixing. To extract the labeled lipids, 400 μl of chloroform and 600 μl of 0.9% NaCl are added with mixing, and the tubes are centrifuged in a tabletop centrifuge at medium speed for 15 min. The organic phase is transferred to a 12 × 75-mm test tube, and 80 μl is transferred to a vial, evaporated, and combined with 5 ml of cocktail for liquid scintillation counting to calculate the total incorporation of [^{14}C]glucose into lipid products. The remainder of the lipid sample is dried under nitrogen. The lipids are resuspended in 40 μl of chloroform/methanol (6 : 1, v/v), applied to a silica gel TLC plate, and developed in a solvent system consisting of chloroform/methanol/acetic acid/water (85 : 15 : 15 : 3, v/v). The distribution of [^{14}C]glucose among glucosylceramide (R_f = 0.33), steryl glucoside (R_f = 0.4), and acylated steryl glucoside (R_f = 0.6), the only labeled lipids, is quantified by radiometric scanning of the TLC plate. Because glucosylceramide and steryl glucoside do not migrate very differently in this (or any other tested) TLC solvent system, an instrument for radiometric scanning of the plates is practically essential. To determine the total radioactivity recovered in glucosylceramide, the proportion of radioactivity comigrating with glucosylceramide is multiplied by the value for total [^{14}C]glucose incorporation into lipid products.

Although labeling of steryl glucoside and acylated steryl glucoside is observed readily using this or other assay protocols, the enzyme catalyzing ceramide glucosylation is apparently not very robust. Thus, it is not uncommon for an occasional microsomal membrane preparation to exhibit poor enzyme activity. It should also be noted that the time course for ceramide glucosylation is sigmoidal, with a negligible incorporation of label detected before 20 min. This kinetic behavior is expected for such a consecutive reaction and points to the need for long incubation times (90–150 min) in the assay.

Substrate Specificity and Properties of the Reaction

The apparent K_m for UDP-glucose is approximately 50 μM (for both sterol glucosyltransferase and glucosylceramide synthase, as expected for the proposed reaction scheme). The endogenous steryl glucoside content of the membrane preparation used as the source of the enzyme precludes any accurate measurement of enzyme specific activity. Although various types and concentrations of ceramide have been tested in the assay, none stimulates glucosylceramide formation. This is consistent with studies of ceramide glucosyltransferase[23] and suggests that endogenous ceramide is the preferred substrate. For this reason, ceramide is not added to the assay. Ceramide glucosylation in bean microsomes is not inhibited by D-L-*threo*-1-phenyl-2-decanoylamino-3-morpholino-1-propanol (PDMP), an effective inhibitor of ceramide glucosyltransferase in animal tissues.[26] Omitting UDP-[^{14}C]glucose and including steryl [^{14}C]glucoside in the assay result in the formation of labeled glucosylceramide. Labeled steryl glucoside is generated using the assay described earlier but decreasing the reaction time and then purifying the labeled sterol derivative using a silica Sep-Pak cartridge.[25]

Inositolphosphorylceramide Synthase

The first step in the pathway leading to complex InsPCers is the formation of inositolphosphorylceramide. Becker and Lester[27] demonstrated in yeast that inositolphosphorylceramide formation occurs by the transfer of inositol phosphate from phosphatidylinositol to ceramide. A similar mechanism of inositolphosphorylceramide synthesis has been demonstrated in microsomal membranes isolated from plant tissues.[28] Assays monitoring the incorporation of either [^{3}H]inositol-phosphate (from [^{3}H]phosphatidylinositol) or fluorescent ceramide into inositolphosphorylceramide were developed and optimized for the plant enzyme.

Assay

The assay contains 50 mM bis–Tris propane buffer (pH 8.0), 50 μg ceramide, 600 μM phosphatidyl-[[3]H(N)]inositol (from NEN; adjusted to a specific activity of 0.85 μCi/μmol, 0.05 μCi added to each tube), and 100–250 μg of membrane protein in a total volume of 100 μl.

To perform the assay, appropriate amounts of ceramide and [^{3}H]phosphatidylinositol (stored in chloroform) are added to each 13 × 100-mm test tube and evaporated to dryness under nitrogen. Immediately on removal from the nitrogen stream, all aqueous components except membrane are added. The assay tubes are then mixed and sonicated briefly to disperse the lipids and are then equilibrated at 30° in a water bath. The reaction is

[26] N. S. Radin, J. A. Shayman, and J.-I. Inokuchi, *Adv. Lipid. Res.* **26**, 183 (1993).
[27] G. W. Becker and R. L. Lester, *J. Bacteriol.* **142**, 747 (1980).
[28] P. E. Bromley, Y. Li, and D. V. Lynch, unpublished results (1998).

initiated by the addition of membrane, with mixing, and then incubated for 60 min at 30° with mild shaking. The reaction is terminated by adding 1.0 ml of chloroform/methanol (1:9, v/v) and mixing. Following a 10-min centrifugation in a tabletop centrifuge at medium speed to pellet any precipitating material, 800 μl of the supernatant is transferred to a 12 × 75-mm tube and the solvent is evaporated under nitrogen. The residue is dissolved in 40 μl chloroform/methanol/water (16:16:5, v/v), and the sample is applied to a silica gel TLC plate and developed in chloroform/methanol/4.2 M NH$_4$OH (90:70:20, v/v). In this solvent system, inositolphosphorylceramide migrates 0.5–1.0 cm below phosphatidylinositol. The distribution of radioactivity in substrate and product is determined by quantitative radiometric scanning. The proximity of inositolphosphorylceramide and phosphatidylinositol on the TLC plate necessitate the use of a radiometric scanner.

To quantify relative enzyme activity, the proportion of radioactivity comigrating with inositolphosphorylceramide is multiplied by the total radioactivity added to the assay. This can be used to estimate enzyme specific activity (nanomoles of product formed per minute per milligram protein) by taking into account the specific radioactivity of the [³H]phosphatidylinositol, the assay duration, and the amount of protein added. However, this calculation does not take into account the contribution of endogenous phosphatidylinositol and so is only an approximation. To account for any degradation of [³H]phosphatidylinositol, an assay performed using heat-inactivated microsomes is run in parallel with the others, and the radioactivity present in the region of the chromatogram corresponding to inositolphosphorylceramide is subtracted as background.

Alternatively, inositolphosphorylceramide synthase activity can be assayed using fluorescent NBD C$_6$-ceramide as the labeled substrate. The assay is the same as described earlier, with the following changes: Ceramide is replaced by 4 μg of NBD C$_6$-ceramide (nitrobenzoxadiazolhexanoyl-sphingosine; from Molecular Probes), and [³H]phosphatidylinositol is replaced by 600 μM soy phosphatidylinositol (from Avanti Polar Lipids). Following TLC, the fluorescent product is detected by exposing the plate to UV light, photographing the plate, and quantifying the amount of fluorescence corresponding to inositolphosphorylceramide using a digital imaging system. The amount of fluorescent product formed in an assay with heat-inactivated microsomes is subtracted as background.

Substrate Specificity and Properties of the Reaction

Ceramides containing hydroxy or nonhydroxy fatty acyl chains or C$_6$-NBD as the acyl chain all serve as substrate, i.e., they stimulate product formation. In the absence of exogenous ceramides, some product formation is detected, suggesting that endogenous ceramide is the substrate for the enzyme. Similarly, endogenous phosphatidylinositol may serve as the inositol phosphate donor to NBD C_6-ceramide. However, the product formation is stimulated by exogenous phosphatidylinositol, with half-maximum activity observed at a concentration of 180 μM. Aureobasidin A, a specific inhibitor of fungal inositolphosphorylceramide synthase,[29] is a potent inhibitor of the plant enzyme, exhibiting an IC_{50} of 0.8 nM. Enzyme activity has been detected in microsomal preparations from a variety of plant tissues and species using the assay procedure described earlier.

Ceramidase and Glucosylceramidase

Ceramide and glucosylceramide degradation in membrane preparations from bean and squash have been described only briefly.[12,30] The assays for these two enzymes catalyzing the breakdown of sphingolipids in plants were developed from a radiometric sphingomyelinase assay[31] but use fluorescent substrates. Ceramidase and glucosylceramidase assays differ only in the type of labeled substrate employed, so the ceramidase assay procedure is described in full.

Assay

The assay contains 150 mM potassium phosphate (pH 5.4), 0.1% Triton X-100, 5 μg NBD C_6-ceramide (nitrobenzoxadiazolhexanoylsphingosine; from Molecular Probes), and 50 μg of membrane protein in a total volume of 200 μl. To perform the assay, NBD C_6-ceramide and Triton X-100 are combined in a 13 × 100-mm test tube with 200 μl of excess chloroform. After mixing, the solvent is evaporated under a nitrogen stream. Phosphate buffer is added and each tube is mixed and sonicated briefly.

Membrane is added to each test tube with mixing and is kept on ice prior to initiating the reaction by transfer of the tubes to a 37° shaking water bath. After incubation for 60 min the reaction is terminated by the addition of 1.5 ml of chloroform/methanol (1:2, v/v) and 300 μl of water, mixing after each addition. To complete the lipid extraction, 20 μl of ceramide carrier (1 mg/ml in ethanol), 20 μl of palmitic acid carrier (1 mg/ml in ethanol), 500 μl of chloroform, and 300 μl of 0.1 M KH_2PO_4 (pH 2)

[29] M. M. Nagiec, E. E. Nagiec, J. A. Baltisberger, G. B. Wells, R. L. Lester, and R. C. Dickson, *J. Biol. Chem.* **272,** 9809 (1997).

[30] M. T. Lee and D. V. Lynch, unpublished results (1993).

[31] J. B. Carre, O. Morrand, P. Homayoun, F. Roux, J. M. Bourre, and N. Baumann, *J. Neurochem.* **52,** 1294 (1989).

are added sequentially to each test tube, mixing between each addition. The tubes are centrifuged in a tabletop centrifuge at medium speed for 5–8 min, and a 700-μl aliquot of the lower phase is transferred to a 12 × 75-mm test tube and the solvent evaporated under a nitrogen stream. The dried lipids are redissolved in 40 μl of chloroform/methanol (6 : 1, v/v) and are applied quantitatively to a silica gel TLC plate. The plate is developed in chloroform/methanol/acetic acid (90 : 2 : 8, v/v). In this solvent system, values of R_f for ceramide and fatty acid are 0.4 and 0.7, respectively. Following TLC, the fluorescent product is detected by exposing the plate to UV light, photographing the plate, and quantifying the amount of fluorescence corresponding to free fatty acid using a digital imaging system. The amount of fluorescent product formed in an assay with heat-inactivated microsomes is subtracted as background.

The assay of glucosylceramidase activity is performed in the same way, except that (1) 10 μg NBD C_6-glucosylceramide is added in place of NBD C_6-ceramide and (2) two products, fluorescent ceramide and fluorescent fatty acid, are quantified on the TLC plate using a digital imaging system. Presumably, the fluorescent fatty acid is generated by the action of ceramidase on the fluorescent ceramide produced by glucosylceramidase activity; however, the enzymatic formation of glucopsychosine and C_6-NBD acyl chain cannot be ruled out.

To optimize assay conditions when using different plant tissue membrane preparations as the source of ceramidase (or glucosylceramidase), variables such as pH and the type and concentration of detergent should be manipulated. The potential effects of divalent cations (see later) should also be assessed.

Substrate Specificity and Properties of the Reaction

In the membrane preparations tested to date, ceramidase activity is greatest between pH 5.2 and pH 5.6 and in the presence of 0.1% Triton X-100. Substrate specificity has not been investigated. Although a divalent cation is not included in the assay procedure described earlier, we have found that enzyme activity increases 100% in the presence of 1 mM calcium or magnesium. Glucosylceramidase activity is also detected in plant tissues and exhibits properties and assay requirements similar to those of ceramidase. Higher levels of activity were found for both enzymes in microsomes from terminal tissues (mature squash fruit or cotyledons of etiolated beans) as compared to actively growing tissues (bean hypocotyls). Nevertheless, the results provide evidence of sphingolipid turnover in both types of tissue, which presents possibilities for glucosylceramidase and ceramidase involvement in modulating the concentrations of ceramide and free long chain base in the plant cell.

Sphinganine Kinase

The activity of sphinganine kinase in microsomes isolated from corn shoots has been characterized.[32] Activity is assayed by monitoring the conversion of [³H]sphinganine to [³H]sphinganine-1-phosphate, which is recovered in the aqueous phase following lipid extraction. This assay for sphinganine kinase activity was adapted from the method of Louie et al.[33]

Assay

The assay contains 100 mM potassium phosphate (pH 7.4), 5 mM ATP, 10 mM MgCl$_2$, 10 mM KF, 10% (v/v) dimethyl sulfoxide (DMSO), 80 μM D-erythro-[³H]sphinganine (adjusted to a specific activity of 10–20 μCi/μmol, 0.08–0.16 μCi added to each tube), and 40 μg membrane protein. Labeled sphingosine and sphinganine are available commercially, and a procedure for reducing the double bond of sphingosine with tritium to yield D-erythro-[³H]sphinganine is given at the end of the article.

To perform the assay, [³H]sphinganine in methanol is added to each 13 × 100-mm test tube. The solvent is evaporated under nitrogen and the long chain base is redissolved in 10 μl of DMSO. Phosphate buffer, ATP, KF, and MgCl$_2$ are added to each tube, and the tubes are mixed, sonicated, and equilibrated in a 37° shaking water bath. The reaction is initiated by the addition of membrane to each tube, mixing, and returning the tube to the water bath. The tubes are incubated in the water bath for 30 min before terminating the reaction by the addition of 500 μl of 1.5 M NH$_4$OH to each tube. Two milliliters of chloroform/methanol (2:1, v/v) is added to each tube, mixing thoroughly. Tubes are centrifuged at high speed in a tabletop centrifuge for 10 min and are then allowed to sit for 10 min. The upper aqueous phase of each tube (approximately 725 μl) containing the [³H]sphinganine-1-phosphate product is transferred to another tube, combined with 300 μl of chloroform, mixed, centrifuged again for 10 min, and allowed to sit for 10 min. This wash with chloroform serves to remove any remaining [³H]sphinganine from the upper phase. Three hundred microliters of the upper phase is transferred to a scintillation vial and is combined with 5 ml of scintillation cocktail for liquid scintillation counting.

The radioactivity of each sample is proportional to the amount of sphinganine-1-phosphate produced, based on thin-layer chromatographic analyses of the radioactivity present as substrate and product in the upper and lower phases of the lipid extract.[32] Background radioactivity from assays using heat-inactivated membranes is subtracted from the radioactivity re-

---

[32] G. J. Crowther and D. V. Lynch, *Arch. Biochem. Biophys.* **337,** 284 (1997).

[33] D. D. Louie, A. Kisic, and G. J. Schroepfer, Jr., *J. Biol. Chem.* **251,** 4557 (1976).

covered using native microsomes. Enzyme activity is calculated from the scintillation counting after subtracting background and taking into account the proportion of the total volume of upper phase taken for counting, counting efficiency, the specific radioactivity of the labeled substrate, assay duration, and amount of protein added.

Substrate Specificity and Properties of the Reaction

Sphinganine kinase from corn utilizes D-*erythro*-sphinganine and ATP as substrates. The enzyme exhibits a preference for ATP (with an apparent K_m of 0.81 mM) but GTP can also serve as the phosphate donor. The optimum sphinganine concentration is approximately 100 μM. Results of competition experiments suggest that the enzyme can also phosphorylate sphingosine but not DL-*threo*-sphinganine or phytosphingosine. However, the possible existence of a kinase specific for phytosphingosine cannot be ruled out. Of all the enzyme activities described in this article, only sphinganine kinase exhibits any significant activity in the supernatant following ultracentrifugation. The specific activity of the enzyme in corn microsomes is 25-fold higher than that reported for preparations from brain tissue.[33] Bean microsomes exhibit lower kinase activity than corn membrane preparations. Our results suggest that the properties of the kinase from corn microsomes are distinct from those of the mammalian and protistan enzymes in some respects.[33–36]

Related Procedures

Membrane Isolation Procedures

All of the enzyme assays described in this article have been developed using membrane preparations from plant tissues. Three different methods of preparing membrane fractions are described briefly. In all cases, all steps are done on ice and the membranes are stored in aliquots at −80°.

Method 1. Microsomes from bean hypocotyl (and other plant tissues) used for assaying serine palmitoyltransferase,[9] sphinganine *N*-acyltransferase,[12] ceramide glucosylation,[25] inositolphosphorylceramide synthase,[28] and ceramidase[12,30] activities are prepared by first homogenizing the diced tissues in 1–2 volume equivalents of cold homogenizing medium containing 0.5 M sucrose, 50 mM HEPES (pH 7.8), 5 mM EDTA, 2 mM DTT, 0.5% polyvinylpyrrolidone, and 0.1% bovine serum albumin (fatty acid free).

[34] C. B. Hirschberg, A. Kisic, and G. J. Schroepfer, Jr., *J. Biol. Chem.* **245,** 3084 (1970).
[35] W. Stoffel, E. Bauer, and J. Stahl, *Z. Physiol. Chem.* **355,** 61 (1974).
[36] B. M. Buehrer and R. M. Bell, *J. Biol. Chem.* **267,** 3154 (1992).

After filtering through cheesecloth and centrifuging at 12,000g for 30 min, the supernatant is centrifuged at 105,000g for 1 hr. The resulting pellet is resuspended in wash medium [10 mM HEPES (pH 7.8), 2 mM EDTA, and 4 mM DTT] and centrifuged as before at 105,000g. The pellet is resuspended in wash medium or wash medium containing 20% glycerol prior to freezing.

Method 2. To isolate microsomes from corn tissue for use in assays of sphinganine kinase activity,[32] the same processing and centrifugation steps as described earlier are used, but the homogenizing medium contains 0.25 M sucrose, 100 mM potassium phosphate/50 mM HEPES (pH 7.4), 20% glycerol, 2 mM EDTA, 4 mM DTT, 1 mM mercaptoethanol, 0.5% polyvinylpyrrolidone, and 0.1% bovine serum albumin, and the wash/storage medium contains 100 mM potassium phosphate/10 mM HEPES (pH 7.4), 20% glycerol, 2 mM EDTA, 4 mM DTT, and 1 mM mercaptoethanol.

Method 3. The postnuclear membrane fraction used in assays of acyl-CoA-independent ceramide formation (ceramide synthase)[12] is prepared by homogenizing diced tissue in 2 volume equivalents of cold homogenizing medium containing 0.1 M KCl and 50 mM HEPES (pH 7.8), filtering through cheesecloth, and centrifuging at 2000g for 10 min. The resulting supernatant is centrifuged at 105,000g for 1 hr, resuspended in homogenizing medium, washed by centrifugation, and resuspended again in homogenizing medium prior to freezing.

Production of D-erythro-[³H]Sphinganine

This procedure, similar to others described previously,[11] reduces the double bond of sphingosine using tritium, so the predicted product is D-*erythro*-[4,5-³H]sphinganine. The labeled product formed from this reaction exhibits chromatographic properties and reactivity in enzyme assays consistent with it being D-*erythro*-sphinganine.

In a stoppered vial flushed with nitrogen gas, 525 μg NaB[³H]$_4$ (from NEN; approximately 360 mCi/mmol, 5 mCi added to vial) is combined with 400 μl of 2 mg/ml D-*erythro*-sphingosine (from Biomol) in degassed methanol, 2 mg of platinum oxide suspended in 500 μl of degassed methanol, and 20 μl of 1 M hydrochloric acid. Components are added with a syringe through the stopper in the order just given. The vial is placed in a 50° water bath and shaken. After 2 hr, 2 mg of NaBH$_4$ in 200 μl of degassed methanol is added to the vial, which is allowed to sit in the water bath for another hour with continued periodic shaking. The contents of the vial are combined with 2 ml of chloroform and 2 ml of 0.1 M NH$_4$OH in a glass centrifuge tube. The mixture is centrifuged at 5000g for 10 min, and the upper (aqueous) phase is discarded. Sodium sulfate (approximately 200–500 mg) is added to the remaining solution to absorb any water-soluble contaminants. Thin-layer chromatography is used to verify the purity of the tritiated sphinganine product. Labeled sphinganine is not very stable and is particularly susceptible to decomposition when dried on glass surfaces or on TLC plates, so every effort should be made to minimize the duration of such exposure. The labeled sphinganine may be purified by applying the long chain base (in chloroform) to a silica Sep-Pak cartridge and eluting with chloroform/methanol mixtures containing increasing amounts of methanol. The concentration of sphinganine in the tritiated stock is calculated based on parallel experiments to determine the recovery of long chain base following tritiation or by high-performance liquid chromatography analysis[37] of an aliquot of the labeled base.

[37] A. H. Merrill, Jr., E. Wang, R. E. Mullins, W. C. L. Jamison, A. Nimkar, and D. C. Liotta, *Anal. Biochem.* **171,** 373 (1988).

[16] Purification and Characterization of Recombinant Human Acid Sphingomyelinase Expressed in Insect *Sf*21 Cells

By S. LANSMANN, O. BARTELSEN, and K. SANDHOFF

Introduction

Human acid sphingomyelinase (haSMase, EC 3.1.4.12) is a lysosomal enzyme catalyzing the hydrolysis of sphingomyelin to ceramide and phosphorylcholine. An inherited deficiency of the enzyme activity results in Niemann-Pick disease type A and B, an autosomal recessive lipid storage disorder accompanied by a massive accumulation of sphingomyelin in lysosomes.[1] In 1987 we isolated acid sphingomyelinase from human urine.[2] The enzyme was shown to be a monomeric 72-kDa glycoprotein containing a protein core of about 61 kDa. The purified urinary haSMase had a specific enzyme activity of about 2.5 mmol sphingomyelin cleaved per hour and milligram protein using a micellar assay system. Several alternatively spliced haSMase cDNAs were isolated and the genomic sequence encoding

[1] E. H. Schuchman and R. J. Desnick, *in* "The Metabolic Basis of Inherited Disease" (C. R. Scriver, A. L. Beaudet, W. S. Sly, and D. Valle, eds.), p. 2601. McGraw Hill, New York, 1995.
[2] L. E. Quintern, G. Weitz, H. Nehrkorn, J. M. Trager, A. W. Schram, and K. Sandhoff, *Biochim. Biophys. Acta* **922,** 323 (1987).

haSMase was characterized.[3-5] Ceramide, the product of sphingomyelin degradation, was found to be an important cellular signaling molecule involved in apoptosis, cell differentiation, and proliferation.[6] Recent studies indicate a direct involvement of aSMase in certain receptor-mediated cell signaling events.[7]

Studies on the biosynthesis and processing of haSMase in fibroblasts and transfected COS-1 cells revealed extensive posttranslational processing during transport to lysosomes.[8,9]

Many functional and structural studies on haSMase require milligram amounts of enzymatically active protein. However, human tissue is not a suitable enzyme source due to the low aSMase expression level.[2,10] In order to obtain essential amounts of the enzyme, we established a baculovirus-mediated expression of recombinant haSMase (r-haSMase) in *Spodoptera frugiperda* 21 (*Sf*21) cells and developed an efficient purification procedure.[11]

Acid Sphingomyelinase Assay

The assay is carried out following previous descriptions.[2]

Reagents

Enzyme activity is measured with [³H]sphingomyelin (³H labeled in the choline moiety) from bovine brain as substrate in the presence of the nonionic detergent Nonidet P-40 (NP-40).

Each assay contains a total volume of 50 μl: 10 μl diluted enzyme solution (10% substrate degradation during the time of incubation should not be exceeded) in 30 mM Tris–HCl, pH 7.2, and 40 μl of 100 μM [³H]sphingomyelin (2.08×10^{11} Bq/mol) in 250 mM sodium acetate buffer,

[3] L. E. Quintern, E. H. Schuchman, O. Levran, M. Suchi, K. Ferlinz, H. Reinke, K. Sandhoff, and R. J. Desnick, *EMBO J.* **8,** 2469 (1989).

[4] E. H. Schuchman, M. Suchi, T. Takahashi, K. Sandhoff, and R. J. Desnick, *J. Biol. Chem.* **266,** 8531 (1991).

[5] E. H. Schuchman, O. Levran, L. Pereira, and R. J. Desnick, *Genomics* **12,** 197 (1992).

[6] R. Kolesnick, *Trends Cell Biol.* **2,** 232 (1992).

[7] L. A. Peña, Z. Fuks, and R. Kolesnick, *Biochem. Pharmacol.* **53,** 615 (1997).

[8] R. Hurwitz, K. Ferlinz, G. Vielhaber, H. Moczall, and K. Sandhoff, *J. Biol. Chem.* **269,** 5440 (1994).

[9] K. Ferlinz, R. Hurwitz, G. Vielhaber, K. Suzuki, and K. Sandhoff, *Biochem. J.* **301,** 855 (1994).

[10] S. Lansmann, K. Ferlinz, R. Hurwitz, O. Bartelsen, G. Glombitza, and K. Sandhoff, *FEBS Lett.* **399,** 227 (1996).

[11] O. Bartelsen, S. Lansmann, M. Nettersheim, T. Lemm, K. Ferlinz, and K. Sandhoff, *J. Biotech.* **63,** 29 (1998).

pH 4.8, containing 0.1% (w/v) NP-40. The substrate solution is obtained as follows: 50 μl [^{3}H]sphingomyelin (8.52 × 10^{12} Bq/mol, 3.34 × 10^9 Bq/liter) in toluene/ethanol (1:1, v/v) is mixed with 80 μl of 10 mM unlabeled sphingomyelin in toluene/ethanol (1:1, v/v) and evaporated under a stream of nitrogen. After the addition of 80 μl 10% (w/v) NP-40, the mixture is evaporated under a stream of nitrogen at 40°. Following the addition of 8 ml of 250 mM sodium acetate buffer, pH 4.8, the solubilized lipids are sonicated for 3 min at 10° and 200 W in a water-cooled cup horn of the sonifier 250 (Branson, CT).

Assay Procedure

After addition of the ice-cooled substrate mixture to the enzyme solution the enzyme assays are incubated for 30–60 min at 37°. The reaction is stopped by the addition of 800 μl chloroform/methanol (2:1, v/v) and 250 μl distilled water. Enzyme activity is quantified by mixing 200 μl of the upper methanol/water phase containing [^{3}H]phosphorylcholine, the water-soluble product of sphingomyelin degradation, with scintillation cocktail and counting the radioactivity. Blanks are obtained by incubating buffer solution in the absence of enzyme.

Expression and Purification Procedure

In order to generate milligram amounts of glycosylated and enzymatically active haSMase necessary for further structural analysis, the baculovirus/insect *Sf*21 cell expression system was established and optimized successfully by Bartelsen *et al.*[11] Full-length haSMase cDNA was inserted into the *Eco*RI site of the baculovirus transfer vector pVL1393 downstream of the polyhedrin promotor. Recombinant virus was generated by contransfection of *Sf*21 cells with recombinant transfer vector and baculovirus DNA. Infection of *Sf*21 cells with recombinant virus leads to the expression of r-haSMase under transcriptional control of the viral polyhedrin promotor. Nearly 80% of the total r-haSMase activity is secreted into the medium.[11] In order to avoid contamination of the final enzyme preparation with hydrophobic bovine serum albumin, enzyme expression should be performed in the absence of fetal calf serum. For that reason, insect cells have to be adapted to serum-free medium: IPL41 medium (GIBCO BRL) is supplemented with 20 ml/liter yeast extract ultrafiltrate (Sigma), 1 ml/liter lipid mix concentrate (Sigma), 1% (w/v) Pluronic-F68 (Sigma), and 50 mg/liter gentamycin. In order to achieve large quantities of recombinant enzyme, the expression is performed in a 20-liter impeller-stirred Biolafitte reactor as detailed previously.[11]

The following purification protocol might serve as a general instruction for the purification of r-haSMase obtained by an eukaryotic expression system. All chromatographic steps are carried out at 4° using a flow rate of about 30 ml/hr during loading, washing, and elution, unless stated otherwise. Elution steps are generally carried out in reverse direction to minimize loss of enzyme. Eluate fractions containing 25% or more of the maximal value obtained for specific r-haSMase activity are pooled and used for the following chromatographic step. Unless stated otherwise, all buffers contain 0.02% (w/v) NaN$_3$ as preservative and 0.1% (w/v) nonionic detergent to prevent precipitation and adsorption of the hydrophobic enzyme.

Table I summarizes the results of a typical purification of r-haSMase from 6-liter cell- and serum-free conditioned medium using the baculovirus expression vector system. The purification procedure starts with an (NH$_4$)$_2$SO$_4$ precipitation followed by sequential chromatography on concanavalin A (Con A)-Sepharose CL-4B (Sigma), Octyl-Sepharose CL-4B (Sigma), Matrex Gel Red A (Amicon), and Fractogel EMD DEAE-650(S) [DEAE-650(S), particle size: 0.02–0.04 mm, Merck].

(NH$_4$)$_2$SO$_4$ Precipitation

Concentration of the cell culture supernatant, as well as separation from the shear protectant Pluronic-F68, is achieved by a 60% (NH$_4$)$_2$SO$_4$ precipitation. After the addition of 0.1% (w/v) NP-40 and 0.02% (w/v) NaN$_3$, the serum-free expression supernatant is brought to 60% saturation

TABLE I
PURIFICATION OF r-haSMase FROM 6 LITERS OF SERUM-FREE CELL
CULTURE SUPERNATANT[a]

Purification step	Total protein (mg)	Total activity (μmol hr^{-1})	Specific activity (μmol hr^{-1} mg^{-1})	Yield (%)	Enrichment (fold)
Conditioned medium	2500	2940	1.2	100	1
60% (NH$_4$)$_2$SO$_4$ precipitation	1480	2210	1.5	75	1.3
Con A-Sepharose	105	1540	15	52	13
Octyl-Sepharose	25.4	1190	47	40	40
Matrex Gel Red A[b]	6.38	910	157	31	130
DEAE-650(S)[b]	1.36	600	440	20	370

[a] Data are from O. Bartelsen, S. Lansmann, M. Nettersheim, T. Lemm, K. Ferlinz, and K. Sandhoff, *J. Biotech.* **63**, 29 (1998).

[b] Total enzyme activity and total amount of protein of the concentrated Matrex Gel Red A eluate and of the concentrated final enzyme preparation are shown.

with fine-powdered $(NH_4)_2SO_4$ and stirred overnight at 4°. Following ultra-centrifugation (235,000g, 30 min, 4°, Beckman J2-21M), the collected precipitate is solubilized in 200 ml of 30 mM Tris–HCl, pH 7.2, containing 0.5 M NaCl, 0.1% (w/v) NP-40, 1 mM Ca/Mg/MnCl$_2$, and 1 μM leupeptin (buffer A). After additional ultracentrifugation (235,000g, 30 min, 4°, Beckman J2-21M) and subsequent steril filtration (0.2 μm), the clarified enzyme solution can be stored at −80° for several months without significant loss of enzyme activity.

Con A-Sepharose Chromatography

The dissolved $(NH_4)_2SO_4$ precipitate is applied to a Con A-Sepharose column (1.6 × 11 cm) equilibrated with buffer A. Unspecifically bound proteins are removed by washing with 10 column volumes of buffer A. Bound glycoproteins are eluted at room temperature with a linear gradient of 5–25% (w/v) α-D-methylglucopyranoside in buffer A (2 × 90 ml) containing 0.05% (w/v) NP-40. This reduced concentration of hydrophobic NP-40 in the Con A-Sepharose eluate allows efficient washing during the following Octyl-Sepharose chromatography in the presence of NP-40.

Octyl-Sepharose Chromatography

Fractions of the Con A-Sepharose eluate with the highest specific r-haSMase activity are pooled and loaded onto an Octyl-Sepharose column (1.6 × 10 cm) equilibrated with 30 mM Tris–HCl, pH 7.2 (buffer B). After washing with 10 column volumes of 0.005% (w/v) NP-40 in buffer B, bound proteins are eluted with 150 ml of 1% (w/v) NP-40 in buffer B. Fractions containing the highest specific aSMase activity are identical to those with the highest protein concentration. Therefore, gradient elution at this purification step is not advisable.

Matrex Gel Red A Chromatography

Pooled fractions of the Octyl-Sepharose eluate with the highest specific r-haSMase activity are loaded onto a Matrex Gel Red A column (1 × 8 cm, flow rate: 20 ml/hr) equilibrated with buffer B containing 0.1% (w/v) NP-40 (buffer C). During the following washing step the detergent NP-40 [critical micellar concentration (CMC): 0.25×10^{-3} mol/liter] is replaced with dialyzable β-D-octylglucopyranoside (CMC: 14.5×10^{-3} mol/liter, 25°). Exchange of NP-40 for a dialyzable detergent is necessary to enable >20-fold concentration of the enzyme preparation in the final purification step. Therefore the column is washed with 10 column volumes of equilibration buffer followed by 5 column volumes of 0.15% (w/v) β-D-octylglucopyranoside in buffer B (buffer D). Bound proteins are eluted with a linear gradient of 0–1 M NaCl in buffer D (2 × 30 ml).

DEAE-650(S) Chromatography

Desalting and concentration of the Matrex Gel Red A eluate up to a final volume of about 0.5 ml are performed using Centricon 50 microconcentrators (Amicon). By this step an additional enrichment of specific r-haSMase activity is achieved due to the removal of most proteins smaller than 50 kDa. Pooled, desalted, and concentrated fractions of the Matrex Gel Red A eluate with the highest specific r-haSMase activity are mixed with 9 volumes of 50 mM Tris–HCl, pH 8 (at 4°), containing 0.15% (w/v) β-D-octylglucopyranoside (buffer E) and applied immediately to a DEAE-650(S) column (1 × 1 cm, flow rate: 30 ml/hr) equilibrated with buffer E. Under these conditions, about 65% of r-haSMase activity is collected in the flowthrough and the washing solution (washing is performed with buffer E). Most contaminating proteins (pI < 6) specifically bind to the anion exchanger. Fractions of the flowthrough as well as of the washing solution should be neutralized immediately with 1 M sodium acetate buffer, pH 6.5, containing 0.15% (w/v) β-D-octylglucopyranoside. For reason of enzyme stability, combined fractions containing r-haSMase activity are concentrated up to 1–2 mg enzyme/ml, and buffer is exchanged for 0.15% (w/v) β-D-octylglucopyranoside in 30 mM Tris–HCl, pH 7.2, using Centricon 50 microconcentrators. The purified r-haSMase is aliquoted and frozen in liquid N$_2$ for storage at −80° until use. Under these conditions, loss of enzyme activity is less than 10% per year.

Additional Remarks

At higher r-haSMase expression levels (i.e., the specific enzyme activity of the cell culture supernatant should be at least threefold higher than mentioned in Table I), Matrex Gel Red A chromatography can be omitted. To avoid contamination of the Octyl-Sepharose eluate with hydrophobic NP-40, detergent exchange has to be performed during Con A-Sepharose chromatography. Washing of the Con A-Sepharose column with 10 column volumes of buffer A is followed by washing with 5 column volumes of buffer A containing 0.1% (w/v) β-D-octylglucopyranoside instead of NP-40 (buffer F). Elution of the Con A-Sepharose column can be carried out with 20% (w/v) α-D-methylglucopyranoside in buffer F and the eluate should be fractionized. The following Octyl-Sepharose column is eluted with 1% (w/v) β-D-octylglucopyranoside in buffer B. Centricon 50 microconcentrators should be used to concentrate the Octyl-Sepharose eluate with the highest specific r-haSMase activity to minimize contaminating proteins smaller than 50 kDa. This modified purification procedure yields about 30% of the starting r-haSMase activity.

Properties of Recombinant Human Acid Sphingomyelinase

The described purification procedure results in an apparently homogeneous enzyme preparation as judged by silver staining following Tris–Tricine sodium dodecyl sulfate–polyacrylamide gel electrophoresis (SDS–PAGE). The specific r-haSMase activity is about 0.44 mmol hr^{-1} mg^{-1} using a detergent containing assay system.[11] The purified recombinant enzyme has an apparent molecular mass of 72 kDa determined by Tricine SDS–PAGE under nonreducing conditions. Complete deglycosylation with peptide-N^4-(N-acetyl-β-glucosaminyl)asparagine amidase F (PNGase F) leads to a molecular mass reduction of about 8–9 kDa, giving a diffuse protein band of about 63–64 kDa on a Tricine SDS gel, whereas almost complete resistance to endoglycosidase endo-β-N-acetylglucosaminidase H (Endo H) is observed.[11] It can be assumed that N-glycosylation sites of r-haSMase mainly bear Endo H-resistant trimannosyl core structures. Partial resistance of r-haSMase to PNGase F might be explained by partial α1-3-fucosylation of proximal GlcNAc residues.

Contrary to the results of previous biosynthesis and processing studies on haSMase in human fibroblasts and transfected COS-1 cells, proteolytic

TABLE II
FUNCTIONAL AND STRUCTURAL PROPERTIES OF RECOMBINANT AND PLACENTAL haSMase[a]

	Recombinant haSMase	Placental haSMase
M_r glycosylated (kDa)[b]	72	75
M_r deglycosylated (kDa)[b,c]	63	61
N terminus	His[60]	Gly[83]
N-Glycans[d]	Truncated oligomannose type/partially fucosylated	Hybrid/high mannose type
K_m (μM)	32	25
V_{max} (mmol hr^{-1} mg^{-1})	0.56	1.3
Specific activity (mmol hr^{-1} mg^{-1})	0.44	1.01

[a] Data are from O. Bartelsen, S. Lansmann, M. Nettersheim, T. Lemm, K. Ferlinz, and K. Sandhoff, *J. Biotech.* **63,** 29 (1998) and from S. Lansmann, K. Ferlinz, R. Hurwitz, O. Bartelsen, G. Glombitza, and K. Sandhoff, *FEBS Lett.* **399,** 227 (1996).
[b] Estimated by Tricine SDS–PAGE.
[c] After PNGase F treatment.
[d] Carbohydrate structure of N-glycans is assumed on the basis of the results of Endo H and PNGase F treatment and subsequent Tricine SDS–PAGE.

maturation cannot be observed in infected insect $Sf21$ cells.[8,9,11] Complete deglycosylation of both, intracellular 75/72-kDa r-haSMase precursor and secreted functionally active 72-kDa form, results in peptide cores with the same molecular mass of about 63 kDa.

In comparison with native haSMase (n-haSMase) from human placenta, the recombinant enzyme expressed in infected insect $Sf21$ cells undergoes different proteolytic processing as shown by N-terminal amino acid sequencing of the purified enzymes.[10,11] The recombinant enzyme starts with amino acid His[60] compared to Gly[83] for the placental enzyme (amino acid positions refer to the open reading frame of full-length haSMase-cDNA). This indicates that all six potential N-glycosylation sites are present on both enzymes.[10,11] Under micellar assay conditions, r-haSMase shows maximal activity at pH 4.5–5 and an apparent K_m and V_{max} value of 32 μM and 0.56 mmol/hr per mg protein, respectively, with sphingomyelin as the substrate.[11] Contrary to placental haSMase, the recombinant enzyme can be stimulated two- to threefold in the presence of 0.1 mM Zn^{2+} and of 0.1 mM Cu^{2+}, respectively, using cell culture supernatant as the enzyme source.[11]

The most important functional and structural differences between purified secretory r-haSMase expressed in infected insect $Sf21$ cells and purified n-haSMase from human placenta are presented in Table II.

Acknowledgments

We thank H. Moczall and J. Weisgerber for excellent technical assistance and the Deutsche Forschungsgemeinschaft (SFB 400, A5) for financial support.

[17] Purification of Rat Brain Membrane Neutral Sphingomyelinase

By BIN LIU and YUSUF A. HANNUN

Introduction

Ever since the initial discovery of ceramide-induced apoptosis,[1] it has become critical to understand the cellular mechanisms responsible for the regulation of its biosynthesis. A major route of ceramide generation is the sphingomyelinase (SMase)-catalyzed hydrolysis of cell membrane sphingomyelin (SM). Of the multiple forms of SMases discovered thus far, neutral

[1] L. M. Obeid, C. M. Linardic, L. A. Karolak, and Y. A. Hannun, Science 259, 1769 (1993).

SMase appears to be the best candidate for mediating apoptosis-related sphingomyelin hydrolysis.[2] Hence, purification and biochemical characterization of the membrane neutral SMase is key to delineation of the regulation of this enzyme in cells. This article describes our protocol for purification and characterization of a membrane-bound neutral SMase from rat brain.

Purification Strategy

Rat brains are particularly rich in SMases; however, cerebella and brain stem are particularly enriched in acid SMase[3] and are removed. A homogenate is prepared from the remaining portions of rat brain, and the peripheral membrane proteins, along with a significant portion of acid SMase, are removed by washing with high salt and chelator. The neutral SMase, among other highly hydrophobic membrane proteins, is extracted with a neutral detergent and purified using a succession of chromatographic columns that include ion exchangers, heparin, hydroxyapatite, hydrophobic interaction, and gel filtration. Neutral magnesium-dependent sphingomyelinase (N-SMase) activity is assayed as described in this volume.[4] The complex interplay among highly hydrophobic membrane proteins (neutral SMase), column matrix, detergent concentration, and buffer ionic strength has been analyzed systematically to select the best column sequences and buffer conditions. Furthermore partially purified enzymes are characterized for optimum assay conditions in terms of assay buffer, detergent concentrations, and potential activators and inhibitors.

Reagents

Frozen stripped rat brains, from Pel Freez Biologicals (Rogers, AK)
  Homogenizing buffer (50 mM Tris–HCl, pH 7.4, 2 mM EDTA, 5 mM EGTA, 1 mM sodium orthovanadate, 10 mM β-glycerolphosphate, 1 mM sodium fluoride, 1 mM sodium molybdate, 5 mM dithiothreitol, 5 mM 2-mercaptoethanol, 1 mM phenylmethylsulfonyl fluoride, 20 μg/ml each of chymostatin, leupeptin, antipain, and pepstatin A)
Triton X-100, 10% (w/v), aqueous solution, from Boehringer Mannheim
DEAE-Sephacel column (Pharmacia HR 10/10, 8 ml)
Heparin-Sepharose CL-6B column (Pharmacia HR16/10, 20 ml)

[2] Y. A. Hannun, *Science* **274,** 1855 (1996).
[3] E. R. Sperker and M. W. Spence, *J. Neurochem.* **40,** 1182 (1983).
[4] B. Liu and Y. A. Hannun, *Methods Enzymol.* **311** [18] 1999 (this volume).

Ceramic hydroxyapatite column (American International Chemical, Natick, MA, Pharmacia HR 10/10 column, 8 ml)
Mono Q column (1 ml)
Phenyl Superose column (Pharmacia, 1 ml)
Superose-12 column (Pharmacia, HR 10/30, 24 ml)

Preparation of Detergent Extract of Rat Brain

The entire purification process is performed at 4°. Seven rat brains, without cerebella and brainstem, are homogenized (10 passes, three rounds) with a motor-driven Teflon glass homogenizer in 5 volumes of homogenizing buffer. The crude homogenate is centrifuged for 15 min at 1000g, and the supernatant is incubated for 30 min with 0.5 M NaCl and 5 mM EDTA (final concentrations). After centrifugation for 90 min at 105,000g, the resulting pellet is resuspended in 50 ml of 1% Triton X-100 in homogenizing buffer and rocked for 60 min. Detergent-insoluble materials are removed by centrifugation for 90 min at 105,000g, and the Triton X-100-solubilized membrane proteins are used for subsequent column chromatography.

Purification Scheme

All column runs are performed with a Pharmacia Automated FPLC system (Pharmacia Biotech, Piscataway, NJ) kept at 4°. Before loading onto the DEAE column, detergent-extracted proteins are mixed with 1 volume of buffer A [20 mM Tris–HCl, pH 7.4, 0.005% Triton X-100, 1 mM EDTA, 1 mM EGTA, and an inhibitor cocktail (1 mM β-glycerolphosphate, 0.2 mM sodium fluoride, 0.2 mM sodium molybdate, 0.5 mM dithiothreitol, 10 mM 2-mercaptoethanol, 1 mM phenylmethylsulfonyl fluoride, 1 μg/ml each of chymostatin, leupeptin, antipain, and pepstatin A)]. After washing with 100 ml of buffer A, the column is eluted with a 15-ml linear gradient of 0–100% buffer B (20 mM Tris–HCl, pH 7.4, 1 M NaCl, 0.005% Triton X-100, 1 mM EDTA, 1 mM EGTA, and the inhibitor cocktail) and maintained for 100 ml at 100% buffer B. N-SMase is eluted with a linear gradient (100 ml) from 0 to 100% of buffer C (20 mM Tris–HCl, pH 7.4, 1 M NaCl, 0.5% Triton X-100, 1 mM EDTA, 1 mM EGTA, and the inhibitor cocktail). The flow rate is 0.5 ml/ml and 5-ml fractions are collected. Active fractions are pooled, buffer exchanged to buffer A using PD-10 columns (Pharmacia), and loaded at 0.25 ml/min onto the Heparin-Sepharose CL-6B column equilibrated in buffer D (20 mM Tris–HCl, pH 7.4, 0.1% Triton X-100, 1 mM EDTA, 1 mM EGTA, and the inhibitor cocktail). N-SMase (4-ml fractions) elutes as a flow-through on this column. Active fractions are concentrated with Macrosep concentrators (10-kDa cutoff, Filtron, Northborough, MA), buffer exchanged using the PD-10 columns to buffer E (10 mM potassium phosphate, pH 7.2, 0.1% Triton X-100, 10% glycerol, and the inhibitor cocktail), and loaded onto the hydroxyapatite column. After washing (0.25 ml/min) with 40 ml of buffer E, N-SMase is eluted by a gradient (140 ml) of 0–25% of buffer F (500 mM potassium phosphate, pH 7.2, 0.1% Triton X-100, 10% glycerol, and the inhibitor cocktail). Fractions (4 ml) are collected into tubes preloaded with 8 μl each of 0.5 M EDTA and EGTA (1 mM final concentration). Active fractions are pooled, concentrated with Macrosep concentrators (10K), and loaded onto the Mono Q column equilibrated in buffer G (20 mM Tris–HCl, pH 7.4, 0.1% Triton X-100, 1 mM EDTA, 1 mM EGTA, 10% glycerol, and the inhibitor cocktail). After washing with 10 ml of buffer G, N-SMase is eluted at 0.2 ml/min by a linear gradient (40 ml) of 0–50% of buffer H (20 mM Tris–HCl, pH 7.4, 1 M NaCl, 0.1% Triton X-100, 1 mM EDTA, 1 mM EGTA, 10% glycerol, and the inhibitor cocktail). Active fractions are pooled, concentrated, adjusted to 2 M NaCl with 5 M NaCl (in 20 mM Tris–HCl, pH 7.4, 1 mM EDTA, 1 mM EGTA), and loaded onto the phenyl Superose column equilibrated in buffer I (20 mM Tris–HCl, pH 7.4, 2 M NaCl, 1 mM EDTA, 1 mM EGTA, and the inhibitor cocktail). After washing with 15 ml of buffer I, the column is eluted at 0.25 ml/min by a linear gradient of 0–100% buffer J (20 mM Tris–HCl, pH 7.4, 1 mM EDTA, 1 mM EGTA, and the inhibitor cocktail) over 15 ml and maintained at 100% buffer J for 20 ml. The column is finally eluted with a gradient of 0–0.5% Triton X-100 over 15 ml (Fig. 1). Active fractions are concentrated to 0.2 ml and loaded onto a Superose-12 column equilibrated in buffer L (20 mM Tris–HCl, pH 7.4, 0.15 M NaCl, 0.1% Triton X-100, 1 mM EDTA, 1 mM EGTA, and the inhibitor cocktail). The column is eluted at 0.2 ml/min with 30 ml of buffer L. As shown in Table I, through the combination of detergent extraction and six chromatographic columns, N-SMase is enriched by more than 2000-fold over the homogenate.

Properties of Purified N-SMase

Stability

Detergent extracts are stable for several weeks at 4° and for several months at $-70°$. However, the purified enzyme is unstable at both temperatures, which makes it very difficult to employ more steps to purify the enzyme further.

Optimum pH

The purified enzyme has a pH optimum of 7.5. Some activity is detected at pH up to 9.0 using Tris–HCl buffer. No activity is detected in the

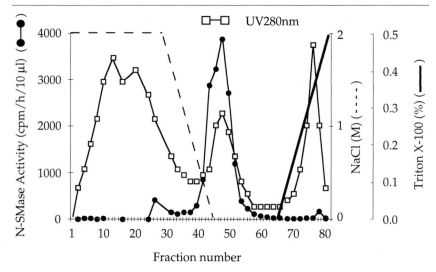

Fig. 1. Chromatography of N-SMase on phenyl Superose column.

acidic pH (4–6) or alkaline pH range (10–11) (Fig. 2). Similar activities are obtained with Tris–HCl, HEPES, or PIPES buffer in the neutral pH range (7–8); HEPES gives slightly higher activity at pH 7.5 than Tris–HCl or PIPES.

Lipid Cofactors

N-SMase is stimulated significantly by phosphatidylserine (PS) and only modestly by other lipids tested. The PS stimulation of N-SMase is not

TABLE I
PURIFICATION OF N-SMase FROM RAT BRAIN

Steps	Total activity (unit)	Total protein (mg)	Specific activity (unit/mg)	Purification (fold)	Yield (%)
Homogenate	36.2	510	0.071	1	100.0
Triton extract	21.6	61	0.355	5	59.7
Cytosol	0.4	75	0.005	0	1.1
DEAE-Sephacel	11.5	13.5	0.85	12	31.8
Heparin-Sepharose	9.0	2.5	3.60	51	24.9
Hydroxyapatite	4.0	0.3	13.33	188	11.0
MonoQ	5.3	0.1	53.00	746	14.6
Phenyl Superose	3.5	0.035	100.00	1408	9.6
Superose 12	2.2	0.015	146.66	2037	6.1

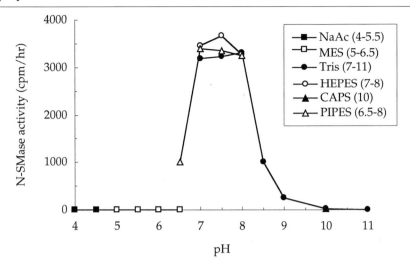

Fig. 2. pH optimum of purified N-SMase. The pH spectrum of N-SMase is determined using the indicated buffers. Numbers in parentheses indicate the pH ranges covered by the individual buffers.

affected by phosphatidic acid (PA), phosphatidylcholine (PC), phosphatidylethanolamine (PE), and phosphatidylinositol (PI). Diolein and short chain C_2- and C_6-ceramide have no effect on PS-stimulated N-SMase activity.

In addition to a high dependence on PS for activity, the amount of coexisting lipids, especially PS, in Triton X-100-extracted N-SMase preparations from lipid-rich rat brain tissues appears to affect the responsiveness of N-SMase to PS stimulation. Several lines of evidence support this notion. First, when examining the effects of PS systematically at all steps of purification, minimum stimulation is detected with enzyme in the postnuclear homogenate fraction, and an increasing degree of PS stimulation is observed with enzyme from subsequent steps such that an almost 20-fold stimulation is seen with enzyme from the last step (Table II). Second, N-SMase is lipidated when extracted from rat brain membranes by Triton X-100 and then delipidated on further purification. The Triton X-100 extract contains a considerable amount of lipids as judged by phospholipid phosphate content. Upon purification of N-SMase by the DEAE-Sephacel column, nearly half (49.5%) of the lipids in the loading material is removed as flow-through, whereas another 24.9% is further removed by elution with 1 M NaCl. Over one-fourth of the lipids (25.6%) is still associated with the main N-SMase peak eluted with Triton X-100, which is used for the next step of purification. Quantitatively, the detergent-eluted N-SMase peak from the DEAE column has a lipid phosphate content of 148.4 nmol/ml. In contrast, the N-SMase peak eluted by the NaCl gradient off the Mono Q column contains

TABLE II
EFFECT OF PS ON N-SMase ACTIVITY

Sample	Average fold stimulation
PNS	1.1
Triton X-100 extract of membranes	1.4
DEAE peak	2.5
Heparin peak	4.7
hydroxyapatite peak	4.5
Mono Q peak	9.9
Phenyl peak	13.2
Superose 12 peak	17.5

only 6.0 n mol lipid phosphate/ml. Third, when the Mono Q column is used earlier in the purification scheme to resolve the Triton X-100 extract of membrane proteins, N-SMase requires detergent for the elution, and little activity comes off the column in 1 M NaCl (Fig. 3), similar to what is observed with the DEAE column (Fig. 4). However, when the Mono Q column is incorporated into the later stage of the purification (Table I) and an N-SMase preparation of much higher purity is applied, N-SMase no longer requires detergent and is eluted in the presence of 0.2–0.3 M NaCl (Fig. 5).

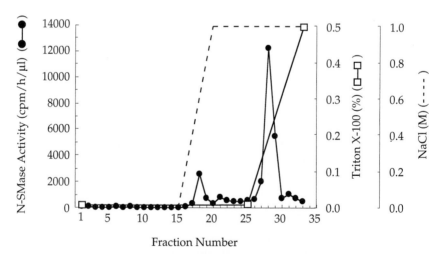

FIG. 3. Chromatography of Mono Q column used directly for resolving the detergent extract of rat brain membrane proteins. Membrane proteins are extracted from one rat brain and resolved on a Mono Q column (1 ml) according to the conditions used for the DEAE column.

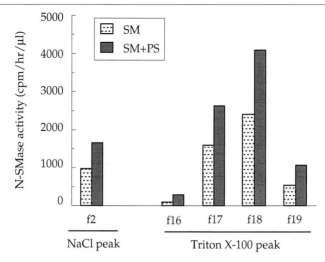

FIG. 4. Comparison of basal- and PS-stimulated N-SMase activity separated by DEAE-Sephacel column chromatography. Peak fractions are assayed for basal (5 mol% SM) and PS (5 mol% SM and 5 mol% PS) stimulated N-SMase activity.

Concluding Remarks

The availability of a purification method for the membrane-associated, magnesium-dependent sphingomyelinase should allow detailed biochemical investigation of the properties of this enzyme. This should enhance our

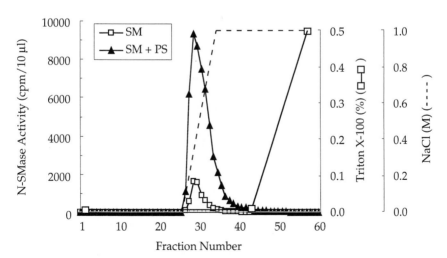

FIG. 5. Basal and PS-stimulated N-SMase activity separated by Mono Q column. Peak fractions are assayed as described in Fig. 4.

ability to determine *in vitro* mechanisms of regulation with possible insight into cellular mechanisms of regulation. A gene has been cloned that enhanced the activity of sphingomyelinase when expressed in cells, although it was not determined whether the gene product is a bona fide sphingomyelinase.[5] Efforts in our laboratory are aimed at determining the relationship of this clone to the purified enzyme. In any case, these studies have enabled a biochemical and molecular approach to the study of sphingomyelin/ceramide metabolism and signaling.

[5] S. Tomiuk, K. Hofmann, M. Nix, M. Zumbansen, and W. Stoffel, *Proc. Natl. Acad. Sci. U.S.A.* **95,** 3638 (1998).

[18] Sphingomyelinase Assay Using Radiolabeled Substrate

By BIN LIU and YUSUF A. HANNUN

Introduction

Ceramide is a product of membrane sphingolipid breakdown and has been proposed as an important mediator of apoptosis. A major source of ceramide is the hydrolysis of membrane sphingomyelin (SM) to generate ceramide and phosphorylcholine, a enzymatic reaction catalyzed by sphingomyelinase (SMase). To date, five types of SMases have been described, including the lysosomal acidic SMase (A-SMase), the cytosolic Zn^{2+}-dependent acidic SMase, the membrane neutral magenesium-dependent SMase (N-SMase), the cytosolic magnesium-independent N-SMase, and the alkaline SMase.[1] Activation of SMases, especially A-SMase and N-SMase, in cells in response to growth factors, cytokines, chemotherapeutic agents, irradiation, nutrient removal, and stress conditions is believed to be involved in the regulation of cell growth, differentiation, cell cycle arrest, and programmed cell death.[1-4] Therefore, determination of the activity of SMases has been a powerful tool for studying the role of ceramide in the death of cells in response to various stimuli. This article provides protocols for

[1] B. Liu, L. M. Obeid, and Y. A. Hannun, *Semin. Cell Dev. Biol.* **8,** 311 (1997).
[2] Y. A. Hannun, *Science* **274,** 1855 (1996).
[3] R. Kolesnick and Z. Fuks, *J. Exp. Med.* **181,** 1949 (1995).
[4] A. H. Merrill, Jr., E. M. Schmelz, D. L. Dillehay, S. Spiegel, J. A. Shayman, R. T. Riley, K. A. Voss, and E. Wang, *Toxicol. Appl. Pharmacol.* **142,** 208 (1997).

determining the activities of neutral and acidic SMases using radiolabeled substrates.

Sample Preparation

Cells grown in suspension are washed twice with ice-cold phosphate-buffered saline (PBS). Adherent cells are detached with trypsin (0.01%)–EDTA (0.5 mM) followed by washing twice with ice-cold PBS. Pelleted cells (1×10^7) are disrupted by three cycles of freezing and thawing in a methanol–dry ice bath in 200 μl of lysis buffer. For neutral SMase, the lysis buffer consists of 50 mM Tris–HCl, pH 7.4, 5 mM EDTA, 2 mM EGTA, 5 mM β-glycerolphosphate, 1 mM sodium fluoride, 1 mM sodium molybdate, 1 mM dithiothreitol (DTT), 1 mM phenylmethylsulfonyl fluoride, 1 μg/ml each of chymostatin, leupeptin, antipain, and pepstatin A. For acidic SMase, the lysis buffer is 50 mM sodium acetate, pH 5.0. The lysate is centrifuged for 10 min at 1000g at 4° and the supernatant is saved for SMase assay. Protein content is determined by the Bio-Rad dye-binding assay with bovine serum albumin as standard.

Assay Protocol

Principle

Sphingomyelin labeled in the choline moiety and presented in Triton X-100-mixed micelles is hydrolyzed by the action of SMase to yield radioactive choline phosphate and ceramide. After extraction and phase separation by the method of Folch,[5a] the radiolabeled product in the aqueous phase is quantified by liquid scintillation counting.

Reagents

[N-methyl-^{14}C]Sphingomyelin is synthesized as described,[5b] and stored at −20° as a stock solution of 0.1 mCi/ml, 20–50 mCi/mmol in chloroform–methanol (9:1)

Bovine brain sphingomyelin, powder obtained from Avanti Polar Lipids (Alabaster, AL), made as a stock solution at 5 μmol/ml, in chloroform–methanol (9:1)

Phosphatidylserine(PS), from Avanti Polar

[5a] J. Folch, M. Lees, and G. Stanley, *J. Biol. Chem.* **226,** 497 (1957).

[5b] T. Okazaki, A. Bielawska, N. Domae, R. M. Bell, and Y. A. Hannun, *J. Biol. Chem.* **269,** 4070 (1994).

Lipids, diluted to 5 μmol/ml in chloroform–methanol (9:1)
Triton X-100, 10% (w/v) aqueous solution, from Boehringer Mannheim
1 M dithiothreitol
200 mM Tris–HCl, pH 7.4
1 M MgCl$_2$
200 mM sodium acetate, pH 5.0
Chloroform–methanol (2:1)

Assay Conditions

For N-SMase, the reaction mixture contains 100 mM Tris–HCl, pH 7.4, 5 nmol of [^{14}C]sphingomyelin (100,000 dpm per reaction), 5 nmol PS, 5 mM DTT, 0.1% Triton X-100 (1.54 mM), 5 mM magnesium chloride and enzyme for a final volume of 100 μl.

A-SMase activity is measured in a 100-μl reaction mixture consisting of enzyme preparation in 100 mM sodium acetate, pH 5.0, 5 nmol [^{14}C]sphingomyelin (100,000 dpm), and 0.1% Triton X-100.

Preparation of Substrate/Detergent-Mixed Micelles

[^{14}C]Sphingomyelin, bovine brain sphingomyelin (in amounts to achieve the desired specific activity) and PS are added to a 13 × 100-mm borosilicate glass tube and dried under nitrogen. SMase assay ingredients are added to the dried lipids (50 μl/assay tube) to yield 200 mM Tris–HCl, pH 7.4, 0.2% Triton X-100, 10 mM magnesium chloride, and 10 mM DTT (for N-SMase) or 200 mM sodium acetate, pH 5.0, and 0.2% Triton X-100 (A-SMase). The tube is then sonicated for 1 min in a bath sonicator followed by incubation for 1 min at 37°. Additional sonication may be applied if the solution appears turbid after incubation at 37°. The substrate solution is always prepared within 10 min before use.

Assay Procedure

The cell homogenate (10–50 μg protein) is pipetted into borosilicate glass tubes (12 × 75 mm) in an ice–water bath and diluted to 50 μl with 20 mM Tris–HCl, pH 7.4 (for N-SMase), or 20 mM sodium acetate, pH 5.0 (for A-SMase). Blank tubes contain no enzyme preparation. The enzyme reaction is initiated by adding 50 μl of substrate and transferring the tubes to a 37° water bath. The reaction is stopped by adding 1.5 ml of chloroform:methanol (2:1, v/v) and 0.2 ml of water. After vortexing and phase separation by centrifugation at 1500g for 5 min, a portion (0.2 ml) of the upper aqueous phase is removed and mixed with 5 ml of scintillation cocktail for counting. The total radioactivity of the product is calculated based on the assumed total volume of the upper aqueous of 0.8 ml. The specific activity of the enzyme can be calculated based on the specific activity of the substrate and the protein content of the enzyme preparation and is usually expressed as nanomoles of substrate hydrolyzed per milligram protein per hour.

Comments

1. The N-SMase appears to be extremely prone to proteolysis and other factors. Inclusion of protease inhibitors, phosphatase inhibitors, and reducing agents stabilizes activity significantly during extraction.[6,7] DTT stimulates the activity of N-SMase, more so for the purer N-SMase[7] than the crude homogenate.[8] In contrast, DTT inhibits the activity of A-SMase potently.[8] In addition to DTT, the activity of N-SMase is enhanced greatly by PS.[7] Therefore, both DTT and PS are included in the assay buffer for N-SMase.

2. All reactions should be carried out with quantities of enzyme preparations and for the length of incubation time that give less than 10% hydrolysis of substrate. Typically, the amounts of enzyme preparation required are in the range of 25–100 μg protein, and the reaction is linear with incubation time up to 120 min. Because the quantities of either N-SMase or A-SMase vary considerably from cell type to cell type,[1] exact conditions as to the amount of enzyme and the length of incubation should be determined for each application.

3. In some cases, Triton X-100 is used for the extraction of total SMase. Hence, these enzyme preparations will add various amounts of Triton X-100 to the assay. The amount of detergent added to the substrate preparation should be adjusted so that the final concentration of Triton X-100 in the reaction mixture is maintained at 0.1%.

[6] S. Adam-Klages, D. Adam, K. Wiegmann, S. Struve, W. Kolanus, J. Schneider-Mergener, and M. Krönke, *Cell* **86,** 937 (1996).
[7] B. Liu, D. F. Hassler, G. K. Smith, K. Weaver, and Y. A. Hannun, *J. Biol. Chem.* **273,** 34472 (1998).
[8] B. Liu and Y. A. Hannun, *J. Biol. Chem.* **272,** 16281 (1997).

[19] Robotic Assay of Sphingomyelinase Activity for High Throughput Screening

By Arminda G. Barbone, Alisa C. Jackson, David M. Ritchie, and Dennis C. Argentieri

Introduction

Neutral sphingomyelinase is an integral membrane protein that hydrolyzes membrane sphingomyelin (SM) to yield ceramide and phosphocholine. The activity of this enzyme has been shown to be upregulated by a number of stimuli, which include tumor necrosis factor-α (TNF-α),[1,2] interleukin-1β (IL-1β),[3] Fas L,[4] and serum deprivation.[5] The hydrolysis product, ceramide, has been associated with signaling cascades leading to apoptosis, proliferation, and inflammation (reviewed in Refs. 6 and 7).

It has been reported that on engagement of TNF-α to the TNFr1-p55, a signaling protein, known as a "factor that activates neutral sphingomyelinase activity" (FAN), binds to the neutral sphingomyelinase activation domain (NSD) of the TNF receptor (TNFr).[8] This enables FAN to activate neutral sphingomyelinase by an unknown mechanism. An acid form of sphingomyelinase, also stimulated by the TNF receptor–ligand interaction, is primarily a lysosomal enzyme activated by different signaling proteins that bind to the TNFr death domain.[9] Acid sphingomyelinase is involved in apoptosis and it has been suggested that the pathways of the two sphingomyelinases are not overlapping.[10]

Ceramide generated by neutral sphingomyelinase activates several downstream targets, which include ceramide-activated protein kinase

---

[1] K. Dressler, S. Mathias, and R. N. Kolesnick, *Science* **255,** 1715 (1992).

[2] M. Kim, C. Linardic, L. Obeid, and Y. Hannun, *J. Biol. Chem.* **266,** 484 (1991).

[3] S. Mathias, A. Younes, C. Kan, I. Orlow, C. Joseph, and R. Kolesnick, *Science* **259,** 519 (1993).

[4] C. Tepper, S. Jayadev, B. Liu, A. Bielawska, R. Wolff, S. Yonehara, Y. Hannun, and M. Seldin, *Proc. Natl. Acad. Sci. U.S.A.* **92,** 8443 (1995).

[5] S. Jayadev, B. Liu, A. Bielawska, J. Lee, F. Nazire, M. Y. Pushkareva, L. Obeid, and Y. Hannun, *J. Biol. Chem.* **270,** 2047 (1995).

[6] B. Liu, L. Obeid and Y. Hannun, *Sem. Cell Dev. Biol.* **8,** 311 (1997).

[7] R. A. Heller and M. Kronke, *J. Cell Biol.* **126,** 5 (1994).

[8] S. Adam-Klages, D. Adam, K. Wiegmann, S. Struve, W. Kolanus, J. Schneider-Mergener, and M. Kronke, *Cell* **86,** 937 (1996).

[9] S. Adam-Klages, R. Schwandner, D. Adam, D. Kreder, K. Bernardo, and M. Kronke, *J. Leukocyte Biol.* **63,** 678 (1998).

[10] K. Wiegmann, S. Schutze, T. Machleidt, D. Witte, and M. Kronke, *Cell* **78,** 1005 (1994).

(CAPK)[11] and ceramide-activated protein phosphatase (CAPP).[12] CAPP is thought to be involved in cell cycle arrest and CAPK activates mitogen-activated protein kinase (MAPK), which in turn phosphorylates and activates cytosolic phospholipase A_2 (cPLA$_2$).[10] Arachidonic acid (AA), released as a result of PLA$_2$ activity, leads to the generation of a variety of inflammatory mediators such as prostaglandins and leukotrienes. Arachidonic acid may also be responsible for activating neutral sphinomyelinase when generated through TNF-α stimulation.[13] These aspects of neutral sphingomyelinase signaling make this enzyme a possible target of inhibition for inflammatory diseases.

Sphingomyelinase activity can be measured by several methods; however, the simplest involves an extraction assay using radiolabeled sphingomyelin, with a ^{14}C label on the phosphocholine, as substrate for the enzyme.[14] After incubation of the substrate with enzyme, the labeled phosphocholine can be separated from the remaining, unhydrolyzed sphingomyelin by extraction with a mixture of chloroform and methanol. The intact sphingomyelin and ceramide are captured in the solvent phase, whereas the labeled product, phosphocholine, partitions to the upper aqueous phase. Radioactivity in an aliquot of the upper phase can be counted in a liquid scintillation counter as a relative measure of sphingomyelin hydrolysis. Because this manual assay was not amenable to high throughput screening, we developed a robotic method for an extraction assay in a 96-well format that allows for the high throughput screening of inhibitor compounds of neutral sphingomyelinase. This assay retains the sensitivity and reproducibility of the manual extraction assay while providing the ability to perform the assay in a microplate.

Method

Preparation of Cell Lysate

The enzyme source for this assay was a whole cell lysate prepared from U937 cells (ATCC). These cells were chosen because they provided an abundant source of human enzyme with high sphingomyelinase activity. Other sources of enzyme that had been considered for the assay were human brain tissue and human neutrophil membranes. The human brain

[11] J. Liu, S. Mathias, Z. Yang, and R. Kolesnick, *J. Biol. Chem.* **269,** 3047 (1994).
[12] R. Wolff, R. Dobrowsky, A. Bielawska, L. Obeid, and Y. Hannun, *J. Biol. Chem.* **269,** 19605 (1994).
[13] S. Jayadev, H. Hayter, N. Andrieu, C. J. Gamard, B. Liu, R. Balu, M. Hayakawa, F. Ito, and Y. Hannun, *J. Biol. Chem.* **272,** 17196 (1997).
[14] M. Lister, C. Crawford-Redick, and C. Loomis, *Biochim. Biophys. Acta* **1165,** 314 (1993).

is a source of high neutral sphingomyelinase activity but was not considered for the assay due to the cost and difficulty in obtaining these tissues. Neutrophils, however, did not yield sufficient activity for high throughput screening. Bacterial neutral sphingomyelinase from *Bacillus cereus* (Sigma), a commercially available source of high enzyme activity, was used for some of the assay development.

U937 cells are cultured in Triplex flasks (Nunc) in RPMI 1640 with 10% fetal bovine serum (GIBCO) at 37° and 5% CO_2 and are harvested at a cell density of $1-1.5 \times 10^6$ cells/ml. The cells are centrifuged at 300g for 10 min at 5° and then washed with cold Dulbecco's phosphate-buffered saline (GIBCO). The cell pellet is resuspended in buffer A with protease inhibitors [20 mM Tris, pH 7.4, 250 mM sucrose, 10 mM EGTA, 0.1 mM sodium vanadate, 100 μM sodium molybdate, 1 mM phenylmethylsulfonyl fluoride (PMSF), 10 μg/ml leupeptin, 2 μM pepstatin] (Sigma) at 2×10^8 cells/ml and the cells are disrupted in a nitrogen cavitator (PARR Instrument Co., Moline, IL) at 500 psi for 45 min on ice. Aliquots of lysate are flash frozen and stored at $-40°$. The U937 lysate is diluted 1:3 in buffer A with protease inhibitors for manual or robotic assays for an equivalent concentration of 6×10^7 cells/ml (30–50 μg/assay).

Preparation of Substrate

Bovine [*N*-methyl-^{14}C]sphingomyelin (specific activity 47 mCi/mM, Amersham) and phosphatidylinositol (PI)(Avanti) are evaporated under a stream of nitrogen to dryness and are resuspended at an equimolar concentration of 0.1 mM in water with 0.5% Triton X-100 (Sigma). The solution is vortexed for 30 min and then sonicated for 30 min. The substrate can be prepared 24 hr in advance and stored at room temperature. PI is included in the reaction mixture as it was found to increase the activity of neutral sphingomyelinase.

Manual Assay

The reaction mixture, containing 25 μl 2× Tris–Mg buffer (0.2 M Tris–HCl, pH 7.5, 0.16 M MgCl), 10 μl prepared substrate (0.1 mM ^{14}C-SM), 10 μl enzyme, and 5 μl test compound or test compound vehicle, in a final volume of 50 μl, is incubated for 1 hr at 37° in 1.5-ml microcentrifuge tubes. At the end of incubation, 30 μl water is added along with 175 μl chloroform/ methanol (2:1, v:v) and the tubes are vortexed vigorously. They are centrifuged in a microfuge for 5 min at 10,000g, and 50 μl of the upper aqueous phase is removed, added to scintillant, and counted in a liquid scintillation counter. Enzyme activity is demonstrated by an increase in cpm resulting from the generation of ^{14}C]phosphocholine.

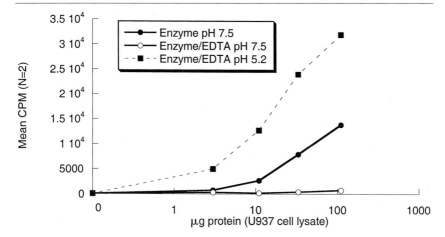

FIG. 1. The U937 lysate was tested at various concentrations and with three different buffer conditions. At pH 7.5 and in the absence of magnesium, the magnesium-dependent sphingomyelinase activity is decreased to background levels as compared to the activity seen in the presence of magnesium. Acid sphingomyelinase activity is present at pH 5.2. Activity is expressed as mean cpm measured in duplicate assays.

Because the membrane neutral sphingomyelinase is magnesium dependent, and activity is diminished in the absence of magnesium, controls are run in a reaction mixture containing 0.1 M Tris–HCl, pH 7.5, 0.05 M EDTA. This inhibition of neutral sphingomyelinase in the absence of magnesium at pH 7.5 is demonstrated in Fig. 1. There is a magnesium-independent neutral sphingomyelinase that has been indentified,[15] although very little of this activity was found in the U937 cell lysates under these conditions. The U937 cell lysates also contain acid sphingomyelinase activity at pH 5.2 as seen in Fig. 1.

Robotic Assay

The first factor considered in the conversion of the manual extraction assay to a robotic assay was to find an appropriate multiwell plate in which the assay could be run. Polypropylene 96-well polymerase chain reaction (PCR) plate inserts (Denville Scientific) with MicroAmp bases (Perkin Elmer) (Fig. 2) were selected for the assay because of their solvent resistance (needed due to chloroform/methanol extraction) and conical well configuration. These plates were compared to round-bottom polypropylene plates

[15] T. Okazaki, A. Bielawska, N. Domae, R. Bell, and Y. Hannun, *J. Biol. Chem.* **269**, 4070 (1994).

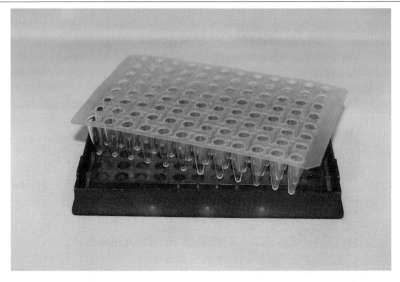

FIG. 2. Solvent-resistant PCR plate inserts and Microamp bases used for the robotic assay.

(Costar) and it was found that the robot could mix with greater turbulence in a conical bottom, providing a more complete extraction. This was evident by the signal-to-noise ratios in the comparison of the two plate types with the manual assay (Fig. 3).

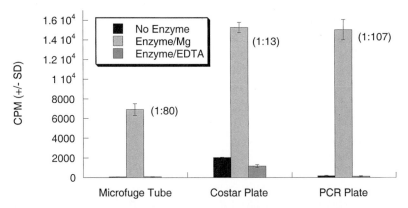

FIG. 3. Round-bottom polypropylene plates and conical-bottom PCR plates were compared to the manual extraction assay using *Bacillus cereus* sphingomyelinase. PCR plates produced a greater signal-to-noise ratio than the round-bottom plates, although it is unclear as to why the total counts were higher than those in the manual assay. Values for signal to noise (shown in parentheses) represent the ratio of enzyme with Tris–Mg buffer to enzyme with Tris–EDTA buffer.

The robot used is a Biomek 2000 (Beckman) (Fig. 4) with 37 and 4° incubators and a tip rack carousel (Brandel). The plates are preloaded into 37° incubators at the start of the assay and, for each addition to the plate, the plates are removed from and returned to the incubators. The first addition is 25 μl 2× Tris–Mg buffer (0.2 M Tris, pH 7.5, 0.16 M MgCl$_2$), 5 μl test sample (compound) in a DMSO–HEPES solvent (26% DMSO, 0.4 M HEPES, pH 7.4), and 10 μl [^{14}C]sphingomyelin substrate. The second addition is 10 μl of diluted U937 lysate, with mixing. The enzyme trough can be preloaded into the 4° incubator, where it remains until enzyme addition begins. It is kept chilled on the deck with refrigerant gel, which surrounds the enzyme trough, during addition to the 96-well plates. The lysate is mixed between plate additions to maintain even distribution of the enzyme. After a 1-hr incubation at 37°, 30 μl of water is added to each well. The final addition to the plates is 175 μl chloroform/methanol (2:1, v:v), with vigorous mixing. It is necessary for the robot to prewet the tips with the solvent to prevent the solvent from dripping from the tips during transfer from the solvent trough to the plates. The solvent-containing plates are removed manually from the incubators as they are generated, covered with solvent-resistant adhesive covers (TiterTops, USA Scientific), and centrifuged at 1500g at 4°. The plates are uncovered and placed on stacks, rather than the incubators, from which they are now accessed and the robot transfers 50 μl of the upper phase to Flashplates (NEN). The process of

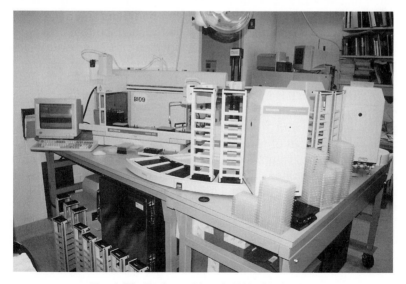

Fig. 4. The Beckman Biomek 2000 with side loader.

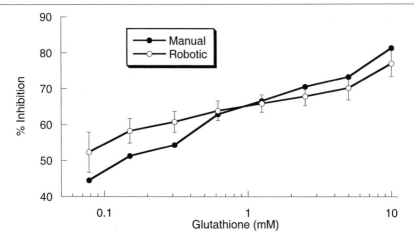

FIG. 5. Comparison of the glutathione dose–response curve in manual and robotic extraction assays shows similar inhibition of neutral sphingomyelinase activity achieved for the same concentrations of glutathione. The percentage inhibition values for the robotic assay are the mean of five assays and those for the manual assay are the mean of duplicate samples.

upper phase aspiration is performed very slowly in order to avoid mixing of the two phases. The Flashplates are then placed into a fume hood overnight to evaporate all liquid and on the following day the plates are counted in a 96-well scintillation counter (Topcount, Packard). It is necessary to evaporate the transferred upper phase in order to bring the signal into close proximity with the scintillant on the plate bottom.

A column of control conditions is included on each plate, which include two wells without enzyme, four wells with enzyme, and two wells with 25 μl 2× Tris–EDTA buffer (0.2 M Tris, pH 7.5, 0.1 M EDTA) in place of Tris–Mg buffer for the inhibited enzyme control. The enzyme wells are averaged and used as the control reaction reference to calculate percentage inhibition in wells with test compounds. A quality control plate is also included with each set of plates and contains a dose–response curve of the enzyme with oxidized glutathione (Sigma), an inhibitor of neutral sphingomyelinase.[16] A glutathione dose–response curve in the robotic and manual assay demonstrates that the similar level of inhibition of the U937 lysate could be measured at equivalent concentrations in both assays (Fig. 5).

The assay on the Biomek has a run time of 5.5 hr and is capable of processing seven test plates and one control plate in that time. Several factors limit the number of plates included per run. One of these factors

[16] B. Liu and Y. Hannun, _J. Biol. Chem._ **272,** 16281 (1997).

is the length of time needed for addition of the various assay components. For example, the length of time required for addition of enzyme to the eight plates, with mixing, is 1 hr, which is the same length of time allowed for the reaction to proceed in the manual assay. The robot, however, requires an extra hour for the addition of water (normally a quick step in the manual assay), to keep the incubation times equal for each plate. Another factor is the length of time that the enzyme remains on the deck. Although the enzyme is chilled with refrigerant, it remains cold for approximately 1 hr, after which enzyme activity begins to deteriorate. One of the practical issues was to keep the time required for hands-on work to a minimum, as the removal, centrifugation, and replacement of plates onto the robot require manual intervention.

The assay is validated by examination of the variation across the plate to ensure that variability would not obstruct detection of a potential inhibitor. Table I shows the variation across the plate with an average % CV (coefficient of variation) for the entire plate of 10.2% and an acceptable signal-to-noise ratio of 1:8, demonstrating a strong reproducibility from well to well. The enzyme activity obtained in the manual assay (1962.9 cpm) is in the same range as that observed in the robotic assay.

TABLE I
REPRODUCIBILITY OF NSMASE ASSAY ENZYME COUNTS ACROSS PCR PLATES[a]

Plate column no.	Mean cpm	St Dev	cpm with EDTA
1	1949.5	242.7	
2	2468.6	225.8	
3	2372.4	147.2	
4	2402.0	143.2	
5	2235.6	215.0	
6	2118.1	141.2	
7	2153.3	186.5	
8	2112.9	156.8	
9	2153.7	162.0	
10	2191.6	188.1	
11	2274.2	204.5	
12			283.9 ($N = 2$)
Manual	1962.9	63.0	141.0 ± 67.0 ($N = 5$)

[a] The neutral sphingomyelinase assay was performed as described in Methods with the entire plate run as a control enzymatic assay reaction with the exception of column 12. Column 12 consisted of four wells as background with no enzyme (cpm = 569.8 ± 47.8), two wells with EDTA (shown above, and represent complete inhibition of the enzyme), and two additional wells with enzyme (cpm = 2321.6). Column 12 serves as a quality control column to assure assay integrity from plate to plate. %CV values ranged from 6.0 to 12.5 for the 11 columns shown above, with the entire plate %CV = 10.2.

This robotic extraction assay for high throughput screening of inhibitors of neutral sphingomyelinase is as reproducible and as sensitive as a manual assay. With this method, it is possible to test a total of 616 compounds in seven plates with a run time of 5.5 hr. This assay also has the potential to be used in the screening of other lipases normally tested by extraction by varying the enzymatic conditions.

[20] A High Throughput Sphingomyelinase Assay

By Daniel F. Hassler, Ronald M. Laethem, and Gary K. Smith

Introduction

Sphingomyelinases (SMases) are phophodiesterases that catalyze hydrolysis of the phosphodiester linkage of sphingomyelin (SM) to yield ceramide or ceramide-1-phosphate as the lipidic product and phosphorylcholine or choline as the headgroup product.[1] SMases have been assayed using a variety of substrates. Radiolabeled sphingomyelin, prepared via incorporation of ^{3}H or ^{14}C into the headgroup moiety,[2] is the most common substrate. Colorimetric and fluorescent assays have been developed that utilize SM analogs containing chromophores or fluors.[3] However, these classes of substrates differ significantly from authentic SM and consequently are not widely used.

Assays employing radiolabeled substrate require physical separation of product from substrate. Usually this separation is accomplished with liquid–liquid phase extractions[4] that effectively separate lipophilic substrate and hydrophilic product into organic and aqueous phases, respectively. The reaction components can also be separated chromatographically. These methods are tedious and are not suitable for large numbers of samples. An alternate method employs headgroup radiolabeled SM with separation of the unreacted substrate from the labeled product by precipitation.[5,6] In this strategy, the labeled substrate is selectively precipitated while the la-

[1] M. W. Spence, *Adv. Lipid Res.* **26**, 3 (1993).
[2] W. Stoffel, D. LeKim, and T. S. Tschung, *Hoppe-Seyler's Z. Physiol. Chem.* **352**, 1058 (1971).
[3] J. J. Gaudino, K. Bjergarde, P. Chan-Hui, C. D. Wright, and D. S. Thomson, *Bioorg. Med. Chem. Lett.* **7**, 1127 (1997).
[4] E. G. Bligh and W. J. Dyer, *Can. J. Biochem. Physiol.* **37**, 911 (1959).
[5] J. N. Kanfer, O. M. Young, D. Shapiro, and R. O. Brady, *J. Biol. Chem.* **240**, 1081 (1966).
[6] R. M. Laethem, Y. A. Hannun, S. Jayadev, C. J. Sexton, J. C. Strum, R. Sundseth, and G. K. Smith, *Blood* **91**, 4350 (1998)

beled headgroup product remains soluble. The latter can then be transferred easily for quantitation by liquid scintillation counting. The advantage of this method is the ease with which multiple samples can be processed simultaneously. This article describes methods of exploiting this separation technique to greatly increase the efficiency and throughput of SMase assays.

Enzyme Preparation

This strategy has been applied to the assay of both crude cell or tissue extracts and partially purified preparations of both neutral and acidic pH optima SMases. Crude extracts are valuable for screening tissue sources[6] or for determining activity changes elicited by various treatments. Purified preparations are preferred for inhibitor screening and kinetic and mechanistic studies. The preparation of partially purified neutral SMase and use of this enzyme in high-throughput screening are described.

Reagents

Homogenization buffer: 25 mM Tris (pH 7.5), 0.25 M sucrose, 1 mM EDTA, 1 mM dithiothreitol (DTT), 1 μM pepstatin A, 33 μM antipain, 33 μM chymostatin, and 10 μM leupeptin

Membrane solubilization buffer: 20 mM Tris (pH 7.5), 6% (w/v) CHAPS, 0.15 M NaCl, 1 mM DTT, 1 mM EDTA, 1 μM pepstatin A, 33 μM antipain, 33 μM chymostatin, and 10 μM leupeptin

Gel filtration buffer: 20 mM Tris (pH 7.5), 1.5% (w/v) CHAPS, 15% (w/v) glycerol, 0.15 M NaCl, 1 mM DTT, 1 mM EDTA, and 10 μM leupeptin

Anion-exchange buffer A: 20 mM Tris (pH 8.0), 1% (w/v) Triton X-100, 1 mM DTT, and 25% (w/v) glycerol

Anion-exchange buffer B: 20 mM Tris (pH 8.0), 1% (w/v) Triton X-100, 1 mM DTT, 1 M NaCl, and 25% (w/v) glycerol

Preparation of Rat Brain Membranes

Neutral pH SMases (N-SMase) are usually associated with membranes and can be enriched from cells or tissues by preparing a total membrane fraction. Rat brain is a rich source of membrane-associated N-SMase. Homogenize whole adult rat brains (3–4 g) in 40 ml of homogenization buffer and ultrasonically disrupt for 5 × 30 sec on ice using a microtip probe sonicator (Heat Systems-Ultrasonics, Inc.) on medium setting. Centrifuge at 1000g for 10 min at 4° to remove unbroken material. Pellet the total membrane fraction containing most of the N-SMase by centrifugation of the supernatant at 200,000g for 30 min at 4°.

Delipidation

The membrane fraction containing N-SMase is also enriched in lipids, including SM, ceramide, and other lipids that must be removed before reliable N-SMase measurements can be obtained. Removal of endogenous lipid from N-SMase requires the disruption of membranous protein/lipid interactions followed by separation of the protein and lipid components while maintaining enzyme activity. This is accomplished by solubilizing the total membrane fraction with 6% (w/v) CHAPS buffer followed by gel filtration.

Solubilize the total membrane pellet in 20 ml of solubilization buffer at 4° by repipetting until particulates are dissolved. Remove insoluble material by centrifugation at 200,000g for 30 min at 4° and add glycerol to the resulting supernatant to a final concentration of 15% (w/v). Clarify, if necessary, by passage through a 0.2-μm syringe tip filter. Load the clarified CHAPS extract onto a Sephacryl S-300 column (2.6 cm i.d. $\times$ 1 m) at 20 ml/hr and elute with gel filtration buffer at 4°. The enzyme (Fig. 1) elutes

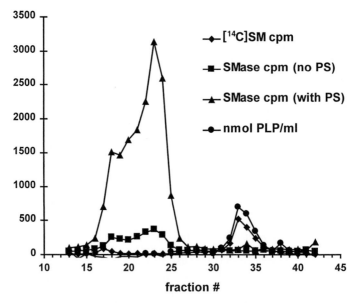

FIG. 1. Fractions from the gel filtration column (10 ml) were assayed for N-SMase activity using a substrate mixture with or without PS. To monitor phospholipids, the enzyme sample was spiked with about 1 μCi of [14C]SM and its elution monitored by direct scintillation counting of a portion of each fraction. Phospholipid (phospholipid phosphate) was measured directly by performing a Bligh and Dyer extraction on a portion of each fraction followed by phosphorus analysis of the dried organic phase as described in B. N. Ames and D. T. Dubin, *J. Biol. Chem.* **235,** 769 (1960).

well ahead of the endogenous lipid and shows strong dependence on phosphatidylserine for activity.

Concentration and Detergent Exchange

Because rat brain N-SMase is inhibited by CHAPS, this detergent must be removed. Pool the active gel filtration fractions and dilute threefold with anion-exchange buffer A. Load this pool onto a Mono Q HR 10/10 column (Pharmacia) equilibrated in anion-exchange buffer containing 50 mM NaCl. Wash the column with 100 ml of equilibration buffer and elute active N-SMase with a linear salt gradient (0.05–0.5 M NaCl). The enzyme obtained following this process contains little or no CHAPS or dissociable endogenous lipid and is ready for further purification or assays. Store this material frozen and dilute it appropriately into freshly prepared 20 mM Tris–HCl (pH 7.5), 0.1% (w/v) Triton X-100, and 10 mM DTT for assays.

Assay Method

Reagents

Store Triton X-100 (SigmaUltra) as a 10% (w/v) stock in water at 4°, phosphatidylserine (PS) (Avanti Polar Lipids) in chloroform at −20°, and bovine [14C]methylcholine SM (New England Nuclear or from Dr. A. Bielawska, Medical University of South Carolina, 54.5 Ci/mol) at −20° in either 1:1 methanol:toluene or 9:1 chloroform:methanol. Store 1 M DTT as small aliquots in water at −20°. Prepare 5% (w/v) fatty acid-free bovine serum albumin (BSA) (Sigma) in water with 0.02% sodium azide and store at 4°. Dissolve trichloroacetic acid (TCA) in water at 8% (w/v) and store at room temperature. Store other assay reagents as concentrated stocks at room temperature or 4°. A scintillation cocktail with high capacity for aqueous samples is necessary to keep volumes low so that counting of product can be accomplished in multiwell plates (Wallac Optiphase "SuperMix" works well).

Procedure

Final N-SMase reaction components (50 μl reaction volume) include
0.1 M Tris–HCl (pH 7.5)
10 mM MgCl$_2$
2–5 mM DTT
0.2–0.25% (w/v) Triton X-100
0.0217% (w/v) synthetic dioleoylphosphatidylserine (diC18:1 PS) (8 mole %)

44,400 cpm [^{14}C]SM (0.21 mole %)
Compound being tested for SMase inhibition
Enzyme sample

Prepare a twofold concentrated assay mixture by combining appropriate volumes of PS, SM, and Triton X-100 stocks. If the SM stock is in toluene : methanol, dry the PS and SM stock mixture completely under a nitrogen stream and redissolve in chloroform prior to adding the Triton X-100 stock. Thorough mixing is necessary and results in a cloudy emulsion. Evaporate the chloroform by blowing a stream of nitrogen over the mixture until it is clear and viscous. Add Tris–HCl (pH 7.5) and MgCl$_2$ stocks and mix. Add water to bring reagents to a final twofold concentration and mix to achieve a clear solution. Store at −20°. This stock can undergo several freeze/thaw cycles and is stable for at least several months.

Dispense 25 μl of appropriately diluted rat brain N-SMase enzyme preparation into wells of Costar 96-well Serocluster U-bottom polypropylene plates. If compounds are being tested for inhibition, adjust enzyme volume and add compound so that the total volume is 25 μl.

Add 25 μl of twofold concentrated assay mixture to wells containing enzyme or enzyme dilution buffer for no enzyme controls. Mix by gentle agitation or by repipetting. Incubate the plate in a 37° humidified incubator for 0.5–2 hr, then stop the reaction by adding 50 μl of 5% BSA and 100 μl of 8% TCA to all wells. Mix by mechanical agitation or repeated inversion of a sealed plate (Wallac sealing tape). Sealing is recommended to avoid loss of liquid from wells (many sealers perform poorly so be sure to test using the nonradioactive reaction components). Centrifuge the sealed plate (1500g for 5 min at room temperature) to pellet the precipitated substrate and leave the soluble radiolabeled headgroup in the supernatant.

Quantitate SMase activity by direct liquid scintillation counting of the headgroup product. Transfer 100 μl of supernatant (about half of the total volume) containing the labeled headgroup to a separate counting plate [Wallac (PET) or Costar] containing 175 μl of Wallac Optiphase "SuperMix" scintillation cocktail per well. Use care not to transfer any of the white precipitate containing the radiolabeled SM. Cover with a plate sealer, mix thoroughly by inversion, and count in a multiwell scintillation counter. Alternatively, aliquots of the supernatants can be counted in scintillation vials.

Assay Parameters

Product Characterization

Reactions are carried out with SMase from *S. aureus* (Sigma) and crude Triton X-100 extracts of Molt-4 human T-cell leukemia cells for 45 min at

37°. Lipid is extracted using the method of Bligh and Dyer, and the resulting aqueous phase is lyophilized. Reactions are run in duplicate with one sample from the set resuspended in H_2O and the other is resuspended in 10% (v/v) triethanolamine (pH 9.6). To the tubes at pH 9.6, 10 U of alkaline phosphatase (Sigma) is added and incubated at 37° for 30 min. Labeled phosphorylcholine is visualized using thin-layer chromatography (TLC) on silica plates (Whatman Silica Gel 60, 20 cm) developed in an equilibrated TLC chamber using a 50:50:5 mixture of 0.9% (w/v) aqueous $NaCl$:methanol:concentrated NH_4OH. The developed plate is air dried and exposed to a phosphorimager screen. Only reactions containing Mg^{2+} produce a radioactive spot comigrating with unlabeled phosphorylcholine standard. Its mobility is shifted after treatment with alkaline phosphatase. The results confirm that phosphorylcholine is formed with both purified *S. aureus* SMase and crude lysates of Molt-4 cells.

Enzyme, Time, and Phosphatidylserine Dependence

Reaction progress should be linear with time and enzyme (Figs. 2 and 3) until 10% of the substrate is consumed. Prolonged incubations with a low turnover may not be linear due to enzyme instability.

PS has a marked stimulatory effect on rat brain N-SMase. PS stimulation of N-SMase is more pronounced following delipidation of the enzyme. Presumably, endogenous PS or other lipids stimulate the activity and tend to minimize additional stimulation by exogenous lipid. Stimulation by PS

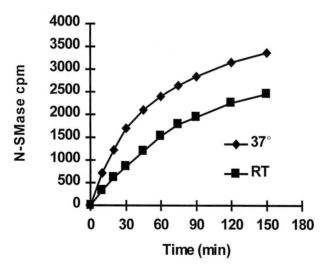

Fig. 2. Reaction progress is time dependent. Linearity is lost when greater than 10% of the substrate is turned over or with prolonged incubation.

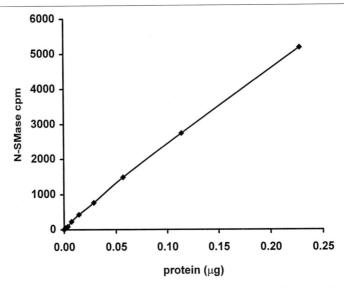

FIG. 3. Samples for protein dependence were incubated at 37° for 30 min.

is saturable and decreases at high mole fractions of PS due to substrate dilution (Fig. 4). Activity in the absence of added PS may be PS independent or due to residual PS present in the enzyme preparation.

Kinetic Parameters

Affinities of SM and cosubstrates for the enzyme were determined by measuring reaction velocities as functions of the concentrations of the respective components. The K_m values for SM and Mg^{2+} were determined to be 0.6 mole % (20 μM under these assay conditions) and 0.2 mM, respectively. Turnover of radiolabeled SM was competitively inhibited by cold SM.

Discussion

The method described here offers researchers a practical means of assaying multiple samples simultaneously. Increases in productivity and efficiency afforded by this method make it useful for screening large numbers of compounds to discover inhibitors. Radiolabeled material that is not precipitated in the assay accumulates slowly in stock solutions of SM, resulting in increased background. A background of up to 2% of the total counts is acceptable, but 1% or less is preferred. Substrate stocks yielding

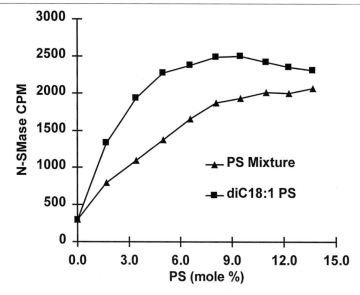

FIG. 4. The delipidated enzyme is stimulated by phosphatidylserine. Synthetic dioleoylphosphatidylserine is more potent than a mixture of bovine brain phosphatidylserines.

high backgrounds can be purified by performing a Bligh and Dyer liquid/liquid phase extraction to remove headgroup contaminants from the organic stock solution.

The assay can be adapted easily for assaying A-SMase by altering the assay components as desired. An acetate buffer at pH 5.0 with 1 mM zinc acetate replaces the Tris and magnesium used for N-SMase. Both DTT and PS can be omitted for A-SMase assays. Triton X-100 should be included for A-SMase measurements.

The assay allows for some compositional flexibility. The rat brain N-SMase can tolerate up to 10% of the reaction volume as dimethyl sulfoxide with minimal loss of activity. The critical separation of product and substrate may be affected, however, if significantly larger amounts of detergent are included in the assay.

Manual Assays

Although the assay protocol described here is performed most efficiently in multiwell plates, it can be adapted to tubes of many types with good results. Those lacking equipment for performing multiple simultaneous liquid transfers, multiwell plate centrifugation, or multiwell plate-based liquid scintillation counting can nonetheless perform this assay (with adaptations) using more common techniques and equipment.

Automated Assays

Multiple simultaneous liquid handlers can be utilized to greatly increase the efficiency of assaying large numbers of samples. Dispensing test compounds, enzyme, assay mixture, BSA, TCA, scintillation cocktail, and stopped reaction supernatants to multiwell plates via automated liquid handlers greatly reduces time, effort, and errors when assaying many samples. Automation of all aspects of the assay may be possible but is not necessarily practical or efficient. Sealing and mixing plates must be performed carefully to minimize sample loss before quantitation and thus are best performed manually. In our laboratory, an assay automated at all but the sealing and mixing steps allowed screening of several hundred thousand compounds in the search for inhibitors of the N-SMase.

Acknowledgment

The authors thank Dr. Robert M. Bell for valuable suggestions in the development of the assay and for his support of this work.

[21] Analysis of Sphingomyelin Hydrolysis in Caveolar Membranes

By Rick T. Dobrowsky and Valeswara Rao Gazula

Introduction

Caveolae and caveolae-related domains (CRDs) are distinct, specialized regions of the plasma membrane that share a somewhat similar lipid constitution enriched in cholesterol, sphingomyelin (SM), phosphatidylinositols, and glycosphingolipids.[1–4] CRDs are likely to be ubiquitous entities in most if not all cells,[5,6] whereas caveolae are not necessarily present in all cell types.[7] Caveolae are nonclathrin-coated invaginations of the plasma membrane that possess a distinct flask-like morphology that arises from the

[1] D. A. Brown and J. K. Rose, *Cell* **68**, 533 (1992).

[2] J. Liu, P. Oh, T. Horner, R. A. Rogers, and J. E. Schnitzer, *J. Biol. Chem.* **272**, 7211 (1997).

[3] L. J. Pike and L. Casey, *J. Biol. Chem.* **271**, 26453 (1996).

[4] M. P. Lisanti, P. E. Scherer, Z. Tang, and M. Sargiacomo, *Trends Cell Biol.* **4**, 213 (1994).

[5] D. A. Brown and E. London, *Biochem. Biophys. Res. Commun.* **240**, 1 (1997).

[6] R. J. Schroeder, S. N. Ahmed, Y. Zhu, E. London, and D. A. Brown, *J. Biol. Chem.* **273**, 1150 (1998).

[7] A. Gorodinsky and D. A. Harris, *J. Cell Biol.* **129**, 619 (1995).

association of lipids within these domains with the structural protein of caveolae, caveolin.[8] Indeed, expression of caveolin in cells lacking caveolae, but possessing CRDs, leads to the formation of morphologic caveolae. These results suggest that CRDs may be viewed as precaveolae.[9]

Many experimental approaches to the isolation of caveolae and CRDs have exploited the detergent insolubility of these domains, which is in large part due to the sphingolipid composition of these membranes. Relative to phospholipids, sphingolipids are more hydrophobic, undergo more hydrogen bonding, and tend to cluster within cell membranes.[10] Consequently, these regions and their associated proteins tend to be insoluble in nonionic detergent at low temperature and display a low buoyant density in sucrose gradients. It is the lipid composition of these regions and not the presence of caveolin or other proteins that is primarily responsible for their detergent insolubility.[5] This is exemplified by the detergent solubility of caveolin prior to its interaction with lipids within CRDs, which then renders caveolin detergent insoluble.[11]

Caveolae and CRDs have emerged as potential sites for the sequestering and integration of signal transduction pathways.[9] Numerous signaling molecules such as G-proteins, ras, protein kinase C isoforms, nitric oxide synthase, and src-related tyrosine kinases have been demonstrated to localize to caveolae or CRDs.[9] Further, several receptor systems have also been demonstrated to signal from caveolae, including the epidermal growth factor receptor,[12] platelet-derived growth factor receptor,[13] and the low-affinity neurotrophin receptor, $p75^{NTR}$.[14]

Evidence has implicated caveolae and CRDs as sites for ligand-induced SM hydrolysis.[14,15] Interestingly, although the Mg^{2+}-dependent neutral sphingomyelinase (SMase) is localized primarily to the plasma membrane,[16] an acid SMase activity has been reported to localize to caveolae.[15] Thus, caveolae and CRDs are emerging as potential sites for organizing and sequestering molecular components involved in regulating receptor-linked SM metabolism. Further, it is likely that caveolae may be critical sites for

[8] S. Li, K. S. Song, S. S. Koh, A. Kikuchi, and M. P. Lisanti, *J. Biol. Chem.* **271**, 28647 (1996).

[9] T. Okamoto, A. Schlegel, P. E. Scherer, and M. P. Lisanti, *J. Biol. Chem.* **273**, 5419 (1998).

[10] T. Harder and K. Simons, *Curr. Opin. Cell Biol.* **9**, 534 (1997).

[11] M. Murata, J. Peranen, R. Schreiner, F. Wieland, T. V. Kurzchalia, and K. Simons, *Proc. Natl. Acad. Sci. U.S.A* **92**, 10339 (1995).

[12] P. Liu, Y. Ying, Y. G. Ko, and R. G. W. Anderson, *J. Biol. Chem.* **271**, 10299 (1996).

[13] C. Mineo, G. L. James, E. J. Smart, and R. G. W. Anderson, *J. Biol. Chem.* **271**, 11930 (1996).

[14] T. R. Bilderback, R. J. Grigsby, and R. T. Dobrowsky, *J. Biol. Chem.* **272**, 10922 (1997).

[15] P. Liu and R. G. W. Anderson, *J. Biol. Chem.* **270**, 27179 (1995).

[16] M. W. Spence, J. Wakkary, and H. W. Cook, *Biochem. Biophys. Acta* **719**, 162 (1982).

interactions between tyrosine kinase and sphingolipid-signaling pathways.[17] Understanding the biochemical mechanisms affecting interactions between sphingolipid and tyrosine kinase signaling pathways in caveolae represents a novel and emerging area of cell regulation.

This article describes some of the commonly used procedures for isolating caveolar membranes and examining ligand-dependent SM turnover. Additionally, we provide some cautionary notes on the use of these procedures in the identification of caveolae/CRDs as localized sites for signal transduction and for the enrichment of various signaling molecules.

Measurement of Sphingomyelin Hydrolysis in Caveolin-Enriched Membranes from Cultured Cells

Detergent Extraction

This procedure exploits the detergent insolubility of caveolae in Triton X-100 at low temperature and the low buoyant density of this membrane fraction in sucrose gradients.[18] The purification of caveolae is monitored by assaying for the presence of caveolin, the marker protein for caveolae. Although this procedure readily separates caveolin-enriched membranes (CEMs) from noncaveolar membranes (NCMs), it does not necessarily produce pure caveolar vesicles. Therefore, it is premature to call this membrane fraction "caveolae" without morphologic documentation of the homogeneity of the membrane vesicles. Similarly, this procedure will also purify CRDs in cells lacking caveolae. It is important to note that CRDs and caveolae may coexist in some cells and that these domains will likely copurify in the following procedure. Although these domains may be separated using cationic silica beads,[19] the current lack of availability of these beads renders this separation problematic.

Buffers

Phosphate-buffered saline (PBS)
MES-buffered saline (MBS): 25 mM MES, pH 6.5, 5 mM EDTA, and 150 mM NaCl

---

[17] C. Wu, S. Butz, Y. Ying, and R. G. W. Anderson, *J. Biol. Chem.* **272,** 3554 (1997).

[18] M. P. Lisanti, P. E. Scherer, J. Vidugririene, Z. Tang, A. Hermanowski-Vosatka, Y.-H. Tu, R. F. Cook, and M. Sargiacomo, *J. Cell Biol.* **126,** 111 (1994).

[19] J. E. Schnitzer, D. P. McIntosh, A. M. Dvorak, J. Liu, and P. Oh, *Science* **269,** 1435 (1995).

MBST: MBS plus 1% Triton X-100, 1 mM Pefabloc or phenylmethylsulfonyl fluoride (PMSF), 10 μg/ml leupeptin, aprotinin, and bestatin each

80, 35, and 5% (w/v) sucrose in MBS

Procedure

Metabolic Labeling

Prepare two 15-cm dishes of cells and allow them to reach at least 70% confluency. Four or more 10-cm dishes may also be substituted. If ligand-dependent SM hydrolysis is to be measured, label the cells for 3 days in medium containing 0.5 μCi/ml [^{3}H]choline chloride. Other isotopes such as [^{3}H]palmitate or [^{14}C]serine (with serine-deficient medium) may also be used if desired. Because previous work has suggested that the ligand-sensitive pool of SM is the last to incorporate the radiolabel,[20] it is important to incubate the cells for at least 48 to 72 hr with the desired radioisotope to reach metabolic equilibrium. Although these times are influenced by the growth rate of particular cell lines, 48 hr is usually sufficient to reach metabolic equilibrium for cells with a rapid doubling time, whereas longer incubation times may be required for cells with prolonged population doublings. After the end of the incubation, dispose of the radioactive medium, wash the monolayers twice with fresh serum-free medium or PBS, and place the cells in serum-free medium for at least 4 hr. This resting period is important as simple medium changes can dramatically affect sphingolipid levels in cultured cells.[21]

Isolation of CEMs

Treat the cells with the ligand of choice and, following treatment, aspirate the medium rapidly and wash the cells twice with ice-cold PBS. After aspirating the PBS, add 2 ml of ice-cold MBST and quickly scrape the cells from the plate with a cell scraper. Care should be exercised that the pH of the MBST is adjusted to 6.5 because neutral pH buffers give lower yields of caveolar membranes.[22] Transfer the lysate to a 10-ml homogenization tube and set on ice for at least 15–20 min. Homogenize the samples using a tight-fitting Teflon pestle (20 strokes), maintaining the tube on ice at all times. Transfer 2 ml of the sample to a 12-ml ultracentrifuge tube that can be accommodated within the buckets of a SW41 or comparable rotor.

[20] C. M. Linardic and Y. A. Hannun, *J. Biol. Chem.* **269,** 23530 (1994).
[21] E. R. Smith, P. L. Jones, J. M. Boss, and A. H., Merrill, Jr., *J. Biol. Chem.* **272,** 5640 (1997).
[22] M. Sargiacomo, M. Sudol, Z. Tang, and M. P. Lisanti, *J. Cell Biol.* **122,** 789 (1993).

Adjust the extract to 40% sucrose by the addition of 2 ml of 80% sucrose in MBS lacking Triton X-100. Vortex the tube to mix the solutions thoroughly. Form a discontinuous sucrose gradient by carefully overlaying this solution with 4 ml of the 35% sucrose in MBS. A clear interface should be evident between the 35 and 40% sucrose layers. Next, carefully overlay the 35% sucrose layer with 4 ml of the 5% sucrose in MBS. The difference in density results in a readily visible interface. A linear 5–35% sucrose gradient may also be substituted for the discontinuous gradient. Prepare a balance tube if necessary using the same solutions (substitute 2 ml of MBS for the lysate) and centrifuge the samples at 39,000 rpm in a SW41 rotor for 16–18 hr at 4°. Alternatively, a minimal centrifugation time of 3 hr may be used because it also gives satisfactory recovery of CEMs.[3] We typically collect 15 × 0.8-ml fractions beginning from the top of the gradient. An aliquot of each fraction may be used to determine the distribution of total radiolabel over the gradient (Fig. 1A). Additional aliquots of each fraction are used for the determination of protein content. Because total cellular protein distributes primarily to high density regions of the gradient (Fig. 2, closed circles), reliable protein measurements can be obtained using 0.1–0.2 ml of fractions 3–8 and 0.02–0.05 ml of fractions 9–15.

CEMs are typically recovered as a visible band of material located at the interface of the 5 and 35% sucrose layers, fractions 4–6.[14] Alternatively, CEMs may be located by light scattering (600 nm). The presence of CEMs is indicated by comigration of caveolin with these gradient fractions. This is performed easily by SDS–PAGE and immunoblot analysis using any of several commercially available caveolin antibodies. If desired, the CEMs can be concentrated by pooling together the fractions, diluting in MBS, and concentrating the membranes by centrifugation at 100,000g for 30 min. However, some investigators concentrate the membranes by centrifugation in a microfuge.[22]

Caveolin is well resolved from the bulk of cellular protein as over 95% of the total cellular protein is typically recovered in NCMs (fractions 10–15) (Fig. 2, closed circles). However, depending on the cell type, about 50–70% of the cellular SM is recovered in the CEMs,[14] whereas the remainder distributes primarily to fractions 12–15 representing bulk plasma membrane (Fig. 1A). Thus, relative to cellular protein, SM is highly enriched in CEMs (Fig. 2, closed squares).

Advantages and Disadvantages

A major advantage of preparing CEMs using detergent insolubility as a criterion is the relative ease of the procedure. However, a cautionary note is that slight alterations in the protein-to-detergent ratio may alter

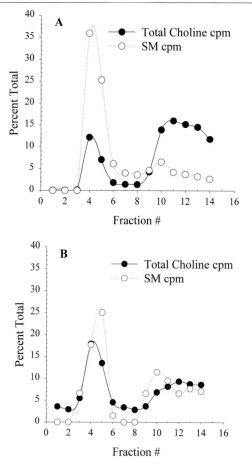

FIG. 1. Distribution of [³H]SM following detergent (A) and nondetergent extraction (B) and centrifugation through discontinuous sucrose gradients. Cells were fractionated as described in the text and the amount of [³H]SM in each fraction was plotted as a percentage of the total [³H]SM recovered in the gradient. The distribution of [³H]SM relative to the total amount of radioactivity present is also shown.

the solubility of caveolar components.[23] Because the mechanisms whereby receptors couple to both neutral or acid SMases are poorly understood, loss of critical components due to changes in this ratio may cause inconsistent results. In this regard, detergent extraction results in the selective loss

[23] A. Uittenbogaard, Y. Ying, and E. J. Smart, *J. Biol. Chem.* **273,** 6525 (1998).

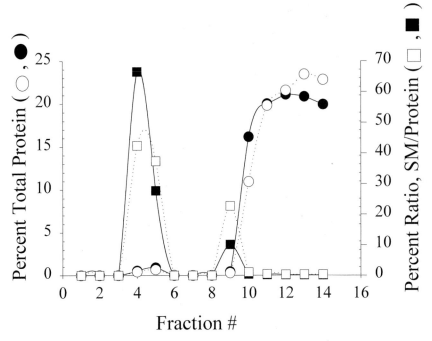

Fraction #

Fig. 2. Enrichment of [³H]SM in CEMs relative to cellular protein following detergent (●, ■) or nondetergent extraction (○, □) and centrifugation through discontinuous sucrose gradients. Cells were fractionated as described in the text and the amount of protein (●, ○) in each fraction was plotted as a percentage of the total recovered in the gradient. The amount of SM in the gradient fractions was determined in the same manner and the ratio of SM/protein plotted (■, □).

of lipid-modified proteins from caveolae.[24] An additional concern is that this procedure does not result in a homogeneous preparation of caveolar vesicles. Whereas isolated caveolae are composed primarily of vesicles ranging in diameter from about 50 to 80 nm, detergent extraction likely results in a very heterogeneous population of vesicles, many of which may be derived from the plasma membrane proper.[25] Thus, although useful for indicating whether ligand-induced SM hydrolysis may localize to detergent-insoluble membrane domains, it is not a definitive demonstration of a caveolar compartmentalization of this response.

[24] K. S. Song, S. Li, T. Okamoto, L. A. Quilliam, M. Sargiacomo, and M. P. Lisanti, *J. Biol. Chem.* **271,** 9690 (1996).

[25] R. V. Stan, W. G. Roberts, D. Predescu, K. Ihida, L. Saucan, L. Ghitescu, and G. E. Palade, *Mol. Biol. Cell* **8,** 595 (1997).

Detergent-Free Extraction

Because detergent can preferentially solubilize lipid-modified caveolae-associated proteins, several detergent-free methods have been developed to avoid this problem. The sodium carbonate procedure is a variation to that described earlier.[24]

Buffers

MBS containing 0.5 M sodium carbonate, pH 11
90% sucrose in MBS
5 and 35% sucrose in MBS containing 250 mM sodium carbonate

Procedure

Isolation of CEMs

Cells are scraped into 2 ml of ice-cold MBS containing 0.5 M sodium carbonate, pH 11, producing a very viscous cell lysate. The cells are then homogenized using 20 strokes in a Dounce homogenizer, three 10-sec pulses with a Polytron tissue grinder (half-maximal speed), and three 20-sec bursts with a microtip sonicator to break up the plasma membrane and detach the caveolae. The lysate is then brought to 45% sucrose containing 250 mM sodium carbonate by the addition of 2 ml of 90% sucrose in MBS (prepare the 90% sucrose by dissolving the sugar with mild heating). The 45% sucrose is overlaid with 4 ml of 35% sucrose in MBS followed by 4 ml of 5% sucrose in MBS (both containing 250 mM sodium carbonate). The tubes are then centrifuged as described earlier. Under this protocol, CEMs also migrate in fractions 4–6 and noncaveolar membranes (NCMs) comprise the bulk of fractions 10–15.

Advantages and Disadvantages

Similar to detergent extraction, the sodium carbonate extraction procedure is easy and reliably separates CEMs from NCMs and results in a similar protein profile (Fig. 2, open circles). Additionally, SM is similarly distributed over the gradient (Fig. 1B) and enriched in CEMs relative to protein (Fig. 2, open squares). Importantly, the recovery of CEMs is somewhat improved over the detergent extraction method and does not remove lipid-modified proteins from these membrane domains. However, sodium carbonate has been used extensively to determine if proteins strongly attach to membranes. Thus, this procedure may remove loosely attached protein components that may impair the development of *in vitro* approaches to

Control	100 ng/ml NGF	10 ng/ml NT-3	10 ng/ml IL-1β	
4 5 6 7	4 5 6 7	4 5 6 7	4 5 6 7	Frac. #

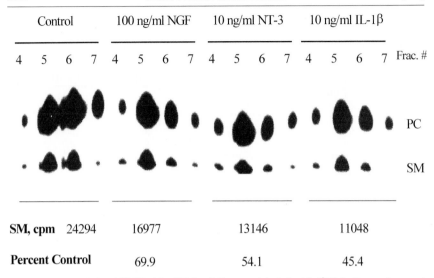

	PC
	SM

	Control	100 ng/ml NGF	10 ng/ml NT-3	10 ng/ml IL-1β
SM, cpm	24294	16977	13146	11048
Percent Control		69.9	54.1	45.4

FIG. 3. Hydrolysis of [³H]SM in CEMs. Cells were labeled with [³H]choline and treated with the indicated ligands. CEMs were isolated and [³H]SM content determined by TLC. After normalization to total protein in each fraction the percentage SM hydrolysis was determined. PC, phosphatidylcholine.

determine how receptors may couple to enzymes regulating SM metabolism in CEMs. Moreover, because the extraction is performed typically in the pH range of 9–11, the direct measurement of enzymes involved in sphingolipid metabolism is problematic. The membranes may be dialyzed against buffer to remove the sodium carbonate and enzyme activity assessed.

Assay of Sphingomyelin Content of Caveolin-Enriched Membranes

Once CEMs have been isolated, we measure SM levels in the membranes by thin-layer chromatography (TLC) or by using the bacterial SMase assay.[14]

Bacterial SMase Assay

Transfer a 0.4-ml aliquot of each fraction to a screw-cap glass tube and perform a Bligh and Dyer lipid extraction.[26] Quantitation of SM is determined as described.[27] Evaporate one-half of the organic layer and

[26] E. G. Bligh and W. J. Dyer, *Can. J. Biochem. Physiol.* **37,** 911 (1959).
[27] S. Jayadev, C. M. Linardic, and Y. A. Hannum, *J. Biol. Chem.* **269,** 5757 (1994).

solubilize the residue in 0.05 ml of 200 mM Tris–HCl, pH 7.5, 20 mM MgCl containing 1% Triton X-100. Add 0.05 ml of a 2 U/ml solution of *Streptomyces* sp. SMase (diluted in 10 mM Tris–HCl, pH 7.5) and incubate the tubes for 2 hr at 37°. Terminate the reaction by adding 1.5 ml of chloroform:methanol (2:1). Add 0.2 ml of water and recover the released [³H]choline in the aqueous layer after centrifugation. Normalize the amount of SM hydrolyzed to the protein content of each fraction. Because the bacterial SMase quantitatively hydrolyzes any SM present in the sample,[27] ligand-induced SM hydrolysis is revealed by a decrease in the amount of SM present in CEMs from treated cells versus control cells.

Thin-Layer Chromatography of SM

Alternatively, SM hydrolysis may be measured by TLC. Numerous solvent systems exist for the separation of SM from other lipids, but two common systems used in our laboratory are chloroform:methanol:glacial acetic acid:water (65:35:8:4) or chloroform:methanol:2 N NH₄OH (65:35:8). Fit a TLC chamber with a paper wick (Whatman filter paper) and place enough solvent into the chamber to cover about the bottom 1–2 cm of the TLC plate. Do not individually add the solvent components to the chamber, mix them in a separate flask and add the solution to the chamber. Next, allow the chamber to equilibrate for at least 3–4 hr prior to placing the TLC plate in the chamber. If two TLC plates are to be developed, a single chamber may be used. If this is the case, place both plates into the chamber together and ensure that they do not touch one another and are adequately immersed in the solvent.

Following the Bligh and Dyer lipid extraction, evaporate 1.5 ml of the chloroform extract and resuspend the residue in 0.03 ml of chloroform. Apply 0.02 ml to the silica gel and dry the spot thoroughly. Develop the plate to about 2 cm from the top and allow the solvent to evaporate in a fume hood. Spray the silica gel with an autoradiography enhancer such as En³Hance (spray liberally but do not create runs), wrap the plate in plastic wrap, and expose to film for 2–3 days at −80°. Figure 3 shows a typical result for the analysis of SM hydrolysis from CEMs by this method from p75^{NTR}-NIH 3T3 cells labeled with [³H]choline and treated with several agonists.

[22] Ceramidases

By Mariana Nikolova-Karakashian *and* Alfred H. Merrill, Jr.

Introduction

Ceramidase (*N*-acylsphingosine deacylase) hydrolyzes ceramide to a free sphingoid base and fatty acid. Because ceramide can induce cell death,[1,2] whereas the phosphorylated product of sphingosine, sphingosine-1-phosphate, is a potent inducer of cell proliferation and survival signals, ceramidase(s) play a key role in regulating the amounts of the bioactive sphingolipids. This article summarizes methods for assaying ceramidases as well as factors known to participate in the regulation of these enzymes.

Overview of Mammalian Ceramidases

Ceramidase activity was first characterized as an enzyme with an acidic pH optimum present in all rat tissues examined.[3,4] This enzyme was later shown to account for a rare genetic disorder called Farber's disease where there is accumulation of ceramide in the lysosomes of spleen, cerebellum, fibroblasts, and kidney.[5] An activity with alkaline pH optimum, however, was retained in Farber's patients, indicating that at least two ceramidases hydrolyze ceramide.[6] Both enzymes are membrane associated; however, the acidic enzyme is localized primarily in the lysosomes and activated by Saposine D[7] which inhibits the alkaline enzyme (IC_{50} 1–5 μM).[8] The purified acid ceramidase[9] shows little specificity with respect to the fatty acid chain length of ceramidases, but prefers sphingosine over sphinganine-containing ceramides. In addition to the acidic and alkaline activities, hydrolysis of ceramide is also found at neutral pH in rat liver, small intestine, and

[1] S. Spiegel and A. H. Merrill, Jr., *FASEB J.* **10**, 1388 (1996)
[2] Y. A. Hannun, *Science* **274**, 1855 (1996)
[3] S. Gatt, *J. Biol. Chem.* **241**, 3724 (1966)
[4] E. Yavin and S. Gatt, *Biochemistry* **8**, 1692 (1969)
[5] H. Moser, *in* "The Metabolic Base of Inherited Disease" (C. S. Serwer, A. L. Beaudet, W. S. Sly, and D. Wall, eds.), 7th ed., p. 2589. McGraw Hill, New York, 1995.
[6] M. Sugita, J. T. Dulaney, and H. W. Moser, *Science* **178**, 1100 (1972)
[7] N. Azuma, J. C. O'Brian, H. W. Moser, and Y. Kishimoto, *Arch. Biochem. Biophys.* **311**, 354 (1994)
[8] A. Bielawska, M. S. Greenberg, D. Perry, S. Jayadaev, J. A. Shayman, C. McKay, and Y. A. Hannun, *J. Biol. Chem.* **271**, 12646 (1996)
[9] K. Bernardo, R. Hurvitz, T. Zenk, R. Desnick, K. Ferlinz, E. Schuchman, and K. Sandhoff, *J. Biol. Chem.* **270**, 11098 (1995)

epithelial cells.[10,11] Studies with isolated plasma membranes from rat liver[12] have shown a Mg^{2+}-dependent generation of sphingosine, presumably from ceramide. Ceramidase activity with a neutral pH optimum has also been found in isolated liver microsomes.[13] In intact cells or tissues, however, the neutral and alkaline activities do not have distinct pH maxima and often appear as a broad plateau in the range of pH 6.5 to pH 9.0.

These enzymes appear to metabolize ceramides from different sources, because studies with fibroblasts from Farber's patients have shown that acid ceramidase hydrolyzes ceramides that enter the cell via the LDL receptor, whereas endocytotic vesicle-derived ceramide is hydrolyzed by a nonlysosomal ceramidase.[14] Multiple ceramidases appear to generate sphingosine as a "second messenger," according to studies of PDGF signaling in glomular mesangial cells[15] and IL-1β signaling in primary hepatocytes.[16]

In Vitro Assay of Ceramidase Activity

Using a Radiolabeled Substrate

Traditionally, ceramidase has been measured using [³H] or [¹⁴C]oleoyl sphingosine as an exogenous substrate[3,17] with either tissue homogenates or cell fractions as enzyme sources. A disadvantage of this method is that ceramides are hydrophobic molecules and therefore high concentrations of detergent are required for delivery of the substrate. Furthermore, detection of the product often requires TLC separation. Nonetheless, this method has been used successfully in a variety of cell systems. A protocol adapted from the original references[3,17] is given later; more recent modifications of the method can be found in Refs. 8 and 15.

Preparation of Enzyme Source. If cultured cells are used, the cells are harvested in ice-cold phosphate-buffered saline (PBS), pelleted, and resus-

---

[10] M. Spence, S. Beed, and H. W. Cook, *Biochem. Cell. Biol.* **64,** 400 (1985)

[11] A. Nilson, *Biochim. Biophys. Acta* **176,** 339 (1969)

[12] C. W. Slife, E. Wang, R. Hunter, S. Wang, C. Burgess, D. C. Liotta, and A. H. Merrill, Jr., *J. Biol. Chem.* **264,** 10371 (1989)

[13] W. Stoffel and I. Melzner, *Hoppe-Seylers Z. Physiol. Chem.* **361,** 755 (1980)

[14] S. L. Sutrina and W. W. Chen, *J. Biol. Chem.* **257,** 3039 (1982)

[15] C. Martinez, M. McKenna, and M. Kester, *J. Biol. Chem.* **270,** 23305 (1995)

[16] M. N. Nikolova-Karakashian, E. T. Morgan, C. Alexander, D. C. Liotta, and A. H. Merrill, Jr., *J. Biol. Chem.* **272,** 18718 (1997)

[17] M. Sugita, M. Williams, J. T. Dulaney, and H. T. Moser, *Biochim. Biophys. Acta* **398,** 125 (1975)

pended in the same 0.25 M sucrose, 0.1 M EDTA-containing buffer. Application of 0.1 to 0.4 mg of tissue or cell homogenate per incubation has been shown to give a linear response. Cells are disrupted by sonication in the presence of a protease inhibitor cocktail [1 mM CaCl$_2$, 1 mM phenylmethylsulfonyl fluoride (PMSF), 1 mM benzamidine, and 20 μg/ml freshly added leupeptin, pepstatin, aprotinin, and 1 mM sodium vanadate). If subcellular fractionation is desired, the usual fractionation methods are used. For assays of animal tissues, they are homogenized in the same buffers.

Preparation of Exogenous Substrate. The method used most often for the solubilization of exogenous long-chain ceramide is that of Sugita *et al.*[17] For 10 assays: 500 nmol of N-[14C-oleoyl]sphingosine (specific activity of 2000 dpm/nmol), 1 mg of Tween 20, and 2.5 mg of Triton X-100 (all as stock solution in chloroform:methanol, 2:1, by volume) are mixed in a glass tube, the organic solvent is evaporated under nitrogen, and 200 μl of a 20-mg/ml aqueous sodium cholate solution is added. The suspension is sonicated in five short bursts. Other radiolabeled ceramides, such as N-[14C-lauroyl]sphingosine, can also be used as substrate; however, ceramides labeled in the sphingoid moiety via reduction of the 4-*trans* double bond appear to be less effective.[9,12] The efficacy of the solubilization procedure can be verified by counting the amount of radioactivity in a small aliquot.

Assay Mixture. Ceramidase activity is assayed in 0.125 M sodium acetate buffer, pH 4.5 (for acidic enzyme), 0.125 M HEPES, pH 8.0 (for the alkaline enzyme), or 0.1 M Tris, pH 7.2 (for the neutral activity), 0.125 M sucrose, and 0.01 mM EDTA, with initiation of the assay by the addition of 20 μl of substrate suspension and 0.1–0.5 mg of enzyme preparation (cell or tissue homogenate) for a final volume of 0.2 ml. These are mixed in borosilicate glass tubes with continuous shaking at 37° for the desired assay time (usually 1 hr).

Detection of Product. For quantitation of the amount of fatty acid released during incubation, 10 μl of 10 mg/ml ice-cold oleic acid (in chloroform:methanol, 2:1, by volume) is added to each tube at the end of the incubation to serve as a carrier during the analysis and then one of the following is done.

Phase Partitioning of Product. The reaction mixture is stopped by the addition of 2 ml of basic Dole's solution (isopropanol:heptane:1 N NaOH, 40:10:1, by volume). Heptane (1.8 ml) and water (1.6 ml) are added, and the mixture is vortexed and then centrifuged for 10 min at 2000g. The upper phase is aspirated and the lower phase is washed twice with heptane and then 1 ml of 1 N H$_2$SO$_4$ and 2.4 ml of heptane are added to the lower phase. After vortexing for 1 min and centrifugation at 2000g for 10 min, 1-ml aliquots from the upper phase are analyzed by a scintillation counter.

Thin-Layer Chromatography. One milliliter of chloroform:methanol (2:1, by volume) and 1 ml of acidic Dole's solution (isopropanol:heptane: 2 N H$_2$SO$_4$, 40:10:1 by volume) are added to the assay mixture. After 10 min, 0.4 ml of heptane and 0.6 ml of water are added, the samples are vortexed, and the phases are separated by centrifugation for 10 min at 2000g. The upper phase is transferred to new tubes and the heptane is dried *in vacuo*. The dried lipids are resuspended in 15 μl of chloroform:methanol (2:1, by volume), applied on a silica gel G TLC plate, and developed in chloroform:methanol:acidic acid (94:1:5, by volume). The free fatty acid is identified by exposure to iodine vapors and comparison with standard. The respective spots are scraped and the radioactivity is counted by a scintillation counter with correction for quenching by silica.

Phase partitioning is a faster and easier method for quantitating the amount of fatty acid released during the reaction; however, copartitioning of ceramide and the fatty acid is possible and, therefore, initial analysis of the last heptane phase by TLC is recommended.

Using a Fluorescent Substrate

The use of NBD-hexanoic-ceramide (available commercially from Molecular Probes and Matreya) instead of radiolabeled long-chain ceramide simplifies the separation and quantitation of the released fatty acid. The NBD ceramide is not a natural substrate; however, assays with purified acid ceramidase have shown it to be as effective as natural ceramides.

The method described here uses cultured cells as an enzyme source; however, it can be adapted easily for tissue homogenates or isolated membrane fractions.

Preparation of Enzyme Source. The cells (ca. 1 mg of protein) are harvested and washed twice with ice-cold PBS, then lysed in 0.5 ml of 0.2% Triton X-100 in 10 mM Tris, pH 7.4, supplemented with 10 μg/ml leupeptin and aprotinin, 1 mM sodium vanadate, 1 mM 2-mercaptoethanol, 1 mM EDTA, and 15 mM NaCl. The lysate is homogenized with three passes through a 25-gauge needle, transferred to an Eppendorf "bullet" tube, and a 10-μl aliquot is saved for protein assay.

Delivery of Substrate. One microliter of NBD-ceramide (from a 10 mM stock solution in ethanol) is added to the cell lysate, and the mixture is vortexed and left on ice for 10 min to allow the substrate to equilibrate among the membranes.

Enzyme Assay. Aliquots (ca. 50 μl) corresponding to 0.1 mg of cell protein and approximately 1 nmol of NBD-ceramide, respectively, are added to borosilicate tubes containing 0.2% Triton X-100 and 10 mM Tris, pH 7.4 (for neutral ceramidase), 0.5 M acetate buffer, pH 4.5 (for acid ceramidase), or 10 mM HEPES, pH 9.0 (for alkaline ceramidase) with a final assay volume of 0.3 ml. The mixture is incubated at 37° for 1 hr in a reciprocal water bath and then stopped by the addition of 1 ml of mobile phase (methanol:water:85% phosphoric acid, 850:150:1.5, by volume).

Separation and Detection of Product. After adding the mobile phase, the samples are incubated at 37° for 1 hr to allow the proteins to precipitate, and the insoluble material is removed by centrifugation. Aliquots of the clear supernatant are injected onto a reversed-phase column (Nova-Pack, C18, Waters Corp.) and eluted with the mobile phase (methanol:water:85% phosphoric acid, 850:150:1.5, by volume)[16] at a flow rate of 2 ml/min, with detection of NBD fluorescence (excitation 455 nm; emission 530 nm).[18] In this system, the elution time for NBD-fatty acid is 1.6 min and that of the NBD-ceramide is 10.3 min. A typical HPLC profile is shown in Fig. 1. The mass of the fatty acid released is calculated by comparison with the fluorescence of the NBD-lipid standard.

As with *in vitro* assays in general, assays with each enzyme should be tested for (i) linearity with time and amount of protein and (ii) the concentration of exogenous substrate compared to the endogenous ceramide.

In Situ Assays of Ceramidase

One of the major disadvantages of an *in vitro* assay of enzyme activity is that the protein of interest does not always retain full enzymatic activity due to proteolytic cleavage, loss of cofactors during preparation of the cell extracts, change in the active conformation during solubilization of the enzyme, and so on. These problems can be bypassed by *in situ* assays, two of which are described in this article.

Use of Fluorescent Labeling

In this *in situ* assay, cells are incubated with NBD-ceramide and the formation of NBD-fatty acid is followed. NBD-fatty acid is not reutilized and can be measured easily. This approach allows comparison of the rate of ceramide degradation on agonist addition to the intact cells.

To deliver NBD-ceramide to the cells, 25 ml of tissue culture medium is transferred to a sterile 50-ml tissue culture tube and NBD-ceramide (10 μl of a 10 mM stock solution in ethanol) is injected rapidly into the medium. After 1 min of mixing, the medium is added to the cells in culture, replacing the old medium.

---

[18] O. Martin and R. Pagano, *Anal. Biochem.* **159,** 101 (1986)

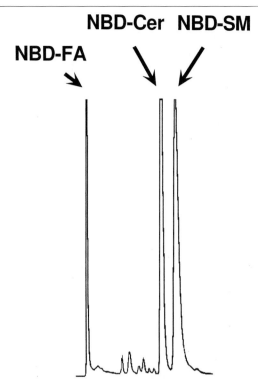

Fig. 1. A typical HPLC profile of lipid extract from hepatocytes cultured in the presence of NBD-ceramide for 16 hr. NBD-FA, NBD-fatty acid; NBD-Cer, NBD-ceramide; NBD-SM, NBD-sphingomyelin.

After the desired incubation period, the medium is removed, the cells are washed and harvested by standard procedures, 1 ml of mobile phase (methanol : water : 85% phosphoric acid, 850 : 150 : 1.5, by volume) is added, and the tubes are incubated for 1 hr at 37°. The insoluble material is pelleted and the amounts of NBD-fatty acid in the supernatant are analyzed by HPLC as described earlier. This assay is especially useful for measuring the activation of ceramidase by agonists.[16] The cells are "prelabeled" with NBD-ceramide for a period optimized for each cell type (e.g., 6 to 12 hr for hepatocytes)[16] and then the agonist is added and the fluorescent lipids are analyzed over time. The following factors should be kept in mind: (i) Under the condition of this assay, part of the NBD-ceramide will be converted to more complex sphingolipids, such as NBD-sphingomyelin or NBD-glucosphingolipids. During the HPLC analysis, these lipids are separated readily from NBD-ceramide and can be identified by comparison with the respective fluorescent standards. (ii) The amount of NBD-fatty acid released may represent a subcellular pool of ceramide and ceramidases. (iii) The time for NBD-ceramide "prelabeling" must be optimized for each cell system. The goal is to allow the uptake of NBD-ceramide to an apparent "steady state" that is perturbed on agonist addition.

A substantial portion of the NBD-ceramide is usually localized in the plasma membrane and Golgi, whereas the amount in cells can be increased when the NBD-ceramide is loaded onto LDL particles.[19]

Measuring the Mass of Sphingosine

Another assay of ceramidase with minimal perturbation of the system is to incubate the membranes at 37° and measure the release of free sphingoid bases[20] from endogenous ceramide.[12] The interpretation of these results is complex because the mass of endogenous ceramide that is available for hydrolysis may also vary; however, it can be made less limiting by adding an exogenous sphingomyelinase.

Regulation of Ceramidase Activity

The following factors affect ceramidase activity.

Substrate Specificity

As mentioned earlier for acid ceramidase, there is no apparent specificity in respect to the length of the fatty acid chain. It is intriguing, however, that acid ceramidase (measured *in vitro* using purified enzyme) and neutral ceramidase (measured by the generation of free sphingoid bases by plasma membrane) both show very little activity with the sphinganine-containing ceramides (dihydroceramides). This may have interesting biological consequences because it is now accepted that only unsaturated ceramides (but not dihydroceramides) are effective in inducing growth arrest and cell death. Selectivity may indicate that dihydroceramides (which are biosynthetic intermediates for complex sphingolipids) may be spared from "unnecessary turnover."

Biological Regulators

Studies with cells that are pretreated with sodium vanadate demonstrate an increase in both basal and agonist-stimulated ceramidases.[15,16] In con-

---

[19] T. Levade, M. Lernth, D. Graber, A. Moisand, S. Vermeersch, R. Salvayre, and P. J. Courtnoy, *J. Lipid Res.* **37,** 2525 (1996)

[20] A. H. Merrill, Jr., E. Wang, R. Mullins, W. Jamison, S. Nimkar, and D. C. Liotta, *Anal. Biochem.* **171,** 337 (1988)

trast, treatment of the cells with genistein, an inhibitor of tyrosine-specific protein kinases, decreases the activity and blocks agonist-induced stimulation. It is likely that tyrosine phosphorylation is essential for ceramidase activity, although it is not clear at this point whether the ceramidase molecule per se or an upstream effector is phosphorylated.

Inhibitors

Two inhibitors of ceramidase, D-e-MAPP[21] and *N*-oleoylethanolamine,[22] have been reported. D-e-MAPP is a potent and specific inhibitor of alkaline enzyme and is ineffective toward acid ceramidase. Its close stereoisomer, L-*threo*-MAPP, does not inhibit the enzyme, but rather serves as a substrate for it. D-e-MAPP is equally effective when supplied to intact cells or to isolated membrane fraction. When supplied to HL-60 cells, it induces an increase in the endogenous ceramide and suppresses cell growth.

Although not as potent an inhibitor of ceramidase (the IC_{50} is 500–700 μM), *N*-oleoylethanolamine, nevertheless, inhibits growth factor-induced mutagenesis in rat mesangial cells, most likely by preventing sphingosine and sphingosine-phosphate generation.

[21] A. Bielawska, M. S. Greenberg, D. Perry, S. Jayadev, J. A. Shayman, C. McKay, and Y. A. Hannun, *J. Biol. Chem.* **271,** 12646 (1996).
[22] J. Quintains, J. Kilkus, C. L. McShan, A. R. Gottschalk, and G. Dawson, *Biochem. Biophys. Res. Commun.* **202,** 710 (1994).

[23] Purification of Acid Ceramidase from Human Placenta

By T. Linke, S. Lansmann, and K. Sandhoff

Introduction

Human acid ceramidase (haCerase, EC 3.5.1.23, *N*-acylsphingosine amidohydrolase) is a lysosomal enzyme that catalyzes the hydrolysis of ceramide into sphingosine and free fatty acid.[1] *In vivo* and *in vitro* studies on sphingolipid metabolism showed that degradation of ceramide by haCerase is stimulated by a nonenzymic glycoprotein called sphingolipid activator

[1] D. F. Hassler and R. M. Bell *in* "Advances in Lipid Research," Vol. 26, p. 49. Academic Press, San Diego, 1993.

protein D (SAP-D).[2,3] A deficiency in haCerase activity leads to the autosomal recessive sphingolipid storage disorder called Farber disease, caused by accumulation of ceramide in lysosomes. Seven different phenotypes of this rare disorder are known to date.[4]

A purification procedure using human urine from patients with peritonitis as an enzyme source yielding an apparently homogeneous enzyme preparation and providing peptide sequence data that subsequently led to the isolation and characterization of a full-length cDNA encoding haCerase has been described.[5,6] Since copurification of sphingolipid activator protein C (SAP-C) and glucocerebrosidase has been reported by Aerts *et al.*,[7] we tested each purification step for the presence of SAP-D in haCerase-containing fractions. Western blot analysis indicated that SAP-D copurified with haCerase.[8] In order to allow further functional and structural studies on haCerase, a purification procedure yielding SAP-D free, enzymatically active haCerase was established and optimized.

Acid Ceramidase Assay

The haCerase assay is essentially performed as described previously.[5]

Reagents

In a final volume of 100 μl, the incubation mixture contains 10–50 μl undiluted enzyme solution, 17 nmol N-lauroylsphingosine (synthesized according to Okazaki *et al.*[9]) solubilized with 0.5% (w/v) Triton X-100, 0.2% (w/v) Tween 20, 0.2% (w/v) Nonidet P-40 (NP-40), and 0.8% (w/v) sodium cholate, 250 mM sodium acetate buffer (pH 4.0), 5 mM EDTA. Sphingosine has been described to be a weak competitive inhibitor of haCerase.[1] Therefore, the amount of enzyme added should not hydrolyze more than 10% of the total substrate.

[2] A. Klein, M. Henseler, C. Klein, K. Suzuki, K. Harzer, and K. Sandhoff, *Biochem. Biophys. Res. Commun.* **200,** 1440 (1994).

[3] N. Azuma, J. S. O'Brien, H. W. Moser, and Y. Kishimoto, *Arch. Biochem. Biophys.* **311,** 354 (1994).

[4] H. W. Moser, in "The Metabolic and Molecular Basis of Inherited Disease" (C. Scriver, A. L. Beaudet, W. S. Sly, and D. Valle, eds.), p. 2589. McGraw Hill, New York, 1995.

[5] K. Bernardo, R. Hurwitz, T. Zenk, R. J. Desnick, K. Ferlinz, E. H. Schuchman, and K. Sandhoff, *J. Biol. Chem.* **279,** 11098 (1995).

[6] J. Koch, S. Gärtner, C. M. Li, L. Quintern, K. Bernardo, O. Levran, D. Schnabel, R. J. Desnick, E. H. Schuchman, and K. Sandhoff, *J. Biol. Chem.* **271,** 33110 (1996).

[7] J. M. Aerts, W. E. Donker-Koopman, C. van-Laar, S. Brul, G. J. Murray, D. A Wenger, J. A. A. A. Barranger, J. M. Tager, and A. W. Schram, *Eur. J. Biochem.* **163,** 583 (1987).

[8] T. Linke, unpublished results, 1997.

[9] T. Okazaki, A. Bielaswka, R. M. Bell, and Y. A. Hannun, *J. Biol. Chem.* **265,** 15823 (1990).

*Incubation Procedure and Preparation of Enzyme Assay for
 HPLC Analysis*

The enzyme assays are incubated for 30 min at 37°. The reaction is stopped by adding 800 μl HPLC-grade (2:1, v/v) chloroform/methanol and 200 μl of 100 mM NH$_4$HCO$_3$ solution. D-*erythro*-1,3-dihydroxy-2-aminotetradecane (C$_{14}$-sphinganine) and D-*erythro*-1,3-dihydroxy-2-amino-hexadecane (C$_{16}$-sphinganine), solubilized in HPLC-grade methanol, are added to each incubation mixture as internal standards (500 pmol each). The short chain sphingosine derivatives are available commercially from Matreya Lipids. Sphinganine derivatives can be prepared by reduction of the double bond.[10] The test tubes are mixed for 10 min and then centrifuged for 5 min at 10,000g. The chloroform phase contains the liberated sphingosine as well as both internal standards. The lower phase is removed carefully and collected in a fresh vial. For quantitative recovery of sphingosine and the internal standards, the extraction is repeated by adding 600 μl (2:1, v/v) chloroform/methanol to each incubation vial. The extraction steps are performed as described above. The combined organic phases are evaporated to dryness under a stream of nitrogen.

Alkaline Hydrolysis Procedure

During tissue homogenization, the added detergent will solubilize large amounts of phospholipids, which interfere with the reversed-phase HPLC analysis of the liberated sphingosine. To determine haCerase activity of the homogenate, supernatant and solubilized precipitates of the (NH$_4$)$_2$SO$_4$ precipitation, the enzyme assays should be treated as follows. The reaction is stopped with 300 μl HPLC-grade methanol and 50 μl internal standard (500 pmol of C$_{14}$- and C$_{16}$-sphinganine). The mixtures are evaporated to dryness under a stream of nitrogen. The dried enzyme assays are then taken up in 300 μl of 100 mM methanolic KOH solution and incubated for 2 hr at 37°. The liberated sphingosine and the internal standards are extracted by the addition of 600 μl chloroform and 300 μl of 100 mM NH$_4$HCO$_3$ solution. The extraction procedure is repeated as described earlier and pooled organic phases are evaporated under a stream of nitrogen.

Quantification of Liberated Sphingosine

The dried extracts from the enzyme assays are solubilized in 50 μl HPLC-grade methanol and mixed for 5 min, followed by addition of 50 μl

[10] G. Schwarzmann, *Biochim Biophys. Acta* **529,** 106 (1978).

of the commercially available o-phthalaldehyde reagent solution Fluoroaldehyde (Pierce). Alternatively, o-phthalaldehyde reagent solution can be prepared as follows: 100 μl of an ethanol solution containing 5 mg of o-phtalaldehyde and 5 μl 2-mercaptoethanol is added to 9.9 ml of a 3% (w/v) boric acid solution, adjusted to pH 10.5 with KOH pellets.[11] The incubation assays are mixed for 5 min at room temperature and then diluted with 0.9 ml HPLC buffer [methanol/5 mM potassium phosphate, pH 7.0, 9/1 (v/v)] to a final volume of 1 ml. An aliquot of each sample (100–200 μl) is then injected onto a Purosphere RP18-e column (125 $\times$ 4 mm, Merck), equilibrated in HPLC buffer. The derivatized sphinganine and sphingosine bases are eluted isocratically (flow rate: 1 ml/min, Shimadzu LC-10AT solvent delivery system) and detected by a fluorescence detector (Shimadzu RF-535, excitation wavelength: 340 nm, emission wavelength: 435 nm).[12,13] Quantification of liberated sphingosine is performed with Windows-based standard liquid chromatography software (Class CR-10, Shimadzu).

Purification Procedure

Human placenta was chosen as the enzyme source due to the difficulty of obtaining sufficient amounts of postmortem human tissue. All purification steps are carried out at 4° unless stated otherwise. The flow rate during loading, washing, and elution is maintained at 30 ml/hr unless stated otherwise. All buffers contain 0.02% (w/v) NaN$_3$ as preservative. All fractions with at least 30% or more of maximum peak activity are pooled. Table I summarizes the results of the purification of haCerase using 15 human placentas as starting material.

Homogenization of Human Placenta and (NH$_4$)$_2$SO$_4$ Precipitation

Fresh human placentas (not older than 2 days) are collected from a local hospital and stored at $-80°$ until used. Human placentas (2.5–3 kg) are freed from the umbilical cord, adipose, and skin tissue, cut into small pieces, and homogenized briefly together with water and ice in a blender (final volume $\sim$8 liters). Further homogenization is performed with an Ultra Turrax in the presence of 0.1% (w/v) Nonidet P-40 (or the equivalent Igepal C-630, Sigma) for the additional solubilization of protein. After centrifugation (45 min, 10,000g, Sorvall RC-5B) fine-powdered (NH$_4$)$_2$SO$_4$ is added to the supernatant to 30% saturation and stirred over-

[11] M. Roth, *Anal. Biochem.* **43,** 880 (1971).
[12] H. W. Jarrett, K. D. Cooksy, B. Ellis, and J. M. Anderson, *Anal. Biochem.* **153,** 189 (1986).
[13] A. H. Merrill, E. Wang, R. E. Wang, W. C. L. Jamisson, S. Nimkar, and D. C. Liotta, *Anal. Biochem.* **171,** 373 (1988).

TABLE I
PURIFICATION OF ACID CERAMIDASE FROM HUMAN PLACENTA

Purification step	Protein (mg)	Volume (ml)	Activity (nmol hr^{-1})	Specific activity (nmol mg^{-1} hr^{-1})	Yield (%)	Enrichment (fold)
Homogenate[a]	228,224	8600	221,396	0.97	100	1
30–60% (NH$_4$)$_2$SO$_4$ precipitation[a]	47,300	860	155,060	3.27	79	3.3
Con A–Sepharose	1,370	250	52,334	38.2	24	39
Octyl–Sepharose	340	86	38,080	112	17	115
Matrex gel red A[b]	9.6	24	12,316	1,283	5.5	1,322
DEAE–Fractogel[b,c]	2.8	4	1,363	486	0.6	501

[a] In order to determine acid ceramidase activity in the homogenate and solubilized pellet of 30–60% (NH$_4$)$_2$SO$_4$ precipitation, an alkaline hydrolysis step (see text) was included in the preparation of the incubation mixture for HPLC analysis.
[b] Before concentration with a Centricon 50 microconcentrator (Amicon).
[c] HaCerase containing fractions of DEAE chromatography flow through.

night. Following centrifugation (45 min, 10,000g, Sorvall RC-5B), (NH$_4$)$_2$SO$_4$ is added to the supernatant to 60% saturation and the suspension is stirred overnight. The precipitate, which contains haCerase activity, is collected by centrifugation (45 min, 10,000g, Sorvall RC-5B) and is dissolved in approximately 800 ml buffer A (30 mM Tris–HCl, pH 7.2, 500 mM NaCl, 1 mM Mg/Ca/MnCl$_2$, 1 μM leupeptin, 0.1% w/v NP-40). The solubilized precipitate is clarified by ultracentrifugation (30 min, 235,000g, Beckmann J2-21M) and sterile filtration (0.2 μm).

Step 1: Concanavalin A (Con A)–Sepharose Chromatography. The dissolved (NH$_4$)$_2$SO$_4$ precipitate is loaded onto a Con A–Sepharose column (2.6 × 20 cm, Sigma) equilibrated with buffer A. The column is washed with at least 10 column volumes of buffer A to remove nonspecifically bound proteins. Bound glycoproteins are eluted in reverse direction at room temperature with a linear gradient of 0–20% (w/v) α-D-methylglucopyranoside in buffer A (2 × 200 ml).

Step 2: Octyl–Sepharose Chromatography. Pooled fractions of the Con A–Sepharose eluate are loaded onto an Octyl–Sepharose column (2.6 × 16 cm, Pharmacia) equilibrated in buffer B (30 mM Tris–HCl, pH 7.2, 100 mM NaCl). The column is washed with at least 10 column volumes of buffer B. Proteins are eluted in the reverse direction with a linear gradient of 0–1% w/v NP-40 in buffer B (2 × 120 ml).

Step 3: Matrex Gel Red A Chromatography. Combined fractions of the Octyl–Sepharose eluate are loaded onto a Matrex gel red A column (1.6 × 10 cm, Amicon) equilibrated in buffer C (30 mM Tris–HCl, pH 7.2,

0.1% w/v NP-40). The column is first washed with at least 5 column volumes of buffer C and is then equilibrated with five column volumes of buffer D (30 m*M* Tris–HCl, pH 7.2, 0.1% w/v β-D-octylglucopyranoside). Detergent exchange is necessary to facilitate subsequent concentration and desalting steps using Centricon 50 microconcentrators (Amicon). Bound proteins are then eluted in reverse direction with a linear gradient of 0–1.5 *M* NaCl in buffer D (2 × 75 ml).

Step 4: Anion-Exchange Chromatography. Pooled fractions from the Matrex gel red A eluate are concentrated ~10-fold, desalted, and equilibrated with buffer D using Centricon 50 microconcentrators (Amicon). This step will result in a further increase in specific haCerase activity due to the removal of most proteins smaller than 50 kDa. The concentrated protein solution is applied to a Fractogel EMD DEAE-650 (S) column (1 × 2 cm, Merck) equilibrated with buffer D. Purified haCerase can be collected in the first fractions of the flow through under these conditions. Residual bound haCerase activity is eluted with a linear gradient of 0–1 *M* NaCl (w/v) in buffer D (2 × 30 ml). HaCerase-containing fractions are pooled, concentrated (final concentration ~1 mg/ml) in a Centricon 50 microconcentrator (Amicon), and stored at −80° until needed.

Additional Remarks

This purification protocol will result in a copurification of acid ceramidase, acid sphingomyelinase (haSMase), and SAP-D in the early chromatographic steps. Separation of haCerase from haSMase is achieved during Matrex gel red A chromatography. The early fractions that contain haCerase are nearly haSMase free, whereas haSMase elutes at higher salt concentrations. Additional purification of haSMase is achieved by anion exchange or by immunoaffinity chromatography as described by Lansmann *et al.*[14]

Copurification of SAP-D with acid ceramidase was detected by Western blot analysis of pooled fractions from Octyl–Sepharose and Matrex gel red A chromatography with anti-SAP-D antibodies.[8] Separation of SAP-D and haCerase is achieved by anion-exchange chromatography (step 4). Because of the low isoelectric point (p*I*) value of SAP-D of 4.8,[15] SAP-D binds strongly to the anion exchanger under the conditions used for haCerase purification. The separation of SAP-D from the haCerase preparation is accompanied by a decrease in the specific activity of haCerase, as seen in Table I.

[14] S. Lansmann, K. Ferlinz, R. Hurwitz, O. Bartelsen, G. Glombitza, and K. Sandhoff, *FEBS Lett.* **399,** 227 (1996).
[15] W. Fürst and K. Sandhoff, *Biochim. Biophys. Acta* **1126,** 1 (1992).

Properties of Placental Acid Ceramidase

Purified placental haCerase is a heterodimeric glycoprotein with an apparent molecular mass of 50 kDa as judged by nonreducing, Tris–Tricine SDS–PAGE. Under reducing conditions, haCerase separates into two subunits, designated α ($M_r \sim 13{,}000$) and β ($M_r \sim 40{,}000$). Complete deglycosylation with peptide N-glycanase F reduces the apparent molecular mass of the β subunit to 28 kDa. The α subunit is not glycosylated.

Placental haCerase has a pI value of 6.8 and shows the highest catalytic activity in the pH range of 3.7–4.2. In the described detergent-based assay system, haCerase displays the highest catalytic activity toward the substrate N-lauroylsphingosine. Decreasing degradation rates are observed with shorter or longer acyl chain ceramide derivatives. Stimulation of enzymatic activity by the addition of nonionic detergents such as NP-40 or Triton X-100 and cholates is caused by increased solubilization of the highly hydrophobic substrate ceramide in aqueous solution. The purification procedure must be performed in the presence of detergents for the complete extraction of haCerase from tissue and to minimize adsorptive losses of the hydrophobic enzyme. HaCerase can be stored at $-80°$ for months without significant loss of enzymatic activity.

Acknowledgments

This work was supported by a grant from the Graduiertenkolleg "Krankheiten des Zentralen Nervensystems" to Thomas Linke and a research grant from the Deutsche Forschungsgemeinschaft (SFB 400). We thank Frau Heidi Moczall for excellent technical assistance.

[24] Ceramide Kinase

By SANDRA BAJJALIEH and ROBERT BATCHELOR

The generation of ceramide from sphingomyelin constitutes a unique signaling pathway that, like the glycerol lipid signaling pathways, regulates many cellular processes.[1] Ceramide has at least two effectors: It activates protein phosphatase 2A[2] and induces the activation of a proline-directed protein kinase.[3,4] Ceramide is likely to have other actions as well. For

[1] Y. A. Hannun, J. Biol. Chem. 269, 3125 (1994).
[2] R. T. Dobrowsky, C. Kamibayashi, M. C. Mumby, and Y. A. Hannun, J. Biol. Chem. 268, 15523 (1993).
[3] R. N. Kolesnick and M. Kronke, Annu. Rev. Physiol. 60, 643 (1998).
[4] Y. A. Hannun, Science 274, 1855 (1996).

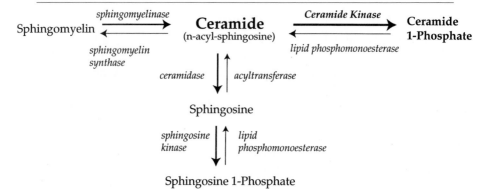

FIG. 1. Position of ceramide kinase in the sphingolipid signaling cascade. The hydrolysis of sphingomyelin generates the second messenger ceramide, which can be modified further to produce two other lipids that act as intracellular signals, sphingosine and sphingosine 1-phosphate. The phosphorylation of ceramide by ceramide kinase produces the novel lipid ceramide 1-phosphate, which is hypothesized to act as a unique intracellular signal.

example, its role in regulating membrane charge and fusability is just beginning to be explored.

In addition to acting as a signaling molecule, ceramide is also a substrate for modifying enzymes that generate other lipid messengers (Fig. 1). Ceramidases remove the amide-acyl moiety of ceramide producing sphingosine, an inhibitor of several protein kinases, including protein kinase C and CAM kinase II.[5,6] Phosphorylation of sphingosine generates sphingosine phosphate which, in addition to inhibiting protein kinases, also has unique actions including the regulation of intracellular calcium channels[7] and the inhibition of apoptotic processes.[8]

Ceramide undergoes a cycle of phosphorylation and dephosphorylation, the function of which is unknown. It has been hypothesized that phosphorylation of ceramide inactivates it as a signaling molecule in the same way that the phosphorylation of diacylglycerol stops the activation of protein kinase C.[9] However, it is likely that ceramide phosphate acts as a unique signaling molecule. The calcium-stimulated production of ceramide 1-phosphate in brain synaptic vesicle membranes suggests that the generation of ceramide phosphate plays a role in regulating the secretion of neurotransmitters.[10] One hypothesized action of ceramide 1-phosphate in the synapse

[5] Y. A. Hannun and R. M. Bell, *Science* **243**, 500 (1989).

[6] A. B. Jefferson and H. Schulman, *J. Biol. Chem.* **263**, 15241 (1988).

[7] T. K. Ghosh, J. Bian, and D. L. Gill, *Science* **248**, 1653 (1990).

[8] O. Cuvillier, G. Pirianov, B. Kleuser, P. G. Vanek, O. A. Coso, J. S. Gutkind, and S. Spiegel, *Nature* **381**, 800 (1996).

[9] K. A. Dressler and R. N. Kolesnick, *J. Biol. Chem.* **265**, 14917 (1990).

[10] S. M. Bajjalieh, T. F. Martin, and E. Floor, *J. Biol. Chem.* **264**, 14354 (1989).

is to promote secretion by increasing the fusability of vesicle membranes.[11] This hypothesis is supported by the observation that the addition of ceramide 1-phosphate to liposomes shifts the rate and extent of their calcium-dependent fusion.[12] Ceramide 1-phosphate levels may also regulate cell proliferation, a hypothesis suggested by the observation that exogenous ceramide 1-phosphate increases DNA synthesis in fibroblasts.[13] While this effect could result from conversion of ceramide 1-phosphate to the mitogen sphingosine 1-phosphate, the observation that ceramide 1-phosphate is not a substrate for brain ceramidases[11] suggests that it acts directly to stimulate proliferation.

Ceramide kinase was initially described as a calcium-stimulated lipid kinase that copurifies with brain synaptic vesicles.[10] Purification of synaptic vesicles leads to an enrichment of ceramide kinase activity and a loss of diacylglycerol kinase (DGK) activity, suggesting that the enzyme associated with synaptic vesicles is specific for ceramide. Ceramide-specific kinase activities have also been reported in HL-60 cells[9] and in neutrophils.[12] As with the synaptic vesicle enzyme, enrichment of ceramide kinase from these cells results in a reduction of DGK activity.

The ceramide kinase activities reported to date share a number of features. Both the synaptic vesicle and HL-60 cell enzymes catalyze the generation of ceramide 1-phosphate (vs ceramide 3-phosphate).[10,14] All three are membrane-associated and demonstrate neutral pH optima[12,14] (R. Batchelor and S. Bajjalieh, unpublished observations). Finally, all ceramide kinases described are stimulated by calcium.

However, several differences between the reported ceramide kinase activities suggest that they represent distinct isoforms. First, the apparent K_m for ceramide differs between the synaptic vesicle and HL-60 forms (4 μM for the synaptic vesicle enzyme and 45 μM for the HL-60 cell enzyme),[10,14] although this may reflect variations in assay conditions such as differences in lipid substrate presentation. Second, while all ceramide kinases are stimulated by calcium, the extent and EC_{50} of calcium stimulation differ between neutrophil and synaptic vesicle/HL-60 cell enzymes. Third, the activities demonstrate differential sensitivities to magnesium. Synaptic vesicle ceramide kinase is stimulated by both calcium and magnesium.[10] HL-60 cells appear to contain separate calcium- and magnesium-stimulated activities[14] whereas neutrophil ceramide kinase is insensitive to magnesium.[12] In addi-

[11] R. Shinghal, R. H. Scheller, and S. M. Bajjalieh, *J. Neurochem.* **61,** 2279 (1993).
[12] V. T. Hinkovska-Galcheva, L. A. Boxer, P. J. Mansfield, D. Harsh, A. Blackwood, and J. A. Shayman, *J. Biol. Chem.* **273,** 33203 (1998).
[13] A. Gomez-Munoz, P. A. Duffy, A. Martin, L. O'Brien, H.-S. Byun, R. Bittman, and D. N. Brindley, *Mol. Pharm.* **47,** 883 (1995).
[14] R. N. Kolesnick and M. R. Hemer, *J. Biol. Chem.* **265,** 18803 (1990).

tion, the presence of ceramide kinase activity in brain cytosol (S. Bajjalieh and R. Batchelor, unpublished observations) suggests either the existence of a soluble form of the enzyme or one that translocates between cytosol and the membrane compartments. Resolution of these apparent differences awaits the purification of ceramide kinases and cloning of the cDNAs that encode them.

Assaying Ceramide Kinase Activity

Ceramide kinase activity is assayed by measuring the phosphorylation of exogenous ceramide presented in mixed detergent/lipid micelles. The assay is a variation of the diacylglycerol kinase assay developed by Priess et al.[15] to quantitate diacylglycerol.

1. To generate substrate micelles, transfer 50 μg/reaction of brain ceramides to a glass test tube and dry under nitrogen. Add 20 μl/reaction of cardiolipin/β-octylglucoside micelles (see next paragraph), vortex the mixture, and sonicate in a bath sonicator until clear. We have found that the time required to fully incorporate ceramide into micelles varies with the sonicator used. Substrate micelles are incubated on ice until use.

Cardiolipin/β-octylglucoside micelles [5 mM cardiolipin, 7.5% β-octylglucoside in 1 mM diethylenetriaminepentaacetic acid (DETAPAC), pH 6.6] are generated by mixing 15 mg dry cardiolipin, 150 mg β-octylglucoside, 20 μl 0.1 M DETAPAC, pH 6.6, and 1980 μl water. This mixture is vortexed and then sonicated in a bath sonicator until the cardiolipin goes into solution (5–50 min depending on the energy and heat used in sonication). Micelles are stored under nitrogen at $-20°$. We have found that the source of cardiolipin is crucial, as minor contaminants in some preparations inhibit enzyme activity. Preparations of cardiolipin suitable for the assay can be obtained from Avanti Polar Lipids (Alabaster, AL).

2. Ceramide kinase reactions (100 μl final volume) are assembled by mixing 20 μl substrate micelles with 20 μl 5× reaction buffer [75 mM MOPS, pH 7.2, 250 mM NaCl, 15 mM CaCl$_2$, 2.5 mM dithiothreitol (DTT), 5 mM EGTA, pH 7.5] and 50 μl of enzyme (in 10 mM MOPS, 2 mM EGTA, 1 mM DTT) to yield reactions containing 880 μM ceramides, 1 mM cardiolipin, 1.5% β-octylglucoside, 0.2 mM DETAPAC, 20 mM MOPS, pH 7.2, 50 mM NaCl, 1 mM DTT, 2 mM EGTA, and 3 mM CaCl$_2$ yielding 1 mM free calcium as calculated using the MaxChelator program. The reaction is initiated with the addition of 10 μl 5 mM magnesium, 5

[15] J. Preiss, C. R. Loomis, W. R. Bishop, R. Stein, J. E. Niedel, and R. M. Bell, J. Biol. Chem. **261**, 8597 (1986).

mM γ-^{32}P-ATP (500 μM final concentration, 40–100 mCi/mmol, 2–5 μCi/ reaction) and is incubated at 30° for 15 min.

3. Reactions are stopped with the addition of 600 μl chloroform and 600 μl methanol. Lipids are extracted by vortexing for 20 sec. Phases are broken by adding 530 μl 1 M KCl in 20 mM MOPS, pH 7.2, vortexing for 30 sec, and centrifuging for 5 min. Four hundred microliters of the organic (lower) phase is removed with a Hamilton syringe and dried under nitrogen.

4. Lipids are resolved by thin-layer chromatography. Dried lipids are resuspended in 25 μl chloroform : methanol (95 : 5) and spotted onto silica gel 60 plates. Plates are resolved in chloroform : methanol : acetic acid (65 : 15 : 10). In this system, ceramide phosphate runs with an R_f of 0.5. Radiolabeled ceramide 1-phosphate is located by autoradiography and its identity confirmed by comparison to radiolabeled standards. The amount generated is quantitated by scraping and scintillation counting.

Ceramide 1-phosphate standards are generated with *Escherichia coli* diacylglycerol kinase, which phosphorylates both diacylglycerol and cer-amide.[16] One to 5 μg of recombinant enzyme (Calbiochem) is incubated in reactions identical to those used to assay ceramide with the following exceptions. EGTA is replaced with EDTA, and calcium is replaced with 10 mM MgCl$_2$ to yield 8 mM free magnesium as calculated using the MaxChelator program. It should be noted that some preparations of *E. coli* diacylglycerol kinase contain significant amounts of diacylglycerol and will therefore produce measurable amounts of phosphatidic acid, which migrates with an R_f of 0.73 in the solvent system used to resolve ceramide 1-phosphate.

The assay, as described, is linear for 20 min when used to measure ceramide kinase activity in both crude and partially purified samples. However, given the low specific activity of the ATP, the assay is not highly sensitive. For a more sensitive assay, the specific activity of ATP can be increased by decreasing the magnesium ATP concentration to 10 μM. Although this assay produces a robust signal, it is not linear with time beyond 2 min at 30° when high activity fractions are assayed and so is not appropriate for comparing activity between samples.

Factors Affecting Enzyme Activity

Endogenous Inhibitors

Some tissues may contain a lipidic inhibitor of ceramide kinase. This was first suggested by the observation that some preparations of cardiolipin

---

[16] E. G. Schneider and E. P. Kennedy, *J. Biol. Chem.* **248**, 3739 (1973).

do not support the assay and therefore might contain an inhibitory contaminant. Two other findings support this interpretation: (1) total measurable ceramide kinase activity is higher in partially purified membrane fractions than in crude tissue extracts and (2) higher concentrations of crude membrane fractions produce less measurable activity than lower concentrations, yielding an inverted U-shaped dose curve. This latter effect can be partially reversed by treating samples with low concentrations of detergent. Because of these potential inhibitors, partial purification may be required to measure ceramide kinase activity in tissues or cell lines of interest.

Salt

The assay will tolerate the addition of KCl to 0.4 M without a significant effect on activity. Higher concentrations inhibit enzyme activity.

Reducing Agents

Ceramide kinase activity is extremely sensitive to oxidation. To protect enzyme activity, a minimum of 1 mM DTT should be added fresh to all solutions.

Detergents

As stated earlier, measurable ceramide kinase activity in crude membrane fractions is increased by the addition of low concentrations of detergent. We have found 0.1% sodium cholate to be effective. However, the activity of partially purified ceramide kinase is inhibited by Triton X-100, Triton X-114, and sodium cholate at concentrations above the critical micelle concentration, perhaps due to effects on the mixed micelle substrate.

Storage

Brain ceramide kinase is stable for ≤6 mo. in tissue that is frozen rapidly and kept at −70°. It does not need to be prepared from fresh tissue. Once in solution, the enzyme is stable at 4° for a minimum of a week. Enzyme stability is facilitated by the periodic replenishing of reducing agents. Microsome preparations and crude extracts can be stored at −20° for weeks with little loss of activity. However, enzyme stability decreases with increasing purification.

Extraction of Ceramide Kinase from Membranes

All reported ceramide kinases are membrane associated.[10,12,14] Brain ceramide kinase is an extrinsic membrane protein. It can be extracted from membranes with 0.5 M KCl (S. Bajjalieh, unpublished) and partitions into the aqueous phase of Triton X-114 extracts (S. Bajjalieh, unpublished). We have found that the recovery in salt extracts is variable and can be affected by calcium concentrations during membrane isolation and extraction (R. Batchelor and S. Bajjalieh, unpublished), a finding that suggests calcium-regulated enzyme translocation.

Consistent extractions are obtained with Triton X-114, and this method is therefore suggested. To extract ceramide kinase from crude synaptic vesicle preparations, membranes are resuspended in 10 mM MOPS, pH 7.2, 2 mM EGTA, 150 mM KCl, 1 mM DTT, and 2% Triton X-114. The mixture is agitated at 4° for a minimum of 2 hr after which insoluble material is removed by centrifugation at 100,000g for 1 hr. The supernatant is layered over a cushion of 0.25 M sucrose, 10 mM MOPS, pH 7.2, 2 mM EGTA, and 1 mM DTT (20% of extract volume) in a conical culture tube and is heated to 30° for 15 min to induce clouding of the Triton X-114. The heated extract is centrifuged at 600g for 6 min. The aggregated detergent collects at the bottom of the tube beneath the sucrose cushion. The aqueous (top) phase contains the majority of the ceramide kinase activity. It should be noted that total recovered activity is reduced due to the inhibitory effects of residual detergent in the aqueous phase.

Separation of Ceramide Kinase from Diacylglycerol Kinase and Sphingosine Kinase Activities

Methods

Extracted ceramide kinase can be purified further by anion-exchange chromatography (Fig. 2). Interestingly, the ability of the enzyme to bind anion exchange resins varies with the method used to extract it from membranes. For example, the majority of ceramide kinase activity in salt extracts of synaptic vesicles does not bind quaternary-amine (Q) resins (S. Bajjalieh, unpublished). Likewise, Kolesnick and Hemer[14] found that HL-60 cell ceramide kinase extracted from membranes with β-octylglucoside does not bind DEAE resin. However, brain ceramide kinase in Triton X-114 extracts binds both DEAE and Mono-Q resins (S. Bajjalieh, unpublished). These observations suggest that ceramide kinase is part of a protein complex whose other members affect its surface charge.

Enrichment of ceramide kinase activity produces a loss of both diacylglycerol kinase and sphingosine kinase activities (Fig. 2). These results, together with earlier observations that ceramide kinase and diacylglycerol kinase activities do not cofractionate[10,14] and with the characterization of

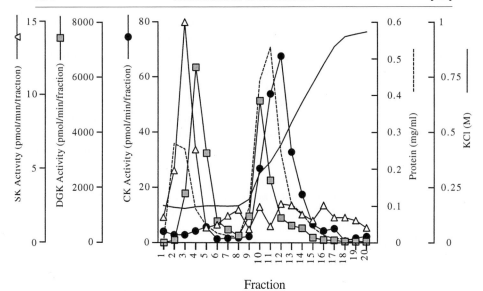

Fig. 2. Separation of ceramide kinase (CK) from diacylglycerol kinase (DGK) and sphingosine kinase (SK). Mono Q anion-exchange chromatography was performed on the aqueous
phase of a Triton X-114 extract of crude synaptic vesicles. Fractions were assayed for the
indicated kinase activity.

a sphingosine-specific kinase,[17] indicate that each of these lipid kinases is
specific. The specificity of lipid kinases is interesting in light of the nonspe-
cificity of two lipid phosphomonoesterases that have been purified.[18] This
suggests that specificity in lipid phosphorylation is regulated in a single direction.

Approximately 15 mg of protein is loaded onto a 5-ml HiTrap Mono
Q anion-exchange column (Pharmacia, Uppasala, Sweden) in MED buffer
(10 mM MOPS, pH 7.2, 2 mM EGTA, pH 7.5, 1 mM DTT) containing
150 mM KCl. The column is washed with 25 ml MED, 150 mM KCl and
then eluted with a 50-ml MED gradient containing 0.15 to 1 M KCl. The
column is eluted further with 25 ml MED, 1 M KCl. The flow rate at all
stages is 2 ml/min. Five-milliliter fractions are collected. Fifty μl of each
is assayed for ceramide kinase, diacylglycerol kinase, and sphingosine kinase
activities. All three enzymes are assayed as described for ceramide kinase,

[17] T. Kohama, A. Olivera, L. Edsall, M. M. Nagiec, R. Dickson, and S. Spiegel, *J. Biol. Chem.*
273, 23722 (1998).
[18] D. W. Waggoner, A. Gomez-Munoz, J. Dewald, and D. N. Brindley, *J. Biol. Chem.* **271**,
16506 (1996).

with the following modifications. The diacylglycerol kinase assay contains 50 μg of diacylglycerol as the lipid substrate, along with 2 mM EDTA, pH 8.0, and 5 mM MgCl$_2$ to yield 3 mM free MgCl$_2$, as calculated using the MaxChelator program. No calcium is added to the assay and the incubation time is decreased to 10 min at 30°. The sphingosine kinase assay contains 50 μg of sphingosine as the lipid substrate and is incubated for 20 min at 30°. Phosphatidic acid standards are made using *E. coli* diacylglycerol kinase as the enzyme source in a DGK assay. Sphingosine phosphate standards are generated by treatment of sphingomyelin with phospholipase D using the method of Zhang *et al.*[19] Protein was assayed with the Bradford protein assay (Bio-Rad) using bovine serum albumin as a standard.

Acknowledgment

We thank Dr. William Catterall for reviewing the manuscript. This work was funded by grants from the University of Washington Research Royalty Fund and the Sloan Foundation.

[19] H. Zhang, N. N. Desai, A. Olivera, T. Seki, G. Brooker, and S. Spiegel, *J. Cell. Biol.* **114,** 155 (1991).

[25] Assaying Sphingosine Kinase Activity

By ANA OLIVERA, KEITH D. BARLOW, and SARAH SPIEGEL

Introduction

Sphingosine kinase is a unique lipid kinase that catalyzes the phosphorylation of long chain sphingoid bases on their primary hydroxyl group. While this enzyme was originally characterized for its role in the catabolism of sphingoid bases to palmitaldehyde and phosphoethanolamine,[1] the potential physiological roles of sphingosine kinase and its product, sphingosine-1-phosphate (SPP), are receiving renewed attention with the discovery that SPP and other products of sphingolipid catabolism are biologically active compounds that mediate diverse cellular effects. Cellular processes that have been reported to be modulated by SPP include proliferation,[2-4] sur-

[1] W. Stoffel, G. Sticht, and D. LeKim, *Hoppe-Seyler's Z. Physiol. Chem.* **349,** 1149 (1968).
[2] H. Zhang, N. N. Desai, A. Olivera, T. Seki, G. Brooker, and S. Spiegel, *J. Cell Biol.* **114,** 155 (1991).
[3] A. Olivera and S. Spiegel, *Nature* **365,** 557 (1993).

vival,[5-7] organization of the cytoskeleton,[8,9] motility,[8,10,11] neurite retraction and cell rounding,[12,13] and differentiation.[14-16] Accumulation of SPP has also been shown to modulate several signaling pathways, including the activation of phospholipase D leading to the formation of phosphatidic acid,[17] activation of the Raf/MEK/ERK signaling cascade,[18] and mobilization of Ca^{2+} from internal stores via a mechanism that is independent of inositol lipid hydrolysis and arachidonic acid release.[19,20] Moreover, it has been shown that SPP has dual actions, acting intracellularly as a second messenger and extracellularly as a ligand for cell surface receptors.[21]

Several biological stimuli have been shown to activate sphingosine kinase and increase cellular levels of SPP, including platelet-derived growth factor (PDGF),[3,8] nerve growth factor (NGF),[14] protein kinase C (PKC) activators,[22,23] the B subunit of cholera toxin,[24] vitamin D_3,[6] activation of m2 and m3

[4] C. S. Rani, A. Berger, J. Wu, T. W. Sturgill, D. Beitner-Johnson, D. LeRoith, L. Varticovski, and S. Spiegel, *J. Biol. Chem.* **272**, 10777 (1997).

[5] O. Cuvillier, G. Pirianov, B. Kleuser, P. G. Vanek, O. A. Coso, S. Gutkind, and S. Spiegel, *Nature* **381**, 800 (1996).

[6] B. Kleuser, O. Cuvillier, and S. Spiegel, *Cancer Res.* **58**, 1817 (1998).

[7] O. Cuvillier, D. S. Rosenthal, M. E. Smulson, and S. Spiegel, *J. Biol. Chem.* **273**, 2910 (1998).

[8] K. E. Bornfeldt, L. M. Graves, E. W. Raines, Y. Igarashi, G. Wayman, S. Yamamura, Y. Yatomi, J. S. Sidhu, E. G. Krebs, S. Hakomori, and R. Ross, *J. Cell Biol.* **130**, 193 (1995).

[9] F. Wang, C. D. Nobes, A. Hall, and S. Spiegel, *Biochem. J.* **324**, 481 (1997).

[10] Y. Sadhira, F. Ruan, S. Hakomori, and Y. Igarashi, *Proc. Natl. Acad. Sci. U.S.A* **89**, 9686 (1992).

[11] F. Wang, K. Nohara, O. Olivera, E. W. Thompson, and S. Spiegel, *Exp. Cell Res.* **247**, 17 (1999).

[12] F. R. Postma, K. Jalink, T. Hengeveld, and W. H. Moolenaar, *EMBO J.* **15**, 2388 (1996).

[13] J. R. Van Brocklyn, Z. Tu, L. Edsall, R. R. Schmidt, and S. Spiegel, *J. Biol. Chem.* **274**, 4626 (1999).

[14] L. C. Edsall, G. G. Pirianov, and S. Spiegel, *J. Neurosci.* **17**, 6952 (1997).

[15] R. A. Rius, L. C. Edsall, and S. Spiegel, *FEBS Lett.* **417**, 173 (1997).

[16] F. S. Lee, J. Hagler, Z. J. Chen, and T. Maniatis, *Cell* **88**, 213 (1997).

[17] N. N. Desai, H. Zhang, A. Olivera, T. Seki, G. Brooker, and S. Spiegel, *J. Biol. Chem.* **267**, 23122 (1992).

[18] J. Wu, S. Spiegel, and T. W. Sturgill, *J. Biol. Chem.* **270**, 11484 (1995).

[19] M. Mattie, G. Brooker, and S. Spiegel, *J. Biol. Chem.* **269**, 3181 (1994).

[20] A. Melendez, R. A. Floto, A. J. Cameron, D. J. Gilooly, M. M. Harnett, and J. M. Allen, *Curr. Biol.* **8**, 210 (1998).

[21] J. R. Van Brocklyn, M. J. Lee, R. Menzeleev, A. Olivera, L. Edsall, O. Cuvillier, D. M. Thomas, P. J. P. Coopman, S. Thangada, T. Hla, and S. Spiegel, *J. Cell Biol.* **142**, 229 (1998).

[22] N. Mazurek, T. Megidish, S.-I. Hakomori, and Y. Igarashi, *Biochem. Biophys. Res. Commun.* **198**, 1 (1994).

[23] B. M. Buehrer, E. S. Bardes, and R. M. Bell, *Biochim. Biophys. Acta* **1303**, 233 (1996).

[24] F. Wang, N. E. Buckley, A. Olivera, K. A. Goodemote, Y. Su, and S. Spiegel, *Glycoconj. J.* **13**, 937 (1996).

muscarinic cholinergic receptors by carbachol,[25] and cross-linking of FcεR1[26] and FcγR1 antigen receptors.[27] Based on our previous study of the activation of sphingosine kinase by growth- and survival-promoting agents, we have proposed that SPP is a novel intracellular second messenger.[28] Intracellular levels of SPP in unstimulated cells typically range from 0.5 to 20 pmol per million cells, depending on the cell type and the method used for quantitation,[29,30] and are determined by the balance of sphingosine kinase-mediated formation and its degradation by a membrane-associated phosphatase[31,32] and a pyridoxal phosphate-dependent lyase in the endoplasmic reticulum.[33,34] The potential role for SPP as a second messenger induced by growth factor signaling was first described in quiescent Swiss 3T3 fibroblasts. Exposing these cells to PDGF or serum produced a transient increase in sphingosine kinase activity and a rapid increase in SPP levels that was not seen with other mitogens, such as epidermal growth factor (EGF).[3] PDGF has subsequently been shown to increase SPP levels in both arterial[8] and airway smooth muscle cells,[35] further implicating a role for sphingosine kinase in the mitogenic signaling pathways induced by this growth factor. Moreover, competitive inhibitors of sphingosine kinase, such as DL-*threo*-dihydrosphingosine (DHS) and *N,N*-dimethylsphingosine (DMS), prevent the formation of SPP and block cellular proliferation induced by either PDGF or serum, whereas proliferation induced by EGF is unaffected.[3,4]

We have also shown that NGF can stimulate sphingosine kinase in PC12 cells by signaling through the TrkA receptor.[14] DMS blocks the cytoprotective effects of NGF,[30] indicating that endogenous SPP may have an important role in mediating survival as well as mitogenesis. Given that SPP can antagonize apoptosis mediated by ceramide, which is known to accumulate as a

[25] D. Meyer zu Heringdorf, H. Lass, R. Alemany, K. T. Laser, E. Neumann, C. Zhang, M. Schmidt, U. Rauen, K. H. Jakobs, and C. J. van Koppen, *EMBO J.* **17,** 2830 (1998).

[26] O. Choi, J.-H. Kim, and J.-P. Kinet, *Nature* **380,** 634 (1996).

[27] A. Melendez, R. A. Floto, D. J. Gillooly, M. M. Harnett, and J. M. Allen, *J. Biol. Chem.* **273,** 9393 (1998).

[28] S. Spiegel, O. Cuvillier, L. C. Edsall, T. Kohama, R. Menzeleev, Z. Olah, A. Olivera, G. Pirianov, D. M. Thomas, Z. Tu, J. R. Van Brocklyn, and F. Wang, *Ann. N.Y. Acad. Sci.* **845,** 11 (1998).

[29] T. Yatomi, F. Ruan, H. Ohta, R. J. Welch, S. I. Hakomori, and Y. Igarashi, *Anal. Biochem.* **230,** 315 (1995).

[30] L. C. Edsall, J. R. Van Brocklyn, O. Cuvillier, B. Kleuser, and S. Spiegel, *Biochemistry* **37,** 12892 (1998).

[31] P. P. Van Veldhoven and G. P. Mannaerts, *Biochem. J.* **299,** 597 (1994).

[32] S. Mandala, R. Thornton, Z. Tu, M. Kurtz, J. Nickels, J. Broach, R. Menzeleev, and S. Spiegel, *Proc. Natl. Acad. Sci. U.S.A* **95,** 150 (1998).

[33] P. P. Van Veldhoven and G. P. Mannaerts, *J. Biol. Chem.* **266,** 12502 (1991).

[34] J. Zhou and J. D. Saba, *Biochem. Biophys. Res. Commun.* **242,** 502 (1998).

[35] S. Pyne, J. Chapman, L. Steele, and N. J. Pyne, *Eur. J. Biochem.* **237,** 819 (1996).

result of a variety of stress responses,[36,37] we have also proposed that the intracellular ratio of these two sphingolipid metabolites may be a sensor that regulates the life/death decision of cells.[5] According to this hypothesis, stress stimuli induce the accumulation of ceramide and sphingosine, whereas growth and survival signals activate sphingosine kinase to increase levels of SPP. In support of this concept, it has been shown that unfertilized mouse oocytes exposed to therapeutic levels of the antitumor drug doxorubicin undergo ceramide-mediated apoptosis that is inhibited by SPP.[38] Several laboratories have also reported that sphingolipid metabolites modulate stress responses in yeast, suggesting that the sphingolipid biostat functions as an evolutionarily conserved regulatory mechanism. In particular, two genes have been identified in *Saccharomyces cerevisiae,* LBP1 and its homolog LBP2, that encode hydrophobic proteins that act as specific phosphatases to phosphorylated sphingoid bases.[32] Deletion of LBP1 and LBP2 results in increased levels of long chain phosphates, in decreased levels of ceramide, and in dramatically enhanced survival on heat shock.[32] The discovery of spontaneous yeast mutants with diminished sphingosine kinase activity that have reduced growth rates and increased sensitivity to heat stress further indicates that sphingosine kinase has an important role in cell growth and survival.[39]

Purification and cloning of mammalian sphingosine kinases should further facilitate the study of the physiological functions of SPP and its potential roles as a second messenger. Rat sphingosine kinase, purified 6×10^5-fold to apparent homogeneity, has an apparent molecular weight of approximately 49,000 and K_m values of 5 and 93 μM for sphingosine and ATP, respectively.[40] Based on peptide sequences derived from the purified rat protein, the first mammalian sphingosine kinases (SPHK1a and SPHK1b) were subsequently cloned from the mouse.[41] Sequence analyses indicate that SPHK1a and SPHK1b are unrelated to other known kinases, yet comparison of the predicted murine sequences to sphingosine kinase sequences cloned from *S. cerevisiae*[42] and *Caenorhabditis elegans* reveals several domains that are highly conserved in all of these proteins.[41] Northern blot analysis of mouse tissues showed that sphingosine kinase mRNA was most abundant in the adult lung

[36] P. Santana, L. A. Peña, A. Haimovitz-Friedman, S. Martin, D. Green, M. McLoughlin, C. Cordon-Cardo, E. H. Schuchman, Z. Fuks, and R. Kolesnick, *Cell* **86,** 189 (1996).

[37] Y. Hannum, *Science* **274,** 1855 (1996).

[38] G. I. Perez, C. M. Knudson, L. Leykin, S. J. Korsmeyer, and J. L. Tilly, *Nature Med.* **3,** 1228 (1997).

[39] M. M. Lanterman and J. D. Saba, *Biochem. J.* **332,** 525 (1998).

[40] A. Olivera, T. Kohama, Z. Tu, S. Milstien, and S. Spiegel, *J. Biol. Chem.* **273,** 12576 (1998).

[41] T. Kohama, A. Olivera, L. Edsall, M. M. Nagiec, R. Dickson, and S. Spiegel, *J. Biol. Chem.* **273,** 23722 (1998).

[42] R. C. Dickson, E. E. Nagiec, M. Skrzypek, P. Tillman, G. B. Wells, and R. L. Lester, *J. Biol. Chem.* **272,** 30196 (1997).

and spleen, while being barely detectable in skeletal muscle and liver. Transient transfections of either human embryonic kidney cells or NIH 3T3 fibroblasts with expression vectors encoding either of the murine sphingosine kinase isozymes increased cellular sphingosine kinase activity more than 100-fold. Transfected cells demonstrated a marked increase in the mass of SPP, with a concomitant decrease in the amounts of sphingosine and, to a lesser extent, ceramide.[41] Crude cell and tissue preparations, purified rat sphingosine kinase, and the cloned mouse enzymes demonstrated a similar degree of substrate specificity by phosphorylating D-*erythro*-sphingosine, but not phosphatidylinositol, diacylglycerol, ceramide, or the competitive inhibitors, DMS and DHS.

This article describes a simple and accurate method by which the activity of sphingosine kinase can be measured. We have successfully used this method to establish the responsiveness of sphingosine kinase to several biologically relevant factors, including PDGF,[3,4] NGF,[14,15] the B subunit of cholera toxin,[24] and vitamin D_3,[6] as well as to determine the specific activity of sphingosine kinase at different steps during purification.[40]

Sphingosine Kinase Assay

The recent accumulation of experimental evidence implicating SPP as an important mediator of several cellular processes has increased the need for a method that can accurately and sensitively measure sphingosine kinase activity and its activation by diverse stimuli. Several assays of varying complexity and reproducibility have previously been used to determine sphingosine kinase activity.[43-45] The major problem with these methods has been the separation of product from substrate, as SPP is to some extent soluble in both polar and nonpolar solvents. In addition, some assays have utilized [^{3}H]sphingosine as substrate, which reduces the sensitivity of the assay and, consequently, its versatility for different applications, given the relatively low specific activity of the commercially available substrate. Moreover, the synthesis of [^{3}H]sphingosine is not a simple task and commercially available material is costly.

As an alternative, we have developed a simple, sensitive, reproducible, and relatively rapid method by which sphingosine kinase activity can be measured using [γ-^{32}P]ATP and sphingosine as substrates. In brief, the samples to be assayed are incubated at 37° with sphingosine and [γ-^{32}P]ATP. The ^{32}P-labeled lipids are then solvent extracted in acidic conditions. Seventy to 80% of the labeled SPP partitions into the organic phase, whereas most of the

[43] W. Stoffel and G. Assmann, *Hoppe-Seyler's Z. Physiol. Chem.* **351,** 1041 (1970).
[44] D. D. Louie, A. Kisic, and G. J. Schroepfer, *J. Biol. Chem.* **251,** 4557 (1976).
[45] B. M. Buehrer and R. M. Bell, *J. Biol. Chem.* **267,** 3154 (1992).

unreacted [γ-^{32}P]ATP remains in the aqueous phase. Extracted phospholipids are then resolved by thin-layer chromatography (TLC), visualized by autoradiography, and quantified using either a scintillation counter or a phosphoimager.

Materials

D-*erythro* sphingosine from Matreya or Sigma is stored as a 50 mM solution in ethanol. This solution is stable for months when stored at −20° in a screw-top glass tube under N$_2$ to prevent oxidation. Tissue culture-grade, fatty acid-free bovine serum albumin (Sigma) is prepared as a 4-mg/ml stock solution. High purity adenosine trisphosphate is available from Boerhinger Mannheim. High specific activity (10 mCi/ml) [γ-^{32}P]ATP is obtained from NEN or Amersham.

The buffer for the sphingosine kinase assay is typically prepared in 10-ml aliquots and stored at −70° for 1–2 weeks. The assay buffer is prepared by mixing 200 μl 1 M Tris–HCl, pH 7.4, 100 μl 0.1 M EDTA, 50 μl 100 mM deoxypyridoxine, 150 μl 1 M sodium fluoride, 0.7 μl 2-mercaptoethanol, 50 μl 0.2 M sodium orthovanadate, 10 μl 10 mg/ml leupeptin and aprotinin, 10 μl 10 mg/ml trypsin inhibitor, 86.4 mg β-glycerolphosphate, 20 μl 0.2 M phenylmethylsulfonyl fluoride (PMSF), 4 ml 50% glycerol, and 5.36 ml of water.

The assay is conducted in 10- to 15-ml disposable conical glass centrifuge tubes with screw caps. [^{32}P]SPP is analyzed on 10 × 20-cm silica TLC plates from Merck.

Preparation of Cell Lysates

Cells may be cultured in either 10-cm plates or 6-well cluster plates. To test the effect of specific extracellular factors on sphingosine kinase, cells are either grown to confluency if they arrest in G$_0$ by contact inhibition or serum starved overnight (or both). If the extracellular factor to be tested is a ligand for a tyrosine kinase receptor, it is helpful to incubate the cells in the presence of 50 μM sodium orthovanadate, a phosphatase inhibitor, both for 30 min prior to the addition of the factor to be tested and during the incubation period. After stimulation with the factor, the cells are placed on ice and washed twice with an appropriate volume of cold phosphate-buffered saline (PBS). Next, a maximum of 500 μl of sphingosine kinase assay buffer is added to the cells, the cells are harvested by scraping, and transferred to a microfuge tube. The cells are lysed by repeated freezing and thawing and lysates are then transferred to prechilled ultracentrifuge tubes and spun at 105,000g for 90 min at 4° to obtain the cytosolic fraction. Supernatants can be stored at −70° for several weeks, but longer term storage results in loss of activity.

In the majority of the cell types that we have tested, most of the sphingosine kinase activity is present in the cytosolic fraction. However, some of the activity, ranging from 10 to 30%, can also be detected in the particulate fraction. The putative amino acid sequence of mouse sphingosine kinase indicates that this enzyme should be localized in the cytosol. In addition, when sphingosine kinase is overexpressed in various cell lines, most of the activity is associated with the cytosol. The sequence of yeast sphingosine kinase also suggests that the enzyme is a cytosolic protein, but surprisingly, most of the yeast sphingosine kinase activity is associated with membrane fractions.

Preparation of Sphingosine Substrate

Sphingosine can be solubilized as either a sphingosine–BSA complex or in mixed micelles with Triton X-100. Sphingosine–BSA complexes are preferred because Triton X-100 can affect sphingosine kinase activity from some types of cells. Sphingosine kinase purified from rat kidneys and sphingosine kinase cloned from mouse are both activated by Triton X-100.

A 1 mM solution of sphingosine complexed with BSA is prepared as follows: (i) an appropriate volume of a 4-mg/ml solution of BSA is transferred to a glass tube; (ii) the appropriate volume of 50 mM stock solution of sphingosine is added to the tube drop by drop, with vortexing, to make a 1 mM solution; and (iii) the resulting sphingosine–BSA solution is then sonicated in a bath sonicator for 1–2 min. The solution may appear to be slightly cloudy, but no particulate matter should be present.

To prepare a 1 mM solution of sphingosine in Triton X-100 micelles, the appropriate volume of 50 mM sphingosine is added to 5% Triton X-100 and vortexed briefly. A clear solution should result. Both types of preparation of sphingosine are stable at $-20°$ for several months.

Preparation of ^{32}P-ATP/Mg^{2+} Substrate

The ATP substrate is freshly prepared daily and contains 20 mM ATP and 200 mM MgCl$_2$. Immediately before initiating the sphingosine kinase reaction, nine parts of unlabeled ATP/Mg^{2+} solution are mixed with one part of [γ-^{32}P]ATP (approximately 10 μCi/sample), vortexed, and kept on ice until used.

Sphingosine Kinase Reaction

Glass conical tubes are placed on ice and suitable volumes of the cell extract to be tested are added. The optimal protein concentration for assaying Swiss 3T3 fibroblast extracts is 40–120 μg, but this must be determined independently for each cell type and experimental conditions. A sufficient volume of sphingosine kinase assay buffer is then added to bring the total volume to 180 μl. Ten microliters of 1 mM sphingosine (final concentration 50 μM) is then added as either a sphingosine–BSA complex or as sphingosine–Triton X-100 micelles. The tubes are vortexed gently and the reactions are started by the addition of 10 μl of [^{32}P]ATP substrate solution and the tubes are placed in a 37° water bath. For most cell types, 30-min incubations are sufficient, but the linearity of the reaction must be established for each cell type.

Separation of [^{32}P]SPP and Quantitation of SPHK Activity

The sphingosine kinase reaction is terminated by placing the tubes on ice and adding 20 μl of 1 M HCl to each sample, followed by the addition of 0.8 ml of chloroform/methanol/HCl (100 : 200 : 1, v/v). The tubes are then vortexed vigorously and after 5–10 min at room temperature, 240 μl of chloroform and 240 μl of 2 M KCl are added for phase separation. The tubes are vortexed vigorously, and after 5–10 min at room temperature, centrifuged again at 1000g for 5–10 min. The aqueous (upper) layer containing unreacted [^{32}P]ATP is removed by aspiration. An aliquot (50–100 μl) of the organic (lower) phase is applied to a TLC plate at a distance of 2 cm from the bottom and with a separation between samples of at least 0.5 cm. Drying of the applied sample while spotting can be facilitated either by gently applying heat using a hair dryer on its lowest setting or by warming on a hot plate. Standard SPP is applied in one lane. The TLC plate is developed in 1-butanol/methanol/acetic acid/water (80 : 20 : 10 : 20, v/v) by ascending chromatography until the solvent front almost reaches the top of the plate. The plate is then removed dried in a fume hood. The SPP standard is visualized by spraying the plate with ninhydrin, and labeled SPP is quantified using either autoradiography and a scintillation counting of scraped spots or with a phosphoimager. Exposing autoradiography film to the TLC plate overnight or exposing a phosphor screen for at least 1 hr is usually sufficient.

An aliquot of the [^{32}P]ATP substrate solution is counted to determine the specific activity, and the sphingosine kinase activity can then be calculated in picomoles SPP formed per minute from the amount of radioactivity measured in the SPP spot.

Identification and Characterization of the Product

While the comigration of radiolabeled products with the SPP standard and the absence of a spot corresponding to SPP in samples to which sphingosine is not added are necessary routine controls, it is important to unequivocally characterize the reaction product when first measuring sphingosine kinase activity from a novel source. Additional characterization of [^{32}P]SPP can be accomplished by comparing its TLC migration with standard SPP in several different solvent systems, such as chloroform/ethanol/water

(65:35:8, v/v; $R_{f_{SPP}}$ = 0.31); chloroform/methanol/ammonium hydroxide (13:7:1, v/v; $R_{f_{SPP}}$ = 0.0) in which most phospholipids, but not SPP, migrate from the origin; chloroform/methanol/acetic acid (30:30:2:5, v/v; $R_{f_{SPP}}$ = 0.35); and chloroform/methanol/ammonium hydroxide (4:1:0.1, v/v).

The labeled SPP product may also be characterized based on the known resistance of SPP to alkaline hydrolysis. In this method, the organic phase containing either the SPP standard or the radiolabeled SPP product is evaporated, resuspended in 0.1 M methanolic KOH, and incubated at 37° for 1 hr. After neutralization, the lipids are extracted with chloroform and water and resolved by TLC. SPP can also be cleaved by periodate followed by borohydride reduction to yield [^{32}P]ethylene glycol monophosphate, which can be resolved and identified by paper chromatography.[2]

Acknowledgments

This work was supported by research grants from the National Institutes of Health (GM43880 and CA61774) and by a grant from the American Cancer Society (BE-275). We thank Dr. S. Milstien and the members of the Spiegel laboratory for helpful suggestions and discussions.

[26] Yeast Sphingosine-1-Phosphate Phosphatases: Assay, Expression, Deletion, Purification, and Cellular Localization by GFP Tagging

By CUNGUI MAO and LINA M. OBEID

Introduction

Dihydrosphingosine (sphinganine)-1-phosphate (DHS-1-P) phosphatase dephosphorylates DHS-1-P, sphingosine-1-P, and phytosphingosine-1-P to dihydrosphingosine (DHS), sphingosine and phytosphingosine, respectively. DHS-1-P phosphatase was first identified biochemically in the membranes of rat liver cells by De Ceuster et al.[1] and was cloned from the yeast *Saccharomyces cerevisiae* by Mao et al.[2] and Mandala et al.[3]

[1] P. De Ceuster, G. P. Mannaerts, and P. P. Van Veldhoven, *Biochem. J.* **311,** 139 (1995).

[2] C. Mao, M. Wadleigh, G. M. Jenkins, Y. A. Hannun, and L. M. Obeid, *J. Biol. Chem.* **272,** 28690 (1997).

[3] S. M. Mandala, R. Thornton, Z. Tu, M. B. Kurtz, J. Nickels, J. Broach, R. Menzeleev, and S. Spiegel, *Proc. Natl. Acad. Sci. U.S.A.* **95,** 150 (1998).

In the liver cells of rats, the activity of the DHS-1-P phosphatase with high affinity and specificity toward DHS-1-P or sphingosine-1-P is associated with plasma membrane. This enzyme has a pH optimum of 7.5 and is sensitive to Zn^{2+} and detergents, but not to the chelator EDTA. In fact, it is activated by EDTA; however, its physiological functions are unknown.

In the yeast *S. cerevisiae,* there are two DHS-1-P phosphatases, YSR2 and YSR3. YSR2 is a 409 amino acid protein with a p*I* of 9.2, whereas YSR3 is a 404 amino acid protein with a p*I* of 7.8. Both YSR2 and YSR3 are integral membrane proteins with multiple transmembrane domains. YSR2 has 53% overall identity to YSR3. They both have a novel phosphatase motif consisting of three conserved domains as first identified by Stukey *et al.*[4] Both YSR2 and YSR3 have *in vitro* activity consisting of dephosphorylation of DHS-1-P, sphingosine-1-P, and phytosphingosine-1-P. The yeast DHS-1-P phosphatases have restricted substrate specificities. They only catalyze dephosphorylation of phosphorylated long chain sphingoid bases rather than ceramide-1-phosphate and phospholipids. Like the mammalian DHS-1-P phosphatase, YSR2 and YSR3 are not inhibited by divalent cations and are activated by EDTA.

YSR2 and YSR3 differ in at least three aspects of their physiological functions. YSR2 is localized to the endoplasmic reticulum, whereas YSR3 is localized to vacuoles. Also, YSR2 has higher transcription levels, whereas the mRNA of YSR3 is barely detectable. As to their physiological functions, YSR2, but not YSR3, is critical for the incorporation of exogenous DHS into ceramide. *YSR2* deletion causes accumulation of DHS-1-P, decreasing sphingolipid formation but increasing glycerolipid formation from dihydrosphingosine in yeast cells. However, *YSR3* deletion does not affect the incorporation of DHS, although DHS-1-P is accumulated. Therefore, YSR2 is an important mediator of sphingolipid metabolism in the yeast *S. cerevisiae.* YSR2 appears to act as a metabolic switch, controlling the direction of the DHS flux between the biosynthetic pathway and the catabolic pathway. Second, YSR2 is involved in heat stress responses in the yeast cells, as deletion of *YSR2* confers heat tolerance. Third, overexpression of YSR2 suppresses yeast cell growth more significantly than that of YSR3.

Clearly, the DHS-1-P phosphatase YSR2 is a novel lipid phosphatase that mediates the regulation of cellular levels of sphingoid bases DHS and sphingosine, the immediate products; and phosphorylated forms DHS-1-P and sphingosine-1-P, the substrates. It also regulates the concentration of ceramide, the product that occurs further downstream of DHS. Sphingosine and ceramide induce apoptosis in mammalian cells and inhibit the growth of both mammalian and yeast cells. DHS is involved in the heat stress

[4] J. Stukey and G. Carman, *Prot. Sci.* **6,** 469 (1997).

responses in the yeast. However, sphingosine-1-P has a mitogenic effect on certain mammalian cells. Sphingosine-1-P also antagonizes ceramide-mediated apoptosis. Therefore, the DHS-1-P phosphatase has important roles in the control of cellular events mediated by these lipid messengers.

To determine these specific roles, the genes had to be cloned. Cloning of the phosphatase genes made it possible to characterize the enzyme both biochemically and genetically. This article describes enzyme epitope tagging, as well as expression and purification of the tagged protein (YSR2). The *in vivo* and *in vitro* enzyme assays are also included. Finally, a method of cellular localization of the enzymes using GFP tagging will be discussed.

Methods

Cloning of the Open Reading Frame of the DHS-1-P Phosphatase Gene YSR2

DHS-1-P phosphatase gene *YSR2* was cloned using the DHS-1-P lyase-deficient strain JS16, which is hypersensitive to sphingosine due to the accumulation of sphingosine-1-P. An increase in the dosage of the *YSR2* gene in a high-copy plasmid (YEp24) in JS16 cells conferred sphingosine resistance due to the attenuation of the toxicity of sphingosine-1-P. *YSR2* was selected by sphingosine resistance. By searching the *Saccharomyces* genomic database (SGD, http://genome-www.stanford.edu/Saccharomyces/), a homolog of YSR2 has been found, designated YSR3. Sequences of the genes encoding DHS-1-P phosphatases have been registered in SGD.

To facilitate the purification of the protein encoded by the gene *YSR2* the protein is tagged by a peptide tag, Flag. The Flag tag is an eight amino acid peptide (DYKDDDDK) that has been used widely due to the high specificity of its monoclonal antibody across species. To tag the protein, the open reading frame (ORF) of the gene *YSR2* is fused with the sequence (GACTACAAGGACGACGATGATAAG) encoding the Flag peptide using polymerase chain reaction (PCR). The details of the procedures are as follows. First, the following primers for PCR amplification of YSR2 are synthesized: the forward primer, 5'-CGGGGTACCATGG<u>ACTACAAG GACGACGATGATAAG</u>GTAGATGGACTGAATACCTCGAACAT TAG G-3', and the reverse primer, 5'-CGGGAATTCTTATGCTATATT TAGAGGGAAAATAGGACGGGGC-3'. Then, to a 0.5-ml Eppendorf thin-wall tube, 75 ng of the yeast genomic DNA as the template, 0.8 μl of 25 mM dNTPs, 10 μl of the PCR buffer, and 1 μl of each primer (100 μM) and enough water are added to make 98 μl of the PCR reaction. The tube is vortexed briefly. The PCR reaction is initiated by adding 2 μl (5 units) of DNA polymerase (Pfu, Stratagene) to the reaction mixture, which has been preheated to 94°. The PCR reaction conditions are 1 cycle of 2 min at 94°, 30 cycles of 30 sec at 94°, 45 sec at 55°, and 2 min at 72° followed by 1 cycle of 10 min at 72°.

The PCR products are purified by a QIAquick PCR purification kit (Qiagen) and digested with *Kpn*I followed by *Eco*RI. The digested DNA fragment is purified from the agarose gel using a QIAquick gel extraction kit (Qiagen) and subcloned into the sites *Kpn*I and *Eco*RI of the yeast expression vector pYES2 (Invitrogen) by following the standard procedures of DNA cloning. The expression construct (pYSR2-Flag), which encodes the Flag-tagged YSR2, is amplified in *Escherichia coli* and purified using a plasmid midi kit (Qiagen).

Expression of DHS-1-P Phosphatase in Yeast Cells

There are several commercially available vectors for gene expression in yeast. Some have constitutive promoters whereas others have inducible promoters. pYES2 (Invitrogen) is one of most used vectors with an inducible promoter, *Gal*1, which is activated by galactose but suppressed by glucose. pYES2 is used to control YSR2 expression because overexpression of YSR2 arrests yeast cell growth. To express YSR2-Flag, the pYSR2-Flag plasmid is introduced into the yeast strain JK9-3d α, which is cultured and maintained in YPD medium, using the lithium acetate method.[5] To obtain sufficient cells without expression of YSR2, the pYSR2-Flag-containing cells are first cultured to midlog phase in the SC-ura medium with 2% glucose at 30°. The cells are then washed twice with sterile water and transferred to the SC-ura medium with 2% galactose. The gene expression is induced maximally after 16 hr in the galactose medium. Expression levels of the tagged YSR2 are detected by Western blotting analysis using the anti-Flag monoclonal antibody (M2, Kodak).

Purification of the Flag-Tagged Protein

The YSR2 is an integral membrane protein. To purify the Flag-tagged protein, microsomes must be obtained from yeast cells. Most cytosolic proteins are removed in preparing the microsomes. Proteins are extracted out of the membranes by Triton X-100, and the tagged protein YSR2-Flag is enriched by an affinity column coupled with the monoclonal antibody against the Flag peptide. The detailed procedures are as follows: cells (3×10^9) expressing the Flag-tagged protein are washed twice with ice-cold water, suspended in 2 ml of an ice-cold extraction buffer [20 m*M*

[5] R. D. Gietz and R. Woods, *in* "Molecular Genetics of Yeast: A Practical Approach." Oxford University Press, 1994.

Tris–HCl buffer, pH 7.4, containing 5 mM EDTA, 0.1 mM phenylmethyl-sulfonyl fluoride (PMSF), 20 μg/ml protease inhibitor mixture CLAP (chy-mostatin, leupeptin, pepstatin, and aprotinin), and 5 mM dithiothreitol (DTT)] in a 50-ml tube. The acid-washed glass beads (425–600 μm, Sigma) are added to the tube to the meniscus. The tube is vortexed vigorously for 30 sec and then chilled on ice for 45 sec. Vortexing and chilling are repeated alternatively for 15 cycles. The glass beads and cell debris are removed by centrifugation at 6000g for 5 min at 4°. The microsome-con-taining supernatant is transferred to a new tube. The glass beads are washed with 0.5 ml of the extraction buffer. The supernatant is transferred to the same tube after centrifugation as described earlier. Membranes are precipitated by centrifugation at 100,000g for 40 min at 4°, washed, and resuspended in the extraction buffer. To isolate the tagged protein from the microsomes, total membrane proteins are solubilized in a lysis buffer (20 mM Tris–HCl, pH 7.4, containing 150 mM NaCl, 1% Tween 20, 0.1 mM PMSF, 1 mM EDTA, and 1 mM EGTA). After membrane debris is removed by centrifugation under the same conditions, 1 ml of agarose beads coupled with the anti-Flag monoclonal antibody M2 is added to the protein solution and incubated overnight at 4°. The beads are loaded onto a column (5 ml, Bio-Rad) and washed three times with 15 ml of TBS buffer (50 mM Tris–HCl, pH 7.4, containing 150 mM NaCl). The tagged protein is eluted with 0.1 M glycine (pH 3.5). The eluant is neutralized immediately by adding 0.05 volume of 1 M Tris–HCl (pH 7.4). The protein concentration is measured using silver staining (Bio-Rad) on SDS–PAGE with bovine serum albumin (BSA) as a standard. The purified protein can be frozen at −80°. It is stable for several cycles of freezing and thawing.

Activity Assay of DHS-1-P Phosphatase

Activity of the DHS-1-P phosphatase can be assayed two different ways using a different radiolabeled substrate: D-*erythro*-[4,5-^{3}H]dihydrosphingo-sine-1-phosphate or D-*erythro*-dihydrosphingosine-1-^{32}P. Using the first substrate, the assay is based on the formation of tritiated DHS. Using the latter substrate, the assay is based on a decrease in DHS-1-^{32}P due to release of ^{32}P from the substrate.

Using D-erythro-[4,5-^{3}H]dihydrosphingosine-1-phosphate as a Substrate

D-*erythro*-[4,5-^{3}H]dihydrosphingosine-1-phosphate   (ARC,   Inc.)   in methanol is dried under a N$_2$ stream and is resuspended in the assay buffer (100 mM Tris–HCl, pH 7.4, containing 5 mM EDTA, 0.1 mM PMSF, 20 μg/ml CLAP, and 0.3% fatty acid-free BSA) to a final concentration of 20 μM and a specific activity of 1 μCi/mmol for D-*erythro*-[4,5-^{3}H]dihydrosphingosine. In 50 μl of reaction, 25 μl of microsomes (~20–30 μg of proteins) or the affinity purified protein is added to 25 μl of the substrate in 5-ml glass tubes to initiate the enzymatic reaction at 30° for 20 min. Reactions are stopped by adding 1.5 ml of chloroform:methanol (2:1) and 0.35 ml of water. The tubes are vortexed. The organic and aqueous phases are separated by centrifugation at 3000 rpm for 5 min.

In the organic phase, the product D-*erythro*-[4,5-^{3}H]dihydrosphingosine is collected into a new tube and dried under a N$_2$ stream. The lipids are dissolved in 50 μl of chloroform:methanol:water (2:1:0.2) and applied to a silicon gel 60 TLC plate (Merck) along with the standard D-*erythro*-[4,5-^{3}H]dihydrosphingosine and D-*erythro*-[4,5-^{3}H]dihydrosphingosine-1-phosphate. The plate is developed in a solvent system consisting of chloroform:methanol:4.2 N ammonium hydroxide (9:7:2). The plate is dried and exposed to a ^{3}H screen of PhosphorImager (Molecular Dynamics, Inc.) for at least 6 hr and scanned by PhosphorImager (Storm, Molecular Dynamics). Using the software ImageQuant, the [^{3}H]dihydrosphingosine band as well as the bands of other labeled lipids are detected and quantitated. If a PhosphorImager is not available, the TLC plate can be sprayed with En^{3}Hance (NEN, Inc.) and radioautographed using X-ray films (Kodak). Longer exposure time (2–3 days) is required to detect the labeled lipids by film. The dihydrosphingsine spots are identified using the dihydrosphingosine standard and are scraped into 0.5 ml of 1% (w/v) SDS and counted on a scintillation counter. The corresponding region from the blank control is also scraped and counted. Absolute counts for the enzymatic reaction can be obtained by subtracting the blank background counts, contributed by autodegradation of the substrate in the absence of the enzyme.

Using Dihydrosphingosine-1-^{32}P as a Substrate

Dihydrosphingosine can be phosphorylated to DHS-1-^{32}P by sphingosine kinase in the presence of [^{32}P]ATP following the procedure by Mandala et al.[3] The activity of the enzyme is detected by converting DHS-1-^{32}P to DHS due to the release of ^{32}P from DHS-1-^{32}P. Briefly, in the presence of ^{32}P-ATP, dihydrosphingosine is phosphorylated by sphingosine kinase in a yeast crude extract or the partially purified kinase from murine cell lines as described.[3] ^{32}P-labeled dihydrosphingosine is purified by TLC separation. Enzymatic reactions are assembled by adding 50 μl of DHS-1-^{32}P (1 nmol, 25,000 cpm, 0.3% BSA) to 50 μl of the microsomes containing DHS-1-P phosphatase in the assay buffer. After incubation at 30° for 20 min, the reactions are stopped by adding 20 ml of 1 N HCl and are extracted with 0.8 ml of chloroform:methanol:concentrated HCl (100:200:1). The samples are vortexed vigorously. In order to separate phases, 240 ml of chloroform and 240 ml of 2 M KCl are added. After vortexing, the samples are centrifuged. After the aqueous phase is aspirated, lipids in the organic phase are dried under a N_2 stream and resolved under the same conditions as described earlier. The remaining DHS-1-^{32}P in the reactions is identified and measured by the PhosphorImager.

An advantage of this method is that it is not very expensive. However, it is time-consuming and laborious. Another disadvantage is that the substrate cannot be kept for long due to the rapid decay of ^{32}P. In addition, this method utilizes the crude or the partially purified sphingosine kinase, which will compromise substrate purity. Because sphingosine kinase has been cloned from both yeast and mammalian cells, recombinant sphingosine kinase expressed in *E. coli* is preferred in order to eliminate the complexity of the substrates.

Creation of DHS-1-P Phosphatase-Deficient Mutants

Wild-type *S. cerevisiae* cells maintain low levels of free sphingoid bases and their phosphorylated forms. Mutant cells with deletion of either the DHS-1-P phosphatase or the DHS-1-P lyase accumulate DHS-1-P. Therefore, the mutant deficient either in the phosphatase or in the lyase is important for studying the role of DHS-1-P in cells.

In *Saccharomyces,* the one-step transplacement method, which is based on homologous integration, is the one most often used to delete a gene. Knockout constructs, containing a marker gene flanked with left and right portions of the gene of interest, are made by PCR using a plasmid as a template, which carries a selectable marker such as the G418-resistant gene marker, or autotrophic markers *URA3, LEU2,* and *HIS3*. After constructs are introduced into the yeast, homologous recombination between the flanking fragments integrates the marker gene into a chromosome and deletes the gene of interest simultaneously.

To delete *YSR2* in yeast, the following primers are used: the forward primer, 5′-GAGCCAGGACTCTCTAACCCCAATGACTTTCAAGAG CCCAGCTGAAGCTTCGTACGC-3′, and the reverse primer, 5′-AG GACGGGGCTGCACATTACAACGGTGAATGGGCATAGGCCAC TAGTGGATCTG-3′. The gene disruption construct is amplified by PCR using the plasmid pFA6[2] as the template, which contains the G418-resistant gene. PCR reaction conditions are the same as described in the amplification of the YSR2 ORF. The construct is introduced into strain JK9-3d α using the lithium acetate method, and G418-resistant clones are selected on YPD plates with 220 μg/ml G418. Gene disruption is confirmed by two PCRs. The first PCR is conducted using the same primers described earlier. The second PCR is carried out by using one of the primers described earlier and a primer upstream or downstream of the gene. Genomic DNAs of the G418-resistant clones are isolated as described[2] and are used as DNA templates for both PCR reactions. Both PCRs should result in products if the correct integration occurs.

Studies of Sphingolipid Metabolism Using the Phosphatase-Deficient Mutant

In vivo labeling using the radioactive precursors is an important method to study the *de novo* synthesis and metabolism of sphingolipids. Sphingolipids can be labeled from tritiated serine, palmitic acid, DHS, or inositol. All of these precursors are incorporated into both sphingolipids and phospholipids. The radiolabeled phospholipids can be deacylated by base hydrolysis using monomethylamide and removed by extraction.[6] Base hydrolysis is essential for labelings with precursors other than DHS and inositol as other precursors are incorporated inadequately into sphingolipids.

Based on *in vivo* labeling with different precursors, it has been found that DHS-1-P phosphatase was essential for incorporating exogenous DHS into ceramides and other sphingolipids containing ceramide moiety.[2] In the YSR2-deficient mutant, degradation of DHS through phosphorylation by DHS kinase and subsequent cleavage by DHS-1-P lyase dominates DHS metabolism. The degradative products are then incorporated into glycerolipids. Therefore, the mutant has significantly increased labeled glycerolipids.

Tritiated DHS Labeling of Yeast Cells

A single colony of the wild-type or the mutant strain is inoculated into 5 ml of YPD medium in a glass culture tube that is rotated on a drum roller at 30° overnight (approximately 14 hr). The next morning, the overnight culture is added to 50 ml of fresh YPD medium in a 250-ml flask and shaken for 6 hr at 250 rpm at 30°. The cells are collected and washed twice with sterile water and suspended in SC medium at a density of 3×10^7 cells/ml. One milliliter of the cells is aliquoted to a Falcon plastic culture tube, and 3 μCi DHS in ethanol is added. The cells are labeled from 30 min to 2 hr at 30°. After labeling, the cells are collected, washed twice with water, and suspended in the extraction solvent (ethanol:water:diether:pyridine:NH$_4$OH, 15:15:5:1:0.018) in a glass tube with a cap. The total lipids are extracted by incubating at 60° for 15 min. Supernatants are transferred to a new tube after centrifugation at 3500 rpm for 5

[6] C.-S. Oh, D. A. Toke, S. M. Mandala, and C. E. Martin, *J. Biol. Chem.* **272**, 17376 (1997).

min. Lipids in the supernatants are dried under a N_2 stream. Seventy-five microliters of chloroform : methanol : water (2 : 1 : 0.1) is added to dissolve the lipids. Fifty microliters of lipids along with the lipid standards are spotted onto TLC plates. The plates are then developed in the following solvent system: chloroform : methanol : 4.2 N NH_4OH (9 : 7 : 2). The plates are dried and exposed to the tritium screen of PhosphorImager for at least 4 hr. The plates are scanned by the Storm PhosphorImage Scanner. The profiles of labeled lipids are analyzed with ImageQuant. In mutant cells, sphingolipids, including phytosphingosine, ceramide, IPC, MIPC, and $M(IP)_2C$, are not synthesized from exogenous DHS or DHS-1-P, but DHS-1-P is accumulated.

Tritiated Serine and Palmitic Acid Labeling of Yeast Cells

Procedures of labeling of sphingolipids with serine or palmitic acid in yeast cells are the same as with DHS, except for the labeling period and the radioactivity of precursors. Longer incubation (more than 4 hr) and more radioactivity (at least 10 μCi per 3 $\times$ 10^7 cells) are required to achieve sphingolipid labeling with serine or palmitic acid to the same extent as with DHS. In addition, base hydrolysis of glycerolipids is highly recommended in order to resolve sphingolipids on TLC plates better.

Involvement of DHS-1-P Phosphatase in Heat Shock Responses

Mutant cells deficient in DHS-1-P phosphatase grow normally as that of the wild-type do at a temperature less than $37°$. However, the mutant cells are more resistant to heat treatment as high as 50 to $55°$. To test the heat resistance of the mutant, exponential cells at a density of 10^6 cells/ml in SC or YPD medium are heat treated for 5–15 min at 50 or $55°$. The nonheat-stressed cells are incubated at room temperature. The cells are spread onto YPD plates (200–300 cells per plate) after they are diluted serially with fresh YPD medium at room temperature. The plates are incubated at $30°$ for at least 2 days. Colonies on each plate are counted. The postheat shock viability is measured by comparing colony numbers before and after heat shock. The mutant has a higher postheat shock viability.

Localization of DHS-1-P Phosphatase

To understand how the enzyme functions in cells, it is important to know where the enzyme is localized in cells. Although fluorescent immunostaining is widely used for localizing proteins in mammalian cells, its application in yeast cells is restricted because yeast cells are small and have rigid cell walls. The emergence of living color protein, GFP, as a labeling probe makes it easier to study the localization of proteins in the living yeast cells. A GFP mutant, which has the yeast and fungal codon preference, has been developed.[7] It can be fused to the carboxyl- or amino-terminals of any yeast proteins to make fusion proteins. The fusion proteins are well expressed in the yeast cells. The wild-type or other versions of GFP such as GFPuv can also be expressed in yeast. They should be preferably fused to carboxyl-terminals of the yeast proteins to achieve higher expression of the fusion proteins.

To fuse GFP to the YSR2, the DNA sequence encoding GFP is amplified by PCR using GFP containing plasmid pBAD-GFPuv [constructed by cloning GFPuv cDNA (Clontech) into the bacterial expression vector pBAD (Invitrogen)] as the template. The following primers are used: the forward primer, 5'-CCC AAG CTT ATG GCT AGC AAA GGA GAA GAA C, and the reverse primer, 5'-CCC GGT ACC TTT GTA GAG CTC ATC CAT GCC. The PCR conditions are the same as described previously. The GFP sequence is cloned into the sites *Hin*dIII and *Kpn*I of the plasmid pYSR2, which is constructed by cloning the ORF of YSR2 into the yeast plasmid pYES2 (Invitrogen, Inc.). pYES2 contains the *Gal*1 promoter, which is induced by galactose, but suppressed by glucose. The construct for expression of the GFP-YSR2 fusion protein is introduced into the yeast cells using the lithium acetate method. The expression of the GFP-YSR2 is induced by galactose, which is added to SC-ura medium. The fluorescence of the yeast cells can be visualized by a confocal microscope (Zeiss) with a 100× objective lens. Neither fixation nor washing is needed to visualize the fluoresence of the yeast cells. YSR2 is localized to the endoplasmic reticulum. Data suggest that dephosphorylation of DHS-1-P might occur in the endoplasmic reticulum.

Summary

DHS-1-P phosphatases cloned from yeast represent novel lipid phosphatases, which were not thought to exist in yeast. Identification and characterization of YSR2 and YSR3 have demonstrated that the DHS-1-P phosphatase is an important mediator in the biosynthesis of sphingolipids and in the maintenance of the balance of signaling lipid molecules ceramide, sphingosine, and sphingosine-1-P. Methods introduced here for purification, activity assay, *in vivo* labeling, and cellular localization using GFP tagging are expected to facilitate our understanding of this enzyme.

---

[7] B. P. Cormack, G. Bertram, M. Egerton, N. A. Gow, S. Falkow, and A. Brown, *Microbiology* **143**, 303 (1997).

Reagents

Molecular biology enzymes including T4 DNA ligase, calf intestine alkaline phosphatase, and restriction enzymes are purchased from Boehringer Mannheim. DNA polymerase Pfu is purchased from Stratagene. [³H]Dihydrosphingosine and 3,4-[³H]dihydrosphingosine-1-P are purchased from ARC, Inc. Sphingosine is prepared by this laboratory. Other chemicals are purchased from Sigma.

[27] Analysis of Ceramide 1-phosphate and Sphingosine-1-phosphate Phosphatase Activities

By David N. Brindley, Jim Xu, Renata Jasinska, and David W. Waggoner

Introduction

This article is concerned with the measurements of ceramide 1-phosphate and sphingosine-1-phosphate phosphatase activities in mammalian tissues, although the same type of assays should be appropriate for general use with samples prepared from other organisms. Studying the dephosphorylation of these lipid phosphate esters is important as both of these substrates and their dephosphorylation products are bioactive. Removal of the phosphate group could therefore attenuate the signal from the sphingolipid phosphates and, at the same time, generate alternative signals from the products, ceramide and sphingosine. Consequently, the phosphatases occupy an important position in cell activation by lipid mediators by being able to switch the signal from one mediator to another.

Sphingosine-1-phosphate is an important extracellular signaling molecule that stimulates specific external receptors.[1–3] This interaction increases phospholipase D activity, the production of phosphatidate and diacylglycerol, and the consequent signaling cascades that are generated by these bioactive glycerolipids. Exogenous sphingosine-1-phosphate also causes the

[1] S. Spiegel, O. Cuvillier, E. Fuior, and S. Milstien, *in* "Sphingolipid-Mediated Signal Transduction" (Y. A. Hannun, ed.), Chap. 9. Landes Co., Austin, TX, 1997.
[2] K. Gonda, H. Okamoto, N. Takuwa, Y. Yatomi, H. Okazaki, T. Sahurai, S. Kimura, R. Sillard, K. Harii, and Y. Takuwa, *Biochem. J.* **331,** 67 (1998).
[3] A. Gómez-Muñoz, A. Abousalham, Y. Kikuchi, D. W. Waggoner and D. N. Brindley, *in* "Sphingolipid-Mediated Signal Transduction" (Y. A. Hannun, ed.), Chap. 8. Landes Co., Austin, TX, 1997.

0076-6879/99 $30.00

mobilization of intracellular Ca^{2+} and generates a potent mitogenic signal that counteracts ceramide-induced apoptosis.[4] Sphingosine-1-phosphate is also produced as part of the signal that is generated by platelet-derived growth factor (PDGF), but not by epidermal growth factor (EGF).[4] This effect of PDGF is mediated by the activation of sphingosine kinase. Presumably the sphingosine-1-phosphate is produced inside the cell and then activates internal cell targets, including the stimulation of mitogen-activated protein (MAP) kinase and AP1.[5] Sphingosine-1-phosphate phosphatase activity could therefore destroy exogenous or endogenous sphingosine-1-phosphate and attenuate its signal. The production of sphingosine can initiate several signaling events, including the mobilization of intracellular Ca^{2+} and the stimulation of cell division.[1-4] Sphingosine inhibits protein kinase C (PKC) activities,[6] which can occur either directly, or indirectly through inhibition of Mg^{2+}-dependent phosphatidate phosphohydrolase (PAP-1) or Mg^{2+}-independent phosphatidate phosphohydrolase (PAP-2) activities.[7] Sphingosine can also increase phosphatidate concentrations by stimulating diacylglycerol kinase and phospholipase D activities in some cells.[7]

The role of ceramide 1-phosphate in cell signaling is less clear. Dressler and Kolesnick[8] identified a pathway in which ceramide originating from the action of neutral sphingomyelinase, but not glucosylceramidase, is converted to ceramide 1-phosphate by a Ca^{2+}-dependent kinase. Shingal et al.[9] identified a ceramide 1-phosphate phosphatase in rat brain, and Boudker and Futerman[10] characterized a phosphatase that specifically hydrolyzes ceramide 1-phosphate in plasma membranes of rat liver. Exogenous ceramide 1-phosphate stimulates cell division in fibroblasts, but unlike PA, lysoPA, and sphingosine 1-phosphate, it does not activate PLD and MAP kinase, nor inhibits adenylate cyclase.[11] Little is known about the physiological effects of ceramide 1-phosphate, although it may regulate some aspects of synaptic vesicle function.[9] Ceramide 1-phosphate also has pathological and toxic effects since a sphingomyelinase D has been identified as the active principle of the venom of the brown recluse spider, *Loxosceles reclusa.*[12]

---

[4] O. Cuvillier, G. Pirianov, B. Kleuser, P. G. Vanek, O. A. Coso, S. Gutkind, and S. Spiegel, *Nature* **381,** 800 (1996).

[5] A. Olivera and S. Spiegel, *Nature* **365,** 557 (1993).

[6] A. H. Merrill, Jr., and V. L. Stevens, *Biochim. Biophys. Acta* **1010,** 131 (1989).

[7] D. N. Brindley and D. W. Waggoner, *Chem. Phys. Lipids* **80,** 45 (1996).

[8] K. A. Dressler and R. N. Kolesnick, *J. Biol. Chem.* **265,** 14917 (1990).

[9] R. Shinghal, R. H. Scheller, and S. M. Bajjalieh, *J. Neurochem.* **61,** 2279 (1993).

[10] O. Boudker and A. H. Futerman, *J. Biol. Chem.* **268,** 22150 (1993).

[11] A. Gómez-Muñoz, P. A. Duffy, A. Martin, L. O'Brien, H.-S. Byun, R. Bittman, and D. N. Brindley, *Mol. Pharmacol.* **47,** 883 (1995).

[12] A. P. Truett and L. E. King, Jr., *Adv. Lipid Res.* **26,** 275 (1993).

Sphingomyelinase D is also produced by some bacteria, including *Coryne-bacterium pseudotuberculosis* and *Vibrio damsela*.[12] Although ceramide 1-phosphate can be produced directly from sphingomyelin by the toxins of these organisms, there is no evidence for this reaction occurring as a result of a mammalian enzyme.

Dephosphorylation of ceramide 1-phosphate generates ceramide. This compound is produced by the activation of sphingomyelinases through the interaction of specific agonists such as tumor necrosis factor-α (TNF-α), interleukin-1, interferon-γ, fas ligand, and nerve growth factor (by the p75 receptor) with cell surface receptors.[13–16] Ceramide production is often associated with decreased cell growth and in producing apoptosis.[13–16] Ceramides stimulate specific serine/threonine kinases[13] and phosphatases[15] and they inhibit the activation of phospholipase D.[3,17] In confluent rat2 fibroblasts, ceramides can activate tyrosine kinase activity, resulting in the formation of Ras-GTP, the stimulation of phosphatidylinositol 3-kinase, and the activation of MAP kinase.[18] Such an effect can explain why ceramides and TNF-α stimulate cell division, rather than cause apoptosis, in confluent fibroblasts.[19–23]

In light of these observations, it is important to understand the regulation and biological importance of enzymes that dephosphorylate ceramide 1-phosphate and sphingosine-1-phosphate and how these enzymes interact with the respective kinases that synthesize these compounds. The interplay of the kinases and phosphatases controls the relative balance between the concentrations of important sphingolipid mediators, which regulate the aggregate signal that the cell receives in terms of structure, growth, and death.

Characterization of Mammalian Ceramide 1-phosphate and Sphingosine-1-phosphate Phosphatases (Lipid Phosphate Phosphatases)

So far, dephosphorylation of ceramide 1-phosphate and sphingosine-1-phosphate in mammals appears to be catalyzed by type 2 phosphatidate

[13] R. Kolesnick and D. W. Golde, *Cell* **77,** 325 (1994).
[14] R. A. Heller and M. Krönke, *J. Cell Biol.* **126,** 5 (1994).
[15] Y. A. Hannun, *J. Biol. Chem.* **269,** 3125 (1994).
[16] Y. A. Hannun, *Science* **274,** 1855 (1996).
[17] A. Abousalham, C. Liossis, L. O'Brien, and D. N. Brindley, *J. Biol. Chem.* **272,** 1069 (1997).
[18] A. N. Hanna, E. Y. W. Chan, J. Xu, J. C. Stone, and D. N. Brindley, *J. Biol. Chem.* **274,** 12722 (1999).
[19] B. J. Sugarman, *Science* **230,** 943 (1985).
[20] A. Olivera, N. E. Buckley, and S. Spiegel, *J. Biol. Chem.* **267,** 26121 (1992).
[21] B. B. Aggarwal, S. Singh, R. LaPushin, and K. Totpal, *FEBS Lett.* **364,** 5 (1995).
[22] E. J. Battegay, E. W. Raines, T. Colbert, and R. Ross, *J. Immunol.* **154,** 6040 (1995).
[23] R. Kolesnick and Z. Fuks, *J. Exp. Med.* **181,** 1949 (1995).

phosphohydrolase activity. This activity was originally characterized in mammalian cells based on a lack of requirement for bivalent cations and insensitivity to inhibition by N-ethylmaleimide.[24] We now know that this activity is catalyzed by a family of at least three mammalian lipid phosphate phosphatases, which were redesignated LPP-1, LPP-2, and LPP-3.[25] There is a human splice variant of LPP-1, called LPP-1a. The reason for the new nomenclature was to produce a more accurate name that reflected the activity and physiological functions of the phosphatases. LPP activity isolated from rat liver dephosphorylates lysophosphatidate, ceramide 1-phosphate, sphingosine 1-phosphate,[26] and diacylglycerol pyrophosphate[27] with efficiencies similar to phosphatidate. These substrates are also mutually competitive,[26] indicating that the same active site is used by the different phosphate esters. LPP does not hydrolyze phosphatidylcholine, phosphatidylethanolamine, phosphatidylserine, phosphatidylinositol, or diphosphatidylglycerol, thus demonstrating that it is not a general phospholipase C. LPP will not dephosphorylate, nor is it inhibited by water-soluble phosphate monoesters such as glycerol 3-phosphate.[24] The reaction catalyzed by LPP also obeys a surface dilution kinetic model[26] that confirms that this enzyme is a lipid phosphate phosphohydrolase.

Details of the structures and properties of the mammalian LPPs have been reviewed previously.[25] The LPPs belong to a phosphatase superfamily that was identified by Stukey and Carman[28] and which was expanded upon further by Hemricka et al.[29] and Neuwald.[30] This superfamily includes Wunen (a Drosophila protein), yeast, and bacterial DGPPases, yeast LPP-1 and phytosphingosine phosphatases, human glucose 6-phosphatases, and fungal chloroperoxidase.[25,28–30] These phosphatases contain three highly conserved domains within a larger motif. The LPPs possess six putative membrane-spanning regions, which is compatible with their being integral membrane proteins.[25,28–30]

Recombinant LPP-1 dephosphorylates phosphatidate, lysophosphatidate, diacylglycerol pyrophosphate, ceramide 1-phosphate, and sphingosine-1-phosphate with relatively similar efficiencies (unpublished work) as

[24] Z. Jamal, A. Martin, A. Gómez-Muñoz, and D. N. Brindley, J. Biol. Chem. 266, 2988 (1991).
[25] D. N. Brindley and D. W. Waggoner, J. Biol. Chem. 273, 24281 (1998).
[26] D. W. Waggoner, A. Gómez-Muñoz, J. Dewald, and D. N. Brindley, J. Biol. Chem. 271, 16506 (1996).
[27] D. A. Dillon, X. Chen, G. M. Zeimetz, W.-I. Wu, D. W. Waggoner, J. Dewald, D. N. Brindley, and G. M. Carman, J. Biol. Chem. 272, 10361 (1997).
[28] J. Stukey and G. M. Carman, Prot. Sci. 6, 469 (1997).
[29] W. Hemricka, R. Renirie, H. L. Dekker, P. Bernett, and R. Wever, Proc. Natl. Acad. Sci. U.S.A. 94, 2145 (1997).
[30] A. F. Neuwald, Prot. Sci. 6, 1764 (1997).

does purified LPP from rat liver.[26] In contrast, Kai *et al.*[31] showed that human LPP-1 hydrolyzed phosphatidate and lysophosphatidate but had relatively little activity toward ceramide 1-phosphate and sphingosine-1-phosphate. Both sphingosine-1-phosphate and phosphatidate were good substrates for LPP-3.[31] Hooks *et al.*[32] reported that LPP-1 and LPP-2 had about 30% higher activity against phosphatidate compared to lysophosphatidate, but that LPP-2 had higher activity against lysophosphatidate. All three isoforms also dephosphorylated *N*-oleoylethanolamine phosphate, but this activity was relatively lower for LPP-2. The most rigorous kinetic analysis using lipids presented in Triton X-100 micelles comes from work on LPPs overexpressed in Sf9 cells.[33] These results indicate that LPP-1 and LPP-3 may have a greater catalytic efficiency (V_{max}/K_m) for glycerolipid substrates and that LPP-2 dephosphorylates glycerolipid and sphingolipid phosphates with similar efficiencies.[33]

Methods

A convenient way to measure ceramide 1-phosphate and sphingosine-1-phosphate phosphatase activities is to employ ^{32}P-labeled substrates and to measure the release of ^{32}P$_i$. This gives greater sensitivity and a low background to the assay compared to determining the production of P$_i$ chemically.

Synthesis of Ceramide 1-[^{32}P]phosphate

Nonradioactive ceramide 1-phosphate can be obtained commercially or can be synthesized by the phosphorylation of the desired ceramide by *Escherichia coli* diacylglycerol kinase. Essentially any ceramide can be used but the reaction rate of the kinase is much lower with C_2-ceramide (*N*-acetylsphingosine) compared to longer chain ceramides.[17] D-*erythro*-C_8-ceramide 1-phosphate is synthesized using C_8-ceramide (2 mg) solubilized in 5 m*M* cardiolipin, 7.5% octyl-β-glucopyranoside, and 1 m*M* diethylenetriaminepentaacetic acid by sonication and resuspended in a mixture containing 50 m*M* imidazole, pH 6.6, 50 m*M* NaCl, 100 m*M* MgCl$_2$, 1 m*M* EGTA, and 0.38 units of diacylglycerol kinase/ml.[11] The reaction is started with 25 m*M* ATP. After 12 hr at 37°, the incubation mixture is supplemented with 0.38 units of diacylglycerol kinase/ml and the reaction is continued for another 2 hr. The reaction is stopped by extracting the lipids according

[31] M. Kai, I. Wada, S.-I. Imai, F. Sakana, and H. Kanoh, *J. Biol. Chem.* **272,** 24572 (1997).
[32] S. B. Hooks, S. P. Ragan, and K. R. Lynch, *FEBS Lett.* **427,** 188 (1998).
[33] R. Roberts, V. A. Sciorra, and A. J. Morris, *J. Biol. Chem.* **273,** 22059 (1998).

to Bligh and Dyer,[34] except that the phases are separated by adding 2 M KCl in 0.2 M H_3PO_4 instead of water. The organic phase is dried under N_2, and the lipid is applied to thin-layer plates in chloroform/methanol (1:1, v/v). The plates are developed sequentially with chloroform/methanol/ NH_4OH (65:35:7.5, by volume) and chloroform/acetone/acetic acid/ methanol/water (50:20:15:10:5, by volume). To identify C_8-ceramide 1-phosphate on the thin-layer plate, [γ-^{32}P]ATP can be included in one of the reaction tubes so that the product can be detected by autoradiography with X-ray film or a radioactive scanner. Nonradioactive C_8-ceramide 1-phosphate can also be identified by spraying the plates lightly with distilled water when lipids are detected as white area whereas the background is gray. In this solvent system, C_8-ceramide 1-phosphate is identified as a single spot at R_f 0.59. The distance of migration of ceramide phosphates increases with increasing chain length of the fatty acid linked to sphingosine. The ceramide 1-phosphates are eluted with three washes of 3 ml of chloroform/methanol/acetic acid/water (50:39:1:1, by volume). Silica is removed by centrifugation, and 1 ml of water is added to the combined supernatants to separate the phases. The organic phase is washed once with 4 ml of methanol/water (1:1, v/v). C_8-ceramide 1-phosphate is standardized by phosphate analysis[8] and is stored in chloroform at $-20°$. To characterize C_8-ceramide 1-phosphate further, it is deacylated in 6 M HCl/ butan-1-ol (1:1, v/v) for 1 hr at $100°$, and the sphingosine-1-phosphate is resolved by thin-layer chromatography using butan-1-ol/acetic acid/water (3:1:1, by volume). Recovery of C_8-ceramide 1-phosphate as sphingosine-1 phosphate is about 65% as reported previously for long-chain ceramide 1-phosphate.[8] Sphingosine-1-phosphate is identified by phosphate- and amino-specific sprays and by using an authentic standard that is prepared essentially as described by Van Veldhoven et al.[35] The purity of C_8-ceramide 1-phosphate is confirmed by thin-layer chromatography with four additional solvent systems: (A) chloroform/methanol/water (65:35:8, v/v/v), (B) chloroform/methanol/acetic acid (9:1:1, v/v/v), (C) butan-1-ol/acetic acid/ water (3:1:1, v/v/v), and (D) chloroform/methanol/NH_4OH (65:35:7.5, v/v/v). In all cases there is a single spot that is positive for phosphate sprays. R_f values are 0.26, 0.11, 0.50, and 0.05, respectively, for the four solvent systems.

Synthesis of Sphingosine-1-[^{32}P]phosphate

Nonradioactive preparations of sphingosine-1-phosphate can be obtained commercially or can be synthesized by the enzymatic digestion of

[34] E. G. Bligh and W. J. Dyer, *Can. J. Biochem. Physiol.* **37,** 911 (1959).
[35] P. P. Van Veldhoven, R. J. Fogelsong, and R. M. Bell, *J. Lipid Res.* **30,** 611 (1989).

sphingosylphosphorylcholine with phospholipase D essentially as described by Van Veldhoven et al.[35] Our group modified the procedure slightly.[36] The first incubation is for 2 hr followed by a second incubation that lasts 3–4 hr. After each of these incubations the samples are frozen, instead of being maintained at 4°. This helps improve the precipitation of sphingosine-1-phosphate. Tubes are centrifuged, and the pellets are pooled and then dispersed in water by sonication. After freezing-thawing and centrifuging, the final pellet is dispersed in 1 ml acetone by sonication. This sample is centrifuged, and the final pellet is dissolved in methanol. The sphingosine-1-phosphate is then purified by thin-layer chromatography according to Zhang et al.[37] Sphingosine-1-phosphate has an R_f of 0.47–0.49 when plates are developed with butan-1-ol/acetic acid/water (3:1:1, by volume). The purity of sphingosine-1-phosphate is confirmed by analysis of the purified product in three additional solvent systems, which include solvents I and V from Desai et al.[38] and a solvent system that consists of chloroform/acetone/acetic acid/methanol/water (50:20:15:10:5, by volume). The R_f for sphingosine-1-phosphate in the latter solvent is 0.39. In each system there is a single spot that is positive with ninhydrin and phosphate sprays.[39] Sphingosine-1-phosphate is eluted from the silica in three washes with methanol/chloroform/HCl (9:1:0.05, by volume).

^{3}H- or ^{32}P-labeled sphingosine-1-phosphate can be prepared by deacylation of ceramide-1-phosphate, essentially according to Desai et al.[38] The preparation of labeled ceramide 1-phosphate is essentially as described previously using E. coli diacylglycerol kinase. Ceramide 1-[^{32}P]phosphate is treated with 6 M HCl/butan-1-ol (1:1, v/v) for 60 min at 100° to produce sphingosine-1-[^{32}P]phosphate. This compound is purified by thin-layer chromatography[37] using 1-butanol/acetic acid/water (3:1:1, by volume). The product also cochromatographs with the nonradioactive sphingosine 1-phosphate in the solvent systems described earlier. As an alternative sphingosine-1-[^{32}P]phosphate can be obtained directly by using sphingosine kinase and [γ-^{32}P]ATP.[40]

[36] A. Gómez-Muñoz, D. W. Waggoner, L. O'Brien, and D. N. Brindley, *J. Biol. Chem.* **270,** 26318 (1995).

[37] H. Zhang, N. N. Desai, A. Olivera, T. Seki, G. Broker, and S. Spiegel, *J. Cell Biol.* **114,** 155 (1991).

[38] N. N. Desai, H. Zhang, A. Olivera, T. Seki, M. E. Mattie, and S. Spiegel, *J. Biol. Chem.* **267,** 23122 (1992).

[39] R. M. C. Dawson, D. C. Elliott, W. H. Elliott, and K. M. Jones, *in* "Data for Biochemical Research," p. 480. Clarendon Press, Oxford, 1987.

[40] A. Olivera, T. Kohama, Z. Tu, S. Milstien, and S. Spiegel, *J. Biol. Chem.* **273,** 12576 (1998).

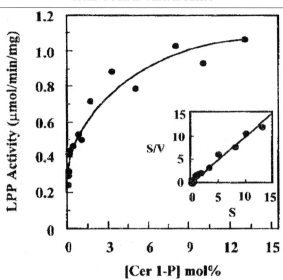

FIG. 1. Kinetics of long-chain ceramide 1-phosphate dephosphorylation by lipid phosphate phosphatase in the Triton micelle assay. The activity of LPP purified from rat liver[26] was measured at a saturating bulk concentration (100 μM) of ceramide 1-phosphate. The appropriate mol% of ceramide 1-phosphate in the micelles was adjusted by varying the concentration of Triton X-100.[42]

Measurement of Ceramide 1-phosphate and Sphingosine 1-phosphate Phosphatase Activities in Cell Lysate, Isolated Membranes, or after Purification

Assays with isolated cell membranes are performed in a volume of 100 μl containing 100 mM Tris/maleate buffer, pH 6.5, and 1 mM N-ethylmaleimide. Reactions are started by adding ^{32}P-labeled sphingosine-1-phosphate or ceramide 1-phosphate in Triton X-100 mixed micelles. Routinely, assay reactions contain a final concentration of 0.6 mM substrate in 8 mM Triton X-100.[26,41] Incubations are performed at 37° and for a time such that <12% of the substrate is hydrolyzed. A detailed kinetic analysis of the reaction can be performed using the surface dilution kinetic model proposed by Carman *et al.*[42] in which the concentration of sphingosine-1-phosphate or ceramide 1-phosphate is expressed as a percentage of the lipid in the Triton micelle (Fig. 1). In the assay of ceramide 1-phosphatase, the reaction is terminated by adding 0.5 ml of 0.1 M HCl in methanol and

[41] D. W. Waggoner, A. Martin, J. Dewald, A. Gómez-Muñoz, and D. N. Brindley, *J. Biol. Chem.* **270,** 19422 (1995).

[42] G. M. Carman, R. A. Deems, and E. A. Dennis, *J. Biol. Chem.* **270,** 18711 (1995).

1.5 ml of chloroform/methanol (2:1). Phases are separated by adding 0.9 ml of 2 M KCl and 0.3 M H_3PO_4, and the tubes are centrifuged to isolate a clear top phase that contains the released $^{32}P_i$. Radiolabeled $^{32}P_i$ in the aqueous phase is quantitated by scintillation counting.[26] For the hydrolysis of sphingosine-1-[^{32}P]phosphate, the main problem is that this compound cannot be removed efficiently from $^{32}P_i$ by chloroform extraction. Originally, we stopped the reactions with 10 μl of concentrated formic acid, and $^{32}P_i$ was separated from the sphingosine phosphate by thin-layer chromatography in butan-1-ol/acetic acid/water (3:1:1, by volume).[26] $^{32}P_i$ was then quantitated by scintillation counting.[26] This method proved to be tedious and we now separate the $^{32}P_i$ by extraction of a phosphomolybdate complex as described next.

Measurement of Phosphatase Activities Using Exogenous Ceramide 1-phosphate and Sphingosine 1-phosphate with Intact Cells

These assays were designed primarily to determine the dephosphorylation of sphingosine-1-phosphate and ceramide 1-phosphate by intact cells. Sphingosine-1-phosphate is believed to be an important extracellular activator.[1-3] The likely mode of its natural presentation to cells is as an albumin complex. Short-chain ceramide 1-phosphates can also be presented in this way and they can also cause cell activation.[11] However, long-chain ceramide phosphates do not activate cells effectively unless they are added in a mixture of methanol/dodecane (49:1, v/v).[43] It is not known whether exogenous ceramide 1-phosphate is a physiological activator of cells and, if so, whether it becomes accessible to target cells from the membranes of adjacent cells. For studying the ability of cells to dephosphorylate exogenous lipid phosphate esters we therefore employed cultured ECV 304 cells and rat2 fibroblasts. Cells are rinsed twice with phosphate-buffered saline and incubated in DMEM containing 0.1% fatty acid-free bovine serum albumin for 2 hr. Unless indicated to the contrary, the medium is replaced with 0.8 ml of the same medium containing the desired concentration of ceramide 1-[^{32}P]phosphate and sphingosine-1-[^{32}P]phosphate (about 1 Ci/mol), which are dispersed by sonication. In the case of C_8-ceramide 1-[^{32}P]phosphate, this substrate might be converted to sphingosine-1-[^{32}P]phosphate by ceramidase, but subsequent metabolism to $^{32}P_i$ would be catalyzed by LPP-1 itself and this side reaction would be minimized by the excess of ceramide 1-phosphate.[26] Medium (0.5 ml) is transferred into 0.5 ml of 1 M HClO$_4$ to precipitate protein and the majority of the labeled lipid. After centrifugation, 0.8 ml of supernatant is extracted twice with 0.8 ml of butan-1-ol to

[43] A. Gómez-Muñoz, L. M. Frago, L. Alvarez, and I. Varela-Nieto, *Biochem J.* **325,** 435 (1997).

remove any remaining lipid. A portion (0.5 ml) of the aqueous phase is recovered, 50 μl of 125 mM ammonium molybdate is added, and the mixture is extracted with 0.6 ml of isobutanol/benzene (1 : 1; v/v). $^{32}P_i$ is recovered quantitatively in the organic phase as a phosphomolybdate complex,[17] and radioactivity is measured in 0.5-ml samples by scintillation counting. This method distinguishes P_i from other water-soluble phosphates that do not partition into the organic phase. Medium from dishes that do not contain cells or a zero time control is extracted to measure blank values.

An example of such an assay for the degradation of exogenous C_8-ceramide 1-phosphate is shown in Fig. 2. Maximum rates of dephosphorylation are maintained for about 10 min, even though only about 10% of the C_8-ceramide 1-phosphate is converted. The initial rates of dephosphorylation for C_8-ceramide 1-phosphate can be determined at the 5- and 10-min points, and these initial rates are similar in magnitude to those obtained with lysoPA. The reason for the nonlinear time course with C_8-ceramide

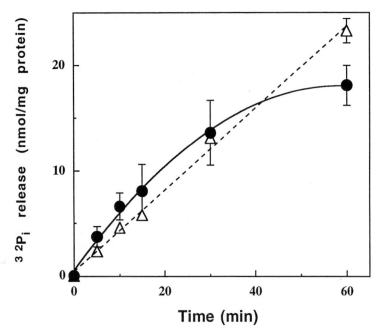

Fig. 2. Time course for the dephosphorylation of lysophosphatidate and C_8-ceramide 1-phosphate. Intact rat2 fibroblasts were incubated at 37° with 0.1% BSA and 50 μM ^{32}P-labeled lysoPA ($\triangle$) or C_8-ceramide 1-phosphate ($\bullet$) for the time indicated and the production of $^{32}P_i$ in the medium was determined. Results are means ± SD (where large enough to be shown) for triplicate assays from three experiments.

1-phosphate is not yet established. Figure 3 shows the dephosphorylation of exogenous sphingosine-1-phosphate as a function of its concentration by intact ECV 304 cells. These results are used to calculate an apparent K_m value for sphingosine-1-phosphate that is approximately 62 μM. ECV cells that overexpress LPP-1, and particularly LPP-3, exhibit enhanced dephosphorylations of sphingosine-1-phosphate.

Concluding Remarks

So far the phosphatase activities that have been detected toward ceramide 1-phosphate and sphingosine-1-phosphate appear to be catalyzed by

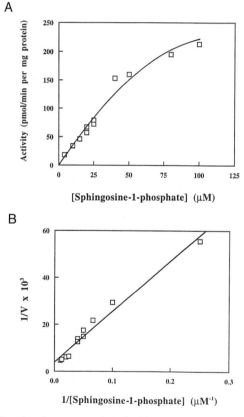

FIG. 3. Effect of changing the concentration of sphingosine-1-phosphate on its dephosphorylation by ECV 304 cells. Intact ECV cells were incubated at 37° for 20 min with 0.1% BSA and various concentrations of ^{32}P-labeled sphingosine-1-phosphate and the production of [32]P$_i$ in the medium was determined. The double reciprocal plot was used to calculate apparent K_m and V_{max} values.

the LPP family of enzymes. An enzyme preparation can be incubated with excess N-ethylmaleimide (1 mM unless thiol-protecting agents are used in the homogenizing buffer) to check if the activity is resistant to this treatment as expected for LPPs. Also, if dephosphorylation is catalyzed by LPPs, then there should be no Mg^{2+} or Ca^{2+} requirement. Lipid phosphate phosphatases may be found in the future that will have different characteristics. Two different types of assays have been described in this article. The first employed the ceramide or sphingosine phosphates in micelles of Triton X-100. This assay can be used to determine the total phosphatase activity against these substrates in a lysate or homogenate. The second assay employs intact cells with exogenous ceramide or sphingosine phosphates. Determining the release of P_i under these conditions measures the effective activity of the enzyme that can degrade external substrates. Our results indicate that this portion of the LPP activity is able to attenuate cell activation by these exogenous lipids, especially when their concentrations are low.[44] Measurement of the various phosphatases that degrade ceramide 1-phosphate and sphingosine-1-phosphate is necessary to elucidate the biological functions and regulations of these enzymes.

[44] R. Jasinska, Q-X. Zhang, C. Pilquil, I. Singh, J. Xu, J. Dewald, D. A. Dillon, L. G. Berthiaume, G. M. Carman, D. W. Waggoner, and D. N. Brindley, *Biochem. J.* **340,** 677 (1999).

[28] Sphingosine-1-phosphate Lyase

By PAUL P. VAN VELDHOVEN

Sphingoid bases are a group of related aliphatic 2-amino-1,3-diols and characteristic constituents of sphingolipids. The most prevalent base in mammals is sphingenine [D(+)-*erythro*-2-amino-4-*trans*-1,3-octadeca-decenediol; sphingosine]. The saturated analog, sphinganine (dihydrosphingosine), and the 4-hydroxyderivative of it, 4D-hydroxysphinganine (phytosphingosine), are less common but are major sphingoid bases in other organisms such as plants, invertebrata, fungi, and protozoa.[1] All sphingoid bases found in nature possess the 2D,3D-configuration, better known as D-*erythro* (in the case of sphingenine or sphinganine), and generally contain 18 carbon atoms.[2] The catabolic removal of sphingoid bases involves a phosphorylation of the primary hydroxyl group, followed by cleavage of

[1] K. A. Karlsson, *Chem. Phys. Lipids* **5,** 6 (1970).
[2] In this article, the substituents at C-2, C-3, and C-4 of the sphingoid bases are specified by the prefix D or L. This prefix refers to the orientation of the substituents to the right (D) or left (L) of the carbon chain when the sphingoid base is drawn vertically in a Fischer projection with C-1 at the top [Nomenclature of Lipids, *Eur. J. Biochem.* **79,** 11 (1977)].

the generated phosphorylated sphingoid base (PSB)[3] between the carbon atoms carrying the amino and the secondary hydroxyl groups. The phosphorylation is catalyzed by sphingosine kinase (for reviews on this enzyme, see Refs. 4 and 5).[6] Mammalian tissues contain a cytosolic and a (presumably plasma) membrane-associated sphingosine kinase, the latter one being responsible for the major portion of the overall tissue activity.[7,8] The cleavage reaction is catalyzed by sphingosine-1-phosphate lyase (sphingosine-phosphate lyase), also named dihydrosphingosine-1-phosphate aldolase or sphinganine-1-phosphate alkanal lyase (EC 4.1.2.27) (for a review, see Ref. 9). In yeast, this reaction would be the rate-limiting step in sphingolipid breakdown.[10]

Interest in PSBs has increased substantially as these zwitterionic lipids appear to evoke different cellular responses. Hence, in addition to being catabolic intermediates, PSBs (especially sphingenine phosphate) are now considered to be inter- and/or intracellular messengers.[11] Whether sphingosine-phosphate lyase plays a role in the attenuation of bioactive sphingenine phosphate has not been critically evaluated, but, most likely, in mammalian cells a plasma membrane-bound sphingosine phosphatase activity[12–14] would be involved in the inactivation process.

Sphingosine-phosphate Lyase Characteristics

Sphingosine-phosphate lyase belongs to the class of pyridoxal phosphate-dependent carbon-carbon lyases and acts on the 1-phosphorylated

---

[3] DTT, dithiothreitol; PSB, phosphorylated sphingoid base; SaPC, sphinganyl-1-phosphorylcholine; SPE, solid-phase extraction; sphinganine-, sphingenine-, and sphingosine-1-phosphate, sphinganine, sphingenine, and sphingosine phosphate; TLC, thin-layer chromatography.

[4] B. Buehrer and R. M. Bell, *Adv. Lipid Res.* **26,** 59 (1993).

[5] A. Olivera, K. D. Barlow, and S. Spiegel, *Methods Enzymol.* **311** [25] 1999 (this volume).

[6] Although sphingosine is sometimes used to refer to sphingenine, in this article the name is used as a generic term to refer to 2D,3D-sphingoid bases without specification of the substituents or degree of unsaturation. In this sense, the names sphingosine kinase, sphingosine-phosphate lyase, and sphingosine phosphatase are used because these enzymes act on sphingenine, sphinganine, or 4D-hydroxysphinganine or their phosphorylated derivatives.

[7] S. Gijsbers, C. Causeret, and P. P. Van Veldhoven, *Lipids* **34,** A066, in press.

[8] A. Olivera, T. Kohama, Z. Tu, S. Milstien, and S. Spiegel, *J. Biol. Chem.* **273,** 12576 (1998).

[9] P. P. Van Veldhoven and G. P. Mannaerts, *Adv. Lipid Res.* **26,** 69 (1993).

[10] J. D. Saba, F. Nara, A. Bielawska, S. Garrett, and Y. A. Hannun, *J. Biol. Chem.* **272,** 26087 (1997).

[11] S. Spiegel and A. H. Merrill, Jr., *FASEB J.* **10,** 1388 (1996).

[12] P. De Ceuster, G. P. Mannaerts, and P. P. Van Veldhoven, *Biochem. J.* **311,** 139 (1995).

[13] D. W. Waggoner, A. Gómez-Muñoz, and D. N. Brindley, *J. Biol. Chem.* **271,** 165056 (1996).

[14] M. Kai, I. Wada, S.-I. Imai, F. Sakane, and H. Kanoh, *J. Biol. Chem.* **272,** 24572 (1997).

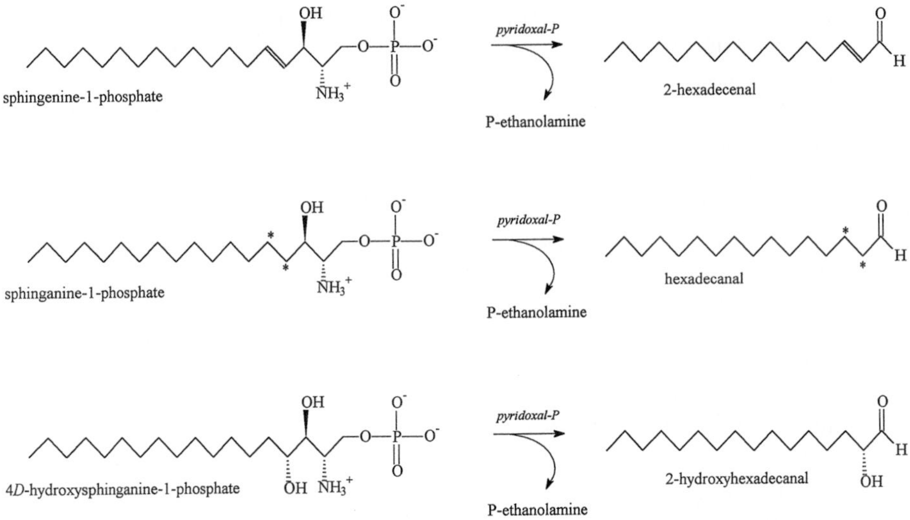

FIG. 1. Structures of the most common naturally occurring phosphorylated sphingoid bases are shown. Their degradation by sphingosine-phosphate lyase results in the formation of a fatty aldehyde and phosphoethanolamine. Asterisks denote the position of the tritium label in [4,5-³H]sphinganine phosphate and in the generated hexadecanal.

derivatives of sphingoid bases. The cleavage products are an aliphatic fatty aldehyde and phosphoethanolamine (Fig. 1). Most of the properties of the enzyme have been treated comprehensively in a previous review.[9] Therefore, only those characteristics of direct importance for the activity measurements will be discussed briefly, together with new relevant data. For more details and the complete list of original references, the interested reader is referred to this review.

Sphingosine-phosphate lyase is an ubiquitous enzyme present in all tissues in vertebrates, in invertebrates, plants, fungi, and unicellular protozoa. One apparent exception are platelets, which are devoid of lyase activity,[15,16] as has been confirmed.[17] The protein is firmly associated with the endoplasmic reticulum and with the catalytic site facing the cytosol. Detergent solubilization causes inactivation of the enzyme,[18] and attempts to purify the enzyme from rat liver microsomes failed.[19] The yeast lyase cDNA

[15] W. Stoffel, G. Assmann, and E. Binczek, *Hoppe Seyler's Z. Physiol. Chem.* **351,** 635 (1970).
[16] W. Stoffel, G. Heimann, and B. Hellenbroich, *Hoppe Seyler's Z. Physiol. Chem.* **354,** 562 (1973).
[17] Y. Yatomi, S. Yamamura, F. Ruan, and Y. Igarashi, *J. Biol. Chem.* **272,** 5291 (1997).
[18] P. P. Van Veldhoven and G. P. Mannaerts, *J. Biol. Chem.* **266,** 12502 (1991).
[19] P. P. Van Veldhoven, unpublished data.

has been cloned,[10] however, and subsequently, by homology, the murine[20] and human[19] counterparts. The primary structure of the lyase of the nematode *Caenorhabditis elegans* is also known.[20] The respective mRNAs encode a protein with a molecular mass of approximately 61 to 65 kDa. Only one membrane span, located near the amino terminus of these proteins, appears to be present.

PSBs, like all sphingolipids, possess two optical centers. Although sphingosine kinase phosphorylate(s) all four isomers of sphingoid bases, only the phosphate esters of the naturally occurring 2D,3D-isomers are recognized by the lyase. Substrates mostly studied are the 1-phosphorylated derivatives of sphingenine, sphinganine and 4D-hydroxysphinganine. The cleavage of these PSBs results in the formation of 2-*trans*-hexadecenal, hexadecanal, and 2D-hydroxyhexadecanal, respectively. The other reaction product is phosphoethanolamine (Fig. 1). In intact systems, both cleavage products are actively further metabolized as discussed in detail in Ref. 9. Interestingly, fatty aldehydes derived from sphingolipid breakdown are, after reduction to fatty alcohols, incorporated adequately in ether glycerolipids.

In contrast to this stereospecificity, the lyase does not seem to be very discriminative toward the chain length and possible substituents of the sphingoid base moiety. The shorter (down to 7 carbons)[21] and longer (20 carbons) analogs of sphinganine phosphate are cleaved,[22] as well as the phosphate ester of sphingenine with a 4-*cis* unsaturation.[23] Other sphingenine phosphate analogs carrying a methyl group at carbon 4 or 5, either with a 4-*trans* or 4-*cis* double bond, are also recognized; 4-methyl-4-*cis*-sphingenine phosphate, however, is a poor substrate.[23] For most of these substrates, no detailed kinetic analysis is available. Sphingenine phosphate and sphinganine phosphate are degraded at similar rates. The K_m values for sphinganine phosphate determined in rat liver microsomes range from 9 to 16 μM,[18,24] whereas a value of 110 μM was reported for the lyase of *Tetrahymena pyriformis*.[25]

An analog with a modified head group, sphinganine-1-phosphonate, is also cleaved but is slower compared to sphinganine phosphate.[24] Sphinganine, however, is not a substrate.[26] Modifications of the amino group of

[20] J. Zhou and J. D. Saba, *Biochem. Biophys. Res. Commun.* **242**, 502 (1998).

[21] W. Stoffel, G. Sticht, and D. LeKim, *Hoppe Seyler's Z. Physiol. Chem.* **350**, 63 (1969).

[22] W. Stoffel and A. Scheid, *Hoppe Seyler's Z. Physiol. Chem.* **350**, 1593 (1969).

[23] G. van Echten-Deckert, A. Zsoche, T. Bar, R. R. Schmidt, A. Raths, T. Heinemann, and K. Sandhoff, *J. Biol. Chem.* **272**, 15825 (1998).

[24] W. Stoffel and M. Grol, *Chem. Phys. Lipids* **13**, 372 (1974).

[25] W. Stoffel, E. Bauer, and J. Stahl, *Hoppe Seyler's Z. Physiol. Chem.* **355**, 61 (1974).

[26] C. Causeret, S. Gijsbers, and P. P. Van Veldhoven, unpublished data.

the substrate are not well tolerated. *N*-Acetylsphinganine phosphate and (most likely) *N*,*N*-dimethylsphingenine phosphate are not cleaved.[19]

Pyridoxal phosphate is used as cofactor, which appears loosely associated with the lyase. Hence, carbonyl-reactive compounds (cyanide, semicarbazide, hydrazines, amines) as well as pyridoxal phosphate analogs (deoxypyridoxine phosphate) will inhibit the lyase reaction. The cofactor is thought to form a Schiff base with the free amino group of the PSB and to remain associated with the two-carbon fragment during the cleavage reaction.[27] Amino acids involved in the binding of the cofactor have not been established. Although lyases display a high homology to glutamate decarboxylases, the signature motif for the pyridoxal phosphate binding sites present in glutamate decarboxylases is not found in the deduced amino acid sequences of the already cloned lyases.[19] Motifs found in other pyridoxal phosphate-dependent enzymes do not seem to be present in the cloned lyases, unless with some mismatches.[19]

The lyase is sensitive toward sulfhydryl reagents such as *N*-ethylmaleimide, iodoacetamide, and *p*-chloromercuribenzoate. The inhibitory effects of these agents are due to the alkylation of a cysteine residue, the SH group of which is postulated to be involved in a nucleophilic attack of the pyridoxal phosphate complexed substrate at the third carbon.[28] Comparison of the primary structures of the cloned lyases reveals the presence of two conserved cysteine residues located within a conserved stretch of amino acids in the middle part of the protein.[19]

Heavy metal ions are potent inhibitors, with Ca^{2+} and Zn^{2+} being the most potent ones, and chelators stimulate the activity.[18]

Only a few (specific) inhibitors are known. Sphinganine-1-phosphonate acts as a competive inhibitor (K_i 5 μM; in rat liver microsomes).[24] The 2*D*,3*L*-isomer of sphinganine phosphate is also a competitive inhibitor (K_i 9.7 μM; in *T. pyriformis*), but the 2*L*,3*L*-isomer does not influence the lyase.[25] 4-Methyl-4-*cis*-sphingenine phosphate might be inhibitory.[23] The aminopentol prepared from fumonisin B1 by alkaline hydrolysis and resembling a sphingoid base (but with a 2*D*,3*L* configuration) affects the lyase (IC_{50} 20 μM).[26] The different isomers of sphingoid bases are weak inhibitors, with the 2*D*,3*D*-isomer being the most potent (IC_{50} 75 μM).[26] The effects of *N*,*N*,*N*-trimethyl-3-D,L-sphingenine phosphate (IC_{50} 150 μM) and 3-D,L-sphingenyl-1-phosphorylcholine (IC_{50} 600 μM) are negligible.[29] Various detergents are deleterious to the enzyme, with *N*,*N*-dimethyldodecylamine

[27] T. Shimojo, T. Akino, Y. Miura, and G. J. Schroepfer, Jr., *J. Biol. Chem.* **251,** 4448 (1976).
[28] W. Stoffel, *Chem. Phys. Lipids* **5,** 139 (1970).
[29] S. Su and P. P. Van Veldhoven, unpublished data.

N-oxide (LDAO) being very potent (IC$_{50}$ 50 μM).[18,29,30] A selective inhibitor is a 2-vinyl analog of sphinganine phosphate (IC$_{50}$ 2.4 μM; in rat liver microsomes).[31] The action of the latter analog is likely caused by the presence of the allylic amino group.

Sphingosine-phosphate Lyase Activity Measurements

Considerations

Because of the low activity of the lyase, mainly radiometric assays have been developed and/or used to measure the enzyme. Labeled substrates that have been prepared include [1-^{14}C]sphinganine phosphate,[32] [3-^{14}C]sphinganine phosphate,[32] [1-^{3}H]sphinganine phosphate,[32] [4,5-^{3}H]sphinganine-phosphate,[14] [5-^{14}C]sphingenine phosphate,[33] [1-^{3}H]sphingenine phosphate,[34] (3-D,L)-[3-^{3}H]sphingenine phosphate,[17,35] and sphingenine [^{32}P]phosphate.[13,36–38] The preparation of some of these substrates will be described further. The tritiated compounds, [4,5-^{3}H]sphinganine phosphate and [3-^{3}H]sphingenine phosphate, are the most convenient to work with because they are relatively easy to obtain or prepare and the label is retained in the produced aldehyde, facilitating the measurements (see Fig. 1). The use of [4,5-^{3}H]sphinganine phosphate is preferred because [3-^{3}H]sphingenine phosphate, unless specified, is racemic at carbon 3 and the 2D,3L isomer has been reported to be inhibitory,[25] and during the possible oxidation of the produced fatty aldehyde the tritium label is lost.

The following assay is based on the quantitation of tritiated hexadecanal generated from [4,5-^{3}H]sphinganine phosphate[18] and has been employed in extracts of rat tissues,[9] cultured mammalian cells,[9] and yeast.[10] In crude cellular or tissue extracts, fatty aldehydes can be metabolized further, mainly oxidized to fatty acids. In order to recover the eventually formed labeled palmitic acid, the assay mixture is subjected to an extraction and

[30] IC$_{50}$ values determined in rat liver microsomes were 4.4 mM for 3-[(3-cholamidopropyl)dimethylammonio]-1-propanesulfonate (CHAPS),[18] 2 mM for octanoyl-N-methylglucamide (MEGA-8),[19] 1.5 mM for dodecylmaltoside,[29] and 0.7 mM for octylglucoside.[18]

[31] A. Boumendjel and S. P. F. Miller, *Tetrahed. Lett.* **35**, 819 (1994).

[32] W. Stoffel, G. Sticht, and D. LeKim, *Hoppe Seyler's Z. Physiol. Chem.* **349**, 1745 (1968).

[33] C. B. Hirschberg, A. Kisic, and G. J. Schroepfer, *J. Biol. Chem.* **245**, 3084 (1970).

[34] W. Stoffel and G. Assmann, *Hoppe Seyler's Z. Physiol. Chem.* **351**, 1041 (1970).

[35] A. Gómez-Muñoz, D. W. Waggoner, L. O'Brien, and D. N. Brindley, *J. Biol. Chem.* **270**, 26318 (1995).

[36] K. A. Dressler and R. N. Kolesnick, *J. Biol. Chem.* **265**, 14917 (1990).

[37] N. N. Desai, H. Zhang, A. Olivera, M. E. Mattie, and S. Spiegel, *J. Biol. Chem.* **267**, 23122 (1992).

[38] A. Olivera, J. Rosenthal, and S. Spiegel, *Anal. Biochem.* **223**, 306 (1994).

phase separated under acidic conditions. Under these conditions, however, the substrate will distribute between both phases and a subsequent separation of the reaction products from the substrate by TLC (or SPE) is required.

To limit the hydrolysis of the cofactor, the reaction medium is buffered with a high concentration of phosphate and fortified with NaF. Hydrolysis of the substrate, however, cannot be completely counteracted under the assay conditions. One should keep in mind that in some tissues, e.g., nervous tissue, very high sphingosine phosphatase activities have been found.[12] Hence, such samples should be sufficiently diluted.

Despite their zwitterionic character, PSBs display a peculiar solubility, being poorly soluble in aqueous as well as in organic solvents. In the assay, the substrate is solubilized in Triton X-100. This detergent, in contrast to many others,[18,30] does not interfere with (mammalian) sphingosine-phosphate lyases.

Finally, sphingosine-phosphate lyase is a rather stable enzyme. The activity is well preserved when mammalian tissues or cells are stored at $-80°$. The lyase activity in mammalian tissues and cultured cells ranges between 20 and 100 pmol/min mg protein.[9]

Procedure

For enzyme incubations, the following stock solutions are required:

0.5 M potassium phosphate buffer, pH 7.4, store at $4°$

0.1 M EDTA, pH 7.4; store at $4°$

1% (w/v) Triton X-100; store at $4°$

1 M DTT; store at $4°$ in a well-closed plastic tube

0.5 M NaF; store frozen in aliquots

5 mM pyridoxal phosphate; store frozen in aliquots protected from light

[4,5-^{3}H]Sphinganine phosphate (specific activity approximately 80,000 dpm/nmol); approximately 0.5 to 1.0 mM methanolic solution; store at $-20°$; before use, warm up and place in a sonication bath; hazy solution

1% (w/v) HClO$_4$; store at room temperature

1% (w/v) HClO$_4$/methanol (8/2, v/v); keep at room temperature, most conveniently in a 2-ml dispenser

Chloroform/methanol (1/2, v/v); keep at room temperature; most conveniently in a 2-ml dispenser

Chloroform; keep at room temperature; most conveniently in a 1-ml dispenser

For the analysis of the reaction products, the following solutions/materials are used:

Chloroform/methanol (8/2, v/v) containing 5 mM palmitic acid (or oleic acid), hexadecanol, and hexadecanal. The synthesis of the latter compound is described in Ref. 18. Store solution at $-20°$ in well-closed screw cap tubes.

TLC tanks lined with thin chromatography paper and containing solvent A (chloroform/methanol/acetic acid; 50/50/1, v/v) or solvent B (hexane/diethylether/acetic acid; 70/30/1, v/v)

Silica G 60A plates (20 × 20 cm; 0.25-mm-thickness glass plates; Merck or comparable TLC plates)

Aldehyde staining solution: dissolve 0.25 g of 4-amino-3-hydrazino-5-mercapto-1,2,4-triazole (Purpald; Aldrich Chemical Company, Milwaukee, WI) in 5 ml of 2.5 N HCl and adjust volume to 50 ml with methanol; stable at room temperature[18]

1% (w/v) sodium dodecyl sulfate (SDS)

The reaction mixture, sufficient for 25 assays, is prepared as follows. Dry 200 nmol of tritiated substrate under nitrogen in a 10-ml plastic or disposable glass tube and dissolve white residue in 500 μl of Triton X-100 solution. Add subsequently 1 ml of buffer, 50 μl of EDTA, 250 μl of NaF, 5 μl of DTT, 250 μl of pyridoxal phosphate, and 1945 μl of water (total volume of 4 ml). Upon addition of the pyridoxal phosphate solution, one will notice a color change from yellow to greenish-yellow, which is due to formation of the Schiff base between cofactor and substrate (if not observed, the solubization of the substrate is questionable). Keep reaction mixture at $37°$ and place, before use, in a sonication bath for several seconds to ensure complete solubilization of the substrate (by preparing some excess of reaction mixture, aliquots can be removed for the determination of the radioactivity in order to check the substrate solubility).

The assay, modified slightly from the original procedure described in Ref. 18, is performed as follows. Dilute samples with a suitable medium (e.g., 0.25 M sucrose, 5 mM MOPS, pH 7.5, 1 mM EDTA, and 1 mM DTT for mammalian tissue homogenates) and place 40 μl of the diluted sample at the bottom of 5-ml (disposable) screw-cap glass tubes. Include at least one blank containing only dilution medium. Start reaction by adding 160 μl of reaction mixture and place tubes at $37°$ in a shaking water bath. Stop reaction after 60 min by adding 0.2 ml of HClO$_4$, followed by 1.5 ml of chloroform/methanol and vortex tubes thoroughly. Induce phase separation by adding 0.5 ml of chloroform and 0.5 ml of HClO$_4$, mix again by vortexing. Centrifuge tubes at low speed to facilitate phase separation and remove upper phase. Wash lower phase once with 1 ml of HClO$_4$/methanol. Transfer a known volume of lower phase (0.6 ml/0.85 ml total) to a new series of tubes, dry solvent under nitrogen, and dissolve residue in 50 μl of chloroform/methanol (8/2, v/v) containing 5 mM of appropriate carriers (palmitic acid, hexadecanol, and hexadecanal). Spot 20 μl on TLC plates (that were previously developed with acetone and dried) and develop plates in solvent A. In this system, hexadecanal (R_f 0.87) and its further metabolites, palmitic acid (R_f 0.80) and hexadecanol (R_f 0.84), migrate close to the front, well separated from sphinganine (R_f 0.30) and the substrate, which remains at the origin. Allow plates to dry and place in a closed chamber containing some iodine crystals. The most intense spot corresponds to Triton-X 100 (R_f 0.70). Selective visualization of aldehydes can be obtained by spraying the dried plates with an acidic solution of 4-amino-3-hydrazino-5-mercapto-1,2,4-triazole, followed by exposure to ammonia fumes.[18] Aldehydes will stain pink-violet on a clear background.[39] If exposed too long, background will also stain. Scrape silica in the region with R_f between 0.75 and 0.95 in scintillation vials containing 0.5 ml of 1% SDS, add suitable scintillation fluid, and determine radioactivity.[40] If analysis of secondary metabolites is desirable, develop plates in solvent B. R_f values for hexadecanal, palmitic acid, and hexadecanol are 0.48, 0.28, and 0.16, respectively. If a radioscanner is available, plates can be scanned for radioactivity; however, in general, the counting efficiency of tritium labeled compounds absorbed in silica layers is very low.

Synthesis of Substrates

One major obstacle in the determination of sphingosine-phosphate lyase is the limited availability of cold and suitably labeled substrate. Presently, suppliers of (racemic) sphingenine phosphate are manyfold (Sigma, St. Louis, MO; Biomol, Plymouth Meeting, PA; Toronto Research Chemicals, Ontario, Canada; ICN, Costa Mesa, CA; Calbiochem-Novabiochem, San Diego, CA; LC Laboratories, Woburn, MA; Universal Biologicals Ltd., Gloucestershire, UK). Sphinganine phosphate can be obtained from Biomol or LC Laboratories. [4,5-^{3}H]Sphinganine phosphate has also become available commercially (American Radiolabeled Chemicals, Inc., St. Louis, MO). Because of the considerably high cost, especially of the labeled substrate, synthesis in the laboratory is still worth considering. Although differ-

---

[39] Solutions of hexadecanal, like that of other aldehydes, are unstable. On prolonged storage, two additional spots, staining less intense with Purpald and migrating somewhat faster than the main component, are seen.

[40] In order to limit the risk of exposure to silica dust, TLC scrapings are performed in a hood and the silica is directly collected in an SDS solution. The detergent helps elute the lipids from the silica, resulting in higher initial counts for most tritiated lipids tested. This might depend, however, on the kind of scintillation fluid. Therefore, it is recommended to count the vials at a later time point to ensure that there was no further rise in the counts.

ent methods have appeared for the total synthesis of PSBs (see Ref. 41), these have not been applied to prepare labeled compounds. Given the increasing commercial availability of labeled sphingoid bases, (racemic) [3-^{3}H]sphingenine (ICN; Dupont-NEN, Boston, MA) and [4,5-^{3}H]sphinganine (American Radiolabeled Chemicals, Inc.), chemical phosphorylation of these lipids, as described for nonlabeled sphingoid bases,[42] becomes an interesting possibility. In the absence of facilities for organic synthesis, however, the simplest way to obtain labeled PSB is via biochemical approaches, sometimes combined with straightforward and simple organic chemistry. A first method consists of the N-acylation of labeled sphingenine or sphinganine with a suitable activated short chain fatty acid (e.g., hexanoylanhydride or hexanoylchloride), followed by phosphorylation of the generated ceramide by means of *Escherichia coli* DAG kinase,[43,44] and acidic hydrolysis of the ceramide 1-phosphate to destroy the amide bond.[13,35–37,45] The *E. coli* kinase is available commercially (see Ref. 46) and this method allows also for the preparation of ^{32}P-labeled PSBs.[13,36,37] In a second method, sphingoid bases are phosphorylated enzymically. Different sources of sphingosine kinase have been described: platelets,[15] cytosolic fractions from rat liver,[33] and partially purified preparations from Swiss 3T3 cells,[38] from BALB/c 3T3 fibroblasts,[47] or from rat kidney cytosol,[48] and purified rat renal enzyme.[8] Similar to the first method, PSBs labeled either in the sphingoid base or in the phosphate group can be generated. In a last method, [4,5-^{3}H]sphinganyl-1-phosphorylcholine (SaPC) is treated with phospholipase D from *Streptomyces chromofuscus*.[18,49] Tritiated SaPC can be obtained by acidic hydrolysis of tritiated sphingomyelin,[18] the latter being generated by the catalytic reduction of sphingomyelin in the presence of [^{3}H]Na- or KBH$_4$.[18,50]

Regardless of the method employed to prepare PSBs, one should be aware that these lipids are sparingly soluble in polar and organic solvents, which is a severe drawback for their purification/isolation. A convenient

[41] A. Bielawska, Z. Szulc, and Y. A. Hannun, *Methods Enzymol.* **311** [44] 1999 (this volume).

[42] A. Boumendjel and S. P. F. Miller, *J. Lipid Res.* **35**, 2305 (1994).

[43] P. P. Van Veldhoven, R. M. Bishop, and R. M. Bell, *Anal. Biochem.* **183**, 177 (1989).

[44] P. P. Van Veldhoven, R. M. Bishop, D. A. Yurivich, and R. M. Bell, *Biochem. Mol. Biol. Intern.* **36**, 21 (1995).

[45] Due to the acidic conditions, sphingenine-phosphate prepared in this way might be racemic at carbon 3 (see Ref. 1).

[46] D. Perry, A. Bielawska, and Y. A. Hannun, *Methods Enzymol.* **311** [3] 1999 (this volume).

[47] N. Mazurek, T. Megidish, S. Hakomori, and Y. Igarashi, *Biochem. Biophys. Res. Commun.* **198**, 1 (1994).

[48] S. Gijsbers and P. P. Van Veldhoven, unpublished data.

[49] P. P. Van Veldhoven, R. Foglesong, and R. M. Bell, *J. Lipid Res.* **30**, 611 (1989).

[50] G. Schwarzmann, *Biochim. Biophys. Acta* **529**, 106 (1978).

way worked out during our studies on tissue levels of PSBs and their metabolism[12,51] appears to be the use of reversed-phase columns. The following protocol is currently used to obtain (tritiated) sphinganine phosphate. Dissolve approximately 30 μmol of (tritiated) SaPC in 10 ml of 0.2 M ammonium phosphate buffer, pH 8.0,[52] add approximately 100 μg (or 2 U) of phospholipase D (*Streptomyces chomofuscus;* Boehringer, Mannheim, Germany), and allow to incubate at 37°. Collect the precipitating PSBs after 1 hr by centrifugation at 4° at 10,000 g for 10 min. Allow the supernatant to incubate further overnight to obtain a second precipitate. Combine the copious pellets in 10 ml of 0.1 M NH$_4$OH and apply the foamy solution (depending on the actual PSB concentration, some haziness might be present) on an activated 500-mg C$_{18}$-SPE column (Bond-Elut C$_{18}$; Varian, Harbor City, CA or Sep-Pak Vac RC-C$_{18}$; Millipore, Milford, MA). Wash the SPE column with 2 $\times$ 5 ml of 0.1 M NH$_4$OH and with 2 $\times$ 5 ml of 0.1 M NH$_4$OH containing 30% (v/v) methanol. Elute PSBs with repetitive (up to six times) additions of 5-ml aliquots of 0.1 M NH$_4$OH containing 60% (v/v) methanol and analyze the eluates for the presence of PSBs by TLC (solvent chloroform/methanol/water 60/35/8, v/v, R_f approximately 0.25).[53] Possible contaminants, such as sphingoid bases and SaPC, are retained on the columns under these conditions. Dry fractions containing PSBs under nitrogen and dissolve the ammonium salts in methanol. Store at −20°. Alternatively, instead of relying on their insolubility, PSBs can be recovered in the upper phase during extractions according to Bligh and Dyer when alkaline solutions (e.g., 1 M NH$_4$OH) are used for phase separation.[51] Prior to application of such upper phase on C$_{18}$-SPE columns, remove organic solvents by rotovap or dilute twofold with water.

Acknowledgments

The author thanks Dr. G. P. Mannaerts for his helpful editorial comments. Research was supported by grants from the "Fonds voor Wetenschappelijk Onderzoek—Vlaanderen; Project G.0240.98" and the "Interuniversitaire Attractiepolen; Project P4/23."

[51] P. P. Van Veldhoven, P. de Ceuster, R. Rozenberg, G. P. Mannaerts, and E. de Hoffmann, *FEBS Lett.* **350,** 91 (1994).

[52] In the original procedure (Ref. 49), ammonium acetate was used as buffer. Apparently phospholipase D preparations contain a phosphatase that cleaves the generated PSBs. This phosphatase activity is suppressed by phosphate ions.[19]

[53] The volume of medium required to elute the PSBs increases with the load of PSBs. This is presumably related to the limited solubility of PSBs, approximately 1 mM in the elution medium. Hence the specified volumes are adequate for the isolation of approximately 10 mg of PSBs.

[29] Sphingolipid Hydrolases and Activator Proteins

By Uwe Bierfreund, Thomas Kolter, and Konrad Sandhoff

Introduction

Glycosphingolipids (GSLs)[1] are amphiphilic constituents of eukaryotic plasma membranes. They contain a hydrophobic ceramide moiety that anchors them in the lipid bilayer and a hydrophilic, extracellular oligosaccharide chain. GSLs are involved in cell-type specific adhesion processes[2] and can influence the activity of membrane-bound receptors and enzymes.[3] Inherited defects within the GSL degradation pathway give rise to various human diseases, the sphingolipidoses.[4] The constitutive degradation of GSLs occurs in the acidic compartments of the cell: endosomes and lysosomes. Cell surface-derived GSLs reach the lysosomal compartment by endocytic membrane flow, presumably on the surface of intraendosomal and intralysosomal vesicles.[5] Here, the GSLs are cleaved into their building blocks by the stepwise action of hydrolytic enzymes. Glycosidases cleave off the sugar residues from the nonreducing end of the GSLs. In the case of GSLs with short oligosaccharide chains of less than four sugar residues, the degrading enzymes need the assistance of protein cofactors, the so-called sphingolipid activator proteins. The sphingolipid activator proteins known so far are derived from two genes. One gene on chromosome 5 carries the information for the GM2 activator[6] and the other gene on chromosome 10 for the SAP precursor, also called prosaposin,[7–9] which is processed to four homologous proteins (also called saposins): SAP-A, SAP-B, SAP-C, and SAP-D.[10] Mechanistically, the GM2 activator and SAP-B were shown to function by solubilizing the lipid substrate, thus

---

[1] C. L. M. Stults, C. C. Sweeley, and B. A. Macher, *Methods Enzymol.* **179**, 167 (1989).

[2] A. Varki, *Glycobiology* **3**, 97 (1993).

[3] R. L. Schnaar, *Glycobiology* **1**, 477 (1991).

[4] T. Kolter and K. Sandhoff, *Brain Pathol.* **8**, 79 (1998).

[5] K. Sandhoff and T. Kolter, *Trends Cell Biol.* **6**, 98 (1996).

[6] H. Klima, A. Tanaka, D. Schnabel, T. Nakano, M. Schröder, K. Suzuki, and K. Sandhoff, *FEBS Lett.* **289**, 260 (1991).

[7] W. Fürst, W. Machleidt, and K. Sandhoff, *Biol. Chem. Hoppe-Seyler* **369**, 317 (1988).

[8] J. S. O'Brien, K. A. Kretz, N. Dewji, D. A. Wenger, F. Esch, and A. L. Fluharty, *Science* **241**, 1098 (1988).

[9] T. Nakano, K. Sandhoff, J. Stümper, H. Christomanou, and K. Suzuki, *J. Biochem.* **105**, 152 (1989).

[10] G. Vielhaber, R. Hurwitz, and K. Sandhoff, *J. Biol. Chem.* **271**, 32438 (1996).

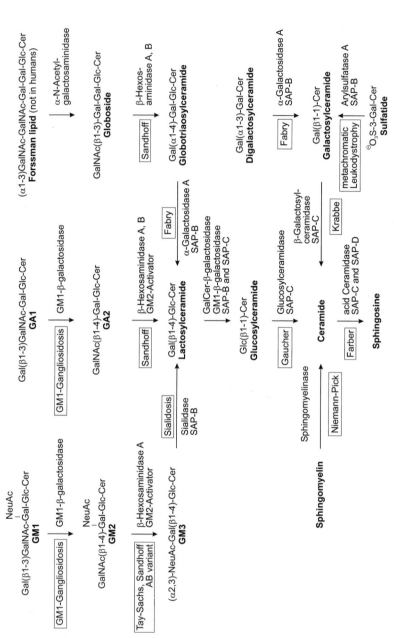

Fig. 1. Degradation of selected sphingolipids in human lysosomes. The eponyms of individual inherited diseases (boxes) are given. Individual metabolites show considerable heterogeneity within the lipid portion, which is not indicated. Those activator proteins that are required for the respective degradation step *in vivo* are indicated. Variant AB, AB variant of GM2 gangliosidosis (deficiency of GM2 activator protein). SAP, sphingolipid activator protein (modified from Ref. 118).

making it accessible to the water-soluble enzyme, whereas the other SAPs seem to facilitate interaction of the respective hydrolase with the lipid in or at the membrane in an as yet less well-characterized way.[4] While approaches for assaying sphingolipid hydrolases[11] and sphingolipid activator proteins[12] have been described in detail in former volumes of this series, this article focuses on experimental approaches to mimic the *in vivo* degradation of glycolipids as closely as possible. This is required to obtain reliable data on activator protein specificity and function.

GM2 Activator

The *in vivo* degradation of ganglioside GM2, the major storage compound in GM2 gangliosidoses, requires the water-soluble enzyme β-hexosaminidase A and a lysosomal glycolipid-binding protein, the GM2 activator.[13,14] *N*-Acetylglucosamine- and *N*-acetylgalactosamine residues on glycoconjugates, which extend far enough into the lysosol, are cleaved by the β-hexosaminidases without the requirement for an activator protein. The *in vivo* degradation of glycolipid substrates of hexosaminidase with short carbohydrate head groups such as ganglioside GM2 or glycolipid GA2 needs the assistance of the GM2 activator protein. A range of experimental data[15] suggests the mechanism of action of this protein. The surface-active GM2 activator binds and lifts ganglioside GM2 from a lipid aggregate or lipid bilayer and forms a water-soluble, 1:1 protein–lipid complex. β-Hexosaminidase A binds to this complex, cleaves the ganglioside to yield ganglioside GM3, and releases the activator–lipid complex. The activator may reinsert GM3 into the membrane and bind another GM2 molecule. In addition to the protein–lipid interaction among the GM2 activator, β-hexosaminidase A, and ganglioside GM2, a protein–protein interaction between the GM2 activator and β-hexosaminidase A is involved in this process.[16] Homoallelic mutations within the structural gene of the GM2 activator have been identified in four patients with the AB variant of GM2 gangliosidosis, where the GM2 activator is missing. These mutations lead

---

[11] K. Suzuki, *Methods Enzymol.* **138,** 727 (1987).

[12] E. Conzelmann and K. Sandhoff, *Methods Enzymol.* **138,** 792 (1987).

[13] R. A. Gravel, J. T. R. Clarke, M. M. Kaback, D. Mahuran, K. Sandhoff, and K. Suzuki, in "The Metabolic and Molecular Bases of Inherited Disease" (C. Scriver, A. L. Beaudet, W. S. Sly, and D. Valle, eds.), 7th ed., Chapter 92, p. 2839. McGraw-Hill, New York, 1995.

[14] K. Sandhoff, K. Harzer, and W. Fürst, in: The Metabolic and Molecular Bases of Inherited Disease (C. Scriver, A. L. Beaudet, W. S. Sly, and D. Valle, eds.), 7th ed., Chapter 76, p. 2427. McGraw Hill, New York, 1995.

[15] E. M. Meier, G. Schwarzmann, W. Fürst, and K. Sandhoff, *J. Biol. Chem.* **266,** 1879 (1991).

[16] H. J. Kytzia and K. Sandhoff, *J. Biol. Chem.* **260,** 7568 (1985).

to premature degradation of the gene products.[17–19] This results in a block within the GM2 degradation in the lysosomes that can be circumvented by feeding native GM2 activator to the culture medium of the mutant fibroblasts.[20] In contrast to isolated deficiencies of SAP-B or SAP-C, a complete loss of GM2 activator function leads to a storage disease with infantile onset. This indicates that no significant GM2 degradation occurs *in vivo* in the absence of the activator. In addition to this function of the GM2 activator, the lipid storage pattern observed in GM2 activator-deficient mice[21] and in AB variant brain tissue[22] indicates that it also facilitates the degradation of GA2. Sufficient amounts of pure GM2 activator are available from different expression systems in *Escherichia coli*[20,23] and in insect cells.[24]

SAP Precursor-Derived Activator Proteins

The first sphingolipid activator protein was discovered by H. Jatzkewitz in 1964.[25] It is required for the hydrolytic degradation of GSLs carrying a sulfuric ester group, the sulfatides. This sphingolipid activator protein, SAP-B or saposin B, is a small glycoprotein of 80 amino acids bearing three disulfide bridges and an N-linked carbohydrate chain.[26] It is able to bind GSLs with broader specificity than the GM2 activator and may be regarded as a physiological detergent that presents GSLs to water-soluble enzymes as substrates.[27] The inherited deficiency of SAP-B leads to a lysosomal storage disease that resembles metachromatic leukodystrophy. In addition to SAP-B, three other sphingolipid activator proteins (SAP-A, SAP-C, and SAP-D) are derived proteolytically from the SAP precursor

[17] U. Schepers, G. J. Glombitza, T. Lemm, A. Hoffmann, A. Chabàs, P. Ozand, and K. Sandhoff, *Am. J. Hum. Genet.* **59**, 1048 (1996).

[18] M. Schröder, D. Schnabel, R. Hurwitz, E. Young, K. Suzuki, and K. Sandhoff, *Hum. Genet.* **92**, 437 (1993).

[19] M. Schröder, D. Schnabel, K. Suzuki, and K. Sandhoff, *FEBS Lett.* **290**, 1 (1991).

[20] H. Klima, A. Klein, G. van Echten, G. Schwarzmann, K. Suzuki, and K. Sandhoff, *Biochem. J.* **292**, 571 (1993).

[21] Y. Liu, A. Hoffmann, A. Grinberg, H. Westphal, M. P. McDonald, K. M. Miller, J. N. Crawley, K. Sandhoff, K. Suzuki, and R. L. Proia, *Proc. Natl. Acad. Sci. USA* **94**, 8138 (1997).

[22] K. Sandhoff, K. Harzer, W. Wässle, and H. Jatzkewitz, *J. Neurochem.* **18**, 2469 (1971).

[23] Y. Y. Wu, J. M. Lockyer, E. Sugiyama, N. V. Pavlova, Y.-T. Li, and S.-C. Li, *J. Biol. Chem.* **269**, 16276 (1994).

[24] C. G. Schütte, T. Lemm, G. J. Glombitza, and K. Sandhoff, *Prot. Sci.* **7**, 1039 (1998).

[25] E. Mehl and H. Jatzkewitz, *Hoppe-Seyler's Z. Physiol. Chem.* **339**, 260 (1964).

[26] W. Fürst, J. Schubert, W. Machleidt, H. E. Meyer, and K. Sandhoff, *Eur. J. Biochem.* **192**, 709 (1990).

[27] S. C. Li, S. Sonnino, G. Tettamanti, and Y. T. Li, *J. Biol. Chem.* **263**, 6588 (1988).

(pSAP). They are homologous to each other and have similar properties, but differ in their specificity and in the mechanism of action. SAP-C deficiency leads to an atypical form of Gaucher disease.[28,29] Isolated deficiencies of SAP-A and SAP-D have not been reported to date. A human patient, however, has been identified who lacks the whole SAP precursor protein due to a homoallelic mutation within the start codon of pSAP.[30] pSAP deficiency leads to storage of various sphingolipids with no or short oligosaccharide chains, such as ceramide, glucosylceramide, sulfatides, digalactosylceramide, and globotriaosylceramide.[31] Loading studies with cells from this patient clarified the *in vivo* function of SAP-D.[32] The accumulation of ceramide in pSAP-deficient cells can be overcome by the addition of SAP-D to the culture medium. This indicates a role of SAP-D for the lysosomal degradation of ceramide by acid ceramidase. It was also shown that SAP-C is able to reduce ceramide storage in pSAP-deficient cells (A. Klein, unpublished results). These findings were confirmed in an *in vitro* liposomal assay system for acid ceramidase (T. Linke, personal communication). In pSAP-deficient cells the impaired galactosylceramide degradation could be partially restored by the addition of SAP-A, SAP-C, or a mixture of both activators to the culture medium. Because galactosylceramide degradation is apparently normal in cells derived from patients with SAP-C deficiency, SAP-A seems to be involved in galactosylceramide degradation in living cells.[33] An animal model of SAP precursor deficiency has been developed in mice.[34]

Assays for Sphingolipid Activator Proteins, Lysosomal Hydrolases, and Their GSL Substrates

The activity of sphingolipid activator proteins is quantified by measuring their capability to stimulate GSL hydrolysis by lysosomal hydrolases *in vitro*. Extrapolation of these *in vitro* results to the *in vivo* situation is

---

[28] H. Christomanou, A. Aignesberger, and R. P. Linke, *Biol. Chem. Hoppe-Seyler* **367,** 879 (1986).

[29] D. Schnabel, M. Schröder, and K. Sandhoff, *FEBS Lett.* **284,** 57 (1991).

[30] D. Schnabel, M. Schröder, W. Fürst, A. Klein, R. Hurwitz, T. Zenk, J. Weber, K. Harzer, B. C. Paton, A. Poulos, K. Suzuki, and K. Sandhoff, *J. Biol. Chem.* **267,** 3312 (1992).

[31] V. Bradova, F. Smid, B. Ulrich-Bott, W. Roggendorf, B. C. Paton, and K. Harzer, *Hum. Genet.* **92,** 143 (1993).

[32] A. Klein, M. Henseler, C. Klein, K. Suzuki, K. Harzer, and K. Sandhoff, *Biochem. Biophys. Res. Commun.* **200,** 1440 (1994).

[33] K. Harzer, B. C. Paton, H. Christomanou, M. Chatelut, T. Levade, M. Hiraiwa, and J. S. O'Brien, *FEBS Lett.* **417,** 270 (1997).

[34] N. Fujita, K. Suzuki, M. T. Vanier, B. Popko, N. Maeda, A. Klein, M. Henseler, K. Sandhoff, H. Nakayasu, and K. Suzuki, *Hum. Mol. Genet.* **5,** 711 (1996).

not necessarily valid. For example, the *in vitro* degradation of micellar ganglioside GM1 by GM1-β-galactosidase is dependent on SAP-B.[35] There is, however, no accumulation of GM1 in patients with a pSAP deficiency.[36] To date, it is not clear whether SAP-B contributes to the GM1 degradation in living cells or whether this only occurs in artificial assay systems. A related case is the stimulation of acid sphingomyelinase by SAP-D (see later).

For the investigation of activator-dependent sphingolipid degradation, *in vitro* assays that mimic the *in vivo* situation as closely as possible are essential. One common approach for the determination of lysosomal hydrolase activity is the use of artificial substrates that liberate a fluorescent leaving group on enzymatic hydrolysis. For example, a 4-methylumbelliferyl residue can be coupled with the sugar substrates of the respective hydrolase. In such a system, the different hydrolases are usually able to generate fluorescent 4-methylumbelliferone without the assistance of sphingolipid activator proteins. For the determination of activator activities, endogenous substrates are required that bear a radioactive label in an appropriate position.[37] *In vitro,* the presence of detergents or activator proteins is required for the hydrolysis of these substrates. For the analysis of function and specificity of activator proteins, liposomal assay systems are preferably used that avoid the addition of detergents. Moreover, this approach makes demands on specific vesicle properties such as lipid composition and curvature. The *in vivo* function and the activity of the sphingolipid activator proteins can be determined by loading studies in cultured cells. The degradation of exogenously added radiolabeled gangliosides[38] is monitored by thin-layer chromatography. An alternative was the metabolic labeling of sphingolipids with [3-^{14}C]serine. Both experimental approaches are described here for cultured skin fibroblasts derived from patients of sphingolipid activator protein deficiencies.

GM2 Degradation

The activity of the GM2 activator can be measured through degradation of ganglioside GM2. GM2 is usually isolated by the method of Svennerholm[39] from brain tissue of patients who died from GM2 gangliosidosis.

[35] S.-C. Li, H. Kihara, S. Serizawa, Y.-T. Li, A. L. Fluharty, J. S. Mayes, and L. J. Shapiro, *J. Biol. Chem.* **260,** 1867 (1985).
[36] B. Schmid, B. C. Paton, K. Sandhoff, and K. Harzer, *Hum. Genet.* **89,** 513 (1992).
[37] S. Sonnino, M. Nicolini, and V. Chigorno, *Glycobiology* **6,** 479 (1996).
[38] G. Schwarzmann, *Biochim. Biophys. Acta* **529,** 106 (1978).
[39] L. Svennerholm, *Methods Carbohydr. Chem.* **6,** 464 (1972).

Another purification protocol is described by Li and co-workers[40] who observed that GM2 constitutes the major ganglioside of striped mullet (*Mugil cephalus*) roe. The terminal *N*-acetylgalactosamine moiety is radiolabeled by the galactose oxidase method.[41–43] β-Hexosaminidase A is purified from human liver,[44] placenta,[45,46] kidney,[47,48] brain,[49] or urine.[50]

In the following assay, ganglioside GM2 is degraded in a detergent-free micellar assay system. β-Hexosaminidase A alone is not able to degrade GM2 in this system and, as *in vivo*, the GM2 activator is needed for the reaction. The described assay could be used to determine the GM2-activator content during the purification procedure after the first ion-exchange chromatography[12] and also for the pure GM2 activator.

Incubation mixtures contain the following components in a final volume of 40 μl: sodium citrate buffer, 100 mM, pH 4.0; bovine serum albumin (BSA), 2.5 μg; β-hexosaminidase A, 100 mU; [^{3}H-GalNAc]GM2 (5–10 Ci/mol), 10 nmol; and GM2 activator (0.5–3 μg) in 10 μl water.

Preparation of Assay

Prepare a [^{3}H-GalNAc]GM2 stock solution in buffer containing [^{3}H-GalNAc]GM2, β-hexosaminidase A, and BSA.

1. Transfer an appropriate amount of [^{3}H-GalNAc]GM2 in organic solution, e.g., chloroform–methanol (1:1, v/v) and remove the solvent under a stream of nitrogen. Add an appropriate amount of sodium citrate buffer and BSA. If not indicated otherwise, the following steps are carried out on an ice bath. Sonify the [^{3}H-GalNAc]GM2 stock solution for 5 min and then add the appropriate amount of β-hexosaminidase A.

[40] Y.-T. Li, Y. Hirabayashi, R. De Gasperi, R. K. Yu, T. Ariga, T. A. W. Koerner, and S.-C. Li, *J. Biol. Chem.* **259**, 8980 (1984).

[41] R. M. Bradley and J. N. Kanfer, *Biochim. Biophys. Acta* **84**, 210 (1964).

[42] Y. Suzuki and K. Suzuki, *J. Lipid Res.* **13**, 687 (1972).

[43] A. Novak, J. A. Lowden, Y. L. Gravel, and L. S. Wolfe, *J. Lipid Res.* **20**, 678 (1979).

[44] J. Burg, A. Banerjee, E. Conzelmann, and K. Sandhoff, *Hoppe-Seyler's Z. Physiol. Chem.* **364**, 821 (1983).

[45] J. E. S. Lee and A. Yoshida, *Biochem. J.* **159**, 535 (1976).

[46] B. Geiger and R. Arnon, *Methods Enzymol.* **50**, 547 (1978).

[47] J. E. Wiktorowicz, Y. C. Awasthi, A. Kurosky, and S. K. Srivastava, *Biochem. J.* **165**, 49 (1977).

[48] S. K. Srivastava, *in* "Practical Enzymology of the Sphingolipidoses" (R. H. Glew and S. P. Peters, eds.), p. 217. A. R. Liss, New York, 1977.

[49] R. A. Aruna and D. Basu, *J. Neurochem.* **27**, 337 (1976).

[50] D. V. Marinkovic and J. N. Marinkovic, *Biochem. Med.* **20**, 422 (1978).

2. Add 30 μl of the [^{3}H-GalNAc]GM2 stock solution to 10 μl of the GM2 activator samples in a sealed vessel (Eppendorf cap).

4. Incubate the samples for 1–3 hr at 37° and then stop the reaction with 40 μl methanol.

Separation of N-[^{3}H]Acetylgalactosamine from [^{3}H-GalNAc]GM2

1. Introduce small pieces of silanized glass fiber wadding (from Macherey-Nagel, Düren, Germany) into glass Pasteur pipettes and add 0.5 ml of silica gel RP18 (1 : 1 suspension in methanol).

2. Rinse the columns once with 1 ml methanol and twice with 1 ml chloroform–methanol–0.1 M potassium chloride (3 : 48 : 47, v/v/v) and apply the samples to the columns. Wash the reaction vials with 80 μl chloroform–methanol–0.1 M potassium chloride (3 : 48 : 47, v/v/v).

3. Rinse the columns with 2 ml water (doubly distilled quality) to elute and collect N-[^{3}H]acetylgalactosamine.

4. Add 10 ml scintillation fluid (Ultima Gold, Packard) to the flow-through fraction and measure the radioactivity in a scintillation counter (with correction for quenching).

5. Subtract blank values from samples without addition of GM2 activator.

In the absence of GM2 activator, the reaction rate is very low. In the presence of the GM2 activator it decreases rapidly with increasing ionic strength.[51] Therefore, samples have to be dialyzed against water before the activator content is determined. An activator unit (AU) is defined by the amount of [^{3}H-GalNAc]GM2 cleaved under standard conditions; 1 AU = 1 nmol [^{3}H-GalNAc]GM2/(unit β-hexosaminidase A per min).[52] The activity of the β-hexosaminidase A is assayed with the fluorogenic substrate 4-methylumbelliferyl-β-D-N-acetylglucosaminide[53–56] (1 mM in 50 mM citrate buffer, pH 4.5). One enzyme unit is defined as the amount of enzyme that splits 1 μmol of the substrate per minute under standard conditions.

[51] E. Conzelmann, J. Burg, G. Stephan, and K. Sandhoff, *Eur. J. Biochem.* **123,** 455 (1982).

[52] E. Conzelmann and K. Sandhoff, *Hoppe-Seyler's Z. Physiol. Chem.* **360,** 1837 (1979).

[53] K. Sandhoff, E. Conzelmann, and H. Nehrkorn, *Hoppe-Seyler's Z. Physiol. Chem.* **358,** 779 (1977).

[54] D. H. Leaback and P. G. Walker, *Biochem. J.* **78,** 151 (1961).

[55] N. Dance, R. G. Price, D. Robinson, and J. L. Stirling, *Clin. Chim. Acta* **24,** 189 (1969).

[56] S. K. Srivastava, A. Yoshida, Y. C. Awasthi, and E. Beutler, *J. Biol. Chem.* **249,** 2046 (1974).

Purification of SAP-A, SAP-B, SAP-C, and SAP-D

SAPs are isolated and purified from human Gaucher spleen as described by Sano and co-workers[57] and modified by Klein et al.[32] In comparison to normal spleen there is a 20- to 22-fold increase of SAP contents in Gaucher spleen.[58] Gaucher spleen (200 g) is homogenized with 600 ml water and heated to 70° for 1 hr. After centrifugation (17700g, 60 min, 4°), the supernatant is adjusted to pH 3.0 with trichloroacetic acid and the precipitate is removed by further centrifugation. The supernatant is dialyzed for 2 days against 50 liters 20 mM phosphate buffer (pH 6.0) and fractionated by column chromatography on Q-Sepharose (Hiload 16/10 Q-Sepharose, Pharmacia). SAPs are eluted with a 0–0.3 M NaCl linear gradient in 20 mM phosphate buffer, (pH 6.0) and detected on Western blots with the different anti-SAP antibodies.[32] Fractions containing the different (but still not completely separated) SAPs are pooled, concentrated by ultrafiltration, and fractionated by gel filtration (Superdex 75 HR 10/30, Pharmacia). Fractions containing the different SAPs are pooled and fractionated again by reversed-phase HPLC on a Nucleosil C4 column (250 × 4 mm i.d., Nucleosil-300 C4, 5 μm, Knauer, Germany). SAPs are eluted in 30 min by a linear gradient of 0–90% acetonitrile in water containing 0.05% trifluoroacetic acid with a flow rate of 1 ml/min. Effluents containing the different SAPs are evaporated to dryness and dissolved in water (1 mg/ml).

Globotriaosylceramide Degradation by α-Galactosidase A and SAP-B

Globotriaosylceramide can be isolated from the neutral lipid fraction of many tissues, e.g., liver, kidney, and spleen. Erythrocytes[59] and porcine intestine[60,61] have been recommended as convenient sources. The terminal galactose residue is labeled radioactively by the galactose oxidase method.[42] α-Galactosidase A can be purified from human liver or placenta by standard chromatographic techniques[62–64] or by affinity chromatography.[65] Sufficient

[57] A. Sano, N. S. Radin, L. L. Johnson, and G. E. Tarr, J. Biol. Chem. 263, 19597 (1988).
[58] S. Morimoto, Y. Yamamoto, J. S. O'Brien, and Y. Kishimoto, Proc. Natl. Acad. Sci. U.S.A 87, 3493 (1990).
[59] T. Taketomi and N. Kawamura, J. Biochem. 72, 791 (1972).
[60] C. Suzuki, A. Makita, and Z. Yoshizawa, Arch. Biochem. Biophys. 127, 140 (1968).
[61] K. J. Dean and C. C. Sweeley, in "Practical Enzymology of Sphingolipidoses" (R. H. Glew and S. P. Peters, eds.), p. 173. A. R. Liss, New York, 1977.
[62] E. Beutler and W. Kuhl, J. Biol. Chem. 247, 7195 (1972).
[63] W. G. Johnson and R. O. Brady, Methods Enzymol. 28, 849 (1972).
[64] K. J. Dean and C. C. Sweeley, J. Biol. Chem. 254, 10006 (1979).
[65] D. F. Bishop and R. J. Desnick, J. Biol. Chem. 256, 1307 (1981).

amounts of pure enzyme are available from different expression systems.[66,67] In the following assay, globotriaosylceramide is incorporated into unilamellar liposomes as substrate. Degradation depends on the presence of both α-galactosidase A and SAP-B.[68]

Incubation mixtures contain the following components in a final volume of 50 μl: citrate buffer, 50 mM sodium, pH 4.6; BSA, 5 μg; α-galactosidase (recombinant enzyme[66]), 2.4 mU; liposomes, 10 μl; and SAP-B (0.5–5 μg).

Preparation of Liposomes

Lipid concentration, 2 mM
Egg phosphatidylcholine, 93% (mol/mol)
dl-α-Tocopherol, 2% (mol/mol)
[^{3}H-Gal]globotriaosylceramide (5–10 Ci/mol), 5% (mol/mol)

The assay is carried out according to Kase *et al.*[68] Prepare a liposome suspension in water:

1. Transfer the different lipids (egg phosphatidylcholine, *dl*-α-tocopherol, [^{3}H-Gal]globotriaosylceramide) in organic solution, e.g., chloroform–methanol, (1:1, v/v), remove the organic solvent under a stream of nitrogen, and keep the samples under vacuum for at least 1 hr.
2. Add an appropriate volume of water (doubly distilled quality) and suspend the lipids by vigorous shaking and sonication for 1 min in a water-cooled cup horn at 100 W (Branson sonifier B12; Branson, Danbury).
3. Freeze-thaw the solution 4 to 10 times in liquid nitrogen to ensure solute equilibration between trapped and bulk solutions.
4. Press 19 times through an extrusion device (Liposo Fast-Basic; Avestin Inc., Ottawa, Canada) using two stacked polycarbonate membranes (pore size diameter ≈100 nm).[69]

Preparation of Assay

1. Mix 50 mM buffer, 5 μg BSA, 2.4 mU α-galactosidase A and SAP-B (total volume 40 μl) in a test vial. Add 10 μl of the liposome suspension. Carry out all steps on an ice bath if not indicated otherwise.

[66] S. Ishii, R. Kase, H. Sakuraba, S. Fujita, M. Sugimoto, K. Tomito, T. Semba, and Y. Suzuki, *Biochim. Biophys. Acta* **1204,** 265 (1994).
[67] F. Matsuura, M. Ohta, Y. A. Ioannou, and R. J. Desnick, *Glycobiology* **8,** 329 (1998).
[68] R. Kase, U. Bierfreund, A. Klein, T. Kolter, K. Itoh, M. Suzuki, Y. Hashimoto, K. Sandhoff, and H. Sakuraba, *FEBS Lett.* **393,** 74 (1996).
[69] R. C. MacDonald, R. I. MacDonald, B. P. M. Menco, K. Takeshita, N. K. Subbarao, and L. Hu, *Biochim. Biophys. Acta* **1061,** 297 (1991).

2. Incubate the samples for 1 hr at 37° and stop the reaction with 50 μl methanol.
3. Separate the liberated tritiated galactose from [^{3}H]globotriaosylceramide with small reversed-phase column RP-18 as described in the GM2 activator assay.
4. Add 10 ml scintillation fluid (Ultima Gold, Packard) to the flowthrough fraction and measure the radioactivity in a scintillation counter.
5. In control experiments, SAP-B and α-galactosidase A are added after termination of the incubation of liposomes.

The reaction rate decreases with increasing ionic strength and is very low in the absence of SAP-B. Highly purified preparation of SAP-A, SAP-C, and SAP-D at concentrations up to 5 μg/assay do not enhance the hydrolysis of globotriaosylceramide by α-galactosidase significantly. In the presence of 2.4 mU of α-galactosidase A and 1 μg SAP-B, about 7% of globotriaosylceramide present on the outer surface of the liposome is degraded within 1 hr. There is a linear relationship between enzyme activity and reaction time until 1 hr. Alternatively, the assay could also be run with globotriaosylceramide micelles, produced by ultrasonic irradiation. Under this condition the α-galactosidase is able to cleave the substrate without SAP-B. Adding SAP-B will increase the substrate hydrolysis (U. Bierfreund, unpublished results). α-Galactosidase A activity is assayed with the artificial substrate 4-methylumbelliferyl-α-D-galactopyranoside.[70] One enzyme unit is defined as the amount of enzyme that splits 1 μmol of substrate per minute under standard conditions.

Degradation of Ganglioside GM1 by GM1-β-Galactosidase and SAP-B

SAP-B is able to stimulate several reactions *in vitro*.[68,71,72] In principle, any one of them can be used to assay for this cofactor, but the degradation of ganglioside GM1 by acid β-galactosidase is most conveniently used because this reaction proceeds fast enough to permit sensitive measurements of the activator content and is the least sensitive to disturbances by impurities. Globotriaosylceramide is much more difficult to prepare in pure form than ganglioside GM1. Degradation of sulfatide by arylsulfatase A is comparatively slow and may be strongly inhibited by contaminating proteins

[70] R. J. Desnick, K. Y. Allen, S. J. Desnick, M. K. Raman, R. W. Bernlohr, and W. Krivit, *J. Lab. Clin. Med.* **81,** 157 (1973).
[71] A. Zschoche, W. Fürst, G. Schwarzmann, and K. Sandhoff, *Eur. J. Biochem.* **222,** 83 (1994).
[72] A. M. Vaccaro, F. Ciaffoni, M. Tatti, R. Salvioli, A. Barca, D. Tognozzi, and C. Scerch, *J. Biol. Chem.* **270,** 30576 (1995).

and other compounds so that the assay requires the use of highly purified enzyme and activator.[12]

In this assay the stimulation of the enzymatic degradation of ganglioside GM1 to GM2 by acid β-galactosidase is measured. Terminal [^{3}H]galactose-labeled ganglioside GM1 is available commercially (ARC) and can be diluted to the desired specific activity of around 100 Ci/mol with nonradioactive GM1 (BioCarb or Sigma). It can also be synthesized by the galactose oxidase/boro[^{3}H]hydride method.[73]

Lysosomal β-galactosidase can be purified from human liver or placenta by affinity chromatography on immobilized p-aminophenyl- or 6-amino-hexyl-β-D-thiogalactopyranoside.[74,75]

The affinity gel is available commercially. A β-galactosidase with the same specification and pH optimum is also available from Boehringer Mannheim, Germany (β-galactosidase from bovine testes).

Incubation mixtures contain the following components in a final volume of 50 μl: sodium citrate buffer, 50 mM, pH 4.5; GM1-β-galactosidase, 5 mU; [^{3}H-Gal]GM1 (100 Ci/mol), 1 nmol; and SAP-B (0.5–5 μg) in 10 μl water.

Preparation of Assay

Prepare a [^{3}H-Gal]GM1 stock solution in buffer:

1. Transfer an appropriate amount of the [^{3}H-Gal]GM1 from organic solution, e.g., chloroform–methanol (1:1, v/v) into a sealed vessel (Eppendorf cap) and remove the solvent in a stream of nitrogen. Add an appropriate amount of sodium citrate buffer. Prepare all steps on an ice bath if not indicated otherwise.
2. Sonicate the [^{3}H-Gal]GM1 stock solution for 5 min and then add the appropriate amount of GM1-β-galactosidase.
3. Add 25 μl of the [^{3}H-Gal]GM1 stock solution to 25 μl of the SAP-B samples.
4. Incubate the samples for 1 hr at 37° in a sealed vessel (Eppendorf cap) and stop the reaction with 50 μl methanol.

[73] K. Suzuki, *in* "Practical Enzymology of the Sphingolipidoses" (R. H. Glew and S. P. Peters, eds.), p. 101. A. R. Liss, New York, 1977.
[74] A. L. Miller, R. G. Frost, and J. S. O'Brien, *Biochem. J.* **165,** 591 (1977).
[75] J. Lo, K. Mukerji, Y. C. Awasthi, E. Hanada, K. Suzuki, and S. K. Srivastava, *J. Biol. Chem.* **254,** 6710 (1979).

Separation of [³H]Galactose from [³H-Gal]GM1

1. Introduce small pieces of silanized glass fiber wadding (from Macherey-Nagel, Düren, Germany) into glass Pasteur pipettes and add 0.5 ml of silica gel RPl8 (1:1 suspension in methanol).
2. Rinse the columns once with 1 ml methanol and twice with 1 ml methanol–2 mM aqueous galactose solution (1:1, v/v) and apply the samples to the columns.
3. Rinse the columns with 2 ml methanol–2 mM aqueous galactose solution (1:1, v/v) to elute the liberated [³H]galactose.
4. Add 10 ml scintillation fluid (Ultima Gold, Packard) to the flow-through fraction and measure the radioactivity in a scintillation counter.
5. Blank runs with water instead of SAP-B solution are subtracted.

The reaction depends almost linearly on the SAP-B concentration up to about 10 μM (5 μg SAP-B per assay). The activity of β-galactosidase is usually measured with the fluorogenic substrate 4-methylumbelliferyl-β-D-galactoside, e.g., 1 mM in 50 mM citrate/0.1 M phosphate buffer, pH 4.0, with 0.1 M NaCl.[73] One unit of GM1-β-galactosidase is defined as the amount of enzyme that splits 1 μmol of this substrate per minute under the previous conditions at 37°.

Sulfatide Degradation by Arylsulfatase A and SAP-B

Substrate, enzyme preparation, and the assay conditions are described in a previous volume of this series.[12]

Glucosylceramide Degradation by Glucocerebrosidase and SAP-C

In nonpathological tissues, glucosylceramide is only present in low concentrations. In liver and spleen of patients with Gaucher's disease, glucosylceramide is stored in large amounts. Gaucher spleen is the most common source of glucosylceramide used in enzymatic and other studies. Extraction and purification can be performed as described elsewhere.[76,77] [¹⁴C-Glc] glucosylceramide is prepared as described briefly here: D-[U-¹⁴C]glucose is peracetylated and subsequently activated as the 1-bromotetraacetyl-D-[U-¹⁴C]glucose according to the procedure of Fischer,[78] followed by cou-

[76] N. S. Radin, *J. Lipid Res.* **17**, 290 (1976).
[77] S. P. Peters, R. H. Glew, and R. E. Lee, *in* "Practical Enzymology of the Sphingolipidoses" (R. H. Glew and S. P. Peters, eds.), p. 71. A. R. Liss, New York, 1977.
[78] E. Fischer, *Chem. Ber.* **44**, 1898 (1911).

pling with 3-O-benzoyl-2-N-dichloroacetyl-D-erythro-sphingosine,[79] subsequent deprotection and acylation according to the method of Shapiro et al.,[80] with minor modifications of Sarmientos et al.[81] Glucocerebrosidase can be purified from human placenta.[82] Glucocerebrosidase activity can be assayed with the fluorogenic substrate 4-methylumbelliferyl-β-D-glucoside [2 mM, 0.2 M acetate buffer (pH 5.5)[83,84] or 2 mM, 50 mM acetate buffer (pH 4.5)].[85–87] We used the procedure of Vaccaro et al.[72] [4-methylumbelliferyl-β-D-glucoside 2.5 mM, 0.1 M citrate, 0.2 M phosphate buffer, pH 5.6, 0.1% (v/v) Triton X-100, and 0.25% (w/v) sodium taurocholate]. One enzyme unit is defined as the amount of enzyme that splits 1 μmol of substrate per minute under standard conditions. It should be mentioned that the glucocerebrosidase is stimulated by acidic phospholipids[83,88–90] or detergents[88,91,92] and that the measured activity is strongly dependent on the amount and nature of such additions. Very few experiments are reported that make use of acid lipids occurring in the lysosomes, such as bis(monoacylglycero)phosphate, phosphatidylinositol, or dolicholphosphate.[93]

Incubation mixtures contain the following components in a final volume of 40 μl: citrate buffer, 50 mM sodium, pH 4.5; SAP-C, 0.5–3 μg; liposomes, 20 μl; and glucocerebrosidase, 50 μU.

Preparation of Liposomes

Lipid concentration, 6.25 mM
[^{14}C-Glc]GlcCer, 3% (mol/mol) (2.4 Ci/mol, 20,000 dpm)
Cholesterol, 23% (mol/mol)

[79] F. Sarmientos, G. Schwarzmann, and K. Sandhoff, *Eur. J. Biochem.* **146**, 59 (1985).
[80] D. Shapiro, E. S. Rachaman, and T. Sheradsky, *J. Am. Chem. Soc.* **86**, 4472 (1964).
[81] F. Sarmientos, G. Schwarzmann, and K. Sandhoff, *Eur. J. Biochem.* **160**, 527 (1986).
[82] F. S. Furbish, H. E. Blair, J. Shiloach, P. G. Pentchev, and R. O. Brady, *Methods Enzymol.* **50**, 529 (1978).
[83] S. P. Peters, P. Coyle, C. J. Coffee, and R. H. Glew, *J. Biol. Chem.* **252**, 563 (1977).
[84] S. P. Peters, C. J. Coffee, R. H. Glew, R. E. Lee, D. A. Wenger, S.-C. Li, and Y.-T. Li, *Arch. Biochem. Biophys.* **183**, 290 (1977).
[85] S. L. Berent and N. S. Radin, *Biochim. Biophys. Acta* **664**, 572 (1981).
[86] D. A. Wenger and S. Roth, *Biochem. Int.* **5**, 705 (1982).
[87] W. T. Norton and S. E. Poduslo, *J. Neurochem.* **21**, 759 (1973).
[88] G. L. Dale, D. G. Villacorte, and E. Beutler, *Biochem. Biophys. Res. Commun.* **71**, 1048 (1976).
[89] F. Choy and R. G. Davidson, *Pediatr. Res.* **14**, 54 (1980).
[90] M. W. Ho and D. Light, *Biochem. J.* **136**, 821 (1973).
[91] M. W. Ho, *Biochem. J.* **136**, 721 (1973).
[92] S. P. Peters, P. Coyle, and R. H. Glew, *Arch. Biochem. Biophys.* **175**, 569 (1976).
[93] G. Wilkening, T. Linke, and K. Sandhoff, *J. Biol. Chem.* **273**, 30271 (1998).

Acidic lipids, 10% (mol/mol) (phosphatidylserine, phosphatidylinositol)
Phosphatidylcholine, 64% (mol/mol)
   The assay method described is that used by Wilkening et al.[93] The
liposomes should be prepared like the liposomes in the assay with
globotriaosylceramide/SAP-B/α-galactosidase.

Preparation of Assay

   Prepare a stock solution containing buffer, SAP-C, and glucocerebrosidase.

   1. Add 30 μl of the stock solution to 10 μl of the liposome solution on ice.
   2. Incubate the samples for 1 hr at 37° and stop the reaction with 50
μl methanol.
   3. Separate the liberated [^{14}C]glucose from [^{14}C-Glc]GlcCer by a small
reversed-phase column RP-18, which is described in the assay with the
GM2 activator.
   4. Add 10 ml scintillation fluid (Ultima Gold, Packard) to the flow-
through fraction and measure the radioactivity in a scintillation counter.
   5. Controls are run by adding SAP-C and glucocerebrosidase after
termination of the incubation of liposomes.

Degradation of Galactosylceramide by β-Galactosylceramidase
   and SAP-C

   Substrate, enzyme preparation and assay conditions are described in a
previous volume of this series.[12]

Ceramide Degradation by Acid Ceramidase and SAP-D

   SAP-D is one of four activator proteins that is derived from the SAP
precursor protein by proteolytic processing.[94] Since its isolation, the exact
function of SAP-D remained unclear for a long time. In first reports it was
shown that SAP-D is able to stimulate acid sphingomyelinase *in vitro* in
both a detergent-based micellar and liposomal assay system,[95,96] It was then
demonstrated that purified SAP-D is able to reduce ceramide storage in a

[94] W. Fürst and K. Sandhoff, *Biochim. Biophys. Acta* **1126,** 1 (1992).
[95] S. Morimoto, B. M. Martin, Y. Kishimoto, and J. S. O'Brien, *Biochem. Biophys. Res. Commun.* **156,** 403 (1988).
[96] M. Tayama, S. Soeda, Y. Kishimoto, B. M. Martin, J. W. Callahan, M. Hiraiwa, and J. S. O'Brien, *Biochem. J.* **290,** 401 (1993).

cell line that is deficient of all four SAPs due to a mutation in the initiation codon of the prosaposin gene.[32] Just prior to these *in vivo* results it was reported that SAP-D is able to stimulate acid ceramidase activity in a detergent-based *in vitro* assay system.[97] A different approach for an *in vitro* assay system for SAP-D circumvents the use of detergents for the solubilization of the highly hydrophobic ceramide in aqueous solution. The purpose of the liposomal assay system is to mimic the *in vivo* situation more closely through the use of phospholipids as solubilizing agents.

Ceramide Degradation

The assay can be divided into four different steps: (a) purification of acid ceramidase, (b) preparation of liposomes, (c) ceramidase assay, and (d) quantification of sphingosine liberated in the enzymatic reaction. Of these four steps, only the preparation of the liposomes and the procedure of the assay for ceramide degradation in the presence of SAP-D will be described here. The purification of acid ceramidase and quantification of the liberated sphingosine are described elsewhere in this volume.[97a]

Incubation mixtures contain the following components in a final volume of 100 μl: 1 M sodium acetate buffer, 25 μl (pH 4.0, final concentration 250 mM); EDTA, 5μl, 100 mM (final concentration 5 mM); liposomes, 40 μl; SAP-D containing samples in up to 20 μl; acid ceramidase, 0.5–2.5μg in up to 10 μl; and add water to a final volume of 100 μl.

Preparation of Liposomes

Lipid concentration, 2.5 mM
Dioleoyl phosphatidylglycerol, 83% (mol/mol)
N-lauroyl sphingosine (C_{12}-ceramide), 17% (mol/mol)

Liposomes are prepared similar to the liposomes described in the assay with globotriaosylceramide/SAP-B/α-galactosidase A.

1. Prepare the incubation mixture in the order described earlier on ice.
2. Vortex the mixtures shortly and incubate at 37° for 30 min.
3. Stop the reaction by adding 800 μl chloroform/methanol (2/1) and 300 μl 100 mM NH$_4$HCO$_3$ buffer.
4. After vigorous shaking, centrifuge the samples for 10 min (10,000 rpm) in a small bench-top centrifuge.

---

[97] N. Azuma, J. S. O'Brien, H. W. Moser, and Y. Kishimoto, *Arch. Biochem. Biophys.* **311,** 354 (1994).

[97a] T. Linke, S. Lansmann, and K. Sandhoff, *Methods Enzymol.* **311** [23] 1999 (this volume).

5. Carefully remove and collect the lower organic phase into a new vial.

6. Add 600 μl chloroform/methanol (2/1) to the aqueous phase and repeat the extraction procedure.

7. The pooled lower organic phases contain the liberated sphingosine and are evaporated to dryness under a stream of nitrogen.

The liberated sphingosine is quantified according to the procedure by Roth[98] and Merrill et al.,[99] which is described in detail elsewhere in this volume.[97a] Stimulation of acid ceramidase activity is detected by comparing the amount of liberated sphingosine in SAP-D containing samples with samples in which SAP-D was omitted.

In order to obtain a high level of sensitivity, ceramidase samples should be free of SAP-D. This is accomplished by the purification procedure described elsewhere in this volume.[97a]

The presence of SAP-D in the sample can also be confirmed by Western blot analysis with specific anti-SAP-D antibodies.[32]

Determination of Sphingolipid Activator Proteins
in Cultured Fibroblasts

Compared to several in vitro assays, the specificity of activator proteins is determined more reliably in metabolic studies with cultured skin fibroblasts. Currently, two different in vivo test systems are suitable. In one of these, sphingolipids radiolabeled in the lipid moiety are added exogenously to cultured skin fibroblasts from a patient with the AB variant or with the SAP precursor deficiency. The added lipid should be a catabolic precursor of the lipid degraded in the reaction of interest, e.g., ganglioside GM1 in the case of GM2 degradation. The exogenously added activator protein restores the normal catabolism of the affected cells, indicating the function of the added activator protein. In the other system, cellular sphingolipids of cultured skin fibroblasts are labeled with [3-^{14}C]serine. Cultured skin fibroblasts derived from a patient with activator protein deficiency may be profitably employed to study the influence of the added activator proteins on the cellular sphingolipid pattern. As an example, the following assay is described for the GM2 activator. Ganglioside GM1 radiolabeled in the lipid moiety is added to GM2 activator-deficient skin fibroblasts (variant AB of GM2 gangliosidosis) and the lipid patterns of these cells are analyzed. Biosynthetic labeling of sphingolipids is described for SAP precursor-deficient cultured skin fibroblasts.

[98] M. Roth, Anal. Chem. 43, 880 (1971).
[99] A. H. Merrill, Jr., E. Wang, R. E. Mullins, W. C. L. Jamisson, S. Nimkar, and D. C. Liotta, Anal. Biochem. 171, 373 (1988).

GM1 Loading Studies in GM2 Activator-Deficient Fibroblasts

Loading tests with sphingolipids radiolabeled in the lipid moiety have been developed for the analysis of sphingolipid metabolism in cultured cells and for the diagnosis of lysosomal storage diseases. For example, all variants of GM2 gangliosidosis can be diagnosed by combining loading tests with the determination of β-hexosaminidase A activity.[100] Despite normal hexosaminidase A activity, almost no GM2 degradation is observed in GM2 activator-deficient skin fibroblasts.[101] When GM2 activator protein is added to the culture medium of skin fibroblasts from such cells, it is taken up by the cells and the catabolism of ganglioside GM2 is normalized.[102] More detailed information about the investigation of sphingolipid metabolism and trafficking using radiolabeled compounds can be found elsewhere in this volume.[102a]

Fibroblasts Loading with Ganglioside GM1 Radiolabeled in the Lipid Portion

Normal human skin fibroblasts and GM2 activator-deficient skin fibroblasts from a patient with the AB variant of GM2 gangliosidosis are cultured in Dulbecco's modification of Eagle's medium (DMEM) containing 10% (v/v) fetal calf serum (heat inactivated, 30 min at 56°), streptomycin (100 μg/ml), and penicillin (100 units/ml).[103] At 24 hr after reaching confluency, the fibroblasts are washed as described earlier and then incubated with medium (1 ml/10 cm^2 culture surface) containing 3–50 nmol [^{3}H]GM1 labeled in the sphingoid base (specific activity at least 100 Ci/mol). After a time period of 24 hr up to 120 hr, the fibroblasts are harvested. This should be done by trypsination (0.25% trypsin for 15 min at 37°), as it has been shown by electron spin resonance (ESR) studies that the trypsination and washing procedure effectively removes all glycolipids adsorbed to the cell surface but not glycolipids inserted into cellular membranes.[104]

---

[100] P. Leinekugel, S. Michel, E. Conzelmann, and K. Sandhoff, *Hum. Genet.* **88**, 513 (1992).

[101] K. Sandhoff, E. Conzelmann, E. F. Neufeld, M. M. Kaback, and K. Suzuki, *in* "The Metabolic Bases of Inherited Disease" (C. Scriver, A. L. Beaudet, W. S. Sly, and D. Valle, eds.), 6th ed., Chapter 72, p. 1807. McGraw-Hill, New York, 1989.

[102] S. Sonderfeld, E. Conzelmann, G. Schwarzmann, J. Burg, U. Hinrichs, and K. Sandhoff, *Eur. J. Biochem.* **149**, 247 (1985).

[102a] L. Riboni, P. Viani, and G. Tettamanti, *Methods Enzymol.* **311** [51] 1999 (this volume).

[103] G. Weitz, T. Lindl, U. Hinrichs, and K. Sandhoff, *Hoppe-Seyler's Z. Physiol. Chem.* **364**, 863 (1983).

[104] G. Schwarzmann, P. Hoffmann-Bleihauer, J. Schubert, K. Sandhoff, and D. Marsh, *Biochemistry* **22**, 5041 (1983).

Uptake of GM2 Activator into Cultured Fibroblasts

Recombinant GM2 activator or GM2 activator purified from human tissues is added to the [³H]GM1-containing medium (10 μg GM2 activator/ ml medium). Alternatively, fibroblasts are labeled with [³H]GM1 for 48 hr in a pulse period. After the culture medium is exchanged, the cells are chased for 48 hr with medium containing the GM2 activator protein. Isolation and thin-layer separation of sphingolipids are described elsewhere in this volume.[104a] The solvent system used for development of TLC is chloroform/methanol/0.22% aqueous $CaCl_2$ (60:35:8, by volume).[105] Radioactive bands are visualized by fluorography[102] or by phosphoimaging (phosphoimager Fuji/Raytest BAS 1000). Spots can be identified by various analytical methods, including immunostaining, comigration with appropriate standard lipids, mass spectrometric analysis, or enzymatic digestion. In normal fibroblasts, GM2 degradation is stimulated 1.3-fold in the presence of the GM2 activator, indicating that the availability of the GM2 activator rather than that of β-hexosaminidase A is the limiting factor for GM2 degradation in human fibroblasts.[17]

[³H]Sulfatide-Loading Studies in pSAP- or
SAP-B-Deficient Fibroblasts

Loading tests with [³H]sulfatide radiolabeled in the lipid moiety have been developed for the diagnosis of metachromatic leukodystrophy (MLD), where arylsulfatase A is deficient.[106–108] The loading test with [³H]sulfatide is also suitable for the diagnosis of SAP-B deficiency.[109,110] Almost no sulfatide degradation is observed in SAP-B-deficient fibroblasts.[109,110] When SAP-B is added to the culture medium of skin fibroblasts from such patients it is taken up by the cells and sulfatide is degraded to galactosylceramide.[111]

[104a] H. Schulze, C. Michel, and G. van Échten Deckert, *Methods Enzymol.* **311** [4] 1999 (this volume).

[105] D. H. van den Eijnden, *Hoppe-Seyler's Z. Physiol. Chem.* **352,** 1601 (1971).

[106] H. Kihara, C. K. Ho, A. L. Fluharty, K. K. Tsay, and P. L. Hartlage, *Pediatr. Res.* **14,** 224 (1980).

[107] T. Tonnesen, P. V. Bro, K. Brondum Nielsen, and C. Lykkelund, *Acta Paediatr. Scand.* **72,** 175 (1983).

[108] T. Kudoh and D. A. Wenger, *J. Clin. Invest.* **70,** 89 (1982).

[109] W. Schlote, K. Harzer, H. Christomanou, B. C. Paton, B. Kustermann-Kuhn, B. Schmid, J. Seeger, U. Beudt, I. Schuster, and U. Langenbeck, *Eur. J. Pediatr.* **150,** 584 (1991).

[110] D. A. Wenger, *APMIS Suppl.* **40,** 81 (1993).

[111] A. Klein, Ph.D. Thesis Bonn, p. 45 (1994).

Loading of Fibroblasts with [³H]Sulfatide Radiolabeled in the Lipid Portion

The fibroblasts to be analyzed (e.g., with pSAP or SAP-B deficiency) are cultured in DMEM containing 10% (v/v) fetal calf serum (heat inactivated, 30 min at 56°), streptomycin (100 μg/ml), and penicillin (100 units/ml).[103] After reaching confluency (24 hr), the fibroblasts are washed as described earlier and cultured in minimal essential medium containing Glutamax (MEM) containing 10% (v/v) heat-inactivated fetal calf serum with SAP-B (25 μg/ml medium) for 24 hr. The cells are washed with medium and pulse labeled with [³H]sulfatide (e.g., 0.3 μCi/ml, 12 nmol/ml, 24 Ci/Mol) in MEM containing 10% heat-inactivated fetal calf serum. [³H]Sulfatide is dissolved in ethanol and added to the medium at 37° with a final concentration of less than 5 μl ethanol/ml medium.

After a 48-hr pulse the cells are chased with MEM containing 10% fetal calf serum and SAP-B (25 μg/ml) for 120 hr. Fibroblasts are harvested like the cells in the [³H]GM1 loading experiment and the lipid pattern is visualized as described earlier.

Biosynthetic Labeling of GSLs in Fibroblasts with [3-¹⁴C]Serine

Serine is the substrate for the enzyme serine palmitoyl transferase, which catalyzes the committed step of sphingolipid biosynthesis, condensation of serine with palmitoylcoenzyme A to form 3-ketosphinganine.[112] The carbon in position one of serine is liberated during this reaction,[113] thus [3-¹⁴C]serine can be used as a metabolic precursor of sphingolipids in loading experiments. Labeling of cultured fibroblasts with [3-¹⁴C]serine for 24 hr followed by a chase of 120 hr allows the pronounced labeling of endogenously synthesized sphingolipids. For analyzing the effects of the sphingolipid activator proteins SAP-A to SAP-D, cultured fibroblasts from a patient with deficiency of the SAP-precursor[30] are available for biosynthetic labeling with [3-¹⁴C]serine.

One hour before the labeling experiment, the culture media of normal human fibroblasts and SAP precursor-deficient fibroblasts[30,31,114] are changed to MEM containing Glutamax containing 5% (v/v) heat-inactivated fetal calf serum. Fibroblasts are labeled for 24 hr with [3-¹⁴C]serine (54 Ci/mol; 1 μCi/ml medium) in MEM containing 0.3% fetal calf serum.

[112] A. H. Merrill, Jr., and E. Wang, *J. Biol. Chem.* **261**, 3764 (1986).

[113] T. Kolter and K. Sandhoff, *Chem. Soc. Rev.* **25**, 371 (1996).

[114] K. Harzer, B. C. Paton, A. Poulos, B. Kustermann-Kuhn, W. Roggendorf, T. Grisar, and M. Popp, *Eur. J. Pediadr.* **149**, 31 (1989).

The medium is removed and replaced by a medium containing L-serine (185 nmol/ml), 0.6% fetal calf serum, and the different SAPs (25 μg/ml). The surplus of L-serine should displace the [3-^{14}C]serine. Fibroblasts are harvested like the cells in the [^{3}H]GM1 loading experiment.

Compared to normal fibroblasts, cells from patients with SAP precursor deficiency contain highly increased levels of labeled ceramide, glucosylceramide, lactosylceramide, and ganglioside GM3.[32] Most of the sphingolipids migrate in TLC as double bands due to their heterogeneous ceramide moiety. The addition of SAP-B to the chase medium of SAP precursor-deficient fibroblasts decreases the amount of labeled lactosylceramide significantly and to a lesser extent than that of ganglioside GM3, whereas other stored sphingolipids are not affected. The addition of SAP-D to the chase medium of SAP precursor-deficient fibroblasts decreases exclusively the labeling of ceramide down to almost normal levels.[32] With such pulse-chase studies, the *in vivo* activity of the SAPs can be analyzed. Furthermore, [3-^{14}C]serine labeling tests with different pulse or chase times have been used for the diagnosis of Farber disease and the detailed study of ceramide metabolism.[115]

Comments

We reported on the different detergent-free assay systems for sphingolipid activator proteins, hydrolases, and sphingolipids that are currently used in our laboratory to test the activity of the different sphingolipid activator proteins. In the future, the use of unilamellar vesicles containing negatively charged lysosomal lipids will become increasingly important for the study of this metabolic pathway at phase boundaries. Bis(monoacylglycero)phosphate is synthesized within the acidic compartments[116] and is therefore specifically found in lysosomes and also in endosomes.[117] Other lipids, such as phosphatidylinositol and dolicholphosphate, together with their degradation products, such as fatty acids and dolichol, also occur in lysosomes. These components might influence lipid degradation and should be incorporated into assay systems to mimic the lysosomal conditions and to understand the selective degradation of glycolipids.

Several radiolabeled substrates, however, are difficult to obtain together with the sphingolipid activator protein-deficient fibroblasts for *in vivo* ex-

---

[115] G. van Echten-Deckert, A. Klein, T. Linke, T. Heinemann, J. Weisgerber, and K. Sandhoff, *J. Lipid Res.* **38,** 2569 (1997).

[116] B. Amidon, A. Brown, and M. Waite, *Biochemistry* **35,** 13995 (1996).

[117] T. Kobayashi, E. Stang, K. S. Fang, P. de Moerloose, R. G. Parton, and J. Gruenberg, *Nature* **392,** 193 (1998).

[118] T. Kolter and K. Sandhoff, *J. Inher. Metab. Dis.* **21,** 548 (1998).

periments. However, the sometimes enormous differences between *in vivo* and *in vitro* observations are a strong motivation to establish assay systems that mimic the lysosomal environment much closer. The assays presented can be used to study the function of a single component and also their interaction within the whole system.

Acknowledgments

We thank Dr. Gerhild van Echten-Deckert, Michaela Wendeler, Thomas Linke, Gunther Uhlhorn-Dierks, and Gundo Wilkening (Bonn) for critically reading the manuscript. We gratefully acknowledge the critical comments of Ernst Conzelmann (Theodor-Boveri-Institut für Biowissenschaften der Universität, Würzburg, Germany), who carefully read this manuscript. Work done in the authors' laboratory was supported by the Deutsche Forschungsgemeinschaft (SFB 284).

[30] Sphingolipid Hydrolyzing Enzymes in the Gastrointestinal Tract

By Rui-Dong Duan and Åke Nilsson

In the intestinal tract, there are enzymes that hydrolyze both endogenous and exogenous sphingolipids. The alkaline sphingomyelinases (SMase) of the gut and human bile have been most studied, and a major part of this article is focused on these enzymes. It also summarizes studies of ceramidase, glycosylceramidase, and the digestion of dietary sphingolipids.

Enzymes Hydrolyzing Dietary Sphingomyelin

Occurrence of Alkaline SMase in Tissues of Different Species

Of different types of phosphosphingolipids, sphingomyelin (SM) is the most prevalent. Several types of SMases have been identified, and acid and neutral SMase are discussed elsewhere in this volume. In the intestinal tract, a distinct enzyme that hydrolyzes SM was discovered in 1969 by Nilsson[1] and named alkaline SMase. The enzyme has an alkaline pH optimum around 9, and subcellular fractionation demonstrates an enrichment in brush border preparations. The alkaline SMase activity is localized specifically in the intestinal tract and is not detectable in other organs, including

---

[1] Å. Nilsson, *Biochim. Biophys. Acta* **164,** 575 (1968).

brain, kidney, lung, spleen, testis, pancreas, and stomach, or in milk and urine.[2] The intestinal alkaline SMase is not bacterial in origin as similar activity is found in ordinary mice and germ-free mice and in meconium of human fetus as early as 26 weeks of gestation. In the intestines of different experimental animals, the highest activities were in rat, mouse, pig, and baboon, lower activity in rabbit, and little activity in the guinea pig.[2] Whether these species differences indicate genetic variation or an influence of diet composition on the expression of the enzyme in the intestine is unknown.

Nyberg et al.[3] found a rather high alkaline SMase activity in both human gallbladder bile and hepatic bile. Similar activities are present in both bacterial positive and negative bile samples. Interestingly, bile alkaline SMase is only found in human bile and not in bile of other species such as rat, guinea pig, hamster, pig, sheep, and baboon.[2] Chen et al.[4] reported SMase activity in human pancreatic juice obtained during endoscopic cannulation of the pancreatic and bile ducts. The authors remarked that the samples with highest activity were bile colored and thus may be of biliary origin as others were not able to demonstrate SMase activity in pure human pancreatic juice.[1,2] Homogenates of rat pancreas contain no alkaline SMase activity, but like other tissues, it contains Mg^{2+}-dependent neutral SMase and acid SMase.[1,5]

Distribution of Alkaline SMase in the Intestinal Tract

In rats, this activity is essentially absent in stomach and upper duodenum, but increases in the small intestine and reaches maximal levels in the distal part of jejunum, followed by a rapid decline in the ileum. The enzyme is also present in the colonic mucosa with a decreasing activity gradient from ascending colon down to the rectum.[6] This distribution pattern is similar for other animals and humans, except in human duodenum where considerable alkaline SMase activity is found.[2] This activity may be contributed by human bile SMase.

Although alkaline SMase is enriched in brush border, high activity has been identified in the intestinal lumen with a longitudinal distribution pattern parallel to that of the mucosa.[6] The lumenal activity may originate from the sloughing of mucosal cells and from dissociation of the enzyme

[2] R.-D. Duan, E. Hertervig, L. Nyberg, T. Hauge, B. Sternby, J. Lillienau, A. Farooqi, and Å Nilsson, Dig. Dis. Sci. **41,** 1801 (1996).

[3] L. Nyberg, R.-D. Duan, J. Axelsson, and Å. Nilsson, Biochim. Biophys. Acta **1300,** 42 (1996).

[4] H. Chen, E. Born, F. C. Johlin, and F. J. Field, Biochem. J. **286,** 771 (1992).

[5] R.-D. Duan, L. Nyberg, and Å. Nilsson, Int. J Panceratol. **18,** 288 (1995).

[6] R.-D. Duan, L. Nyberg, and Å. Nilsson, Biochim. Biophys. Acta **1259,** 49 (1995).

from the mucosal membrane caused by lumenal factors, e.g., bile and bile salts.[7]

Method of Alkaline SMase Assay

Alkaline SMase can be determined by the method of Gatt et al.[8] with modifications. The assay buffer is 30 mM Tris–HCl (pH 9.0) containing 0.15 M NaCl, 2 mM EDTA, and 0.16% bile salt mixture [42% taurocholate, 26% taurodeoxycholate, 22% glycocholate, and 10% glycochenodeoxycholate (w/w)]. The bile salt mixture can be replaced by taurocholate alone at a final concentration of 10 mM. EDTA in the buffer serves as an inhibitor of Mg^{2+}-dependent neutral SMase. The substrate [^{14}C]choline-labeled bovine SM is dissolved in ethanol and stored at $-20°$. Before assay, it is suspended in 0.15 M NaCl containing 0.16% bile salt mixture with a final ethanol concentration of $\leq 1\%$. At this concentration, ethanol has no influence on alkaline SMase activity.

Because alkaline SMase activity varies greatly with different locations of the intestine or different bile samples, two protocols can be used for alkaline SMase assay: one for samples with low activity and one for those with high activity. When low alkaline SMase is assayed, a 5-μl sample is added to 75 μl of assay buffer followed by the addition of 20 μl of [^{14}C]SM suspension (40,000 dpm, 400 pmol). The mixture is incubated at 37° for 30 min and the reaction is terminated by adding 0.4 ml of chloroform/methanol (2 : 1). The tubes are shaken vigorously and centrifuged for 5 sec at 10,000 rpm or at 1500 rpm for 10 min. After phase separation, a 100-μl sample is taken from the upper phase for radioactivity counting. Enzyme activity can be expressed either as percentage hydrolysis of the total radioactivity or as nmol/hr/mg protein according to the specific activity of [^{14}C]SM.

When alkaline SMase activity is high (e.g., by the preceding assay), a second protocol is used in which unlabeled SM is used to dilute the [^{14}C]SM. Note, however, that this changes the concentration of the SM in the assay.

Isolation of Intestinal Alkaline SMase

Intestinal alkaline SMase can be isolated from rat intestinal mucosa as follows.

Elution of the Enzyme from the Brush Border by Bile Salt. Based on the findings that bile salts dissociate the enzyme from the mucosal membrane, the intestinal segment (50 cm long) is cannulated and rinsed with 25 ml of 0.15 M NaCl twice, then removed, opened longitudinally, and

[7] R.-D. Duan, Y. Cheng, H. D. Tauschel, and Å. Nilsson, *Dig. Dis. Sci.* **43,** 26 (1998).
[8] S. Gatt, *Biochem. Biophys. Res. Commun.* **68,** 235 (1976).

soaked in 10 ml of 0.15 M NaCl containing 1 mM phenylmethyl sulfonyl fluoride, 1 mM benzamidine, and 6 mM taurodeoxycholate with shaking at 100 cycle/min for 30 min at room temperature. This procedure avoids homogenizing the whole mucosal membrane.

Precipitation of Protein by Acetone. The solubilized protein is decanted into a beaker that is put in an ice-methanol bath (about −10°). Acetone, which has been prechilled at −20°, is added slowly into the medium until 50% (v/v) to precipitate the protein. More than 95% of the alkaline SMase is recovered by centrifugation of the precipitated proteins. After centrifugation, the proteins are dissolved in a small volume of 20 mM Tris–HCl buffer, pH 8.2, containing 1 mM benzamidine.

Anion-Exchange Chromatography. The sample is then loaded at room temperature on a high Q anion-exchange column (Bio-Rad) that has been equilibrated with 20 mM Tris–HCl buffer containing 1 mM benzamidine, pH 8.2, followed by elution with a NaCl gradient from 0 to 0.5 M NaCl. The intestinal alkaline SMase is eluted by 0.05 M NaCl.

Phenyl Sepharose Hydrophobic Interaction Chromatography (HIC, Pharmacia). Fractions from HQ with high SMase activity are pooled and ammonium sulfate is added to 1.0 M followed by loading at room temperature on an HIC column that has been equilibrated with 10 mM Tris–maleate buffer, pH 7.0, containing 1 M ammonium sulfate and 1 mM benzamidine. The column is eluted stepwise with the same buffer containing 0.75, 0.5, and 0 M ammonium sulfate, respectively. Alkaline SMase is eluted by 0.75 M ammonium sulfate.

Sephacryl S-200 Gel (Pharmacia) Chromatography. The most active fractions from HIC are pooled and dialyzed at 4° against 30 mM Tris–HCl, pH 7.5, containing 1 mM benzamidine. After dialysis, the sample is concentrated by ultrafiltration (YM 10 membrane, from Micron) and loaded on a Sephacryl S-200 column (1.5 × 100 cm) followed by elution with 30 mM Tris buffer containing 1 mM benzamidine and 0.15 M NaCl at 4°. Alkaline SMase activity is eluted at the volume corresponding to a protein band of 70 kDa as checked by 10% SDS–polyacrylamide gel electrophoresis (PAGE).

The identity of alkaline SMase can be checked by nondenaturing (10%) PAGE at 4° (mercaptoethanol or dithiothreitol should not be added to sample or running buffers). After electrophoresis, the whole lane is cut into 10 pieces and each piece is minced, soaked in 30 mM Tris buffer containing 3 mM bile salt mixture, 2 mM EDTA, and 0.15 M NaCl overnight at 4°, and SMase activity is determined.

Isolation of Human Bile Alkaline SMase

Precipitation of Lipoproteins by Dextran Sulfate. To each 10 ml of human bile, 35 μl of 10% dextran sulfate (MW 50,000) and 100 μl of 1 M CaCl$_2$

are added, followed by stirring for 45 min at 4°. This procedure removes lipoproteins and other proteins that have no SMase activity.

Protein Precipitation by Acetone. The supernatant is decanted and the protein is precipitated by adding acetone to 50% as described for intestinal SMase isolation. More than 95% of the alkaline SMase activity is recovered on centrifugation. The precipitated protein is dissolved in a small volume of 20 mM Tris buffer containing 1 mM benzamidine, pH 8.2.

DEAE Anion-Exchange Chromatography. The dissolved proteins are loaded on a DEAE Sepharose column that has been equilibrated with 20 mM Tris buffer, pH 8.2, containing 1 mM benzamidine followed by elution with NaCl gradient from 0 to 1 M NaCl in the same buffer. Bile alkaline SMase is eluted by NaCl at a concentration of 0.15 M NaCl.

Phenyl Sepharose and Sephacryl S-200 Chromatography. After this procedure, the sample is subjected to phenyl Sepharose hydrophobic interaction chromatography and Sephacryl S-200 gel chromatography as described for intestinal alkaline SMase.

Sphingosylphosphocholine (SPC) Affinity Chromatography. SPC (available commercially or prepared by acid hydrolysis of bovine SM) can be coupled to activated CH-Sepharose 4B according to the instructions of the manufacturer (Pharmacia, Sweden). The SMase fraction(s) from the phenyl Sepharose and Sephacryl S-200 chromatography step described earlier is desalted and loaded onto a SPC Sepharose column that has been equilibrated with 30 mM Tris–HCl containing 1 mM benzamidine, pH 7.4, followed by elution with a NaCl gradient from 0 to 0.4 M NaCl in the same buffer. The bile alkaline SMase is eluted at 0.18 M NaCl.

Tips for Isolation of Alkaline SMase

Acidic pH rapidly and irreversibly inactivates both intestinal and bile alkaline SMase. In addition, ammonium sulfate can reduce alkaline SMase activity. Thus, for HIC chromatography, the concentration of ammonium sulfate supplemented to the sample should not be above 1 M, and salt should be removed by dialysis as soon as possible after chromatography.

Properties of Alkaline SMase

pH Optima. The optimal pH of both intestinal and bile alkaline SMase is about 9.0. The enzymes have no activity at acid pH but exhibit some activity at pH 7.4 (about 10% of that at pH 9.0). When a crude material such as bile or intestinal mucosal homogenate is assayed at neutral pH, it may be difficult to distinguish whether the activity is derived from neutral SMase or alkaline SMase. Measuring the activity both at pH 7.5 in the presence of 2 mM Mg^{2+} and at pH 9.0 in the presence of 2 mM EDTA might help clarify which enzyme is responsible.

Substrate Specificity. Isolated alkaline SMases from rat intestinal mucosa and bile hydrolyze SM to ceramide and phosphocholine. At alkaline pH, they have no activity against phosphatidylcholine (PC).[3,6,9] However, at neutral pH and in the presence of Ca^{2+}, bile alkaline SMase has weak activity with PC. There is no activity with either phosphatidylethanolamine or *p*-nitrophenyl phosphate. When assayed in 30 mM Tris–HCl buffer, pH 9.0, containing 0.16% bile salt mixture, 2 mM EDTA, and 0.15 M NaCl, the V_{max} of purified intestinal and human bile alkaline SMases are 150 and 46 μmol/hr/mg, respectively[9] (and unpublished data).

Influence of Divalent Ions. Generally speaking, the activity of alkaline SMase is not dependent on either Mg^{2+} or Ca^{2+}.[2,3,9] The activities of the purified enzyme measured in the presence of 4 mM EDTA and EGTA are not different from those measured in the presence of 2 mM Mg^{2+} or Ca^{2+}. For purified rat intestinal alkaline SMase, 4 mM Mg^{2+} does increase the activity by 33%, whereas 8 mM Mg^{2+} reduces activity by 75%. The reason for this biphasic effect is unknown.

Bile Salt Dependence. In the absence of bile salts, little alkaline SMase activity is detected.[1,3,9] Activity increases with the concentration of bile salt up to the critical micellar concentration (CMC). The effects of bile salts vary greatly with different bile salts, and among the bile salts tested, taurocholate is the most effective, followed by taurodeoxycholate. Cholate, glycocholate, and deoxycholate only slightly increase alkaline SMase activity. These data indicate that SMase may have specific binding sites for trihydroxy bile salts.

Stability of Alkaline SMase. Alkaline SMase is relatively stable. Heating the isolated alkaline SMase from human bile or rat intestine at 50° for 30 min causes little reduction of activity.[10] Activity is stable at pH ranging from 7.0 to 10.0, but is lost in acidic conditions, e.g., incubation of enzyme at pH 6.0 reduces the activity by 50% and at pH 5 all activity is lost.[10] The inactivation seems irreversible. In addition, trypsin and chymotrypsin, the major proteinases in the intestine, have no inhibitory effect on bile or intestinal alkaline SMase.[10]

Factors That May Reduce Alkaline SMase Activity

In addition to acid pH, several other factors decrease alkaline SMase.

CHAPS. Although it has the same steroid nucleus as taurocholate, CHAPS has no stimulatory effect on alkaline SMase and inhibits taurocho-

[9] R.-D. Duan and Å. Nilsson, *Hepatology* **26,** 823 (1997).
[10] R.-D. Duan, *Scand. J. Gastroenterol.* **33,** 673 (1998).

late-induced activation of alkaline SMase,[9] with 50% inhibition occurring at a ratio of CHAPS/taurocholate of 1/5. The inhibition is reversible by dilution of CHAPS and may be due to a competition of CHAPS with taurocholate or with SM to bind to the enzyme because the head group structure of CHAPS shares some similarities with SM.

Triton X-100. Triton X-100 has been used in the determination of acid and neutral SMase, however, it significantly inhibits TC-induced activation of alkaline SMase.[6]

Bile Salts. Although the alkaline SMase is bile salt dependent, above the CMC, the stimulatory effect declines rapidly. Taurocholate is the most effective bile salt to stimulate alkaline SMase; other bile salts, such as taurodeoxycholate, may suppress the effect of TC.

Phosphatidylcholine. PC inhibits bile alkaline SMase, with 50% inhibition at a PC/SM ratio of 28.[9]

Ceramidases in the Gut

Studies by Nilsson in 1969 indicated that little or no ceramide is absorbed intact, but is hydrolyzed to sphingosine and fatty acid before absorption.[11] Ceramide hydrolysis could be demonstrated with human duodenal contents and the mucosa of rat and pig small intestine.[1] Like alkaline SMase, ceramidase activity is enriched in brush border fractions. The enzymes have not been fully characterized so far, but three types of enzymes may hydrolyze ceramide in the gut: (a) the pancreatic and human milk bile salt stimulated lipase (BSSL), (b) ceramidase(s) with a neutral or slightly alkaline pH optimum that may be identical to the ceramidase in other tissues, and (c) the acid lysosomal ceramidase that is generally found in tissues.

Bile Salt-Stimulated Lipase (BSSL)

Hui and co-workers[12] reported that pancreatic BSSL has lipoamidase activity with an artificial substrate. Nyberg *et al.*[13] showed that human milk whey and pure human milk BSSL have ceramidase activity. To measure BSSL-induced ceramidase activity, 47.2 nmol of a mixture of labeled [1-^{14}C]palmitoylsphingosine and unlabeled ceramide is dissolved in CHCl$_3$ and evaporated under nitrogen. Tris–HCl buffer (30 mM, pH 8.5) containing 4 mM taurocholate is added for a final volume of 0.5 ml, and the mixture is sonicated for 5 min. The enzyme sample is then added and

[11] Å. Nilsson, *Biochim. Biophys. Acta* **164,** 575 (1968).
[12] D. Y. Hui, K. Hayakawa, and J. Oizumi, *Biochem. J.* **291,** 65 (1993).
[13] L. Nyberg, A. Farooqi, L. Bläckberg, R.-D. Duan, Å. Nilsson, and O. Hormel, *J. Pediatr. Gastroenterol. Nutr.* **27,** 560 (1998).

incubated for 1 hr at 37°. The liberated fatty acids are extracted by adding 3 ml $CH_3OH:CHCl_3:$heptane (28:25:20, v/v/v) and 1 ml 0.05 M ($K_2CO_3 + K_2B_2O_4$), pH 10.0.[14] After centrifugation, an aliquot of the aqueous phase is taken for scintillation counting. Under these conditions, 80% of the liberated fatty acids are recovered in the upper phase, whereas all of the ceramide partitions in the lower phase.

When the effects of different bile salts are examined at pH 8.5, optimal activity is obtained with 2–6 mM glycocholate or TC. Conjugated dihydroxy bile salts give less stimulation and have maximal effects at lower concentration. When ceramide is included in a triglyceride–PC emulsion or a dispersion with PC or SM, the activity toward ceramide is low. It is therefore difficult to know to what extent BSSL hydrolyzes ceramide *in vivo*.

Glycosylceramidases in the Gastrointestinal Tract

Brady and co-workers[15] conducted a 2330-fold purification of an enzyme from the intestinal tissue of 14-day-old rats that hydrolyzed glucosylceramide (glucocerebroside). The particulate fraction sedimenting between 600 and 8400g for 12 min was solubilized with deoxycholate and then precipitated between 35 and 50% saturation with ammonium sulfate. Additional purification was obtained by stepwise elution from a triethylaminoethyl cellulose column. The pH optimum was 6.0 with a small inflection at pH 5.0. The activity was enhanced by the addition of 10 mg of sodium cholate/ml. Addition of 1.3 mg of Cutscum/ml caused a further 12% stimulation and was routinely included in the assay. Leese and Semenza[16] purified a phlorizine hydrolase, and glucosylceramide may be a physiological substrate for the phlorizine hydrolase. Kobayashi and Suzuki[17] found a highly active galactosylceramidase in murine intestine, but this intestinal enzyme was not activated by taurocholate and was activated by taurodeoxycholate with a pH optimum at 5.2. The taurodeoxycholate-activated enzyme could be separated from the taurocholate-activated one and from GM1-ganglioside β-galactosidase by octyl-Sepharose hydrophobic chromatography. The purified enzyme[18] was named intestinal glucosylceramidase due to the higher activity with glucosylceramidase than galactosylceramide. The purification procedure included deoxycholate solubilization, chromatography on Con A-Sepharose, DEAE-cellulose, hydroxyapatite, octyl-Sepharose,

[14] P. Belfrage and M. Vaughan, *J. Lipid Res.* **10,** 341 (1969).
[15] R. O. Brady, A. E. Gal, J. N. Kanfer, and R. M. Bradley, *J. Biol. Chem.* **240,** 3766 (1965).
[16] H. J. Leese and G. Semenza, *J. Biol. Chem.* **248,** 8170 (1973).
[17] T. Kobayashi and K. Suzuki, *J. Biol. Chem.* **256,** 1133 (1981).
[18] T. Kobayashi and K. Suzuki, *J. Biol. Chem.* **256,** 7768 (1981).

Sepharose 4B, Decyl agarose, and Sephadex G-200 and led to a ca. 200-fold increase in the specific activity. The enzyme had an apparent molecular mass of 130 kDa on SDS–PAGE and 290–300 kDa by gel chromatography on Sepharose 4B or Sephadex G-220. Antibodies against the enzyme did not cross-react with brain or kidney galactosylceramidase. The enzyme was active with glucosylceramide, galactosylceramide, lactosylceramide, galactosylsphingosine, and glucosylsphingosine, but was inactive with GM1, asialo GM1, desialylated fetuin, and desialylated transferrin. Among disaccharides, the enzyme showed the highest catalytic activity toward lactose and hydrolyzed phlorizine.

The developmental pattern of lactase and several glycosylceramidase activities has been studied.[19] To assay glucosylceramidase, these authors used the assay of Kobayashi and Suzuki,[18] i.e., citrate-phosphate buffer, pH 6.0, containing 0.1 M citrate, 0.2% sodium TDC, 30–40 nmol of the respective glycolipid substrate in a total volume of 100 μl, and a reaction time of 2 hr. The enzyme activities were studied in immunoprecipitates from homogenates using a monoclonal antibody. The developmental patterns of the enzyme activities for the glycolipid substrates were similar to that found for lactase. Galactosyl- and lactosylceramide inhibited lactose hydrolysis by 38% without a competitive pattern, suggesting two different active sites for lactose and glycolipid hydrolysis, respectively. The authors conclude that the protein immunoprecipitate is a multifunctional enzyme and that the lactase-phlorizin hydrolase may play an important role in the digestion of glycolipids.

The Course of Sphingomyelin and Glycosylceramide Absorption

Generally, the results of studies of sphingolipid absorption *in vivo* are compatible with a physiological role of the alkaline SMase and the brush border glucosylceramidase in sphingolipid digestion, whereas the contribution of the different ceramidase activities is difficult to evaluate at present. The absorption of intact sphingolipids and ceramide is low, whereas released free sphingoid bases are rapidly absorbed and degraded in the mucosal cells. When fatty acid or sphingosine-labeled SM is fed to lymphatic duct cannulated rats, little or no intact SM is transported by the chyle.[11] The appearance of ceramide or other sphingolipids is also very low. With ^{3}H labeling at position 3 of sphingosine, little radioactivity is recovered in chyle when free [^{3}H]sphingosine, [^{3}H]ceramide, or [^{3}H]SM is given orally. With dihydrosphingosine labeled at position 11–12, the recovery of radioac-

[19] H. A. Büller, A. G. van Wassenaer, S. Raghavan, R. K. Montgomery, M. A. Sybicki, and R. J. Grand, *Am. J. Physiol.* **257**, G616 (1989).

tivity in chyle is higher, mainly in triacylglycerols. An explanation is that the sphingosine bases have been converted to palmitic acid in the mucosa and incorporated into the chylomicron TG.

When [14C]stearic acid-labeled SM is given, 30–40% of the radioactivity is recovered in chyle, indicating that hydrolysis and subsequent absorption of the stearic acid are incomplete. Some intact SM and ceramide are recovered in feces[3]; in experiments with [3-3H]sphingosine-labeled SM, about 30% of the radioactivity is recovered in feces, mainly as ceramide.[3] Absorption is extended over the whole length of the intestine, with some reincorporation of labeled sphingosine into ceramide and glycosphingolipids of the mucosa.[20,21] Hydrolysis and uptake are highest where the highest alkaline SMase activities are observed, i.e., in the middle of the small intestine. The capacity to digest SM seems, however, to be limited. When 25 mg was fed in milk, substantial amounts of undigested SM were found in the content of the ileum. At lower doses a rather high proportion of radioactive ceramide was found in the lumen, whereas little radioactive sphingoid bases accumulated.

The course of absorption of [3H]sphingosine-labeled glucosylceramide is similar to that of SM.[22] Glucosylceramide absorption is incomplete, with some intact substrate and ceramide appearing in the feces.

In accordance with the finding that sphingoid bases are oxidized rapidly after absorption, a mucosal homogenate from guinea pig small intestine was able to convert [3H]dihydrosphingosine to palmitic acid, with palmitaldehyde as an intermediate.[23] Sphingoid bases are degraded in two steps[24]: they are first phosphorylated at the 1-hydroxyl group and then cleaved to a long chain aldehyde and ethanolamine phosphate by sphingosine-1-phosphate lyase. In comparisons of lyase activities of different tissues, van Veldhoven and Mannaerts[25] found that levels expressed per milligram protein in intestinal mucosa were two- to threefold higher than in liver.

Physiological Role and Pathological Implications

Hydrolysis of sphingolipids in the intestinal tract may have important roles both physiologically and pathologically. Ceramide is now known as

[20] E. M. Schmelz, K. Crall, R. La Roque, D. L. Dillemay, and A. H. Merrill, Jr., *J. Nutr.* **124,** 702 (1994).

[21] L. Nyberg, Å. Nilsson, P. Lundgren, and R.-D. Duan, *J. Nutr. Biochem.* **124,** 702 (1997).

[22] Å. Nilsson, *Biochim. Biophys. Acta* **176,** 339 (1969).

[23] Å. Nilsson, *Acta Chem. Scand.* **24,** 598 (1970).

[24] W. Stoffel, *Chem. Phys. Lipid.* **11,** 318 (1973).

[25] P. P. van Veldhoven, and G. P. Mannaerts, *Adv. Lipid Res.* **26,** 69 (1993).

a potent inducer of apoptosis.[26-28] Generation of ceramide from hydrolysis of SM and other sphingolipids in the intestine could be implicated in the gastrointestinal development and proliferation of epithelial cells. Human milk has been reported to stimulate the maturation of the intestinal tract[29] and the effect may well be attributed to the SM content in the milk. Dudeja et al.[30] showed that the administration of 1,2-dimethylhydrazine induced an increased level of SM and decreased activity of neutral SMase in colonic mucosa. The group of Merrill found that supplement of purified bovine SM[31-33] or chemically synthesized SM reduced the number of aberrant colonic crypt foci in mice and increased the proportion of benign adenomas of the total tumors induced by 1,2-dimethylhydrazine. When three types of SMase were assayed in human biopsy samples or surgical dissected samples, alkaline SMase activity was decreased by 50% in colorectal adenomas compared with control, by 75% in colorectal carcinomas compared with the adjacent mucosa, and by 90% in both adenomatous tissue and the flat mucosa of familial adenomatous polyposis, as compared with control,[10,34,35] which indicate that the downregulation of alkaline SMase may be an early event in the development of colorectal cancer. In addition, because familial adenomatous polyposis and colon cancer result from mutations of several tumor suppresser genes, particularly the APC gene, further investigation of whether the expression of alkaline SMase is related to the APC gene protein is of high interest.

[26] Y. A. Hannun and L. M. Obeid, Adv. Exp. Biol. Med. **407,** 145 (1997).
[27] R. N. Kolesnick and M. Kronke, Annu. Rev. Physiol. **60,** 643 (1998).
[28] A. H. Merrill, Jr., E. M. Schmelz, D. L. Dillehay, S. Spiegel, J. A. Shayman, J. J. Schroeder, R. T. Riley, K. A. Voss, and E. Wang, Toxicol. Appl. Pharmacol. **142,** 208 (1997).
[29] M. B. Yellis, Gastroenterol. Nurs. **18,** 11 (1995).
[30] P. K. Dudeja, R. Dahiya, and T. A. Brasitus, Biochim. Biophys. Acta **863,** 309 (1986).
[31] D. L. Dillehay, S. J. Webb, E. M. Schmeiz, and A. H. Merrill, Jr., J. Nutr. **124,** 615 (1994).
[32] E. M. Schmelz, D. L. Dillehay, S. K. Webb, A. Reiter, J. Adams, and A. H. Merrill, Jr., Cancer Res. **56,** 4936 (1996).
[33] E. M. Schmelz, A. S. Bushnev, D. L. Dillehay, D. C. Liotta, and A. H. Merrill, Jr., Nutr. Cancer **28,** 81 (1997).
[34] E. Hertervig, Å. Nilsson, L. Nyberg, and R.-D. Duan, Cancer **79,** 448 (1997).
[35] E. Hertervig, Å. Nilsson, R.-D. Duan, J. Björk, and R. Hultkrantz, Gastroenterology **114,** A610 (1998).

[31] Properties of Animal Ceramide Glycanases

By Manju Basu, Patrick Kelly, Mark Girzadas, Zhixiong Li,
and Subhash Basu

Introduction

Ceramide glycanase (CGase), the endoglycosidase that cleaves various glycosphingolipids in a one-step process liberating ceramide and the corresponding oligosaccharides, has been characterized from different annelids,[1–4] bacteria,[5,6] and mollusk.[7,8] The same CGase activity has also been detected and purified from mammalian sources.[9,10] The discovery of the mammalian CGases raises the possibility of glycosphingolipids (GSLs) being involved in the signal transduction and/or apoptotic processes as a secondary messenger. The ceramide and its breakdown product sphingosine, which are very much in the limelight of apoptosis and signal transduction cascades,[11–14] have so far been predicted to come from the breakdown of the major membrane lipid sphingomyelin by the action of sphingomyelinase.[15,16] Substantial CGase activities have been observed in different mammalian organs, although the highest activity was found in the mammary tissues. A parallel between CGase activity and the gestation as well as lactation has also been observed.[9] However, the significance of this observa-

[1] Y.-T. Li and S.-C. Li, *Methods Enzymol.* **179,** 479 (1989).
[2] Y.-T. Li, Y. Ishikawa, and S.-C. Li, *Biochem. Biophys. Res. Commun.* **141,** 167 (1987).
[3] B. Zhou, S.-C. Li, R. A. Laine, R. T. C. Huang, and Y.-T. Li, *J. Biol. Chem.* **264,** 12272 (1989).
[4] Y.-T. Li and S.-C. Li, *Methods Enzymol.* **242,** 146 (1994).
[5] M. Ito and T. Yamagata, *Methods Enzymol.* **179,** 488 (1989).
[6] H. Ashida, K. Yamamato, H. Kumagai, and T. Tochikura, *Eur. J. Biochem.* **205,** 729 (1993).
[7] S. S. Basu, S. Dastghieb-Hosseini, G. Hoover, Z. Li, and S. Basu, *Anal. Biochem.* **222,** 270 (1994).
[8] S. Dastghieb, Z. Li, M. Basu, N. Radin, and S. Basu, *FASEB J.* **10,** A1240 (1996).
[9] M. Basu, M. Girzadas, S. Dastgheib, J. Baker, F. Rossi, N. S. Radin, and S. Basu, *Ind. J. Biochem. Biophys.* **34,** 142 (1997).
[10] M. Basu, S. Dastgheib, M. Girzadas, P. H. O'Donnell, C. W. Westervelt, Z. Li, J.-I. Inokuchi, and S. Basu, *Acta Pol. Biochim.* **42,** 327 (1998).
[11] Y. A. Hannun and R. M. Bell, *Science* **243,** 500 (1989).
[12] Y. A. Hannun, *J. Biol. Chem.* **269,** 3125 (1994).
[13] Y. A. Hannun and L. M. Obeid, *Trends Biochem. Sci.* **20,** 73 (1994).
[14] L. M. Obeid, C. M. Linardic, L. A. Karolak, and Y. A. Hannun, *Science* **259,** 1769 (1993).
[15] S. Chatterjee, *Adv. Lipid Res.* **26,** 27 (1993).
[16] S. L. Schissel, X. Jiang, J. Tweedie-Hardman, T. Jeong, E. Hurt Camejo, J. Najib, J. H. Rapp, K. J. Williams, and Ira Tabas, *J. Biol. Chem.* **273,** 2738 (1998).

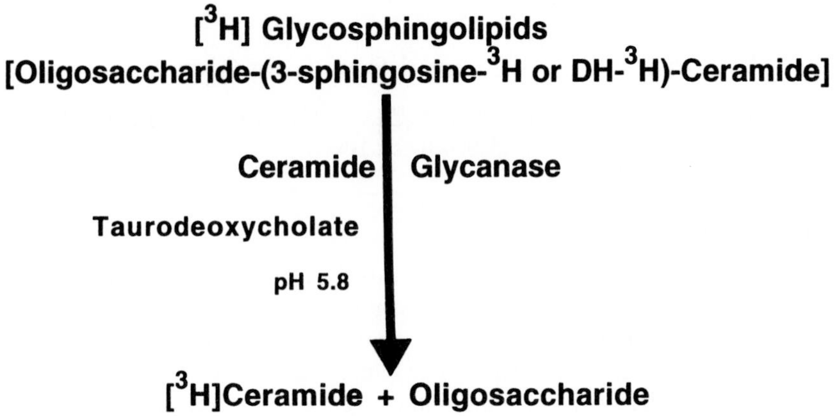

[³H] Glycosphingolipids
[Oligosaccharide-(3-sphingosine-³H or DH-³H)-Ceramide]

Ceramide | Glycanase

Taurodeoxycholate

pH 5.8

[³H]Ceramide + Oligosaccharide

FIG. 1. General reaction catalyzed by ceramide glycanase.

tion has not been established as it was for bovine mammary glucosidase.[17] CGase activities in the kidneys and the mammary tissues of newborn rats parallel that found in the lactating mammary tissue. This article deals with the study of CGase characterized from mammalian sources.

Figure 1 shows the reaction catalyzed by the ceramide glycanase.

Methods

Preparation of Radiolabeled Substrates

The principle of the assay method for the detection of ceramide glycanase activity is based on the use of ceramide-labeled glycosphingolipids. Two different methods for labeling have been followed.

Substrates with Label at the Sphingosine Double Bond of Ceramide ([DH-³H]GSL). Tritiation is performed by labeling the glycosphingolipids with sodium borotritide in the presence of paladium chloride catalyst according to the method of Scwarzmann and Sandhoff.[18] In brief, 1 mg of GSL is suspended in tetrahydrofuran and radiolabeled with $NaBT_4$ in the presence of $PdCl_2$ as catalyst at room temperature for 6 to 8 hr with shaking followed by incubation with 2 mg of $NaBH_4$ for 2 hr. The unreacted $NaBT_4$ is then removed using a C_{18} Sep-Pak column. The labeled GSL is purified further by a Biosil column and specific activity is determined by densitometric scanning before using as substrate.

Substrates with 3-Sphingosine-Labeled Ceramide ([3-Sph-³H]GSL). The method of Chigorno et al.[19] has been adapted for the labeling of substrates

[17] K. Shailubahi, E. S. Saxena, A. K. Balapure, and I. K. Vijay, *J. Biol. Chem.* **265,** 9701 (1990).

[18] G. Scwarzmann and K. Sandhoff, *Methods Enzymol.* **138,** 138 (1987).

[19] V. Chigorno, M. Valesecchi, M. Nicolini, and S. Sonnino, *Ind. J. Biochem. Biophys.* **34,** 150 (1997).

with little modification. In this method, the 3-hydroxyl group of the sphingosine is oxidized to form a ketone by reacting with dichlorodicyanobenzoquinone in anhydrous toluene. The ketone is then subsequently reduced using $NaBT_4$ followed by chase with $NaBH_4$ in a propanol water solution, (7:3, v/v). Unreacted sodium borotritide is first removed by using a C_{18} Sep-Pak column followed by purification of the labeled GSL by silica gel 200–400 mesh column chromatography. The purity of the labeled substrates is checked by thin-layer chromatography (TLC) followed by autoradiography.

Assay of CGase

The assay of CGase activity is based on the quantification of the cleaved ceramide from labeled GSL in the presence of the respective enzyme sources as described previously.[7–10] Authenticity of the reaction is checked by running the unlabeled GSLs with the enzyme and identifying the cleaved ceramide by Coomassie brilliant blue stain[20] on TLC using developing solvent chloroform:methanol (95:5). For the assay with radiolabeled substrates, the incubation mixture contained the labeled substrate (20–30 × 10^6 cpm/μmol), taurodeoxycholate detergent, 50 to 100 μg; sodium acetate buffer, 10 μmol, pH 5.5–5.8; and the enzyme protein, 50–100 μg. After a 2- to 4-hr incubation at 37°, the cleaved ceramide is partitioned from the uncleaved substrate and the oligosaccharide by the addition of 50 μl each of *n*-propanol and heptane. The upper layer, which contains the cleaved radiolabeled ceramide, is then spotted on SG-81 paper and chromatographed in chloroform:methanol (9:1) in a descending fashion. The cleaved labeled ceramide, which moves ca. 3 cm behind the solvent front, is quantitated using the liquid scintillation technique.

Preparation of Ceramide Glycanase

All the steps for enzyme preparation have been conducted at 4°. Mammary tissues from rat and rabbit are homogenized in 20 mM HEPES buffer, pH 7.0, containing 0.32 M sucrose and 0.1% 2-mercaptoethanol using a polytron 10ST homogenizer. The homogenate is then centrifuged at 100,000g, resulting in a pellet and supernatant, both of which have CGase activities in the amounts of 35 and 65%, respectively. We have so far characterized the CGase activity present in the soluble supernatant fraction. Partial heat stability has been noted for CGase proteins from both sources previously,[9,10] which had been utilized as a purification step in the case of rabbit CGase.[10] These enzymes are also found to be hydrophobic in nature as they bind to the hydrophobic column (phenyl Sepharose) in the presence of salt (2 M sodium chloride) and are eluted either by washing with buffer

---

[20] K. Nakamura and S. Handa, *Anal. Biochem.* **142,** 406 (1984).

TABLE I
KINETIC PARAMETERS OF RAT AND RABBIT CERAMIDE GLYCANASES

Parameter	Rat	Rabbit
pH optimum	5.0–6.0	5.0–6.0
Heat stability (50%)	50° × 2 min	50° × 15 min
Cation requirement	None	None
Cation inhibition	$Hg^{2+} > Cu^{2+} > Zn^{2+}$	$Hg^{2+} > Zn^{2+} > Cu^{2+}$
Approximate molecular weight	62,000	54,000

without salt as found for rabbit CGase[10] or with a high concentration of detergent (1% octyl glucoside) in the absence of salt as observed for rat CGase,[9] respectively. CGase proteins from annelids and the mollusk are also found to be of the same hydrophobic nature, and the octyl Sepharose column has also been used for the purification of leech and clam CGase proteins.[4,8] About 1400-fold purification has been achieved for the rabbit CGase[10]; however, functionally active rat CGase has been found to be about 300-fold purified after phenyl Sepharose column chromatography. Further studies have been conducted with these purified proteins. Exoglycosidases present in very high concentrations in the starting soluble supernatants have been removed by hydrophobic column chromatography as shown in the case of rabbit CGase.[10] A similar result has been observed for rat CGase (unpublished). Substantial levels of CGase activities have also been detected in various human tumor cells[10,21] using substrates labeled in either way as indicated earlier.

Properties of Mamammalian Ceramide Glycanase

Stability

The ceramide glycanase in general is a very stable enzyme. Purified as well as crude enzyme preparations are stable for at least 2 years at 4° and for at least 6 months at −18°. The presence of octyl glucoside in the purified sample further increases its stability by at least 3–4 months.

Animal CGase proteins are found to be somewhat heat stable. Both rat and rabbit CGase proteins are found to retain more than 80% enzymatic activity after heating at 55° for 2 min[9,10] (Table I). However, the clam CGase protein is stable even after boiling for 5 min at 100°.[8,22] This unique heat stability of CGase has not been reported for other sources.

[21] M. Basu, P. Kelly, P. O'Donnell, M. Miguel, M. Bradley, S. Sonnino, S. Banerjee, and S. Basu, *Bioscience Rep.* in press (1999).

[22] S. Basu, B. Dartgheib, P. Kelly, Z. Li, and M. Basu, *Methods Enzymol.* in press (1999).

Molecular Weight

The purified rabbit CGase protein exhibited a molecular weight of around 54,000 on SDS–PAGE,[10] whereas the same from rat was visualized at 62,000 (Table I). Both are immunostained at the respective areas with antibody produced against purified clam CGase.[8] This immuno cross-reactivity indicates probable structural similarities between CGase proteins from different sources. In one of the purified rabbit enzyme fractions, however, another high molecular weight band at 94,000 has been observed with positive immunostain against the same antibody,[10] the reason for which is not clearly understood. When CGase proteins are run on SLS–PAGE (sodium lauryl sarcosine) under native condition, the activities from both rat and rabbit enzyme bands at approximately 64,000 and 66,000, respectively. This molecular weight diversity of the rabbit CGase protein is poorly understood at the present time.

Detergent Requirement

The universal requirement for the optimal activity of CGase isolated from various sources is the presence of a detergent, although the nature of the detergent might vary. For all the mammalian CGase activities tested so far, among various detergents, taurodeoxycholate is found to be the best for the hydrolysis of ganglio-, lacto-, and globo-series glycosphingolipids. This is in contrast with the observation found for leech CGase where sodium taurocholate contributed to optimal activity.[4] The CGase activity is almost negligible without added detergent.

pH

Another distinctive difference between mammalian CGase and the same from lower organisms is the optimum pH. In all the mammalian CGase activities tested so far, the optimum pH has been found to be between pH 5.0 and 5.8 in sodium acetate buffer[9,10] for rat and rabbit enzymes (Table I). With tumor cell CGases, the pH between 5.5 and 5.8 has been found to be optimum in acetate buffer[21] with radiolabeled GSLs as substrates. However, in clam and annelids the pH ranges between 4 and 4.5,[1,7,8] although in leech macrobdella the pH optimum is at 5.0 in sodium acetate buffer.[4] The higher pH optimum (5.5–5.8) for the animal CGase activity reduces the probability of the cleavage of the substrate by any endo- or exoglycosidases present in the enzyme preparation as their optimum pH most of the time is between 4.0 and 4.5.

Cation Dependence

All the CGases reported until now have been found to be cation independent for their optimal activities and, consequently, are not inhibited by the addition of exogenous EDTA (Table I). However, mercury, zinc, and copper inhibit the CGase activities from annelids,[1,4] mollusk,[7,8] and mammalian[9,10] sources between 80 and 95% (Table I). Manganese, calcium, and cobalt do not show any effect whereas iron and nickel are about 30% inhibitory to all mammalian CGase activities.

Substrate Specificity

Various glycosphingolipids labeled both at the 3-sphingosine and at the ceramide double bond have been used as substrates for CGase activities. Substrate specificity studies, using substrates labeled at the ceramide double bond, are found to be comparable in the hydrolysis of both acidic and neutral GSLs of ganglio series for CGase from both rat and rabbit mammary tissues. Similar trends are also observed with all three classes of GSLs.[10] However, when 3-sphingosine-tritiated substrates are used, neutral GSLs are found to be better substrates than acidic GSLs. Also, when unlabeled GSLs are used, the substrate as well as detergent specificities seem to alter for purified mammalian CGase fractions. For easier availability, the dihydrosphingosine-labeled substrates have been used until now. With the method of Chigorno *et al.,*[19] a whole array of GSLs have been labeled, and substrate specificity studies are being conducted on a more wide scale. Unlike mammalian CGase, the mollusk CGase shows highest specificity toward both neutral and acidic GSLs of the ganglio series GSLs (discussed in detail elsewhere in this volume).[22] Ganglio series GSLs are also found to be the best substrates for the majority of the annelid CGases, although diversions are noted.[1] It would be of interest on the view of substrate specificities of CGases of different origins to specify their respective physiological roles, if any. The higher specificities of mammalian CGases toward globo series GSL could be significant as globoside is the most abundant GSL in humans.

The CGase activity from leech macrobdella has been shown to be associated with transglycosylation activity, which transfers the whole oligosaccharide chain from GM1 (Galβ1-3GalNAcβ1-4(NeuAcα2-3)Galβ1-4Glc-Cer) to different 1-alkanols, to 1,8-octanediol, and to 4-phenylbutanol.[1] This transferring ability of the mammalian CGases has not been tested. However, no enzyme has been reported until now for the one-step biosynthesis of GSL by transferring an oligosaccharide chain to the ceramide.[23] The classification of CGases on the basis of hydrophobicity of the acceptor substrate could be attempted as it was proposed for glycolipid:glycosyltransferases (GSL:GLTs) into CARS (carbohydrate recognition enzyme) and HY-CARS (hydrophobic and carbohydrate recognition enzymes).[23]

[23] S. Basu, *Glycobiology* **1**, 469 (1991).

The specificity of the CGases has been studied according to their ability for cleaving different classes of glycolipids with acidic and neutral carbohydrate chains as discussed earlier. The major emphasis for the characterization of ceramide glycanase in mammalian systems is to relate the CGase activity that produces ceramide by the one-step cleavage of glycosphingolipids to some kind of physiological relevance; at present, natural GSLs have been tested as substrates. More work is needed before achieving that goal, and experiments are under way in that direction using human tumor cells in culture.[21]

Hydrophobic Nature of CGase

Effect of Ceramide Structural Analogs

The CGase cleaves the glycosidic bond between hydrophobic ceramide and the oligosaccharide chain of a GSL. Previously, glucosyltransferase (GlcT-1;[24]), which catalyzes the transfer of glucose from UDP-glucose to the hydrophobic ceramide with the formation of Glc-Cer (glucosylceramide), had been shown to be inhibited by a structural analog of ceramide, PDMP   (1-phenyl-2-decanoylamino-3-morpholino-1-propanol · HCl).[25,26] Because both the synthetic enzyme GlcT-1 and the hydrolytic enzyme CGase interact between the hydrophobic ceramide and the carbohydrate moiety, the effect of PDMP and its other structural analogs toward CGase activity was also tested. PPMP (1-phenyl-2-hexadecanoylamino-3-morpholino-1-propanol · HCl), a higher analog of PDMP, showed stronger inhibition for both rat and rabbit mammary CGase activity using substrate labeled at the sphingosine double bond at the concentration of 0.75 μM and higher.[9,10] As shown in Table II, inhibition with hydrophobic analogs is observed when 3-sphingosine-labeled substrates are used (Table II). Although the effect of PDMP and its homologs has not been tested with either annelid or bacterial CGase activities. A detailed study with purified clam CGase indicates substantial inhibition of the activity with an increasing concentration of PPMP, and the inhibition is of a mixed nature,[22] also observed for the rat CGase activity (unpublished). These inhibitions are indicative of the hydrophobic nature of the enzyme. MAPP (myristoylaminophenylpropanol), another homolog of PDMP, has been shown to inhibit

[24] S. Basu, B. Kaufman, and S. Roseman, *J. Biol. Chem.* **248**, 1388 (1973).
[25] N. S. Radin and J.-I. Inokuchi, *Trends Glycosci. Glycotech.* **3**, 200 (1991).
[26] N. S. Radin, J. A. Shyaman, and J.-I. Inokuchi, *Adv. Lipid Res.* **26**, 183 (1993).

TABLE II
EFFECT OF SUBSTRATE ANALOGS ON CERAMIDE GLYCANASE[a]

Addition	Concentration (mM)	Rat	Rabbit
		% inhibition of CGase activity	
Sphingosine (D-*erythro*-C$_{18}$)	1.0	75	81
Ceramide	1.0	76	56
L-/D-PDMP	1.0	50	60
L-/D-PPMP	1.0	70	80

[a] Reaction conditions remain the same as described in the text except that different substrate analogs are added in the indicated concentrations in the incubation mixtures.

ceramidase activity.[27] Chatterjee *et al.*[28] also showed inhibition of other GSL:GLTs in cultured cells using PDMP *in vivo*. Apoptosis of human colon carcinoma cells by L-PPMP (1.0–8.0 μM) has also been observed.[29] However, effects of low L-PPMP concentration (*in vivo*) on CGase activities in these cells are not known. It should be noted here that PPMP, a higher homolog of PDMP, is more inhibitory with all mammalian CGases tested so far.[10]

Effects of Alkyl and Acyl Chains and of Substrate Analogs

It was observed previously that alkyl amines and fatty acids with hydrocarbon side chains of C$_{12}$ or higher inhibit CGase activities *in vitro* in the mammalian systems[10] and also in mollusk.[8] The substrate analog sphingosine and the reaction product ceramide have also been tested for the inhibition of CGase activities. Both ceramide and sphingosine are found to be inhibitory to all mammalian CGases (Table II). These observations suggest the involvement of either the hydrocarbon chain or the fatty acyl chain of the substrate in the enzymatic reaction. To elaborate this observation further, studies have been conducted using synthetic sphingosines containing varying chain lengths, which indicate the linear correlation between the CGase inhibition and the sphingosine chain length as seen in Fig. 2. Studies with synthetic lactosylceramide containing various acyl chain lengths also indicated similar results (unpublished). From these studies it is further anticipated that one of the long side chains of the ceramide moiety might contribute significantly in enzyme–substrate recognition.

The strong binding of the CGase enzyme protein to the hydrophobic

[27] A. Bielawska, M. S. Greenberg, D. Perry, S. Jayadev, J. Shyaman, C. McKay, and Y. A. Hannun, *J. Biol. Chem.* **271**, 12646 (1996).
[28] S. Chatterjee, T. Cleveland, W.-Y. Shi, J.-I. Inokuchi, and N. S. Radin, *Glycoconj. J.* **13**, 481 (1996).

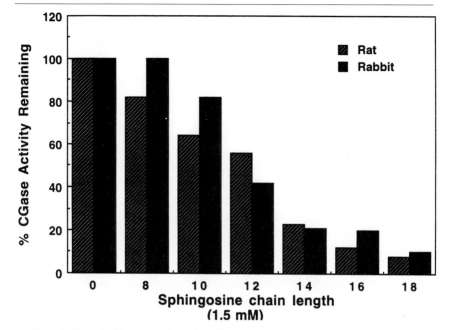

Fig. 2. Effect of sphingosine of varying chain length on ceramide glycanase activity. The reaction conditions are the same as described in the text except that sphingosines containing various chain lengths are added in indicated concentrations.

column, as well as the inhibition of the enzymatic activity by hydrophobic substrate analogs, clearly indicates a strong hydrophobic nature of the protein. However, recognition of both ceramide (hydrophobic moiety) and the oligosaccharide moieties by the mammalian ceramide glycanases has been noted. This particular endoglycosidase perhaps can be classified according to Hy-CARS classification[23] as discussed for GSL:GLTs. In addition, studies with clam CGase show that the inhibition by PPMP is of mixed nature,[22] stressing this point further.

Effect of Chemotherapeutic Drugs

Studies indicate that CGase activities in various human cancer cells in culture are quite high.[21] It has also been noted previously that CGase activity changes with development as the activity was found to increase with embryonic age in developing chicken brains.[29] Another interesting observation shows that CGase activity was high in newborn rat tissues,

[29] P. Kelly, C. Flanagan, M. Basu, M. Miguel, M. Bradley, E. Ahmad, and S. Basu, *Proc. Int. Glyco. Meeting,* Aug 22–26, Tokyo, in press, 1999.

TABLE III
EFFECT OF CHEMOTHERAPEUTIC AGENT ON CGase ACTIVITIES[a]

Addition (1.0 mM)	Rat		Rabbit	
	% inhibition of CGase activity			
	GSL substrate		GSL substrate	
	DH-³H	3-Sph-³H	DH-³H	3-Sph-³H
None	0	0	0	0
Tamoxifen	91	91	66	67
4-Hydroxytamoxifen	77	99	63	87
cis-platin	0	85	1	44

[a] Indicated drugs were added in the incubation mixtures as described and assayed following the method given in the text.

particularly in spleen.[9] In view of these observations it could be speculated that ceramide glycanase may play some significant role in the carcinoembryonic process.[29] It has also been suggested that apoptosis, a normal process during development and morphogenesis, may contribute to pathological conditions with improper regulation. Embryogenesis and carcinogenesis processes are considered to follow parallel growth patterns but differ significantly in regulation and differentiation. The presence of CGase in both developing tissues and carcinoma cells indicates a housekeeping nature of the enzyme with possible involvement in other cellular processes. Studies with human colon carcinoma and human neuroblastoma cells indicate induction of apoptosis in these cells after treatment with cis-platin (0.25 μM) as well as tamoxifen (0.25 μM) for 20 hr as observed by the appearance of DNA ladder formation in comparison to control cells.[30] The CGase activity in these drug-treated cells does seem to increase a little in contrast to the inhibition found in vitro.[30] However, the exogenous addition of tamoxifen, 4-hydroxyltamoxifen, and cis-platin to the incubation mixture seems to inhibit the CGase activity significantly in both rat and rabbit mammary tissue preparation when either the 3-sphingosine or the dihydrosphingosine-labeled substrate is used (Table III). However, cis-platin failed

[30] S. Basu, M. Bradley, M. Miguel, B. Mikrilla, M. Nakajima, L. Davis, J. Ingold, and M. Basu, Proc. New Front. Glyco. Lipid Biol. 20th Century, Aug 28–30, Takushima, Japan, in press, 1999.

to inhibit the activity in either enzyme preparation when the dihydrosphin-gosine-labeled substrate was used (Table III). If apoptosis is a general physiological process and if CGase activity does really increase with the drug treatment, a more definite role of CGase in the cellular events can be anticipated with more production of ceramide from GSLs as the sphingo-myelinase does from sphingomyelin.[15,16] More definitive experiments are in progress with human carcinoma cells in culture to establish the role of mammalian CGase in cellular processes.

[32] Enzymatic *N*-Deacylation of Sphingolipids

By MAKOTO ITO, KATSUHIRO KITA, TOYOHISA KURITA,
NORIYUKI SUEYOSHI, and HIROYUKI IZU

Introduction

Lysosphingolipids, which are sphingolipids with an *N*-deacylated cer-amide moiety, are present at low levels in normal tissues, but accumulate abnormally in cells in various lysosomal storage diseases.[1] For example, in Krabbe's disease, which is caused by a deficiency of β-galactosylceramidase, abnormal accumulation of galactosylceramide as well as its lyso form is observed.[2] Lyso-GM2 and lysosphingomyelin have been detected in the brain of patients with Tay-Sachs[3] and Niemann-Pick type A[4] disease, respec-tively, whereas they are barely detectable in the normal brain. Lysosphingo-lipids inhibit protein kinase C, which could be responsible for the pathogen-esis of sphingolipidoses.[5]

Several lines of evidence have suggested the biological significance of lysosphingolipids in various cell activities. Tyrosine-specific autophosphory-lation of the epidermal growth factor receptor of A431 cells is inhibited by lyso-GM3 as well as by GM3, and both of these can be detected in the cells.[6] It has been reported that lysosphingomyelin is a potent mitogen in

[1] B. Rosengren, J.-E. Mansson, and L. Svennerholm, *J. Neurochem.* **49,** 834 (1987).
[2] T. Miyatake and K. Suzuki, *Biochem. Biophys. Res. Commun.* **48,** 538 (1972).
[3] S. Neuenhofer, E. Conzelmann, G. Schwarzmann, H. Egge, and K. Sandhoff, *Biol. Chem. Hoppe-Seyler* **367,** 241 (1986).
[4] C. Rodriguez-Lafrasse and M. T. Vanier, *J. Neurochem.* **61,** 101 (1993).
[5] Y. A. Hannun and R. M. Bell, *Science* **235,** 670 (1987).
[6] N. Hanai, G. A. Nores, C. MacLeod, C. R. Torres-Mendez, and S. Hakomori, *J. Biol. Chem.* **263,** 10915 (1988).

a variety of cell types[7] and stimulates DNA-binding activity of the transcription activator protein AP-1,[8] being recognized as a mediator of intracellular calcium release.[9] The gene for the putative receptor of lysosphingomyelin responsible for the release of Ca^{2+} from the endoplasmic reticulum has been cloned.[10]

Lysosphingolipids are useful for preparing sphingolipid derivatives containing appropriately labeled fatty acids and can also be coupled with either appropriate proteins or gel matrix for affinity columns utilizing the amino groups newly generated in lysosphingolipids.

This article describes a novel enzyme, tentatively designated sphingolipid ceramide N-deacylase (SCDase), which is capable of cleaving the N-acyl linkage of ceramides in various glycosphingolipids as well as sphingomyelin to produce their lyso forms.[11] To date, the preparation of lysosphingolipids has been performed using purely chemical procedures,[12] which are somewhat troublesome, time-consuming, and give a low yield. Using the SCDase we were able to obtain easily the lyso forms of all species of glycosphingolipids and sphingomyelin without any alternation of their polar portions and sphingoid moieties.

Materials

SCDase is purified from the culture fluid of *Pseudomonas* sp. TK4 by the method described previously[11] or obtained from Takara Shuzo Co., Japan. Glycosphingolipids are purchased from Iatron Lab. Inc., Japan. Sphingomyelin and ceramide from bovine brain are from Matreya.

Enzyme Assay

Principle

SCDase catalyzes the following reaction: glycosphingolipids (or sphingomyelin) + H_2O → lysoglycosphingolipids (or lysosphingomyelin) + fatty acids. Quantitative assay of SCDase activity is performed using cold or

[7] N. N. Desai and S. Spiegel, *Biochem. Biophys. Res. Commun.* **181,** 361 (1991).
[8] A. Beger, D. Rosenthal, and S. Spiegel, *Proc. Natl. Acad. Sci. U.S.A.* **92,** 5885 (1995).
[9] L. A. Kindman, S. Kim, T. V. McDonald, and P. Gardner, *J. Biol. Chem.* **269,** 13088 (1994).
[10] C. Mao, S. H. Kim, J. S. Almenoff, X. L. Rudner, D. M. Kearney, and L. A. Kindman, *Proc. Natl. Acad. Sci. U.S.A.* **93,** 1993 (1996).
[11] M. Ito, T. Kurita, and K. Kita, *J. Biol. Chem.* **270,** 24370 (1995).
[12] S. Neuenhofer, G. Schwarzmann, H. Egge, and K. Sandhoff, *Biochemistry* **24,** 525 (1985).

radioactive GM1 as the substrate to determine the lyso-GM1 generated (Assay I) or [^{14}C]fatty acids released (Assay II).

Preparation of Substrates

A mixture of gangliosides is prepared from bovine brain using a method described previously.[13] GM1 is prepared from crude gangliosides by digestion with neuraminidases isolated from *Clostridium perfringens* (Sigma) or by using sialidase-producing marine *Pseudomonas* sp. YF-2 as a microbial biocatalyst,[14] followed by C$_{18}$ reverse-phased chromatography and DEAE-Sepharose A25 anion-exchange chromatography. [^{14}C]GM1 is prepared by the condensation of [^{14}C]stearic acid to lyso-GM1 by SCDase.

Procedure: Assay I

The reaction mixture contains 10 nmol of GM1 and an appropriate amount of the enzyme in 20 μl of 25 mM sodium acetate buffer, pH 6.0, containing 0.8% Triton X-100. Following incubation at 37° for 30 min, the reaction is stopped by heating in a boiling water bath for 5 min. The reaction products are freeze-dried by a Speed-Vac concentrator (Savant Instruments, Inc.), redissolved in 15 μl of chloroform/methanol (1/2, v/v), and analyzed by thin-layer chromatography (TLC) using chloroform/methanol/10% acetic acid (5/4/1, v/v) as the developing solvent.

Glycosphingolipids and lysoglycosphingolipids are visualized by spraying the TLC plates with orcinol-H$_2$SO$_4$ reagent and scanning them with a Shimadzu CS-9300 chromatoscanner with the reflectance mode set at 540 nm. The extent of hydrolysis is calculated as follows: hydrolysis (%) = (peak area for lyso-GM1 generated) × 100/(peak area for remaining GM1 + peak area for lyso-GM1 generated).

Assay II

Radioactive GM1, of which stearic acid is labeled with ^{14}C, is used for the assay. The reaction is performed as described in Assay I, but the reaction mixture contains 0.1 nmol [^{14}C]GM1 (55 mCi/mmol) and 9.9 nmol cold GM1 instead of 10 nmol of cold GM1. The digestion products are analyzed by TLC and determined by a BAS1000 image analyzer (Fuji Film Co., Japan). The extent of hydrolysis is calculated as follows: hydrolysis (%) = (PSL for fatty acid generated) × 100/(PSL for remaining GM1 + PSL for fatty acid generated). Here, PSL is photostimulated luminescence.

[13] R. W. Ledeen and R. K. Yu, *Methods Enzymol.* **83,** 139 (1982).
[14] Y. Fukano and M. Ito, *Appl. Environ. Microbiol.* **63,** 1861 (1997).

Unit Definition

One enzyme unit (U) is defined as the amount capable of catalyzing the hydrolysis of 1 μmol of GM1 per minute under the conditions shown in Assay I. Values of 10^{-3} U enzyme are expressed as 1 mU.

Properties of SCDase[11]

Enzymatic Purity

The purified enzyme preparation is completely free from the following enzyme activities: α- and β-galactosidases, β-N-acetylhexosaminidase, α-N-acetylgalactosaminidase, α-L-fucosidase, α- and β-mannosidases, sialidase, sphingomyelinase, endoglycoceramidase, and proteases. Confirmation of the absence of these activities is made using 20 mU of SCDase for each assay and incubation with the appropriate substrate for 16 hr.

General Properties

The general properties of the enzyme are as follows: optimal activity at pH 5.0–6.0 and stable between pH 4.0 and 9.0; potently inhibited by Hg^{2+}, Cu^{2+}, and Zn^{2+} (2 mM), but not by Ca^{2+}, Mn^{2+}, Mg^{2+}, and EDTA, all at the same concentration. The enzyme retains 80% of its activity when kept at 60° for 30 min and can be kept at $-85°$ for 2 months without any loss of activity.

The addition of Triton X-100 or taurodeoxycholate at a concentration of 0.8% (w/v) increases the enzyme activity about 10-fold in comparison with that in the absence of the detergent.

Substrate Specificity

The extent of hydrolysis of various sphingolipids after exhaustive digestion with the enzyme is summarized in Table I. This enzyme shows quite wide specificity, i.e., it hydrolyzes both neutral and acidic glycosphingolipids, including sulfatide, and also a range from simple glycosphingolipids (cerebrosides) to complex polysialogangliosides (GQ1b).

Furthermore, the enzyme hydrolyzes not only glycosphingolipids, but also sphingomyelin. It is notable, however, that the enzyme does not hydrolyze completely all sphingolipid substrates tested even after prolonged incubation.

TABLE I

Specificity of Sphingolipid Ceramide N-Deacylase[a]

Name	Structure	Hydrolysis (%)
Glucosylceramide (GlcCer)	Glcβ1-1'Cer	39
Galactosylceramide (GalCer)	Galβ1-1'Cer	36
Sulfatide	HSO₃-3Galβ1-1'Cer	54
Lactosylceramide	Galβ1-4Glcβ1-1'Cer	48
Asialo GM1	Galβ1-3GalNAcβ1-4Galβ1-4Glcβ1-1'Cer	65
Globotetraosylceramide (Gb4Cer)	GalNAcβ1-3Galα1-4Galβ1-4Glcβ1-1'Cer	49
Globopentaosylceramide (Gb5Cer)	GalNAcα1-3GalNAcβ1-3Galα1-4Galβ1-4Glcβ1-1'Cer	56
GM3	NeuAcα2-3Galβ1-4Glcβ1-1'Cer	51
GM2	GalNAcβ1-4(NeuAcα2-3)Galβ1-4Glcβ1-1'Cer	72
GM1a	Galβ1-3GalNAcβ1-4(NeuAcα2-3)Galβ1-4Glcβ1-1'Cer	60
GD3	NeuAcα2-8NeuAcα2-3Galβ1-4Glcβ1-1'Cer	73
GD1a	NeuAcα2-3Galβ1-3GalNAcβ1-4(NeuAcα2-3)Galβ1-4Glcβ1-1'Cer	52
GQ1b	NeuAcα2-8NeuAcα2-3Galβ1-3GalNAcβ1-4(NeuAcα2-8NeuAcα2-3)Galβ1-4Glcβ1-1'Cer	53
Sphingomyelin	Choline phosphate-Cer	28

[a] Sphingolipids (10 nmol each) were incubated at 37° for 16 hr with 2 mU of the enzyme in 20 μl of 20 mM sodium acetate buffer, pH 6.0, containing 0.8% (w/v) Triton X-100. The extent of hydrolysis was determined by TLC and calculated as described in the text. Glycosphingolipids and sphingomyelin were visualized by orcinol-H₂SO₄ and Coomassie brilliant blue reagents, respectively. Each value is the mean of four independent experiments.

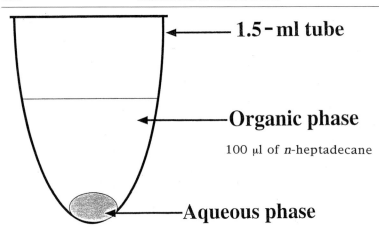

1.5 - ml tube

Organic phase

100 µl of n-heptadecane

Aqueous phase

10 µl of 50 mM sodium acetate buffer, pH 6.0,
containing 10 nmol sphingolipids,
2 milliunits SCDase, 0.8% taurodeoxycholate
and 0.1% Triton X-100

FIG. 1. N-Deacylation of various sphingolipids by sphingolipid ceramide N-deacylase (SCDase) in the biphasic system. The reaction mixture was incubated at 37° for 16 hr without shaking. The extent of hydrolysis of various sphingolipids was increased by 15–30% in comparison with that in the normal aqueous system (Table I).

Effects of Organic Solvents

The addition of n-heptadecane to the reaction mixture increases the hydrolysis rate of glycosphingolipids by SCDase.[15] Addition of the organic solvent seems to be effective for removing fatty acids, which may cause feedback inhibition, from the reaction system. Actually, fatty acids are detected in the organic phase, but hardly in the aqueous phase, after reaction in this biphasic system.[15] It should also be noted that SCDase catalyzes the reverse hydrolysis (condensation) reaction very efficiently.[16]

Preparation of Lysosphingolipids by SCDase in the Biphasic System

A 100-µl portion of n-heptadecane is added to 10 µl of 50 mM sodium acetate buffer, pH 6.0, containing 10 nmol of glycosphingolipids, 2 mU of SCDase, 0.8% (w/v) taurodeoxycholate, and 0.1% (w/v) Triton X-100 (Fig. 1). The mixture is then incubated at 37° for 16 hr without shaking. Lysoglycosphingolipids are retained in the aquatic phase, whereas fatty acids move

[15] H. Izu, T. Kurita, M. Sano, I. Kato, and M. Ito, Abstract of XIXth International Carbohydrate Symposium DP 066, San Diego, 1998.
[16] S. Mitsutake, K. Kita, N. Okino, and M. Ito, *Anal. Biochem.* **247,** 52 (1997).

to the organic phase. The extent of hydrolysis of GM1, asialo GM1, GD1a, and sphingomyelin in the biphasic system is increased by 15–30% in comparison with that in the normal aqueous system (Table I).

Comments

Lysosphingomyelin is usually prepared by acid hydrolysis of sphingomyelin.[16] As a consequence of C-3 epimerization of the sphingosine base during acid hydrolysis, the resulting lysosphingomyelin consists of a mixture of D-*erythro*-(2S,3R)- and L-*threo*-(2S,3S)-stereoisomers.[17] However, lysosphingomyelin prepared by SCDase was demonstrated by ^{1}H nuclear magnetic resonance (NMR) and ^{13}C NMR to be of the naturally occurring D-*erythro*-(2S,3R)-isomer.[18]

Acknowledgments

This work was supported in part by a Grants-in-Aid for Scientific Research Priority Areas (09240101) and Scientific Research (B) (09460051) from the Ministry of Education, Science, and Culture of Japan.

[17] R. C. Gaver and C. C. Sweeley, *J. Am. Oil Chem. Soc.* **42,** 294 (1965).
[18] N. Sueyoshi, H. Izu, and M. Ito, *J. Lipid Res.* **38,** 1923 (1997).

[33] Genetic Approaches for Studies of Glycolipid Synthetic Enzymes

By SHINICHI ICHIKAWA and YOSHIO HIRABAYASHI

Introduction

The synthesis of sphingolipids starts from the condensation of L-serine and palmitoyl-CoA. After several reactions, ceramide, a pivotal sphingolipid, is synthesized as the core structure of all sphingolipids, including sphingomyelin (SM) and over 300 species of glycosphingolipids (GSLs). The synthesis of most GSLs begins with the glucosylation of ceramide catalyzed by GlcCer synthase (ceramide glucosyltransferase, GlcT-1, EC 2.4.1.80) and its product glucosylceramide (GlcCer)[1] (Fig. 1). These GSLs

[1] S. Basu, B. Kaufman, and S. Roseman, *J. Biol. Chem.* **243,** 5802 (1968).

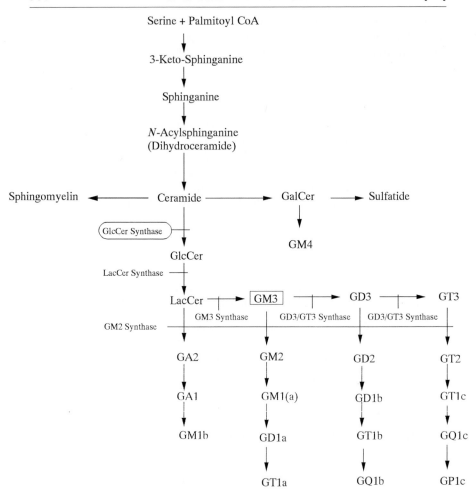

FIG. 1. Synthetic pathways of sphingolipids and GSLs. The synthesis of sphingolipids starts with the condensation of L-serine and palmitoyl-CoA. After ceramide is synthesized, it serves as the core structure of all sphingolipids, including sphingomyelin and over 300 species of GSLs. Most GSLs are derived from GlcCer synthesized by GlcCer synthase (EC 2.4.1.80). The steps catalyzed by LacCer, GM3, and GM2 synthases are also indicated.

are believed to play important roles in a variety of cellular processes such as cell recognition, growth, development, and differentiation.[2] Ceramide glucosylation is also important as a regulatory factor for the intracellular level of ceramide, which is now regarded as a second messenger.[3,4]

[2] A. Varki, *Glycobiology* **3**, 97 (1993).
[3] S. Ichikawa and Y. Hirabayashi, *Trends Cell Biol.* **8**, 198 (1998).
[4] M. Ito and H. Komori, *J. Biol. Chem.* **271**, 12655 (1996).

Enzymes involved in the initial steps of sphingomyelin and glycosphin-golipid syntheses are usually difficult to isolate because they are tightly attached to membranes. The development of expression cloning by Seed and Aruffo[5] offered an excellent system for the isolation of sphingolipid-metabolizing enzymes. Theoretically, the method can be applied for the isolation of most of the cDNAs encoding sphingolipid synthetic enzymes. We have cloned GlcCer synthase by this expression cloning method.[6] This article describes the methods used for the cloning and analyses of GlcCer synthase and also discusses the possible application of this approach to the isolation of other sphingolipid synthetic enzymes.

Assay

General Considerations

Enzyme activities of glycosphingolipid synthetic enzymes are usually measured by the incorporation of an isotope-labeled sugar in the lipid fraction. However, a much simpler method utilizing fluorescent probe-labeled sphingolipids (NBD-sphingolipids) has been developed. NBD-lipids can be used for the assay of various sphingolipid-metabolizing enzymes, including GlcCer, LacCer, GM3, GM2, and sphingomyelin synthases. Examples of GlcCer synthase assays follow. Assay conditions for other enzymes differ somewhat as some glycosyltransferases require divalent cations and other phospholipids for activity.

Assay Based on a Ratioisotope-Labeled Nucleotide Sugar

This protocol is based on Basu et al.[1] with a slight modification. The cell lysate for the enzyme assay is prepared as follows: Cells are harvested by scraping and are washed with phosphate-buffered saline (PBS). The cells are resuspended in 20 mM Tris–HCl buffer (pH 7.5) containing 0.25 M sucrose. Cells are lysed by sonication or by freeze-thawing. The incubation mixture (50 μl) consists of 0.5% Triton X-100 [because Triton X-100 is sometimes inhibitory, a mixture of Tritor X-100 and Cutscum (2:1) may be used], 20 mM Tris–HCl buffer (pH 7.5), 500 μM [^{14}C]UDP-Glc [glucose-^{14}C(U) specific activity 11 GBq/mmol purchased from New England Nuclear Products (Boston, MA) (3×10^5 dpm/reaction is used)], 0.3 mM ceramide, and 600 μg protein of cell lysate as an enzyme source. The mixture is incubated for 2 hr at 30°. The reaction is stopped with 500 μl of chloroform–methanol (2:1, v/v). After mixing and centrifugation, the

[5] B. Seed and A. Aruffo, Proc. Natl. Acad. Sci. U.S.A. **84,** 3365 (1987).
[6] S. Ichikawa H. Sakiyama, G. Suzuki, I.-P. J. K. Hidari, and Y. Hirabayashi, Proc. Natl. Acad. Sci. U.S.A. **93,** 4638 (1996).

lower phase is washed with chloroform–methanol–0.1 M KCl (3:48:47, v/v), dried, and subjected to scintillation counting or analyzed by thin-layer chromatography (TLC) followed by autoradiography. The position of GlcCer on TLC is identified by visualization of cold GlcCer with orcinol-H_2SO_4 reagent[7] prepared as follows: 200 mg of orcinol is dissolved in 11.4 ml of H_2SO_4 and is brought to 100 ml with H_2O. The TLC plate is sprayed with the reagent and heated at 100° for 5 to 10 min.

Assay Based on a Fluorescent-Labeled Sphingolipid

This is a modification of the method developed by Lipsky and Pagano.[8] The enzyme activities of GlcCer and SM synthases can be measured simultaneously using C_6-NBD-Cer as a substrate. For other enzymes, including LacCer⁻, GM3-, and GM2 synthases, C_{12}-NBD-substrates should be used and the conditions for product formation and extraction optimized for each enzyme. Only C_{12}-NBD-GlcCer is available commercially from Sigma Chemical Co. (St. Louis, MO), but other substrates can be prepared enzymatically from C_{12}-NBD-GlcCer using crude cell lysates as glycosyltransferase sources. The following assay conditions are used for GlcCer and SM synthases. C_6-NBD-Cer (Molecular Probes) (50 μg) and lecithin (500 μg) are mixed in 100 μl ethanol and the solvent evaporated; 1 ml of water is added and the mixture is sonicated in a water bath type sonicator to form liposomes. A standard reaction mixture (100 μl) is composed of 20 mM Tris–HCl (pH 7.5), 500 μM UDP-Glc, 20 μl liposomes, and 50 μg cell protein. The mixture is incubated at 30° for 4 hr and then 100 μl of chloroform–methanol (2:1, v/v) is added to stop the reaction. After mixing and centrifugation, the lower phase is transferred to a new tube, evaporated, and applied to silica gel 60 plates (E. Merck). NBD-lipids are separated by the solvent system of $CHCl_3$-CH_3OH-12 mM $MgCl_2$ in H_2O (65:25:4, v/v) and visualized by UV illumination (middle wavelength). The fluorescent intensities of C_6-NBD-GlcCer and C_6-NBD-SM are measured by a TLC scanner (Shimadzu CS-9300 or equivalent). The R_f values for C_6-NBD-Cer, C_6-NBD-GlcCer, and C_6-NBD-SM are 0.75, 0.53, and 0.07, respectively. A silica gel paper SG-81 (Whatman) may be used instead of the plate for more sensitive measurement, in which case the bands corresponding to C_6-NBD-GlcCer are cut out, dried, the lipids extracted from the paper by 500 μl of chloroform–methanol (2:1, v/v), and fluorescent intensity measured with a fluorometer (λ_{ex} = 475 nm and λ_{em} = 525 nm).

[7] L. Svennerholm, *J. Neurochem.* **1,** 42 (1956).
[8] N. G. Lipsky and R. E. Pagano, *J. Cell Biol.* **100,** 27 (1986).

cDNA Cloning

General Considerations

The cDNA cloning of the glycosyltransferases in the early steps of GSL synthesis was achieved either by the classical method using a DNA probe[9–11] or by expression cloning.[12–15] These enzymes are hydrophobic proteins and are tightly attached to membranes, which makes their purification very difficult. Expression cloning is advantageous because it does not require enzyme purification. Theoretically, most of the sphingolipid synthetic enzymes could be cloned using this strategy. If the target enzyme is expressed in all types of cells, the establishment of a mutant cell line deficient in the enzyme is necessary, as is the case for GlcCer synthase.[16,17] Many procedures for the isolation of mutant cells have been reported so far.[18–20] If the lack of an enzyme is lethal, the cloning may be achieved by selecting surviving cells after transfection with a library DNA. A target cDNA can be isolated by immunochemical methods using an antibody against the enzyme product. The method, so-called "panning"[21] or cell sorting, is commonly used and will be described in a later section. If the antibody against an immediate enzyme product is not available or if the product is not expressed strongly on the cell surface, an antibody against a distal product of the metabolic

[9] S. Schulte and W. Stoffel, *Proc. Natl. Acad. Sci. U.S.A.* **90,** 10265 (1993).

[10] T. Nomura, M. Takizawa, J. Aoki, H. Arai, K. Inoue, E. Wakisaka, N. Yoshizuka, G. Imokawa, N. Dohmae, K. Takio, M. Hattori, and N. Matsuo, *J. Biol. Chem.* **273,** 13570 (1998).

[11] A. Ishii, M. Ohta, Y. Watanabe, K. Sakoe, M. Nakamura, J. Inokuchi, Y. Sanai, and M. Saito, *Glycoconj. J.* **14,** S49 (1998).

[12] Y. Nagata, S. Yamashiro, J. Yodoi, K. O. Lloyd, H. Shiku, and K. Furukawa, *J. Biol. Chem.* **267,** 12082 (1992).

[13] K. Sasaki, K. Kurata, N. Kojima, N. Kurosawa, S. Ohta, N, Hanai, S. Tsuji, and T. Nishi, *J. Biol. Chem.* **269,** 15950 (1994).

[14] K. Nara, Y. Watanabe, K. Maruyama, K. Kasahara, Y. Nagai, and Y. Sanai, *Proc. Natl. Acad. Sci. U.S.A.* **91,** 7952 (1994).

[15] M. Haraguchi, S. Yamashiro, A. Yamamoto, and K. Furukawa, *Proc. Natl. Acad. Sci. U.S.A.* **91,** 10455 (1994).

[16] M. Nozue, H. Sakiyama, K. Tsuchiya, Y. Hirabayashi, and M. Taniguchi, *Int. J. Cancer* **42,** 734 (1988).

[17] S. Ichikawa, N. Nakajo, H. Sakiyama, and Y. Hirabayashi, *Proc. Natl. Acad. Sci. U.S.A.* **91,** 2703 (1994).

[18] P. Stanley, *Annu. Rev. Genet.* **18,** 525 (1984).

[19] T. Tsuruoka, T. Tsuji, H. Nojiri, E. H. Holmes, and S. Hakomori, *J. Biol. Chem.* **268,** 2211 (1993).

[20] K. Hanada, M. Nishijima, and Y. Akamatsu, *J. Biol. Chem.* **265,** 22137 (1990).

[21] L. J. Wysocki and V. L. Sato, *Proc. Natl. Acad. Sci. U.S.A.* **75,** 2844 (1978).

pathway may be used. Usually an antibody against the end product is recommended.

Expression Cloning of GlcCer Synthase cDNA

Cells used in this experiment were routinely maintained in Dulbecco's modified Eagle medium (DMEM) (GIBCO Laboratories, New York) supplemented with 10% fetal calf serum (FCS) under 5% CO_2. Because GlcCer synthase is expressed in all types of cells, the establishment of a mutant lacking enzyme activity was necessary. A mouse melanoma B16 (MEB-4) was treated with a mutagen, N-methyl-N'-nitro-N-nitrosoguanidine (MNNG) (1.5 μg/ml) and cells not expressing sialyllactosylceramide (GM3) were selected in combination with an anti-GM3 antibody (M2590)[22] and complement (Giblatlar Laboratories, Ohio). Working concentrations of the antibody and complement must be determined empirically for each experiment. After 10 times selection of the mutagenized cells with M2590, 100 clones were isolated by limiting dilution. From these GM3-deficient cell lines, a mutant deficient in GlcCer synthase, GM-95, was identified. GM-95-PyT, a recipient cell line for transient expression, was established by cotransfection of GM-95 with pSV2neo (American Type Culture Collection, Maryland) and pPSVE-PyE (the plasmid carrying the early region containing the largeT of polyoma virus, a gift from Dr. M. Fukuda, The Burnham Institute) plasmids followed by selection with G418 (GIBCO-BRL). The method of stable expression will be described in a later section.

The quality of the cDNA library is very important for expression cloning. Many types of high-quality cDNA libraries are now available commercially from Invitrogen (California) and Clontech (California). We usually use libraries from these companies. These two companies also offer custom cDNA library construction service for expression cloning. Several methods for the construction of expression libraries have been described in detail[23,24] and will not be discussed further here.

Recipient cells must express polyoma virus (for rodent cells) or SV40 virus (for human cells) large T proteins encoding DNA helicases for intranuclear replication of plasmid vectors for strong expression of cDNA inserted in the vector, and this also enables plasmid rescue from the cells recovered by panning or cell sorting. We routinely use pcDNA I vector

---

[22] Y. Hirabayashi, A. Hamaoka, M. Matsumoto, T. Matsubara, T. Tagawa, S. Wakabayashi, and M. Taniguchi, *J. Biol. Chem.* **260**, 13328 (1985).

[23] A. Aruffo, *in* "Cell-Cell Interaction: A Practical Approach" (B. R. Stevenson, W. J. Gallin, and D. L. Paul, eds.), p. 55. IRL Press at Oxford University Press, New York, 1992.

[24] M. Kriegler, *in* "Gene Transfer and Expression: A Laboratory Mannual," p. 114. Stockton Press, New York, 1990.

(Invitrogen) because this vector contains both polyoma and SV40 replication origins. Plasmids of the cDNA library were prepared by the standard cesium chloride method[25] or by a Qiagen column according to the manufacturer's instructions.

We use electroporation for transfection of the cDNA library DNA to cells. The following conditions are optimized for COS7, CHO, and melanoma B16-derived cell lines, including GM-95-PyT. Cells are harvested at 50% confluency and 2×10^7 cells washed with K-PBS$^-$ (30.8 mM NaCl, 120.7 mM KCl, 8.1 mM Na$_2$HPO$_4$, and 1.46 mM KH$_2$PO$_4$) twice and resuspended in 400 μl of K-PBS$^-$ supplemented with 5 mM MgCl$_2$ (KPBS$^+$). To this cell suspension, 100 μg of library DNA dissolved in 400 μl K-PBS$^+$ is added and kept on ice for 10 min. Cells are then transferred to a 0.4-cm cuvette and exposed to a 300-V pulse with a capacitance of 960 μF by Gene Pulsar (Bio-Rad, California) and placed on ice. After 10 min, the cell suspension is diluted with 5 ml of cold serum-free DMEM, incubated at 25° for 30 min, and cultured at 37° in DMEM supplemented with 20% FCS. A total of 1×10^8 cells are transfected as described earlier and cultured in five 15-cm culture dishes.

The cell sorting or panning procedure is used for the enrichment of the positive cells (Fig. 2). We prefer panning because the method can screen a large number of cells quickly. The method is advantageous when the expression of a target mRNA quantity is small. However, if the binding between a cell surface antigen (in this case enzyme products or their metabolites) and the antibody is not strong enough to hold cells on the plate, cell sorting is a better choice. The concentration of the primary antibodies to be used must be determined empirically. We strongly recommend the use of monoclonal antibodies prepared from ascites fluid. If ascites fluid is used without purification, it must be diluted more than 100 times as a high concentration of the fluid usually promotes nonspecific binding of cells to plates. Panning plates are prepared as follows. Falcon 1007 plates (6 cm, noncoated) are treated with 2 ml of 10 mg/ml antimouse IgM goat IgG (Cappel) in 50 mM Tris–HCl, pH 9.5, for 2 hr. The antibody solution is removed and the plates are washed three times with 0.5 M NaCl. The plates are then blocked with 3 ml of 10 mg/ml bovine serum albumin in PBS.

At 48–72 hr after transfection, the medium is removed from the dishes and the dishes are rinsed two or three times with new DMEM without FBS to remove dead cells. This step is very important because dead cells tend to stick to the panning plate nonspecifically. After washing, 5 ml of 5 mM EDTA in PBS is added to each plate. After 30 min, the cells are detached

[25] J. Sambrook, E. F. Fritsch, and T. Maniatis, "Molecular Cloning: A Laboratory Manual." Cold Spring Harbor Laboratory, Cold Spring Harbor, NY, 1989.

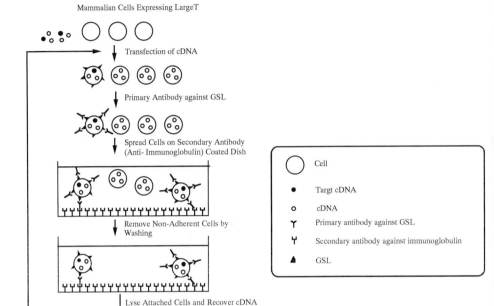

Fig. 2. Cells containing the target cDNA are concentrated by panning as follows. Cells carrying a desirable cDNA express the product (in this case, GM3), are treated with an antibody against the product GM3, and are added to antiantibody-coated plates. Only cells bound by the anti-GM3 antibody are trapped on plates. These cells are lysed and plasmids prepared. After amplification, the rescued plasmids are subjected to another round of transfection and panning for further concentration of the target cDNA.

from the plates by pipetting, washed with PBS twice, and resuspended at a concentration of 5×10^7 cells/ml in PBS-EDTA-NaN$_3$ [PBS containing 5% FBS, 0.02% sodium azide (optional), and 0.5 mM EDTA]. They are incubated with a primary antibody for 2 hr. In case of GlcCer synthase, we used an anti-GM3 monoclonal antibody, M2590 (Wako, Japan) (20 μg/ ml). If the binding between the antigen and the antibody is not strong enough to hold the cells on the plate, a bifunctional cross-linker may be used to fix the bound antibody on the cell surface.[15] For cross-linking, the cells are washed twice with ice-cold PBS, diluted to 1/5 with PBS containing 50 mM HEPES buffer (pH 8.3) and 0.2 mM bis(sulfosuccinimidyl)suberate, and kept on ice for 30 min. The cells are then washed twice with ice-cold PBS, resuspended in 10 ml of PBS-EDTA-NaN$_3$, and distributed to five 6-cm panning plates coated with antimouse IgM goat IgG. After 4 hr incubation at 25°, nonadherent cells are removed by gentle washing with PBS-EDTA-NaN$_3$. The plasmids are rescued from adherent cells by the method of Hirt.[26] To each washed dish, 400 μl 0.6% SDS and 10 mM EDTA are added. After 30 min, the lysate in each dish is transferred to a 1.5-ml Eppendorf tube and 100 μl of 5 M NaCl is added. After incubation on ice for 20 hr, the mixture is cenrifuged at 15,000 rpm for 15 min and the supernatant is transferred to a new tube. This is extracted successively with phenol–chloroform (1:1, v/v) and chloroform. The aqueous phase is transferred to a new tube again and to this is added 10 μl of 1 mg/ml glycogen and 1 ml of ethanol. After centrifugation, the pellet is resuspended in 100 μl of TE (10 mM Tris–HCl, pH 8.0, containing 1 mM EDTA) and ethanol precipitated again. The pellet is washed with 80% ethanol and dried. The pellet (containing plasmids) is redissolved in 5 μl TE and 1–2 μl is transformed into 10 μl of electrocompetent $E.$ $coli$ in H$_2$O, MC1061/ P3 cells (Invitrogen) by electroporation.

Electrocompetent can be prepared by washing midlog phase $E.$ $coli$ cells with 10% ice-cold glycerol. For electroporation, the mixture of DNA and cells is transferred to a 0.1-cm cuvette, cooled on ice for 10 min, and exposed to a 1.6-kV pulse with a capacitance of 25 μF and pulse controller unit to 200 Ω by Gene Pulsar (Bio-Rad). After the pulse, 1 ml of SOC medium[25] is added immediately and incubated for 30 min at 37°. The cells are then inoculated into 500 ml of 2 × YT containing 12.5 μg/ml ampicillin and 7.5 μg/ml tetracycline and are grown overnight. The plasmid is then prepared by the standard cesium chloride method[25] or by a Qiagen column and another round of panning performed as described earlier.

After the second round of panning and electroporation, SIB selection is performed (Fig. 3). By the second transformation, approximately 500 $E.$ $coli$ colonies are formed on a 15-cm dish from 2 μl of the rescued plasmid. They are divided into 32 pools after transfer onto a nylon membrane. Each piece of the membrane is dissected by a razor blade and is transferred to a test tube containing 2 ml 2 × YT culture containing 12.5 μg/ml ampicillin and 7.5 μg/ml tetracycline. After incubation at 37° overnight, plasmids are prepared from each pool by the alkaline lysis method followed by PEG precipitatior.[25] They are introduced into GM-95-PyT cells cultured in 24-well plates by the DEAE dextran method. Conditions of the DEAE dextran transfection differ from cell types and are optimized for mouse melanoma B16 cell lines, including GM-95-PyT. For the transfection of COS cells, refer to Ref. 23. One-third of plasmid prepared from a 2-ml culture (40 μl in PBS) is mixed with 10 μl of 50 mg/ml DEAE dextran (Sigma). GM-95-PyT cells are grown in 6-well plates in DMEM containing 10% FCS until they reach 50% confluency; then the medium is replaced by 1 ml DMEM containing 10% Nu-Serum (Collaborative Research). The

[26] B. Hirt, $J.$ $Mol.$ $Biol.$ **26**, 365 (1967).

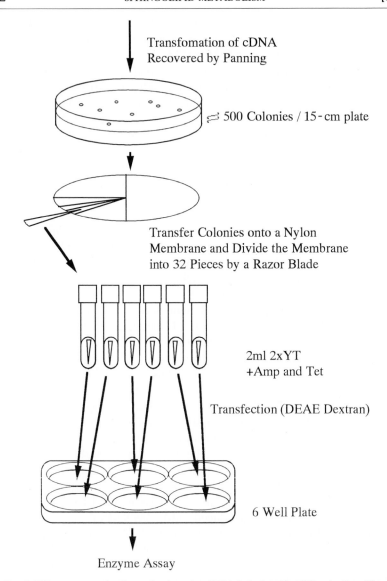

Transfomation of cDNA
Recovered by Panning

⇌ 500 Colonies / 15-cm plate

Transfer Colonies onto a Nylon
Membrane and Divide the Membrane
into 32 Pieces by a Razor Blade

2ml 2xYT
+Amp and Tet

Transfection (DEAE Dextran)

6 Well Plate

Enzyme Assay

FIG. 3. After two rounds of transfection, the cDNA is isolated by SIB selection. Plasmids rescued from the second panning are transformed to *E. coli,* and the colonies that form are transferred onto a nylon membrane and dissected into 32 pieces by a razor blade. Plasmids are prepared from colonies from each piece of membrane and transfected into GM-95-PyT cells cultured in 6-well plates. Two days after the transfection, cells are harvested for enzyme assay. In the case of GlcCer synthase, 62 individual clones from a positive pool are examined similarly and GlcCer synthase cDNA, pCG-1, is isolated.

DNA–DEAE dextran complex is added to each well and incubated for 2 hr. After incubation, the medium is replaced by DMEM containing 20% FCS. Cells are harvested between 48 and 72 hr after transfection for the enzyme assay. GlcCer synthase activity is measured in the cells harvested from each well. Individual *E. coli* clones from the positive pools are examined for activity as described earlier and GlcCer synthase cDNA is isolated. The GlcCer synthase cDNA in the vector is designated as pCG-1. It is advisable to stop panning after the second cycle because further rounds of electroporation and panning do not promote further enrichment of a desired cDNA.

For determination of the nucleotide sequence, we used the cycle sequencing method[27] with Thermo Sequanase (Amersham Life Technologies, Illinois) in combination with the LI-COR 4000L sequencer (Li-Cor, NE) according to the manufacturer's instructions. The system gives excellent results for the determination of G + C-rich sequences and usually approximately 1 kb of nucleotide sequence is determined from a priming site.

Expression of Recombinant Enzymes

Most of the enzymes involved in GSL synthesis are located at the lumenal side of the Golgi apparatus and are anchored via their N-terminal in the Golgi membranes (type II membrane proteins). Large-scale production of these enzymes in *E. coli* is relatively easy when the N-terminal signal-anchor sequence is deleted because this does not cause loss of activity in most cases. In contrast to these glycosyltransferases, GlcCer synthase is located on the cytoplasmic side of the Golgi membranes, anchored via its N-terminal signal-anchor sequence (a type III membrane protein). Because removal of the signal anchor sequence from this enzyme inactivates catalytic activity, GlcCer synthase is extremely difficult to produce in large quantities, and high-level expression is toxic to both *E. coli* and mammalian cells. However, cells expressing the recombinant enzyme are still useful for the studies of the enzyme properties and GSL functions.

Expression in Escherichia coli

Escherichia coli expression is suitable for analyses of cDNA involved in sphingolipid synthetic enzymes because *E. coli* contain neither sphingolipids nor their synthetic enzymes. Although the expression level of GlcCer synthase in *E. coli* is low, the cells can produce more enzyme protein than most mammalian cells. We use a system in which the genes are controlled

---

[27] F. Sanger, S. Nicklen, and A. R. Coulson, *Proc. Natl. Acad. Sci. U.S.A.* **74**, 5463 (1977).

by the T7 promoter (pET series, Novagen),[28] which is inducible by the addition of isopropyl-β-thiogalactopyranoside (IPTG) to the medium. A variety of pET vectors with different cloning sites are now available. A restriction site must be introduced in a cDNA to locate its initiation codon several base pairs downstream of the ribosome-binding sequence (SD sequence) in the vector. This can be achieved easily by the polymerase chain reaction (PCR)[29] using primers with an extra restriction site attached to their 5' ends (Fig. 4A). In addition to the restriction site sequence, a nucleotide sequence may be added to protect the 5' end from nuclease. The DNA fragment with the full-length coding sequence with NdeI and BamHI sites at 5' and 3' ends, respectively, is generated by PCR (Fig. 4A) and is cloned into the SmaI site of the Bluescript KS vector (Strategene, California) (Fig. 4B). The cloning step into Bluescript KS may be omitted if several extra nucleotide sequences are added to 5' ends of primers for efficient cutting by restriction enzymes. After amplification in E. coli DH5α (BRL) cells, the insert is excised with NdeI and BamHI and then cloned into E. coli expression vector pET3a (Fig. 4C). The resulting plasmid pET-CG-1 is transformed into the E. coli strain BL21 (DE3), and the clone is purified by isolation of single colonies, after streaking a liquid culture on a plate. For expression of the cloned GlcCer synthase, cells harboring the plasmid are grown in NZCYM medium containing 400 μg/ml ampicillin at 37°. When the cell density reaches 0.25 OD_{600}, IPTG is added to a final concentration of 1 mM and the cells are incubated for an additional 5 hr. After the incubation, the cells are harvested and disrupted by sonication. For the enzyme assay, 150 μg of cell lysate protein is added to each reaction mixture and incubated for 7 hr. For maximal yield of the protein, the induction is performed at 37°. If most of the protein is expressed as an inactive inclusion body, induction at 30° improves the result.

Stable Expression in Mammalian Cells

Cell lines that express specific GSLs are valuable tools for studying GSL functions. To study biological functions of GSLs, the parental cell line of GM-95 (MEB-4) is not ideal because it may have multiple mutations (GM-95 and MEB-4 show different properties, including their morphology; however, these differences may be due to other mutations in MEB-4 than the GlcCer synthase gene). With this point in mind, we established GM-

[28] F. W. Studier, A. H. Rosenberg, J. J. Dunn, and J. W, Dubendorff, *Methods Enzymol.* **185,** 60 (1990).
[29] R. K. Saiki, S. Sharf, F. Faloona, K. B. Mullis, G. T. Horn, H. A. Erlich, and N. Arnheim, *Science* **230,** 1350 (1985).

95 stably expressing GlcCer synthase and compared the properties with GM-95. The plasmid pCG-1 isolated in the previous section carries the full-length GlcCer synthase cDNA in the pcDNA I vector and was used for transfection. Note that the 5′-untranslated region is necessary for the efficient expression of mammalian cDNA, or one may attach the Kozack sequence[30] preceding an initiation codon artificially. GM-95 cells are plated in 10-cm culture dishes in 10 ml DMEM supplemented with 10% FCS and incubated until they reach 50% confluency. The medium is replaced by 10 ml serum-free DMEM and to this is added the mixture of pCG-1 (60 μg), pSV2neo (6 μg), and Lipofectin (150 μg) in 300 μl of water. As a control, the pcDNA I vector and pSV2neo are transfected. The next day, the cells are subjected selection by G418 (800 μg/ml). After 2 weeks, more than 100 colonies are obtained from each plate. The cells are pooled, replated, maintained for 2 months in medium containing G418, and cloned by limiting dilution. The lipid composition and reactivity of the transfectants with M2590 are examined for each clone.

The lipid pattern is analyzed as follows. The cells are scraped at subconfluency, washed twice with PBS, and lyophilized. Total lipids are extracted from the cells with 20 volumes of $CHCl_3$-CH_3OH (2:1, v/v), filtered, and evaporated to dryness. The lipids are redissolved in a small volume of $CHCl_3$-CH_3OH (2:1, v/v) and chromatographed on precoated silica gel TLC plates (E. Merck) developed with $CHCl_3$-CH_3OH-12 mM $MgCl_2$ in H_2O (65:25:4, v/v). GSLs are visualized with orcinol-H_2SO_4 reagent as described earlier. For immunochemical detection of GSLs, TLC immunostaining is carried out according to Higashi et al.[31] Total lipids are applied to a POLYGRAM SIL G plate (Macherey-Nagel, FRG), developed in $CHCl_3$-CH_3OH-12 mM $MgCl_2$ in H_2O (5:4:1, v/v), and stained as described previously. For the analysis of ceramide, total lipids are spotted on a silica gel HPTLC plate, developed 2 cm from the origin with CH_3Cl-CH_3OH-12 mM $MgCl_2$ in H_2O (60:35:8, v/v), the plate is dried, and is rechromatographed in $CHCl_3$-CH_3OH-CH_3COOH (90:2:8, v/v). The plate is dried completely and sprayed with cupric-acetate-phosphoric acid reagent to visualize lipids.[32,33]

Expression of GSLs is also examined by staining cells with M2590 as follows. Cells are allowed to attach to glass coverslips for 24 hr. The cells

[30] M. Kozack, Cell **44,** 283 (1986).
[31] H. Higashi, Y. Fukui, S. Ueda, S. Kato, Y. Hirabayashi, M. Matsumoto, and M. Naiki, *J. Biochem.* **95,** 1517 (1984).
[32] R. Selvam and N. S. Radin, *Anal. Biochem.* **112,** 338 (1981).
[33] M. E. Fewster, B. J. Burns, and J. F. Mead, *J. Chromatogr.* **43,** 120 (1969).

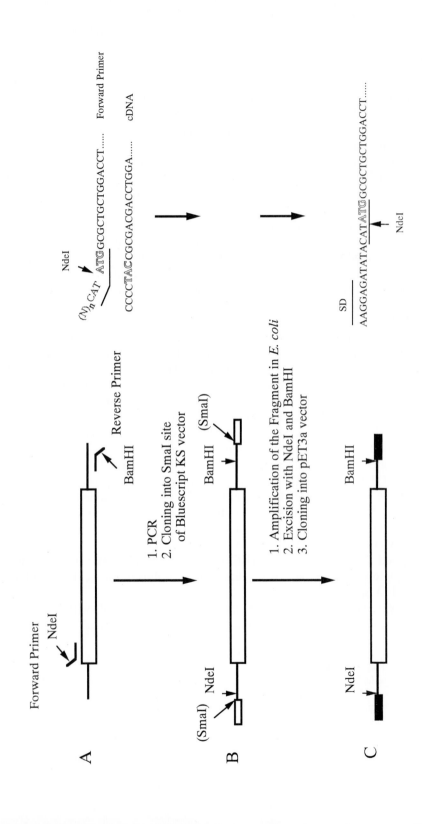

are washed with PBS, fixed with 3% formaldehyde, blocked for 1 hr with 10% fetal calf serum, and reacted with M2590 (10 μg/ml). After washing with PBS, the cells are stained with FITC-labeled anti-mouse IgM (Cappel).

Gene Cloning

Isolation of a Gene

Development of P1[34] and bacterial artificial chromosome vectors (BAC)[35] allow the isolation of large-size genomic DNA fragments. Genome Systems Inc. now offers genomic libraries in P1 and BAC vectors, and nylon membranes spotted with DNA from independent colonies. The membrane is used for quick identification and isolation of a target gene by dot blot hybridization. This company also offers a custom library screening service by PCR of SIB libraries and by the dot blot hybridization. We have isolated the mouse GlcCer synthase genomic DNA fragment in pBeloBAC11 by BAC high-density filter screening (embryonic stem cell genomic library of 129Svs/J mouse) using a ^{32}P-labeled 1.1-kb human GlcCer synthase cDNA fragment as a probe (positions between 298 and 1347).[6]

Analysis of Gene Structure

The isolated large genomic DNA fragment (approximately 120 kb in size, roughly estimated by a reverse field electrophoresis) is digested with

---

[34] N. Sternberg, Proc. Natl. Acad. Sci. U.S.A. **87**, 103 (1990).
[35] H. Shizuya, B. Birren, U.-J. Kim, V. Mancino, T. Slepak, Y. Tachiri, and M. Simon, Proc. Natl. Acad. Sci. U.S.A. **89**, 8794 (1992).

FIG. 4. Construction of the E. coli expression vector. (A) The entire coding region is amplified by PCR with the forward primer around the translation initiation site, reverse primer corresponding to the 3'-untranslated region, and pCG-1 harboring the entire GlcCer synthase cDNA as a template. The extra nucleotides are added to create restriction sites in both forward and reverse primers. The NdeI site of the forward primer is designed to overlap the translation initiation site of GlcCer synthase. In addition to the restriction sites, one or two (n = 1 or 2) nucleotide sequences may be added to protect the 5' end from nuclease. (B) The amplified fragment is cloned into the SmaI site of the Bluescript KS vector by blunt end ligation. In case step B is omitted, more than five ($n > 5$) and three additional nucleotide sequences are required for efficient cutting by NdeI and BamHI, respectively. (C) The cDNA fragment is excised by double digestion of NdeI and BamHI and cloned into the same restriction sites of the pET3a vector. This simply locates the initiation site several nucleotides downstream of the ribosome-binding site (the SD sequence). N stands for A, T, G, or C.

restriction enzymes and subcloned into pZero-1 (Invitrogen) or Bluescript KS vectors. Clones containing exons are isolated by colony hybridization using a 1.1-kb DIG-labeled *Bam*HI–*Eco*RI fragment of mouse GlcT-1 cDNA (positions between 186 and 1340).[36] The DIG DNA labeling kit (Boehringer Manheim, Tokyo, Japan) is used for the probe preparation, according to manufacturer's instructions. The DIG system is convenient and as sensitive as the isotope system, although it sometimes gives a relatively high background. Standard hybridization for DIG-labeled probes is carried out at 42° for 30 hr in 5× SSC containing 50% formaldehyde, 4% SDS, 10% blocking solution (Boehringer Mannheim), 100 μg/ml of salmon sperm DNA, and an appropriate amount of DIG-labeled probe (500 ng/ml for random prime-labeled probe and 30 pmol/ml for oligonucleotide probe). After hybridization, the membrane is washed with 2× SSC/0.5% SDS and 0.1× SSC/0.5% SDS each for 30 min at 50°. Hybridized DNA fragments are reacted with alkaline phosphatase-labeled anti-DIG antibody and visualized with nitroblue tetrazolium salt and 5-bromo-4-chloro-3-indo-lylphosphate as substrate. In some cases, DNA fragments containing exon–intron boundaries are amplified by PCR using exon primers, subcloned, and sequenced. We were successful in amplifying most of the introns of the GlcCer gene, the largest of which was the first intron (12 kb). For this, the Expand High Fidelity PCR system (Boehringer Mannheim) is used according to manufacturer's instructions. PCR is performed in 100 μl of a reaction mixture containing template DNA (1 μg of a BAC DNA), 1× Expand HF buffer, 1.5 mM MgCl$_2$, 200 μM each of dNTPs (dATP, dTTP, dCTP, and dGTP), a pair of primers (40 pmol each), and the enzyme mixture (0.75 U). The Tm-5° of primers is set around 68°, and a two-cycle PCR of 94 and 68° is performed. It is also possible to sequence the BAC clone plasmids directly using the cycle sequencing kit as described earlier. Exon sequences and exon–intron boundaries are defined by comparison with the mouse cDNA sequence.

Acknowledgments

We thank Y. Nagai (Mitsubishi Kagaku Institute of Life Science) for encouraging our work. We also thank K. Ozawa for illustrations. This work was supported by the Frontier Research Program of the Institute of Physical and Chemical Research (RIKEN), the Grants-in Aid for Encouragement of Young Scientists (No. 09780586 to S.I.) and for Scientific Research of Priority Areas (05274106 to Y.H.) from the Ministry of Education, Science, and Culture of Japan, and the Mizutani foundation.

[36] S. Ichikawa, K. Ozawa, and Y. Hirabayashi, *Biochem. Mol. Biol. Int.* **44**, 1193 (1998).

[34] Use of Yeast as a Model System for Studies of Sphingolipid Metabolism and Signaling

By Namjin Chung and Lina M. Obeid

Introduction

Using the yeast *Saccharomyces cerevisiae* as a model system for sphingo-lipid research is not only pertinent but also advantageous. The rapidly increasing number of yeast genes that have been identified to encode enzymes of sphingolipid metabolism and their sequence conservation with mammalian genes verify this idea.[1] The structure and the metabolism of sphingolipids are essentially conserved among eukaryotes, including human and yeast, yet yeast has only a limited number of sphingolipid species and their metabolic pathways are simpler.[1,2] The yeast genome has been completely sequenced, and this has facilitated the identification of many genes involved in sphingolipid metabolism. Needless to say, the power of yeast genetics has proved to be very useful for mapping out complex signal transduction pathways.[3,4]

This article describes basic methods of yeast genetics suitable for analyzing sphingolipid metabolism and signaling for those who are not familiar with yeast genetics, but to some extent familiar with terms and basic methodologies of sphingolipid research. Readers are thus referred to other articles in this volume for general sphingolipid methodologies. First, we will attempt to use the example of growth inhibition by phytosphingosine for describing how to apply yeast genetic techniques to study sphingolipid-mediated signal transduction,[5] and then we will discuss strategies and examples of how to clone genes involved in sphingolipid metabolism.

Getting Started

The pathways for *de novo* synthesis and degradation of yeast sphingo-lipids are illustrated in Fig. 1. Readers are referred to a review for the detailed description of these pathways.[1] Some sphingolipid biosynthetic

[1] R. C. Dickson, *Annu. Rev. Biochem.* **67,** 27 (1998).
[2] S. Hakomori, *in* "Handbook of Lipid Research" (J. N. Kanfer and S. Hakomori, eds.), Vol. 3, p. 1. Plenum, New York, 1983.
[3] D. E. Levin and B. Errede, *Curr. Opin. Cell Biol.* **7,** 197 (1995).
[4] J. Schultz, B. Ferguson, and G. F. Sprague, Jr., *Curr. Opin. Gen. Dev.* **5,** 31 (1995).
[5] N. Chung, C. Mao, G. M. Jenkins, Y. A. Hannun, J. Heitman *et al.,* unpublished results.

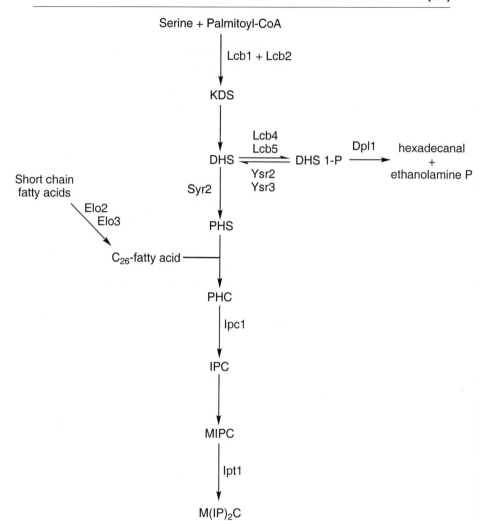

FIG. 1. Metabolic pathways of yeast sphingolipids. Abbreviations for the names of sphingolipids are defined in the text. Enzymes identified to catalyze some of the steps are shown as three-letter gene product names. Other steps remain blank to indicate that the gene products for the corresponding steps are not yet identified.

and degradative intermediates, collectively referred to here as sphingolipid metabolites,* are involved in the regulation of certain biological functions. Their structures are depicted in Fig. 2.

Sources and Preparations of Sphingolipid Metabolites

Many sphingolipids and their metabolites are available commercially, including PHS (Sigma, P2795), DHS (Biomol, SL-125), SPH (Biomol, EI-155), C_2-Cer (Biomol, SL-100), KDS (Matreya, 1876), and STA (Sigma, S9273). PHC is at present not available from any commercial sources, but could be synthesized chemically with the methods described elsewhere in this volume.[5a]

Most lyophilized sphingolipid metabolites can be dissolved in ethanol with brief vortexing and incubation at room temperature. A convenient concentration for stock solutions is 20 mM for most use, and stock solutions should be stored at $-20°$. Precipitates may develop while in storage, but they can be redissolved by incubating briefly at room temperature or at $37°$. Instead of ethanol, dimethyl sulfoxide (DMSO) can be used as a solvent, but in this case solutions should be stored in small aliquots to avoid repeated freezing and thawing.

Media Preparation

Recipes for yeast medium are found in another volume of this series.[6] To add sphingolipids to (1) liquid medium: prewarmed sphingolipid metabolites can be added directly to cell culture (it is not recommended to store liquid medium containing sphingolipid metabolites at any temperature for later use); (2) solid medium: first autoclave at $121°$ for 20 min and cool down to $60°$ before sphingolipids are added. It is customary to include a detergent such as Tergitol (Type NP-40; Sigma) to a final concentration of 0.05% for the purpose of even distribution of sphingolipids in solid agar.[7] The presence of detergent does not affect growth inhibition by PHS (our

---

* Abbreviations for sphingolipid metabolites are used throughout the article as follows: KDS, 3-ketodihydrosphingosine (3-ketosphinganine); DHS, dihydrosphingosine (sphinganine); PHS, phytosphingosine (5-hydroxysphinganine); PHC$_n$, phytoceramide (n indicates the number of carbons in a fatty acyl chain); IPC, inositolphosphoceramide; MIPC, mannosylinositolphosphoceramide; and M(IP)$_2$C, mannosyldiinositolphosphoceramide. Mammalian derivatives D-erythro-sphingosine (SPH) and ceramide (CER) are used for comparison and specificity and are structurally related, but metabolically unrelated stearylamine (STA) is used as a negative control.

[5a] A. Bielawska et al., Methods Enzymol. **311** [42] 1999 (this volume).
[6] C. Guthrie and G. R. Fink, Methods Enzymol. **194** (1991).
[7] G. B. Wells and R. L. Lester, J. Biol. Chem. **258**, 10200 (1983).

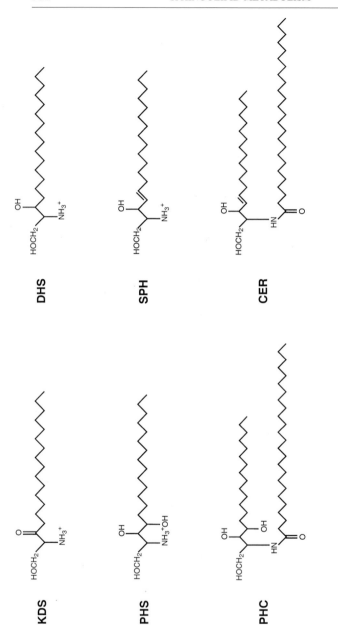

FIG. 2. Structures of sphingolipid metabolites. Abbreviations for the names of sphingolipids are defined in the text. KDS and DHS are sphingolipid metabolites found in both yeast and animals; PHS and PHC are found in yeast; and SPH and CER are found in animals. Note the hydroxyl group at C-4 positions of PHS and PHC, and the double bonds between the C-4 and C-5 positions of SPH and CER. STA is a structurally related, but metabolically unrelated, analog of sphingolipid metabolites, which is often used as a negative control.

unpublished observation), therefore, it is worthwhile to test medium with or without detergent. Solid medium containing sphingolipids can be stored in the dark in a cold room for more than a few months without significant loss of biological activity.

Growth Inhibition by PHS: An Example

It was reported previously that C_2-Cer, a short-chain analog of mammalian ceramides, arrests the yeast cell cycle at the G_1 phase only under limited conditions.[8] However, PHS was shown to inhibit growth in diverse experimental conditions with good reproducibility.[5] We have demonstrated that PHS inhibits growth of auxotrophic, but not prototrophic yeast strains, by inhibiting nutrient uptake activities through accelerating ubiquitin-dependent degradation of nutrient permeases.[5]

Strain Background

Different strains appear to show different degrees of tolerance toward PHS such that some strains appear to be more sensitive to PHS than others.[5] Therefore, it is necessary to determine an effective concentration of PHS for growth inhibition. For example, the W303 strain and some strains with S288C background are sensitive to 20 μM PHS whereas strains with $\Sigma 1278b$ background are sensitive to 40 μM PHS, but resistant to 20 μM PHS.

There are two plausible reasons for the different sensitivities. One reason is that different strains with different parental origins have been significantly diverged genetically and this could result in different sensitivity. Another reason is related to the different combinations of auxotrophic markers found in different strains. PHS inhibits the uptake of various nutrients, and inhibition of the uptake of certain kinds of nutrients is apparently more critical for viability. Among these are tryptophan and leucine, so strains with *trp1 leu2* markers are more sensitive to PHS than those with the single *trp1* or *leu2* marker, and prototrophic strains are resistant to PHS in the concentration range where the action of PHS is specific.

With these factors in mind, a parental strain should be chosen carefully depending on the purpose of one's research.

Liquid Medium Culture

Liquid culture is a convenient and quantitative way to measure growth inhibition of yeast strains by PHS. It usually requires smaller amounts of

---

[8] J. T. Nickels and J. R. Broach, *Genes Dev.* **10**, 382 (1996).

PHS or other sphingolipid metabolites and less time to complete a growth assay in liquid medium than in solid medium. Growth inhibition becomes apparent within a few hours after addition of PHS. This is not the result of cell cycle arrest, as the ratio of budded cells to total cells remains relatively constant throughout the incubation time of more than 18 hr. Control treatment with ethanol or DMSO, which is used as a vehicle for PHS or other sphingolipid metabolites, should be included in all experiments, but it does not affect growth at the concentrations typically used for many experiments.

Growth inhibition by PHS is not affected significantly by cell density. It is conventional to seed cells into fresh medium at a density of one million cells per milliliter, and cell growth can be monitored every 2 to 3 hr by counting cell numbers using a hemocytometer or by reading absorbance at 600 nm (A_{600}). It is convenient to prepare a standard curve to correlate A_{600} to cell density for a particular spectrophotometer and a particular strain.

Solid Medium Culture: Streaking vs Spreading

Streaking cells on agar plates has been a traditional way of assessing yeast cell growth on solid medium. Using the round-head end of a toothpick, streaking with three changes of toothpicks should result in a serial dilution of cells and the formation of single colonies on the third streak. Whether PHS or a particular sphingolipid metabolite inhibits growth should be judged by comparing colony formation in the third streaks. It is recommended to use 100-mm-diameter petri dishes divided into four or six equal sections for unquestionable determination of growth inhibition. Eight-section plates also can be used depending on the potency of growth inhibition by a particular sphingolipid metabolite.

Spreading for single colony formation is another method to assay growth on solid medium. For this assay, yeast cells are grown to log phase, counted using a hemocytometer, diluted with appropriate medium, and plated onto solid medium containing PHS or other sphingolipid. Approximately 300 cells are suitable for counting colonies on a 100-mm-diameter plate after 2 days of incubation. Cells should be spread evenly with a sterile glass rod or with six to eight sterile glass beads (diameter of 4 mm; Fisher Scientific, 11-312B); the latter often results in more even spreading than the former.

Both types of assays are used to test the colony-forming ability of isolated single cells, but each has its advantages and disadvantages. Streaking is convenient, takes less time for an assay, and saves the number of plates and the amount of PHS to be used. Also, more than two strains could be compared visually side by side on a single plate, eliminating plate-to-plate variations. However, the sensitivity of streaking lags behind that of spreading. With streaking, cells could manage to form colonies in an area where cell density is high enough to override the cytotoxicity of PHS. Therefore, especially when eight-section plates are used, it is often necessary to scrutinize the end of streaks for colony-forming ability. However, spreading usually gives all-or-none results: wild-type cells form no single colony on a certain concentration of PHS plate (Fig. 3). Also, with spreading, it is easy to quantify the cytotoxicity of PHS by counting the number of colonies. However, spreading uses more plates and time compared to streaking. In addition, because there could be lot-to-lot variations with different batches of medium, it is recommended to use medium from the same batch for a single set of experiments.

Yeast genetic analysis often involves tetrad dissection of spores. For checking growth phenotypes, dissection should be first performed on regular medium because, in many cases, spores fail to germinate on a medium containing PHS. Then, the sensitivity to PHS should be examined by streaking or spreading germinated spores on a medium containing PHS.

Use of Synthetic Complete or Drop-out Medium

Synthetic medium, including complete medium (SC) and drop-out medium (SC-Ura, e.g.,), are widely used in many genetic experiments. There is an unresolved issue in using synthetic medium for sphingolipid studies:

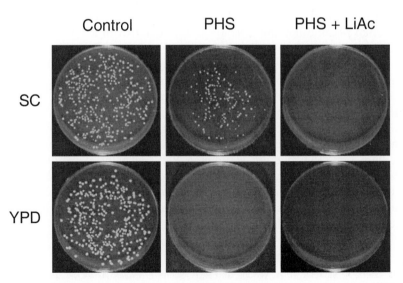

FIG. 3. Effects of medium type on PHS cytotoxicity. Control indicates mock treatment with ethanol vehicle, and LiAc indicates 10 m*M* lithium acetate.

cells are more viable on synthetic medium containing PHS than on YPD medium containing PHS (Fig. 3). Obviously, this is due to differences in the compositions of the two media,[6] which could be attributed to some impurities in YPD medium exaggerating the cytotoxicity of PHS or to additives in synthetic medium that help cells survive on PHS. Data show that PHS inhibits the uptake of various nutrients,[5] suggesting that the discrepancy in PHS cytotoxicity in two different types of medium probably stems from differences in the concentrations of amino acids and/or ammonium sulfate.

It is thought that perhaps PHS permeates into cells with less efficiency in synthetic medium than in YPD medium, maybe because of the long hydrocarbon moiety of PHS. Following a method of transporting DNA into cells,[9] lithium acetate is added to facilitate transport of PHS across the plasma membrane barrier. At 100 mM, (the concentration used for DNA transformation) lithium acetate is quite toxic, but 10 mM does not affect cell growth appreciably. When 10 mM lithium acetate is added together with 20 μM PHS to synthetic medium, growth inhibition is enhanced and becomes comparable to that in YPD medium containing 20 μM PHS in that no single colonies form, thus making this combination suitable for genetic screening (Fig. 3). An interpretation of this observation could be that the enhanced cytotoxicity of PHS in synthetic medium is due to the facilitated uptake of PHS. It is also possible that lithium acetate may inhibit nutrient permeases by either direct or indirect interactions.

In practice, this matter could affect two types of genetic experiments: (1) when plasmid-bearing cells are tested in a growth assay and (2) when cells are screened for viability in certain drop-out medium as an indicator of transformation. The latter is particularly problematic because a false-positive colony could be mistaken as a transformant. Until we fully understand how PHS works in synthetic medium, one could simply include 10 mM lithium acetate in synthetic medium for both types of experiments. For growth assays of plasmid-bearing cells, one could first grow the cells on synthetic drop-out medium and then streak a colony from the drop-out medium on YPD medium containing PHS. Usually, depending on the plasmid type, the plasmid is not easily lost from a population of cells bearing it over 2 days of incubation on YPD medium.[10] Thus, if one predetermines the rate of plasmid loss and includes a proper control strain in each experiment, the results should be reproducible and reliable.

[9] D. Gietz, A. St. Jean, R. A. Woods, and R. H. Schiestl, *Nucleic Acids Res.* **20,** 1425 (1992).
[10] P. Hieter, C. Mann, M. Snyder, and R. W. Davis, *Cell* **40,** 381 (1985).

Use of Positive Selection Markers

It was demonstrated that *TRP1* wild-type strains are resistant to PHS whereas trp1 auxotrophic mutant strains are sensitive to PHS.[5,11] This implies that PHS inhibits tryptophan uptake. As we extended our investigation, it became apparent that PHS inhibits the uptake of various nutrients, including leucine, histidine, uracil, and tryptophan. When strains that have different combinations of auxotrophic markers were tested, only the tryptophan- or leucine-prototrophic strains were resistant to PHS, whereas the histidine- or uracil-prototrophic strains remained sensitive to PHS. Therefore, growth phenotypes against PHS and nutrient uptake are not in perfect unison. However, this can be explained by a hypothesis that PHS inhibits the uptake of a wide range of nutrients, but that only some nutrients, such as tryptophan and leucine, are critical for viability. It is not known at present if and how other auxotrophic markers, such as *LYS2* and *ADE2,* contribute to the growth phenotype with PHS.

For practical purposes, one can safely use the *URA3* or the *HIS4* markers for plasmids and gene disruption when PHS resistance is examined. It is generally not advisable to use the *TRP1* or the *LEU2* markers, but they may be used in biochemical studies not involving phenotypic examinations. A better option is to use a positive selection marker such as the G418^R gene, which is neutral. A polymerase chain reaction-generated G418^R gene cassette could be used efficiently to disrupt a specific target gene. Specific procedures are described in detail elsewhere.[12]

Thin-Layer Liquid Chromatographic Analysis of Yeast Sphingolipids

After the construction of a mutant strain of sphingolipid metabolism, it is often necessary to examine and compare the sphingolipid profiles of the mutant to that of the reference wild-type strain. Thin-layer liquid chromatography (TLC) is a convenient tool for this purpose. A protocol originally devised by Hanson and Lester,[13] improved later by Mandala *et al.*[14] and by Oh *et al.*,[15] provides a simple and easy method to analyze yeast sphingolipids by TLC.

[11] M. S. Skrzypek, M. M. Nagiec, R. L. Lester, and R. C. Dickson, *J. Biol. Chem.* **273,** 2829 (1998).

[12] A. Wach, A. Brachat, R. Pohlmann, and P. Philippsen, *Yeast* **10,** 1793 (1994).

[13] B. A. Hanson and R. L. Lester, *J. Lipid Res.* **21,** 309 (1980).

[14] S. M. Mandala, R. A. Thornton, B. R. Frommer, J. E. Curotto, W. Rozdilsky, M. B. Kurtz, R. A. Giacobbe, G. F. Bills, M. A. Cabello, and I. Martin *et al., J. Antibiot.* **48,** 349 (1995).

[15] C. S. Oh, D. A. Toke, S. Mandala, and C. E. Martin, *J. Biol. Chem.* **272,** 17376 (1997).

Specifically, yeast cells are grown until lag or early log phase before being labeled with [³H]serine (final activity at 5 μCi/ml; American Radiolabelled Chemicals, ART-246) for a few doublings (4 to 6 hr). Five to 10 mm of cells at 1×10^7 cells/ml are then treated for specific experimental purposes, harvested immediately by centrifugation at 1000g at 4°, and washed twice with cold water. Pelleted cells are resuspended with 1 ml of the extraction buffer (ethanol:water:diethyl ether:pyridine:ammonium hydroxide, 15:15:5:1:0.018) and incubated in a 60° water bath with slow agitation for 15 min. When a higher yield is desired, the extraction could be repeated and combined with the first extract. After drying under a nitrogen stream, the [³H]serine-labeled extract is subjected to mild alkaline methanolysis with 0.5 ml monomethylamine reagent (25% monomethylamine in ethanol) at 52° for 30 min. The alkali-resistant extract is again dried under nitrogen and suspended in an appropriate volume (up to 100 μl) of chloroform:methanol:water (16:16:5). A portion of the suspension is applied to Whatman silica gel TLC plates and resolved by chloroform:methanol:4.2 N ammonium hydroxide (9:7:2). Radioactive bands on the TLC plate can be visualized by autoradiography after treatment with En³Hance (NEN Life Science) and quantified using PhosphorImager (Molecular Dynamics) after exposure to a tritium screen.

Strategies for Cloning Genes Involved in Sphingolipid Metabolism

The rapidly increasing number of identified genes involved in sphingolipid metabolism underscores the value of different strategies. One such strategy is based on growth phenotypes against sphingolipid metabolic inhibitors or sphingolipid metabolites and another is based on DNA/protein sequence homology searches that take advantage of the availability of the complete yeast genome sequences (http://genome-www.stanford.edu/Saccharomyces/). By discussing some of these cases, we hope to help develop general ideas for cloning more genes involved in sphingolipid metabolism in the future.

LCB1 and LCB2

Selection of yeast mutants requiring long-chain bases (e.g. KDS, DHS, or PHS) yielded two complementation groups, *lcb1* and *lcb2*.[7] Because growth defects of the mutant strains could be rescued by KDS, it was reasoned that these mutant strains are defective in serine palmitoyltransferase which catalyzes the condensation between L-serine and palmitoyl-CoA to produce KDS, and the genes encoding serine palmitoyltransferase, *LCB1* and *LCB2*, were cloned by complementation (see article [1], this volume).[16,17] Later, the genes for mammalian serine palmitoyltransferase were cloned based on sequence homology.[18]

SYR2

The *SYR2* gene was originally cloned because it was required for the cytotoxicity exerted by syringomycin E, a cyclic lipodepsipeptide.[19] In search for a biological function, the Syr2 protein sequence was compared to protein databases by BLAST algorithm, and significant similarities were found with the Erg3 (C-5 sterol desaturase) and the Erg25 (C-4 sterol methyl oxidase) sequences.[20] Despite sequence homology between the Syr2 and the enzymes involved in sterol metabolism, Syr2 was not shown to have a role in sterol metabolism. The presence of an eight-histidine motif in the Syr2, which is a characteristic signature of desaturases, hydroxylases, oxidases, or decarbonylases involved in lipid metabolism, prompted investigation into a possible role in sphingolipid metabolism. This endeavor finally showed that the Syr2 is essential for DHS hydroxylase, which catalyzes C-4 hydroxylation of DHS to produce PHS.

LCB4, LCB5, YSR2, YSR3, and DPL1

These five genes encode three important enzymes involved in the degradation of sphingoid bases: *LCB4* and *LCB5* for sphingosine kinases,[21] *YSR2* and *YSR3* for DHS (PHS) 1-phosphate phosphatases,[22,23] and *DPL1* for DHS (PHS) 1-phosphate lyase.[24] In the course of degradation, DHS (PHS) is phosphorylated to DHS (PHS) 1-phosphate by *LCB4* or *LCB5,* which is in turn degraded to hexadecanal and phosphoethanolamine by *DPL1*

[16] R. Buede, C. Rinker-Schaffer, W. J. Pinto, R. L. Lester, and R. C. Dickson, *J. Bacteriol.* **173,** 4325 (1991).

[17] M. M. Nagiec, J. A. Baltisberger, G. B. Wells, R. L. Lester, and R. C. Dickson, *Proc. Natl. Acad. Sci. U.S.A.* **91,** 7899 (1994).

[18] K. Hanada, T. Hara, M. Nishijima, O. Kuge, R. C. Dickson, and M. M. Nagiec, *J. Biol. Chem.* **272,** 32108 (1997).

[19] P. Cliften, Y. Wang, D. Mochizuki, T. Miyakawa, R. Wangspa, and J. Hughes, *Microbiology* **142,** 477 (1996).

[20] M. M. Grilley, S. D. Stock, R. C. Dickson, R. L. Lester, and J. Y. Takemoto, *J. Biol. Chem.* **273,** 11062 (1998).

[21] M. M. Nagiec, M. Skrzypek, E. E. Nagiec, R. L. Lester, and R. C. Dickson, *J. Biol. Chem.* **273,** 19437 (1998).

[22] C. Mao, M. Wadleigh, G. M. Jenkins, Y. A. Hannun, and L. M. Obeid, *J. Biol. Chem.* **272,** 28690 (1997).

[23] S. M. Mandala, R. Thornton, Z. Tu, M. B. Kurtz, J. Nickels, J. Broach, and S. Spiegel, *Proc. Natl. Acad. Sci. U.S.A.* **95,** 150 (1998).

[24] J. D. Saba, F. Nara, A. Bielawska, S. Garrett, and Y. A. Hannun, *J. Biol. Chem.* **272,** 26087 (1997).

(Fig. 1). Phosphorylation of DHS (PHS) and subsequent dephosphorylation by Ysr2 or Ysr3 also have important roles in the uptake of exogenous sphingoid bases.[22,25]

DPL1 (formerly known as *BST1* for bestower of sphingosine tolerance) was cloned by virtue of resistance to 1 mM SPH when present as multicopies.[24] In contrast, a *dpl1Δ* mutant strain is more sensitive to SPH (25 μM) than an isogenic wild-type strain, most likely due to the cytotoxic level of accumulation of SPH 1-phosphate. This led to the cloning of another important gene in sphingoid base metabolism, *YSR2*, by selecting a multicopy suppressor of the SPH sensitivity of the *dpl1Δ* mutant strain.[22] The *YSR2* gene encodes DHS (PHS) 1-phosphate phosphatase, as evidenced by the accumulation of DHS 1-phosphate in a *ysr2Δ* mutant strain and by an *in vitro* assay showing the conversion of DHS to DHS 1-phosphate by the purified protein.[26] A homology search of the yeast protein database with Ysr2 amino acid sequences identified another DHS 1-phosphate phosphatase gene, YSR3 (53% identity).

An independent study identified the same DHS 1-phosphate phosphatase gene *LBP1/YSR2* via a different approach.[23] In this case, a suppressor mutant strain of *cho1Δ* phosphatidylserine (PS) synthase mutant strain was selected to clone the gene encoding DHS 1-phosphate phosphatase. The logic behind this selection strategy is that the *cho1Δ* mutant strain is auxotrophic for phosphoethanolamine, which could be provided by another mutation in sphingolipid metabolic pathway. If the DHS 1-phosphate phosphatase gene is mutated, then DHS 1-phosphate would be only destined to phosphoethanolamine.

Whereas the multicopy suppressor analysis of *dpl1Δ* mutant strain identified DHS 1-phosphate phosphatase genes, transposon mutagenesis of DPL1 yielded a SPH-resistant suppressor mutant strain, which enabled the cloning of DHS kinase, *LCB4*.[21] Homology searches revealed another DHS kinase gene, *LCB5* (53% identity).

ELO2 and ELO3

These genes encode fatty acid elongases responsible for the synthesis of C_{26}-fatty acids, which are substrates for PHC synthase.[15] These were identified by a homology search of Elo1, which is involved in the elongation of the fatty acid 14:0 to 16:0. The *ELO2* (56% amino acid sequence identity to *ELO1*) and the *ELO3* (52% identity) genes encode fatty acid elongases responsible for C_{24}- and C_{26}-fatty acid synthesis, respectively.

[25] L. Qie, M. M. Nagiec, J. A. Baltisberger, R. L. Lester, and R. C. Dickson, *J. Biol. Chem.* **272,** 16110 (1997).

[26] C. Mao and L. M. Obeid, *Methods Enzymol.* **311** [26] 1999 (this volume).

General Strategies Emerging from the Examples

These examples suggest that one could clone genes involved in sphingo-lipid metabolism by exploiting phenotypes related to sphingolipid metabolites or inhibitors with a basic understanding of sphingolipid metabolic pathways. In fact, all of the genes mentioned previously, except *ELO2* and *ELO3* were cloned with such strategies. Although the speed of cloning genes in sphingolipid metabolism has been remarkable, there are still some genes to be cloned. An example of such a gene is PHC synthase, which has been elusive for many years. The inhibitors of PHC synthase, fumonisin B1 and australifungin, may be used to clone the PHC synthase gene in the future.

The availability of the complete yeast genome sequence has facilitated the cloning of genes involved in sphingolipid metabolism greatly. *ELO2* and *ELO3* are such examples. Also, mammalian serine palmitoyltransferase genes, *YSR3* and *LCB5* were identified quickly because it was possible to carry out a homology search against the complete genome sequence database.

Acknowledgments

We thank Dr. Joseph Heitman for discussions regarding yeast genetic methods and Dr. Yusuf A. Hannun for discussions regarding sphingolipid methods. N.C. is a recipient of a fellowship from the Korea Foundation of Advanced Studies.

Section II

Inhibitors of Sphingolipid Biosynthesis

[35] Isolation and Characterization of Novel Inhibitors of Sphingolipid Synthesis: Australifungin, Viridiofungins, Rustmicin, and Khafrefungin

By Suzanne M. Mandala and Guy H. Harris

Sphingolipid synthesis is an essential process in yeast and the pathogenic fungi that cause life-threatening human infections such as candidiasis, aspergillosis, and cryptococcosis. Although many steps in the human and fungal sphingolipid biosynthetic pathway are similar, there are several enzymes found uniquely in fungi (Fig. 1) that are potential targets for the development of nontoxic therapeutic antifungals. In our screening program, we have found that natural products are a rich source of structurally diverse inhibitors of sphingolipid synthesis. Natural product inhibitors to four different enzymes that affect sphingolipid synthesis have been discovered: sphingofungins,[1–3] lipoxamycin,[4] myriocin/ISP1,[5] and viridiofungins[6–8] inhibit serine palmitoyltransferase; fumonisin B1[9] and australifungin[10,11] inhibit ceramide synthase; aureobasidins,[12] khafrefungin,[13] and rustmicin[14–16] inhibit inositol phosphoceramide synthase; and minimoidin inhibits the fatty acid elongation pathway (S. M. Mandala and G. H. Harris, unpublished). Most of these compounds have fungicidal activity against a broad spectrum of pathogenic fungi, but only the inhibitors of inositol phosphoceramide are fungal selective; compounds that inhibit early biosynthetic steps show comparable activity against orthologous mammalian enzymes. This article describes a method to identify sphingolipid inhibitors and a detailed proto-

[1] M. M. Zweerink *et al., J. Biol. Chem.* **267**, 25032 (1992).
[2] F. M. VanMiddlesworth *et al., J. Antibiot.* **45**, 861 (1992).
[3] W. S. Horn *et al., J. Antibiot.* **45**, 1692 (1992).
[4] S. M. Mandala *et al., J. Antibiot.* **47**, 376 (1994).
[5] Y. Miyake *et al., Biochem. Biophys. Res. Commun.* **211**, 396 (1995).
[6] G. H. Harris *et al., Tetrahed. Lett.* **34**, 5235 (1993).
[7] S. M. Mandala *et al., J. Antibiot.* **50**, 339 (1997).
[8] T. Esumi *et al., Tetrahed. Lett.* **39**, 877 (1998).
[9] E. Wang *et al., J. Biol. Chem.* **266**, 14486 (1991).
[10] S. M. Mandala *et al., J. Antibiot.* **48**, 349 (1995).
[11] O. D. Hensens *et al., J. Org. Chem.* **60**, 1772 (1995).
[12] M. M. Nagiec *et al., J. Biol. Chem.* **272**, 9809 (1997).
[13] S. M. Mandala *et al., J. Biol. Chem.* **272**, 32709 (1997).
[14] S. M. Mandala *et al., J. Biol. Chem.* **273**, 14942 (1998).
[15] G. H. Harris *et al., J. Antibiot.* **51**, 837 (1998).
[16] B. Tse *et al., J. Org. Chem.* **62**, 3236 (1997).

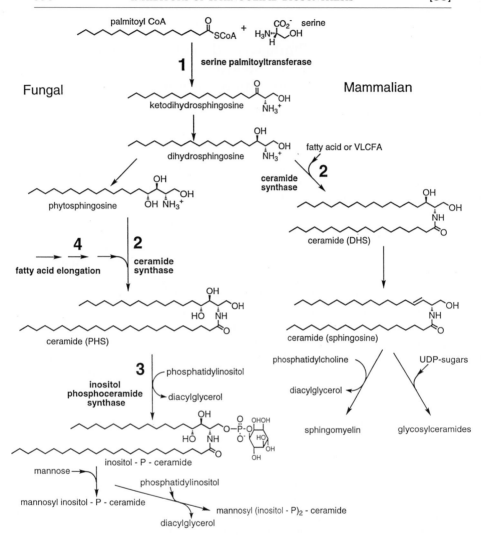

FIG. 1. Pathways for fungal and mammalian sphingolipid biosynthesis. Steps subject to natural product inhibition are indicated: (1) sphingofungins, myriocin, lipoxamycin, and viridiofungin; (2) fumonisins, and australifungin; (3) rustmicin, khafrefungin, and aureobasidin A; and (4) minimoidin. VLCFA, very long chain fatty acid; DHS, dihydrosphingosine; PHS, phytosphingosine.

col for the isolation of australifungin. More abbreviated descriptions of the isolation of viridiofungins, khafrefungin, and rustmicin are also included. Structures of these compounds are shown in Fig. 2.

Bioassay for Sphingolipid Inhibition

Sphingolipid synthesis in the yeasts *Saccharomyces cerevisiae, Candida albicans,* and *Cryptococcus neoformans* can be measured in a 96-well plate using [^{3}H]inositol incorporation into lipids, which accounts for 99% of total inositol found in trichloroacetic acid (TCA) precipitates. The assay relies on chemical hydrolysis to discriminate the alkali-stable sphingolipids from phosphatidylinositol (PI); on alkaline methanolysis, PI is degraded to glycerophosphoinositol, a water-soluble product that can be washed away. Details for the *Candida*-based assay are given.

Assay

Sphingolipid inhibitors identified to date have poor solubility in water and some are unstable in an aqueous environment; they are stored as

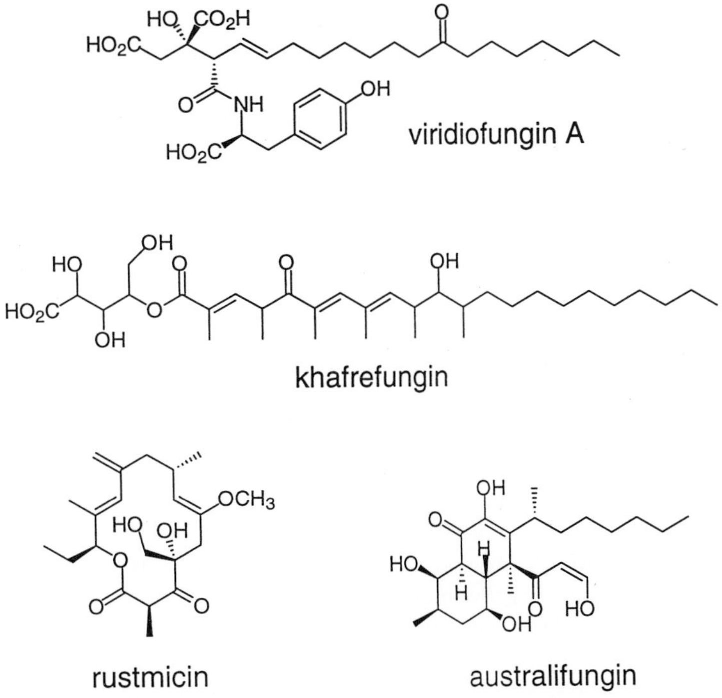

Fig. 2. Structures of natural product inhibitors of sphingolipid synthesis.

concentrated stocks in methanol or dimethyl sulfoxide (DMSO) at −80°. For testing, they are diluted to five times the desired test concentrations in 96-well plates containing medium [Difco yeast nitrogen base with glucose (YNBD)], modified with up to 20% solvent. Forty microliters of each sample is transferred to a 96-well test plate. Cell inoculum is prepared by culturing cells to logarithmic phase in YNBD medium containing 1% casamino acids, buffered to pH 5.2 with 40 mM sodium succinate. Cells are mixed with [^{3}H]inositol and 160 μl is added immediately to drug-containing test plates to give final concentrations of 1.6–2.0 × 10^6 cells/ml and 2.0 μCi [^{3}H]inositol/ml. Plates are incubated at 30° for 3 hr.

The assay is terminated with the addition of 67 μl of 20% TCA to each well and plates are incubated at 4° for 20 min. Cells are collected onto glass fiber filter mats in a 96-well cell harvester that is programmed for two 10-sec H$_2$O washes. Filters are dried thoroughly in a warm air stream and sealed into sample bags with 10 ml liquid scintillant. Total [^{3}H]inositol incorporation is quantitated in an LKB BetaPlate liquid scintillation counter (Wallac). To measure [^{3}H]inositol incorporation in the sphingolipid fraction, filter mats are removed from scintillant, placed into glass dishes containing 200 ml of 0.2 N KOH in methanol:toluene (1:1), and rotated on a platform shaker at room temperature for 60 min. The solution is neutralized by adding 1.25 ml glacial acetic acid, and the filters are sequentially transferred and incubated for 10 min in glass dishes containing 200 ml each of methanol, cold 5% TCA, methanol, and a second methanol wash. Filters are dried and counted again in the BetaPlate.

Data Analysis

Radioactivity remaining on the filter mats after alkaline methanolysis is a measure of sphingolipid synthesis. By subtracting these counts from the total, the amount of incorporation into PI can be calculated. The relative amount of label in PI and sphingolipids will vary depending on the length of assay and growth rate of cells. [^{3}H]Inositol is first converted to PI, which is subsequently metabolized to sphingolipids. Consequently, short labeling periods favor high ratios of PI to sphingolipids, whereas longer labeling periods favor sphingolipid synthesis. Using the conditions described (*C. albicans,* 30°, 3 hr), approximately 50% of counts are found in each fraction. Calculating a ratio of PI to sphingolipids can correct for well-to-well variation in cell inoculum or growth and will also eliminate nonspecific inhibitors that reduce total [^{3}H]inositol incorporation. Dividing the PI/sphingolipid ratio measured in drug-treated cells by the PI/sphingolipid ratio from control cells measured in the absence of inhibitors provides normalized values for the relative reduction in sphingolipid synthesis. Thus, a value of 2 represents 50% inhibition of sphingolipid synthesis (relative to PI synthesis and normalized to control cells).

Inhibitor Analysis

Examples of four inhibitors tested at multiple concentrations provide data on potency of the compounds and the relative extent of sphingolipid inhibition (Fig. 3). Inhibitors of inositol phosphoceramide synthase directly block the transfer of [^{3}H]inositol into sphingolipids and produce the highest ratios of PI to sphingolipid. Ratios obtained from inhibitors that block earlier in the pathway are increased if cells are pretreated with drugs prior to [^{3}H]inositol addition, presumably due to the reduction of intermediates in the pathway. The labeling assay detects inhibitors of serine palmitoyltransferase, ceramide synthase, very long chain fatty acid synthesis, and inositol phosphoceramide synthase (see Fig. 1), which can be discriminated by

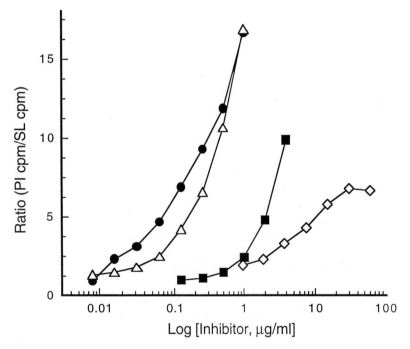

FIG. 3. Inhibition of sphingolipid synthesis as a function of inhibitor concentration. Inhibitor—rustmicin (●), khafrefungin (△), australifungin (■), and viridiofungin A (◇)—were titrated in the sphingolipid inhibition assay. Data are expressed as the ratio of PI over sphingolipid, normalized to controls in the absence of inhibitor.

specific *in vitro* enzyme assays and by labeling cells with precursors to the pathway and analyzing the lipids for the accumulation of intermediates.[7,10,13]

Bioassay-Guided Isolation of Sphingolipid Synthesis Inhibitors

Natural products with a desired biological activity are isolated from plants, microorganisms, or marine invertebrates using a process known as bioassay-guided fractionation. Extracts of the organisms are subjected to repeated fractionation using various forms of partition, adsorption, ion exchange, and size-exclusion chromatography until a pure compound is obtained. Each step of the process is guided by one or more bioassays for the activity of interest. A combination of sphingolipid inhibition assays and agar diffusion antifungal assays have been the most useful for sphingolipid synthesis inhibitors. The structure of the isolated pure natural product is then determined from analysis of nuclear magnetic resonance (NMR), mass spectral (MS), ultraviolet (UV), and infrared (IR) spectroscopic data.

A typical isolation scheme consists of extraction of the biological material, a low-resolution "cleanup" of the crude extract followed by one or two high-resolution purification steps. A crude extract of the organism is prepared by extraction of the material with an organic solvent. The extraction solvent is chosen to maximize the yield of the biological activity of interest while minimizing other components. It is useful to determine a few properties of the activity at this crude extract stage to help choose subsequent fractionation methods. For example: Is the activity water soluble? Is it stable at acidic, basic, or neutral pH? Does the activity appear to be acidic or basic? The initial fractionation of the crude extract is generally scaleable, low resolution, and is chosen to provide as much weight purification of the activity as possible. Gel filtration, ion exchange, or open column adsorption or partition chromatography is typically used for this purpose. The active rich cut from this step is subjected to one or two high-resolution chromatographic separations, such as reversed phase high-performance liquid chromatography (RP-HPLC), to yield a pure natural product. The choice of chromatographic steps requires some experience and is beyond the scope of this article. The reader is referred to the many reviews on the subject.[17,18]

Many of the sphingolipid synthesis inhibitors are unstable and require careful assessment of the recovery of biological activity through the isolation process. This is accomplished by titrating the crude extract in the sphingo-

[17] G. H. Wagman and R. Cooper, "Natural Products Isolation." Elsevier, Amsterdam, 1989.
[18] R. J. P. Cannell, "Natural Products Isolation." Humana, Totowa, NJ, 1998.

lipid inhibition and *in vitro* antifungal assays to establish a baseline level of activity. The combined rich cut from each subsequent fractionation step is similarly titrated and the relative activities compared. Recovery of activity from a fractionation step should be 50% or greater when assay variability and chromatographic yields are considered. More significant loss of biological activity after a fractionation step may indicate compound instability or very poor chromatographic behavior.

The chemical nature of the natural product sphingolipid synthesis inhibitors described to date has caused some specific problems for bioassay-guided fractionation. Many of the inhibitors are structurally similar to sphingolipids, having a polar, ionizable head group combined with a lipophilic tail. As a result, these compounds are strongly retained in both normal and reversed phase chromatographic systems and yield broad chromatographic peaks. To minimize these problems, the ionic form of any ionizable functional group should be controlled by buffering extraction solvents and mobile phases wherever possible. The presence of enolizable or tautomeric functional groups, as in australifungin, can also cause poor chromatographic behavior.

Preparation of Australifungin

Fermentation.[10] Culture MF5672 (ATCC 74157), *Sporormiella australis,* is grown in KF seed medium, 54 ml in 250-ml baffled Erlenmeyer flasks at 25°, 220 rpm, and 50% humidity for 3 days. The seed culture, 2 ml, is used to inoculate 250-ml baffled Erlenmeyer flasks containing 50 ml of production media F1 or MOF. The production flasks are incubated statically for solid medium F1 or are shaken at 220 rpm for liquid medium MOF at 25°, 50% humidity for approximately 14 days.

KF seed medium consists of (per liter): corn steep liquor, 5 g; tomato paste, 40 g; oat flour, 10 g; glucose, 10 g; trace element mix, 10 ml; pH adjusted to 6.8 before autoclaving. [The trace element mix consists of (per liter): $FeSO_4 \cdot 7H_2O$, 1 g; $MnSO_4 \cdot 4H_2O$, 1 g; $CuCl_2 \cdot 2H_2O$, 25 mg; $CaCl_2$, 100 mg; H_3BO_3, 56 mg; $(NH_4)_6Mo_7O_{24} \cdot 4H_2O$, 19 mg; and $ZnSO_4 \cdot 7H_2O$, 200 mg.]

F1 solid fermentation medium consists of (per 250-ml flask): cracked corn, 10 g; Ardamine PH, 2.0 mg; KH_2PO_4, 1.0 mg; $MgSO_4 \cdot 7H_2O$, 1.0 mg; sodium tartrate, 1.0 mg; $FeSO_4 \cdot 7H_2O$, 0.1 mg; $ZnSO_4 \cdot 7H_2O$, 0.1 mg; distilled water 10 ml.

MOF liquid fermentation medium consists of (per liter): D-mannitol, 75 g; oat flour, 15 g; Fidco yeast extract, 5 g; L-glutamic acid, 4 g; [2-(*N*-

morpholino)ethanesulfonic acid] monohydrate (MES), 16.2 g; pH adjusted to 6.0 with NaOH.

Isolation. Australifungin contains two enolizable functional groups: an α-diketone and a β-ketoaldehyde. The tautomeric forms of these functionalities cause severe peak broadening on RP-HPLC columns, making preparative separations poor. Countercurrent chromatography uses immiscible liquids as both stationary and mobiles phases[19,20] and provides an effective final purification. Typical titers of australifungin in fermentations of MF5672 should be 50–70 mg/liter in solid fermentation medium F1. A closely related compound, australifunginol, is also produced at titers of 500–600 mg/liter, but is separated easily from australifungin.[10] Isolation of australifungin, and any of the other inhibitors described, should be monitored with thin-layer chromatography and analytical RP-HPLC as described in Table I.

1. Extraction. One liter of a MF5672 liquid culture is adjusted to pH 3 with concentrated H_3PO_4. Ethyl acetate, 1 liter, is added and the solution is agitated for 1 hr. The ethyl acetate layer is removed and concentrated *in vacuo*. Ethyl acetate, 50 ml/250-ml flask, is added directly to solid fermentations in F1 medium, and the solids are gently disrupted and agitated for 1 hr. The solids are removed by filtration, and the ethyl acetate layer is concentrated to dryness *in vacuo*. Excess oily material is removed from the ethyl acetate by dissolving in 50 ml MeOH and extracting the solution with 150 ml of hexanes. The MeOH layer contains australifungin and is concentrated *in vacuo*.

2. First isolation step:

a. Silica gel 60 (E. Merck, 0.040–0.063 mm, 230–400 mesh), 90 g, is slurried in approximately 500 ml of hexane : ethyl acetate (3 : 2) containing 1% glacial acetic acid. The slurry is poured into a glass chromatography column with approximate dimensions of 2.5 × 50 cm. The liquid is drained to the top of the silica gel bed in preparation for loading the column.

b. The concentrated MeOH layer from step 1 is dissolved in 5 mL of the above hexane : ethyl acetate solution and applied to the column. The column is eluted with the same solution at 10 ml/min collecting 20-ml/fractions. Australifungin-containing fractions are combined and concentrated to dryness. (Note: australifunginol can be recovered by further elution of the silica gel column with ethyl acetate con-

---

[19] Y. Ito and W. D. Conway, "High-Speed Countercurrent Chromatography." Wiley, New York, 1996.

[20] W. D. Conway, "High-Speed Countercurrent Chromatography." VCH, New York, 1993.

TABLE I
CHROMATOGRAPHIC PROPERTIES OF KHAFREFUNGIN, VIRIDIOFUNGIN A, RUSTMICIN,
AND AUSTRALIFUNGIN

| Compound | Thin-layer chromatography[a] | | Reversed-phase HPLC[b] | |
	Solvent	R_f	Mobile phase	t_R (min)
Australifungin	Hexane:EtOAc:HOAc (5:5:0.1)	0.53	CH_3CN:25 mM K_2HPO_4 adjust to pH 6.9 with concentrated H_3PO_4 (55:45)	8.3
Viridiofungin A	Not useful	Not useful	CH_3CN:0.1% aqueous H_3PO_4 (58:42)	10.3
Khafrefungin	Not useful	Not useful	CH_3CN:0.1% aqueous H_3PO_4 (80:20)	11.8
Rustmicin	Hexane:EtOAc (1:1)	0.58	MeOH:25 mM NH_4OAc, pH 4.5 (75:25)	14.8

[a] Silica gel 60, F-254, E. Merck. Detection (TLC): Australifungin and rustmicin can be visualized by spraying with $CeSO_4$ in dilute H_2SO_4 followed by heating. Rustmicin and related macrolides can also be visualized by spraying with 0.5% p-anisaldehyde in a solution of 10% H_2SO_4 in 50% EtOH followed by heating.
[b] Phenomenex Primesphere C8, 5 μm, 4.6 × 250 mm, 1.0 ml/min at 40° for viridiofungin A, khafrefungin, and rustmicin; Phenomenex Ultracarb 30, 4.6 × 150 mm, 1.0 ml/min at 55° for australifungin. UV detection (HPLC): australifungin, 275 nm; viridiofungin, 210 nm; khafrefungin, 290 nm; and rustmicin, 235 nm.

taining 1% glacial acetic acid. The final purification of australifunginol using preparative RP-HPLC is straightforward.)

3. Final purification. The following separation is based on a HSCCC instrument manufactured by P.C. Inc. (Potomac, MD) but should extrapolate to other instruments.

a. One liter of a solvent system consisting of hexane:ethyl acetate:MeOH:25 mM K_2HPO_4 adjusted to pH 6.9 (7:3:5:5) is shaken for 5 min in a separatory funnel and the layers are separated. The lower aqueous layer is pumped into the tail of 300 ml, #14 coil of P.C. Inc. multilayer coil, until the coil is completely filled.

b. The silica gel australifungin rich cut from step 2b is dissolved in 2 ml of each phase of the solvent system and is injected onto the tail of the column. Rotation of the unit is started at 800 rpm in the forward direction, and the mobile phase, upper layer, is pumped tail to head at 3 ml/min collecting 7.5-ml fractions. The peak of australifungin typically elutes with 400–480 ml of mobile phase.

Preparation of Khafrefungin, Viridiofungins, and Rustmicin

Khafrefungin

Fermentation.[21] Culture MF6020 (ATCC 74305), a sterile fungal mycelium, is grown in KF seed medium as described earlier for australifungin. The seed culture, 2 ml, is used to inoculate 250-ml baffled Erlenmeyer flasks containing 50 ml of production media CYS80. The production flasks are incubated at 220 rpm, 25°, 50% humidity for approximately 21–28 days.

CYS80 production medium consists of (per liter): sucrose, 80 g; corn meal (yellow), 50 g; yeast extract, 1 g; no pH adjustment.

Isolation. Khafrefungin is a 22 carbon linear polyketide esterified to the C-4 hydroxyl of an aldonic acid. The lipophilic character of the fatty chain, combined with the polar aldonic acid end, results in strong retention in both normal and reversed-phase chromatographic systems. This behavior is exploited by using low-pressure silica gel chromatography, as the initial isolation step applied to the crude culture extract, to achieve approximately 100-fold weight purification of khafrefungin. Titers of khafrefungin in cultures of MF 6020 are relatively low (10 mg/liter).

A solid fermentation culture MF6020, equivalent to 1 liter of liquid broth, is extracted with methyl ethyl ketone (MEK), 1 liter, filtered and the extract is concentrated to dryness. The residue is dissolved in 20% aqueous MeOH, adjusted to pH 2.5 with concentrated H_3PO_4, and partitioned sequentially with heptane and ethyl acetate. The concentrated ethyl acetate fraction is then applied to a column of silica gel 60 (230–400 mesh, bed volume = 250 ml) preequilibrated with CH_2Cl_2:MeOH:HOAc (90:10:1). The column is then eluted sequentially with CH_2Cl_2:MeOH: HOAc (90:10:1), 7× 100-ml fractions; CH_2Cl_2:MeOH:HOAc:H_2O (70:30:1:2), 5× 200-ml fractions; and MeOH:HOAc (100:1), 3× 200-ml fractions. Khafrefungin is found in the last two eluates. The fractions, typically 10–14, are combined and concentrated to dryness *in vacuo*.

Khafrefungin is purified from this residue using preparative RP-HPLC on Phenomenex Primesphere C8, 5 μm, 9.4 × 250 mm, at a flow rate of 4.0 ml/min using the mobile phase described for analytical RP-HPLC (Table I).

Viridiofungins

Fermentation.[22] Culture MF5628 (ATCC 74084), *Trichoderma viride,* is grown in KF seed medium as described for australifungin. This culture

[21] G. H. Harris *et al., Bioorg. Med. Chem. Lett.,* in preparation.
[22] J. C. Onishi *et al., J. Antibiot.* **50,** 334 (1997).

is used to inoculate 250-ml Erlenmeyer flasks containing 50 ml of BRF production medium. The production flasks are incubated statically for approximately 21 days at 25°, 85% humidity.

BRF solid fermentation medium consists of (per 250-ml flask): brown rice, 5 g; base liquid, 20 ml. [Base liquid consists of (per liter): yeast extract, 1 g; sodium tartrate, 0.5 g; KH_2PO_4, 0.5 g; no pH adjustment.]

Isolation. The citric acid moiety of the viridiofungins is exploited to achieve a functional group selective first isolation step. Individual viridio-fungins are then purified from this mixture using preparative RP-HPLC, or a combination of HSCCC, as described earlier for australifungin, and RP HPLC.[23] Viridiofungin A is the major component of the mixture and is produced at a titer of 0.5–2 g/liter. The titer of viridiofungins B and C is typically 0.05–0.5 g/liter, and numerous additional minor components can be isolated at titers of approximately 1 mg/liter.

A methyl ethyl ketone extract of a BRF medium solid fermentation of culture MF5628, 1 liter, is prepared as described earlier for khafrefungin. The dried MEK extract is dissolved in 250 ml of a buffer solution consisting of 800 ml H_2O : 1200 ml CH_3CN : 4.28 ml formic acid, which is adjusted to pH 4.5 with 2.0 N NaOH. This solution is then extracted twice with an equal volume of hexane, and the aqueous portion is applied to a column of Bio-Rad AG4 × 4 (formate cycle, 100–200 mesh, bed volume = 100 ml), which is preequilibrated in the buffer solution described previously. The viridiofungin mixture is eluted with a solution consisting of 800 ml H_2O : 1200 ml CH_3CN : 11.2 ml concentrated H_2SO_4. The elution of viridio-fungins is monitored using analytical RP-HPLC and by measuring the pH of the eluant; viridiofungins elute at the pH transition. Viridiofungin-containing fractions are combined and desalted by removal of the CH_3CN *in vacuo* followed by extraction of the remaining aqueous solution with ethyl acetate. Purification of viridiofungin A and related minor components is accomplished easily using preparative RP-HPLC as described for khafre-fungin and with scale-up of the conditions shown in Table I.

Rustmicin (Galbonolide A)

Fermentation.[24] Culture MA7086, a *Micromonospora* sp., is grown in 50 ml of seed medium consisting of (per liter): glucose, 10 g; soluble starch, 20 g; yeast extract, 5 g; N-Z amine A, 5 g; beef extract, 3 g; bacto-peptone, 5 g; $CaCO_3$, 1 g; the pH is adjusted to 7.0 prior to the addition of $CaCO_3$. The culture is incubated at 28° and shaken at 220 rpm for 72 hr. A 2-ml aliquot of the seed culture is transferred to 250-ml nonbaffled flasks con-

---

[23] G. H. Harris *et al.,* U.S. patent 5,364,948 (1994).
[24] J. M. Sigmund and C. F. Hirsch, *J. Antibiot.* **51,** 829 (1998).

taining 44 ml of production medium and is incubated at 28°, 220 rpm for 5–6 days.

The production medium consists of (per liter): dextrin, 25 g; β-cyclodextrin 10 g; primary yeast, 14 g; tomato paste, 4 g; $CoCl_2 \cdot 6H_2O$, 0.005 g; UCON-LB625, 4 ml; the pH is adjusted to 7.2 with NaOH.

Isolation. Rustmicin (galbonolide A) was originally discovered because of its potent antifungal activity against phytopathogenic fungi.[25,26] Rustmicin exhibits the best stability at pH 5.5, making it important to buffer aqueous solutions at this pH.[27] The half-life is significantly less than 1 hr at a pH less than 4 or greater than 7. Rustmicin is, however, stable indefinitely in methanol solution at $-20°$. Rustmicin titers are typically 40–60 mg/liter. The predominant related minor component produced by MA7086, at titers comparable to those of rustimicin, is galbonolide B.

Culture MA7086, 1 liter, is extracted by adding an equal volume of methanol and stirring or shaking vigorously for 1 hr. The solids are then removed by filtration or centrifugation, and the resulting filtrate/supernatant is adjusted to pH 5.5. The solution is passed through a column of Mitsubishi SP207 (bed volume = 50 ml), which was equibrated previously in $MeOH : 25$ mM NH_4OAc, pH 4.5 (1 : 1). After washing the column with 100 ml of the equilibration solution, rustmicin and related macrolides are eluted with 250 ml MeOH. Rustmicin is purified further using silica gel chromatography, as described for australifungin, except that a mobile phase of hexane : EtOAc : H_2O (75 : 25 : 0.2) is used. Rustmicin can also be purified from the SP207 rich cut using preparative RP-HPLC (scale-up of the analytical separation described in Table I). Rustmicin can be crystallized following silica gel or RP-HPLC by dissolving in MeOH at a concentration of 20 mg/liter and, with vigorous stirring, diluting with 2 volumes of H_2O. Rustmicin crystallizes on standing at 25°.

Biological Activity of Inhibitors

Ceramide Synthase Inhibitors

The first known inhibitors of ceramide synthase were the fumonisins,[9] which were initially isolated as tumor-promoting agents associated with severe toxicological effects in animals.[28] The sphingolipid inhibition, toxicity, and metabolism of the fumonisins have been characterized extensively

[25] H. Achenbach et al., Tetrahed. Lett. **26**, 6167 (1985).

[26] T. Takatsu et al., J. Antibiot. **38**, 1806 (1985).

[27] H. Achenbach et al., Ann. N.Y. Acad. Sci. **544**, 128 (1988).

[28] W. C. A. Gelderblom et al., Appl. Environ. Microbiol. **54**, 1806 (1988).

in animals and cultured mammalian cells.[29,30] Unfortunately, fumonisins have very poor activity against whole cell fungal sphingolipid synthesis or growth. Fumonisin B1 inhibits fungal ceramide synthase *in vitro*, and the weak whole cell activity may be due to limited penetration. In contrast, australifungin is very potent against fungi and is the preferred inhibitor for studies on fungal ceramide synthesis, but its activity in mammalian systems has had only limited characterization. The major drawback to using australifungin arises from the reactivity of the β-ketoaldehyde moiety. Australifungin reacts with free amines, including the sphingoid base components of ceramide and forms stable keto-enamine complexes (G. H. Harris and S. M. Mandala, unpublished data). The dihydrosphingosine : australifungin conjugate is at least 100-fold less potent at inhibiting ceramide synthase than free australifungin. Formation of the complex can be monitored by UV absorbance with an increase in absorbance at 344 nm, and similar spectral shifts are obtained when australifungin is incubated with other sphingoid bases or amino acids. The formation of inactive complexes is consistent with the 10-fold increase in potency that is measured when casamino acids are omitted from the sphingolipid inhibition assay and the comparatively poor activity that australifungin demonstrates in medium that contains serum or peptone. In addition to the loss in potency, caution is also advised when using australifungin in assays that employ trace amounts of sphingoid bases; high concentrations of australifungin can result in nonspecific inhibition due to sequestration of the sphingoid base.

Serine Palmitolytransferase Inhibitors

Viridiofungins are at least five-fold less potent and are not as specific for serine palmitoyltransferase inhibition as the sphingofungin/myriocin class of compounds. Viridiofungins are known to inhibit squalene synthase, geranyl geranyltransferase, and other enzymes that are sensitive to di- and tricarboxylic acids, albeit at higher concentrations than required for serine palmitoyltransferase inhibition. Because of this difference in enzyme sensitivity, viridiofungin A showed selective inhibition of sphingolipid synthesis in labeling studies in *C. albicans,* with no evidence for inhibition of sterol synthesis at growth inhibitory concentrations of drug.[7,22] For unknown reasons, viridiofungins lack *in vitro* activity against *Saccharomyces* serine palmitoyltransferase and do not inhibit sphingolipid synthesis in this yeast, although they do inhibit the mammalian enzyme and appear to be selective for sphingolipid inhibition in HepG2 cells.

[29] A. H. Merrill, Jr., D. C. Liotta, and R. T. Riley, *Trends Cell Biol.* **6,** 218 (1996).
[30] R. T. Riley *et al., J. Food Protect.* **57,** 638 (1994).

Inositol Phosphoceramide Synthase Inhibitors

Unlike inhibitors to earlier steps in sphingolipid synthesis, khafrefungin and rustmicin do not have any detectable effect on lipid synthesis in mammalian cells (unpublished data and Ref. 13), and rustmicin and aureobasidin A have been nontoxic in animal studies,[14,31] supporting the idea that inositol phosphoceramide synthase is a fungal selective target. Khafrefungin was lytic to washed red blood cells at 12.5 to 25 μg/ml. However, this toxicity may be due to the detergent-like properties of the compound. Although all three inositol phosphoceramide synthase inhibitors are very potent *in vitro* (picomolar to low nanomolar), substantially higher concentrations are required to kill fungi. In the case of rustmicin, two factors that limit whole cell activity were identified: efflux via multidrug transporters and chemical instability.[14] Even at its optimal pH of 5.5, rustmicin degrades relatively rapidly in aqueous media. Despite these limitations, concentrations of rustmicin that are required to completely inhibit phosphosphingolipid synthesis and accumulate ceramide in fungi are achieved easily.

Acknowledgments

We thank M. Zweerink and A. Edison for assay development; R. Giacobbe, M. Nallain, and J. Sigmund for fermentation studies; G. Bills, J. Polishook, and F. Pelaez for mycology; R. Thornton, M. Rosenbach, and M. A. Cabello for assay support; and D. Zink for mass spectrometry.

[31] K. Takesako *et al., J. Antibiot.* **46,** 1414 (1993).

[36] Fermentation, Partial Purification, and Use of Serine Palmitoyltransferase Inhibitors from *Isaria* (=*Cordyceps*) *sinclairii*

By RONALD T. RILEY and RONALD D. PLATTNER

Introduction

There are currently several potent fungal inhibitors of serine palmitoyltransferase (3-ketosphinganine synthase) [palmitoyl-CoA: L-serine C-palmitoyltransferase (decarboxylating); EC 2.31.50], the first enzyme in the *de novo* biosynthesis of ceramides and more complex sphingolipids.[1,2] In

[1] A. H. Merrill, Jr. and E. Wang, *Methods Enzymol.* **51,** 427 (1992).
[2] R. C. Dickson, R. L. Lester, and M. Marek, *Methods Enzymol.* **311** [1] 1999 (this volume).

addition to fungal serine palmitoyltransferase (SPT) inhibitors, there are several suicide inhibitors (cycloserine and haloalanines) of pyridoxal phosphate-dependent enzymes that inhibit SPT in the micromolar range.[3] Fungal SPT inhibitors (Fig. 1) are all structurally similar to the more common mammalian sphingoid bases: sphingosine, sphinganine, and phytosphingosine. Typically, fungal SPT inhibitors are active in the nanomolar range and are competitive with respect to both serine and palmitate.[3,4] For example, myriocin (sometimes referred to as ISP-I or thermozymocidin[5-7]) inhibited SPT with an IC_{50} of approximately 0.3–1 nM using sonicated CTLL-2 cells as the enzyme source.[8] There is some confusion in the literature about the common name (ISP-I = myriocin = thermozymocidin) for the fungal metabolite with the chemical name 2-amino-3,4-dihydroxy-2 hydroxy-methyl-14-oxo-eicos-6-enoic acid. Common names reflect the fungal genera in the cases of ISP-I and myriocin and the more general classification of thermophilic eumycetes in the case of themozymocidin. Other SPT inhibitors (lipoxamycin, viridiofungins, and sphingofungins) have been shown to have IC_{50} values in the nanomolar range in microsomal preparations from HeLa cells.[4,9,10] The adverse *in vitro* cellular effects of fungal SPT inhibitors can be reversed by the addition of free sphingoid bases.[4,11,12] Interest in SPT inhibitors stems from their potential as broad-spectrum antifungal agents and potent immunosuppressants, although studies indicate that the potent immunosuppressive ability of ISP-I may not be due to it potent inhibition of SPT.[13] Fungal SPT inhibitors have proved useful as biochemical tools for exploring the role of sphingolipids in cellular regulation. In

[3] T. Kolter and K. Sandoff, *Chem. Soc. Rev.* **26**, 371 (1996).

[4] S. M. Mandala, R. A. Thornton, B. R. Frommer, S. Driekorn, and M. B. Kurtz, *J. Antibiot.* **50**, 339 (1997).

[5] F. Aragozzini, P. L. Manachini, R. Craveri, B. Rindone, and C. Scolastico, *Tetrahedron* **28**, 5493 (1972).

[6] D. Kluepfel, J. Bagli, H. Baker, M.-P. Charest, A. Kudelski, S. N. Sehgal, and C. Vézina, *J. Antibiot.* **25**, 109 (1972).

[7] T. Fujita, K. Inoue, S. Yamamoto, T. Ikumoto, S. Sasaki, R. Toyama, K. Chiba, Y. Hoshino, and T. Okumoto, *J. Antibiot.* **47**, 208 (1994).

[8] Y. Miyake, Y. Kozutsumi, S. Nakamura, T. Fujita, and T. Kawasaki, *Biochem. Biophys. Res. Commun.* **211**, 396 (1995).

[9] S. M. Mandala, B. R. Frommer, R. A. Thornton, M. B. Kurtz, N. M. Young, M. A. Cabello, O. Genilloud, J. M. Liesch, J. L. Smith, and W. S. Horn, *J. Antibiot.* **47**, 376 (1994).

[10] S. M. Mandala and G. H. Hasres, *Methods Enzymol.* **311** [35] 1999 (this volume).

[11] M. M. Zweerink, A. M. Edison, G. B. Wells, W. Pinto, and R. L. Lester, *J. Biol. Chem.* **267**, 25032 (1992).

[12] S. Nakamura, Y. Kozutsumi, Y. Sun, Y. Miyake, T. Fugita, and T. Kawasaki, *J. Biol. Chem.* **271**, 1255 (1996).

[13] T. Fujita, R. Hirose, M. Yoneta, S. Sasaki, K. Inoue, M. Kiuchi, S. Hirase, K. Chiba, H. Sakamoto, and M. Arita, *J. Med. Chem.* **39**, 4451 (1996).

particular SPT inhibitors are being used to reveal the role of ceramide and sphingolipid metabolities in the regulation of cell growth, differentiation, and cell death.[14] Unfortunately, to the best of our knowledge, none of the fungal SPT inhibitors are currently available commercially and thus their use is not widespread. The purpose of this article is to (i) present a simple and inexpensive method for the fermentation and partial purification of a potent SPT inhibiting activity from cultures of *Isaria* (=*Cordyceps*) *sinclairii* and (ii) provide a few examples of how purified and partially purified SPT inhibitors could be used in studies to reveal the role of increased free sphingoid bases and decreased *de novo* ceramide biosynthesis in the toxicity associated with ceramide synthase inhibitors (fumonisins, AAL-toxins, and australofungins).[15,16,17]

Production of Fungal SPT Inhibitors

There are currently numerous fungi capable of producing inhibitors of SPT (Fig. 1). The first reported fungal SPT inhibitors were the sphingofungins produced by *Aspergillus fumigatus* (ATCC 20857) and *Paecilomyces variotii* (ATCC 74097; Merck Culture Collection Number MF 5537).[11] Lipoxamycin, another SPT inhibitor, is produced by *Streptomyces* (Merck Culture Collection Number MA 6975), and viridiofungins are produced by *Trichoderma viride* (ATCC 74084; Merck Culture Collection Number MF 5628).[4,9] Myriocin, produced by *Myriococcum albomyces* (ATCC 16425), and thermozymocidin, produced by *Mycelia sterilia* (ATCC 20349), were first described in 1972.[6,18] Shortly thereafter, both compounds were shown to be structurally identical, and the cultivation of microorganisms belonging to the thermophilic eumycetes group for the purpose of producing compounds with the structural formula of thermozymocidin was patented.[5,19,20] *Mycelia sterilia* (ATCC 20349) was isolated from a sample of maize-cultivated soil.[19] In 1994, ISP-I, structurally identical to myriocin and thermozymocidin, was isolated and purified from cultures of *Isaria* (=*Cordyceps*) *sinclairii* (ATCC 24400) and was shown to be a potent inhibitor of SPT.[7,8]

[14] A. H. Merrill, Jr., E.-M. Schmelz, D. L. Dillehay, S. Spiegel, J. A. Shayman, J. J. Schroeder, R. T. Riley, K. A. Voss, and E. Wang, *Toxicol. Appl. Pharmacol.* **142,** 208 (1997).
[15] R. T. Riley, K. A. Voss, W. P. Norred, C. W. Bacon, F. I. Meredith, and R. P. Sharma, *Environ. Toxicol. Pharmacol.* **7,** 109 (1999).
[16] E.-M. Schmelz, M. A. Dombrink-Kurtzman, P. C. Roberts, Y. Kozutsumi, T. Kawasaki, and A. H. Merrill, Jr., *Toxicol. Appl. Pharmacol.* **148,** 252 (1998).
[17] N. Ueda, G. P. Kaushal, X. Hong, and S. V. Shah, *Kidney Int.* **54,** 399 (1998).
[18] R. Craveri, P. L. Manachini, and F. Aragozzini, *Experientia* **28,** 867 (1972).
[19] J. F. Bagli, D. Kluepfel, and M. St-Jacques, *J. Org. Chem.* **38,** 1253 (1973).
[20] R. Craveri, P. L. Manachini, and F. Aragozzini, U.S. Patent 3, 758,529 (1973).

Sphingofungin C

ISP-1 = myriocin = thermozymocidin

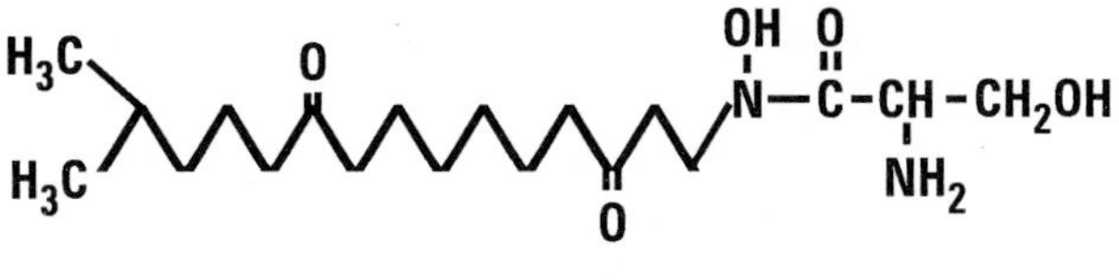

Lipoxamycin

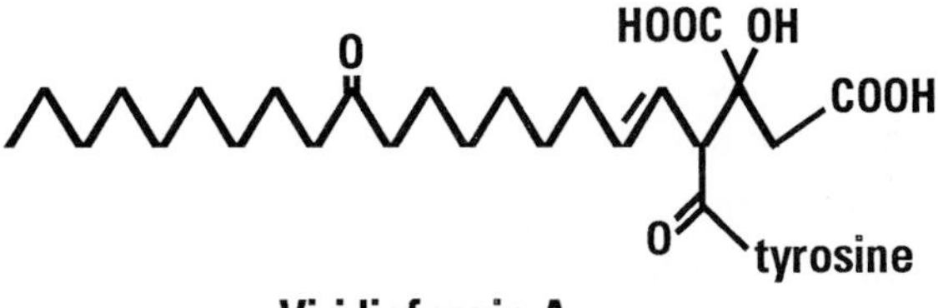

Viridiofungin A$_4$

Fig. 1. Chemical structures of some known fungal serine palmitoyltransferase inhibitors isolated from fermentation. Other fungal metabolites with similar structures that are either proven or suspected inhibitors of serine palmitoyltransferase include sphingofungins A, B, D, E, and F,[1,2] mycestericins A-G,[3] hydroxylipoxamycin,[4] viridiofungins A–C, A1, A2, B2, Z2, A3, A4,[5] fumifungin,[6] and flavovirin.[7] Key to references: [1]W. S. Horn, J. L. Smith, G. F. Bills, S. L. Raghoobar, G. L. Helms, M. B. Kurtz, J. A. Marrinan, B. R. Frommer, R. A. Thornton, and S. M. Mandala, *J. Antibiot.* **45,** 1692 (1992).[2] F. Van Middlesworth, R. Giacobbe, M. Lopez, G. Garrity, J. A. Bland, K. Bartizal, R. A. Fromtling, J. Polishook, M. Zweerink, A. M. Edison, W. Rodilsky, K. E. Wilson, and R. L. Monaghan, *J. Antibiot.* **45,** 861 (1992); [3]T. Fujita, R. Hirose, M. Yoneta, S. Sasaki, K. Inoue, M. Kiuchi, S. Hirase, K. Chiba, H. Sakamoto, and M. Arita, *J. Med. Chem.* **39,** 4451 (1996); [4]S. M. Mandala, B. R. Frommer, R. A. Thornton, M. B. Kurtz, N. M. Young, M. A. Cabello, O. Genilloud, J. M. Liesch, J. L. Smith, and W. S. Horn, *J. Antibiot.* **47,** 376 (1994); [5]S. M. Mandala, R. A. Thornton, B. R. Frommer, S. Driekorn, and M. B. Kurtz, *J. Antibiot.* **50,** 339 (1997); [6]T. Mukhopadhyay, K. Roy, L. Coutinho, R. H. Rupp, B. N. Ganguli, and H. W. Fehlhaber, *J. Antibiot.* **40,** 1050 (1987); [7]M. Sailer, V. Sasek, J. Sejbal, M. Budesinky, and V. Musilek, *J. Basic Microbiol.* **29,** 375 (1989).

Isaria sinclairii occurs in nature as a parasite on moths, is not thermophilic, and is capable of producing about 4–5 mg ISP-I/liter of culture broth at 23 to 25°. Perhaps the best producer of myriocin is *M. sterilia* (ATCC 20349), which has been reported to produce approximately 200 mg myriocin/liter of culture broth at 40°.[7]

Stock Cultures of Isaria (=Cordyceps) sinclairii

The following methods are described under the assumption that the reader is not familar with the basic methods for handling fungi and their use in fermentation. The book "Basic Plant Pathology Methods" is a good source of information on basic methods applicable to the culture of fungi.[21] A common pitfall for the novice is determining whether the organism that is growing in the production medium is in fact the organism of interest. In our laboratory, verification that the fungus in our production medium is identical to that supplied by the American Type Culture Collection is based primarily on colony appearance and morphology and the production of fermentation medium that has potent SPT-inhibiting activity.

The initial stock of *I. sinclairii* (ATCC 24400) was obtained as a slant culture on potato dextrose agar (PDA) from the American Type Culture Collection (Rockville, MD). Stock cultures should not be kept for more than 2 or 3 days at 4°.

Starter Cultures of Isaria (=Cordyceps) sinclairii

Dissolve 39 g of PDA powder (Difco stock No. 0013-01-4, Difco Laboratories, Detroit, MI) in 1 liter of distilled water and heat to boiling. Transfer hot PDA to two 1-liter Erlenmeyer flasks with cotton plugs and autoclave for 20 min. While the PDA is still hot to the touch, pour into several petri dishes (100 × 15 mm, approximately 10 ml/dish) in a laminar flow hood. cover dishes and let stand in the hood until PDA has solidified. Store dishes upside down at room temperature in a plastic bag for several days. Check for microbial contamination.

In our laboratory, the inoculum used for initiation of fermentation media is called the "starter culture." The starter culture is obtained by transferring stock cultures to the PDA dishes. Prepare "starter" *I. sinclairii* cultures from the ATCC stock slant cultures by inoculating three of the PDA dishes using a dissecting needle with a flattened tip and a sharpened edge as follows: flame the needle and cool by touching to sterile agar in

---

[21] O. D. Dhinga and J. B. Sinclair, "Basic Plant Pathology Methods." CRC Press, Boca Raton, FL, 1995.

an extra dish, flame the tube, remove cap and flame the top of the tube, insert the dissecting needle into the tube, and remove small portions of the fungal mat for transfer to the dishes. Tightly wrap the starter culture dishes around the edge with Parafilm M and store upside down at 20–22° on a 12-hr light/dark cycle. The fungus will grow slowly under these conditions. After 10–12 days on the PDA dish there should be a 1- to 2-cm growth of mycelia that appears as a pure white to cream colored, cottony growth forming a colony circular to oval in form (Fig. 2). Microscopic examination at the edge of the colony should reveal branching segmented hyphae. Unfortunately, hyphal morphology is of little value in taxonomy and under the culture conditions described, characteristic condia were not produced. Fortunately, the best test for verifying that the culture is an SPT inhibitor-producing variety is functional and not morphological: the ability to produce ISP-I. Starter cultures are for (i) verifying that the fungus inoculated for fermentation is an *Isaria* sp. (based on colony morphology and toxin production), (ii) checking for contamination (usually colored colonies that do not appear as described earlier), and (iii) maintaining growing cultures for initiating additional fermentations. The starter cultures are slow growing and easily maintained. They should be reinitiated onto fresh PDA dishes

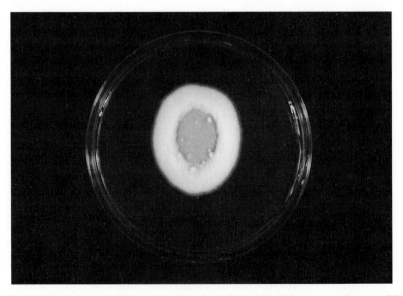

FIG. 2. Morphology of *Isaria* (=*Cordyceps*) *sinclairii*. Mycelial mass growing on a PDA dish (8.5 cm diameter) for 10 days at room temperature and the growing hyphae showing the lack of conidia at 100× using a phase objective.

monthly. As starter cultures age they lose their distinct edge, become grayer in appearance, and become difficult to cut using the dissecting blade.

Fermentation

Once it is established that the starter culture is uncontaminated and the fungus is viable, prepare several liters of fermentation medium (Table I). Allow the fermentation medium to stand for several days as a test for sterility. Cut a small plug of mycelia (0.25–0.5 cm^2) from the starter culture using a dissecting needle and transfer to a 500-ml triple baffled extra deep shake flask with a DeLong style neck and stainless steel closure (Bellco Glass Inc., Vineland, NJ) containing 100 ml of fermentation medium. This flask is incubated at room temperature on a rotary shaker (200 rpm) for 3 weeks. Three to 5 ml of the 3-week-old primary fermentation medium is transferred to a second-stage fermentation consisting of 500 ml of fermentation medium contained in 2.8-liter triple baffled Fernbach flasks (Bellco Glass Inc.). Also transfer 0.5 ml of the primary fermentation medium to two PDA dishes, cover, and allow to dry in the laminar flow hood, seal with Parafilm, and store upside down. This is done to verify that the primary fermentation medium contains only *I. sinclairii* and is free of contamination with other microorganisms. The secondary fermentation is conducted at room temperature on a rotary shaker for 30 days. The secondary fermentation medium should also be checked for contamination as described earlier. Fermentation medium from the 2.8-liter flasks is combined and centrifuged at 18g RCF for 60 min. The clear supernatant is decanted, filtered through

TABLE I
COMPOSITION OF FERMENTATION MEDIUM

Component	Amount (g/liter)
D-Glucose, anhydrous	30
Bacto peptone[b]	5
Yeast extract[c]	3
KH$_2$PO$_4$ (monobasic)	0.3
K$_2$HPO$_4$ (dibasic)	0.3
MgSO$_4 \cdot$7H$_2$O	0.3
pH 5.5 with 1 N HCl	

[a] From T. Fujita, K. Inoue, S. Yamamoto, T. Ikumoto, S. Sasaki, R. Toyama, K. Chiba, Y. Hoshino, and T. Okumoto, *J. Antibiot.* **47,** 208 (1994).

[b] Difco No. 0118-01.

[c] Difco No. 0127-01-7.

0.45-μm, 250-ml Nalgene filter units (Nalge Company, Rochester, NY), and stored at $-20°$.

Bioassay for SPT Inhibition

The bioassay is based on the fact that fumonisin (available from several commercial suppliers) inhibition of sphingosine/sphinganine acyltransferase [(dihydro) ceramide synthase][acyl-CoA:sphingosine N-acyltransferase; EC 2.3.1.24] causes a rapid and very large increase in free sphinganine concentration in cultured cells.[22,23] The fumonisin-induced increase in free sphinganine can be prevented by cotreatment with SPT inhibitors.[15,16,24,25] SPT inhibition is bioassayed in the fermentation medium and other fractions using the inhibition of fumonisin B_1-induced sphinganine accumulation (20 μM, 24 hr) in confluent (growth arrested) cultures of LLC-PK$_1$ cells as an index of SPT inhibition. Other cell types are also suitable if they are resistant to the cytotoxic effects of fumonisin B_1 under the conditions assayed. Typically, growth-arrested epithelial cell lines, primary rat hepatocytes, and rat liver slices are resistant to fumonisin-induced inhibition of cell proliferation and cell death. The ability to inhibit fumonisin-induced free sphinganine accumulation by partially purified material (Fig. 3A) is compared to the activity of highly purified myriocin (Fig. 3B), which was generously provided by T. Fujita, Setsunan University, Osaka, Japan. The results are expressed as myriocin equivalents. The SPT inhibitor-induced decrease in free sphinganine concentration in fumonisin-treated LLC-PK$_1$ cells is determined by high-performance liquid chromatography (HPLC) of base-treated chloroform/methanol extracts utilizing a modification of the extraction method originally described by Merrill *et al.*[26] Sphingoid bases are quantitated based on recovery of the C_{20}-sphinganine internal standard generously provided by A. H. Merrill, Jr. and Dennis Liotta, Emory University, Atlanta, Georgia, which is now also available commercially (Matreya, Inc., Pleasant Gap, PA). A complete description of the HPLC apparatus and derivatization procedure is described in Riley *et al.*[27] This bioassay for

[22] E. Wang, W. P. Norred, C. W. Bacon, R. T. Riley, and A. H. Merrill, Jr., *J. Biol. Chem.* **266,** 14486 (1991).

[23] H.-S. Yoo, W. P. Norred, E. Wang, A. H. Merrill, Jr., and R. T. Riley, *Toxicol. Appl. Pharmacol.* **114,** 9 (1992).

[24] J. J. Schroeder, H. M. Crane, J. Xia, D. C. Liotta, and A. H. Merrill, Jr., *J. Biol. Chem.* **269,** 3475 (1994).

[25] H.-S. Yoo, W. P. Norred, J. Showker, and R. T. Riley, *Toxicol. Appl. Pharmacol.* **138,** 211 (1996).

[26] A. H. Merrill, Jr., E. Wang, R. E. Mullins, W. Charles, L. Jamison, S. Nimkars, and D. C. Liotta, *Anal. Biochem.* **171,** 373 (1988).

[27] R. T. Riley, E. Wang, and A. H. Merrill, Jr., *J. Assoc. Off. Anal. Chem. Int.* **77,** 533 (1994b).

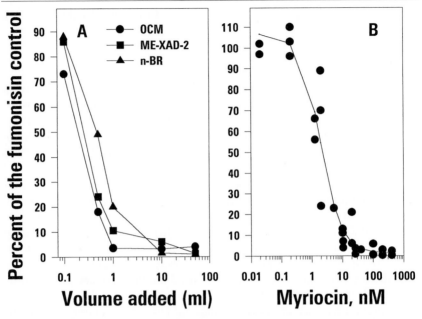

FIG. 3. Reversal of fumonisin B_1-induced free sphinganine accumulation in LLC-PK$_1$ cells by (A) partially purified ISP-I or (B) purified myriocin.[1] The values for pure myriocin are replicate samples from three to four independent experiments expressed as a percentage of the concurrent positive control (positive control was only treated with fumonisin B_1). Cells were grown in 24-well plates until confluent, exposed for 24 hr to pure fumonisin B_1 (20 μM), and various amounts of purified myriocin (nanomolar, nM) were dissolved in 95% ethanol so that the target concentration was achieved by adding a total of 2 μl ethanol/ml of growth medium. For partially purified ISP-I (equivalent volume added, μl), the various fractions of materials isolated from *I. sinclairii* culture material were dried and dissolved in the 95% ethanol vehicle so that the target concentration (μl/ml growth medium) was achieved by adding a total of 2 μl ethanol/ml of growth medium. Cells growing in individual wells were analyzed for free sphinganine by the method of Yoo *et al.*[2] Neither the fumonisin B_1 nor the myriocin showed any visual evidence of being cytotoxic to LLC-PK$_1$ cells. Under these conditions, the concentration of pure myriocin that caused a 50% inhibition of the fumonisin-induced free sphinganine increase was 1.8 $\pm$ 1.0 nM.[3] Based on the EC$_{50}$ for pure myriocin, the total recovery of myriocin equivalents (ISP-I) in each fraction was calculated (Table II). The response curves curves are shown for the original culture material (OCM), the methanol eluate from the XAD-2 column (ME-XAD-2), and the *n*-butanol residue (*n*-BR). Key to reference: [1]T. Fujita, K. Inoue, S. Yamamoto, T. Ikumoto, S. Sasaki, R. Toyama, K. Chiba, Y. Hoshino, and T. Okumoto, *J. Antibiot.* **47,** 208 (1994); [2]H.-S. Yoo, W. P. Norred, J. Showker, and R. T. Riley, *Toxicol. Vitro* **10,** 77 (1996); [3]R. T. Riley, K. A. Voss, W. P. Norred, C. W. Bacon, F. I. Meredith, and R. P. Sharma, *Environ. Toxicol. Pharmacol.* **7,** 109 (1999).

detecting SPT inhibitors has the advantage that it is functionally rather than chemically specific and thus will detect all mammalian SPT inhibitors regardless of structure and without requiring microsomal preparations or radioactive precursors. Using the modified extraction method, approximately 48 samples/day can be extracted; however, a single HPLC analysis requires approximately 20 to 30 min/sample.[28]

Partial Purification of ISP-I

The ISP-I in the fermentation medium (6 liters total) is partially purified using a method based on that described by Fujita *et al.*[7] The filtered fermentation medium (2 liters) is applied to an Amberlite XAD-2 column (2.5 × 40 cm, i.d.), washed with water (4 liters), and eluted with methanol (2 liters). A total of 6 liters of fermentation medium is processed 2 liters at a time. The combined methanol eluate is concentrated on a rotary evaporator under reduced pressure at 50°. The residue is dissolved in water (300 ml) with a little heat and is then extracted first with ethyl acetate and then *n*-butanol. According to Craveri *et al.*[20] thermozymocidin is substantially insoluble in ethyl acetate but somewhat soluble in *n*-butanol (0.05–0.2%). An equal volume of ethyl acetate is added to the water layer in a separatory funnel, the water layer is collected, the upper ethyl acetate layer is extracted with water two more times, and the ethyl acetate layer is discarded. The aqueous layer (970 ml) is extracted twice with *n*-butanol (500 ml). The *n*-butanol layer (1000 ml) is removed and concentrated on a rotary evaporator under reduced pressure at 50°. The *n*-butanol residue is a red oil and is dissolved in approximately 1–2 ml of methanol. The recoveries of SPT inhibitor activity from the various fractions based on the bioassay are shown in Table II and Fig. 3A.

After several days storage in methanol at 4° crystals will begin to form. The supernatant is removed carefully and the crystals are washed gently several times with 0.5 ml of cold methanol. This process is repeated for several days until no more crystals appear. Myriocin is soluble at 2, 0.5, and <0.5 mg/ml in methanol, ethanol, and chloroform, respectively. The melting point of the crystallized material is determined on a Mettler FP5 melting point apparatus (Mettler Instruments, Zurich, Switzerland) and compared to the actual melting point of the pure myriocin that was provided by T. Fujita. The melting point of pure myriocin was 171 to 172° and that of the materials crystallized from *I. sinclairii* (ISP-I) was 169 to 171°. There was a small amount of red color to the crystallized ISP-I. In the original method described by Fugita *et al.*[7] preparative thin-layer chromatography

---

[28] H.-S. Yoo, W. P. Norred, J. Showker, and R. T. Riley, *Toxicol. Vitro* **10**, 77 (1996).

TABLE II
RECOVERIES OF SPT INHIBITORS BASED ON THE BIOASSAY[a]

Fraction	Total recovered
Original fermentation medium	23.1 ± 12.8
Filtrate and water wash of XAD-2 column	<1
Methanol eluate of XAD-2 column	16.6 ± 9.2
n-Butanol residue	11.2 ± 6.2
Precipitated from methanol[b]	4.9

[a] See Fig. 3 and text. Values are the predicted mean values ±95% confidence interval normalized to the total volume of the original fermentation medium (6 liters). The various fractions taken during the extraction and cleanup procedures are described in the text.
[b] Based on the actual dry weight of the material with a melting point of 169° to 171°.

(TLC) was utilized to further purify the crude ISP-I prior to attempting to crystallize the ISP-I in methanol. We did not do this as crystals could be obtained from the n-butanol residues dissolved in methanol directly.

Chromatographic Analysis of ISP-I

The relative purity of ISP-I in the various fractions can be assessed qualitatively by TLC (Fig. 4) and LC/MS (Fig. 5). Briefly, liquid chromatography/mass spectrometry (LC/MS) was done with a Finnigan LCQ mass spectrometer. The liquid chromatograph was a Thermo separations unit composed of a 4000 series pump and an AS 3000 autosampler that were operated by the LCQ software. The LC column was a MetaChem Technologies Intersil 5 μm ODS-3 (150 × 3.0 mm). Flow was 0.3 ml/min of water/methanolic/acetic acid (20/80/0.03 or 25/75/0.03). Flow eluting from the column was directed into the electrospray interface, which was operated with a voltage of 4 kV at a temperature of 220°. Samples were dissolved in 1 ml of methanol. The standard concentration was 10 ng/μl. Aliquots of 10 μl were injected and spectra were recorded from m/z 150 to 800 approximately every second for 20 min. The column void volume was approximately 0.6 ml (2 min).

The LC/MS procedure was able to detect ISP-I (m/z = 402.3) in the culture medium, the butanol extracts, and other crude fractions and is suitable for calculating relative amounts in the nanogram range. In the culture material and butanol fractions, two closely eluting compounds with a m/z = 402 were detected (data not shown). Regarding quantitation, the primary problem when using crude preparations is that matrix effects (pass through of neutral materials) can suppress ionization. Nonetheless, LC/MS

A   B   C

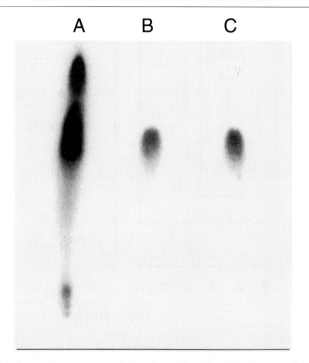

FIG. 4. Thin-layer chromatogram of the *n*-butanol residue (A), the material precipitated from methanol with a melting point of 169 to 171°C (B), and pure myriocin (C). The material in B and C contained 5 μg of material by weight. The origin is shown as a line near the bottom of the plate. Chromatography was conducted using a 5 × 10-cm glass-backed silica gel 60 + F254 high-performance thin-layer chromatography plate with a mobile phase of CHCl₃:methanol:water (60:35:5, v/v). Ninhydrin-positive substances were visualized by spraying the plate with 0.02% ninhydrin in ethanol (spray until saturated and air dry) and heating at 100° for approximately 20 min. Alternatively, materials on the plate can also be visualized nondestructively by spraying the silica gel coating to the point of saturation with distilled water. The mobility of pure myriocin relative to the solvent front was 0.62.

has the potential to be quantitative in nanogram amounts under carefully controlled conditions.

Examples of in Vitro and in Vivo Use of ISP-I

Both pure ISP-I and partially purified fractions (*n*-butanol residues, Fig. 4) have been used successfully in both short-term *in vitro* (≤72 hr) and *in vivo* studies (≤24 hr).[15] For example, the IC₉₅ concentration of ISP-I reversed the antiproliferative effects and prevented fumonisin-induced apoptosis after 48 hr exposure. The partially purified ISP-I was also effective

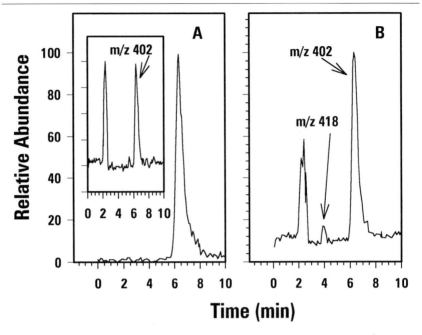

FIG. 5. LC/MS chromatograms. (A) Ion selective chromatogram of myriocin [m/z 402.3 [M + H_2O]; retention time (RT) 6.3 min] provided by T. Fujita, Setsunan University, Osaka, Japan (total ion chromatogram m/z 150 to 800, inset) and (B) the total ion chromatogram of the material (m.p. 169–171°) that precipitated from methanol (Table II). The peak at 2 min is the solvent front from the sample injection. The compound in B with a RT of approximately 4 min is more polar than ISP-I with a m/z of 418 (one more hydroxyl). The total run time was 20 min; however, no peaks (m/z 150–800) were detected after 10 min in A or B.

at reducing free sphinganine *in vivo*. Free sphinganine concentration was reduced 60% in kidney of mice injected i.p. with ISP-I plus fumonisin B_1 when compared to fumonisin B_1 alone. The ability of SPT inhibition to reduce fumonisin B_1-induced sphinganine accumulation *in vivo* may be useful in the development of therapeutic agents for treatment of animals suspected to have been exposed to toxic levels of fumonisin in feeds. For example, herds of horses where one or more horse has shown signs of equine leucoencephalomalacia could be treated with ISP-I to reduce the sphinganine buildup that occurs concurrent to the onset of the liver damage that precedes the onset of clinical signs of ELEM.[29,30] ISP-I will also prove useful in studies to reveal the role of increased ceramide concentration on

[29] E. Wang, P. F. Ross, T. M. Wilson, R. T. Riley, and A. H. Merrill, Jr., *J. Nutr.* **122,** 1706 (1992).
[30] R. T. Riley, J. L. Showker, D. L. Owens, and P. F. Ross, *Environ. Toxicol. Pharmacol.* **3,** 221 (1997).

the inhibition of apoptosis, and other effects, due to either activation of serine palmitoyltransferase or ceramide synthesis.[17,31,32]

Acknowledgments

We would like to acknowledge our appreciation of the hard work, dedication, and creative thinking of Ms. Jency Showker, without whom this work could never have been completed in a timely manner.

[31] T. Weider, C. E. Orfanos, and C. C. Geilen, *J. Biol. Chem.* **273,** 11025 (1998).
[32] J. Xu, C. H. Yeh, S. Chen, L. He, S. L. Sensi, L. M. Canzoniero, D. W. Choi, and C. Y. Hsu, *J. Biol. Chem.* **273,** 16521 (1998).

[37] Isolation and Characterization of Fumonisins

By FILMORE I. MEREDITH

Introduction

Fumonisins B_1 (FB$_1$), B_2 (FB$_2$), B_3 (FB$_3$), and B_4 (FB$_4$) (Fig. 1) are mycotoxins produced by *Fusarium moniliforme* Sheldon, a common fungal contaminate found in corn (*Zea mays* L.) and other cereal grains. A closely related species, *Fusarium proliferatum,* also produces fumonisin mycotoxins on corn. There are many strains found in these two *Fusarium* species that produce varying amounts of each individual fumonisin or combinations of them.[1] Therefore, specific strains should be chosen for isolation that will produce high quantities of the desired fumonisin. This will make the isolation and purification steps less demanding and will ensure that a sufficient level of the required fumonisin is present to justify the time required for the isolation. *Fusarium moniliforme* MRC 826 will produce high concentrations of FB1 and moderate amounts of FB$_2$, whereas *F. proliferatum* M6104 will produce large quantities of FB$_2$ and low levels of the other fumonisins. *Fusarium moniliforme* strain KSU 819 produces FB$_3$ and FB$_4$ but very little FB$_1$ and FB$_2$.[2]

Fumonisin FB$_1$ and FB$_2$ were first isolated and identified in 1988.[3] Purified FB$_1$ isolated from *F. moniliforme* culture material and corn culture

[1] P. E. Nelson, R. D. Plattner, D. D. Shackelford, and A. E. Desjardins, *Appl. Environ. Microbiol.* **57,** 2410 (1991).
[2] S. M. Poling and R. D. Plattner, *J. Agric. Food Chem.* **44,** 2792 (1996).
[3] W. C. A. Gelderblom, K. Jaskiewicz, W. F. O. Marasas, P. G. Thiel, R. M. Horak, R. Veleggaar, and N. P. J. Kriek, *Appl. Environ. Microbiol.* **54,** 1806 (1988).

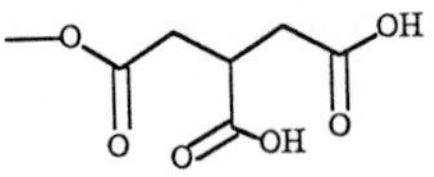

Tricarballylic Acid (TCA)

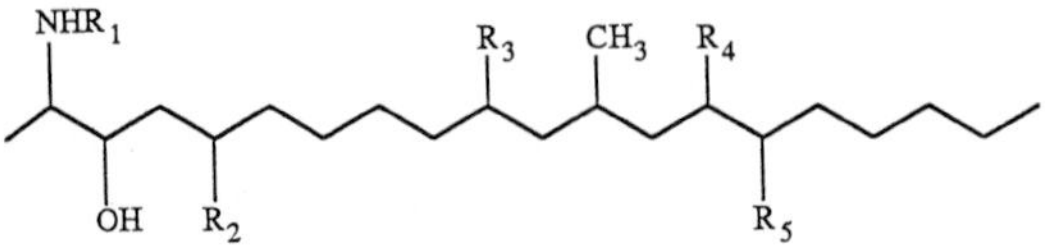

	R1	R2	R3	R4	R5	MW
FB$_1$	H	OH	OH	TCA	TCA	721
FB$_2$	H	OH	H	TCA	TCA	705
FB3	H	H	OH	TCA	TCA	705
FB$_4$	H	H	H	TCA	TCA	689
FA$_1$	COCH$_3$	OH	OH	TCA	TCA	763
FA$_2$	COCH$_3$	OH	H	TCA	TCA	747
FA$_3$	COCH$_3$	H	OH	TCA	TCA	747
PHFB$_1$	H	OH	OH	TCA	OH	563
PHFB$_2$	H	OH	OH	OH	TCA	563
HFB$_1$	H	OH	OH	OH	OH	405

FIG. 1. Chemical structures of fumonisins.

material containing fumonisins have both caused equine leucoencephalo-malacia (ELEM)[4–6] and porcine pulmonary edema (PPE),[7] and both FB$_1$ and FB$_2$ are toxic to broiler chicks, ducklings, and turkey poults.[8–10] *Fu-*

[4] T. S. Kellerman, W. F. O. Marasas, P. G. Thiel. W. C. A. Gelderblom, M. Cawood, and J. A. W. Coetzer, *Onderstepoort J. Vet. Res.* **57,** 269 (1990).

[5] W. F. O. Marasas, T. S. Kellerman, W. C. A. Gelderblom, J. A. W. Coetzer, P. G. Thiel, and J. J. Van der Lugt, *Onderstepoort J. Vet. Res.* **55,** 197 (1988).

[6] P. F. Ross, L. G. Rice, J. C. Reagor, G. D. Osweiler, T. M. Wilson, P. E. Nelson, D. L. Owens, R. D. Plattner, K. A. Harlin, J. L. Richard, B. M. Colvin, and M. I. Banton, *J. Vet. Diagn. Invest.* **3,** 238 (1991).

[7] L. R. Harrison, B. M. Colvin, J. T. Greene, L. E. Newman, and J. R. Cole, *J. Vet. Diagn. Invest.* **2,** 217 (1990).

[8] N. P. Kriek, T. S. Kellerman, and W. F. O. Marasas, *Onderstepoort J. Vet. Res.* **48,** 129 (1981).

sarium moniliforme and the fumonisin mycotoxins are implicated in the etiology of esophageal cancer, which occurs at abnormally high rates in the Tanskei regions of South Africa.[3,11-13] Rats fed corn culture with *F. moniliforme* or naturally contaminated corn exhibit pathological changes in kidney and liver cells.[14] Additional studies with rats fed diets of purified fumonisin B_1 produced similar effects and showed that liver and kidney are the target organs of the mycotoxins.[15] Fumonisins block ceramide synthase, which inhibits the conversion of sphinganine to *N*-acylsphinganine and results in the elevation of free sphinganine and depletion of complex sphingolipids.[16]

The fumonisin backbone is a C_{20} hydroxylated chain with two propane-1,2,3-tricarboxylic side chains called tricarballylic acid (Fig. 1). The four carboxyl groups and the amine group make the FB_1 water soluble whereas FB_2, FB_3, and FB_4 require the addition of an organic polar solvent to affect solubility. Fumonisins are insoluble in nonpolar organic solvents. In polar solvents, FB_1 exists as a zwitterion and with metals interactions can occur. Fumonisins do not have a chromophore and can only be weakly detected by UV at 210 nm. The amine undergoes reactions the same as amino acids forming *N*-acetylated compounds and adducts with sugars. These reactions render fumonisins undetectable as derivatives cannot be formed on the primary NH_2 group.

The structure for FB_1 was first determined by nuclear magnetic resonance.[13,17] Liquid secondary ion mass spectrometry (fast atom bombardment mass spectrometry, FAB/MS) gave the protonated molecular ion for FB_1 as *m/z* 722 and FB_2 as *m/z* 706.[13] The development of deuterium-labeled FB_1 as an internal standard resulted in an improved gas chromatog-

[9] T. Javed, G. A. Bennett, J. L. Richard, M. A. Dombrink-Kurtzman, L. M. Cote, and W. B. Buck, *Mycopathologia* **123,** 171 (1993).

[10] W. L. Brydon, R. J. Love, and L. W. Burgress, *Aust. Vet. J.* **64,** 225 (1987).

[11] W. F. O. Marasas, K. Jaskiewiez, F. S. Venter, and D. J. Van Schalkwyk, *S. Afr. Med. J.* **74,** 110 (1988).

[12] J. P. Rheeder, W. F. O. Marasas, P. G. Thiel, E. W. Sydenham, G. S. Shephard, and D. J. Van Schalkwyk, *Phytopathology* **82,** 353 (1992).

[13] S. C. Bezuidenhout, W. C. A. Gelderblom, R. M. M. Grost-allman, W. F. O. Marasas, G. Spiteller, and R. Vleggaar, *J. Chem. Soc. Chem. Commun.* **11,** 743 (1988).

[14] K. A. Voss, W. P. Norred, R. D. Plattner, C. W. Bacon, and J. K. Porter, *Toxicologist* **9,** 225 (1989).

[15] K. A. Voss, W. J. Chamberlain, C. W. Bacon, and W. P. Norred, *Nat. Toxins* **1,** 222 (1993).

[16] E. Wang, W. P. Norred, C. W. Bacon, R. T. Riley, and A. H. Merrill, Jr., *J. Biol. Chem.* **266,** 14486 (1991).

[17] B. A. Blackwell, O. E. Edwards, A. Fruchier, J. W. ApSimon, and J. D. Miller, in "Fumonisins in Food: Advances in Experimental Medicine and Biology" (L. S. Jackson, J. W. DeVries, and L. B. Bullerman, eds.), p. 75. Plenum Press, New York, 1996.

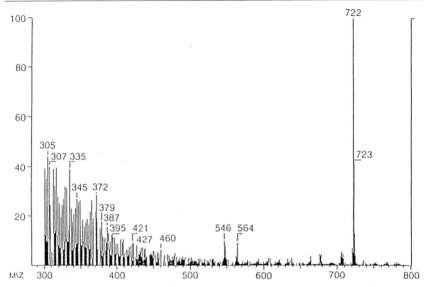

Fig. 2. Fast atom bombardment mass spectroscopy of fumonisin B_1.[18]

raphy-mass spectrometry (GC-MS) method for the derivatized fumonisin B_1 backbone, improved the precision and accuracy of GC/MS for fumonisins, and allowed for development of a quantitative FAB/MS method.[18] The FAB mass spectra of fumonisin B_1 is presented in Fig. 2.[18]

Fumonosin B_1 is available commercially from many suppliers. To prepare it, or other fumonisins, the following procedures would be used.

Fumonisin Production *in Vitro*

Corn Culture

Alberts *et al.*[19] prepares culture material by inoculating autoclaved corn with lyophilized *F. moniliforme* MRC 826 conidia (ATCC No. 52539). Whole yellow corn (400 g) and 400 ml of water are placed in a 2-liter, 11-cm-wide mouth fruit jar, the opening is covered with cotton cloth, and the jar is autoclaved for 1 hr on 2 consecutive days at 121° and 120 kPa. The sterile, hydrated corn is inoculated with lyophilized conidia and is covered with a sterile cotton cloth. The culture material is incubated at 25°. The

[18] R. D. Plattner and B. E. Branham, *J. AOAC. INT.* **77,** 525 (1994).
[19] J. F. Alberts, W. C. A. Gelderblom, P. G. Thiel, W. F. O. Marasas, D. J. Van Schalkwyk, and Y. Behrend, *Appl. Environ. Microbiol.* **56,** 1729 (1990).

surface-to-volume ratio is approximately $2:5$. After 8 weeks, the FB_1 concentration is 11.7 g/kg.

Rice Culture

Rice (Uncle Ben's converted rice, 300 g) and distilled water (300 ml) are placed in 2.8-liter Fernbach flasks. The flasks are held at room temperature for 12 hr and then autoclaved for 30 min.[20,21] The flasks are inoculated with 5 ml of an aqueous suspension of *F. moniliforme* MRC 826 conidia. Conidia have been previously cultured for 7 days on potato dextrose agar and adjusted to 10^9 conidia/ml of water. The inoculated flask is stoppered loosely with a cotton plug and incubated in the dark at 26° for 28–35 days. For the first 10 days of the incubation period the flasks are shaken. After 8 weeks, the yield of FB_1 was 12.2 g/kg. We found that the rice culture does not produce as many small particles and therefore is easier to filter and the colored pigments are at a lower level, making purification easier.

Liquid Culture

At the present time fumonisins are obtained usually from the solid culture technique.[22] This is due in part to the high fumonisin concentrations that can be obtained from solid culture methods. Liquid culture production has definite advantages over the use of solid culture. These advantages are that large amounts of organic solvents are not needed for extractions, purification of individual compounds are easier, and determination and analysis of biosynthetic pathways and control mechanisms are less difficult.[23] The low yield of fumonisins from liquid culture, however, limits this technique.

Isolation and Purification

Method A

Gelderblom *et al.*[3] first reported on the isolation of FB_1 and FB_2 from cultures of *F. moniliforme* MRC 826 by extracting with methanol–water $(3:1)$ followed by solvent partitioning with chloroform.

[20] F. I. Meredith, C. W. Bacon, R. D. Plattner, and W. P. Norred, *J. Agric. Food Chem.* **44,** 195 (1996).

[21] F. Meredith, C. Bacon, W. Norred, and R. Plattner, *J. Agric. Food Chem.* **45,** 3143 (1997).

[22] S. E. Keller and T. M. Sullivan, in "Fumonisins in Foods: Advances in Experimental Medicine and Biology" (L. S. Jackson, J. W. DeVries, and L. B. Bullerman, eds.), p. 205. Plenum Press, New York, 1996.

[23] M. E. Cawood, W. C. A. Gelderblom, R. Vleggaar, Y. Behrend, P. G. Thiel, and W. F. O. Marasas, *J. Agric. Food Chem.* **39,** 1958 (1991).

Detection of the solvent fractions that contained tumor-promoting activity was found by a bioassay procedure using rats and eventually led to the discovery of the fumonisins. The aqueous phase was purified on Amberlite XAD-2, silica gel, and reversed-phase chromatographic columns yielding FB_1 and FB_2 with a purity of about 90%.

Method B

The most widely used procedure used for the separation and isolation of the fumonisins is by Cawood *et al.*[23] in which culture material (1 kg) is extracted twice with ethyl acetate (1 liter) by blending and filtering by gravity through Whatman No. 4 filter paper. The residue is extracted with 1 liter methanol–water (3:1) and twice with 1.5 liters of methanol–water (3:1). The combined extracts are taken to dryness under vacuum. Methanol–water (1:3, 200 ml) is added to the dried culture extract, and the residue is dissolved at 50°. The methanol–water extract is partitioned with chloroform three times (100 ml). The aqueous phase is evaporated to dryness, and the residue is dissolved in methanol–water (3:1) and applied to an XAD-2 column (sample:resin, 1:15) that has been conditioned previously with methanol–water (3:1). The column is washed with 1 liter methanol–water (3:1) and 1.5 liter methanol–water (1:1), and the fumonisins are eluted with 1 liter methanol. The methanol is removed and the residue containing the fumonisins is dissolved in 70 ml of the eluent chloroform–methanol–acetic acid (6:3:1) and submitted to silica gel column (5.5 × 85 cm; 1100 g) chromatography containing anhydrous Na_2SO_4 on top of the silica gel. After elution (flow rate 3.5 ml/min) of the first 1.5 liters, 50-ml fractions are collected. Fractions 5–19 contain FA_1 and FA_2, fractions 30–50 (A) contain FB_2 and FB_3, and fractions 51–90 (B) contain FB_1, FB_2, and FB_3. Fractions 51–90 (B) are combined and the solvent removed under vacuum. The residue (B) is dissolved in 20 ml chloroform–methanol–water-acetic acid (55:36:8:1) and submitted to a second silica gel column that does not contain anhydrous Na_2SO_4. The column is washed with the same solvent used to dissolve the sample. The initial flow rate is 2 ml/min for the first 1.2 liters of eluting solvent and is then increased to 3.5 ml/min and 50-ml fractions are collected. Fractions 50–80 (C) contained FB_1. The monomethyl and dimethyl esters are found in fractions 15–29, whereas FB_2, FB_3, and FB_4 are found in fractions 30–50 (D). Fractions 30–50 (D) are combined for later separation.

FB_1 Purification. FB_1-containing fractions [50–80 (C)] are combined and the solvent removed. The residue is taken up in 20 ml of methanol–water (1:1), and the pH is adjusted to 3.5 and applied to a reversed-phase C_{18} column (1.5 × 50 cm, 200 g) that is prepared by the method of Kingston

and Gerhart.[24] The column has been equilibrated previously with methanol–water (1:1) and the sample flow rate is 1.5 ml/min. A linear gradient of 533 min is used for the separation. The starting solvent is methanol–water (1:1) and the ending solvent is methanol–water (4:1). After elution of the first 400 ml of solvent, 15-ml fractions are collected. Fractions 8–15 contain FB_1, which eluted as a single peak. Purity of the FB_1 (>90%) is determined by high-performance liquid chromatography.[19]

FB₂ Purification. The chloroform–methanol–water–acetic acid (55:36:8:1) solvent of fractions 30–50 (D) that eluted from the second silica gel column is removed, the residue is redissolved in methanol–water (1:1), and the pH of the sample is adjusted to 3.5. The combined fractions of D are applied to the C_{18} reversed-phase column. The C_{18} column is equilibrated with methanol–water (1:1), and the sample is applied to the column. A linear gradient of methanol–water (1:1) to 100% methanol is started. The gradient takes 533 min and the flow rate is 1.5 ml/min. Fractions (15 ml) are collected after elution of the first 300 ml. Fractions 5–14 contain FB_3 and FB_4 with a purity of 20%, and fractions 17–30 contain FB_2 at 90% purity. All of the culture material extracts and all of the fractions from the columns are analyzed for the fumonisins by silica gel thin-layer chromatography (TLC), as described later.

Method C

A simplified procedure[20] for the isolation of fumonisin FB_1 uses culture material (corn or rice) extracted with methanol–water (1:1) at the ratio of 5 ml/g of culture material.[20] The culture material is stirred occasionally and, after 4 hr, is filtered under reduced pressure through Whatman No. 4 filter paper. The residue is resuspended in fresh solvent and extracted overnight. The extracting solvents, which are dark red (almost black) solutions from the pigments,[25] are filtered and the filtrates combined. Approximately half the volume of the extract is removed on a 10-liter rotary evaporator with the water bath temperature at 38°. Reducing the methanol in the extract will decrease the volume of water required to dilute the extract prior to preparative C_{18} reversed-phase chromatography. The fumonisin extract is diluted with water to a content of 10–15% methanol. The diluted extract is applied at 30 ml/min with a preparative HPLC equipped with a gradient controller to a Waters Bondapak PrepPak 500/C_{18} reversed-phase cartridge that has been conditioned previously with 5 void volumes (VV) of methanol and 6 VV of water (VV = 501 ml). The cartridge is washed with water for 15 min and then eluted with a water–methanol gradient

[24] G. I. Kingston and B. B. Gerhart, *J. Chromatogr.* **116,** 182 (1976).
[25] P. S. Steyn, P. L. Wessel, and W. F. O. Marasas, *Tetrahedron* **35,** 1551 (1979).

(Fig. 2). This allows the separation of FB_1 from FB_2 and FB_3. Twenty 450-ml fractions are collected. FB_1 is recovered in fractions 7–16, and FB_3 and FB_2 are in fractions 14–17. Fractions 13–15 contain all three mycotoxins.

FB_1 Purification. Fractions 7–13 that contain FB_1 are combined and are prepared for additional purification by removing the methanol from the solvent by rotary evaporation and lyophilizing the sample. The residue is dissolved in water and is submitted to a Waters Bondapak PrepPak 47 × 300-mm CN cartridge connected in series to a second CN cartridge. The two CN cartridges have been conditioned previously with 2 liters of methanol and 8 liters of water containing 0.5% pyridine. The cartridges are eluted isocratically with 0.5% pyridine in water at a flow rate of 50 ml/min. After one VV, FB^1 elutes from the cartridge. Fractions of 250 ml are collected until 5 liters of solvent have passed through the cartridge. Fractions are analyzed by analytical HPLC to determine the presence of FB^1. Purity is >90% for FB_1.

The use of water–acetonitrile (1:1) as the extracting solvent eliminates *N*-acetylation from occurring and is just as efficient as methanol in extracting the fumonisins.[21] Because of this, acetonitrile is the solvent of choice. The use of water–acetonitrile instead of water–methanol for the preparative chromatography gradient has definite advantages.[21] Fumonisin B_1, FB_2, and FB_3 can be separated on the C_{18} cartridge, and the length of time needed for the preparative chromatography can be decreased by using the water–acetonitrile[21] gradient. Using this gradient, 16 fractions (450 ml) are collected from the preparative HPLC with FB_1 eluting in fractions 10–12, FB_3 eluting in fraction 13, and FB_2 eluting in fractions 14 and 15.

FB_2 Purification. Additional purification steps are required to obtain FB_2 at 90% or greater. Fumonisin B_2 is purified on a centrifugal spinning TLC system (Chromatotron, Harrison Scientific, Palo Alto, CA).[21] A round glass plate is coated with 4-mm-thick TLC grade silica gel (silica gel 60 PF-254 with calcium sulfate, E. Merck).

The absorbent is dried under a 100-W light bulb. After drying, the absorbent is scraped to 2 mm thickness and is activated at 70° for 12 hr. A methanol solution containing FB_2 is applied to the spinning TLC plate under N_2 using a 10-ml syringe held in a specially constructed holder.[21] This allows for a very slow rate of application of the sample and results in a colored band of material deposited on the TLC plate that is less than 5 mm wide.

After allowing the methanol to evaporate, the spinning centrifugal TLC plate is eluted with (A) chloroform–acetone (4:3) and (B) methanol–acetone (1:1) applied as a linear gradient for 180 min at a flow rate of 3 ml/min. The starting solution is 90% (A) and 10% (B) and the ending condition is 50% (A) and 50% (B). Fractions (6 ml) are collected throughout the gradient analysis and every fifth fraction is analyzed by HPLC and fluorescent detection for FB_2. Using this system, FB_2 can be purified to >90%.

Method D

FB_3 and FB_4 Purification. *Fusarium moniliforme,* strain KSU 819, produces high levels of FB_3 and FB_4 with very little FB_1 or FB_2.[2] Acetonitrile–water (1:1) extraction, separation, and purification of FB_3 and FB_4 by NH_2 and tC_{18} cartridges (Waters, Milford, MA) are described by Poling and Plattner.[2] The 10-g NH_2 cartridge is conditioned with 50 ml of methanol followed by 50 ml of methanol–water (1:1). The acetonitrile–water extract (250 ml containing FB_3 at approximately 1 mg/ml and FB_4) is applied to the NH_2 cartridge and the cartridge is washed with 250 ml of methanol–water (1:1) and 250 ml methanol, and the fumonisins are eluted with 5% acetic acid–methanol (250 ml). The eluting solvent is diluted with water to 40% methanol and the solution containing FB_3 and FB_4 is submitted to a tC_{18} cartridge that is conditioned with 100 ml of methanol and 40% methanol–water (100 ml). The tC_{18} cartridge is washed with 100 ml of acetonitrile–water at ratios of 15–85, 25–75, 35–65, 35–65, 45–55, and 55–45 acetonitrile–water. The combined 35–65 fractions contain FB_3 and fraction 45–65 contains FB_4. Recovery for the fumonisins is 95% and purity is 80% for FB_3 and 85% for FB_4.

Method E

Poling and Plattner[26] report a method to remove the colored pigments from extracts containing fumonisins. A weak anion-exchange resin, Amberlite IRA-68 (1-0331, Supelco, Supelco Park, Bellefonte, PA) is washed twice with water and twice with acetonitrile–water (1:1). The resin is allowed to stand in each wash solvent for at least 10 to 15 min. The washed IRA-68 resin (15 g) is added to 500 ml of the fumonisin containing acetonitrile–water (1:1) extract and stirred for 2 hr. The IRA-68 resin is removed from the extract by filtering and is rinsed with 100 ml of acetonitrile–water (1:1). The resin is resuspended in 75 ml acetonitrile–water (1:1) containing 5% acetic acid, stirred for 18 to 21 hr to desorb the fumonisins, filtered, resuspended in 30 ml of the same solvent, and stirred for an additional 2 hr. The solution is filtered and the two filtrates are combined. Recovery of the fumonisins is approximately 95%. The fumonisins are then submitted to additional purification (C_{18} chromatography) to remove the acetic acid and other contaminates.

[26] S. M. Poling and R. D. Plattner, *J. Agric. Food Chem.* **47,** 2349 (1999).

Analytical Methods

High-Performance Liquid Chramotography

Fumonisins lack a chromophore and are detected only weakly by UV at 210 nm. Therefore, derivatives must be prepared for detection and quantitation by HPLC or GC-MS. Derivatization with maleic anhydride is the first method developed for the quantitation of fumonisins by analytical HPLC and UV detection.[27,28] Because this procedure cannot detect fumonisins at the low levels found in food products, it is not satisfactory. A more sensitive method is o-phthalaldehyde (OPA) and fluorescent detection.[29] The OPA reagent is prepared by dissolving 40 mg OPA in methanol and adding 5 ml of 0.1 M sodium borate and 50 μl 2-mercaptoethanol.[29] The reagent is stored in the dark at room temperature and is stable for at least a week. The sample (50 μl) is derivatized with 200 μl OPA and is injected into the HPLC after 1 to 2 min. The time between derivatization and injection is critical as the OPA derivatives are not stable.[29] The derivatives are detected by fluorescence at wavelengths of 335 nm excitation and 440 nm emission.[29] The OPA-HPLC analytical method allows the simultaneous separation and quantitation of FB$_1$ and FB$_2$ using a 250 × 4.6-mm Phenomenex Ultracarb 7 ODS 30 reversed-phase column and a Resolve C$_{18}$ reversed-phase guard column. Fumonisins are eluted by isocratic chromatography using methanol:0.1 M sodium dihydrogen phosphate (80:20) adjusted to pH 3.3 with orthophosphoric acid at a flow rate of 1 ml/min. Detection levels are 50 ng/g for fumonisin FB$_1$ and 100 ng/g for FB$_2$.[29]

We have developed a reliable precolumn procedure using the program capabilities of the 1090 HPLC (Hewelett-Packard, Wilmington, DE) and OPA derivatization that consists of successively drawing 10 μl of OPA (fluoraldehyde, Pierce Chemical Co., Rockford, IL), 6–10 μl sample, and 10 μl OPA and mixing the three components in the sample loop prior to injection.[20] After mixing the reagents and sample, the derivatized sample is held for 1 min and then injected. The 1046A Hewlett Packard fluorescent detector wavelengths are determined for maximum excitation and emission.[20] The maximum wavelengths for FB$_1$ with our detector are 304 nm excitation and 440 nm emission. The derivatized fumonisins are separated isocratically on a 3-μm particle size Rainin C$_{18}$ reversed-phase (10 cm × 4.6 mm) column connected to a Rainin C$_{18}$ guard column. Eluting solvents

[27] E. W. Sydenham, W. C. A. Gelderblom, P. G. Thiel, and W. F. O. Marasas, *J. Agric. Food Chem.* **38,** 285 (1990).

[28] J. F. Alberts, W. C. A. Gelderblom, and W. F. O. Marasas, *Mycotoxin Res.* **8,** 2 (1992).

[29] G. S. Shephard, E. W. Sydenham, P. G. Thiel, and W. C. A. Gelderblom, *J. Liquid Chromatogr.* **13,** 2077 (1990).

are 25% water containing 1% orthophosphoric acid and 75% methanol at a flow rate of 0.8 ml/min. This solvent system does not form salt crystals on the pump plungers. Therefore, pump maintenance is reduced as the seals do not wear as fast nor develop leaks as often. Several other OPA-HPLC systems using different solvent-eluting procedures for fumonisins are reported[20,30–32] as well as different reagents used to form a fluorescent derivative on the fumonisin primary amine. These fluorescent-derivatizing reagents are 4-fluor-7-nitrobenzofurazan,[33] fluorescamine,[34] 9-(fluorenyl-methyl) chloroformate,[35] naphthalene-2,3-dicarboxaldehyde,[36] 4-(N,N-di-methylaminosulfonyl)-7-fluoro-2,1,3-benzoxadiazole,[37] and 6-aminoquino-lyl N-hydroxysuccinimidylcarbamate.[38]

A detector for HPLC systems that does not require the formation of derivatives is the evaporative light-scattering detector (ELSD) (Varex, MkIII, Alltech Assoc., Deerfield, IL).[2,21] The ELSD is a universal mass detector with relatively good sensitivity (10–50 ng).[2,21] One problem of the detector is that it is nonlinear in respect to the amount of material detected and its response.[39] However, the ELSD will detect all compounds in the sample injected on the HPLC column provided they are sufficiently concentrated. The ELSD is very useful in providing information on the purity of the fumonisins. This information is similar to that obtained from LC/ES/MS.[2,21,39] The ELSD is also an excellent way to obtain conditions necessary for HPLC/MS analysis.[39]

Thin-Layer Chromatography

Large numbers of samples can be screened rapidly for fumonisins by thin-layer chromatography. Samples are extracted and the fumonisins are partially purified using either a C_{18} or a strong ion-exchange (SAX) column. Several different solvent systems are reported for silica gel TLC plate

[30] E. W. Sydenham, G. S. Shephard, and P. G. Thiel, *J. Assoc. Anal. Chem.* **75,** 313 (1992).

[31] M. E. Stack and R. M. Eppley, *J. Assoc. Anal. Off. Chem.* **75,** 834 (1992).

[32] P. G. Thiel, E. W. Sydenham, G. S. Shephard, and D. J. Van Schalkwyk, *J. Assoc. Anal. Off. Chem.* **76,** 361 (1993).

[33] P. M. Scott and G. A. Lawrence, *J. Assoc. Off. Anal. Chem.* **77,** 541 (1994).

[34] M. Holcomb, J. B. Sutherland, M. P. Chiarelli, W. A. Korfmacher, H. C. Thompson, J. O. Lay, L. J. Hankins, and C. E. Cerniglia, *J. Agric. Food Chem.* **41,** 357 (1993).

[35] M. Holcomb, H. C. Thompson, and L. J. Hankins, *J. Agric. Food Chem.* **41,** 764 (1993).

[36] G. M. Ware, O. Francis, P. U. Kuan, A. Carman, L. Carter, and G. A. Bennett, *Anal. Lett.* **26,** 1751 (1993).

[37] H. Akiyama, M. Miyahara, M. Toyoda, and Y. Saito, *J. Food Hyg. Soc. Jpn.* **36,** 77 (1995).

[38] C. Velazquez, C. Vanbloemendal, V. Sanchis, and R. Canela, *J. Agric. Food Chem.* **43,** 1535 (1995).

[39] J. G. Wilkes, J. B. Sutherland, M. I. Churchwell, and A. J. Williams, *J. Chromatog.* **695,** 319 (1995).

development. These different solvent systems are chloroform–methanol–water–acetic acid $(55:36:8:1)$,[23] chloroform–methanol–acetic acid $(6:3:1)$,[23,40] $(55:35:10)$,[41] and acetonitrile–water $(85:15)$.[42] Solvent systems used to develop reversed-phase C_{18} TLC plates are the following: methanol–water $(80:20)$,[41] methanol–sulfuric acid–acetic acid $(85:5:10)$,[3] or methanol–4% aqueous KCl $(3:2)$.[43] Fumonisins on the TLC plate are visualized by spraying with 0.5% p-anisaldehyde in methanol–sulfuric acid–acetic acid $(85:5:10)$[3,23,40,41] and heated to 100–120° until a reddish color develops. Another procedure[43] for visualization of fumonisins is spraying the TLC plate with a 0.1 M sodium borate buffer, pH 8–9, and allowing the plate to air dry. The plate is then sprayed with fluorescamine (0.4 mg/ml in acetonitrile). After 1 min, it is sprayed with 0.01 boric acid : acetonitrile $(40:60)$. The plate is air dried and then examined with long wavelength UV light.

Enzyme-Linked Immunosorbent Assay

Enzyme-linked immunosorbent assay (ELISA) methods are an alternative procedure for screening large numbers of samples. This method is fast, fairly reliable, simple, and relatively inexpensive.[44] Correlation values (r value) for ELISA versus GC-MS and HPLC methods are 0.478 and 0.512 ($p < 0.05$), respectively; whereas GC-MS compared to HPLC is 0.946 ($p < 0.01$).[45] ELISA gave higher values for fumonisin content than GC-MS and HPLC, especially in highly contaminated food samples.[45] Variations in sample preparation, extraction procedures, and method differences in GC-MS and HPLC may also affect the resulting values.[45] An additional study[46] compared direct enzyme-linked immunosorbent assay (CD-ELISA) with HPLC analysis. CD-ELISA determinations are consistently higher in fumonisin levels in naturally contaminated corn than in those obtained by HPLC. Differences between the two methods were reduced by decreasing the organic solvent in the extract and by adding a hexane partition step.[46]

[40] T. Ackermann, *J. Appl. Toxic.* **11**, 451 (1991).
[41] G. Rottinghaus, C. E. Coatney, and H. C. Minor, *J. Vet. Invest. Diagn.* **4**, 326 (1992).
[42] R. D. Plattner, D. Weisleder, D. D. Shackelford, R. Peterson, and R. G. Powell, *Mycopathologia* **117**, 23 (1992).
[43] M. M. Abouzied, S. D. Askegard, C. B. Bird, and B. M. Miller, *in* "Fumonisins in Food: Advances in Experimental Medicine and Biology" (L. S. Jackson, J. W. DeVries, and L. B. Bullerman, eds.), p. 135. Plenum Press, New York, 1996.
[44] J. J. Pestka, J. I. Azcona-Olivera, R. D. Plattner, F. Minervini, M. B. Doko, and A. Visconti, *J. Food Prot.* **57**, 169 (1994).
[45] E. W. Sydenham, G. S. Shephard, P. G. Thiel, C. Bird, and B. M. Miller, *J. Agric. Food Chem.* **44**, 159 (1996).
[46] R. A. Shelby, G. E. Rottinghaus, and H. C. Minor, *J. Agric. Food Chem.* **42**, 2064 (1994).

Investigators suggest that structurally related fumonisin-like compounds in the naturally contaminated corn contribute to the differences between ELISA and HPLC procedures.[46] However, the ELISA procedure can still be used as a fast semiquantitative screening method.[46] Concentration data for fumonisin in maize obtained from a competitive indirect immunoassay (CI-ELISA) and from TLC showed that CI-ELISA gave higher values than TLC.[46] Both CI-ELISA and TLC, however, are suitable for rapid screening of maize samples.

Acknowledgment

Mention of firm names or trade products does not imply the endorsement by the U.S. Department of Agriculture over firms or other commercial products.

[38] Inhibitors of Glucosylceramide Synthase

By James A. Shayman, Lihsueh Lee, Akira Abe, and Liming Shu

Introduction

Glycosphingolipids are ubiquitous but poorly understood membrane components. They amass as a result of defective glycosidases or activator proteins for these glycosidases in several sphingolipid storage disorders. These disorders include Gaucher, Fabry, and Tay-Sachs disease where glucosylceramide, globotriaosylceramide (Gb3), and ganglioside GM2, respectively, accumulate within the affected tissues. Glucosylceramide is the base cerebroside for the majority of glycosphingolipids, including Gb3 and ganglioside GM2.

Whereas the roles of these glycosphingolipids in the pathogenesis of sphingolipidoses are well appreciated, the normal physiological roles of glucosylceramide-based glycolipids are less well understood. One means for better understanding these functions is through the selective inhibition of cellular glucosylceramide formation. The search and identification by Radin and co-workers[1] of specific inhibitors of glucosylceramide synthase has provided a widely used tool for understanding the cellular biology of glycosphingolipids. The parent compound, D-*threo*-1-phenyl-2-decanoyl-amino-3-morpholino-1-propanol (PDMP), has found wide application in this regard.[1] More recently, a series of structurally related homologs that

[1] N. S. Radin, J. A. Shayman, and J. Inokuchi, *Adv. Lipid Res.* **26,** 183 (1993).

 0076-6879/99 $30.00

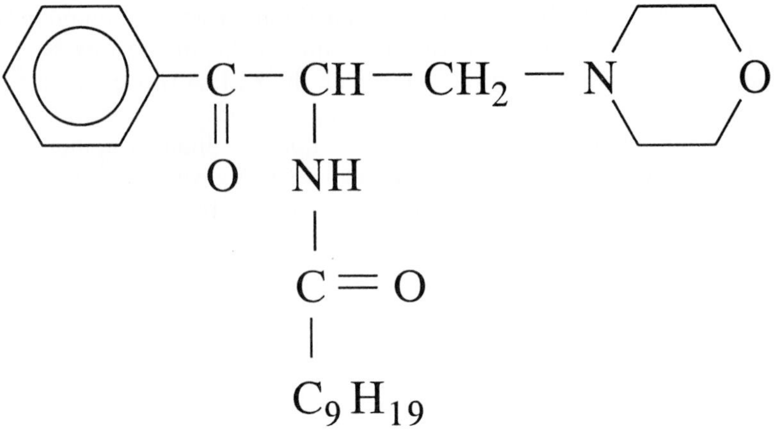

FIG. 1. DL-2-Decanoylamino-3-morpholinopropiophenone.

exhibit greater sensitivity and specificity toward the cerebroside synthase have been designed and tested. This article describes the discovery, synthesis, and characterization of PDMP and these newer homologs.

Discovery of PDMP

In 1977 a search began for an inhibitor of glucosylceramide synthase [ceramide : UDP-glucose glucosyltransferase (EC 2.4.1.80)] with the goal of identifying a therapeutic agent for the treatment of Gaucher disease. A keto amine, DL-2-decanoylamino-3-morpholinopropiophenone-1, was found to inhibit cerebroside synthase, with complete inhibition being observed at ~300 μM (Fig. 1).[2,3] This compound produced complete inactivation of the cerebroside synthase, probably secondary to covalent binding at the active site of the enzyme. The reduction of the ketone produced a compound, PDMP, that was more potent in inhibiting the cerebroside synthase and which acted reversibly.[2] A consequence of the reduction of the propiophenone was the formation of four isomers due to the presence of two asymmetric carbons (Fig. 2).

A mixture of the enantiomers formed by the borohydride reduction of the ketone was subjected to separation by thin-layer chromatography (TLC). Two spots were observed and identified as the DL-*threo* and DL-*erythro* diastereomers of PDMP. The DL-*threo*-PDMP diastereomer was

[2] R. R. Vunnam and N. S. Radin, *Chem. Phys. Lipids* **26**, 265 (1980).
[3] R. R. Vunnam, D. Bond, R. A. Schatz *et al., J. Neurochem.* **34**, 410 (1980).

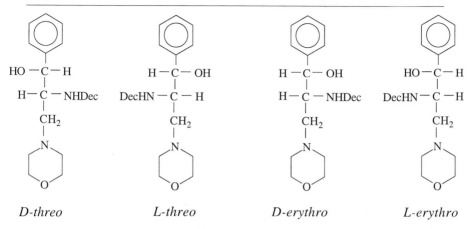

| D-threo | L-threo | D-erythro | L-erythro |

FIG. 2. PDMP enantiomers formed by reduction of DL-2-decanoylamino-3-morpholinopropiophenone. Dec denotes a decanoyl group.

found to be the active mixture. The optical enantiomers of the *threo* mixture were resolved by crystallization of the salts formed with optically active dibenzoyltartaric acid followed by the recovery of the free base and reconversion to the hydrochloride salt. Only the D-*threo* enantiomer was identified as active in inhibiting glucosylceramide synthase.

Improved Homologs of PDMP

The three functional groups comprising PDMP, the phenyl group, fatty acid in amide linkage, and cyclic amine, provide potential sites to alter the potency and specificity of this inhibitor. Attempts to develop improved inhibitors of glucosylceramide synthase were initiated using PDMP as the lead inhibitor with substitutions at the morpholino, decanoyl, or phenyl groups. These new compounds were evaluated in both isolated enzyme preparations and intact cells in a number of assays, including UDP-glucosyltransferase. These compounds were also screened for effects on cell proliferation.

N-acyl chain length substitutions affected glucosylceramide synthase inhibitory activity markedly. An increase in the chain length from 10 (decanoyl) to 16 (palmitoyl) carbons distinctly improved the inhibitory effect.[4] As a result of these early studies, most of the new compounds were synthesized with the *N*-palmitoyl substitution.

[4] A. Abe, J. Inokuchi, M. Jimbo et al., J. Biochem. 111, 191 (1992).

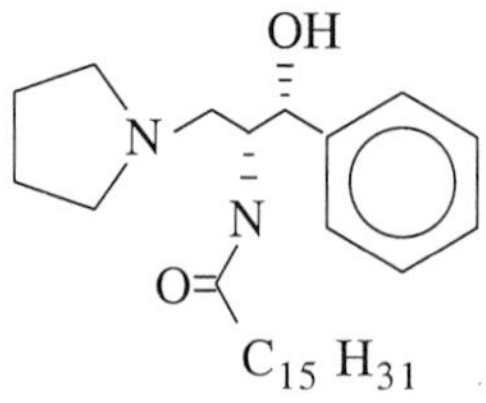

D-*threo*- P4 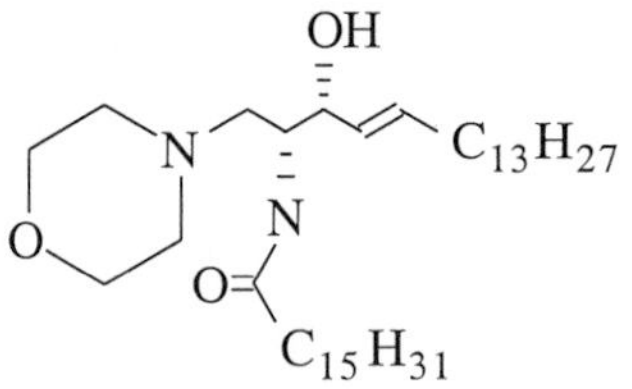   D-*threo*- 1-morpholino-1-deoxyceramide

FIG. 3. Homologs of PDMP with improved activity and increased specificity toward gluco-sylceramide synthase. D-*threo*-1-Phenyl-2-palmitoyl-3-pyrrolidinopropanol (D-*threo*-P4) inhibits glucosylceramide synthase with an IC$_{50}$ of 1.4 μM. D-*threo*-1-Morpholino-1-deoxyceramide inhibits glucosylceramide synthase with minimal secondary changes in cell ceramide content.

Homologs with substitutions for the morpholino function were synthesized.[5] Several of these compounds are more potent in their inhibitory action against glucosylceramide synthase. Replacing the oxygen in the morpholine ring of the palmitoyl homolog of PDMP with a methylene group resulted in an improvement of ~1.4-fold. Replacing the piperidine group with a seven-membered ring decreased activity greatly, whereas use of a five-membered ring quadrupled the effectiveness (50% vs 12.4% inhibition). The next smaller ring analog, the four-membered ring analog, was about as effective as the piperidino compound. Structurally variant six-membered ring analogs containing N-phenylpiperazinyl, N-methylpiperazinyl, and 4-dimethylaminopiperidinyl groups had little or no inhibitory activity. An analog in which the six-membered ring was removed and only a free NH$_2$ remained was not an inhibitor but slightly stimulated the enzyme. An analog in which the ring was severed resulted in a conformationally unrestrained diethylamino group that was a weak inhibitor.

The most potent analog resulting from the "amine analog" phase of homolog development was the pyrrolidino analog (P4), a mixture of four isomers, with an IC$_{50}$ of 5 μM (Fig. 3).[5] In strict analogy to the observations made with the four PDMP isomers, it was thought most likely that only one diastereomer of P4 would contain the active isomer. Support for the validity of this analogy was obtained by separation of P4 into DL-*threo* and DL-*erythro* isomers by reversed-phase high-performance liquid chromatography (HPLC). The separated isomers were designated DL-*threo*-P4 (more polar by TLC) and DL-*erythro*-P4 (less polar by TLC). Assay of glucosylcer-

[5] A. Abe, N. S. Radin, J. A. Shayman et al., J. Lipid Res. **36,** 611 (1995).

amide synthase with each diastereomer at 5 μM showed 15% inhibition by designated DL-*erythro*-P4 and 79% inhibition by DL-*threo*-P4. Further chromatographic analysis of the isomers showed there was a small amount of contamination of each with the other, which could explain the finding of some inhibition by DL-*erythro*-P4. When the two isomers were purified further, the difference in effectiveness was found to be somewhat higher. This supported the analogy with the PDMP isomers where the less polar *threo* species and not the *erythro* species is the active inhibitor of glucosylceramide synthase. For P4, when the actual content of active isomer is taken into account, the IC_{50} becomes 1.4 μM.

Dissociation of Glucosylceramide Synthesis from Cell Growth Inhibition

During this phase of homolog development, the best glucosylceramide synthase inhibitors were studied for their effects on cell growth in several tumor lines. Cell lines used included MCF-7 breast carcinoma cells, HT-29 colon adenocarcinoma cells, H-460 lung large cell carcinoma cells, and 9L brain gliosarcoma cells. Exposure of the cell lines to inhibitors at different concentrations showed that the six new compounds most active against glucosylceramide synthase were very effective growth inhibitors, with IC_{50} values of 0.7 to 2.6 μM.[5] In general, the ability of these new compounds to inhibit glucosylceramide synthase paralleled growth inhibition, providing partial pharmacological support to the hypothesis that glucosylceramide metabolism is important in the regulation of cell proliferation.

This interpretation was prove to be incorrect. When the *erythro* and *threo* diastereomeric forms of P4 were compared, they were found to have very similar effects on MDCK cell growth (Table I). The diastereomers were compared in greater detail. Examination of the alkali-stable lipids by TLC showed that DL-*threo*-P4 was more effective than DL-*erythro* in lowering glucosylceramide levels, as expected from its greater effectiveness *in vitro* as a glucosylceramide synthase inhibitor. The levels of the other lipids (mainly sphingomyelin, cholesterol, and fatty acids) were changed little or not at all. The P4 diastereomers were detected readily at the various levels used, showing that they were taken up by the cells during the incubation period at concentration-dependent rates.

Chromatographic determination of ceramide levels showed that ceramide accumulation was similar for both *erythro* and *threo* forms of P4. Elevated levels of ceramide only appeared at inhibitor concentrations (4 μM), which were more than sufficient to lower glucosylceramide levels. An important finding was that both active (DL-*threo*-P4) and inactive (DL-*erythro*-P4) diastereomers (with respect to glucosylceramide synthase) in-

TABLE I

EFFECTS OF SELECTED CERAMIDE ANALOGS ON GLUCOSYLCERAMIDE SYNTHASE, CERAMIDE, AND
CELL PROLIFERATION PARAMETERS IN MDCK CELLS[a]

Compound	Glucosylceramide synthase activity (% control)	Ceramide (% control)	[³H]Thymidine incorporation (% control)	Cell protein (% control)
DL-erythro-P4 (4 μM)	86.5 ± 0.4	368	45.7 ± 5.5	34.8 ± 6.1
DL-threo-P4 (4 μM)	15.4 ± 1.4	438	51.5 ± 0.9	31.2 ± 5.7
1-Pyrrolidino-1-deoxyceramide (2 μM)	79.1 ± 0.2	100	100	101 ± 2.4
N-Octanoylsphingosine (32 μM)	Not determined	190	65.9 ± 4.9	50.3 ± 2.4

[a] Glucosylceramide synthase inhibitors were used as a complex with BSA. Octanoylsphingosine was incorporated into liposomes consisting of PC and sulfatide. Glucosylceramide synthase measurements were performed at 5 mM for each inhibitor.

duced ceramide accumulation and were also potent inhibitors of cell growth. Thus, ceramide accumulation may be a common mechanism of action of these compounds. Data showing ceramide accumulation induced by DL-erythro-P4 (a diastereomer inactive toward glucosylceramide synthase) was a key finding. It indicated that DL-erythro-P4 must inhibit another pathway of ceramide metabolism, the result of which is ceramide accumulation.

In addition, stereospecific deoxyceramides were synthesized with either morpholino- or pyrrolidino substitutions (Fig. 3).[6] The *threo* form of these aliphatic analogs of PDMP and P4 inhibited glucosylceramide synthase. Unlike PDMP and its *threo* homologs, however, glucosylceramide synthase inhibition was neither associated with an increase in ceramide nor with an inhibition of cell growth. Finally, effects of the short chain ceramides N-octanoylsphingosine and N-acetylsphingosine were evaluated for their effects on endogenous ceramide and glucosylceramide levels. These compounds were found to increase glucosylceramide levels and to increase the levels of endogenous ceramides in addition to inhibiting cell growth.

Thus stereospecific PDMP homologs and short chain ceramides have varying and surprisingly different effects on the levels of endogenous ceramides and cerebrosides in cultured cells. Specific glucosylceramide synthase inhibitors (D-*threo*-PDMP and DL-*threo*-P4) decrease glucosylceramide and increase ceramide. *Erythro* homologs of the pyrrolidino compounds increase ceramide without decreasing glucosylceramide significantly. Deoxyceramide analogs inhibit glucosylceramide formation

[6] K. G. Carson, B. Ganem, N. S. Radin, A. Abe, and J. A. Shayman, *Tetrah. Lett.* **35,** 2659 (1994).

without significantly changing ceramide levels or cell growth rate during a 24-hr exposure. Finally, short chain ceramides increase both endogenous ceramide and glucosylceramide. These compounds provide highly useful tools for discerning the relative importance of ceramide and glucosylceramide in regulating cellular processes. These findings, taken together with other current discoveries regarding ceramide, are strong evidence that the pyrrolidino homologs of PDMP exert their antiproliferative effect by causing the accumulation of ceramide, which may induce cell differentiation and growth arrest. Compounds that raise ceramide independent of inhibiting GSL formation represent a unique class of new antiproliferative agents. Importantly, these studies also indicate that it was possible to create PDMP homologs that blocked glucosylceramide synthase and glycosphingolipid formation without raising ceramide and inhibiting cell growth. Such homologs may therefore be candidate drugs for the treatment of glycosphingolipidoses.

Basis for Inhibitor-Mediated Ceramide Accumulation

An interesting problem emerged from these studies: if ceramide accumulation was comparable for both DL-*threo*- and DL-*erythro*-P4, then what pathway other than glucosylceramide synthase was being inhibited by these homologs? Several ceramide-utilizing enzymes were assayed in an attempt to identify the basis for the ceramide accumulation with the pyrrolidine homologs. Sphingomyelinase, sphingomyelin synthase, ceramidase, and ceramide synthase activities were not inhibited or stimulated substantially by the P4 diastereomers. Although not every ceramide-utilizing enzyme was assayed, the failure to identify a second site of action quickly for the P4 compounds raised the possibility that a novel ceramide pathway was involved. In order to assess this possibility, MDCK cells were incubated with tritiated *N*-acetylsphingosine.[7] Previous studies utilizing radiolabeled *N*-acetylsphingosine were labeled in the acetyl group. If deacetylated by a ceramidase, the acetyl group would be metabolized only minimally to other sphingolipids. MDCK cells metabolized *N*-acetyl[³H]sphingosine rapidly to free tritiated sphingosine, C2-sphingomyelin, and C2-glucosylceramide and normal chain length sphingolipids and GSLs. However, when the cellular lipids were extracted without alkaline methanolysis, a very nonpolar product was identified by TLC. The product formed *N*-acetylsphingosine when subjected to alkaline hydrolysis. This product was subsequently characterized as 1-*O*-acylceramide. Determining the substrate for the 1-*O*-acylceramide synthase was problematic. Eventually it was determined that the

[7] A. Abe, J. A. Shayman, and N. S. Radin, *J. Biol. Chem.* **271,** 14383 (1996).

fatty acid donor was not free fatty acid or fatty acyl CoA. Rather, the 1-
O-acylceramide was formed as a result of a transacylase that utilized the
sn-2 fatty acid of phosphatidylethanolamine or phosphatidylcholine.

Chemical Synthesis of P4 (D,L-*threo* and *erythro* 1-Phenyl-2-palmitoylamino-3-pyrrolidino-1-propanol)

General Synthesis of Inhibitors

The aromatic inhibitors are synthesized by the Mannich reaction from
2-*N*-acylaminoacetophenone, paraformaldehyde, and pyrrolidine, followed
by reduction with sodium borohydride. The reaction produces a mixture
of four isomers due to the presence of two asymmetric carbons. For syntheses in which phenyl-substituted starting materials are used, the chloro,
methoxy, and methyl groups in the acetophenone structure are brominated
and converted to the primary amine. Brominations of the substituted aceto-
phenones (0.08 mol) are performed in chloroform (40 ml) by the addition
of bromine (0.08 mol) in chloroform (10 ml). After stirring for 10 min, the
reaction is quenched by the addition of sodium bicarbonate (saturated in
water) until a pH of 7 is obtained. The organic layer is dried over magnesium
sulfate and concentrated to dryness and the products are recrystallized
from ethyl acetate and hexane.

The synthesis of P4 is as follows (Fig. 4).

Synthesis of Palmitoyl Aminoacetophenone

Aminoacetophone · HCl (17.18 g, 100 mmol) and tetrahydrofuran (250
ml) are placed in a 1-liter three-neck round-bottom flask with a large egg-
shaped stir bar. Sodium (50% in water, 10 ml) is added in three portions.
Palmitoyl chloride (36.6 ml, 20% excess, 120 mmol) in tetrahydrofuran (60
ml) is then added dropwise over 20 min, yielding a dark brown solution.
This mixture is then stirred for an additional 2 hr at room temperature.
The resultant mixture is poured into a separating funnel to remove the
aqueous solution. Chloroform/methanol (2 : 1, 320 ml) is added to the organic layer, which is then washed with water (100 ml). The resultant yellow
aqueous layer is extracted once with chloroform (50 ml). The organic
solutions are pooled and roto-evaporated until almost dry. This residue is
redissolved in chloroform (100 ml) and crystallized by the addition of
hexane (400 ml) and cooling to 4° for 2 hr. The crystals are filtered and
washed with cold hexane until they are almost white. The crystals are dried
in vacuo for 1 day. Twenty-seven grams of product is obtained (72% yield).

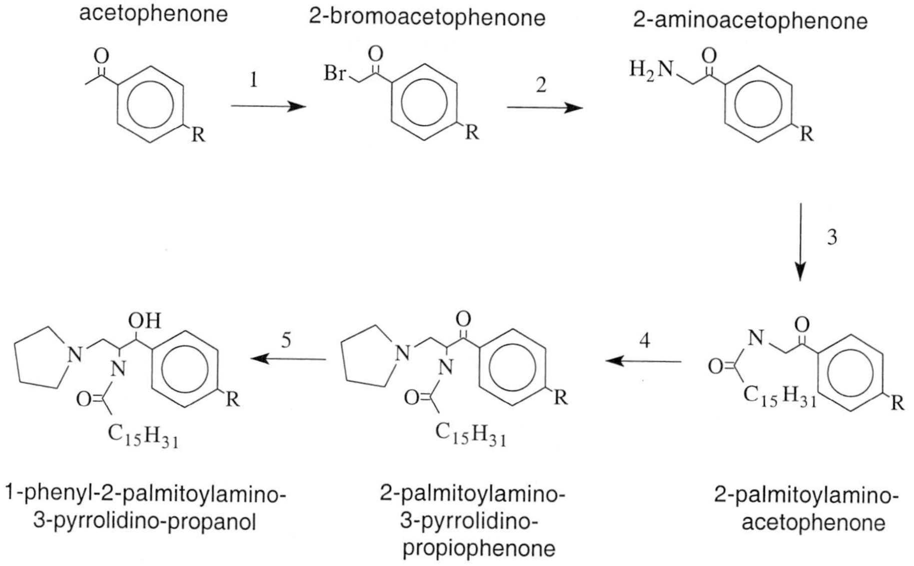

acetophenone   2-bromoacetophenone   2-aminoacetophenone

1-phenyl-2-palmitoylamino-   2-palmitoylamino-   2-palmitoylamino-
3-pyrrolidino-propanol   3-pyrrolidino-   acetophenone
                                   propiophenone

FIG. 4. Synthetic pathway for D-*threo*-1-phenyl-2-palmitoyl-3-pyrrolidinopropanol and phenyl-substituted homologs.

Synthesis of 2-Palmitoylamino-3-pyrrolidinopropiophenone

Palmitoylaminoacetophone (24.4 g, 65.3 mmol), paraformaldehyde (1.8 g, 6.0 mmol), pyrrolidine (7.7 ml, 6.6 g, 93 mmol), and ethanol (400 ml) are added to a 1-liter round flask under a flow of nitrogen. Concentrated HCl (2 ml) is added to this mixture through the condenser and then heated to reflux for 16 hr. The brown solution is then dried completely by use of a solvent evaporator. The residue is redissolved with chloroform (130 ml), washed once with HCl (3 N, 32 ml), twice with H_2O (33 ml), and dried over anhydrous $MgSO_4$. The chloroform solution is roto-evaporated to a viscous oil and then crystallized by the addition of chloroform (20 ml) and hexane (150 ml). Crystals are formed when the solution is kept in a cold room overnight. The crystals are filtered and rinsed with cold hexane (20 ml). The crystals are air dried, yielding 17.7 g (55% yield). An R_f of 0.54 was observed when the product was separated by thin-layer chromatography in a solvent consisting of chloroform/methanol/acetic acid (90:5:10). $[M + H]^+$: 457.7, $C_{29}H_{49}O_2N_2$.

Synthesis of DL-threo and DL-erythro 1-Phenyl-2-palmitoylamino-3-pyrrolidino-1-propanol

2-Palmitoylamino-3-pyrrolidinopropiophenone (17.7 g, 35.9 mmol) in ethanol (100 ml) is sonicated for 20 min and then adjusted to pH 8 with

NaOH (1 M, 15 ml). Sodium borohydride (4.07 g, 108 mmol) is added slowly to the ketoamine mixture on an ice bath. The mixture is then stirred for 4 hr at room temperature, adjusted to pH 4 with HCl (3 N, 45 ml), and extracted with chloroform (200 ml). The aqueous solution is reextracted with chloroform (50 ml). The extracts are combined and then washed with water (50 ml) and saturated sodium chloride (50 ml), dried with anhydrous MgSO$_4$, and roto-evaporated to a viscous oil. The oil is dissolved in ether (150 ml) and chloroform (15 ml) and cooled to 4° overnight. Filtration yields 6.8 g of a pale yellow solid, which is dried in air for 2 days. Further purification is carried out using a 6-mm SiO$_2$ rotor on a chromatotron (Model 8924, Harrison Research, Palo Alto, CA) to remove the erythro enantiomers using a solvent consisting of chloroform/methanol/acetic acid (9:1:1). The eluted diastereomers are dried with a roto-evaporator. The yield of DL-*threo*-P4 is 5.05 g (93–98% of DL-*threo*). In addition, 4.4 g of DL-*threo* and DL-*erythro* in a 1-to-1 ratio is obtained from the filtrate, which is also purified by a 6.5 × 28-cm silica column with 4–9% MeOH in CH$_2$Cl$_2$. [M + H]$^+$: 459.5, C$_{29}$H$_{50}$O$_2$N$_2$.

Resolution of P4 Enantiomers by Chiral Chromatography

Diastereomers of P4 are resolvable by thin-layer chromatography. R_f values of 0.49 and 0.40 for the *erythro* and *threo* diastereomers, respectively, are observed in a solvent system consisting of chloroform/methanol/acetic acid (9:1:1).

Resolution of the DL-*threo* and DL-*erythro* enantiomers of P4 is accomplished using a preparative HPLC column (Chirex 3014: [(S)-val-(R)-1-(α-naphthyl)ethylamine, 20 × 250 mm]). The enantiomers are eluted with hexane/1,2-dichloroethane/ethanol/trifluoroacetic acid (65:30:4.75:0.25) at a flow rate of 7 ml/min. The column eluant is monitored at 254 nm in either the preparative or the analytical mode. Isolated products are reinjected until pure by analytical HPLC analysis, which is determined using an analytical (4.6 × 250 mm) Chirex 3014 column and the same solvent mixture at a flow rate of 1 ml/min. In order to obtain the best separation, each injection contains no greater than 30 mg of P4, and the fractions are pooled to obtain sufficient quantities of isomers of L-*erythro,* D-*erythro,* L-*threo,* and D-*threo* for further biological characterization.

^{1}H nmr (δ, ppm, CDCl$_3$) for D-*threo*-P4: 7.32 (5H, m, Ar-), 7.14 (1H, d, 6.9 Hz, -NH-), 5.03 (1H, d, 3.3 Hz, H-1), 4.43 (1H, m, H-2), 3.76 (2H, m, c-(CH$_2$CH$_2$)$_2$N-), 3.51 (1H, m, H-3), 3.29 (1H, m, H-3), 2.97 (2H, m, c-(CH$_2$CH$_2$)$_2$N-), 2.08 (6H, m, -C(O)CH$_2$(CH$_2$)$_{13}$CH$_3$ and c-(CH$_2$CH$_2$)$_2$N-), 1.40 (2H, m, C(O)CH$_2$CH$_2$(CH$_2$)$_{12}$CH$_3$), 1.25 (24H, m, -C(O)CH$_2$CH$_2$(CH$_2$)$_{12}$CH$_3$), 0.87 (3H, t, 6.7 Hz, C(O)CH$_2$(CH$_2$)$_{13}$CH$_3$). High-resolution mass spectrometry calculated for $[C_{29}H_{50}N_2O_4 + H]^+$: 459.3950, observed 459.3935.

For D-*threo*-P4, the specific rotation $[\alpha]^t\lambda = -5.82$ (6.7 mg/ml); for D-*threo*-PDMP, the specific rotation $[\alpha]^t\lambda = -10.47$ (9.17 mg/ml).

Delivery of Glucosylceramide Synthase Inhibitors into Cultured Cells

PDMP and its structural homologs are lipoidal compounds that spontaneously form aggregates in aqueous solutions. In the absence of suitable dispersants, these inhibitors are absorbed onto the glass and plastic walls of test tubes and cell culture dishes. The dispersion of glucosylceramide synthase inhibitors into culture media can be accomplished through three alternative means: detergents, liposomes, or protein carriers.

Myrj 52

PDMP or a PDMP homolog may be dispersed into a micelle consisting of Myrj 52. The inhibitory effect of the detergent is dependent on both the Myrj 52 concentration and the cell density in the culture dish. In a standard study utilizing Madin Darby canine kidney cells, 5×10^5 cells would be seeded in a 10-cm dish and grown for 24 hr in 8 ml of serum-free defined Dulbeccos modified Eagle's medium (DMEM). After 24 hr the cells would be treated with the glucosylceramide synthase inhibitor. Under these conditions, 15 μg/ml Myrj 52 does not inhibit cell growth.

Fifty milliliters of inhibitor containing DMEM is constituted as follows to yield a final detergent concentration of 15 μg/ml. Myrj 52 (75 μl, 10 mg/ml) in methanol and an appropriate amount of PDMP or PDMP homolog (10 mM in isopropyl alcohol) are mixed in a sterile tube and dried down under a stream of nitrogen. The dried materials are dispersed into 10 ml of medium by use of a temperature-controlled bath sonicator for 30 min at 37°. The mixture is adjusted to 50 ml with additional medium.

Liposomes

PDMP or PDMP homologs may be incorporated into liposomes consisting of phosphatidylcholine and dicetyl phosphate. These liposomes have no effect on MDCK cell growth. Phosphatidylcholine (80 mol%), dicetyl phosphate (10 mol%), and inhibitor (10 mol%) are mixed and dried down under a stream of nitrogen in a glass tube. Phosphate-buffered saline is added to render a final concentration of 2 to 10 μmol of phospholipid per 1 ml of solution. The mixture is dispersed by sonication for 10 min in an ice bath with a probe sonicator. Following sonication, the liposome solution is sterilized by passage through a 0.2-μm membrane filter. Liposomes without inhibitor or with an inactive enantiomer of the inhibitor are used as a control.

Bovine Serum Albumin Complex

A one-to-one complex of inhibitor and bovine serum albumin (BSA) is used as an alternative means for incubating cells with glucosylceramide synthase inhibitors. Two millimolar of fatty acid-free BSA in distilled water is sterilized by passage through a 0.2-μm membrane filter. One part of ethanol alone or of 50 mM PDMP or PDMP homolog in ethanol is injected rapidly into 25 parts of 2 mM BSA and vortexed briefly. The complex is kept at $-20°$. The inhibitor-free complex is used as a control. No effect on MDCK cell growth is observed with 0.1% ethanol.

Effects on Cell Metabolism

Perhaps the greatest utility for glucosylceramide synthase inhibitors has been found in their ability to alter endogenous cell and tissue levels of sphingolipids. Such treatment results in the time-dependent depletion of glucosylceramide and glucosylceramide-based glycosphingolipids and in the accumulation of ceramide.[8] The former effect is due to direct inhibition of the cerebroside synthase; the latter effect is probably the result of inhibition of 1-O-acylceramide synthase. This strategy has been used to implicate glucosylceramide-based glycosphingolipids in cellular growth and differentiation, phospholipase C activity,[9] tyrosine kinase signaling,[10] cell adherence,[11] multidrug resistance,[12] and tumor metastasis.[13] Increased endogenous ceramide concentrations resulting from PDMP and PDMP homolog treatment have been used to test the role of ceramide in NF-κB signaling[14] and cell cycle progression.[15]

Glycosphingolipids

Time- and concentration-dependent decreases in cell glycosphingolipid content are observed on exposure of PDMP or the more potent homologs. The effects are more evident in cells that are actively replicating with little

[8] J. A. Shayman, G. Deshmukh, S. Mahdiyoun et al., J. Biol. Chem. **266,** 22968 (1991).
[9] J. A. Shayman, S. Mahdiyoun, G. Deshmukh et al., J. Biol. Chem. **265,** 12135 (1990).
[10] B. Felding-Habermann, Y. Igarashi, B. Fenderson et al., Biochemistry **29,** 6314 (1990).
[11] J. Inokuchi, K. Momosaki, H. Shimeno et al., J. Cell. Physiol. **141,** 573 (1989).
[12] Y. Lavie, H. Cao, A. Volner et al., J. Biol. Chem. **272,** 1682 (1997).
[13] J. Inokuchi, M. Jimbo, K. Momosaki et al., Cancer Res. **50,** 6731 (1990).
[14] J. C. Betts, A. B. Agranoff, G. J. Nabel, and J. A. Shayman, J. Biol. Chem. **269,** 8455 (1994).
[15] C. S. Sheela Rani, A. Abe, Y. Chang et al., J. Biol. Chem. **270,** 2859 (1995).

change observed in quiescent cells in G_0 arrest. The presence of serum at concentrations greater than 1% while stimulating the growth of many cell lines impedes the activity of the cerebroside synthase inhibitors by allowing them to complex with serum proteins.

Changes in glucosylceramide mass are observable as early as 6 hr following PDMP addition. If cells are labeled with [³H]galactose, a decrease in the radiolabeling of glucosylceramide is observable as early as 1 hr following inhibitor addition. This is consistent with the view that the inhibitors enter the cultured cells rapidly and bind to glucosylceramide synthase. The binding of PDMP to glucosylceramide synthase is reversible. Therefore, if one changes the incubation medium of cultured cells treated with PDMP with inhibitor-free medium, the transcriptionally induced cerebroside synthase will result in a rebound increase of glucosylceramide.[16] The more potent homologs, including P4, do not exhibit such a rebound effect because they bind with a significantly higher affinity. These inhibitors can be recovered in the lipid extracts of treated cells and identified by Coomassie blue staining following separation by thin-layer chromatography.

Although glucosylceramide synthase inhibitors can be documented by radiolabeling cells with [³H]galactose or [³H]palmitate to inhibit ongoing glucosylceramide synthesis, the glucosylceramide mass may often remain at 10–20% of baseline levels after 1 to 2 days of inhibitor treatment. This is likely the result of degradation of more complex glucosylceramide-based glycosphingolipids. The levels of these more complex lipids, including ganglioside GM3 and globotriaosylceramide, also fall in the presence of cerebroside synthase inhibitors.

Ceramide

Cell ceramide content increases over a time course comparable to that of the decrease in glucosylceramide. Initial observations with PDMP suggested that these changes were the result of substrate accumulation due to inhibition of the glycosyltransferase. The observation that *erythro* enantiomers of P4 were equally as effective as *threo* enantiomers in raising ceramide levels led to the identification of the 1-O-acylceramide synthase pathway as a likely source for these changes in ceramide. In addition, PDMP homologs with lower IC_{50} values for inhibition of glucosylceramide synthase display a higher IC_{50} for ceramide accumulation (Fig. 5).

More active glucosylceramide synthase inhibitors, therefore, provide a means for depleting cellular glycosphingolipids in the absence of significant changes in ceramide accumulation. Alternatively, the use of *erythro* enanti-

[16] A. Abe, N. S. Radin, and J. A. Shayman, *Biochim. Biophys. Acta* **1299,** 333 (1996).

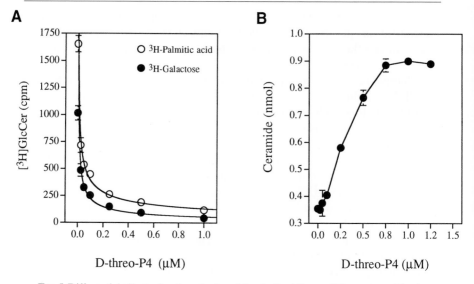

FIG. 5. Differential effects of D-*threo*-1-phenyl-2-palmitoyl-3-pyrrolidinopropanol (D-*threo*-P4) on glucosylceramide depletion (A) and ceramide accumulation (B). In this experiment, glucosylceramide formation in NIH 3T3 cells was determined by radiolabeling with either [³H]galactose or [³H]palmitic acid. The IC_{50} for glucosylceramide formation was ~50 nM. In contrast, 50% of maximal ceramide accumulation was observed at a concentration of ~200 nM. Data are expressed as cpm/50 nmol phospholipid phosphate and nmol ceramide/50 nmol phospholipid phosphate, respectively.

omers of P4 or related inhibitors provides a means for raising endogenous ceramide levels through the 1-O-acylceramide synthase pathway independent of changes in glycosphingolipid content.

Effects of Glucosylceramide Synthase Inhibitors on Other Sphingolipid Pathways

Aside from the inhibition of 1-O-acylceramide synthase at micromolar concentrations, PDMP and some PDMP homologs have been reported to have surprising effects on other sphingolipid-related pathways. L-*threo*-PDMP has been shown to increase the formation of glucosylceramide, lactosylceramide, and ganglioside GM3 by metabolic labeling studies.[17] This stimulation of glycosphingolipid synthesis may be the result of activation of the lactosylceramide synthase (glucosylceramide $\beta 1 \rightarrow$ 4-galactosyltrans-

[17] J. Inokuchi, S. Usuki, and M. Jimbo, *J. Biochem.* (*Tokyo*) **117,** 766 (1995).

ferase).[18] However, the mechanism of this activation has not been elucidated.

Ceramide glycanase cleaves the glycosidic bond between ceramide and the glucosyl group of the intact oligosaccharide on glycosphingolipids. Ceramide glycanase has been reported to be inhibited by high concentrations of DL-*threo*-PDMP.[19]

Sphingomyelin synthase activity is present within the malarial parasite *Plasmodium falciparum*. The parasite induces the formation of a network of tubovesicular membranes when present in the mature erythrocyte independent of its own plasma membrane. This may serve as a transport organelle for the parasite. The tubovesicular membrane network is dependent on the activity of a malarial sphingomyelin synthase. This synthase is reportedly inhibited by the palmitoyl-substituted homolog of PDMP at low micromolar levels.[20] It is probable that this sphingomyelin synthase is distinct from the mammalian synthase, an enzyme that is not inhibited by either *erythro* or *threo* diastereomers of PDMP or P4.

Conclusion

The design, synthesis, and characterization of newer homologs of PDMP have yielded two primary findings. First, more active inhibitors of glucosylceramide synthesis can be developed by simple substitutions at each of the three functional groups. Second, it is possible to dissociate the depletion of glycosphingolipids from secondary metabolic effects such as the accumulation of ceramide. These observations have increased the possibility of developing drugs that may selectively deplete tissues of stored glycosphingolipids and may therefore have some therapeutic potential for the treatment of inherited lipid storage disorders. Short of this goal, glucosylceramide synthase inhibitors should continue to provide useful tools for the further understanding of the biochemistry and cellular biology of sphingolipids.

[18] S. Chatterjee, T. Cleveland, W. Y. Shi *et al., Glycoconj. J.* **13,** 481 (1996).
[19] M. Basu, M. Girzadas, S. Dastgheib *et al., Indian J. Biochem. Biophys.* **34,** 142 (1997).
[20] S. A. Lauer, P. K. Rathod, N. Ghori, and K. Haldar, *Science* **276,** 1122 (1997).

Section III

Chemical and Enzymatic Syntheses

[39] Synthesis of Sphingosine and Sphingoid Bases

By Christopher Curfman *and* Dennis Liotta

Introduction

Sphingosines are a related class of long chain aliphatic compounds possessing a 2-amino-1,3-diol moiety. The most common member of this group found in nature is (2*S*,3*R*)-D-*erythro*-2-amino-1,3-octadec-4*E*-ene-diol **1a** and is frequently referred to as sphingosine. These molecules are the backbone of more complex sphingolipids, including ceramide, sphingo-myelin, cerebrosides, and glycosphingolipids. Ubiquitous to all living organisms, sphingolipids play many biologically important roles. As major constituents of membranes, they contribute to the structural integrity of a cell. Furthermore, sphingolipids facilitate cell–cell signaling, cell recognition, and adhesion.[1,1a]

Given the obvious significance of sphingolipids, efforts for the synthetic preparation of sphingosine and related analogs have generated much interest. There have been many published syntheses of sphingosine and its derivatives in the literature since the first reported sphingosine synthesis by Shapiro,[2,2a] with each synthesis offering something unique. Some syntheses utilize inexpensive starting materials from the chiral pool, whereas others control stereochemistry through asymmetric induction. High yield and few steps may be the hallmark of some preparations, whereas synthetic elegance, stereoselectivity, or the ability to convert to other sphingolipids or natural products may be the attributes of others. Because of this wealth of approaches chemists have a variety of choices for the preparation of synthetic sphingoid bases, but determining which procedure to use depends on the researcher's specific needs. The aim of this article is to illustrate and summarize many protocols, provide insights about each preparation and, in this way, facilitate the selection of sphingolipid synthetic preparations. Sphingosine syntheses can be divided into several categories. Syntheses that start from chiral precursors are divided into those that begin with serine derivatives and those that begin with sugar derivatives. Syntheses that

[1] Y. A. Hunnan and R. M. Bell, *Science* **243,** 500 (1989).
[1a] A. H. Merrill, S. Nimkar, D. Menaldino, Y. A. Hannun, C. Loomis, R. M. Bell, S. R. Tyagi, J. D. Lambeth, V. L. Stevens, R. Hunter, and D. C. Liotta, *Biochemistry* **28,** 3138 (1989).
[2] D. Shapiro and K. Segal, *J. Am. Chem. Soc.* **76,** 5894 (1954).
[2a] D. Shapiro, H. Segal, and H. M. Flowers, *J. Am. Chem. Soc.* **80,** 1194 (1958).

METHODS IN ENZYMOLOGY, VOL. 311

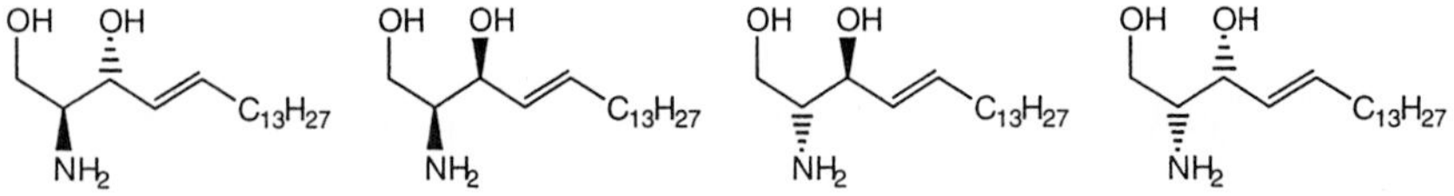

(2S, 3R)-D-erythro-   (2S, 3S)-L-threo-   (2R, 3S)-L-erythro-   (2R, 3R)-D-threo-

1a   **1b**   **1c**   **1d**

SCHEME 1a–1d.

utilize asymmetric epoxidations, aldol reactions, and relevant miscellaneous methods constitute the remaining sections.

Sphingosine has four stereoisomers, **1a–1d**. Whereas most syntheses focus on the preparation of the naturally occurring D-*erythro* analog, **1a**, this article includes several preparations for isomers **1b–1d**, as well as preparations for racemic sphingolipids. Sections discussing similar methods take a historical perspective and, when appropriate, refer to other sources when particular background information may be relevant.

Sphingosines from the Chiral Pool

The use of readily available, inexpensive, chiral starting materials, such as amino acids and sugars, is a major theme in sphingosine preparations. These approaches take advantage of the chirality of the starting material in several ways. A chiral center may be directly conserved in the product or it can influence the setting of another chiral center through a stereoselective process. In general, methods that utilize the chiral pool are beneficial for large-scale and diastereomerically pure sphingolipid preparations.

Sphingosine from Serine

Many syntheses of sphingolipids incorporate inexpensive and commercially available L-serine. These syntheses use serine to construct the 2-amino-1,3-diol head of the sphingoid base. Not only is the stereocenter at the C-2 carbon in serine conserved in the sphingoid product, but through the use of appropriate N-protection strategies, it can influence the stereochemistry at the C-3 position. Thus, a central theme in each synthesis is the elaboration of a configurationally stable, serine-derived synthon with a desired aliphatic chain.

Garner and Parks[3] prepared one of the most commonly used synthons, enantiomerically pure β-hydroxy-α-amino aldehyde **8**, in four steps from

---

[3] P. Garner and J. M. Park, *J. Org. Chem.* **52**, 2361 (1987).

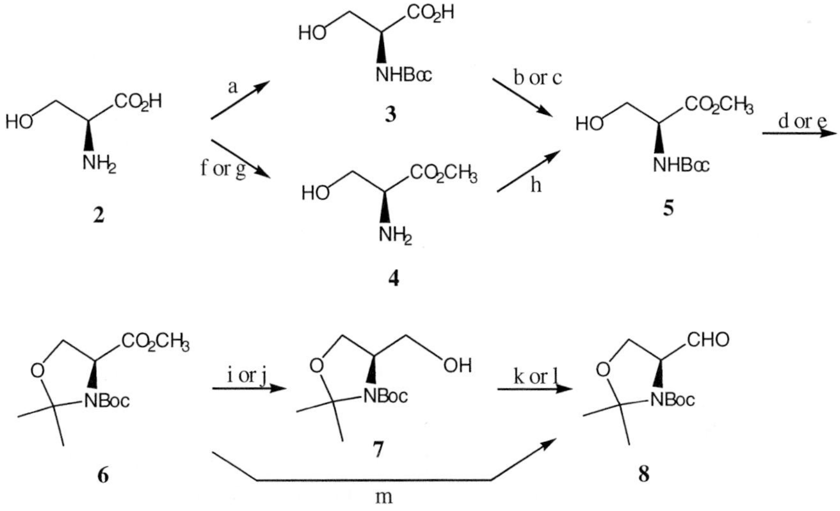

SCHEME 1. Reagents and conditions: (a) (Boc)$_2$O, 1 N NaOH, dioxane, 5°-r.t.; (b) 0.6 M CH$_2$N$_2$, Et$_2$O, 0°; (c) K$_2$CO$_3$, DMF, CH$_3$I, 0°-r.t.; (d) DMP, p-TsOH, benzene, reflux; (e) DMP, BF$_3$-OEt$_2$, acetone, r.t.; (f) acetyl chloride, methanol, 0°-reflux; (g) SOCl$_2$, methanol, r.t.-reflux; (h) (Boc)$_2$O, Et$_3$N, CH$_2$Cl$_2$, r.t.; (i) LAH, THF, r.t.; (j) NaBH$_4$, ethanol, LiCl, 0°; (k) DMSO, (COCl)$_2$, CH$_2$Cl$_2$, −60°, Hunigs base; (l) DMSO, TFAA, CH$_2$Cl$_2$, −78°, Et$_3$N; (m) DIBAL, toluene, −78°.

L-serine 2 (Scheme 1). The amino group of L-serine is initially protected as its N-tert-butoxycarbonyl (N-Boc) derivative by treatment with di-tert-butyl dicarbonate [Boc anhydride or (Boc)$_2$O] in the presence of 1 N NaOH in dioxane. The crude N-Boc-L-serine 3, when treated with a 0.6 M diazomethane solution in ether (Et$_2$O) at 0°, provides methyl ester 5 in 80–90% yield. Treatment with 2,2-dimethoxypropane (DMP) and catalytic p-toluenesulfonic acid (p-TsOH) in refluxing benzene efficiently generates the oxazolidine ring 6. Finally, reduction of the methyl ester with diisobutylaluminum hydride (DIBAL) in toluene at −78° provides the crude target oxazolidine aldehyde 8. Purification of the final product by vacuum distillation yields pure aldehyde 8 in 93–95% e.e. and in >60% yield from L-serine.

Since its initial preparation, the synthesis of this "Garner aldehyde" 8 has undergone several modifications. One is the conversion of N-Boc-L-serine into the methyl ester by treatment with potassium carbonate and methyl iodide in dimethylformamide (DMF).[4] McKillop et al.[5] later published an improved procedure for 8, which avoids the use of toxic and

---

[4] P. Garner and J. M. Park, Org. Synth. 70, 18 (1991).
[5] A. McKillop, R. J. K. Taylor, R. J. Watson, and N. Lewis, Synthesis 31 (1994).

noxious chemicals such as benzene, methyl iodide, and diazomethane. Esterification of L-serine with methanol and acetyl chloride, followed by protection with $(Boc)_2O$ in the presence of triethylamine (Et_3N), in tetrahydrofuran (THF), provides the N-Boc-L-serine methyl ester in 90% overall yield. Subsequent formation of the oxazolidine with DMP in acetone using boron trifluoride etherate $(BF_3 \cdot Et_2O)$ as a catalyst at room temperature avoids the use of benzene as solvent. A DIBAL reduction results in a 60% overall yield of aldehyde **8**.

Later, Dondoni and Perrone[6] reported a method involving a high yielding reduction–oxidation sequence, which avoids typical problems in the DIBAL reduction step, such as overreduction, incomplete reduction, and difficulty in isolating the aldehyde from the reaction mixture. Reduction of ester **6** with lithium aluminum hydride (LAH) in THF at room temperature quickly provides the alcohol **7** in 96% crude yield. Oxidation of crude **7** under Swern conditions [dimethyl sulfoxide (DMSO), oxalyl chloride, CH_2Cl_2, base, $-78°$] gives the aldehyde in 94% yield from the ester and in high enantiomeric purity (96–98% e.e.). The choice of base in the quenching of the Swern oxidation is critical for obtaining high enantiomeric purity. The authors observed that the use of Et_3N causes significant racemization, whereas the use of Hünig's base (diisopropylethylamine), followed by a cold aqueous HCl workup and phosphate buffer wash, avoids such problems. This reduction–oxidation method has the advantages of being suitable for scale-up and requiring little purification of the intermediates.

Williams et al.[7] reported a similar reduction–oxidation sequence that is also useful. Their preparation is particularly suited for the large-scale production of aldehyde **8**. The generation of the methyl ester of L-serine with thionyl chloride in methanol, followed by N-Boc protection with $(Boc)_2O$, affords N-Boc-L-serine methyl ester **5** in high yield. After subsequent preparation of acetonide **6**, treatment with sodium borohydride $(NaBH_4)$ in the presence of LiCl leads to the alcohol **7**. Oxidation of **7** under modified Swern conditions [DMSO, trifluoroacetic acid anhydride (TFAA)] produces aldehyde **8** in 60% yield.

It should be noted that N-Boc-L-serine methyl ester **5** is available commercially. Its use shortens the synthesis of **8** by two steps and is a more economical alternative. Following identical procedures with D-serine provides the antipode of **8** in equally good yields; however, for the preparation of this enantiomer, it is more economical to start with D-serine rather than the expensive N-Boc methyl ester.

[6] A. Dondoni and D. Perrone, *Synthesis* 527 (1997).
[7] L. Williams, Z. Zhang, F. Shao, P. J. Carroll, and M. M. Joullie, *Tetrahedron* **52,** 11673 (1996).

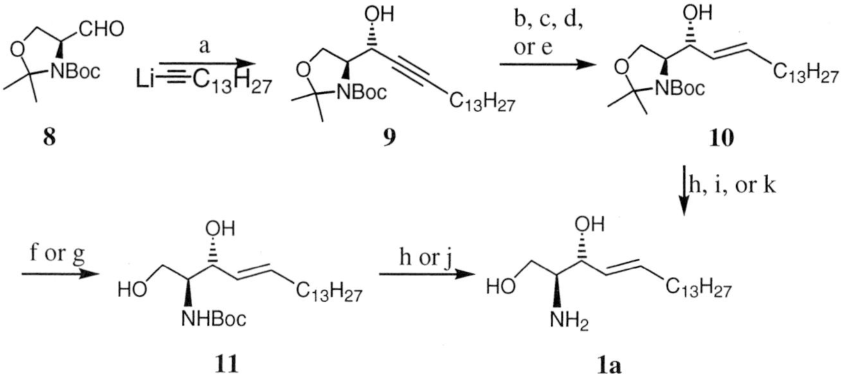

SCHEME 2. Reagents and conditions: (a) THF, HMPA, $-78°$; for *syn* addition use Et$_2$O, ZnBr$_2$; (b) Li/EtNH$_2$, $-78°$; (c) Red-Al, Et$_2$O, $0°$-r.t.; (d) Na/NH$_3$, $-78°$; (e) LAH, DME, reflux; (f) Amberlyst-15, methanol, r.t.; (g) *p*-TsOH methanol, r.t.; (h) 4 *N* HCl, methanol 50:15 v/v, r.t.; (i) 1 *N* HCl, dioxane; (j) 1 *N* HCl, EtOAc, r.t; (k) 1 *N* HCl, THF, 70°.

The next step in the synthesis of sphingoid bases is the addition of a nucleophile to the aldehyde **8**. The Garner aldehyde reacts with lithiated 1-pentadecyne at $-23°$ in THF to produce an 8:1 mixture of propargyl-alcohol diastereomers (Scheme 2).[8] The diastereomers may be separated by flash column chromatography or can be reduced directly with lithium in ethylamine at $-78°$ to give *trans*-olefin **10**.

Removal of the protecting groups with hot aqueous HCl yields D-*erythro*-sphingosine **1a** in 60% overall yield from the aldehyde **8**. The use of the other enantiomer of **8** with identical conditions produces L-*erythro*-sphingosine **1c** as the major diastereomer. It is not necessary to purify the intermediates during the sequence because a final crystallization from hexanes and ethyl acetate gives the pure product. This makes this synthesis attractive for large-scale stereoselective preparations of sphingosine.

Nucleophilic addition to the Garner aldehyde can conceivably occur on either face of the aldehyde, resulting in a mixture of diastereomers. The reaction conditions greatly affect which face of the aldehyde is preferred for nucleophilic attack. Nimkar *et al.*[9] showed that the addition of an alkyne lithiate to this aldehyde proceeds stereoselectively, giving the *erythro* product in a 9:1 ratio over the *threo* product. They rationalized the result as being due to a β chelation-controlled mechanism. Herold[10] studied the addition of metalated alkynes to the aldehyde **8** under different conditions

[8] P. Garner, J. M. Park, and E. Malecki, *J. Org. Chem.* **53**, 4395 (1988).
[9] S. Nimkar, D. Menaldino, A. H. Merrill, and D. Liotta, *Tetrahedron Lett.* **29**, 3037 (1988).
[10] P. Herold, *Helv. Chem. Acta* **71**, 354 (1988).

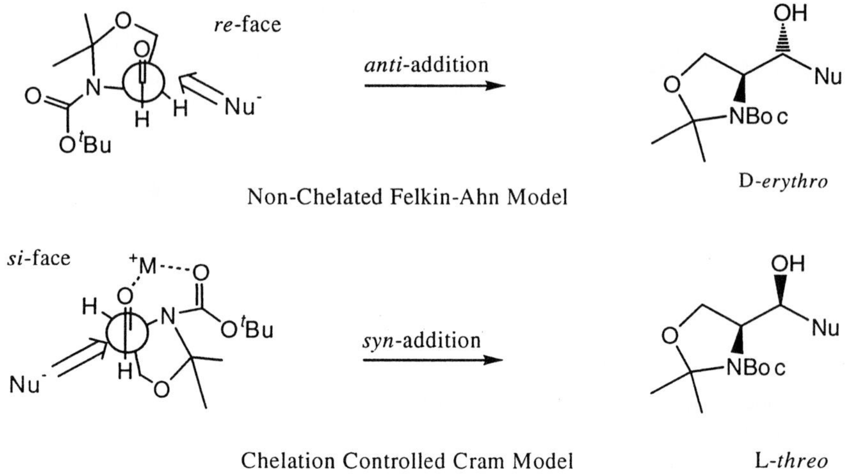

re-face   *anti*-addition   D-*erythro*

Non-Chelated Felkin-Ahn Model

si-face   *syn*-addition   L-*threo*

Chelation Controlled Cram Model

FIG. 1. Diastereoselectivity of nucleophilic additions to Garner aldehyde.

and demonstrated that the solvent and the nature of the organometallic reagent have different effects on diastereoselectivity (Fig. 1). Under strongly nonchelating conditions with lithiated alkynes and the additive hexamethylphosphoramide (HMPA), he observed the D-*erythro* product in a 20:1 ratio over the L-*threo* product. Under strongly chelating conditions with MgBr and CuI, the selectivity reverses to 20:1 L-*threo* over D-*erythro*. A chelation-controlled mechanism can explain the *threo* selectivity. In the absence of chelation, the Conforth model describes the α amino group and the carbonyl group as being antiperiplanar.[11] This configuration allows for a Felkin-Anh type attack at the least hindered *re* face of the aldehyde, which leads to the D-*erythro* product.[12,12a] Williams *et al.*[7] studied the diastereoselectivity of Grignard additions to a Garner-like analog containing a cyclohexyl oxazolidine ring. They also concluded that a Felkin-Anh model predicts an *anti* addition that leads to the D-*erythro* product. Furthermore, X-ray analysis provides evidence of chelation between the urethane carbonyl and the aldehyde. This Cram type chelated intermediate gives rise to the *syn* addition product. Garner and Ramakanth[13] and Karabastos[14] discuss further the selectivity of additions to similar aldehydes.

[11] J. W. Conforth, R. H. Conforth, and K. K. Mathew, *J. Chem. Soc.* 112 (1959).
[12] M. Cherest, H. Felkin, and N. Prudent, *Tetrahedron Lett.* 18, 2199 (1968).
[12a] N. T. Ahn, *Top. Curr. Chem.* 88, 144 (1980).
[13] P. Garner and S. Ramakanth, *J. Org. Chem.* 51, 2609 (1986).
[14] G. J. Karabatsos, *J. Am. Chem. Soc.* 89, 1367 (1967).

Herold[10] finished the syntheses of sphingosine by allowing the Garner aldehyde to react with 1-pentadecynyllithium, prepared by treating 1-pentadecyne with lithium diisopropylamide (LDA) or n-butyllithium at $-78°$. Conducting this reaction in THF with HMPA produces the *erythro*-propargyl alcohol **9**; conducting it in Et_2O in the presence of zinc bromide yields the *threo* isomer. Removal of the acetonide with Amberlyst-15 in methanol, reduction of the triple bond to the *trans* double bond with Red-Al in Et_2O, and Boc deprotection with 1 N HCl in dioxane gives sphingosines **1a** and **1b**. Treating the triple bond with Lindlar's catalyst and H_2 in ethyl acetate affords the *cis* double bond. This procedure is versatile in that it allows one to prepare the 4-*E* or 4-*Z* analogs of (2*S*,3*R*)-D-*erythro*- or (2*S*,3*S*)-D-*threo*-sphingosine. A similar preparation of **1a** utilizes dissolving metal conditions, Na/NH_3, to reduce the propargyl alcohol triple bond. This works well on a small scale; however, for larger scales, LAH reduction in dimethoxyethane (DME) gives better yields.[9]

There are several additional ways to deprotect oxazolidine-carboxylate intermediates. Treating the oxazolidine with *p*-TsOH in methanol allows selective removal of the acetonide. Deprotecting the *N*-Boc-amine with HCl in methanol or ethyl acetate then produces the sphingosine. However, both groups cleave simultaneously by treatment with 15 : 50 (v/v) methanol in 4 N HCl at room temperature, followed by evaporation of the solvent and precipitation of the resulting salt at the interface of a Et_2O and 0.5 N HCl solution.[15]

Another synthesis utilizing the oxazolidine methyl ester **6** involves the formation of a β-ketophosphonate (Scheme 3).[16] Treatment of the ester **6** with dimethyl lithiomethylphosphonate in THF results in the γ-amino-β-ketophosphonate **12**. Under basic conditions, this compound will react with aldehydes in a Horner–Wadsworth–Emmons olefination reaction to yield the enone **13** in 80% yield. The choice of base for this reaction is critical because reports show that strong bases can cause racemization.[17] A mild base for this reaction that does not result in racemization and provides sufficiently clean products is potassium carbonate in acetonitrile.[18] Furthermore, the quality of the potassium carbonate is crucial, requiring activation by drying in the oven overnight. Fortunately, the resulting enone does not require chromatographic separation, which can also cause racemization. A stereoselective reduction of the ketone allows access to either the D-*erythro*

[15] H.-E. Radunz, R. M. Devant, and V. Eiermann, *Liebigs Ann. Chem.* 1103 (1988).
[16] A. M. P. Koskinen and M. J. Krische, *Synlett* 665 (1990).
[17] C. H. Heathcock and T. W. van Geldern, *Heterocycles* **25,** 75 (1987).
[18] A. M. P. Koskinen and P. M. Koskinen, *Synlett* 501 (1993).

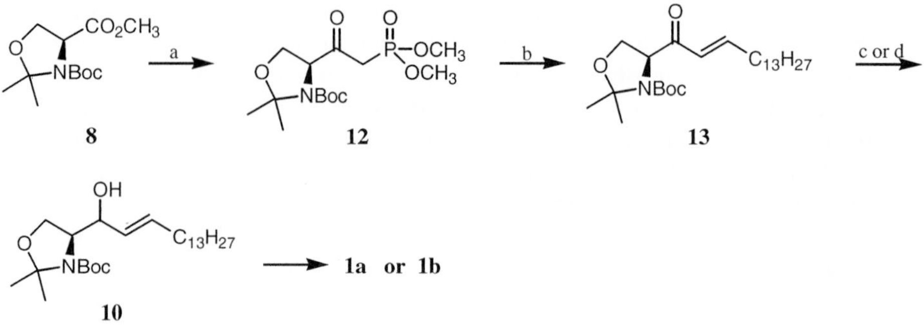

SCHEME 3. Reagents and conditions: (a) $(CH_3O)_2P(O)CH_2Li$, THF; (b) $C_{13}H_{27}CHO$, CH_3CN, K_2CO_3; (c) for *syn* product: L-selectride, THF or nBu_3BHK, THF; (d) for *anti* product: DIBAL, toluene; (e) acid hydrolysis.

or the L-*threo* isomer of **10**.[19] Studies of various reducing conditions show that obtaining the masked *syn* amino alcohol requires sterically bulky reducing agents. Bulky borohydrides, such as potassium tri-*n*-butyl borohydride or L-selectride in THF, give good diastereoselectivity for the L-*threo* product. However, reduction with DIBAL in toluene favors the *anti*-amino alcohol. This procedure is advantageous in that it allows the synthesis of sphingosine **1a** or **1b**.

The procedure of Dondoni *et al.*[20] utilizes a thiazole ring as a formyl anion synthon that adds to aldehyde **8** (Scheme 4). Treatment of thiazole **14** with *n*-butyllithium and then trimethylsilyl chloride (TMSCl) provides 2-trimethylsilyl-thiazole **15** in 93% yield. Desilylation of the thiazole with tetrabutylammonium fluoride (TBAF) in the presence of the Garner aldehyde gives a 92:8 *erythro/threo* ratio of amino-alcohol diastereomers in 85% yield. The major *anti* addition product is purified by crystallization from CH_2Cl_2/hexanes.[21] The excellent *anti* selectivity is consistent with a nonchelated Felkin-Anh model (Fig. 1). A series of reactions unmasks the thiazole group providing the aldehyde **17**.[22] N-methylation with excess methyl iodide, sodium borohydride reduction, and finally hydrolysis in the presence of mercuric chloride affords **17** in 65% overall yield. When subjected to Wittig olefination conditions with hexadecanylidene-triphenylphosphorane and excess lithium bromide, the aldehyde gives the C_{20}-sphingosine intermediate **18** with high *trans* selectivity and in 31% yield. This

[19] A. M. P. Koskinen and P. M. Koskinen, *Tetrahedron Lett.* **34,** 6765 (1993).
[20] A. Dondoni, G. Fantin, M. Fogagnolo, A. Medici, and P. Pedrini, *J. Org. Chem.* **53,** 1748 (1988).
[21] A. Dondoni, G. Fantin, M. Fognolo, and A. Medici, *J. Chem. Soc. Chem. Commun.* 10 (1988).
[22] A. Dondoni, M. Fofafnolo, A. Medici, and P. Pedrini, *Tetrahedron Lett.* **26,** 5477 (1985).

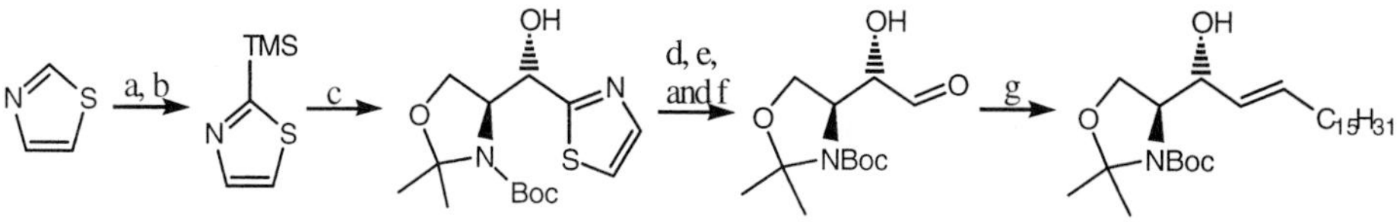

SCHEME 4. Reagents and conditions: (a) n-BuLi, THF, $-78°$; (b) TMSCl; (c) 8, CH_2Cl_2, TBAF, r.t.; (d) MeI, CH_3CN, reflux; (e) $NaBH_4$, methanol, $-10°$; (f) $HgCl_2$, CH_3CN-H_2O, r.t.; (g) $C_{15}H_{31}$ CH$=$PPh$_3$, LiBr, Et$_2$O-toluene, $-30°$.

route is versatile in that the thiazole ring is stable to a variety of conditions. Alternatively, oxidation to the ketone, followed by selective reduction, affords the masked *syn* amino alcohol. Finally, another advantage of this method is that aldehyde **17** is a useful chiron for the synthesis of amino sugars.[23]

The double bond between C-4 and C-5 can be generated by alkyne reduction or by elaborating the Garner aldehyde via Wittig or Horner–Wadsworth–Emmons reactions, as described in the methods described earlier. However, one of the earliest syntheses of sphingolipids provides the double bond directly thorough the addition of vinyl nucleophiles to serine-derived aldehydes.[24] In the first serine-derived synthesis of sphingosine, Newman[25] protected L-serine by *N*-phthaloylation and *O*-acylation with phthalimid-*N*-ethoxycarbonyl and acetic anhydride, respectively. The aldehyde, produced by conversion to the acid chloride with thionyl chloride in benzene, followed by hydrogenation, can be reacted with *trans*-pentadecyldiisobutylalane.[25] The preparation of this vinylalane by treating the corresponding alkyne with DIBAL is a well-established procedure.[26,26a] Chromatographic separation of diastereomers, followed by removal of the protecting groups by methanolysis and treatment with hydrazine, results in sphingosine **1** in good yield.

The second published synthesis of sphingosine using serine, by Tkaczuk and Thorton,[27] also utilizes a vinylalane addition (Scheme 5). Synthesis begins with the protection of L-serine as an oxazoline using a known procedure.[28] Fisher esterification, treatment with benziminoethyl ether, and DI-

[23] A. Dondoni, G. Fantin, M. Fogagnolo, and P. Pedrini, *J. Org. Chem.* **55**, 1439 (1990).
[24] H. Newman, *Tetrahedron Lett.* **47**, 4571 (1971).
[25] H. Newman, *J. Am. Chem. Soc.* **95**, 4098 (1973).
[26] G. Wilke and H. Muller, *Liebigs Ann. Chem.* **629**, 22 (1960).
[26a] G. Zweifel and R. B. Steele, *J. Am. Chem. Soc.* **89**, 2754 (1967).
[27] P. Tkaczuk and E. R. Thornton, *J. Org. Chem.* **46**, 4393 (1981).
[28] D. F. Elliott, *J. Chem. Soc.* 589 (1949).

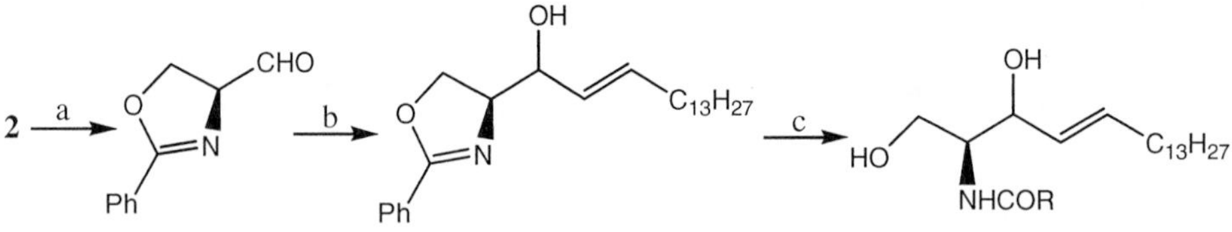

SCHEME 5. Reagents and conditions: (a) HCl, methanol; HN=C(Ph)OEt, CH_2Cl_2, H_2O, r.t.; DIBAL, toluene-hexane, $-78°$; (b) $C_{13}H_{27}CH=CHAl(i\text{-}Bu)_2$; (c) 2 N HCl, THF, r.t; $p\text{-}NO_2\text{-}Ph\text{-}CO_2R$, pyr; $NaOCH_3$, methanol.

BAL reduction of the ester provide the aldehyde **19**. When using the oxazoline aldehyde **19** in their synthesis of a naturally occurring cerebroside, Mori and Funaki[29] found that **19** was unstable and must be used immediately. Therefore, treatment of crude aldehyde **19** with *trans*-pentadecyldiisobutylalane results in **20** as a 1:1 mixture of diastereomers that can be separated by column chromatography. Deprotection with 2 N HCl in THF affords the 1-benzoate hydrochloride salts of sphingosines **1a** and **1b**. This synthesis of sphingosine provides a quick, efficient, high-yielding route to sphingosine and other sphingolipids. Although this method is not diastereoselective, it is possible to separate the diastereomers by column chromatography. Their synthesis proceeds toward the production of ceramide **21** by acylation of the nitrogen and basic deprotection of the C-1 hydroxide with methanol and sodium methoxide.

In his preparation of sphingomyelin analogs, Bruzik[30] utilizes the same procedure. After protection of allylic alcohol **20**, opening the oxazoline ring, acylation of the amine, and removal of the benzoate ester, Bruzik then selectively phosphorylates the C-1 hydroxyl of **21** to produce a series of sphingomyelins. Another illustration of the versatility of this method by Kozikowski and Wu[31] involves the conversion of allyl alcohol **20** into a 5-fluoro derivative utilizing the fluorinating reagent, 2-chloro-1,1,2-trifluoroethylamine via an allylic rearrangement. Removal of the oxazoline ring with 2 N HCl in THF and 5% KOH in methanol/water provides the corresponding 5-fluorosphingosine derivatives.

Sphingosines can also be accessed from L-serine via an α amino acid isoxazolidine derivative. Boutin and Rapoport[32] use commercially available

[29] K. Mori and Y. Funaki, *Tetrahedron* **41**, 2379 (1985).
[30] K. S. Bruzik, *J. Chem. Soc., Perkin Trans. 1* 423 (1988).
[31] A. P. Kozikowski and J.-P. Wu, *Tetrahedron Lett.* **31**, 4309 (1990).
[32] R. H. Boutin and H. Rapoport, *J. Org. Chem.* **51**, 5320 (1986).

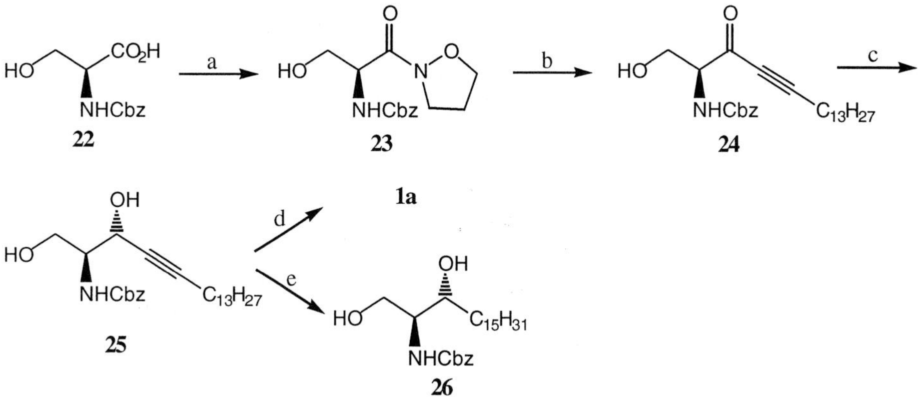

SCHEME 6. Reagents and conditions: (a) 1. isobutyl chloroformate, N-methylmorpholine, THF, $-15°$; 2. isoxazolidine–HCl, THF/H$_2$O, K$_2$CO$_3$, r.t; (b) lithium pentadecyne, THF, $-23°$; (c) NaBH$_4$, methanol, $0°$; (d) Li/NH$_3$, THF, reflux; (e) H$_2$, methanol, Pt/C, H$_2$ (1 atm), r.t.

N-benzyloxycarbonyl-(Cbz)-L-serine in their synthesis (Scheme 6). Earlier work demonstrated the preparation of the isoxazolidine **23** in 76% yield by adding the isoxazolidine hydrochloride and potassium carbonate in THF to a solution of N-Cbz-L-serine, N-methyl morpholine, and isobutyl chloroformate. These types of N-alkylcarbamate amino acid isoxazolidines give good yields with little loss of enantiomeric purity through enolization.[33] Additions to **23** with various alkynyl lithiates provide the ynone **24** in 90% yield. After extensive studies of several reducing agents, sodium borohydride was found to be the most practical choice, providing high yields and good *anti* selectivity. Chromatographic separation of the propargyl-alcohol diastereomers **25** is difficult on a silica gel; however, a sodium borate-impregnated silica gel column does allow for their separation. The removal of the Cbz group and the reduction of the triple bond under dissolving metal conditions occur simultaneously. Reductions of this system under these conditions are problematic due to incomplete reductions, although use of a great excess of lithium and long reaction times improve yields considerably. Reduction of propargyl alcohol **25** with H$_2$ and platinum on carbon provides dihydrosphingosine **26** (sphinganine) in 95% yield. This synthetic route provides sphingosine in 22% overall yield and with >99% e.e. These syntheses involving the α,β-ynone system are flexible in that they allow some control of the C-3 hydroxyl through stereoselective reduction, various alkyl chain lengths, and control of the double bond geometry through different reducing methods.

[33] T. L. Cupps, R. H. Boutin, and H. Rapoport, *J. Org. Chem.* **50,** 3972 (1985).

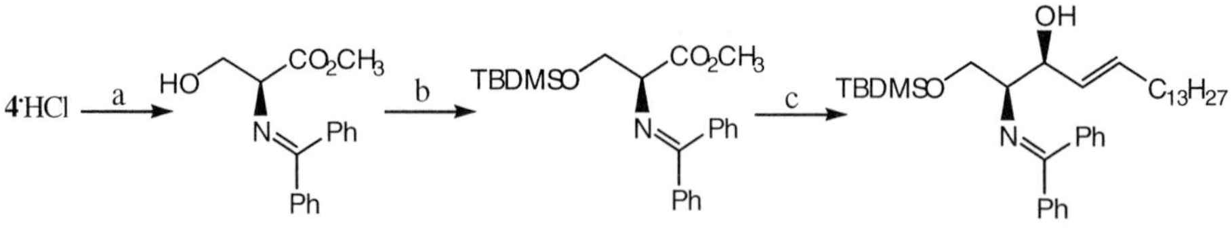

SCHEME 7. Reagents and conditions: (a) 1. diphenylketimine, CH_2Cl_2, r.t; (b) TBDMSCl, imidazole, DMF, r.t; (c) 1:1 DIBAL-TRIBAL, CH_2Cl_2, $-78°$, $C_{13}H_{27}CH=CHLi$; (d) 1 N HCl, THF, r.t.

Polt *et al.*[34] employs the protection of L-serine as its Schiff base derivative in his synthesis of sphingosine (Scheme 7). The commercially available hydrochloride salt of L-serine methyl ester, **4**, is treated with diphenylketimine in CH_2Cl_2 to provide the Schiff base **27**. Reaction of **27** with *tert*-butyldimethylsilyl chloride (TBDMSCl) in the presence of imidazole in DMF protects the primary hydroxyl group as the silyl ether **28**. Treatment of the esters with DIBAL-TRIBAL in CH_2Cl_2 at $-78°$ produces an aluminoxy acetal that can subsequently react with a vinyl lithiate to afford *threo*-sphingosine derivative **29** with exceptionally high diastereoselectivity (20:1). A thorough discussion of the selectivity and mechanistic details of this reaction is given.[34] Hydrolysis then provides sphingosine **1b**. This is a convenient one-pot procedure, which provides (2S,3R)-L-*threo*-sphingosine **1b** in 60% overall yield, and in the absence of racemization. This procedure is also amenable to 1-deoxysphingosine derivatives using alanine as the starting amino acid.

The last example of a serine-derived sphingosine synthesis is a biomimetic approach.[35] It provides access to all four stereoisomers by judicious choice of the N-protecting group and the desired starting serine enantiomer (Scheme 8). The introduction of the double bond is completely *trans* selective. This is advantageous because no separation of isomers is required, as may be the case with methods that utilize Wittig olefinations. The synthesis begins with O-benzyl-L-serine, which is treated with trityl chloride (TrCl) in Et_3N, to provide N-trityl-O-benzyl serine **31**. Then, a THF solution of **31** is allowed to react with the lithium enolate of allyl acetate in the presence of carbonyldiimidazole (CDI) to afford **32**. Alkylation of **32** with 1-tetradecyltriflate gives **33** in 78% yield. The crude product then undergoes deallylation and decarboxylation, on exposure to $Pd(PPh_3)$ under thermodynamic

[34] R. Polt, M. A. Peterson, and L. DeYoung, *J. Org. Chem.* **57**, 5469 (1992).
[35] R. V. Hoffman and J. Tao, *J. Org. Chem.* **63**, 3979 (1998).

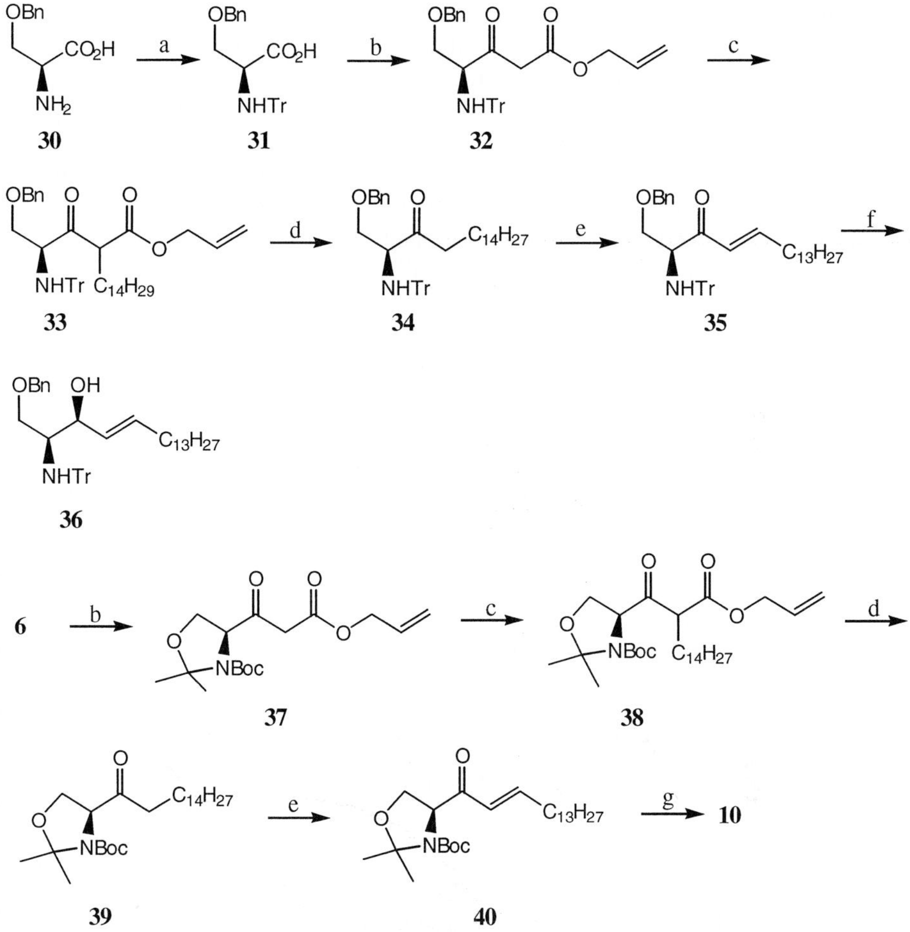

SCHEME 8. Reagents and conditions: (a) TMSCl, TrCl, Et₃N; (b) Im₂CO, LiCH₂CO₂CH₂CH=CH₂; (c) NaH, TfOCH₂C₁₃H₂₇, THF; (d) Pd(PPh₃), morpholine, r.t; (e) NaHMDS, THF, −78°, TMSCl, Pb(OAc)₂, CH₃CN; (f) NaBH₄, CeCl₃ methanol, −20°; (g) NaBH₄, methanol.

conditions, to yield 3-ketosphinganine **34** in 92% yield. Generation of the trimethylsilylenol ether with sodium hexamethyldisilazane (NaHMDS) and TMSCl, followed by treatment with lead acetate, selectively generates *trans*-olefin **35**. Sodium borohydride reduction in the presence of cerium chloride provides the *threo* isomer in 85% yield and with good diastereoselectivity. Moreover, utilizing this protocol with ester **6**, an intermediate in the preparation of Garner aldehyde, allows for a chelated transition state in the sodium borohydride reduction and results in the *erythro* product.[36] Thus, use of the antipodes of the serine starting materials provides access to sphingosines **1c** and **1d**. This synthesis proceeds with 30% overall yield and each diastereomer is available with equal ease and selectivity. Finally, another useful feature of this synthesis and others that involve selective borohydride reductions is that they allow for the preparation of labeled sphingolipids by reducing the ketone with sodium borodeuteride or tritide. Furthermore, the use of ^{14}C-labeled allyl acetate would create a labeled carbon at C-4.

Sphingosine from Sugars

The advantages of preparing sphingosine analogs from sugars are similar to those of amino acids. Sugars are readily available, relatively inexpensive, and optically pure. Typical synthetic protocols involve using the chiral sugar as a template for the sphingosine head group on which alkyl groups are attached in various ways. Most syntheses contain a carbon–carbon bond cleavage that results in an aldehyde that, in turn, can react with ylides or Grignard reagents to produce the aliphatic portion of the sphingosine. These methods, like those from serine, are useful for the preparation of optically pure sphingolipids. Many are amenable to scale-up and often contain intermediates that are useful for preparations of more complex sphingolipids and other natural products. Unfortunately, many of these methods contain multiple protection and deprotection steps, and, furthermore, several intermediates require purification from unwanted side products.

Sphingosines from D-Xylose

The use of D-xylose to prepare sphingoid bases by Kiso *et al.*[37] is a useful example (Scheme 9). Treatment of D-xylose with DMP and catalytic *p*-TsOH in DMF gives a chromatographically separable mixture of acetonated products.[38] The major product, **41**, produced in 35% yield, undergoes a carbon–carbon bond cleavage with sodium metaperiodate in methanol. The resulting D-threose (**42**) reacts with tetradecyltriphenylphosphonium ylide in a Wittig olefination reaction in 35% yield to give olefin **43** as a mixture of *cis* and *trans* isomers. This mixture of isomeric alcohols is activated with methanesulfonyl chloride (MsCl), followed by S$_N$2 substitution

[36] D. A. Evans and K. T. Chapman, *Tetrah. Lett.* **27**, 5939 (1986).
[37] M. Kiso, A. Nakamura, J. Nakamura, Y. Tomita, and A. Hasegawa, *J. Carbohydr. Chem.* **5**, 335 (1986).
[38] M. Kiso and A. Hasegawa, *Carbohydr. Res.* **52**, 95 (1976).

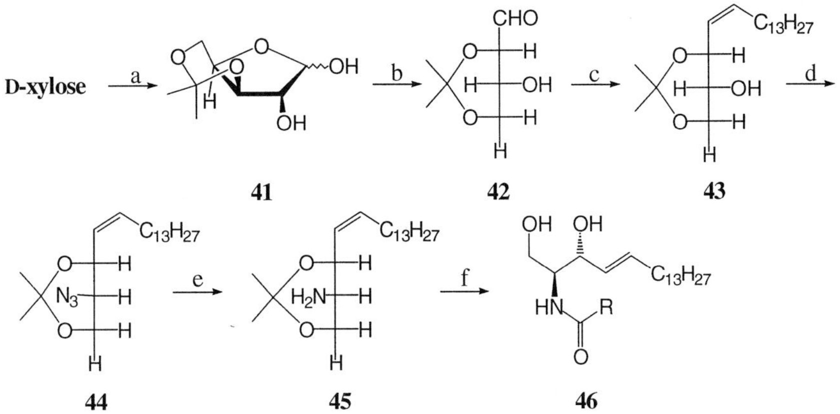

SCHEME 9. Reagents and conditions: (a) DMP, *p*-TsOH, DMF, 45°; (b) NaIO$_4$, methanol; (c) C$_{14}$H$_{29}$PPh$_3$, *n*-BuLi, THF; (d) MsCl, pyr, then NaN$_3$, DMF, 110°; (e) NaBH$_4$, 2-propanol, reflux; (f) RCO$_2$H, DCC, CH$_2$Cl$_2$/dioxane 1:1; HCl.

with sodium azide. (Note: attempts to prepare D-*threo*-sphingosine by inverting the stereochemistry of the free hydroxyl group under Mitsunobu conditions fail due to an acetal group rearrangement.)[39] Purification of azide **44** on silica gel affords the *trans* isomer in 85% yield and the *cis* isomer (shown in Scheme 9) in 83% yield. The azide group is reduced efficiently with sodium borohydride in refluxing 2-propanol. Then, removal of the acetonide of **45** by acid hydrolysis yields **1a**. *N*-acylation *via* dicyclohexyldicarbodiimide (DCC) coupling with a desired fatty acid, followed by deprotection, produces ceramide **46**. This method has the advantage of requiring only a few steps, although several chromatographic separations are required. It also allows for the preparation of azido sphingosines by simple acid hydrolysis of **44**.

Another procedure derived from D-Xylose involves the production of the chloro derivative **47** (Scheme 10).[40] This intermediate undergoes a base-induced elimination reaction with lithium amide in ammonia.[41] The resulting propargyl alcohol **48** can then participate in an alkylation reaction with 1-bromotridecane. A *trans*-selective reduction of subsequently formed **49** with LAH provides olefin **50**, which, when treated with *p*-TsOH, reveals the triol **51**. On treatment with benzaldehyde dimethyl acetal and catalytic *p*-TsOH, the corresponding 1,3-benzylidene derivative **52** is produced as the major product. The C-2 hydroxyl is transformed to azide **53** by conversion to

[39] Y.-L. Li, X.-L. Sun, and Y.-L. Wu, *Tetrahedron* **50**, 10727 (1994).
[40] J. S. Yadev, M. C. Chander, and B. V. Joshi, *Tetrahedron Lett.* **29**, 2737 (1988).
[41] J. S. Yadev, M. C. Chander, and C. S. Rao, *Tetrahedron* **30**, 5455 (1989).

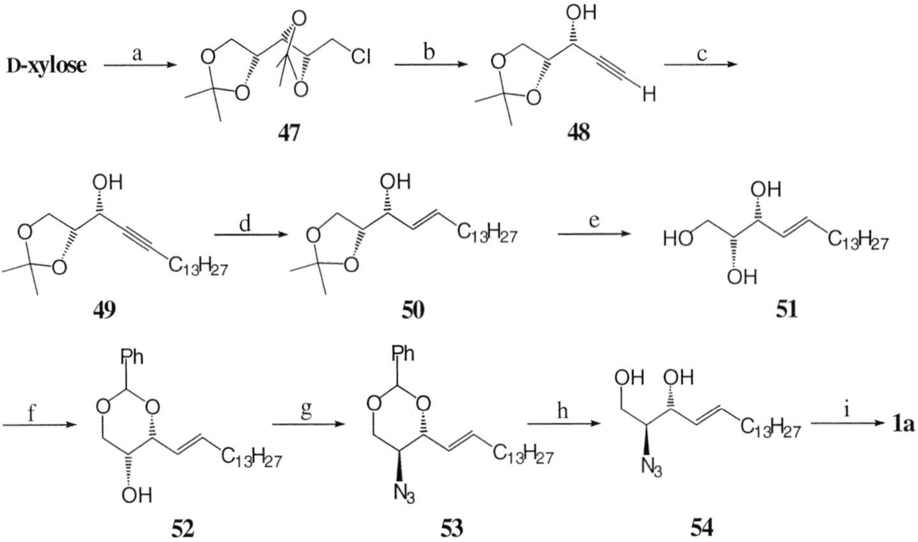

SCHEME 10. Reagents and conditions: (a) DMP, p-TsOH, then PPh$_3$, CCl$_4$, reflux; (b) LiNH$_2$, NH$_3$; (c) n-BuLi, C$_{13}$H$_{27}$Br; (d) LAH; (e) p-TsOH; (f) PhCH(OCH$_3$)$_2$, p-TsOH, CH$_2$Cl$_2$; (g) TsCl, pyr, CH$_2$Cl$_2$, then NaN$_3$, DMF; (h) 3N HCl, THF; (i) LAH.

its toluenesulfonate (tosylate), followed by displacement with sodium azide. Removal of the benzylidene with 3 N HCl followed by LAH reduction of the azide to the amine yields D-*erythro*-sphingosine.[42] Identical procedures starting from D-arabinose provide L-*threo*-sphingosine **1b**.

A useful synthesis using D-xylose by Li *et al.*[39] allows for the large-scale preparation of (2R,3S)-L-*erythro*-sphingosine (Scheme 11). It begins with the preparation of D-xylose diethyl dithioacetal **55** from D-xylose, HCl, and ethyl mercaptan. Treating **55** with copper sulfate and sulfuric acid in acetone results in 2,3-4,5-di-*O*-isopropylidene **56**. Removal of the thioacetal with mercuric oxide and boron trifluoride etherate provides aldehyde **57**.[43] Next, Wittig homologation with tetradecyltriphenylphosphine ylide yields the *cis* olefin as the major product. Isomerization of the *cis* double bond to the *trans* isomer **58** occurs when treated with phenyl sulfide in a cyclohexane/dioxane solution and irradiated by a mercury lamp. Carbon–carbon bond cleavage with periodic acid and subsequent sodium borohydride reduction of the resulting aldehyde give the acetonide **59** in 62% yield. Deprotection of the acetonide with acetic acid produces the triol **60**, which can be converted in a few steps to the (2R,3R)-D-*threo*-sphingosine **1d**.

[42] J. S. Yadev, D. Vidyanand, and D. Rajagopal, *Tetrahedron Lett.* **34**, 1191 (1993).
[43] P. Rollin and J.-R. Pougny, *Tetrahedron* **42**, 3479 (1986).

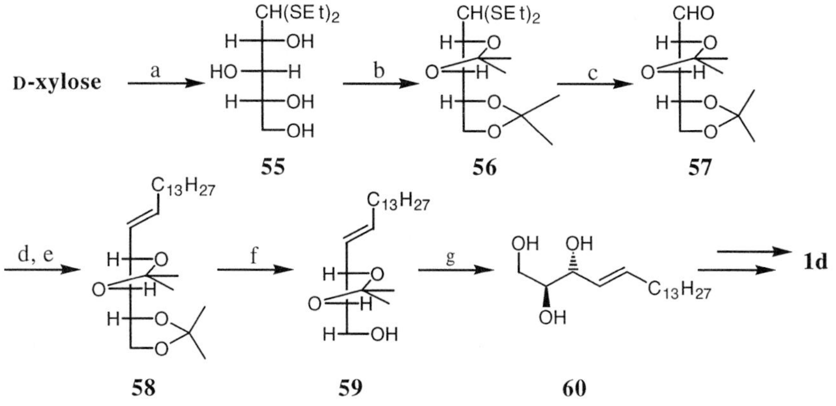

D-xylose →(a)→ **55** →(b)→ **56** →(c)→ **57**

→(d, e)→ **58** →(f)→ **59** →(g)→ **60** → **1d**

SCHEME 11. Reagents and conditions: (a) EtSH, HCl; (b) $CuSO_4$, H_2SO_4, acetone; (c) HgO, $BF_3 \cdot Et_2O$, THF, H_2O, r.t.; (d) $C_{14}H_{29}PPh_3Br$, n-BuLi, THF, $-30°$; (e) PhSSPh, cyclohexane–dioxane 19:1, $h\nu$; (f) H_5IO_6, Et_2O, r.t., then $NaBH_4$, 2-propanol, $0°$; (g) 80% HOAc-H_2O, $60°$.

Sphingosines from D-Glucose

Several groups have prepared sphingosine from D-glucose derivatives. As with the xylose series, these syntheses all involve using the optically pure glucose to serve as a template for asymmetric sphingolipid construction. These protocols also utilize a metaperiodate bond cleavage reaction and a Wittig olefination. Kioke et al.[44] use the readily available diisopropylidene derivative of glucose **61** as their starting material (Scheme 12). Regioselective bond cleavage results in aldehyde **62**. Wittig olefination of this aldehyde with tetradecylidenetriphenylphosphorane in THF gives a mixture of E- and Z-olefins. A photoinduced isomerization of the double bond according to Li's procedure affords olefin **63** as a 15:1 trans/cis mixture in 87% overall yield. After conversion to the corresponding mesylate, acid hydrolysis with acetic acid releases the corresponding diol. Oxidation with sodium metaperiodate in ethanol and subsequent reduction with sodium borohydride give diol **64**. Protection of the diol moiety as its 1-ethoxyethyl derivative allows for the regioselective S_N2 displacement of the mesylate with sodium azide. The azide is reduced to the amine on treatment with sodium borohydride. At this point, amine **66** can be acylated with the corresponding acyl chloride or the ethoxyethyl groups can be removed with Amberlyst-15 to give D-erythro-sphingosine. A similar procedure with the cis-olefin provides 4Z-D-erythro-sphingosine. A nearly identical procedure

[44] K. Koike, M. Numata, M. Sugimoto, Y. Nakahara, and T. Ogawa, Carbohydr. Res. **158**, 113 (1986).

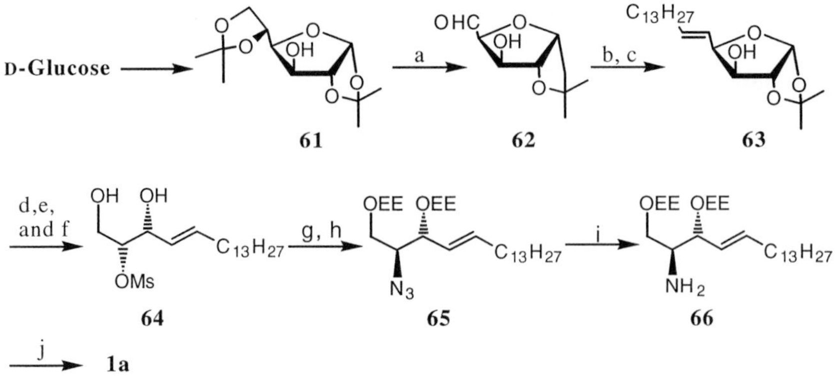

SCHEME 12. Reagents and conditions: (a) H_5IO_6, Et_2O; (b) $C_{14}H_{28}PPh_3Br$, n-BuLi, THF; (c) PhSSPh, cyclohexane–dioxane 19:1, hν, N_2; (d) MsCl, pyr, 0°-r.t; (e) H_2O-HOAc 1:4, 80°; (f) $NaIO_4$, $NaHCO_3$, $EtOH$-H_2O 4:1, $-5°$, then $NaBH_4$, ethanol, 0°; (g) PPTS, ethyl vinyl ether, CH_2Cl_2, r.t.; (h) NaN_3, DMF, 80°; (i) $NaBH_4$, 2-propanol, reflux; (j) Amberlyst-15, 1:1 CH_2Cl_2-MeOH, r.t.

performed by Li et al.[39] also produces the D-*erythro*-sphingosine in good optical purity.

Several related methods utilize the commercially available glucose de-rivative, D-glucosamine. This is a very useful starting material in that the amine functionality is present at the outset, and through appropriate selec-tion of protecting groups, it can be used to control stereochemistry at two stereocenters. One preparation protects the amine functionality as the Cbz derivative, followed by protection of the 1-hydroxyl group as the allyic ether (Scheme 13).[45] Next, treatment with benzaldehyde in the presence of zinc chloride affords 4,6-benzylidene-2-benzyloxycarbonylamino-2-deoxy-α-gluco pyranoside (68). Standard Swern conditions (oxalyl chloride, DMSO, CH_2Cl_2, Et_3N) can be used to oxidize the 3-hydroxyl group to ketone 69. Sodium borohydride then reduces the ketone to alcohol 70, which can be purified by chromatography. Protection of the 3-hydroxyl group takes advantage of the neighboring amine functionality to create a cyclic carbamate tether in the presence of sodium hydride forming 71. The allyl protecting group, once isomerized to a propenyl group with tris(triphe-nylphosphine) rhodium chloride and diazabicyclo[2,2,2]octane (DABCO), can be hydrolyzed simultaneously with the benzylidene group on treatment with HCl releasing 72. After protection of the 6-hydroxyl group with *tert*-butyldiphenylsilyl chloride (TBDPSCl), sodium borohydride reduction of the hemiacetal provides diol 74. Oxidative cleavage of the diol affords

[45] T. Sugawra and M. Narisada, *Carbohydr. Res.* **194,** 125 (1989).

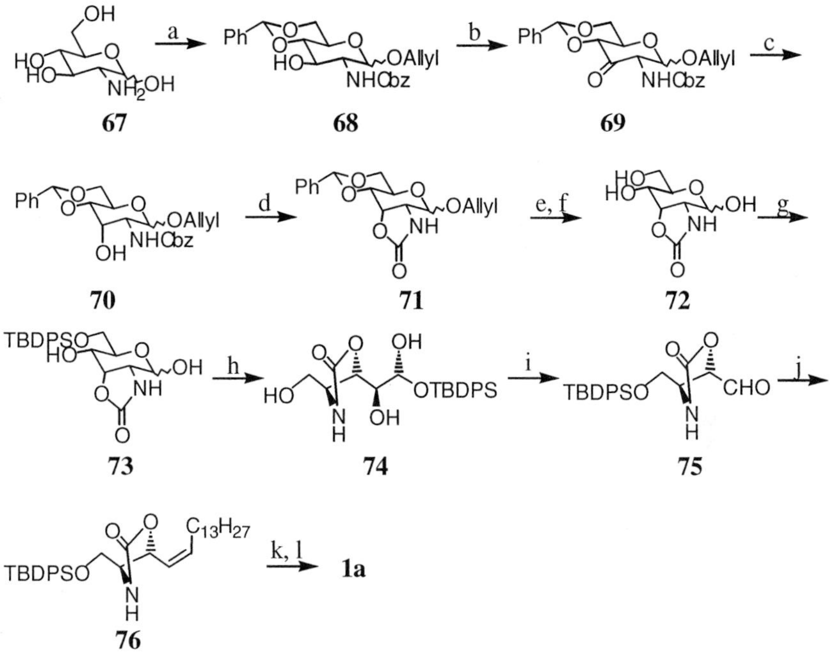

SCHEME 13. Reagents and conditions: (a) Cbz, allyl; (b) DMSO, $(COCl)_2$, CH_2Cl_2, $-78°$; (c) $NaBH_4$, methanol, $0°$; (d) DMF, NaH, $70°$; (e) tris(PPh_3)RhCl, DABCO, EtOH-benzene, H_2O; (f) HCl, methanol; (g) TBDPSCl, imidazole, DMF, r.t.; (h) $NaBH_4$, methanol, $0°$; (i) $NaIO_4$, aq. methanol; (j) $C_{14}H_{29}PPh_3$, n-BuLi, THF, r.t; (k) 1-trimethyltetrazole-5-yl disulfide, benzene, AIBN, reflux, or chromatograph; (l) TBAF, THF, NaOH, ethanol.

aldehyde **75**, which then undergoes a Wittig olefination. This reaction unfortunately proceeds in low yields and results in the formation of the Z-olefin as the major product. These isomers can be separated or isomerized to the E-isomer with 1-trimethyltetrazole-5-yl disulfide and AIBN in refluxing benzene. The synthesis of sphingosine **1a** is complete after removal of the silyl ether with TBAF and base hydrolysis of the carbamate.

The procedure of Murakami et al.[46] uses D-glucosamine in which the amine is protected with $(Cbz)_2O$, and the 5,6-O-ethylidene moiety is formed on treatment with paraldehyde and sulfuric acid (Scheme 14). Reductive opening of the hemiacetal with sodium borohydride in 2-propanol affords the triol **78**. This same reaction does not proceed to completion when carried out in methanol. Protection of the primary alcohol with TBDPSCl allows for the subsequent mesylation of the two secondary alcohols of **79**

[46] T. Murakami, H. Minamikawa, and M. Hato, *Tetrahedron Lett.* **35,** 745 (1994).

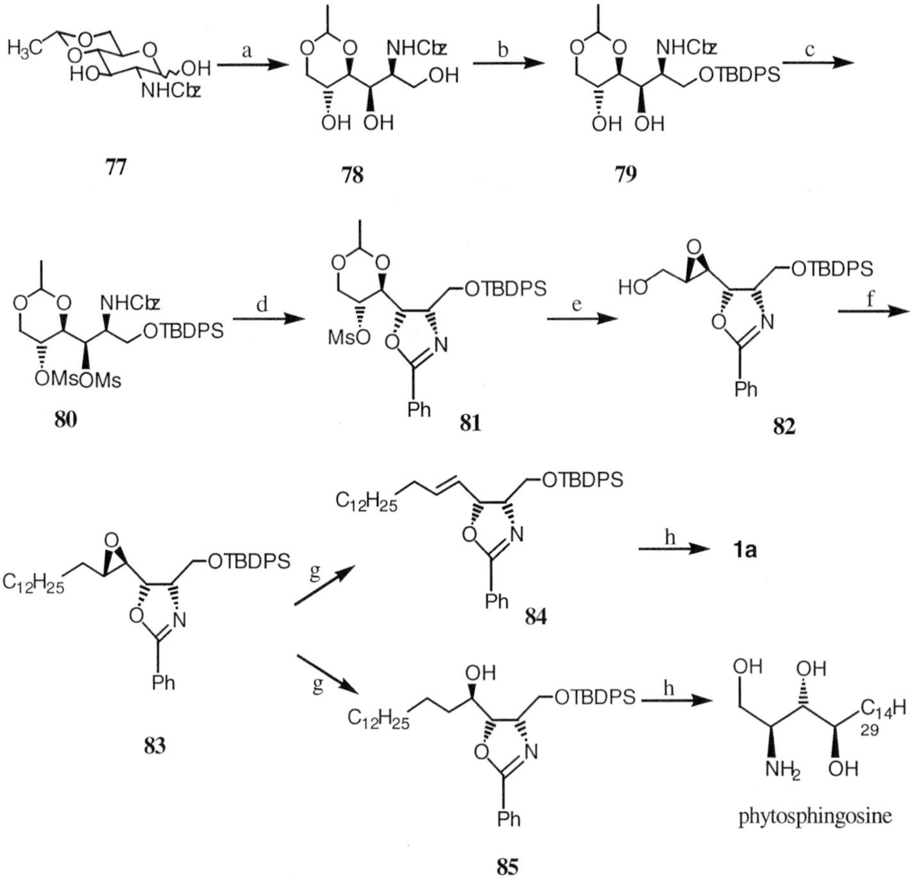

SCHEME 14. Reagents and conditions: (a) NaBH$_4$, 2-propanol H$_2$O 5:1, 0°; (b) TBDPSCl, pyr, CH$_2$Cl$_2$, r.t.; (c) MsCl, Et$_3$N, 0°; (d) pyr, Et$_3$N, toluene, 110°; (e) TiCl$_4$, PhSH, CH$_2$Cl$_2$, 0°, then K$_2$CO$_3$, methanol, 0°; (f) TsCl, DMAP, Et$_3$N, CH$_2$Cl$_2$, 0°, then C$_{12}$H$_{25}$MgBr, CuBr, THF, −30°–0°; (g) NaI, TMSCl, aq. CH$_3$CN, 0°–10°; (h) POCl$_3$, pyr, CH$_3$CN, 0°–15°; (i) n-Bu$_3$SnH, AIBN, toluene, 60°; (j) 4 N HCl, THF, r.t.; then NaOH, aq. ethanol, 95°.

with MsCl. When heated under basic conditions, the dimesylate cyclizes to phenyl oxazoline **81**. Because this compound is unstable to chromatography, it is best to remove the ethylidene group with TiCl$_4$ and thiophenol, which then forms epoxide **82** when treated with potassium carbonate. Grignard homologation of the tosylate derivative of **82** with dodecylmagnesium bromide in the presence of copper bromide results in **83** in 84% yield. A reaction with sodium iodide and TMSCl in aqueous acetonitrile, followed by POCl$_3$ in pyridine, leads to the E-olefin **84** in 85% yield. The protecting

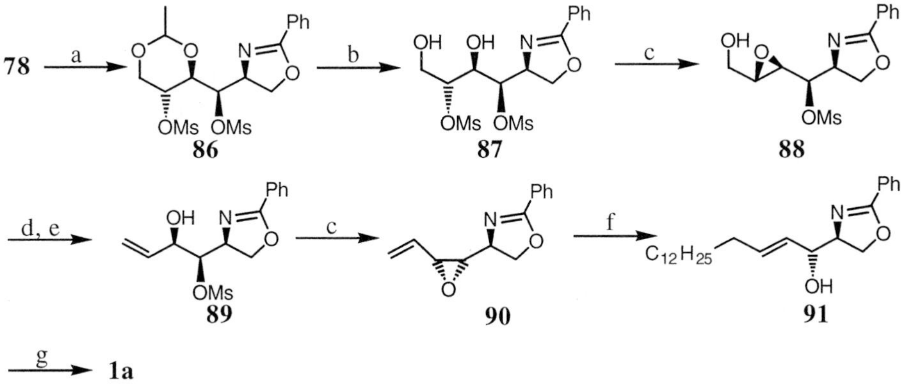

Scheme 15. Reagents and conditions: (a) MsCl, Et₃N, CH₂Cl₂, −10°; (b) TiCl₄, PhSH, CH₂Cl₂, 0°; (c) K₂CO₃, methanol, 0°; (d) I₂, PPh₃, pyr, CH₂Cl₂, 0°; (e) *n*-BuLi, THF, −70°; (f) C₁₂H₂₅MgBr, CuCN, THF, −70° to −20°; (g) 2 *N* HCl, THF, r.t., then NaOH, methanol/ H₂O (1:1), reflux.

group removal involves an initial acid hydrolysis to open the oxazoline ring, followed by a strongly basic hydrolysis, producing D-*erythro*-sphingosine **1a** in a 31% overall yield. The advantages of this method include high yields as well as the preferential formation of *E* double bond formation, thus rendering *Z–E* isomerization unnecessary.

Another procedure from Murakami and Hato[47] differs from the first in that it involves the preparation of a trimesylate from **78** with MsCl, which quickly cyclizes to form the oxazoline (Scheme 15). After the removal of the ethylidene as described earlier base treatment of the diol provides the *trans*-epoxide. The primary alcohol can be converted to the iodide when exposed to iodine and triphenylphosphine. Treatment with *n*-butyl lithium induces a reductive elimination of the epoxy iodide to the corresponding allylic alcohol. This three-step transformation is preferred for large-scale procedures; however, for small scale, a one-pot treatment with iodine and triphenylphosphine affords the allylic alcohol in 70% yield. The *cis*-epoxide can then be formed under basic conditions. Vinyl-epoxide **90** undergoes a 1,4-addition with the organocopper reagent, prepared from dodecylmagnesium bromide and copper cyanide. The resulting *E*-olefin **91** is deprotected with 2 *N* HCl, and subsequent treatment with NaOH yields D-*erythro*-sphingosine **1a**. This particular synthesis is well suited for the preparation of gram quantities of sphingolipids.

[47] T. Murakami and M. Hato, *J. Chem. Soc. Perkin Trans. 1* 823 (1996).

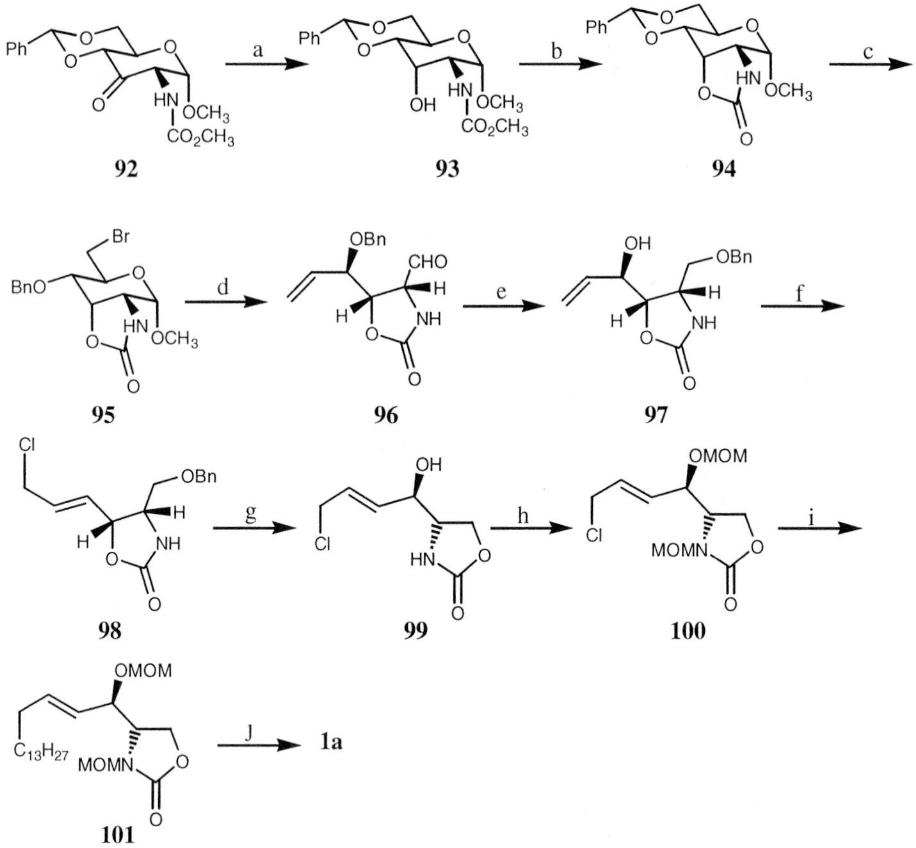

SCHEME 16. Reagents and conditions: (a) NaBH$_4$, CHCl$_3$, EtOH, 0°; (b) NaOCH$_3$, methanol, 60°; (c) NBS, BaCO$_3$, CCl$_4$, 100°; (d) Zn, methanol/H$_2$O, 100°; (e) NaBH$_4$, ethanol, r.t.; (f) SOCl$_2$, DMF, r.t.; (g) K$_2$CO$_3$, methanol, r.t.; (h) p-TsOH, LiBr, CH$_3$OCH$_2$OCH$_3$, r.t.; (i) C$_{12}$H$_{25}$MgBr, CuBr, Et$_2$O, 0°; (j) 1 N NaOH, ethanol, reflux, then 9% HCl, methanol, r.t.

A final synthesis from D-glucosamine utilizes intermediate **92**, which is prepared easily with a series of protection reactions (Scheme 16).[48] Stereoselective reduction of ketone **92**, as mentioned earlier, affords alcohol **93**. Treatment with sodium methoxide in methanol forms oxazolidinone **94**. Oxidative bromination with N-bromosuccinimide (NBS) provides masked bromohydrin **95**. Next, a debrominative ring opening produces a mixture of aldehyde **96** and its methyl hemiacetal. Reduction of the crude product with sodium borohydride affords the oxazolidinone intermediate **97** via a

[48] K. Shinozaki, K. Mizuno, and Y. Masaki, *Chem. Pharm. Bull.* **44,** 927 (1996).

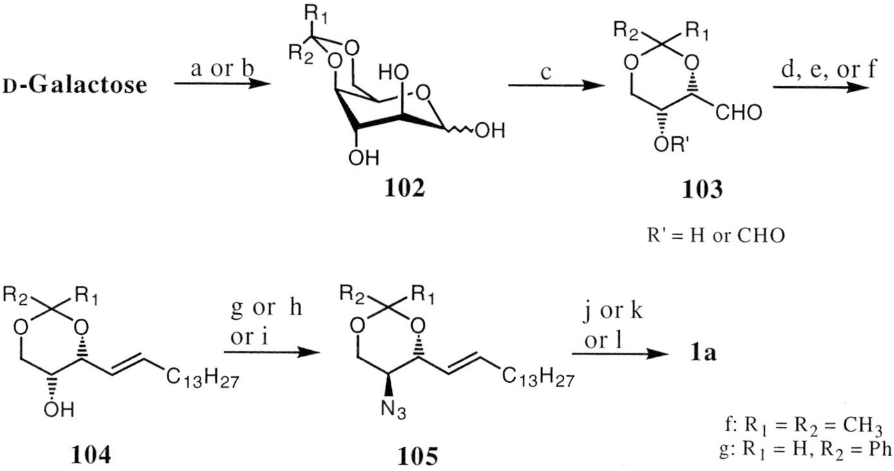

D-Galactose $\xrightarrow{\text{a or b}}$ **102** $\xrightarrow{\text{c}}$ **103** $\xrightarrow{\text{d, e, or f}}$

R' = H or CHO

104 $\xrightarrow[\text{or i}]{\text{g or h}}$ **105** $\xrightarrow[\text{or l}]{\text{j or k}}$ **1a**

f: $R_1 = R_2 = CH_3$
g: $R_1 = H, R_2 = Ph$

SCHEME 17. Reagents and conditions: (a) DMP, DMF, p-TsOH; (b) PhCHO, ZnCl; (c) NaIO$_4$, methanol, r.t.; (d) C$_{14}$H$_{29}$Br, PhLi, CyH-Et$_2$O (7:3), $-60°$-r.t.; (e) C$_{14}$H$_{29}$PPh$_3$, PhLi, LiBr; (f) C$_{14}$H$_{29}$PPh$_3$Br, t-BuOK, THF, 0°, then hν, PhSSPh, CyH-dioxane; (g) MsCl, pyr, $-10°$; (h) MsCl, Et$_3$N, CH$_2$Cl$_2$, 0°; (i) Tf$_2$O; (j) NaBH$_4$, 2-propanol, reflux, then HOAc, 45°; (k) HCl, then H$_2$S, pyr; (l) PPh$_3$, THF-H$_2$O (7:1), r.t., then 1 N HCl, THF, reflux.

1,6-benzyl shift. Treatment with thionyl chloride in DMF transforms the secondary alcohol to the terminal allylic chloride. The intermediate then undergoes an intramolecular rearrangement when treated with potassium pcarbonate. After the protection of the resulting allylic alcohol and the amine as their methoxymethyl ether derivatives, Grignard addition with dodecylmagnesium bromide affords the intermediate **101**. A final deprotection via a sequential base and acid hydrolysis provides D-*erythro*-sphingosine **1a**.

Spingosines from Galactose

Syntheses that produce sphingosines from D-galactose have many similarities to those derived from D-xylose. The compound 4,6-*O*-isopropylidene-D-galactopyranose, which is prepared in one step from D-galactose,[49] undergoes a bond cleavage between C-2 and C-3 when exposed to a solution of sodium metaperiodate in methanol (Scheme 17). The resulting product is a 1:1 mixture of aldehyde **103f** and its formyl derivative. This mixture, when subjected to Wittig olefination, provides olefin **104f** in approximately a 1:1 ratio of E and Z isomers. These isomers can be separated by column

[49] M. Kiso and A. Hasegawa, *Carbohydr. Res.* **52**, 95 (1976).

chromatography and crystallized from aqueous ethanol. Conversion of the *E*-olefin to the mesylate, using MsCl in pyridine, then allows for azide incorporation using sodium azide in DMF. When azide incorporation is attempted using the corresponding trifluoromethanesulfonyl (triflate) derivative, low yields and side products result. Reduction of the azide with sodium borohydride results in the amine, which can either be acylated to provide ceramide derivatives or be deprotected with acetic acid to give D-*erythro*-sphingosine **1a**.[50] Also, intermediate **103f** in this synthesis can be obtained from commercially available 2,4-*O*-isopropylidine-D-xylofuranose by a similar metaperiodate cleavage reaction.[50]

In their preparations of sphingosines and glycosphingolipids, Schmidt and Zimmerman[51,51a] follow a procedure nearly identical to that described earlier (Scheme 17). However, they chose to subject the 4,6-*O*-benzylidine derivative of D-galactose to sodium metaperiodate cleavage conditions. They find that Schlosser's modification of the Wittig reaction, which involves the addition of excess lithium bromide to the reaction mixture, provides the *trans*-olefin **104g** almost exclusively. Introduction of the azide functionality to the *in situ*-generated triflate with sodium azide gives higher yields than those obtained from the mesylate derivative. Opting to first remove the benzylidene by acid hydrolysis allows for the reduction of azide **105g** with hydrogen sulfide to give D-*erythro*-sphingosine **1a**.

Ohashi also finds starting from D-galactose to be a useful starting material in the preparation of l-deoxyhihydroceramide-l-sulfonic acids and other more complex sphingolipids.[52,52a,52b] Unfortunately, in their hands, use of the Schlosser-modified Wittig reaction did not provide the *trans* selectivity reported by Schmidt and Zimmerman. Therefore, they prepared the *E*-olefin by a photoinduced isomerization with phenylsulfide. The azide, prepared via a mesylate intermediate, is reduced to the amine on treatment with triphenylphosphine in aqueous THF. Acid hydrolysis with HCl yields D-*erythro*-sphingosine. (Note: the amine reduction and acid hydrolysis steps can be reversed.) The benzylidene-amine intermediate can be acylated and then deprotected to generate the corresponding ceramides.

One common theme sugar-derived sphingosine synthesis is the installa-

[50] M. Kiso, A. Nakamura, J. Nakamura, Y. Tomita, and A. Hasegawa, *J. Carbohydr. Chem.* **5,** 335 (1986).

[51] R. R. Schmidt and P. Zimmermann, *Tetrahedron Lett.* **27,** 481 (1986).

[51a] P. Zimmermann and R. R. Schmidt, *Liebigs Ann. Chem.* 663 (1988).

[52] K. Ohashi, Y. Yamagiwa, T. Kamikawa, and M. Kates, *Tetrahedron Lett.* **29,** 1185 (1988).

[52a] K. Ohashi, S. Kosai, M. Arizuka, T. Watanabe, M. Fukunaga, K. Monden, T. Uchikoda, Y. Yamagiwa, and T. Kamikawa, *Tetrahedron Lett.* **29,** 1189 (1988).

[52b] K. Ohashi, S. Kosai, M. Arizuka, T. Wtanabe, Y. Yamagiwa, and T. Kamikawa, *Tetrahedron* **45,** 2557 (1989).

tion of the aliphatic olefin through a Wittig reaction. This transformation tends to favor the production of the *cis* double bond, which is a rare structural feature in naturally occurring sphingosines. There are, however, several ways to access the *trans* double bond from *cis/trans* mixtures of Wittig homologation products: photoinduced isomerization of the *cis*-olefin, separation of the *cis* and *trans* isomers by column chromatography, and modifying Wittig conditions to favor the formation of the *trans*-olefin. Several groups have studied this later process in detail and discovered factors that influence the selectivity of this reaction. When preparing sphingosine and dihydrosphingosine from a 3-amino-allofuranose derivative, Reist and Christi[53,53a] found that the *trans* double bond forms preferentially when using the alkyltriphenylphosphonium bromide, excess phenyl lithium, and excess lithium ions in refluxing benzene.

A related procedure from D-galactose by Wild and Schmidt[54] can give rise to phytosphingosine (Scheme 18). Aldehyde intermediate 103, homologated previously using a Wittig reaction above, can be alkylated with the addition of Grignard reagents; however, the diastereoselectivity of such an addition is difficult to control. Oxidation of the crude adducts with pyridinium dichlorochromate (PDC) provides the 4-keto derivative in 70% yield. The ketone reduction proceeds stereospecifically via a samarium iodide-assisted Tishtshenko reaction using acetaldehyde as a reducing agent, followed by treatment with sodium methoxide. The resulting D-arabino derivative 106 then undergoes selective mesylation at the 2-hydroxyl group providing 107. An acid-catalyzed debenzylation with $BF_3 \cdot Et_2O$ provides diol 109 that is acylated with acetic anhydride. Displacement of the mesylate with sodium azide then produces the azide 110. Reduction with LAH quantitatively removes the acetates and reduces the azide providing amine 111. This reduction can also be performed with hydrogen sulfide in pyridine in good yields. To synthesize sphingosine by this route, the azide displacement is performed prior to the removal of the benzylidene. Conversion of resulting azido-alcohol 108 to its phenylselenoxide derivative yields *trans*-105g. An acid hydrolysis of the benzylidene then results in D-*erythro*-sphingosine 1a. This method for preparing sphingosine, by Grignard addition and selenoxide elimination, is more selective for the *E* isomer than earlier methods that depend on the Wittig reaction to introduce unsaturation.

Schmidt[55] developed a novel alternative for the introduction of the 4,5-*trans* double bond using an oxa-Cope rearrangement (Scheme 19). The

[53] E. J. Reist and P. H. Christic, *J. Org. Chem.* **35**, 4127 (1970).
[53a] E. J. Reist and P. H. Christic, *J. Org. Chem.* **35**, 3521 (1970).
[54] R. Wild and R. R. Schmidt, *Tetrahedron: Asymm.* **5**, 2195 (1994).
[55] R. R. Schmidt, T. Bar, and R. Wild, *Synthesis* 868 (1995).

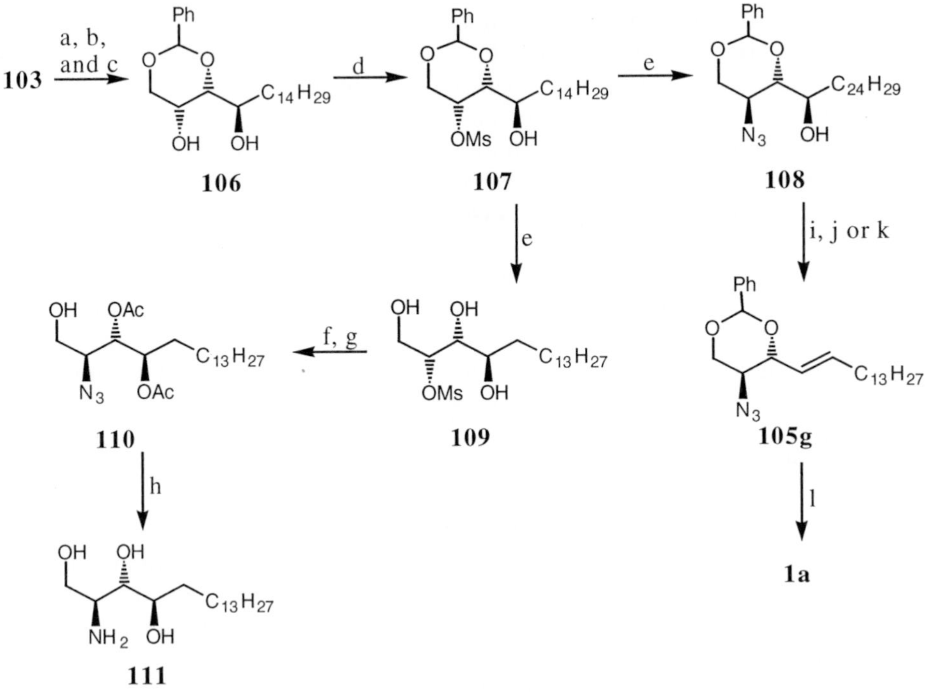

SCHEME 18. Reagents and conditions: (a) $C_{14}H_{29}MgBr$, THF, 60°; (b) PDC, DMF, r.t.; (c) CH_3CHO, SmI_2, THF, −10°, then $NaOCH_3$, methanol; (d) MsCl, pyr, −30°; (e) BF_3OEt_2, methanol, r.t.; (f) Ac_2O, pyr; (g) NaN_3, DMF, 90°; (h) LAH or H_2S, pyr; (i) p-NO_2-TSCl, pyr; (j) Ph_2Se_2, $NaBH_4$, ethanol, 80°, then H_2O_2, THF, 0°-r.t.; (k) DBU, toluene, 80°; (l) HCl.

addition of vinylmagnesium bromide to aldehyde **103g** results in allylic alcohol **112** that is selectively activated with MsCl. The reaction of either isomer of **113** with orthoacetate quantitatively yields oxa-Cope-rearranged product **117**. The authors opted to introduce the azide functionality with tetramethylguanidinium azide in DMF, which provides azide **120** in low yields. To circumvent this situation, a series of protection–deprotection preactions can be employed that give higher overall yields. Specifically, protection of allylic alcohol **112** as its tertbutyldimethylsilyl ether derivative **114**, O-benzoylation of the 2-hydroxy group giving **115**, followed by removal of the silyl ether with triethylammonium fluoride, gives 2-O-benzoyl intermediate **116** that can then undergo the desired oxa-Cope rearrangement. Direct benzoylation of the 2-hydroxy group is unsuccessful. The removal of the benzoyl group with sodium methoxide produces alcohol **119**, which on exposure to triflic anhydride installs the triflate group at C-2. The triflate

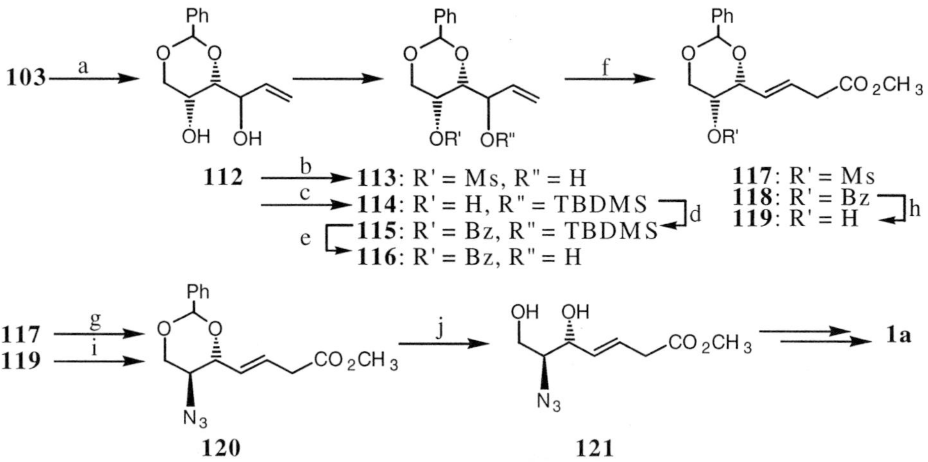

SCHEME 19. Reagents and conditions: (a) $CH_2{=}CHMgBr$, THF, $-20°$; (b) MsCl, CH_2Cl_2, pyr, $-30°$ to $-1°$; (c) TBDMSCl, imidazole, r.t.; (d) BzCl, pyr, r.t; (e) Et_3NHF, THF, r.t.; (f) $CH_3C(OCH_3)_3$, toluene, propanoic acid, $110°$; (g) TMGA, DMF, $95°$; (h) $NaOCH_3$, methanol; (i) Tf_2O, pyr, $-15°$; then NaN_3, DMF, r.t.; (j) p-TsOH, methanol.

can react with sodium azide in DMF in higher yields to afford azide **120**. After the benzylidene group is removed with p-TsOH, the diol ester can either be converted to the acid and removed in a Barton decarboxylation reaction or be reduced to the amino-alcohol with LAH.

The final example of a sugar-derived sphingosine synthesis from D-galactose utilizes the intermediate 3,4,6-tribenzyloxy-D-galactal **122** (Scheme 20). This method also illustrates an alternative for the E selective formation of the 4,5-double bond.[56] When treated with sulfuric acid in the presence of mercuric sulfate, galactal **121** affords enal **123** in 95% yield. A sodium borohydride reduction with cerium chloride gives diol **124**. The authors find that protection of **124** as the diacetate derivative **125** allows for a more efficient coupling with the desired dodecylorganocuprate. This coupling reaction is highly regioselective and provides adduct **126** in 60% yield. Removal of the acetate group with sodium methoxide allows for the introduction of an azide through a standard mesylation-azidotization reaction sequence. Reduction of azide **127** with triphenylphosphine in aqueous THF, followed by a dissolving metal reduction to remove the benzyl groups, completes the synthesis of **1a**.

[56] N. Hirata, Y. Yamagiwa, and T. Kamikawa, *J. Chem. Soc. Perkin Trans. 1* 2279 (1991).

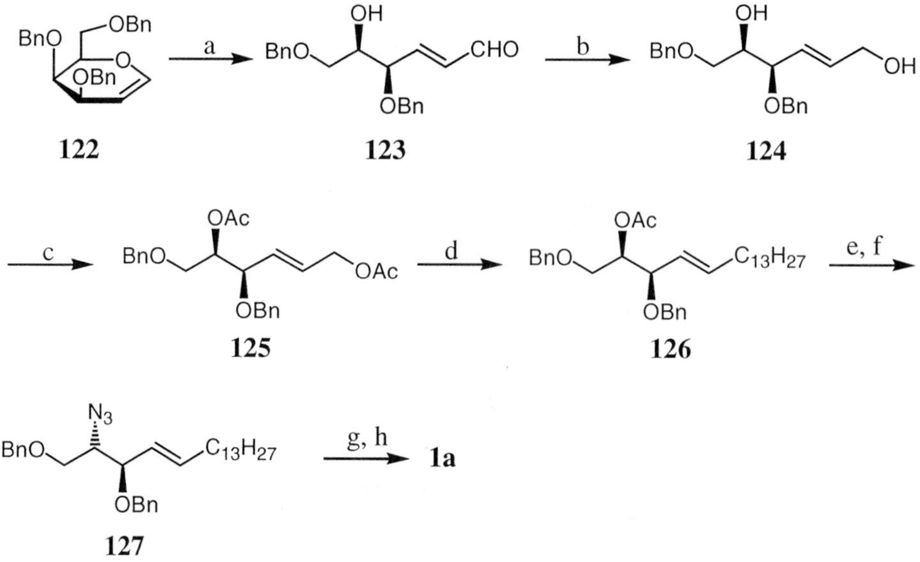

SCHEME 20. Reagents and conditions: (a) 0.01 N H_2SO_4, $HgSO_4$, THF, r.t.; (b) $NaBH_4$, ethanol, $CeCl_3$, r.t.; (c) Ac_2O, DMAP, 0°-r.t.; (d) $C_{12}H_{25}MgBr$, Li_2CuCl_4, THF, $-18°$; (e) $NaOCH_3$, methanol, r.t.; (f) MsCl, pyr, CH_2Cl_2, r.t., then LiN_3, DMF, 90°; (g) PPh_3, THF-H_2O, r.t.; (h) Li/NH_3, THF, $-33°$.

Sphingosines from Ribonolactone

Ribonolactone **128** is a commercially available derivative of L-arabinose and has been used in the synthesis of many natural products.[57] Its use in sphingolipid synthesis allows for the preparation of diastereomerically pure (2S,3S)-L-*threo*-sphingosine in modest yields. Obayashi and Schlosser[58] illustrate its utility in their preparations of sphingosine derivatives (Scheme 21). Protection of **128** as its isopropylidene derivative with acetone and DMP in the presence of an acid, followed by the preparation of the 6-*O*-methoxymethyl ether, provides protected lactone **129** in 74% yield. DIBAL reduction of the lactone to the alcohol, followed by treatment with triphenylphosphine in carbon tetrachloride, produces chloride **130**. Dissolving metal reduction opens the acetonide, which then eliminates to provide the corresponding allylic alcohol. Protection of the alcohol as its methoxymethyl derivative **131**, as before, allows for the selective formation of bromide **132**. When exposed to a strong base, the furane ring opens to obtain acetylene **133**. Generating the alkyne lithiate with *n*-butyllithium enables alkylation

[57] K. L. Bhat. S.-Y. Chen, and M. M. Joullie, *Heterocycles* **23**, 691 (1985).
[58] M. Obayashi and M. Schlosser, *Chem. Lett.* 1715 (1985).

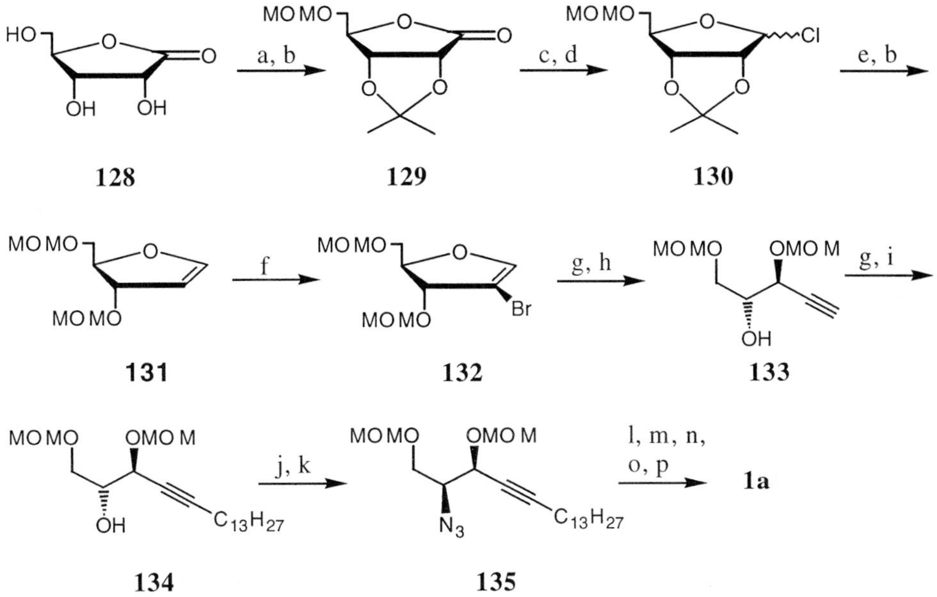

SCHEME 21. Reagents and conditions: (a) DMP, acetone, p-TsOH; (b) ClCH₂OCH₃, Hunig's base; (c) DIBAL, Et₂O; (d) CCl₄, PPh₃; (e) Li/NH₃; (f) Br₂, CCl₄, then DBU; (g) n-BuLi, THF; (h) H₂O; (i) C₁₃H₂₈Br, HMPA; (j) MsCl, Hunig's base; (k) LiN₃, DMF; (l) PPh₃, toluene, 115°; (m) Ac₂O, Et₃N, then H₂O; (n) TMSBr, −30°; (o) NaCO₃, H₂O; (p) NaOEt, ethanol.

by the addition of the appropriate alkyl halide to form **134**. The hydroxyl group serves as the precursor to azide **135** according to standard mesylation-azidotization conditions. Reduction of **135** to the amine with triphenylphosphine and removal of the methoxymethyl ether groups produce the L-*threo*-sphingosine in good diastereomeric purity.

The synthesis of D-*erythro*-sphingosine **1a** is possible from the same protocol as described earlier for L-*threo*-sphingosine **1b**. Using D-mannose, a simple five-step reaction sequence results in the antipode of **129**. A nearly identical reaction sequence yields D-*erythro*-sphingosine in similar yields.

Sphingosine from D-Isoascorbic Acid

This final example of a sugar-derived sphingosine synthesis utilizes D-isoascorbic acid for the production of D-*threo*-sphingosine **1d** (Scheme 22).[59] The homochiral building block arises from D-isoascorbic acid by first gener-

---

[59] A. Tuch, M. Saniere, Y. L. Merrer, and J.-C. Depezay, *Tetrahedron: Asymm.* **7**, 897 (1996).

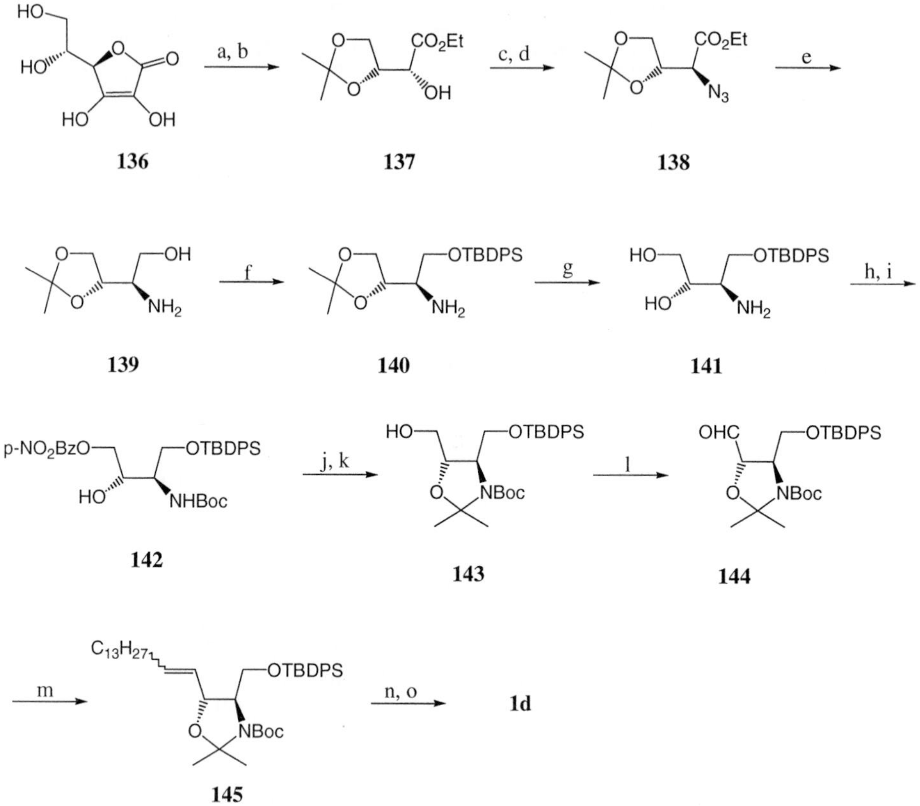

SCHEME 22. Reagents and conditions: (a) DMP, CuSO$_4$, acetone; (b) K$_2$CO$_3$, H$_2$O$_2$, H$_2$O, then EtI, CH$_3$CN, reflux; (c) Tf$_2$O, 2,6-leutidine, CHCl$_3$, −60°; (d) TMGA, −60°-r.t.; (e) LAH, THF, reflux; (f) TBDPSCl, Et$_3$N, DMAP, CH$_2$Cl$_2$, r.t.; (g) TFA, H$_2$O, −5°; (h) Boc-ON, Et$_3$N, dioxane/H$_2$O (1:1); (i)o-NO$_2$C$_6$H$_4$COCl, pyr, CH$_2$Cl$_2$, r.t.; (j) DMP, BF$_3$ · Et$_2$O, r.t.; (k) K$_2$CO$_3$, methanol/THF (1:1), r.t.; (l) (COCl)$_2$, DMSO, Et$_3$N, CH$_2$Cl$_2$; (m) C$_{14}$H$_{29}$PPh$_3$, n-BuLi, THF, −78°-r.t.; (n) PhSSPh, hν, CyH; (o) TFA, H$_2$O (1:1), r.t.

ating the acetonide with DMP and copper(II) sulfate. Treatment with potassium carbonate in the presence of hydrogen peroxide, followed by ethyl iodide, gives alcohol ester **137**. Activation of the alcohol with triflic anhydride allows for azide introduction with guanidinium azide. Reduction of the ester with LAH and protection of the resulting alcohol **139** provide the amino silyl ether **140**. On acid hydrolysis of the acetonide, the diol **141** can be protected in several ways. Investigating various pathways, the authors find that protection of the amino group as its Boc derivative, converting the primary alcohol to its o-nitrobenzoyl derivative, and generating the oxazolidine with DMP, followed by removal of the *o*-nitrobenzoyl group by alkaline hydrolysis, is an efficient strategy. The aldehyde **144**, prepared from the alcohol **143** via Swern oxidation, undergoes Wittig homologation with tetradecyltriphenylphosphonium bromide to provide predominately the *Z* isomer. Photoisomerization of the olefin in the presence of phenyl sulfide produces the easily separable *E*-isomer. Acid hydrolysis of the acetonide reveals D-*threo*-sphingosine with no evidence of epimerization.

Sphingosines from Asymmetric Epoxidations

Several syntheses of sphingolipids begin with achiral starting materials and introduce chirality through asymmetric induction. Perhaps the most significant of these asymmetric reactions is the one developed by Sharpless.[60] The Sharpless asymmetric epoxidation reaction involves the use of titanium tetraisopropoxide, (+)- or (−)-dialkyl tartrate, and an alkyl hydroperoxide. This system epoxidizes allylic alcohols with high enantioselectivity, often above 95% e.e. The epoxidation can be catalytic in titanium and tartrate and the use of molecular sieves can improve efficiency.[61] The ability to synthesize one epoxide enantiomer or the other depends on the choice of the dialkyl tartrate enantiomer. This reaction, as it applies to sphingolipid synthesis, is a key step that sets the stereochemistry of what will later become the sphingosine head group; opening the epoxide, which can be accomplished by several methods, thus sets the stereochemistry of C-2 and C-3 of sphingosine. These methods are advantageous in comparison to those starting from amino acids or sugars in that access to each diastereomer of sphingosine is equally available. Simply switching the dialkyl tartrate enantiomer in the epoxidation allows for the preparation of the sphingosine enantiomers in equal yields using the same reaction procedures. It avoids the need to utilize unnatural and expensive members of the chiral pool.

An early example of this process utilizes the known allylic alcohol, heptadec-2-en-1-ol **146**, to prepare D-*erythro*-dihydrosphingosine (sphinganine) (Scheme 23).[62] The vinyl alcohol results from a modified Horner-Emmons reaction with palmitic aldehyde and ethylacetyl diisopropylphosphonate followed by DIBAL reduction of the ester.[63] Sharpless epoxidation with *tert*-butyl hydroperoxide, titanium isopropoxide, and D-(−)-diethyl tartrate in CH_2Cl_2 efficiently provides (2*R*,3*R*)-epoxy-alcohol **147**. Ammo-

[60] T. Katsuki and K. B. Sharpless, *J. Am. Chem. Soc.* **102**, 5974 (1980).
[61] Y. Gao, R. M. Hanson, J. M. Klunder, H. M. Soo, Y. Ko, and K. B. Sharpless, *J. Am. Chem. Soc.* **109**, 5765 (1987).
[62] K. Mori and T. Umemura, *Tetrahedron Lett.* **22**, 4433 (1981).
[63] W. R. Roush and M. A. Adam, *J. Org. Chem.* **50**, 3752 (1985).

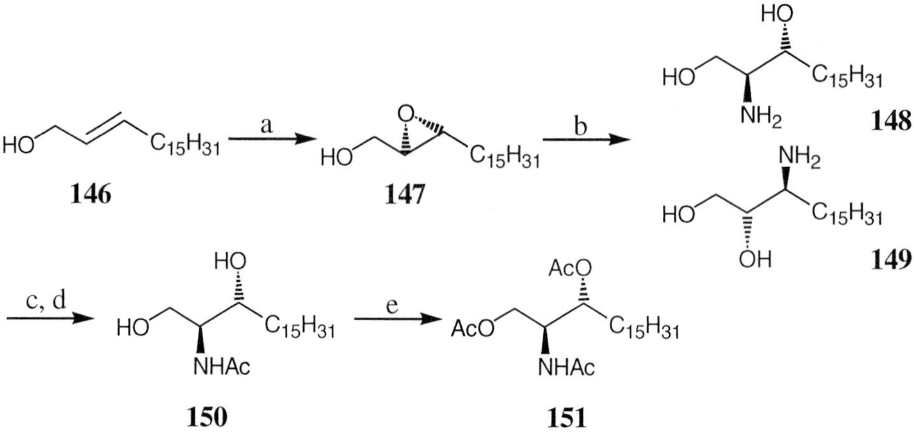

SCHEME 23. Reagents and conditions: (a) t-BuOOH, Ti(i-OPr)$_4$, D-($-$)-diethyl tartrate, CH$_2$Cl$_2$, $-20°$; (b) NH$_3$, methanol, $100°$; (c) Ac$_2$O, methanol; (d) HIO$_4$, methanol, separate; (e) Ac$_2$O, pyr.

nolysis opens the epoxide without regioselective bias, giving a mixture of amino-diols **148** and **149**. These diols are converted to the N-acetates and are then treated with periodic acid. This sequence allows the efficient removal of the undesired 3-amino-1,2-diol from the reaction mixture. The acetate can be removed; although acylation of the alcohols with acetic anhydride results in the triacetate, which is easier to purify and characterize. This chemistry is versatile and has been shown to be amenable to the preparation of other similar natural products.[64]

   An interesting example by Vasella involving the preparation of sphingosine via an asymmetric epoxidation utilizes an intramolecular ring opening.[65] Furthermore, this synthesis is unique in that it represents the first epoxidation on a conjugated dienol system. The synthesis begins with a Sonogashira reaction between 1-pentadecyne and 3-bromoprop-2-en-1-ol (Scheme 24). The resulting enynol **152** is also obtainable from epichlorohydrin and sodium acetylide, followed by protection of the alcohol as its tetrahydropyranyl derivative and subsequent alkylation with butyllithium and tridecylbromide.[66] Enynol **152** undergoes a Sharpless epoxidation reaction with D-($-$)-diethyl tartrate, catalytic titanium tetra-t-butoxide, and $tert$-butyl hydroperoxide. (Note: it is best for **152** to be added slowly to the reaction mixture.) The benzyl urethane **154**, prepared in one pot by first activating

[64] K. Mori and T. Umemura, *Tetrahedron Lett.* **23**, 3391 (1982).
[65] B. Bernet and A. Vasella, *Tetrahedron Lett.* **24**, 5491 (1983).
[66] R. Julina, T. Flerzig, B. Bornet, and A. Vasolla, *Helv. Chim. Acta* **69**, 368 (1986).

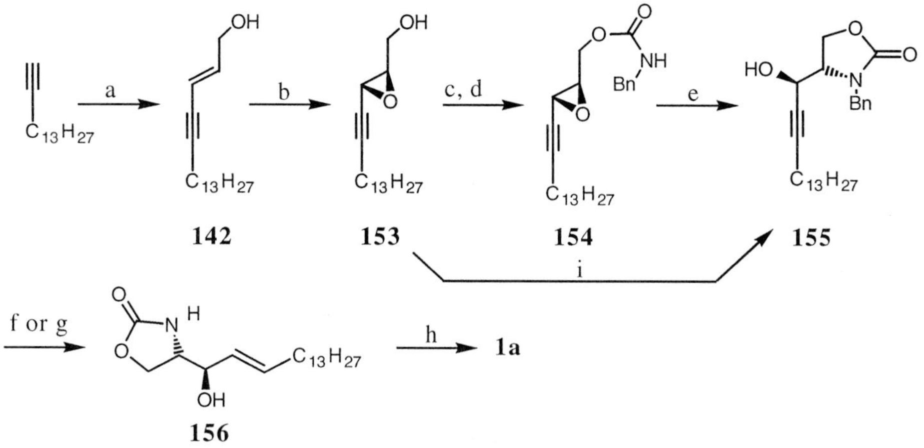

SCHEME 24. Reagents and conditions: (a) BrCH=CHCH₂OH, cat. PdCl₂(PPh₃)₂, CuI, Et₂NH, 40°; (b) Ti(t-OBu)₄, D-(−)-diethyl tartrate, t-BuOOH, 2,3-dimethyl-2-butene, CH₂Cl₂, −20°; (c) ClCO₂C₆H₄(p-NO₂), pyr, CH₂Cl₂, r.t.; (d) BnNH₂, CH₂Cl₂, r.t.; (e) NaHMDS, THF, −20°; (f) Li/NH₃, −20° (2×); (g) Li/EtNH₃, −80°; (h) 2 N NaOH, ethanol, 80°; (i) BnCNO, NaH, THF, 60°.

152 with *p*-nitrophenyl chloroformate and then treating with benzylamine, undergoes an intramolecular epoxide ring opening in the presence of NaHMDS. An alternative for the formation of oxazolidinone **155** is to treat **152** with benzyl isocyanate and sodium hydride in THF. A dissolving metal reduction removes the benzyl group and, although incompletely, reduces the triple bond to form *trans*-olefin **156**. (More efficient conditions for this reaction are lithium in ethylamine at −80°.) Hydrolysis of the carbamate leads to D-*erythro*-sphingosine **1a** in an overall yield of 33%.

Roush and Adam[63] also used the same carbamate intramolecular opening of an epoxide in their synthesis of sphinganine. They form the carbamate with benzyl isocyanate and sodium hydride in THF from epoxy alcohol **147**. (For best results, the reagents should be purified immediately before use and the reaction mixture should be free from moisture and oxygen.) Once the benzyl carbamate forms, it immediately cyclizes to the oxazolidinone. The two diastereoisomers are separable at this stage or can be carried on and later separated as triacetate sphingosine derivatives. Hydrolysis of the oxazolidinone with aqueous lithium hydroxide yields sphinganine **148**. This strategy of using a benzylcarbamate as an ammonia equivalent is a more efficient and practical alternative to ammonolysis.

The Sharpless asymmetric epoxidation is useful not only for the induction of asymmetry into an allylic alcohol, but also for kinetic resolution of

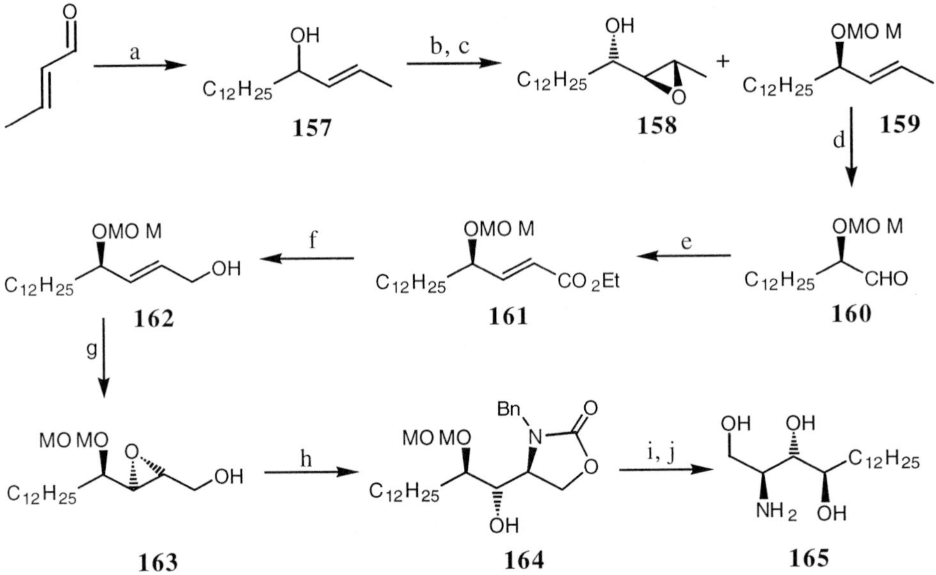

SCHEME 25. Reagents and conditions: (a) $C_{12}H_{25}Br$, Mg, Et_2O, r.t.; (b) (+)-diisopropyl tartrate, t-BuOOH, Ti(i-OPr)$_4$, CH_2Cl_2, $-20°$; (c) MOMCl, Hunigs, THF, r.t.; (d) O_3, methanol, CH_2Cl_2, $-78°$, then $(CH_3)_2S$; (e) $(EtO)_2P(O)CH_2CO_2Et$, NaH, benzene, r.t.; (f) DIBAL, Et_2O, $-10°$; (g) (−)-diisopropyl tartrate, t-BuOOH, Ti(i-OPr)$_4$, CH_2Cl_2, $-20°$; (h) Et_3N, BnCNO, then NaH, r.t.; (i) 2 N NaOH, ethanol, $80°$; (j) Pd/C, H_2, 1 N aq. HCl, cyclohexane.

secondary allylic alcohols. In his preparation of phytosphingosine, Sugiyama[67] demonstrates the elegance and utility of both of these asymmetric processes (Scheme 25). Grignard homologation of crotonaldehyde with dodecylmagnesium bromide affords a racemic mixture of allylic alcohols 157. Kinetic resolution with titanium tetraisopropoxide, *tert*-butyl hydroperoxide, and D-(+)-diisopropyl tartrate gives the 4R-allylic alcohol and 4S-epoxy alcohol 158. The 4R-allylic alcohol, protected as its methoxymethyl ether derivative 159, undergoes ozonolysis to produce the aldehyde 160. (Note: protection of the alcohol alternatively as its TBDMS ether results in an interesting, although undesired, silyl group exchange reaction in the oxazolidinone formation step.) Aldehyde 160 then reacts in a Horner-Emmons reaction with ethylacetyl diisopropylphosphonate to yield *trans*-allylic ester 161. DIBAL reduction of the ester provides masked diol 162. A second Sharpless asymmetric epoxidation with D-(−)-diisopropyl tartrate results in *anti*-epoxy alcohol 163 in 97% yield. Generating the benzylcarbamate with benzyl isocyanate and sodium hydride, as described earlier,

[67] S. Sugiyama, M. Honda, and T. Komori, *Liebigs Ann. Chem.* 619 (1988).

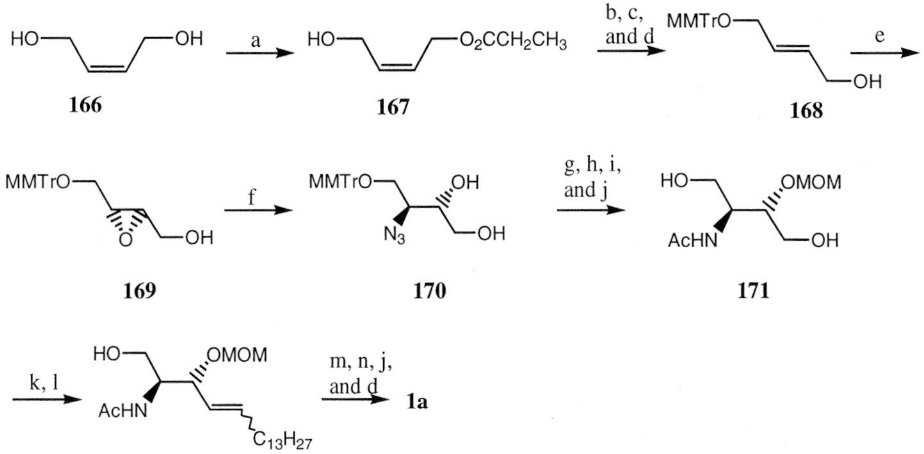

SCHEME 26. Reagents and conditions: (a) $(EtCO)_2O$, acetone; (b) PCC, CH_2Cl_2, then $NaBH_4$; (c) MMTrCl, pyr; (d) 1% KOH, methanol; (e) $(-)$-diethyltartrate, t-BuOOH, Ti(i-OPr)$_4$; (f) Ti(i-OPr)$_2$(N$_3$)$_2$, benzene, reflux; (g) BnCl, Et$_3$N; (h) MOMCl, Hunig's base; (i) LAH; (j) Ac$_2$O, methanol; (k) DMSO, (COCl)$_2$, Hunig's base; (l) $C_{14}H_{29}PPh_3Br$, n-BuLi; (m) 9% HCl, methanol; (n) PhSSPh, hν, CyH-dioxane.

affords only oxazolidinone isomer **164**. Opening of the oxazolidinone ring with sodium hydroxide, debenzylation by hydrogenolysis, and cleavage of the methyoxymethyl ether with HCl, produces $(2S,3S,4R)$phytosphingosine **165**.

An efficient way of stereo- and regioselectively introducing an amino functionality into an epoxide system utilizes the organometallic reagent titanium diisopropoxide diazide, Ti(i-OPr)$_2$(N$_3$)$_2$.[68] This reagent shows a strong preference for opening a *trans*-epoxy alcohol at the C-3 position rather than the C-2 position. The titanium reagent, which is prepared easily, reacts completely in just a few minutes at elevated temperatures. Shibuya et al.[68] demonstrate the utility of this reagent when they prepare all four diastereomers of sphingosine from 1,4-butenediol (Scheme 26). *cis*-Butene-1,4-diol **166**, protected as it monopropanoate **167**, can be isomerized to the *trans*-alkene with pyridinium chlorochromate (PCC) and then reduced with sodium borohydride. Hydroxyl protection with methoxytritylchloride (MMTrCl) in pyridine and basic hydrolysis of the propanoate reveal allylic alcohol **168**. Treatment of *cis*-**167** with this sequence of steps, except for the oxidation step, provides the Z isomer of **168**. Both **168** and its Z isomer

[68] M. Caron, P. R. Carlier, and K. B. Sharpless, *J. Org. Chem.* **53**, 5185 (1988).

can undergo an asymmetric epoxidation with either (+)- or (−)-diethyl tartrate, thus giving rise to the four diastereomers in approximately 75% yield and with 93–97% e.e. [the (2R,3R)-epoxide is shown]. Treatment of **169** with Ti(i-OPr)$_2$(N$_3$)$_2$ produces 3-azido-1,3-diol **170** as the major product. The next series of reactions, benzylation of the primary hydroxyl, protection of the secondary alcohol as the methoxymethyl ether, reduction of the azide, and simultaneous removal of the benzoyl group with LAH, followed by acylation of the amine, afford the masked sphingosine **172** in good yield. The primary alcohol oxidizes readily to the aldehyde under Swern conditions. Wittig olefination with tetradecyltriphenylphosphine results in a 1 : 2 ratio of E and Z isomers. After acid hydrolysis of the methoxymethyl ether and the methoxytrityl groups, the Z isomer can isomerize to the E isomer on irradiation in the presence of phenyl sulfide. The resulting D-*erythro*-sphingosine **1a** is readily purifiable once converted to its triacetate derivative; a basic hydrolysis will yield pure D-*erythro*-sphingosine. In a similar way, utilizing the two-epoxide enantiomers derived from the *cis*-olefin isomer of **168** and the other epoxide enantiomer of **169** accomplishes the production of the three other sphingosine stereoisomers: **1b**, **1c**, and **1d**.

A useful procedure for the preparation of L-*erythro*-sphingosines by Takano *et al.*[69] uses (R,R)-1,2-divinylethylene glycol **173** as the starting material (Scheme 27).[69] A Sharpless asymmetric epoxidation gives a separable mixture of the mono and diepoxides. When treated with DMP, Monoepoxide **174** gives acetonide **175**. The epoxide opens on treatment with potassium *p*-methoxyphenylmethoxide (MPM) and yields the secondary alcohol **176**. The standard mesylation and azidotization reaction sequence provides azide **177**. Exposure of **177** to acid hydrolysis and then DDQ results in oxidative cyclization yielding **178** in good yields. Mesylation of **178** followed by the Grignard addition of dodecylmagnesium bromide in the presence of copper iodide to the allylic mesylate provides the olefin **179**. Acid hydrolysis and azide reduction with LAH results in L-*erythro*-sphingosine **1c**. Following a similar procedure with *meso*-divinylethylene glycol provides D-*threo*-sphingosine **1d**.

One preparation does not involve the Sharpless asymmetric epoxidation but does include a significant epoxide intermediate. It begins with epoxide **180** that is derived from D-(−)-diethyl tartrate (Scheme 28).[70] This epoxide reacts with hydrazoic acid, generated *in situ,* to afford azidohydrin **181**. Reduction of **181** with sodium borohydride provides triol **182** in excellent yields. The 1,2 and 1,3 mixture of acetonides, produced from DMP and *p*-TsOH, equilibrate to the 1,2-O-acetonide on continuous exposure to acid.

[69] S. Takano, Y. Iwabuchi, and K. Ogasawara, *J. Chem. Soc. Chem. Commun.* 820 (1991).
[70] K. Mori and T. Kinsho, *Liebigs Ann. Chem.* 1309 (1991).

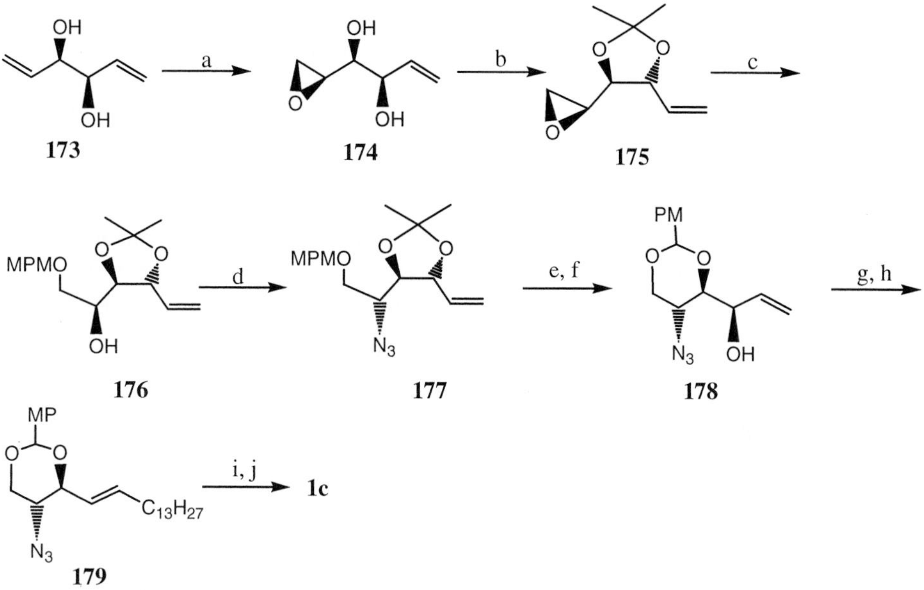

SCHEME 27. Reagents and conditions: (a) D-(−)-diisopropyl tartrate, t-BuOOH, Ti(i-OPr)$_4$, 4°A sieves, CH$_2$Cl$_2$, −20°; (b) DMP, p-TsOH, acetone, r.t.; (c) KH, p-(CH$_3$O)C$_6$H$_4$CH$_2$OH, DMF, 0°; (d) MsCl, pyr, DMAP, CH$_2$Cl$_2$, 0°, then NaN$_3$, DMF, 120°; (e) Amberlyst-15, methanol, r.t.; (f) DDQ, 4°A sieves, CH$_2$Cl$_2$, 0°; (g) MsCl, pyr, DMAP, CH$_2$Cl$_2$, 0°; (h) C$_{12}$H$_{25}$MgBr, CuI, THF, −30°–0°; (i) HCl, methanol, r.t.; (j) LAH, THF.

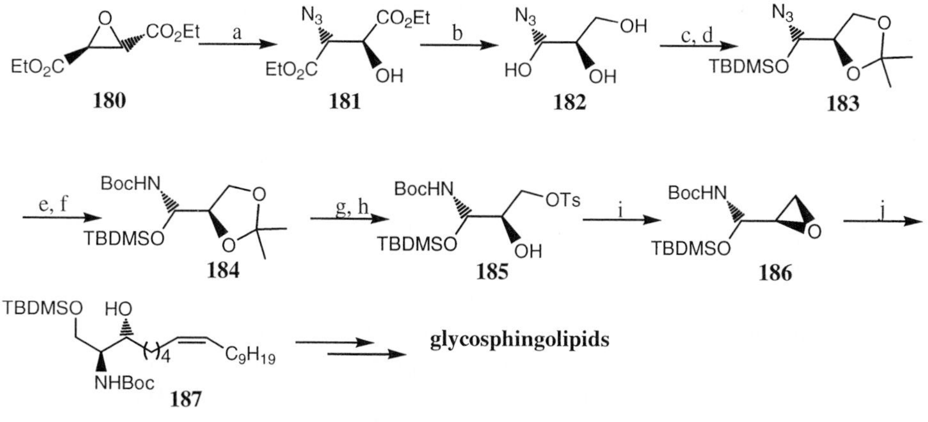

SCHEME 28. Reagents and conditions: (a) TMSN$_3$, methanol, DMF; (b) NaBH$_4$, ethanol, 0°; (c) DMP, acetone, p-TsOH, r.t.; (d) TBDMSCl, imidazole, DMF, r.t.; (e) H$_2$, ethanol, PtO$_2$, r.t.; (f) (Boc)$_2$O, CH$_2$Cl$_2$, r.t.; (g) Amberlyst-15, methanol, r.t; (h) TsCl, pyr, 0°; (i) K$_2$CO$_3$, methanol; (j) alkylBr, Mg, THF, CuI, 0°-r.t.

The primary hydroxyl group is protected with TBDMSCl, which allows the azide to undergo a reduction by catalytic hydrogenation. After protection of the resulting amino group as its Boc derivative **184**, acid hydrolysis removes the acetonide. A selective tosylation of the primary alcohol and treatment with potassium carbonate provides epoxide **186**. This epoxide can now be opened with a variety of alkyl nucleophiles. The authors choose 1-bromo-4E-tetradecene, which was prepared in five steps from 4-butynyl-tetrahydropyranyl ether. The cuprate addition between the alkyl bromide and epoxide affords the masked *erythro* amino alcohol **187**. The compound can be deprotected at this point to reveal the corresponding D-*erythro*-sphingolipid or be carried on to more complex sphingolipids. This method is particularly useful in that a variety of different aliphatic groups can be added to the same epoxide.

Sphingosines from Aldols

There are several valuable examples of sphingosine preparations that utilize aldol reactions. These protocols involve the addition of a stabilized carbanion or enolate to an aliphatic aldehyde. Some methods induce asymmetry in the resulting product through a chiral catalyst or by utilizing a chiral auxiliary. Still other methods are quick, simple, and result in racemic sphingosine adducts. One common feature of sphingosine preparations utilizing aldol reactions is that they contain relatively few steps and are high yielding.

The first examples in this section do not contain classical aldol reactions, although they do involve a stabilized carbanion addition to an aldehyde. These methods utilize 2-nitroethanol as a starting material and are among the earliest reported sphingosine preparations. They are relative simple and would be beneficial for quick and nonstereoselective applications. The deprotonation of 2-nitroethanol is facile due to the electron withdrawing nitro group and the resulting anion can participate in several kinds of addition and substitution reactions. A reduction of the nitro group will produce the amine functionality of the sphingosine head group. However, this compound gives a racemic mixture when added to aldehydes. Methods utilizing 2-nitroethanol are well suited for quick, high-yielding racemic preparations of sphingosine and other sphingolipids (Scheme 29).

One of the earliest syntheses of sphingosine, performed by Grob, couples 2-nitroethanol with 2-penyadecynal.[71] Deprotonation of 2-nitroethanol can occur in the presence of potassium carbonate or sodium hydroxide. The resulting anion adds to the aldehyde to give a 1 : 1 mixture of *erythro* and

---

[71] C. A. Groh and F. Gradient, *Helv. Chim. Acta* **130**, 1145 (1957).

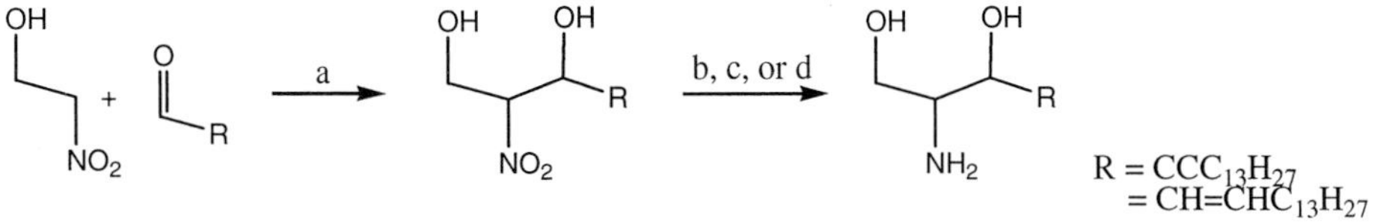

SCHEME 29. Reagents and conditions: (a) K_2CO_3, methanol, r.t.; or NaOH, methanol; (b) Zn, HCl or HOAc; (c) H_2, raney Ni; (d) LAH.

threo diastereomers, which are separable by recrystallization. Reduction of the nitro group to the amine is accomplished with zinc in the presence of HCl or by Raney nickel-catalyzed hydrogenation. The alkyne can be converted to the *trans*-alkene with LAH reduction, the *cis*-alkene with Lindlar's catalyst, or hydrogenated to give sphinganine by catalytic palladium hydrogenation. A more practical route to the preparation of racemic *erythro*-sphinganine entails the coupling of 2-nitroethanol with pentadecanal.[72] Another report using 2-nitroethanol by Mori and Funaki[73] utilizes this procedure to synthesize sphingadienes. They reacted 2-nitroethanol with an aliphatic enynal to obtain a mixture of *erythro* and *threo* isomers. The nitro-propargylic alcohols produced do not separate; readily; however, reducing the nitro group to the amine and subsequent protection as the acetate derivative allow for a more practical separation of diastereomers.

In a preparation of sphingosines that avoids a reduction of a propargylic alcohol, Hino coupled 2-nitroethanol to 2*E*-hexadecenal.[74] Although care must be taken to avoid the competitive 1,4 addition when using such conjugated systems, carrying out the reaction in triethylamine at room temperature gives predominately the 1,2 adduct. The major *threo* isomer can be separated by chromatography from the *erythro* isomer after conversion to the acetonide with DMP and *p*-TsOH. However, the *erythro* isomer can be obtained by thermodynamic equilibration of the *threo* product in refluxing triethylamine. A reduction of the nitro group with LAH results in the amino-acetonide. The problematic formation of dieneamines when deprotecting the amino acetonide requires alternative deprotection strategies. Thus, prior protection of the amino group as the phthalimide, acid hydrolysis of the acetonide, then removal of the phthalimide affords racemic *erythro*-sphingosine.

[72a] C. A. Grob, E. F. Jonny, and H. Ulzinger, *Helv. Chim. Acta* **276–277**, 2249 (1951).
[72b] C. A. Grob and E. F. Jonny, *Helv. Chim. Acta* **263**, 2106 (1952).
[73] K. Mori and Y. Funaki, *Tetrahedron* **41**, 2369 (1985).
[74] T. Hino, K. Nakakyama, M. Taniguchi, and M. Nakagawa, *J. Chem. Soc. Perkin Trans. 1* 1687 (1986).

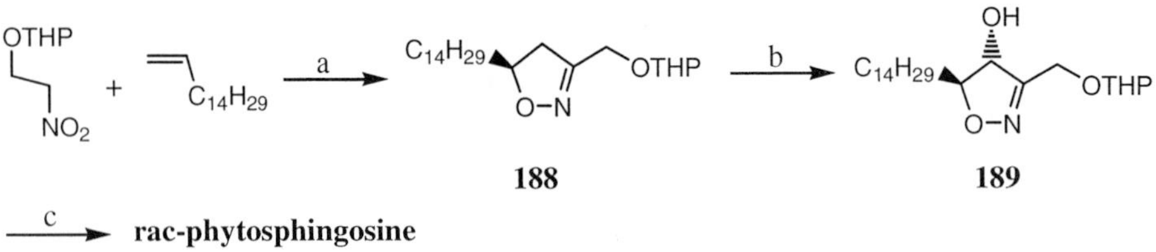

188   189

c → rac-phytosphingosine

SCHEME 30. Reagents and conditions: (a) PhCNO, Et$_3$N; (b) LDA, THF, HMPA, $-78°$; then B(OCH$_3$)$_3$, t-BuOOH, r.t.; (c) LAH, then HCl methanol.

Other starting materials similar to 2-nitroethanol are protected glycines. Schmidt used N,N,O-tri-TMS-protected glycine with LDA to add to $2E$-hexadecenal to give a sphingoacid.[75] Reduction with LAH then provides the racemic *erythro*-sphingosine in excellent yields. The reagent $2E$-hexa-decenal can be prepared easily from dioxene (illustrated in Scheme 31). Treatment of dioxene with *tert*-butyllithium and tridecylbromide results in the substituted dioxene. The conjugated aldehyde forms when exposed to refluxing toluene.[76] These similar reactions are collectively represented in Scheme 29.

An interesting example that utilizes 2-nitroethanol is the formation of 4-hydroxyisoxazolines (Scheme 30).[77] This strategy is useful for the prepara-tion of racemic phytosphingosine. The THP ether derivative of 2-nitroetha-nol, 1-hexadecene, and benzylisocyanate participate in a stereoselective cycloaddition reaction in the presence of triethylamine. When treated with lithium diisopropylamide and trimethoxyborane, followed by an oxidative workup with ammonium hydroxide and hydrogen peroxide, the resulting isoxazoline provides the 4-hydroxyisoxazoline **189**. Ring opening via LAH reduction, followed by removal of the THP group with HCl, provides phytosphingosine in 33% overall yield.

The bislactim ether of L-valineglycine will react with α,β-unsaturated aldehydes to provide the *syn*-aldol addition product nearly exclusively via a lithium enolate (Scheme 31).[78] Exposure of the bislactim to butyllithium and 2-hexadecenal gives a 1:1 mixture of diastereomers. These are easily separable by column chromatography. The undesired *syn* adduct can be converted effectively into the *anti* adduct through a series of oxidation and reduction reactions. It is best at this stage to protect the hydroxyl group as the methoxyphenyl ether. An acid hydrolysis of the *anti* product gives

[75] R. R. Schmidt and R. Klager, *Angew. Chem. Int. Ed. Engl.* **21**, 210 (1982).
[76] R. D. Funk and G. L. Bolton, *J. Am. Chem. Soc.* **110**, 1290 (1988).
[77] W. Schwah and V. Jager, *Angew. Chem. Int. Ed. Engl.* **20**, 603 (1981).
[78] U. Groth, U. Scholkopf, and T. Tiller, *Tetrahedron* **47**, 2835 (1991).

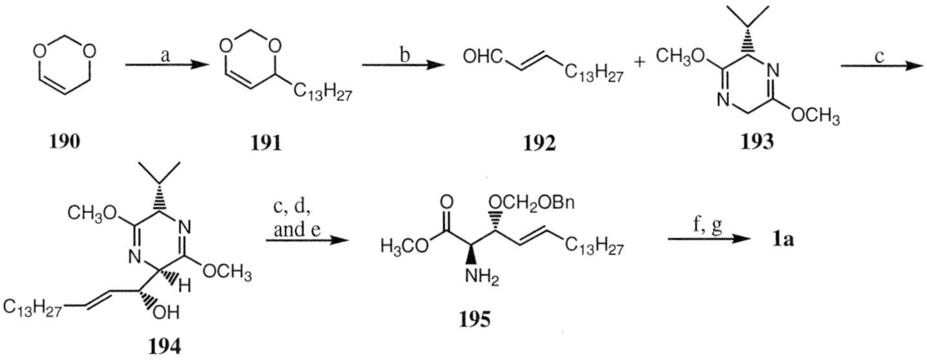

SCHEME 31. Reagents and conditions: (a) t-BuLi, $CH_{13}H_{27}Br$; (b) toluene, reflux; (c) n-BuLi, THF, $-78°$; (d) $BnOCH_2Cl$, Et_3N; (e) 2 N HCl; (f) LAH, Et_2O, reflux; (g) $Li/EtNH_2$.

a sphingosine-ester, which reduces to the protected sphingosine with LAH. Removal of the methoxyphenyl ether under dissolving metal conditions yields the D-*erythro*-sphingosine **1a** in a 21% overall yield. This strategy is particularly suited for large-scale preparations and can give access to all four sphingosine isomers in just five steps.

A novel asymetric-aldol synthesis of D-*erythro*-sphingosine makes use of a chiral gold and iron complex (Scheme 32).[79] In a reaction between 2E-hexadecenal and methyl α-isocyanoacetate catalyzed with an aminoferrocenylphosphine–gold(I) complex, an oxazoline carboxylate results. (The catalyst generates *in situ* with biscyclohexyl isocyanide)gold(I) tetrafluoroborate and (S)-N-methyl-N-[2-(morpholino)ethyl]-1-[(R)-1′,2-bis(diphenylphosphino)ferrocenyl]ethylamine.[80]) The resulting 89 : 11 mixture of *cis* and *trans* isomers is separable by column chromatography. Opening the oxazoline ring by acid hydrolysis gives the resulting β-hydroxyamino acid **196**. A reduction of the *threo* isomer with LAH gives D-*threo*-sphingosine **1d**. Preparation of D-*erythro*-sphingosine by selective acetylation of the amine and primary hydroxyl allows the secondary hydroxyl to be inverted under Mitsunobu conditions.

Nicolaou used chiral auxillaries to induce asymmetry in his preparation of sphingosine (Scheme 33).[81] This synthesis contains few steps, is efficient,

[79] Y. Ito, M. Sawamura, and T. Hayashi, *Tetrahedron Lett.* **29**, 239 (1988).

[80a] Y. Ito, M. Sawamura, and T. Hayashi, *J. Am. Chem. Soc.* **108**, 6405 (1986).

[80b] Y. Ito, M. Sawamura, M. Kohayashi, and T. Hayashi, *Tetrahedron Lett.* **29**, 6321 (1988).

[80c] Y. Ito, M. Sawamura, B. Shirakawa, K. Hayashizaki, and T. Hayashi, *Tetrahedron Lett.* **29**, 235 (1988).

[81] K. C. Nicoluou, T. Canlfield, H. Kataoka, and T. Kunazawa, *J. Am. Chem. Soc.* **110**, 7910 (1988).

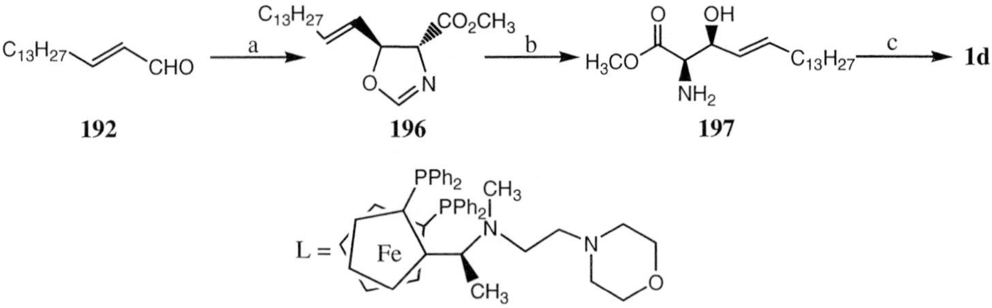

SCHEME 32. Reagents and conditions: (a) CNCH$_2$CO$_2$CH$_3$, AuL; (b) HCl, methanol, 55°; (c) LAH, THF.

and provides enantiomerically pure sphingolipids. When treated with dibutylborontriflate, the oxazolidinone **198** generates the corresponding boron enolate. The enolate then condenses with 2E-pentadecenal to afford **199** in 72% yield. A S$_N$2 displacement of the bromide with sodium hydride gives azide **200**. Allylic alcohol protection with TBDMSCl allows the subsequent removal of the oxazolidinone with lithium borohydride. After desilylation with TBAF and azide reduction with 1,3-propanedithiol in the presence of Et$_3$N, D-*erythro*-sphingosine **1a** results in 81% overall yield. The "sphingosine equivalent" **201** is useful in that it can be derivatized further through

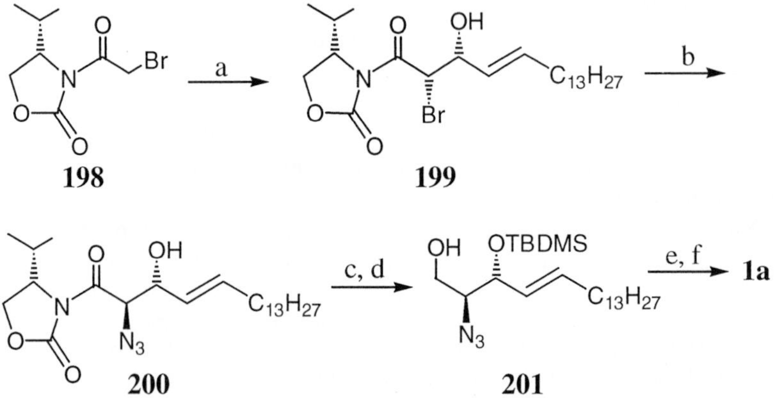

SCHEME 33. Reagents and conditions: (a) (n-Bu)$_2$BOTf, Et$_3$N, −78°-r.t., then 2-pentadecenal, 0°; (b) NaN$_3$, DMSO, r.t.; (c) TBDMSOTf, 2,6-lutidine, CH$_2$Cl$_2$, 0°; (d) LiBH$_4$, THF, 0°; (e) TBAF, THF, r.t.; (f) HS(CH$_2$)$_3$SH, Et$_3$N, methanol, r.t.

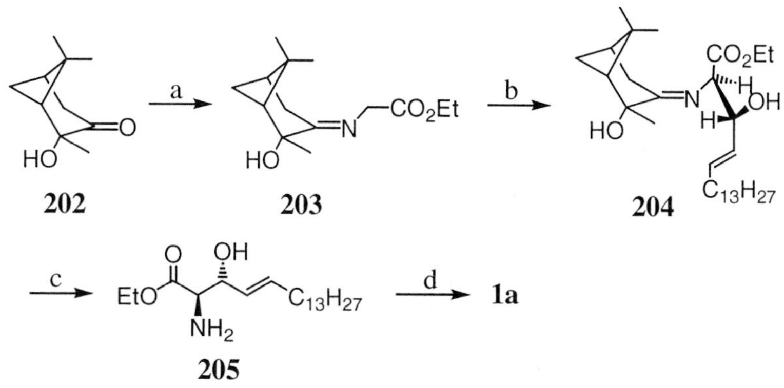

SCHEME 34. Reagents and conditions: (a) $H_2NCH_2CO_2Et$, BF_3Et_2O; (b) $ClTi(OEt)_3$, CH_2Cl_2, 2-hexadecenal, Et_3N, 0°; (c) 2 N HCl, THF, r.t.; (d) $NaBH_4$, ethanol, H_2O.

glycosidation to provide more complex glycosphingolipids. The mechanistic details of this aldol reaction are explained in the Refs. 82 and 82a.

Cavallo later reported an *erythro* and enantioselective aldol involving sphingosine synthesis that contains only four steps (Scheme 34).[83] The synthetic strategy utilizes a titanium enolate derived from an iminoglycinate. The iminoglycinate **202** is prepared from $(+)$-(R,R,R)-hydroxypinanone.[84] The titanium enolate of 202, produced on treatment with one equivalent of $ClTi(OEt)_3$ in the presence of Et_3N, condenses with 2E-hexadecenal to provide the one diastereomer, **203**, in 75% yield. Purification by column chromatography and acid hydrolysis results in a mixture of ethyl and isopropyl esters (9:1). A sodium borohydride reduction of the esters affords D-*erythro*-sphingosine **1a** in 40% overall yield. Following the identical procedure from readily available (S,S,S)-hydroxypinanone provides L-*erythro*-sphingosine **1c**. It should be mentioned that the chiral iminoglycinate **202** can be recovered following acid hydrolysis.

Sphingosines from Miscellaneous Methods

Yamanoi et al.[85] utilized a chiral glycerate in a Horner-Wittig reaction in their preparation of sphingosine (Scheme 35). Starting with dimethyl-

[82] D. A. Evans, E. B. Sjogren, A. E. Weber, and R. E. Conn, *Tetrahedron Lett.* **28**, 39 (1987).
[82a] D. A. Evans and A. E. Weber, *J. Am. Chem. Soc.* **109**, 7151 (1987).
[82b] D. A. Evans. J. V. Nelson, E. Vogol, and T. R. Taber, *J. Am. Chem. Soc.* **103**, 3099 (1981).
[83] A. Solladic-Cavallo and J. Koessler, *J. Org. Chem.* **59**, 3240 (1994).
[84] A. Solladic-Cavallo, M. C. Simon-Wermeister, and J. Schwarz, *Organometallics* **12**, 3743 (1993).
[85] T. Yamunio, T. Akiyama, E. Ishida, H. Abe, M. Amemiya, and T. Innzu, *Chem. Lett.* 335 (1989).

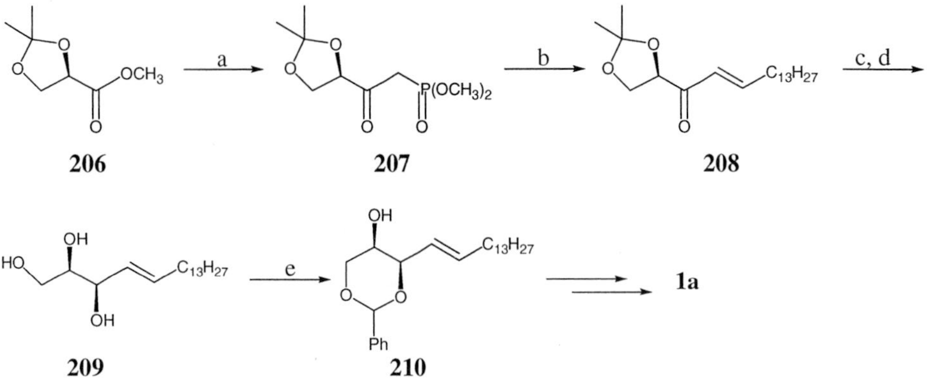

206   **207**   **208**

209   **210**   **1a**

Scheme 35. Reagents and conditions: (a) CH₃P(O)(OCH₃)₂, *n*-BuLi, THF, −78°; (b) C₁₃H₂₇CHO, Cs₂CO₃, 2-propanol, 0°; (c) L-selectride, THF, −78°; (d) H⁺; (e) PhCH(OCH₃)₂, H⁺.

phosphonate, a deprotonation with butyllithium generates a stabilized carbanion that adds to commercially available 2,3-*O*-isopropylidene-D-glycerate (**206**), resulting in the dimethyl 1,2-*O*-isopropylidene-D-glycerolmethylphosphonate **207**. Testing a number of bases in the following Horner-Emmons reaction with tetradecanal, the authors found that one equivalent of cesium carbonate in 2-propanol to be the most successful. The olefination reaction proceeds in high yields (85%) and is preferential for the *E*-olefin. A stereoselective reduction with L-selectride [Li(*s*-Bu)₃BH] results in a 10:1 ratio of *threo-/erythro*-alcohol. Removal of the acetonide by acid hydrolysis, followed by formation of the 1,3-*O*-benzylidene with benzaldehyde dimethyl acetal, allows the C-2 hydroxyl group to be converted to the azide by standard conditions. The reduction of the azide to the amine and removal of the benzylidene-protecting group yields D-*erythro*-sphingosine **1a**.

A novel approach to racemic *erythro*-sphingosine involves a diastereoselective iodocyclization of trichloroacetimidates (Scheme 36).[86] A Horner-Wittig reaction between 4-dimethylphosphono-2-butenoate and tetradecanal produces the dienol **212** after a LAH reduction. Although this olefination reaction favors the formation of the *E* isomer, a significant amount of the *Z* isomer is also present, thus requiring bond isomerization with iodine in benzene. Conversion of **212** to its trichloroacetimidate with trichloroacetonitrile and sodium hydride, and treatment with *N*-iodosuccinimide (NIS), results in an intramolecular cyclization to yield 4,5-dihydro-

[86] G. Cardillo, M. Orena, S. Sandri, and C. Tomasini, *Tetrahedron* **42**, 917 (1986).

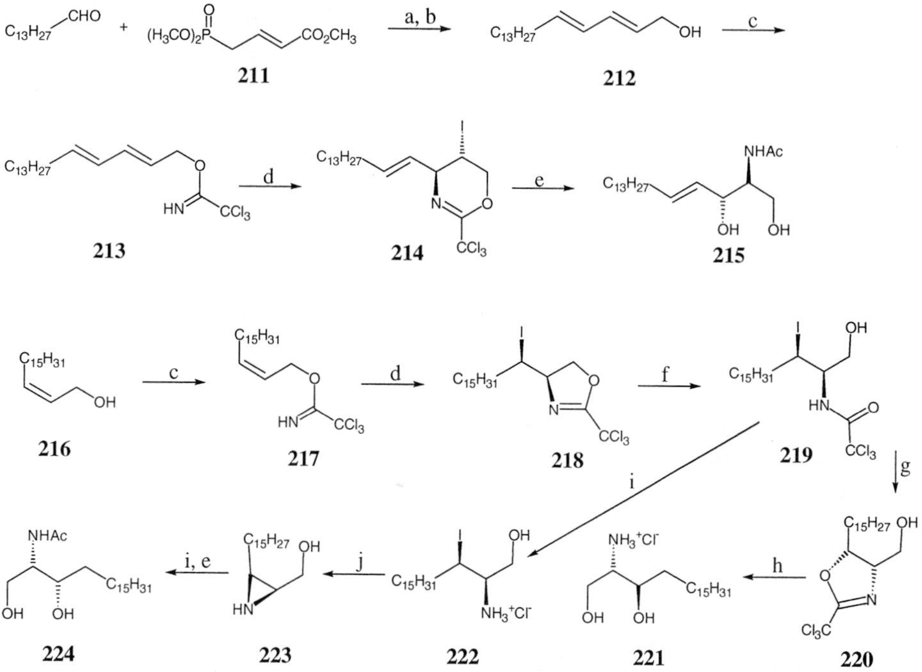

SCHEME 36. Reagents and conditions: (a) LDA, THF, $-40°$; (b) LAH, Et_2O, $-10°$; (c) THF, NaH, CCl_3CN, $0°$; (d) NIS, $CHCl_3$, r.t.; (e) Amberlyst-26 AcO^- form, benzene, $80°$; (f) acetone, H_2O, reflux; (g) Amberlyst-26 CO_3^- form, benzene, reflux; (h) 2 N HCl, methanol; (i) 2 N HCl, acetone, r.t.; (j) Amberlyst-26 CO_3^- form, methanol, r.t.

1,3-oxazine **214** in 95% yield. Acid hydrolysis gives a 3-amino-2-iodo-alcohol, which on exposure to Amberlyst-A26 (in the AcO^- form) in refluxing benzene provides the N-acetyl-($\pm$)-*erythro*-sphingosine **215** via an aziridine intermediate.

In a similar manner, this procedure can be applied for the synthesis of sphinganine by using 2Z-octadecen-1-ol.[87,87a] Using this material, the iodocyclization results in five-membered oxazole ring **218**. Aqueous hydrolysis gives 2-amino-3-iodo-alcohol **219**. The cyclization of amine **219** in the presence of Amberlyst-A26 in benzene gives five-membered oxazole **220**, which provides the ($\pm$)-*erythro*-sphinganine on acid hydrolysis with 2 N HCl. If **219** alcohol is hydrolyzed to the hydrochloride salt prior to treatment

[87] A. Bongini, G. Cardillo, M. Orena, S. Sandri, and C. Tomasini, *J. Chem. Soc. Perkin Trans. 1* 1339 (1986).

[87a] A. Bongini, G. Cardillo, M. Orena, S. Sandri, and C. Tomasini, *J. Chem. Soc., Perkin Trans. 1* 1345 (1986).

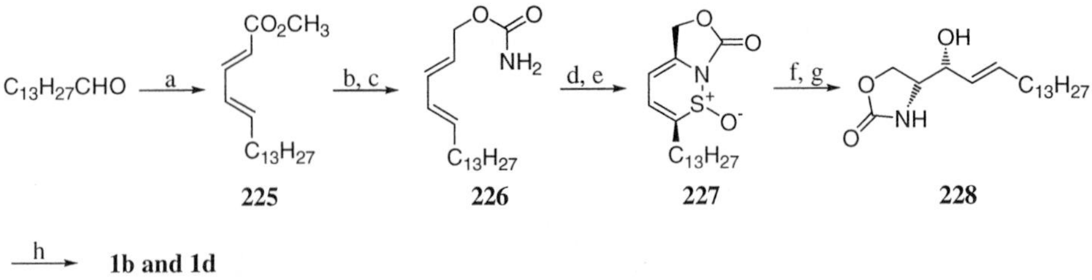

SCHEME 37. Reagents and conditions: (a) $(CH_3O)_2P(O)CH=CHCO_2CH_3$, LDA, THF, $-40°$; (b) LAH, Et_2O, r.t.; (c) NaOCN, TFA/Et_2O; (d) $SOCl_2$, pyr, toluene, 0°; (e) r.t.; (f) PhMgBr, THF, $-60°$; (g) $(CH_3O)_3P$, methanol, 60°; (h) $Ba(OH)_2$, dioxane/H_2O, reflux.

with Amberlyst-A26, then formation of aziridine **223** results. Creating the aziridinium ion with HCl, followed by treatment with Amberlyst-A26 (in the AcO⁻ form), affords a separable mixture of the ($\pm$)-*threo*-N-acetyl-sphinganine **224** and the corresponding 1,2-diol.

Garigipati and Weinreb[88,88a] invoke an intramolecular Diels-Alder cycloaddition of N-sulfinylcarbamates to synthesize sphingosine (Scheme 37). Beginning with tetradecanal, Horner-Emmons olefination with methyl-4-dimethylphosphonate-2-butenoate gives diene **225**. A LAH reduction of the ester, followed by treatment with sodium isocyanate, provides carbamate **226**. After the carbamate is treated with thionyl chloride, an intramolecular Diels-Alder reaction occurs to afford the bicyclic adduct. An allylic sulfoxide intermediate is produced on the addition of phenylmagnesium bromide to **227**. When exposed to trimethylphosphite, this intermediate yields *threo*-*E*-carbamate **228**. Alkaline hydrolysis then provides the racemic *threo*-sphingosine. Utilizing this strategy with (2Z,4E)-octadecadien-1-ol produces racemic *erythro*-sphingosine.

An elegant and noteworthy strategy of sphingolipid synthesis is that of Nugent and Hudlicky (Scheme 38).[89] The key starting material is cyclohexadiene-*cis*-1,2-diol, which is obtained from a biocatalytic oxidation of chlorobenzene with the enzyme toluene dioxygenase from *Pseudomonas putida* 39D.[90] This strategy is useful in that it provides access to all four stereoisomers of sphingosine from one common precursor. For the preparation of D-*erythro*-sphingosine **1a** (shown in Scheme 38), the diol is protected as its acetonide derivative and then converted to the bromohydrin with NBS. After exposure to basic media, which results in the formation of epoxide

[88] R. S. Garigipati and S. Weinreb, *J. Am. Chem. Soc.* **105**, 4499 (1983).
[88a] R. S. Garigipati, A. J. Freyer, R. R. Whittle, and S. M. Weinreb, *J. Am. Chem. Soc.* **106**, 7861 (1984).
[89] T. C. Nugeni and T. Hudlicky, *J. Org. Chem.* **63**, 510 (1998).
[90] T. Budlicky, *Pure App. Chem.* **66**, 2067 (1994).

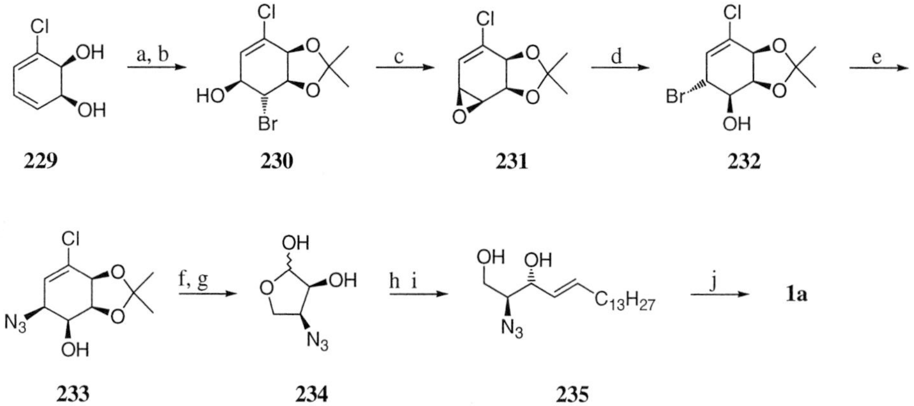

SCHEME 38. Reagents and conditions: (a) DMP, p-TsOH, CH_2Cl_2; (b) NBS, DME/H_2O 3:2, 0°; (c) NaOH, Bu_4HNSO_4, CH_2Cl_2, reflux; (d) LiBr, ethyl acetoacetate, THF, 35°; (e) NaN_3, DMSO; (f) O_3, methanol, −78°, $NaBH_4$, −30°-r.t; (g) Amberlyst-15, H_2O; then $NaIO_4$, H_2O; (h) $C_{14}H_{29}PPh_3Br$, n-BuLi, THF, r.t.; (i) hν, PhSSPh, hexanes/dioxane 4:1; (j) H_2S, pyr, H_2O.

231, treatment with lithium bromide provides bromohydrin **232**. An azide displacement of the bromide affords *cis*-azido alcohol **233**. Ozonolysis of **233**, with subsequent sodium borohydride reduction of the corresponding ozonide, affords lactol **234**. An acid hydrolysis of the acetonide then allows for sodium periodate cleavage of the resulting diol. A direct olefination of the lactol with tetradecyltriphenylphosphonium bromide gives a separable mixture of *cis* and *trans* isomers. The *cis* isomer can be photolyzed to the *trans* isomer with phenyl sulfide. A final azide reduction with hydrogen sulfide provides D-*erythro*-sphingosine **1a**. In a similar manner, the other three stereoisomers of sphingosine can be prepared from diol **229**. The authors also report several optimized conditions for the azide introduction step, ozonolysis, and the Wittig olefination of the lactol. Although this strategy does contain one low yielding step, the Wittig olefination, it is an interesting and elegant synthesis, providing access to all four sphingosine stereoisomers from one common intermediate.

The synthetic strategy of Somfai and Olsson[91] utilizes D-diethyl tartrate in the preparation of D-*erythro-sphingosine* (Scheme 39). A protection of the hydroxyl groups as the p-methoxybenzylidene acetal gives rises to the monoprotected derivative on reduction. A reduction of the esters with LAH, followed by the formation of the acetal **239** with 3-pentanone. The primary alcohol can then be oxidized to the aldehyde under Swern condi-

[91] P. Somfai and R. Olsson, *Tetrahedron* **49**, 6645 (1993).

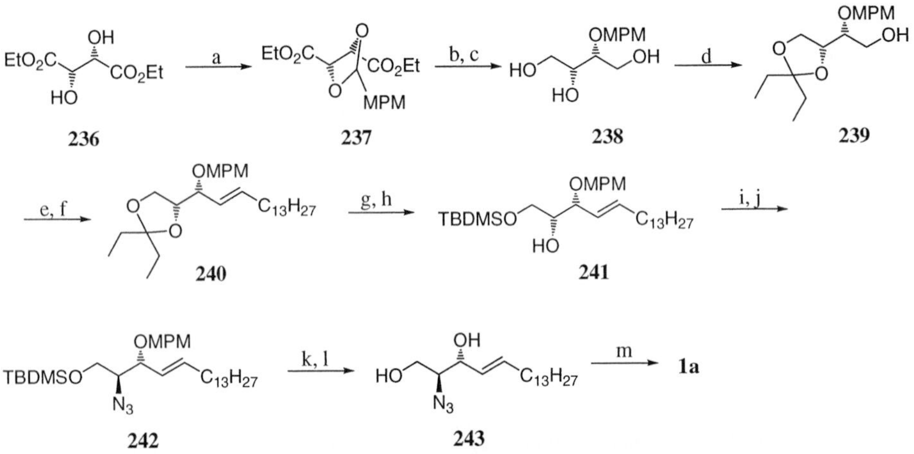

SCHEME 39. Reagents and conditions: (a) p-CH$_3$OPhCH$_2$(OCH$_3$)$_2$, p-TsOH, DMF; (b) NaBH$_4$, LiCl, THF, ethanol; (c) BH$_3$, THF, reflux; (d) 3-pentanone, p-TsOH, THF; (e) DMSO, (COCl)$_2$, Et$_3$N, −78°; (f) C$_{14}$H$_{28}$PPh$_3$Br, PhLi, THF, toluene, −30°, then methanol, H$_2$O; (g) 2% H$_2$SO$_4$, methanol: (h) TBDMSCl, Et$_3$N, DMAP, CH$_2$Cl$_2$; (i) MsCl, pyr, 0°; (j) NaN$_3$, 18-C-6, DMF, 75°; (k) DDQ, CH$_2$Cl$_2$, H$_2$O; (1) TBAF, THF; (m) PPh$_3$, THF, H$_2$O.

tions. (Note: the aldehyde is prone to epimerization and should be used immediately without purification.) Utilizing a Schlosser-modified Wittig reaction provides *trans*-alkene 240 in good yield. Acid hydrolysis of the acetal and selective protection of the primary hydroxyl group as the TBDMS ether results in alcohol 241. A standard mesylation-azidotization sequence yields azide 242, which can be deprotected with DDQ and then TBAF to provide azidosphingosine 243. A final reduction of the azide results in D-*erythro*-sphingosine 1a. This synthetic strategy produces sphingosine stereospecifically in 12 steps with a 31% overall yield.

Solid-Supported Sphingosines

A final and special area of sphingosine synthesis relates to solid-supported sphingosines (Scheme 40). Because sphingosines contain an unreactive ω methyl group on the aliphatic chain, they are not suitable for attachment to a solid support. However, modified sphingosine preparations that put a reactive functionality at the terminus of the aliphatic chain allow for such an attachment. Raun et al.[92] demonstrate this approach in their attempt to prepare a sphingosine kinase affinity matrix. A three-step reaction se-

---

[92] F. Ruan, S. Yamamura, S.-I. Hakomori, and Y. Igarashi, *Tetrahedron Lett.* **36,** 6615 (1995).

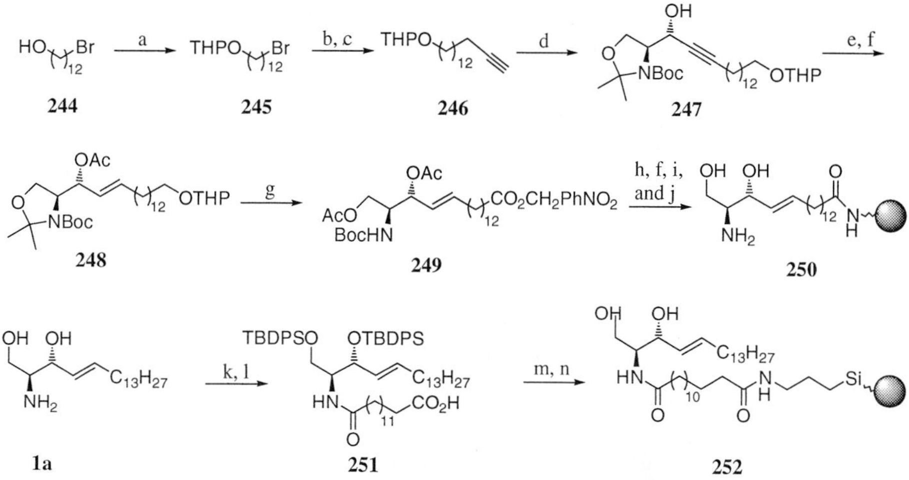

SCHEME 40. Reagents and conditions: (a) DHP, pyr-p-TsOH, CH_2Cl_2; (b) $TMSCH_2CCH$, n-BuLi, TMED, Et_2O; (c) $AgNO_3$, KCN, ethanol/H_2O; (d) n-BuLi, HMPA, THF; (e) Li/NH_3, THF; (f) Ac_2O, pyr; (g) Jones reagent, acetone, then p-NO_2PhCH_2Cl, NaI, $NaHCO_3$, DMF; (h) p-TsOH, methanol; (i) zinc dust, HOAc, H_2O; (j) CPG, HOBt, DIC, CH_2Cl_2, DMF then HOAc; (k) $-O_2CC_{12}H_{24}CO_2-$, DMAP, r.t.; (l) TBDPSCl, imidazole, THF, 36°; (m) CDI, $CHCl_3$, SPA, $C_9H_{19}CO_2H$, $C_2H_5CO_2H$; (n) TBAF, THF.

quence transforms 12-bromo-1-dodecanol to protected alkyne **246**. This alkyne is coupled to Garner aldehyde **8**, as described earlier, resulting in **247**. A dissolving metal reduction of the alkyne, protection of the secondary alcohol, deprotection of the terminal ω-pyranyl ether, an oxidation gives the corresponding acid. The acid is protected as the p-nitrobenzyl ester, which then allows for acid hydrolysis of the acetonide and formation of the diacetate **249**. Removal of the p-nitrobenzyl ester provides the ω acid, which can condense with amino-functionalized controlled pore glass beads.

In a similar procedure, Kozikowski et al.[93] took the ω alcohol and converted it to an azide via a mesylation and azide-displacement reaction sequence. The reduction of the azide with tin chloride, thiophenol, and triethylamine, followed by removal of the protecting groups by methanolysis and acid hydrolysis, leads to an ω-amino-D-erythro-sphingosine. This compound is then coupled to a derivatized agarose gel bead to provide a solid-supported sphingosine.

A related strategy for the preparation of solid-supported ceramides attaches the N-acyl aliphatic portion to a solid support. Yin et al.[94] prepared

[93] A. P. Kozikowski, Q. Ding, and S. Spiegel, *Tetrahedron Lett.* **37**, 3279 (1996).
[94] J. Yin, H. Liu, and C. Pridgeon, *Bioorg. Med. Chem. Lett.* **8** (1998).

sphingosine **1a** in an analogous way to Garner (Scheme 2). Protection of the alcohol groups as their *tert*-butyldiphenylsilyl ethers (TBDPS), followed by treatment with dodecyldicarboxylic acid anhydride, gives the *N*-(13-carboxyl)-tridecanoyl-D-*erythro*-sphingosine **251** in 87% yield. The ω carbonyl is activated as with carbonyldiimidazole (CDI) and coupled to silica propylamine particles. Removal of the TBDPS groups with TBAF affords the solid-supported ceramide.

Conclusions

The wealth of unique synthetic strategies toward sphingosine and related derivatives is apparent. Deciding on which synthesis is largely a matter of preference. Some syntheses require specialized reagents and catalysts. Other syntheses, like some sugar-derived methods, may not be atom economical due to cleavage reactions and may require multiple steps. Still others may not produce the desired stereochemistry or give racemic products. Consequently, a prior understanding of ones needs is essential. Moreover, the synthetic summaries listed earlier constitute a basic outline of well-known strategies, although the represented syntheses are not a comprehensive evaluation of the literature. Furthermore, syntheses that are not mentioned should in no way be considered less important. Several literature reviews can be consulted for more references.[95,95a] The area of sphingosine synthesis is a constantly growing ever-improving field: a testament to the significance of this class of molecules.

[95] H. S. Byun and R. Bittman, "Phospholipid Handbook" (G. Cevc, ed.), p. 97. Dekker, New York, 1993.
[95a] R. M. Devant, *Kontakte* **3,** 11 (1992).

[40] Synthesis of Sphingosine, Radiolabeled-Sphingosine, 4-Methyl-*cis*-sphingosine, and 1-Amino Derivatives of Sphingosine via Their Azido Derivatives

By KARL-HEINZ JUNG and RICHARD R. SCHMIDT

Sphingosine is the major lipid moiety of glycosphingolipids[1] and phos-phosphingolipids[2-4] (sphingomyelins, glycosylphosphatidylinositols). These compounds are expressed on the surface of cell membranes and play an important role in cell–cell interactions,[1-5] in signal transduction,[6-8] and in the anchoring of proteins.[4] Also, sphingosine itself was found to have regulatory properties, and it is of importance, especially as an inhibitor of protein kinase C.[6,7] Because of the large variety of sphingosine moieties found in natural sources and the requirement of labeled and modified sphingosine derivatives for biological and pharmacological investigations, efficient chemical syntheses are of great importance.

Over 100 more or less efficient chemical syntheses of sphingosine and derivatives have been reported since the late 1940s.[9,10] A very convenient and efficient chiral pool synthesis is based on D-galactose as starting material (Scheme 1).[11] This synthesis allows the preparation of large quantities in only a few steps; formation of stereogenic centers is not required and variations in the chain length and other modifications of the long chain are easily accessible.[12] In addition, the azidosphingosine intermediate is a better

---

[1] C. C. Sweely, *in* "Biochemistry of Lipids and Membranes" (D. E. Vance and J. E. Vance, eds.), p. 361. Benjamin Cummings, Menlo Park, CA, 1985.

[2] Y. Barenholz and T. E. Thompson, *Biochim. Biophys. Acta* **604,** 129 (1980).

[3] Y. Barenholz and S. Gatt, *in* "Phospholipids" (J. N. Hawthorne and G. B. Ansell, eds.), p. 129. Elsevier Biomedical Press, Amsterdam, 1982.

[4] V. Eckert, P. Gerold, and R. T. Schwarz, *in* Glycosciences (H.-J. Gabius and S. Gabius, eds.), p. 223. Chapman & Hall, Weinheim, 1997.

[5] S.-I. Hakomori, *Annu. Rev. Biochem.* **50,** 733 (1981).

[6] Y. A. Hannun and R. M. Bell, *Science* **243,** 500 (1989).

[7] A. H. Merill, *J. Bioenerg. Biomembr.* **23,** 83 (1991).

[8] S.-I. Hakomori and Y. Igarashi, *Adv. Lipid Res.* **25,** 147 (1993).

[9] R. M. Devant, *Kontakte* 11 (1992).

[10] K.-H. Jung and R. R. Schmidt, *in* "Lipid Synthesis and Manufacture" (F. D. Gunstone, ed.), p. 208. Sheffield Academic Press, 1998.

[11] R. R. Schmidt and P. Zimmermann, *Tetrahedron Lett.* **27,** 481 (1986); *Liebigs Ann. Chem.* 663 (1988).

[12] R. R. Schmidt, *in* "Synthesis in Lipid Chemistry" (J. H. P. Tyman, ed.), p. 93. The Royal Society of Chemistry, Cambridge, 1996.

METHODS IN ENZYMOLOGY, VOL. 311

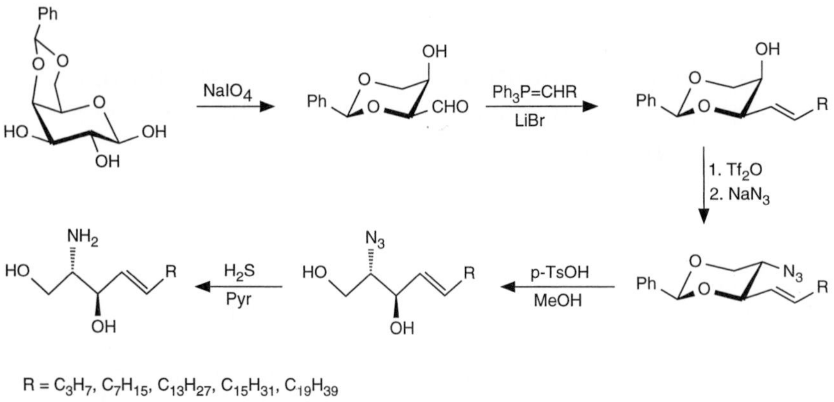

R = C_3H_7, C_7H_{15}, $C_{13}H_{27}$, $C_{15}H_{31}$, $C_{19}H_{39}$

SCHEME 1

glycosyl acceptor in the synthesis of glycosphingolipids than sphingosine itself or other protected derivatives.[12]

From D-galactose, 1,3-O-benzylidene-D-threose is prepared in two steps by benzylidenation[13] and periodate cleavage. The C_4 building block obtained has two stereogenic centers in the required configuration and has suitable protection for the introduction of the nitrogen under inversion of the configuration. The long chain is attached by Wittig reaction with the ylide prepared from an alkyl phosphonium salt. Because of the presence of lithium bromide, the trans-configurated double bond is formed stereoselectively. Conversion of the unprotected 2-OH group into the triflate and substitution by azide are performed in a one-pot reaction. Acid-catalyzed deprotection yields azidosphingosine with the desired D-erythro configuration. By reduction with hydrogen sulfide, natural sphingosine is obtained in only six steps overall.

Application of the described reaction sequence led to C_8-,[14] C_{12}-,[15] C_{18}-,[11] C_{20}-,[11] and C_{24}-sphingosines,[15] which are useful for biochemical studies. The influence of the alkyl chain length on membrane transport and sphingolipid biosynthesis have been investigated.[14,16,17]

[13] E. G. Gros and V. Deulofeu, J. Org. Chem. 29, 3647 (1964).
[14] A. Karrenbauer, D. Jeckel, W. Just, R. Birk, R. R. Schmidt, J. E. Rothmann, and F. Wieland, Cell 63, 259 (1990).
[15] R. Birk, Dissertation, Universität Konstanz, 1991.
[16] D. Jeckel, A. Karrenbauer, R. Birk, R. R. Schmidt, and F. Wieland, FEBS Lett. 261, 155 (1990).
[17] G. van Echten, R. Birk, G. Brenner-Weiß, R. R. Schmidt, and K. Sandhoff, J. Biol. Chem. 65, 9333 (1990).

When radiolabeled compounds are required for biochemical investigations, the 3-[3H]-labeled derivatives of ceramide,[18] cerebrosides,[18,19] gangliosides,[20,21] and sphingomyelin[22] are commonly used. They can be synthesized easily by oxidation/reduction via their 3-keto derivatives. The stereoselectivity of the reduction has not been investigated in detail. The synthesis of [3-[3H]]sphingosine itself, as well as of [3-[3H]]azidosphingosine, is described[23] in Scheme 2 starting from readily accessible azidosphingosine. Selective oxidation of the allylic hydroxy group with DDQ leads to 3-ketoazidosphingosine, which is reduced with sodium boro[3H]hydride to yield a mixture of the 3-[3H]-labeled azidosphingosine with natural D-*erythro* and also L-*threo* configurations. After separation by preparative thin-layer chromatography (TLC), they can be reduced with triphenylphosphine to yield 3-[3H]-labeled sphingosine and L-*threo*-sphingosine, respectively. Thus, the homologous C_{12} and C_{18} derivatives have been prepared[23] and used for the investigation of the sphingolipid biosynthesis regulation by sphingoid bases.[17,24]

Another possible position for tritium labeling has been shown in the synthesis of 1-[3H]-labeled sphingosine[25,26] and ceramide,[26,27] which were obtained from sphingosine, also applying an oxidation/reduction sequence. [7-[3H]]sphingosine has been prepared[28] by a total chemical synthesis using [3-[3H]]tetradecanal according to the synthesis of Shapiro *et al.*[29] Labeling of glycospingolipids and sphingomyelins by catalytic tritiation causes transformation of the sphingosine double bond to [4,5-[3H]]dihydrosphingosine and seems therefore to be less useful.[30] Radiolabeled ceramide containing

[18] Y. Kishimoto and M. T. Mitry, *Arch. Biochem. Biophys.* **161,** 426 (1974).

[19] M. Iwamori, H. W. Moser, and Y. Kishimoto, *J. Lipid Res.* **16,** 332 (1975).

[20] R. Ghidoni, S. Sonnino, M. Masserini, P. Orlando, and G. Tettamanti, *J. Lipid Res.* **22,** 1286 (1981).

[21] V. Chigorno, M. Valsecchi, M. Nicolini, S. Sonnino, *Indian J. Biochem. Biophys.* **34,** 150 (1997).

[22] Y. Stein, K. Oette, Y. Dabach, G. Hollander, M. Ben-Naim, and O. Stein, *Biochem. Biophys. Acta* **1084,** 87 (1991).

[23] R. Birk, G. Brenner-Weiß, A. Giannis, K. Sandhoff, and R. R. Schmidt, *J. Label. Comp. Radiopharm.* **29,** 289 (1991).

[24] E. C. Mandon, G. van Echten, R. Birk, R. R. Schmidt, and K. Sandhoff, *Eur. J. Biochem.* **198,** 667 (1991).

[25] T. Toyokuni, M. Nisar, B. Dean, and S. Hakomori, *J. Label. Comp. Radiopharm.* **29,** 567 (1991).

[26] S. Inooka and T. Toyokuni, *Commun. Appl. Cell Biol.* **11,** 42 (1994) [*Chem. Abst.* **124,** 232894 (1996)].

[27] K. J. Anand, K. K. Sadozai, and S.-I. Hakomori, *Lipids* **31,** 995 (1996).

[28] W. Stoffel and G. Sticht, *Hoppe-Seyler's Z. Physiol. Chem.* **348,** 1561 (1967).

[29] D. Shapiro, K. H. Sega, and H. M. Flowers, *J. Am. Chem. Soc.* **80,** 1194 (1958).

[30] G. Schwarzmann, *Biochim. Biophys. Acta* **529,** 106 (1978).

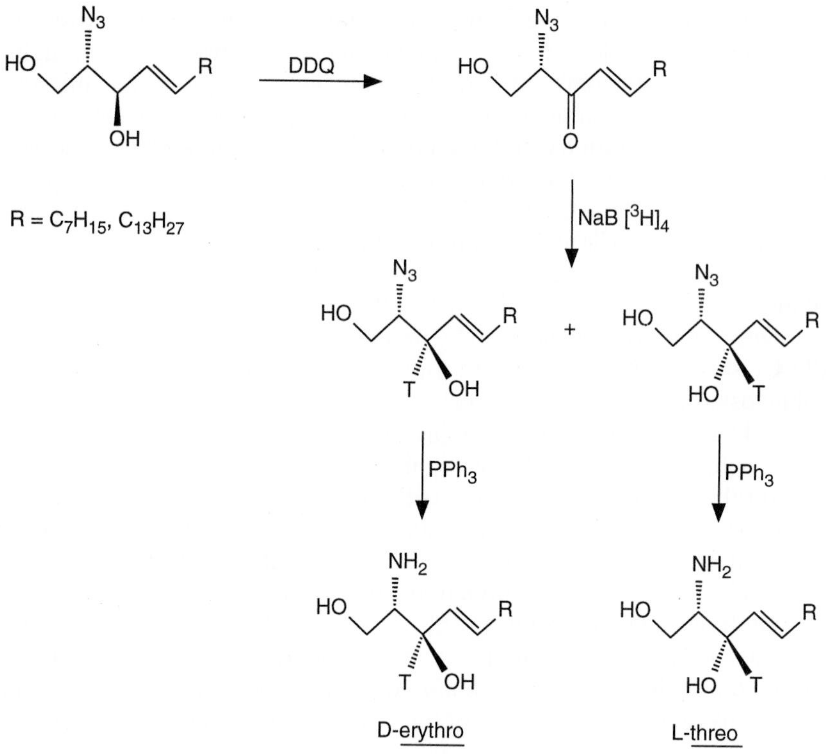

$R = C_7H_{15}, C_{13}H_{27}$

SCHEME 2

a [2,3-^{3}H]fatty acyl residue has been synthesized[14] and used for investigations on sphingomyelin biosynthesis.[14,16]

^{14}C is another suitable isotope commonly used for the radiolabeling of compounds. With [1-^{14}C]-fatty acids, many labeled ceramides,[27,31] glycosphingolipids,[27,31] sphingomyelins,[31,32] and sulfatides[33] have been synthesized and used for biochemical investigations.

In addition to common sphingosine, more than 60 rare derivatives with modified long chain (additional cis-configurated double bonds and/or methyl side chains) have been found in nature.[34–36] According to the strat-

[31] A. H. Futerman and R. E. Pagano, Methods Enzymol. 209, 437 (1992).
[32] Z. Dong and J. A. Butcher, Jr., Chem. Phys. Lipids 66, 41 (1993).
[33] M. Masson, W. X. Li, A. L. Fluharty, J. P. Beaucourt, J. C. Turpin, and N. Baumann, Clin. Chim. Acta 201, 157 (1991).
[34] K.-A. Karlsson, Chem. Phys. Lipids 5, 6 (1970).
[35] N. K. Kochetkov and G. P. Smirnova, Adv. Carbohydr. Chem. Biochem. 44, 387 (1986).
[36] B. Kratzer, Dissertation, Universität Konstanz, 1994.

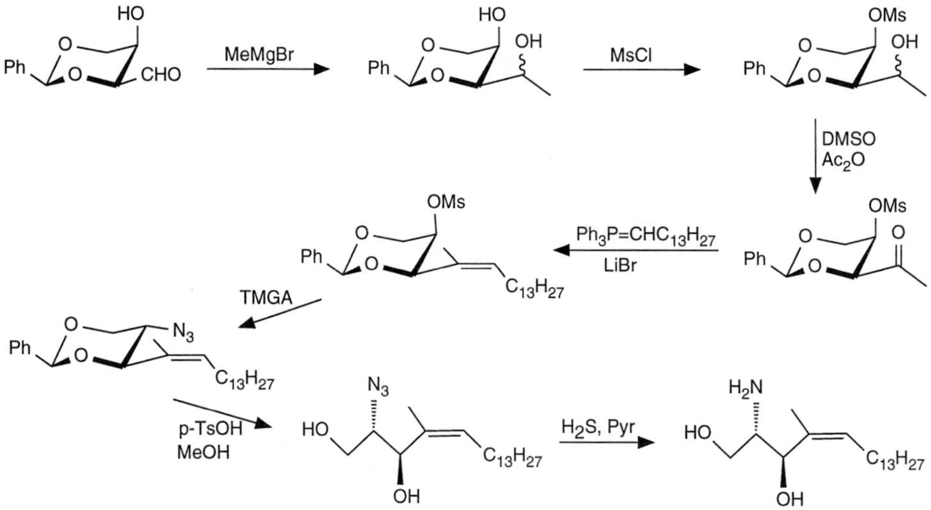

SCHEME 3

egy for the synthesis of azidosphingosine from galactose, *cis*-4-methylsphin-
gosine can be prepared easily (Scheme 3).[37] Again, 1,3-*O*-benzylidene-D-
threose is used as an intermediate, and at first the methyl group is introduced
by the Grignard addition of methylmagnesium bromide. Separation of the
diastereoisomeric products is possible by column chromatography, but not
required for the following steps. After regioselective mesylation of the
2-OH group and oxidation of the 4-OH group to the ketone under Pfitzer–
Moffat conditions, the long chain is attached by the Wittig reaction. The *cis*
configuration of the double bond has been determined by nuclear magnetic
resonance (NMR) spectroscopy with nuclear Overhauser effect (NOE)
experiments. Substitution of the mesylate by azide under inversion of the
configuration and acid-catalyzed deprotection leads to 4-methyl-*cis*-azido-
sphingosine, which can be reduced with hydrogen sulfide as usual to yield
4-methyl-*cis*-sphingosine. Other methyl-branched sphingosines were prepared similarly[37] and used for investigations on their influence on serine
palmitoyltransferase activity in sphingolipid biosynthesis.[38]
    The synthesis of some 1-amino derivatives of sphingosine is described

[37] T. Bär, B. Kratzer, R. Wild, K. Sandhoff, and R. R. Schmidt, *Liebigs Ann. Chem.* 419 (1993).
[38] G. van Echten-Deckert, A. Zschoche, T. Bär, R. R. Schmidt, A. Raths, T. Heinemann, and
    K. Sandhoff, *J. Biol. Chem.* **272,** 15825 (1997).

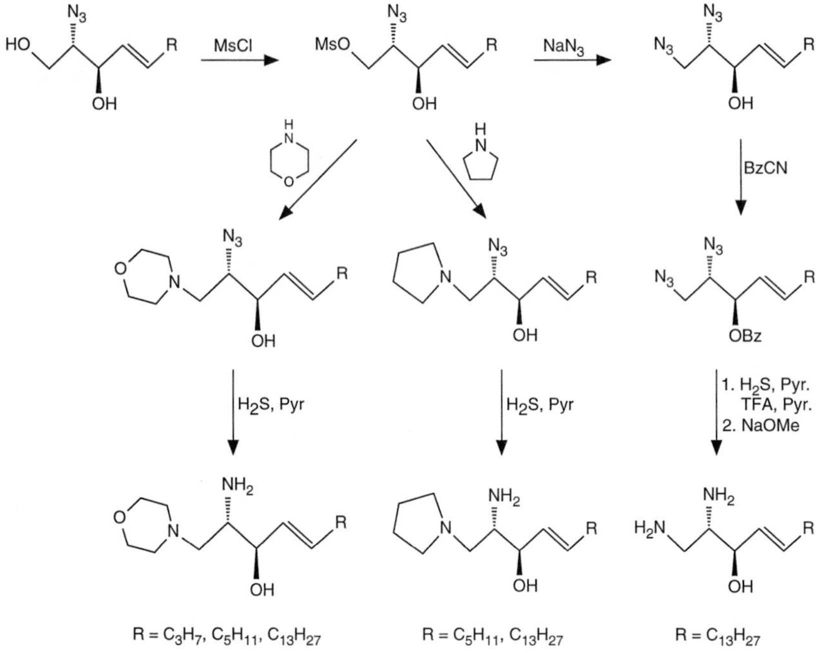

SCHEME 4

(Scheme 4). The readily accessible azidosphingosine is converted into its 1-O-mesyl derivative. By reaction with morpholine, pyrrolidine, and sodium azide the 1-morpholino[39] and 1-pyrrolidino[36,39] derivatives of azidosphingosine, as well as 1,2-diazidosphingosine,[40] are obtained. Reduction with hydrogen sulfide in the usual way yields 1-morpholino-1-deoxy-, 1-pyrrolidino-1-deoxy-, and 1-amino-1-deoxysphingosine, the latter by means of benzoyl protection of the 3-OH group. The influence of 1-morpholino and other derivatives of sphingosine on melanoma cell motility and growth has been investigated,[41] and the 1-pyrrolidino derivative of sphingosine has been shown to be a potent inhibitor of fucosyltransferase,[42] which is involved in glycoprotein synthesis.

[39] T. Bär, Dissertation, Universität Konstanz, 1991.

[40] A. Zaheer, S. M. Khaliq-uz-Zaman, and R. R. Schmidt, unpublished results.

[41] C. Helige, J. Smolle, R. Fink-Puches, R. Hofmann-Wellenhof, E. Hartmann, T. Bär, R. R. Schmidt, and H. A. Tritthart, *Clin. Exp. Metastasis* **14,** 477 (1996).

[42] J. Fischer, R. Zeitler, T. Bär, B. Kratzer, R. R. Schmidt, and W. Reutter, *Fresenius J. Anal. Chem.* **343,** 164 (1992).

General Material and Methods

Solvents are purified in the usual ways, the boiling range of the petroleum ether: 35–60°. The given melting points are uncorrected. Optical rotations are measured with a Perkin-Elmer polarimeter 241 MC; 1-dm cell. The following are used: for (TLC): DC-Plastikfolien Kieselgel 60 F_{254} (Merck; layer thickness 0.2 mm); column chromatography: Kieselgel 60 (Merck; 0.063–0.200 mm); flash chromatography: silica gel (Fa. J. T. Baker, particle size 40 μm); reversed phase chromatography: silica RP18 60 Å (Fa. ICN, particle size 32–63 μm). NMR spectra are taken with the internal standard tetramethylsilane (TMS). Sodium boro[^{3}H]hydride was from Fa. Amersham [specific activity 9.76 (Ci/mmol)].

Preparation of C_{18}-Sphingosine[11]

2,4-O-Benzylidene-D-threose

To a solution of 4,6-O-benzylidene-D-galactose[13] (30.0 g, 111 mmol) in 0.05 M phosphate buffer, pH 7.6 (1.200 ml), sodium periodate (55.0 g, 257 mmol) is added under vigorous stirring. During addition, the pH is adjusted between 7 and 8 by the dropwise addition of 2 N sodium hydroxide solution. After stirring for 1.5 hr the solution is concentrated to dryness *in vacuo*. The solid residue is extracted with tetrahydrofuran (4 × 250 ml). The solution is filtered, dried (MgSO$_4$), and concentrated to yield crude 2,4-O-benzylidene-D-threose (20.0 g, 85%), which is used for the next step without further purification. TLC (toluene/ethanol 3 : 1): R_f = 0.64.

(2R,3R,4E)-1,3-O-Benzylidene-4-octadecene-1,3,4-triol

Tetradecyltriphenylphosphonium bromide (70.0 g, 130 mmol) is suspended in dry, oxygen-free toluene (1000 ml) under nitrogen atmosphere. The mixture is cooled to −30°; phenyllithium, which is prepared from lithium (6.50 g, 940 mmol) and bromobenzene (74.0 g, 470 mmol) in dry diethyl ether (200 ml), is added dropwise. A solution of 4,6-O-benzylidene-D-threose (21.6 g, 104 mmol) in dry tetrahydrofurane (150 ml) is then added dropwise within a 20-min period. After stirring for another 20 min, methanol (150 ml) and then water (250 ml) are added under vigorous stirring. The organic solution is separated and concentrated *in vacuo*. Column chromatography (petroleum ether/ethyl acetate, 9 : 1) of the residue yields the product (27.0 g, 68%); TLC (petroleum ether/ethyl acetate 9 : 1): R_f = 0.21; mp 54–55° (from ethanol); $[\alpha]_D^{22}$ = −3.8 (c = 0.5, CHCl$_3$). ^{1}H NMR (250 MHz, CDCl$_3$): δ = 0.87 (t, 3 H, CH$_3$, J = 6.4 Hz), 1.18–1.50 (m, 22 H, 11 CH$_2$), 2.08 (m, 2 H, 6-H), 2.62 (d, 1 H, OH, $J_{2,OH}$ = 10.4 Hz), 3.53 (dd,

1 H, 2-H, $J_{1a,2}$ = 1.8 Hz, $J_{2,OH}$ = 10.4 Hz), 4.07 (dd, 1 H, 1b-H, $J_{1b,2}$ = 1.2 Hz, J_{gem} = 12 Hz), 4.25 (dd, 1 H, 1a-H, $J_{1a,2}$ = 1.8 Hz, J_{gem} = 12 Hz), 4.42 (d, 1 H, 3-H, $J_{3,4}$ = 6.1 Hz), 5.65 (m, 2 H, 4-H, CHPh), 5.87 (m, 1 H, 5-H), 7.38, 7.52 (2 m, 5 H, C_6H_5).

(2S,3R,4E)-2-Azido-1,3-benzylidene-4-octadecene-1,3-diol

To a solution of (2R,3R,4E)-1,3-O-benzylidene-4-octadecene-1,3-diol (10.0 g, 25.0 mmol) in dry dichloromethane (70 ml) is added dry pyridine (5 ml). Trifluoromethanesulfonic anhydride (8.70 g, 31.0 mmol) is then added at −15°, slowly, dropwise. After 15 min, dry dimethylformamide (250 ml) and sodium azide (7.5 g, 100 mmol) are added. After stirring for 5 h at room temperature, the mixture is poured on water (1000 ml) and extracted with petroleum ether (3 × 100 ml). The organic solution is dried (MgSO$_4$) and concentrated *in vacuo*. Column chromatography (petroleum ether/ethyl acetate, 9:1) of the residue yields the product (8.5 g, 82%) as a colorless oil; TLC (petroleum ether/ethyl acetate 9:1): R_f = 0.80; $[\alpha]_D^{20}$ = −11.7 (c = 3, CHCl$_3$). ^{1}H NMR (250 MHz, CDCl$_3$): δ = 0.87 (t, 3 H, CH$_3$, J = 6.4 Hz), 1.10–1.50 (m, 22 H, 11 CH$_2$), 2.11 (m, 2 H, 6-H), 3.48 (ddd, 1 H, 2-H, $J_{1b,2}$ = 4.8 Hz, $J_{2,3}$ = 9.1 Hz, $J_{1a,2}$ = 11 Hz), 3.61 (dd, 1 H, 1a-H, $J_{1a,2}$ = J_{gem} = 11 Hz), 4.05 (dd, 1 H, 3-H, $J_{3,4}$ = 7.3 Hz, $J_{2,3}$ = 9.1 Hz), 4.33 (dd, 1 H, 1b-H, $J_{1b,2}$ = 4.8 Hz, J_{gem} = 11 Hz), 5.49 (s, 1 H, CHPh), 5.58 (dd, 1 H, 4-H, $J_{3,4}$ = 7.3 Hz, $J_{4,5}$ = 15.6 Hz), 5.97 (m, 1 H, 5-H), 7.37, 7.48 (2 m, 5 H, C_6H_5).

(2S,3R,4E)-2-Azido-4-octadecene-1,3-diol (Azidosphingosine)

To a solution of (2S,3R,4E)-2-azido-1,3-benzylidene-4-octadecene-1,3-diol (8.00 g, 19.3 mmol) in dry methanol (200 ml) are added dry dichloromethane (50 ml) and p-toluenesulfonic acid (170 mg). After stirring for 12 hr the mixture is neutralized with solid NaHCO$_3$. The mixture is filtered and the solution is concentrated. Column chromatography (dichloromethane/methanol, 97.5:2.5) of the residue yields azidosphingosine (4.30 g, 68%) as colorless crystals; mp. 49–50°; TLC (dichloromethane/methanol 95:5): R_f = 0.46; $[\alpha]_D^{22}$ = −32.9 (c = 4, CHCl$_3$). ^{1}H NMR (250 MHz, CDCl$_3$): δ = 0.88 (t, 3 H, CH$_3$, J = 6.4 Hz) 1.15–1.47 (m, 22 H, 11 CH$_2$), 1.95–2.15 (m, 4 H, 6-H, 2 OH), 3.54 (m, 1 H, 2-H), 3.80 (m, 2 H, 1-H), 4.27 (m, 1 H, 3-H), 5.55 (dd, 1 H, 4-H, $J_{3,4}$ = 6.5 Hz, $J_{4,5}$ = 15.5 Hz), 5.83 (m, 1 H, 5-H).

(2R,3R,4E)-2-Amino-4-octadecene-1,3-diol (Sphingosine)

A solution of (2S,3R,4E)-2-azido-4-octadecene-1,3-diol (0.20 g, 0.61 mmol) in pyridine/water (2:1, 15 μl) is saturated with hydrogen sulfide.

After stirring for 48 hr at room temperature the solution is concentrated *in vacuo*. Column chromatography with chloroform/methanol (9:1) followed with chloroform/methanol/water (6.5:3:0.5) yields sphingosine (0.18 g, 95%) as a colorless solid; mp. 79–82° [Ref. 43: mp. 82.5°], $[\alpha]_D^{22} = -2.5$ (c = 6, CHCl$_3$) [Ref. 44: $[\alpha]_D^{22} = -5$ (c = 0.5, CHCl$_3$)].

Homologous C$_8$-,[14] C$_{12}$-,[15] C$_{20}$-,[11] and C$_{24}$-sphingosines[15] are prepared in the same manner.

Preparation of Radiolabeled C$_{18}$-Sphingosine[23]

(2S,4E)-2-Azido-4-octadecene-1-ol-3-on

To a solution of (2S,3R,4E)-2-azido-4-octadecene-1,3-diol (azidosphingosine) (618 mg, 1.9 mmol) in dry benzene (20 ml) is added at room temperature under nitrogen atmosphere 2,3-dichloro-5,6-dicyano-1,4-benzoquinone (DDQ, 522 mg, 2.3 mmol). After stirring for 8–10 days at room temperature the solution is concentrated to dryness *in vacuo*. The residue is dissolved in petroleum ether (200 ml) and washed with 5% aqueous sodium hydroxide solution (30 ml), with a saturated ammonium chloride solution (30 ml), and with brine (30 ml). The organic solution is dried (MgSO$_4$) and concentrated *in vacuo*. The residue is purified by flash chromatography to yield the product (393 mg, 65%) as a colorless solid; mp. 39.5–40.5°; TLC (dichloromethane/methanol, 95:5): $R_f = 0.62$; $[\alpha]_D^{20} = -3.2$ (c = 1, CHCl$_3$). ^{1}H NMR (400 MHz, CDCl$_3$): $\delta = 0.88$ (t, 3 H, CH$_3$, $J = 6.8$ Hz, 1.15–1.37 (m, 20 H, 10 CH$_2$), 1.46–1.49 (m, 2 H, 7-H), 2.24–2.30 (m, 3 H, 6-H, OH), 3.89–4.03 (m, 2 H, 1-H), 4.15 (t, 1 H, 2-H, $J_{1,2} = 4.9$ Hz), 6.36 (d, 1 H, 4-H, $J_{4,5} = 15.6$ Hz), 7.09 (dt, 1 H, 5-H, $J_{5,6a} = 6.8$ Hz, $J_{5,6b} = 6.8$ Hz, $J_{4,5} = 15.9$ Hz).

(2S,3R,4E)-[3-^{3}H]-2-Azido-4-octadecene-1,3-diol ([3-^{3}H] Azidosphingosine) and (2S,3S,4E)-[3-^{3}H]-2-Azido-4-octadecene-1,3-diol ([3-^{3}H]-L-threo-azidosphingsine)

A solution of (2S,4E)-2-azido-4-octadecene-1-ol-3-on (5.5 mg, 17 μmol) in dry diglyme (500 μl) is frozen under argon atmosphere by liquid nitrogen. This freezing procedure is repeated after the addition of more dry diglyme (500 μl). A solution of sodium boro[^{3}H]hydride (50 mCi, 4 μmol) in dry diglyme (100 μl) is then added to the frozen solution. The reaction mixture is allowed to warm to 0° and is stirred vigorously. After 6–8 hr, the reaction

[43] E. Klenk and W. Dibold, *Hoppe-Seyler's Z. Physiol. Chem.* **350**, 1081 (1969).
[44] E. J. Reist and P. H. Christie, *J. Org. Chem.* **35**, 4127 (1970).

mixture is kept for 2 days at 5°. For workup, the reaction mixture is poured into water (20 ml) and methanol (20 ml). Exchangeable tritium is removed by passing through a column with reversed-phase silica gel (RP-18) eluating with water (400 ml). The tritiated product is eluated successively with methanol (250 ml) and methanol/chloroform (1:1, 300 ml). The solution is concentrated to dryness by a nitrogen stream and the residue is purified by preparative TLC (silica gel; petroleum ether/ethyl acetate 1:1) to yield the (2S,3R,4E)-configurated product (1.6 mg, 29%), TLC (petroleum/ether 1:1): $R_f = 0.57$, specific radioactivity: 2900 mCi/mmol, and the (2S,3S,4E)-configurated product (2.7 mg, 49%), TLC (petroleum ether/ethyl acetate 1:1): $R_f = 0.65$, specific radioactivity (3600 mCi/mmol). Analysis by TLC and radiochromatogram scanning indicate the purity. TLC and physical data are identical to those of the corresponding unlabeled compound.

(2S,3R,4E)-[3-³H]-2-Amino-4-octadecene-1,3-diol ([3-³H]Sphingosine)

To a solution of (2S,3R,4E)-[3-³H]-2-azido-4-octadecene-1,3-diol in tetrahydrofurane/water (5:1, 1.5 ml) is added triphenylphosphine (1.5 equivalents) at room temperature. When the reaction is completed (TLC) after a few days, purification is accomplished directly by preparative TLC (silica gel; chloroform/methanol/2 N aqueous NH₃, 40:10:1). The product is extracted from the silica gel with methanol. The solution is filtered through reversed-phase silica gel (RP-18) and then concentrated to dryness by a nitrogen stream to yield the product; TLC (chloroform/methanol/2 N aqueous NH₃, 40:10:1): $R_f = 0.29$; specific radioactivity: 531 mCi/mmol. The purity of the product is checked by TLC and radiochromatogram scanning. Physical data agree with those of the unlabeled compound.

(2S,3S,4E)-[3-³H]-2-Amino-4-octadecene-1,3-diol

([3-³H]-L-*threo*-sphingosine) is prepared from (2S,3S,4E)-[3-³H]-2-azido-4-octadecene-1,3-diol as described for the (2S,3R,4E) isomer. TLC (chloroform/methanol/2 N aqueous NH₃, 40:10:1): $R_f = 0.17$; specific radioactivity: 966 mCi/mmol.

The homologous C_{12} derivatives are prepared in the same manner.[23]

Preparation of 4-Methyl-*cis*-sphingosine[37]

(2S,3R,4R)-1,3-O-Benzylidenepentane-1,2,3,4-tetrol and (2R,3R,4S)-1,3-O-Benzylidenepentane-1,2,3,4-tetrol

A solution of 4,6-O-benzylidene-D-threose (37.0 g, 178 mmol) in tetrahydrofuran (100 ml) is added dropwise under nitrogen to a mixture of 3 M methylmagnesium bromide solution in diethyl ether (180 ml, 540 mmol) and tetrahydrofuran (100 ml). The reaction mixture is then hydrolyzed with an ice-cold saturated ammonium chloride solution; the aqueous layer is saturated with sodium chloride and the mixture is extracted with tetrahydrofuran (3 × 150 ml). The combined extracts are dried (MgSO$_4$) and concentrated. The diastereomeric products can be separated by column chromatography (dichloromethane/methanol, 97.5:2.5). (2R,3R,4R)-1,3-O-benzylidenepentane-1,2,3,4-tetrol: 15.5 g (39%), colorless crystals, mp. 155°; TLC (petroleum ether/ethyl acetate, 25:75): R_f = 0.42; $[\alpha]_D^{20}$ = −0.6 (c = 4, CHCl$_3$). ^{1}H NMR (250 MHz, CDCl$_3$): δ = 1.33 (d, $J_{4,5}$ = 6.4 Hz, 3 H, CH$_3$), 2.54 (br. s, 2 H, 2 OH), 3.64 (dd, $J_{2,3}$ = 1.3 Hz, $J_{3,4}$ = 6.7 Hz, 1 H, 3-H), 3.89 (ddd, $J_{1a,2}$ = 1.5 Hz, $J_{1b,2}$ = 1.8 Hz, $J_{2,3}$ = 1.3 Hz, 1 H, 2-H), 4.05 (dd, $J_{1a,1b}$ = 12.2 Hz, $J_{1a,2}$ = 1.5 Hz, 1 H, 1a-H), 4.10 (dq, $J_{3,4}$ = 6.7 Hz, $J_{4,5}$ = 6.4 Hz, 1 H, 4-H), 4.27 (dd, $J_{1a,1b}$ = 12.2 Hz, $J_{1b,2}$ = 1.8 Hz, 1 H, 1b-H), 5.59 (s, 1 H, C$_6$H$_5$CH), 7.35–7.53 (m, 5 H, C$_6$H$_5$). (2R,3R,4S)-1,3-O-benzylidenepentane-1,2,3,4-tetrol: 14.5 g (36%), colorless oil: TLC (petroleum ether/ethyl acetate, 25:75): R_f = 0.48; $[\alpha]_D^{20}$ = +8.45 (c = 3.0, chloroform).^{1}H NMR (250 MHz, CDCl$_3$): δ = 1.25 (d, $J_{4,5}$ = 6.4 Hz, 3 H, CH$_3$), 2.73 (br. s, 2 H, 2 OH), 3.59–3.64 (m, 2 H, 2-H, 3-H), 4.01 (dd, $J_{1a,1b}$ = 12.2 Hz, $J_{1a,2}$ = 1.2 Hz, 1 H, 1a-H), 4.06 (dq, $J_{3,4}$ = 6.7 Hz, $J_{4,5}$ = 6.4 Hz, 1 H, 4-H), 4.21 (dd, $J_{1a,1b}$ = 12.2 Hz, $J_{1b,2}$ = 1.8 Hz, 1 H, 1b-H), 5.57 (s, 1 H, C$_6$H$_5$CH), 7.35–7.52 (m, 5 H, C$_6$H$_5$).

(2R,3R,4R)-1,3-O-Benzylidene-2-(methylsulfonyloxy)pentane-1,3,4-triol and (2R,3R,4S)-1,3-O-Benzylidene-2-(methylsulfonyloxy)pentane-1,3,4-triol

To a mixture of (2R,3R,4R)-1,3-O-benzylidenepentane-1,2,3,4-tetrol and (2S,3R,4S)-1,3-O-benzylidenepentane-1,2,3,4-tetrol (12.0 g, 53.5 mmol) in dichloromethane (100 ml) and pyridine (20 ml) molecular sieves (4 Å) is added. After cooling to −35°, a solution of methanesulfonyl chloride (10.5 g, 58.9 mmol) in dichloromethane is added dropwise under nitrogen. After 12 hr at −25° the mixture is stirred at −10° for 12 hr (TLC control). Methanol (20 ml) is then added and after 2 hr at −10° the mixture is warmed up to room temperature, washed with water (100 ml), dried (MgSO$_4$), and concentrated to dryness. Filtration of the residue over silica gel (dichloromethane/methanol, 97.5:2.5) and then recrystallization (ethyl acetate/petroleum ether) yields a mixture of the 4R isomer and the 4S isomer (12.3 g, 76%) as a white solid; TLC (petroleum ether/ethyl acetate 3:7): R_f = 0.50. ^{1}H NMR (250 MHz, CDCl$_3$): d = 1.28 (d, $J_{4,5}$ = 5.9 Hz, 3 H, CH$_3^{4S}$), 1.34 (d, $J_{4,5}$ = 6.1 Hz, 3 H, CH$_3^{4R}$), 2.63 (s, 1 H, OH^{4S}), 2.91 (d, J = 5.2 Hz, 1 H, OH^{4R}), 3.13 (s, 3 H, SCH$_3^{4S}$), 3.15 (s, 3 H, SCH$_3^{4R}$),

3.67 (dd, $J_{2,3}$ = 1.2 Hz, $J_{3,4}$ = 8.9 Hz, 1 H, 3-H^{4R}), 3.74 (dd, $J_{2,3}$ = 1.2 Hz, $J_{3,4}$ = 8.3 Hz, 1 H, 3-H^{4S}), 3.92–4.12 (m, 2 H, 4-Ha, 4-H^{4S}), 4.08 (dd, $J_{1a,1b}$ = 13.7 Hz, $J_{1a,2}$ = 1.2 Hz, 1 H, 1a-H^{4S}), 4.13 (dd, $J_{1a,1b}$ = 15.5 Hz, $J_{1a,2}$ = 1.2 Hz, 1 H, 1a-H^{4R}), 4.50 (dd, $J_{1a,1b}$ = 13.7 Hz, $J_{1b,2}$ = 1.2 Hz, 1 H, 1b-H^{4R}), 4.57 (dd, $J_{1a,1b}$ = 15.5 Hz, $J_{1b,2}$ = 1.8 Hz, 1 H, 1b-H^{4S}), 4.72 (ddd, $J_{1a,2}$ = 1.2 Hz, $J_{1b,2}$ = 1.8 Hz, $J_{2,3}$ = 1.2 Hz, 1 H, 2-H^{4S}), 4.95 (ddd, $J_{1a,2}$ = 1.2 Hz, $J_{1b,2}$ = 1.2 Hz, $J_{2,3}$ = 1.2 Hz, 1 H, 2-H^{4R}), 5.58 (s, 1 H, C_6H_5CH^{4R}), 5.60 (s, 1 H, C_6H_5CH^{4S}), 7.36–7.54 (m, 10 H, C_6H_5^{4R}, C_6H_5^{4S}).

(3S,4R)-3,5-O-Benzylidene-3,5-dihydroxy-4-(methylsulfonyloxy)-2-pentanone

A solution of a mixture of (2R,3R,4R)-1,3-O-benzylidene-2-(methylsulfonyloxy)pentane-1,3,4-triol and (2R,3R,4S)-1,3-O-benzylidene-2-(methylsulfonyloxy)pentane-1,3,4-triol (11.4 g, 37.7 mmol) and acetic anhydride (25 ml) in DMSO (70 ml) is stirred at room temperature for 6 hr. After evaporation of the solvent at 0.1 mbar (maximum 35° bath temperature), chromatography of the residue with petroleum ether/ethyl acetate/methanol (60 : 40 : 2) yields the product (10.0 g, 89%) as a colorless oil. An analytically pure product can be obtained by recrystallization (methanol); TLC (petroleum ether/ethyl acetate 3 : 7): R_f = 0.78; $[\alpha]_D^{20}$ = −145 (c = 0.6, CHCl$_3$). ^{1}H NMR (250 MHz, CDCl$_3$): δ = 2.36 (s, 3 H, CH$_3$), 3.02 (s, 3 H, SCH$_3$), 4.15 (dd, $J_{4,5}$ = 1.1 Hz, $J_{5a,5b}$ = 13.6 Hz, 1 H, 5a-H), 4.45 (d, $J_{3,4}$ = 1.8 Hz, 1 H, 3-H), 4.63 (dd, $J_{4,5b}$ = 1.5 Hz, $J_{5a,5b}$ = 13.6 Hz, 1 H, 5b-H), 4.91 (ddd, $J_{3,4}$ = 1.8 Hz, $J_{4,5a}$ = 1.1 Hz, $J_{4,5b}$ = 1.5 Hz, 1 H, 4-H), 5.64 (s, 1 H, C_6H_5CH), 7.39–7.58 (m, 5 H, C_6H_5).

(2R,3R,4Z)-1,3-O-Benzylidene-4-methyl-2-(methylsulfonyloxy)-4-octadecene-1,3-diol

To a suspension of triphenyltetradecylphosphonium bromide (18.4 g, 33.35 mmol) in dry toluene (300 ml, saturated with nitrogen) under nitrogen is added phenyllithium [prepared from lithium (470 mg, 66.70 mmol) and bromobenzene (5.4 g, 33.35 mmol)] in dry diethyl ether (20 ml). The mixture is cooled to −25° and a solution of (3R,4R)-3,5-benzylidene-3,5-dihydroxy-4-(methylsulfonyloxy)-2-pentanone (4.0 g, 13.35 mmol) in dry tetrahydrofuran (80 ml) is added dropwise within 10 min. After 10 min at 0° a saturated ammonium chloride solution (10 ml) is added. The mixture is warmed up to room temperature and the organic layer is separated and concentrated. Chromatography (petroleum ether/ethyl acetate, 9 : 1) of the residue yields the product (2.05 g, 32%) as an oil; TLC (petroleum ether/ethyl acetate 9 : 1): R_f = 0.26; $[\alpha]_D^{20}$ = −35.2 (c = 1.15 CHCl$_3$). ^{1}H NMR (400 MHz, CDCl$_3$): δ = 0.88 (t, J = 6.6 Hz, 3 H, CH$_3$), 1.18–1.42 (m, 22 H, 11 CH$_2$),

1.89 (s, 3 H, CH_3), 2.03–2.10 (m, 2 H, 6a-H, 6b-H), 3.04 (s, 3 H, SCH_3), 4.21 (d, $J_{1a,1b}$ = 13.2 Hz, 1 H, 1a-H), 4.50 (d, $J_{1a,1b}$ = 13.2 Hz, 1 H, 1b-H), 4.59 (s, 1 H, 3-H), 4.89 (s, 1 H, 2-H), 5.39 (m, 1 H, 5-H), 5.66 (s, 1 H, C_6H_5CH), 7.32–7.55 (m, 5 H, C_6H_5).

(2S,3R,4Z)-2-Azido-1,3-O-benzylidene-4-methyl-4-octadecene-1,3-diol

A mixture of (2R,3R,4Z)-1,3-O-benzylidene-4-methyl-2-methylsulfo-nyloxy)-4-octadecene-1,3-diol (520 mg, 1.05 mmol) and tetramethylguani-dinium azide (TMGA) (1.5 g, 9.5 mmol) in dry DMF (30 ml, free of amines) is stirred at 95° for 6 hr. The mixture is then poured into water (300 ml) and extracted with petroleum ether (2 × 100 ml). After the combined extracts have been dried ($MgSO_4$) and concentrated, the residue is purified by column chromatography (petroleum ether/ethyl acetate, 85 : 15) to yield the product (330 mg, 70%) as an oil; TLC (petroleum ether/ethyl acetate 85 : 15: R_f = 0.73, $[\alpha]_D^{20}$ = −27.1 (c = 1, chloroform). ^{1}H NMR (250 MHz, $CDCl_3$): δ = 0.88 (t, J = 6.1 Hz, 3 H, CH_3), 1.25–1.55 (m, 22 H, 11 CH_2), 1.85 (s, 3 H, CH_3), 2.04–2.21 (m, 2 H, 6a-H, 6b-H), 3.65 (dd, $J_{1a,1b}$ = 10.0 Hz, $J_{1a,2}$ = 10.6 Hz, 1 H, 1a-H), 3.72 (ddd, $J_{1a,2}$ = 10.6 Hz, $J_{1b,2}$ = 4.3 Hz, $J_{2,3}$ = 9.2 Hz, 1 H, 2-H), 4.36 (dd, $J_{1a,1b}$ = 10.0 Hz, $J_{1b,2}$ = 4.3 Hz, 1 H, 1b-H), 4.53 (d, $J_{2,3}$ = 9.2 Hz, 1 H, 3-H), 5.52 (s, 1 H, C_6H_5CH), 5.57 (t, $J_{5,6}$ = 6.4 Hz, 1 H, 5-H), 7.32–7.49 (m, 5 H, C_6H_5). Irradiation at δ = 1.85 (4-CH_3) enhances the 5-H and the 2-H signals. Irradiation at δ = 2.04–2.21 (6a/6b-H) enhances the 3-H signal.

(2S,3R,4Z)-2-Azido-4-methyl-4-octadecene-1,3-diol (4-Methyl-cis-azidosphingosine)

To a solution of (2S,3R,4Z)-2-azido-1,3-O-benzylidene-4-methyl-4-octa-decene-1,3-diol (307 mg, 0.718 mmol) in dry methanol (20 ml) and dry dichloromethane (20 ml) is added aqueous hydrochloric acid (20%, 4 ml). After 2 days at 40° the mixture is neutralized with a saturated sodium hydrogen carbonate solution (10 ml) and extracted three times with diethyl ether. The combined organic extracts are dried ($MgSO_4$) and concentrated. Column chromatography (petroleum ether/ethyl acetate, 85 : 15) of the residue yields 4-methyl-cis-azidosphingosine (190 mg, 78%) as a colorless oil; TLC (petroleum ether/ethyl acetate 85 : 15): R_f = 0.79, $[\alpha]_D^{20}$ = −57.3 (c = 1.15, $CHCl_3$). ^{1}H NMR (250 MHz, $CDCl_3$): δ = 0.85 (t, J = 6.6 Hz, 3 H, CH_3), 1.18–1.42 (m, 22 H, 11 CH_2), 1.76 (d, J_{allyl} = 0.8 Hz, 3 H, CH_3), 1.87 (br. s, 1 H, OH), 2.01–2.08 (m, 2 H, 6a-H, 6b-H), 2.16 (br. s, 1 H, OH), 3.54 (ddd, $J_{1a,2}$ = 4.9 Hz, $J_{1b,2}$ = 4.9 Hz, $J_{2,3}$ = 8.5 Hz, 1 H, 2-H), 3.78–3.98 (m, 2 H, 1a-H, 1b-H), 4.64 (d, $J_{2,3}$ = 8.5 Hz, 1 H, 3-H), 5.45 (t, $J_{5,6}$ = 7.6 Hz, 1 H, 5-H).

(2S,3R,4Z)-2-Amino-4-methyl-4-octadecene-1,3-diol
(4-Methyl-cis-sphingosine)

A solution of (2S,3R,4Z)-2-azido-4-methyl-4-octadecene-1,3-diol (190 mg, 0.56 mmol) in pyridine/water (1:1, 10 ml) is saturated with hydrogen sulfide and stirred at room temperature for 24 h (TLC control). The solvent is evaporated (bath temperature max 30°) and the residue is purified by column chromatography. Elution with chloroform/methanol (10:0 → 9:1) (to remove the sulfur) and then with chloroform/methanol (8:2 → 7:3) yields 4-methyl-cis-sphingosine (140 mg, 80%) as an amorphous solid; TLC (chloroform/methanol/2 N ammonia, 80:20:2): R_f = 0.32, $[\alpha]_D^{20}$ = −3.6 (c = 1, $CHCl_3$). ^1H NMR (400 MHz, $CDCl_3$): δ = 0.86 (t, J = 6.6 Hz, 3 H, CH_3), 1.18–1.38 (m, 22 H, 11 CH_2), 1.74 (s, 3 H, CH_3), 1.92–2.13 (m, 4 H, 6a-H, 6b-H, 2 OH), 3.07 (m, 1 H, 2-H), 3.81 (m, 2 H, 1a-H, 1b-H), 4.12 (br. s, 2 H, NH_2), 4.64 (m, 1 H, 3-H), 5.35 (t, $J_{5,6}$ = 6.6 Hz, 1 H, 5-H).

Preparation of 1-Amino Derivatives of Sphingosine

(2S,3R,4E)-2-Azido-1-(methylsulfonyloxy)-4-octadecene-3-ol

To a solution of (2S,3R,4E)-2-azido-4-octadecene-1,3-diol (azidosphingosine) (1.00 g, 3.08 mmol) in dry dichloromethane (60 ml) and pyridine (10 ml) a solution of methanesulfonyl chloride (363 mg, 4.15 mmol) in dichloromethane is added dropwise. After stirring for 15 hr, water (200 ml) is added and the mixture is extracted with dichloromethane (2 × 100 ml). The organic solution is dried ($MgSO_4$) and concentrated. Column chromatography (petroleum ether/ethyl acetate, 75:25) of the residue yields the product (945 g, 76%) as a colorless oil; TLC (petroleum ether/ethyl acetate 7:3): R_f = 0.37; $[\alpha]_D^{22}$ = −23.2 (c = 1.5, $CHCl_3$). ^1H NMR (250 MHz, $CDCl_3$): δ = 0.84 (t, J = 6.6 Hz, 3 H, CH_3), 1.15–1.38 (m, 22 H, 11 CH_2), 2.04 (m, 2 h, 6-Ha, 6-Hb), 2.39 (br. s, 1 H, OH), 3.04 (s, 3 H, S-CH_3), 3.66 (ddd, $J_{1a,2}$ = 7.5 Hz, $J_{1b,2}$ = 3.3 Hz, $J_{2,3}$ = 6.0 Hz, 1 H, 2-H), 4.15 (m, 1 H, 3-H), 4.21 (dd, $J_{1a,1b}$ = 10.8 Hz, $J_{1a,2}$ = 7.5 Hz, 1 H, 1-Ha), 4.35 (dd, $J_{1a,1b}$ = 10.8 Hz, $J_{1b,2}$ = 3.3 Hz, 1 H, 1-Hb), 5.46 (dd, $J_{3,4}$ = 7.4 Hz, $J_{4,5}$ = 15.3 Hz, 1 H, 4-H), 5.76 (dt, $J_{4,5}$ = 15.3 Hz, $J_{5,6}$ = 6.6 Hz, 1 H, 5-H).

(2S,3R,4E)-2-Azido-1-(4-morpholinyl)-4-octadecene-3-ol

A solution of (2S,3R,4E)-2-azido-1-(methylsulfonyloxy)-4-octadecene-3-ol (500 mg, 1.24 mmol) in toluene (50 ml) and morpholine (20 ml) is stirred for 15 hr at 70°. The solution is concentrated and the residue is purified by column chromatography (petroleum ether/ethyl acetate, 6:4) to yield the product (488 mg, 100%) as a colorless oil; TLC (petroleum ether/ethyl acetate 6 : 4): $R_f = 0.27$; $[\alpha]_D^{22} = -8.0$ (c = 1, CHCl$_3$). ^{1}H NMR (250 MHz, CDCl$_3$): $\delta = 0.84$ (t, $J = 6.6$ Hz, 3 H, CH$_3$), 1.15–1.41 (m, 22 H, 11 CH$_2$), 2.03 (m, 2 H, 6a-H, 6b-H), 2.46–2.68 (m, 6 H, 1a-H, 1b-H, CH$_2$NCH$_2$), 3.42 (m, 1 H, 2-H), 3.67 (t, $J = 4.6$ Hz, 4 H, CH$_2$OCH$_2$), 4.10 (m, 1 H, 3-H), 5.44 (dd, $J_{3,4} = 7.0$ Hz, $J_{4,5} = 15.4$ Hz, 1 H, 4-H), 5.77 (dt, $J_{4,5} = 15.4$ Hz, $J_{5,6} = 6.7$ Hz, 1 H, 5-H).

(2S,3R,4E)-2-Amino-1-(4-morpholinyl)-4-octadecene-3-ol[39]

A solution of (2S,3R,4E)-2-azido-1-(4-morpholinyl)-4-octadecene-3-ol (450 mg, 1.14 mmol) in pyridine/water (1 : 1, 30 ml) is saturated with hydrogen sulfide and stirred at room temperature for 15 hr. The solvent is evaporated (bath temperature 30°) and the residue is purified by column chromatography (chloroform/methanol, 95 : 5 → 9 : 1 → 8 : 2 → 7 : 3) to yield the product (399 mg, 95%) as an amorphous solid; TLC (chloroform/methanol/ 2 N ammonia 80 : 20 : 2): $R_f = 0.70$; $[\alpha]_D^{22} = +15.5$ (c = 0.5, CHCl$_3$). ^{1}H NMR (250 MHz, CDCl$_3$): $\delta = 0.84$ (t, $J = 6.6$ Hz, 3 H, CH$_3$), 1.15–1.42 (m, 22 H, 11 CH$_2$), 2.05 (m, 2 H, 6a-H, 6b-H), 2.37–3.05 (m, 10 H, 1a-H, 1b-H, 2-H, CH$_2$NCH$_2$, OH, NH$_2$), 3.66 (t, $J = 4.6$ Hz, 4 H, CH$_2$OCH$_2$), 3.88 (m, 1 H, 3-H), 5.37 (dd, $J_{3,4} = 7.5$ Hz, $J_{4,5} = 15.4$ Hz, 1 H, 4-H), 5.71 (dt, $J_{4,5} = 15.4$ Hz, $J_{5,6} = 6.7$ Hz, 1 H, 5-H).

The homologous C$_8$ and C$_{10}$ derivatives are prepared in the same manner.[39]

(2S,3R,4E)-2-Azido-1-(1-pyrrolidinyl)-4-octadecene-3-ol

A solution of (2S,3R,4E)-2-azido-1-(methylsulfonyloxy)-4-octadecene-3-ol (320 mg, 0.79 mmol) in toluene (30 ml) and pyrrolidine (10 ml) is stirred for 5 hr at 65°. The solution is concentrated and the residue is purified by column chromatography (petroleum ether/ethyl acetate, 3 : 7) to yield the product (291 mg, 97%) as a colorless oil; TLC (petroleum ether/ ethyl acetate 55 : 65): $R_f = 0.25$; $[\alpha]_D^{22} = -12.0$ (c = 1, CHCl$_3$). ^{1}H NMR (250 MHz, CDCl$_3$): $\delta = 0.86$ (t, $J = 6.6$ Hz, 3 H, CH$_3$), 1.15–1.43 (m, 22 H, 11 CH$_2$), 1.76 (m, 4 H, CH$_2$CH$_2$NCH$_2$CH$_2$), 2.06 (m, 2 H, 6a-H, 6b-H), 2.62 (m, 4 H, CH$_2$NCH$_2$), 2.75 (dd, $J_{1a,1b} = 12.5$ Hz, $J_{1a,2} = 4.9$ Hz, 1 H, 1a-H), 2.83 (dd, $J_{1a,1b} = 12.5$ Hz, $J_{1b,2} = 8.3$ Hz, 1 H, 1b-H), 3.33 (ddd, $J_{1a,2} = 4.9$ Hz, $J_{1b,2} = 8.3$ Hz, $J_{2,3} = 7.5$ Hz, 1 H, 2-H), 4.11 (m, 1 H, 3-H), 5.48 (dd, $J_{3,4} = 6.9$ Hz, $J_{4,5} = 15.3$ Hz, 1 H, 4-H), 5.81 (dt, $J_{4,5} = 15.3$ Hz, $J_{5,6} = 7.0$ Hz, 1 H, 5-H), 6.22 (br. s, 1 H, OH).

(2S,3R,4E)-2-Amino-1-(1-pyrrolidinyl)-4-octadecene-3-ol[39]

A solution of (2S,3R,4E)-2-azido-2-(1-pyrrolidinyl)-4-octadecene-3-ol (250 mg, 0.66 mmol) in pyridine/water (2 : 1, 15 ml) is saturated with hydrogen sulfide and stirred at room temperature for 15 hr. The solvent is evaporated (bath temperature 30°) and the residue is purified by column chromatography (chloroform/methanol 9:1 → 8:2 → 1:1) to yield the product (190 mg, 82%) as an amorphous solid; TLC (chloroform/methanol/2 N ammonia 80:20:2): R_f = 40; $[\alpha]_D^{22}$ = +17.3 (c = 1.5, CHCl$_3$). ^{1}H NMR (250 MHz, CDCl$_3$): δ = 0.83 (t, J = 6.6 Hz, 3 H, CH$_3$), 1.15–1.39 (m, 22 H, 11 CH$_2$), 1.74 (m, 4 H, CH$_2$CH$_2$NCH$_2$CH$_2$), 2.01 (m, 2 H, 6a-H, 6b-H), 2.52–2.83 (m, 7 H, 1a-H, 1b-H, 2-H, CH$_2$NCH$_2$), 3.36 (br. s, 3 H, NH$_2$, OH), 3.85 (m, 1 H, 3-H), 5.37 (dd, $J_{3,4}$ = 7.6 Hz, $J_{4,5}$ = 15.4 Hz, 1 H, 4-H), 5.70 (dt, $J_{4,5}$ = 15.4 Hz, $J_{5,6}$ = 6.6 Hz, 1 H, 5-H).

The homologous C$_{10}$ derivative is prepared in the same manner.[36]

(2S,3R,4E)-1,2-Diazido-4-octadecene-3-ol

To a solution of (2S,3R,4E)-2-azido-1-(methylsulfonyloxy)-4-octadecene-3-ol (100 mg, 0.25 mmol) in dry dimethylformamide (1 ml) is added sodium azide (37 mg, 0.57 mmol), and the mixture is stirred at 60° for 24 hr. Water is then added and the mixture is extracted with ethyl acetate. The organic solution is dried (MgSO$_4$) and concentrated *in vacuo*. Flash chromatography (petroleum ether/ethyl acetate, 8:2) of the residue yields the product (66 mg, 76%) as a yellowish oil; TLC (petroleum ether/ethyl acetate, 8:2): R_f = 0.56; $[\alpha]_D^{22}$ = −12.4 (c = 1, CHCl$_3$). ^{1}H NMR (250 MHz, CDCl$_3$): δ = 0.84 (t, J = 6.4 Hz, 3 H, CH$_3$), 1.13–1.40 (m, 22 H, 11 CH$_2$), 2.05 (m, 3 H, OH, 6a-H, 6b-H), 3.39 (dd, $J_{1a,2}$ = 7.4 Hz, J_{gem} = 12.7 Hz, 1 H, 1a-H), 3.43 (dd, $J_{1b,2}$ = 4.2 Hz, J_{gem} 12.7 Hz, 1 H, 1b-H), 3.53 (ddd, $J_{1a,2}$ = 7.4 Hz, $J_{1b,2}$ = 4.2 Hz, $J_{2,3}$ = 5.3 Hz, 1 H, 2-H), 4.17 (dd, $J_{2,3}$ = 5.3 Hz, $J_{3,4}$ = 7.4 Hz, 1 H, 3-H), 5.48 (dd, $J_{3,4}$ = 7.4 Hz, $J_{4,5}$ = 15.4 Hz, 1 H, 4-H), 5.79 (dt, $J_{4,5}$ = 15.4 Hz, $J_{5,6}$ = 6.6 Hz, 1 H, 5-H).

(2S,3R,4E)-3-Benzoyloxy-1,2-diazido-4-octadecene

To a solution of (2S,3R,4E)-1,2-diazido-4-octadecene-3-ol (160 mg, 0.46 mmol) and triethylamine (0.01 ml, 0.07 mmol) in dry dichloromethane (7 ml) is added benzoyl cyanide (7.0 mg, 0.51 mmol). The mixture is stirred at room temperature for 12 hr. Water is then added, and the mixture is extracted with ethyl acetate. The organic solution is dried (MgSO$_4$) and concentrated *in vacuo*. Flash chromatography (petroleum ether/ethyl acetate, 8:2) of the residue yields the product (190 mg, 90%) as a colorless oil; TLC (petroleum ether/ethyl acetate 8:2): R_f = 0.88; $[\alpha]_D^{22}$ = −20.9 (c = 1, CHCl$_3$).^{1}H NMR (250 MHz, CDCl$_3$): δ = 0.85 (t, J = 6.4 Hz, 3 H, CH$_3$), 1.12–1.41 (m, 22 H, 11 CH$_2$), 2.05 (m, 2 H, 6a-H, 6b-H), 3.34 (m, 2 H, 1a-H, 1b-H), 3.87 (ddd, $J_{1a,2}$ = 4.0 Hz, $J_{2,3}$ = 5.4 Hz, $J_{1b,2}$ = 7.4 Hz,

1 H, 2-H), 5.60 (m, 2 H, 4-H, 3-H), 5.95 (m, 1 H, 5-H), 7.48, 7.60, 8.08 (3 m, 5 H, Ph).

(2S,3R,4E)-3-Benzoyloxy-1,2-di-O-trifluoroacetylamino-4-octadecene

A solution of (2S,3R,4E)-3-benzoyloxy-1,2-diazido-4-octadecene (65 mg, 0.14 mmol) in pyridine/water (2:1, 6 ml) is saturated with hydrogen sulfide and stirred at room temperature for 15 hr. The solution is concentrated *in vacuo,* and the residue is dissolved in dry pyridine. Trifluoroacetic acid (190 mg, 0.9 mmol) is added dropwise at $-10°$. After stirring for 2 hr the solution is concentrated *in vacuo.* Flash chromatography (petroleum ether/ethyl acetate, 8:2) of the residue yields the product (150 mg, 69%) as a colorless oil; TLC (petroleum ether/ethyl acetate 8:2): $R_f = 0.42$; $[\alpha]_D^{22} = +6.7$ (c = 0.5, CHCl$_3$). ^{1}H NMR (250 MHz, CDCl$_3$): $\delta = 0.85$ (t, $J = 6.4$ Hz, 3 H, CH$_3$), 1.13–1.40 (m, 22 H, 11 CH$_2$), 2.20 (m, 2 H, 6a-H, 6b-H), 3.75 (dd, $J_{1a,1b} = 10.3$ Hz, $J_{1a,2} = 7.2$ Hz, 1 H, 1a-H), 3.95 (dd, $J_{1a,1b} = 10.3$ Hz, $J_{1b,2} = 3.6$ Hz, 1 H, 1b-H), 4.19 (m, 1 H, 2-H), 4.25 (m, 1 H, 3-H), 5.55 (dd, $J_{3,4} = 7.0$ Hz, $J_{4,5} = 15.3$ Hz, 1 H, 4-H), 5.80 (dt, $J_{4,5} = 15.3$ Hz, $J_{5,6} = 6.5$ Hz, 1 H, 5-H), 6.85 (s, 1 H, NH), 7.77, 7.60, 7.48 (3 m, 5 H, Ph).

(2R,3R,4E)-1,2-Diamino-4-octadecene-3-ol[40]

To a solution of (2S,3R,4E)-3-benzoyloxy-1,2-di-O-trifluoroacetyl-4-octadecene (100 mg, 0.17 mmol) in dry methanol (12 ml) is added 1 N sodium methanolate solution (0.5 ml, 0.5 mmol). After stirring at room temperature for 15 hr the solution is concentrated *in vacuo.* Flash chromatography (chloroform/methanol 95:5 → 9:1 → 8:2) of the residue yields the product as benzoate (45 mg, 90%) as a colorless oil; TLC (chloroform/methanol/2 N ammonia 8:2:0.2): $R_f = 0.43$; $[\alpha]_D^{22} = -1.4$ (c = 0.5, CHCl$_3$). ^{1}H NMR (250 MHz, CDCl$_3$): $\delta = 0.89$ (t, $J = 6.4$ Hz, 3 H, CH$_3$), 1.13–1.41 (m, 22 H, 11 CH$_2$), 2.02–2.34 (m, 5 H, 2 NH$_2$, OH), 3.70 (m, 2 H, 1a-H, 1b-H), 4.29 (m, 1 H, 2-H), 4.48 (m, 1 H, 3-H), 5.54 (dd, $J_{3,4} = 7.0$ Hz, $J_{4,5} = 15.3$ Hz, 1 H, 4-H), 5.85 (dt, $J_{4,5} = 15.3$ Hz, $J_{5,6} = 6.5$ Hz, 1 H, 5-H).

[41] Total Synthesis of Sphingosine and Its Analogs

By Päivi M. Koskinen and Ari M. P. Koskinen

Introduction

The enigmatic nature of the natural membrane lipid sphingosine has intrigued scientists both in medicine and in chemistry for over a century. The revelations of its vast biological importance brought forward via the rapid evolution of enhanced laboratory techniques have increased the pace of research on this essential lipid.

Before 1997, ca. 65 syntheses of sphingosine and its isomers had been published, as well as almost as many syntheses of various analogs and other sphingoid bases. Sphingosine syntheses published before mid-1991 have been reviewed thoroughly by Devant,[1] and syntheses up to 1998 are reviewed by Koskinen and Koskinen.[2] Although many of the strategies behind the published syntheses are very good, weaknesses exist in one or several key steps in the synthetic sequence. For example, it has been difficult to maintain the stereochemistry in the chiral centers and the E/Z selectivity of the double bond. Moreover, syntheses employing relayed or external asymmetric induction are rarely feasible on a larger scale, as neither the chiral auxiliaries nor the catalysts are regenerated and recycled easily. Most importantly, a drawback of almost all published syntheses is that they usually are applicable only to the synthesis of sphingosine itself. These shortcomings prompted us to seek a new route to sphingosine applicable also to its derivatives and various analogs.

General Overview of the Strategy

A starting point for our sphingosine synthesis was to utilize an easily available inexpensive starting material, L-serine, and employ internal asymmetric induction as the key strategy for the selective generation of new asymmetric centers. More specifically, we have developed methodology for the following two reaction types: (1) trans-selective double bond formation with a modified Horner–Wadsworth–Emmons (HWE) reaction starting from a serine-derived β-ketophosphonate[3–6] and (2) diastereoselective

---

[1] R. M. Devant, *Kontakte* 11 (1992).
[2] A. M. P. Koskinen and P. M. Koskinen, *Synthesis* 1075 (1998).
[3] W. S. Wadsworth, Jr. and W. D. Emmons, *J. Am. Chem. Soc.* **83,** 1733 (1961).

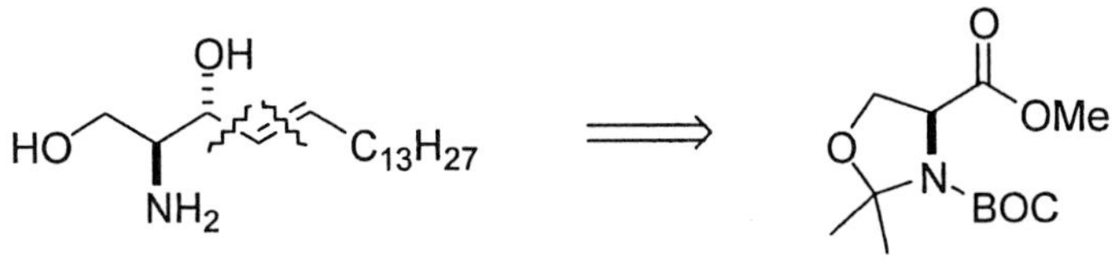

FIG. 1. Strategic bonds in sphingosine leading to L-serine, presented here in its fully protected N-BOC methyl ester form.

enone reductions, where the existing chiral center is used to induce chirality in the new carbinol center.[7,8]

There are two strategic bonds in sphingosine, which, if cleaved retrosynthetically, could eventually lead to L-serine, as can be seen in Fig. 1. L-Serine is presented in its fully protected N-BOC methyl ester form.

L-Serine introduces the C-1–C-3 portion of sphingosine with the desired S-configuration at C-2. The C-4 is introduced with the formation of the β-ketophosphonate. The double bond in E-configuration can then be introduced via the modified HWE reaction using tetradecanal as the required aldehyde, and the resulting enone can be reduced to the corresponding anti or syn allylic alcohol selectively. The synthetic sequence is presented in general form in Scheme 1.

Formation of β-Ketophosphonate and Chiral α,β-Unsaturated Enone

To overcome the problem in E/Z selectivity and to enhance the reactivity of the nonstabilized substrates in the double bond formation, Wadsworth and Emmons[3] used phosphoryl-stabilized carbanions derived from phosphonates, but they were still unable to form the carbanions from nonstabilized phosphonates. Corey and Kwiatkowski[4] reported back in 1966 that such anions can be formed as lithio derivatives through the action of one equivalent of n-BuLi in THF. Heathcock and van Geldern[9] applied this to their rac-norsecurinine synthesis using dimethyl methylphosphonate (DMMP) and reacting it with N-BOC-proline methyl ester. Similarly,

[4] E. J. Corey and G. T. Kwiatkowski, J. Am. Chem. Soc. 88, 5654 (1966).
[5] A. M. P. Koskinen and M. J. Krische, Synlett 665 (1990).
[6] A. M. P. Koskinen and P. M. Koskinen, Synlett 501 (1993).
[7] A. M. P. Koskinen and P. M. Koskinen, Tetrahedron Lett, 34, 6765 (1993).
[8] A. M. P. Koskinen and P. M. Koskinen, U.S. Patent 5,426,228 (1995).
[9] C. H. Heathcock and T. W. van Geldern, Heterocycles 25, 75 (1987).

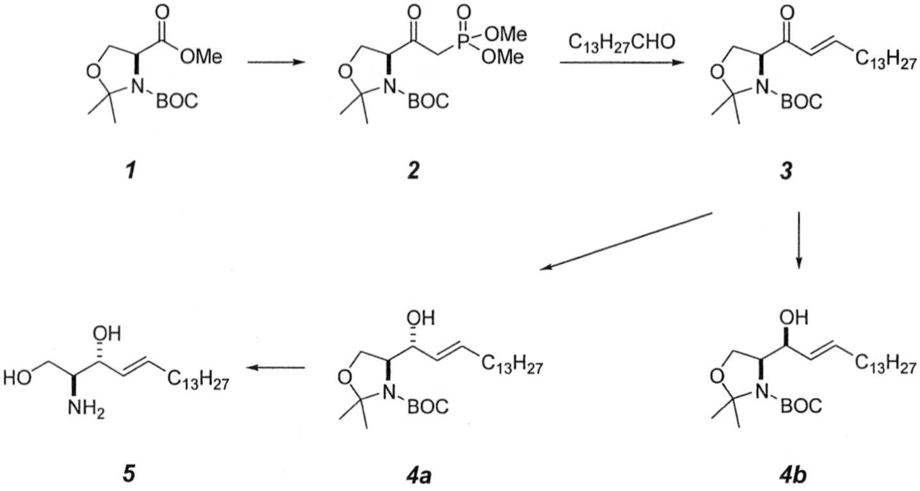

SCHEME 1. General strategy for the sphingosine synthesis.

Yamanoi et al.[10] applied the same methodology to their sphingosine synthesis.

A β-ketophosphonate formed from N-BOC-protected serine methyl ester provides a suitable precursor for a *trans*-selective enone. Accordingly, we chose this as the starting point for our work toward sphingosine and related compounds.

In preliminary work by Koskinen and Krische,[5] the N-BOC serine methyl ester was added to the preformed lithiophosphonate according to the Corey and Kwiatkowski[4] procedure. However, we discovered that one equivalent of DMMP is not sufficient because the yields were consistently below 60%. One reason for this is that the lithiated phosphonate carbanion also forms an adduct with the β-ketophosphonate product, thus consuming an extra equivalent of DMMP. Another possibility could be insufficient generation of the anion, but this explanation was ruled out by performing the reaction with excess n-BuLi with respect to DMMP and varying the time and temperature independently for the formation of the anion. In none of these experiments did the yield of the desired product increase, but the excess base resulted in an increased elimination reaction as a result

[10] T. Yamanoi, T. Akiyama, E. Ishida, H. Abe, M. Amemiya, and T. Inazu, *Chem. Lett.* 335 (1989).

of a direct attack of n-BuLi on serine α-proton, affording what is tentatively assigned as methyl vinyl ether **6**.[11]

6   **7**

Under these conditions, another side product, possibly **7**, began to appear. Unfortunately, this coeluted with the desired β-ketophosphonate and could not be purified for confirmative analysis.[12] However, the formation of **7** could be explained similarly to that of **6** in terms of an abstraction of the α-proton with excess n-BuLi and subsequent elimination. However, the intramolecular HWE is a prerequisite for the reaction and because the desired addition/elimination (A/E) is the kinetic route, performing the reaction below $-70°$ (internal temperature) diminishes the side reaction. After experimenting, the optimal amounts of DMMP and n-BuLi were found to be 210 and 205 mol%, respectively. The best quench conditions proved to be 10% citric acid at 0° adjusting pH to 3.

As demonstrated earlier, the β-ketophosphonate formation with N-BOC serine methyl ester is an equilibrium reaction. This also means that the reaction would not go to completion, and ca. 10% of the starting material remained unreacted. However, the unreacted material was separated very easily from the product and recycled. Chromatographic purification of **2** is necessary anyway because excess DMMP has to be removed. This is done easily by gel filtration with flash silica gel using a two-step gradient: the starting material is eluted with 20% EtOAc in hexanes and the product with 75% EtOAc in hexanes. The alleged methyl vinyl ether **6** is rather nonpolar, but separates from the unreacted starting material. If **7** is formed, an extra gradient step (50% EtOAc in hexanes) has to be added. This purifies the desired product, although with some loss in the yield. The serine-derived β-ketophosphonate **2** is a noncrystalline solid with enantiomeric purity >99%ee determined by chiral high-performance liquid chromatography (HPLC). We have scaled the reaction described earlier up to 100 mmol (ca. 25 g of N-BOC serine methyl ester), and the yields are repeatedly 70–90%.

[11] The tentative structure is assigned based on the ^{1}H NMR δ 1.45 (s, 9H BOC), 3.80 (s, 3H OMe), 5.69 (s, 2H), 6.12 (s, 2H), 7.00 (bs, 1H).

[12] The tentative structure is based on the ^{1}H NMR signals in the mixture of **2** and possibly **7** and comparison with the spectrum of pure **2**.

In the formation of α,β-unsaturated enones, the nature and hence the basicity and reactivity of anions derived from phosphoryl-stabilized reagents have been investigated extensively by Seyden-Penne and co-workers[13,13a] by infrared (IR) and nuclear magnetic resonance (NMR) spectroscopy. The authors concluded that the basicity and reactivity are dependent on both the solvent and the associated cation. They have proposed that metal cations and especially Li^+ most likely form a tight complex with the phosphonate-derived carbanions, thereby enhancing its acidity.

Factors influencing the reactivity per se are now considered to be well established. However, when the phosphonate has a labile chiral center in the γ-position, as is the case with β-ketophosphonate,[2] the classical methods relying on Et_3N, NaH, t-BuO$^-$K$^+$, and even K_2CO_3 in toluene lead to epimerization.[14]

To Masamune and Roush and co-workers[15] these results suggested the possibility that, in the presence of a lithium salt, chiral center bearing phosphonates could be deprotonated easily under mild conditions by amine bases, hence without racemization. The amines they chose were DBU and diisopropylethylamine (DIPEA) with LiCl as the salt. Indeed, they were able to perform HWE reactions without loss of enantiomeric integrity. Another example by Déziel et al.[16] suggests the use of N-methylmorpholine instead of Et_3N. This, however, diminishes the E/Z selectivity considerably.

During earlier studies relating to aminocyclopropanation in our laboratory, we discovered that K_2CO_3 in MeCN is a mild but effective base/solvent system.[17] Potassium cation in MeCN is known to form similar complexes, although weaker than the ones formed by lithium cations.[13] This suggests that a stronger base is needed for HWE. A carbonate base is strong enough to promote the reaction,[14] but under the conditions used earlier, namely with toluene as solvent, it will cause epimerization. Since not only the cation but also the solvent has an effect on reactivity, the question we wanted to answer was: Will MeCN decrease the basicity of K_2CO_3 enough to prevent epimerization, but not enough to interfere with the olefin formation? Achievement of these effects was believed to be

[13] T. Bottin-Strzalko, J. Corset, F. Froment, M.-J. Pouet, J. Seyden-Penne, and M.-P. Simonnin, *J. Org. Chem.* **45,** 1270 (1980).

[13a] T. Strzalko, J. Seyden-Penne, F. Froment, J. Corset, and M.-P. Simonnin, *J. Chem. Soc. Perkin Trans II* 783 (1987).

[14] J. Villiers, M. Rambaud, and B. Kirschleger, *Phosphorus Sulfur* **14,** 385 (1983).

[15] M. A. Blanchette, W. Choy, J. T. Davis, A. P. Essenfeld, S. Masamune, W. R. Roush, and T. Sakai, *Tetrahedron Lett.* **25,** 2183 (1984).

[16] R. Déziel, R. Plante, V. Caron, L. Grenier, M. Llinas-Brunet, J.-S. Duceppe, E. Malenfant, and N. Moss, *J. Org. Chem.* **61,** 2901 (1996).

[17] A. M. P. Koskinen and L. Muñoz, *J. Org. Chem.* **58,** 879 (1993).

possible because K_2CO_3 only dissolves partially in MeCN. Because carbonate is an inorganic base, it would also be attractive from the practical point of view: it is an inexpensive reagent and purification of the products should be easy. This latter is an advantage for γ-chiral α,β-unsaturated ketones because these are labile compounds and all unnecessary handling should be avoided. Preliminary results with 2 and various aldehydes showed that this system can indeed be used as an effective yet mild reagent in HWE olefinations.[5,6]

In the reaction, β-ketophosphonate 2 was dissolved in dry MeCN, and K_2CO_3 was added at ambient temperature. After approximately 15 min the mixture turned pale yellow and formed a heterogeneous suspension. This was taken to indicate formation of the anion. Aldehyde was then added. The optimal amount of base proved to be 200–300 mol%. The reaction time is 25–30 hr when long alkyl chain-containing aldehydes are used. With the just-mentioned conditions there was not decrease in the enantiomeric purity (chiral HPLC).

For comparison, the same reactions were performed using the Masamune–Roush conditions: DBU or DIPEA and LiCl in MeCN.[15] Olefinations performed with K_2CO_3/MeCN gave clearly better yields with all substrates than the Masamune–Roush conditions. In a comparison of optical yields, DIPEA/LiCl/MeCN conditions were clearly better than the other methods, whereas DBU gave disastrous results: The Masamune–Roush conditions utilize organic bases and the product enones must always be purified by chromatography. However, the enones formed with K_2CO_3/ MeCN are usually obtained pure directly from the reaction through careful selection of the correct solvent for extraction. This is not just very convenient but of utmost importance, as some enones epimerize on purification on silica gel![18]

Our K_2CO_3/MeCN method showed its strength most with the actual target compound sphingosine enone 3. The HWE reaction in this case was highly enantioselective, contrary to the Masamune–Roush conditions, which gave both smaller chemical and poorer optical yields (Table I).

Because a small decrease in the optical purity with some substrates was also seen during the HWE reaction,[18] a test with tetradecanal was conducted. When the HWE reaction was allowed to run over 5 days with tetradecanal and 2, some epimerization of the enone was apparent after 3 days. The %ee dropped slowly from >95 to 85, accompanied with decomposition of the reaction mixture. However, it is important to notice that the

---

[18] A. M. P. Koskinen and P. M. Koskinen, unpublished results. Experimental details and rationalization of the modified Horner–Wadsworth–Emmons reaction are communicated separately.

TABLE I

COMPARISON OF CHEMICAL AND OPTICAL YIELDS IN THE HWE REACTION OF THE
β-KETOPHOSPHONATE **2** AND TETRADECANAL, WHEN K_2CO_3/MeCN, DIPEA/LiCl/MeCN,
OR DBU/LiCl/MeCN IS USED AS BASE/SOLVENT SYSTEM

Yield	K_2CO_3/MeCN	DIPEA/LiCl/MeCN	DBU/LiCl/MeCN
Chemical yield (%)	81	75	31
Optical yield (%ee)	>95	68	59

reaction was complete after the first 24 hr (without any loss in optical purity) and was continued for this test purpose only.

Because epimerization was also detected with some enones on purification,[18] the tendency for epimerization of **3** was checked on silica gel. This was done by stirring the enone with the appropriate chromatographic eluent and flash silica gel and analyzing the mixture at intervals with chiral HPLC. Even after 10 days sphingoenone **3** did not show any signs of epimerization and the test was discontinued.

Because tetradecanal dissolves only partially in MeCN, we therefore tried *i*-octane as a cosolvent (1 : 5 ratio with respect to MeCN). There was no significant improvement in either the yield or the reaction time. However, there was no loss of enantiomeric purity either. While this gives reason to believe that the scope of the reaction can be broadened, further studies are required of the nature of carbonate bases in different solvents.[18]

The HWE reaction with serine-derived β-ketophosphonate **2** proved to be at least 98% *E* selective because no *Z* isomer was detected by NMR.

Reduction of the Enone

The two faces of the plane of an unsymmetrically substituted carbonyl compound RCOR′ are enantiotopic, and the reactions of such compounds with chiral reducing agents give rise to asymmetric induction. This method is not preferred if one wishes to induce the selection internally from an existing chiral center in the molecule.

Relative to saturated ketones, enones present one more factor has to be controlled: the 1,4-reduction, which is a serious competitor to the desired 1,2-reduction. Several studies have compared the selectivity of various hydride reducing agents for allylic alcohol formation over saturated ketone or alcohol formation. Most of these studies have been conducted on cyclic enones,[19–19e] however, and results cannot be taken as direct guidelines for

[19] G. R. Meyer, *J. Chem. Educ.* **58,** 628 (1981) (LiAlH$_4$/THF and NaBH$_4$/MeOH).

[19a] A. L. Gernal and J.-L. Luche, *J. Am. Chem. Soc.* **103.** 5454 (1981) (NaBH$_4$/CeCl$_3$).

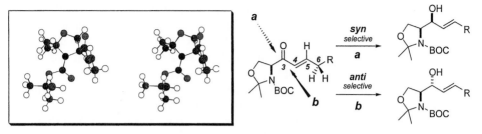

Fig. 2. (Left) A stereo representation and (right) a general structure of a serine-derived enone in an *s-cis* conformation. The possible axes for the approaching hydride are indicated by arrows *a* and *b* leading to *syn* selective or *anti* selective reduction product, respectively.

the reductions of acyclic enones because the conformational factors are much less restricted.

In the following work, our objective was to study which factors are dominating in the reactions between the serine-derived enone (general structure is shown in Fig. 2) and hydride reducing agents and to use this information in developing methods for selective reductions to *syn* and *anti* allylic amino alcohols (Fig. 2).

Syn Selective Reduction

Corey *et al.*[20] had discovered in their prostaglandin synthesis that in the reduction of the acyclic enone the ketone first has to adopt a *trans* coplanar arrangement of hydrogen at the C-5 and C-6 as shown in Fig. 2. They went on to point out, however, that this is not a sufficient prerequisite for the stereoselectivity of the carbonyl reduction, for two other requirements have to be fulfilled as well. First there has to be sufficient steric bulk to block approach along the *b* axis and second the enone has to adopt s-*cis* conformation in order to direct the formation of the 3-S alcohol. If these requirements can be met, an exogenous directing group can control the stereoselectivity.

In a stereo representation (CS Chem3D Pro™) of the general enone

[19b] S. Komiya and O. Tsutsumi, *Bull. Chem. Soc. Jpn.* **60**, 3423 (1987) (NaBH₄/SmCpCl₂ and NaBH₄/ErCpCl₂).

[19c] S. Fukuzawa, T. Fujinami, S. Yamauchi, and S. Sakai, *J. Chem. Soc Perkin. Trans. 1* 1929 (1986) (LiAlH₄/CeCl₃).

[19d] S. Kim and K. A. Ahn, *J. Org, Chem.* **49**, 1717 (1984) (DIBAL-H and DIBAL-H/BuLi).

[19e] S. Krishnamurthy and H. C. Brown, *J. Org. Chem.* **42**, 1197 (1977).

[20] E. J. Corey, K. B. Becker, and R. K. Varma, *J. Am. Chem. Soc.* **94**, 8616 (1972).

structure, also shown in Fig. 2, (R = methyl), it is shown that with our serine-derived enone the requirements for stereoselective reduction are met: the conformation is correct and the BOC group blocks the *b* axis. A noncoordinating hydride reagent should thus favor formation of the *syn* alcohol, which has the 3-S configuration. This selectivity should also increase with the length of the side chain due to the increase in steric bulk and further with a possible backfolding of an alkyl chain.

For our L-*threo*-sphingosine synthesis we chose L-Selectride®, lithium tri-*sec*-butylborohydride,[20] which is a sterically demanding reagent. In preliminary work it was originally used in two different types of solvents: the noncoordinating CH_2Cl_2 and the strongly coordinating THF.[7] The reactions were performed at $-78°$ with 300 mol% of L-Selectride®. Lesser amounts of reducing agent resulted in an incomplete reduction due to the coordination of the first equivalent of the reagent to the carbamate oxygen.

The reaction is complete on addition of the reagent, and the mixture can be warmed to room temperature before quenching the reaction with dilute citric acid or phosphoric acid. The basic and oxidative workup procedures were deliberately avoided because of the sensitive nature of the product. Because the acidic workup procedure is the least effective, the purified yields remained modest (50–70%). However, these are unoptimized conditions and with some experimenting could be improved because there are no impurities present in the reaction mixture (TLC, HPLC).

We have demonstrated with a model compound phenylenone **8** that solvent effect is visible: with the noncoordinating CH_2Cl_2, only reagent control is present and the selectivity drops, whereas with the strongly coordinating THF the steric bulk is the sole factor determining the selectivity and a clear preference is seen for the *syn* diastereomer.[7] In the case of sphingosine enone **3**, only the *syn* diastereomer could be detected by HPLC.

Antiselective Reduction

A simple hydride reducing agent widely known to promote *anti* diastereoselectivity is $NaBH_4$.[19] Normally it is used in alcoholic solvents. In preliminary work with model compounds, practically no diastereoselectivity was achieved regardless of the substrate or solvent.[7] However, no 1,4-reduction, which frequently occurs under these conditions,[19] was detected either. The nonexistent diastereoselectivity can be explained in terms of a partial formation and disproportionation of alkoxyborohydrides with the alcoholic solvent. Hence the reaction starts to resemble the *syn* selective reduction, but because there is more than one reducing species present, no selection is detected.

Gemal and Luche[19a] have studied the use of lanthanides in connection with reducing the portion of 1,4-reduction with BH_4^--reducing agents. They suggest that the lanthanide metal forms a complex with the alcoholic solvent, thus increasing the acidity of the medium and accelerating the reaction. Although we did not observe any 1,4-reduction in the previous experiments, the so-called Luche conditions ($CeCl_3 \times 10 \ H_2O/NaBH_4/ROH$)[19a] were applied next. According to theory, the lanthanide salt preferably complexes with the solvent with the result that no alkoxyborohydride forms. Therefore, $NaBH_4$ would be the only reducing agent improving the *anti* selectivity of the reaction. To our dismay the selectivity with $CeCl_3 \times H_2O$ was of the same order of magnitude with or without it. The selectivity for the reductions of model enone **8** with various $NaBH_4$-based conditions is listed in Table II.

A third approach was then tested. Because *syn* selectivity is achieved under noncoordinating conditions, preference for the formation of the *anti* diastereomer could arise from coordination of the reducing agent, followed by introduction of the hydride along the *b* axis (Fig. 2). This could be the result if a large and strongly coordinating Lewis acid were used. Such a reagent would force the side chain (C-2–C-3 bond) to rotate 180° to prevent collision between the Lewis acid and one of the ketal methyl groups (see the stereorepresentation in Fig. 2). This rotation would expose the *b* axis side to hydride attack more than the *a* axis side. To test this hypothesis, the five different reducing agents listed in Table II were tested. $LiAlH_4/Et_2O$ and RedAl/THF are aluminum hydrides, and hence coordinating reagents. Although they did show a slight preference for the *anti* diastereomer formation, they were clearly too small to give good selectivity. Solladié-Cavallo and Bencheqroun[21] obtained high diastereoselectivity in their *R*-halostatine analog synthesis using s-Bu(i-Bu)$_2$AlHLi/THF and n-Bu$_3$BHK/THF. In our case, however, the *anti* selectivity was poor and *syn* diastereomer formation was slightly preferred. In the absence of a protic solvent the anhydrous $CeCl_3$ could be expected to complex with the carbonyl and we therefore tried this with the large reducing agent L-Selectride®. In this case, the coordinating and steric factors quite clearly canceled each other out, as the selectivity disappeared completely.

Success was finally achieved with DIBAL-H. The first attempt with **8**, with THF as the solvent, in fact gave only a satisfactory result: 4:1 selectivity of the *anti* diastereomer as the major component.[7] However, when the coordinating polar solvent THF was changed to the noncoordinating nonpolar toluene, thus removing the influence of the reaction media as much as possible, the selectivity increased to 6:1.[8] Still better results were achieved

[21] A. Solladié-Cavallo and M. Bencheqroun, *Tetrahedron: Asymmetry* **2,** 1165 (1991).

TABLE II
DIASTEREOSELECTIVITIES IN REDUCTIONS OF MODEL COMPOUNDS USING VARIOUS
COORDINATING HYDRIDE-REDUCING AGENTS

A

8   9

B

10   11

Reaction	Reagent	*syn : anti*
A	NaBH$_4$/MeOH	3 : 1
A	NaBH$_4$/i-PrOH	5 : 6
A	NaBH$_4$/CeCl$_3$ × 10H$_2$O/MeOH	1 : 3
B	NaBH$_4$/CeCl$_3$ × 10H$_2$O/MeOH	1 : 4
A	NaBH$_4$/CeCl$_3$ × 10H$_2$O/i-PrOH	1 : 2
A	LiAlH$_4$/Et$_2$O	1 : 2
A	RedAl/THF	4 : 5
A	s-Bu(i-Bu)$_2$AlHLi/THF	3 : 1
A	n-Bu$_3$BHK/THF	7 : 2
B	L-Selectride®/CeCl$_3$/THF	1 : 1

when the method was applied to enones **3** and **12**. With DIBAL-H in toluene the sphingosine enone **3** gave a single *anti* diastereomer as did its pyrene analog **12**. These results are shown in Table III.

Surprisingly, in the case of the heteroatom-containing enone **10**, only a slight preference for the *anti* diastereomer was achieved. We also detected the formation of considerable amounts of 1,4-reduction product (Table III). This is peculiar, as DIBAL-H was developed precisely to prevent 1,4-reduction and results with other substrates have been very good.[22] Evidently

---

[22] E. Winterfeldt, *Synthesis* 617 (1975).

TABLE III
DIASTEREOSELECTIVITY AND REGIOSELECTIVITY IN THE DIBAL-H REDUCTION OF
PHENYLENONE **8**, SPHINGOSINE, AND ITS PYRENE ANALOG ENONES **3** AND **12**, AND
HETEROATOM CONTAINING ENONE **10**

Substrate[a]	syn:anti in toluene	syn:anti in THF	1,4
8	1:6	1:4	—
10	2:5	—	40%
3	1:>20[b]	—	—
12	1:>20[b]	—	—

[a] ϕ = (pyrene)

[b] No *syn* isomer was detected by HPLC.

a heteroatom in the allylic position of the enone represents a powerful electronic factor, and behavior of these substrates is under study.[23]

[23] Oxidation of optically active allylic alcohols has been studied extensively in our laboratory: O. A. Kallatsa and A. M. P. Koskinen, *Tetrahedron Lett.* **38**, 8895 (1997). A. M. P. Koskinen and P. M. Koskinen, manuscript in preparation.

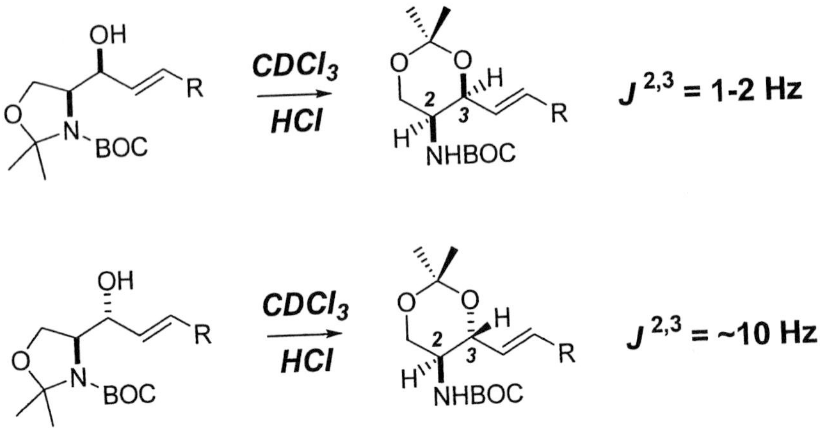

FIG. 3. Proof of the stereochemical assignment for allylic alcohols.

Proof of Stereochemistry

Herold[24] has shown that stereochemical assignment at the newly formed C-3 stereocenter can be made using NMR if the ketal protection is removed and reintroduced. As a result of this manipulation the six-membered ring ketal will preferably form (Fig. 3). In this configuration the C-2 and C-3 protons are in the ring and have a set orientation. Duthaler and co-workers[25] showed that the same effect can be achieved with a catalytic amount of acid and that the amount of HCl present in the normal NMR solvent CDCl₃ is sufficient to promote *trans*-ketalization to the more stable six-membered ring in a single step. Application of the Duthaler method proved the assignments used here for the *syn* and *anti* alcohols to be correct.[7] The *syn* alcohol transketalized completely during 48 hr standing in an NMR solvent. However, the *anti* alcohol resisted the transketalization and only after 2 weeks standing did the reaction take place. This is somewhat surprising because the six-membered ketal in the latter case is the more stable *trans*-substituted product. Evidently the *N*-BOC group interferes with the conformational changes needed for the transketalization, allowing an easier formation of the less stable *cis*-substituted six-membered ring ketal.

[24] P. Herold, *Helv Chim Acta* **71,** 354 (1988).
[25] A. Hafner, R. O. Duthaler, R. Marti, G. Rihs, P. Rothe-Streit, and F. Schwarzenbach, *J. Am. Chem. Soc.* **114,** 2321 (1992).

Final Product

Discovery of *anti*-selective reduction made it possible to complete the synthesis of sphingosine itself. Following the procedures presented earlier, we were able to synthesize alcohols **4a** and **4b** independently from the common precursor enone **3** (see Scheme 1). These are the same intermediates that Garner *et al.*[26] prepared through a less selective route in their sphingosine synthesis. They described their *syn* alcohol **4b** as an enriched 7:1 mixture with **4a**, giving the optical rotation of $[\alpha]_D = -39$ (*c* 0.25, CHCl₃). We were able to obtain the same *syn* diastereomer as a single isomer as shown by NMR and HPLC. The optical rotation for this compound is $[\alpha]_D^{20} = -34.3$ (*c* 0.61, CHCl₃).

The *anti* diastereomer **4a**, which is the natural D-*erythro*-sphingosine in a protected form, was also obtained as a single isomer and gave the optical rotation $[\alpha]_D^{20} = -29.9$ (*c* 0.65, CHCl₃). The same alcohol prepared by Garner *et al.*[26] likewise as a single diastereomer showed an optical rotation of $[\alpha]_D^{20} = -28$ (*c* 0.65, CHCl₃).

The method is thus a formal total synthesis of both D-*erythro*- and L-*threo*-sphingosine. The natural D-*erythro*-sphingosine in protected form can be synthesized in 69% overall yield in just three steps starting from a commercially available *N*-BOC-serine methyl ester (40% unoptimized yield for 10 g of **4a**). The L-*threo*-sphingosine can be achieved in 40% overall yield in three steps from the same starting material.

The isomer **4a** was deprotected by the method described by Garner *et al.*,[26] and analytical data obtained for D-*erythro*-sphingosine **5** were mp. 74° and $[\alpha]_D^{20} = -4.2$ (*c* 0.60, CHCl₃) and are in good accordance with the literature values mp. 72–75°,[26,27] $[\alpha]_D^{20} = -0.58$ (*c* 1.67, CHCl₃),[26] $[\alpha]_D^{20} = -1.3$ (*c* 3.5, CHCl₃),[27] $[\alpha]_D^{20} = -3.4$ (*c* 2, CHCl₃).[28]

Analog Synthesis

As an application of the sphingosine synthesis described earlier we chose to synthesize the pyrene analog **13**, which should be a good candidate for studying the sphingosine metabolism. A six carbon atoms chain length was selected because the total length of the side chain, with the pyrene group included, corresponds best with the total length of the alkyl chain in the natural sphingosine. Also, the hydrophobic properties of the analog

[26] P. Garner, J. M. Park, and E. Malecki, *J. Org. Chem.* **53**, 4395 (1988).
[27] R. H. Boutin and H. Rapoport, *J. Org. Chem.* **51**, 5320 (1986).
[28] Beilstein Vol. IV (1962) Supplement III, p. 854.

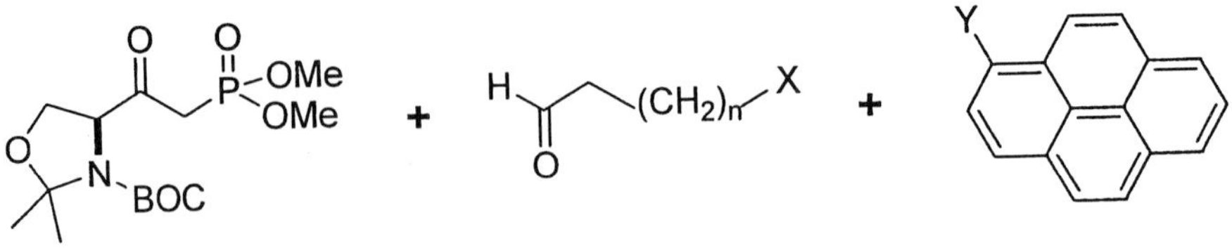

FIG. 4. The three separate building blocks needed in sphingosine analog synthesis.

built with this chain correspond reasonably well with those of sphingosine. Based on modeling, the hydrophobicity of pyrene equals six CH_2 units.[29]

The generality of this overall methodology is well demonstrated in the preparation of this model compound. The synthesis shows how practically any labeled sphingosine analog can be constructed from three separate building blocks.

The single requirement is that the hydrocarbon chain bears an aldehyde at the point of attachment to the β-ketophosphonate. The building blocks are illustrated in Fig. 4. The whole synthetic sequence for this particular model sphingosine analog is presented in Scheme 2.

Conclusions

The Horner–Wadsworth–Emmons reaction offers the synthetic organic chemist an efficient tool for construction of a double bond with high *trans* selectivity. The serine-derived β-ketophosphonate gives access to a large variety of γ-chiral α,β-unsaturated enone structures, which can be prepared with high enantioselectivity by using the modification of the HWE reaction described here.

High diastereoselectivity and regioselectivity in the reduction of enones are highly desired goals. We have demonstrated here how this can be achieved with the right choice of hydride reagent and reaction conditions, which properly maximize the internal asymmetric induction from the existing chiral center in the molecule.

The sequence of reactions—β-ketophosphonate formation, modified HWE reaction, and diastereoselective reduction—enables the synthetic organic chemist to selectively construct molecules with multiple chiral centers. The riddle of sphingosine synthesis has thus been solved. A multigram synthesis of all four isomers of sphingosine can be achieved. D-*erythro* and L-*erythro* isomers are available starting from the natural L-serine, and it can safely be assumed that because D-serine is merely an enantiomer of

---

[29] P. Somerharju, personal communication.

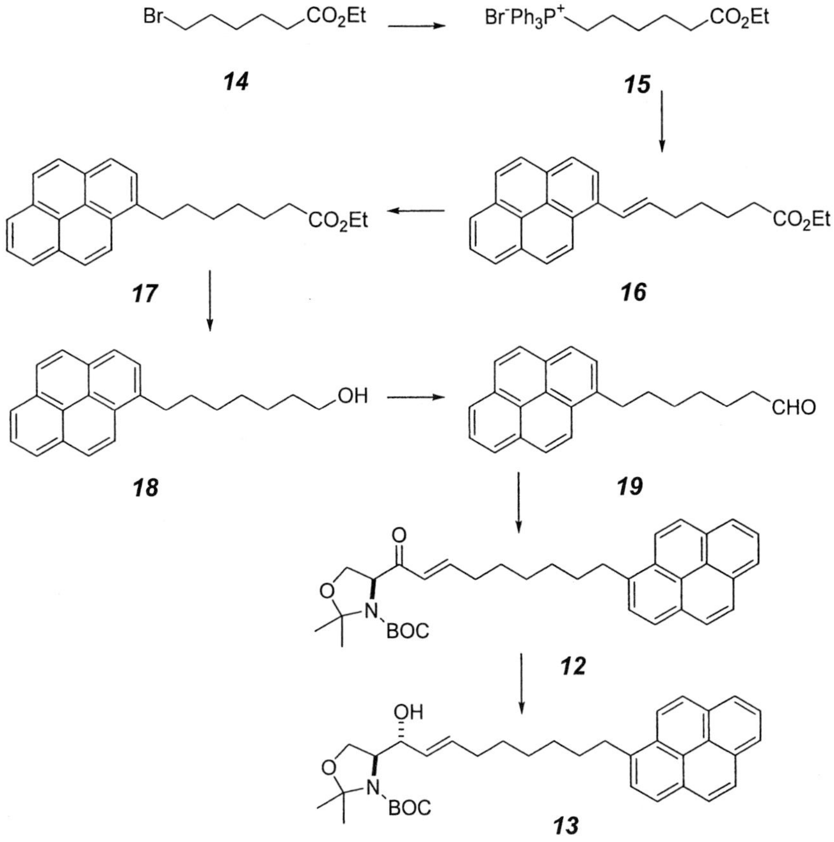

SCHEME 2. Synthetic sequence in a general form for the protected pyrene sphingosine analog.

L-serine that the remaining two isomers, L-*erythro*-and D-*threo*-sphingosines, can be synthesized by precisely the same methodology.

Experimental

General

THF was distilled prior to use from sodium/benzophenone, CH_3CN from phosphorous pentoxide, and CH_2Cl_2 from CaH_2. Toluene was distilled from Na and stored over Na turnings. All the organometallics were transferred by syringe or cannula. All reagents and other solvents, except where noted, were used as obtained from the supplier without further purification.

Melting points are measured with Gallenkamp melting point apparatus MFB-595 and are uncorrected. NMR spectra were determined on Bruker AM200 (^{1}H, 200.13 MHz, ^{13}C 50.32 MHz) and Bruker DX400 (^{1}H, 400.13 MHz, ^{13}C 100.62 MHz). Chemical shifts are reported in ppm (δ) with respect to scale calibrated to the solvent residual signal. Unless aqueous, all reactions were carried out under protective atmosphere (Ar) with magnetic stirring. Temperatures refer to bath temperatures unless otherwise noted. Organic extracts were first treated with brine, dried over Na_2SO_4, filtered, and evaporated with a Büchi rotary evaporator (water aspirator) followed by static evaporation with an oil pump. The Kieselgel 60 F_{254} impregnated aluminum plates were used for analytical TLC. TLC plates were visualized with UV ($\lambda = 254$) and/or either phosphomolybdic acid in 90% EtOH (10 mg/100 ml) or anisaldehyde/glacial acetic acid/H_2SO_4/EtOH (5:1:5:90). Flash chromatography was performed using silica gel 60 (E. Merck) as a stationary phase. HPLC was measured using the following columns: Shandon Hypercil Silica column with Waters Guard-Pak™ precolumn fitted with Resolve™ silica inserts for normal phase chromatography and Daicel chiralcel OD 25 × 0.46 cm with Daicel chiralcel OD 5 × 0.46 cm precolumn for chiral chromatography. Eluents were EtOAc/hexanes or EtOAc/i-octane mixture with flow rates of 3.0–4.5 ml/min (normal phase) and 10% iso-propanol in hexanes; 0.7 ml/min (chiral) detected with 205–254 nm. Mass spectra were measured by the University of Oulu Mass Spectrometry Laboratory on a Kratos 80 mass instrument. Elemental analyses were performed by the University of Oulu Trace Element Laboratory. Optical rotations were determined with Perkin–Elmer 243 B polarimeter (c = g/100 ml).

(S)-1,1-Dimethylethyl 4-[(dimethoxyphosphinyl)acetyl]-2,2-dimethyl-3-oxazolidinecarboxylate (1). Dimethyl methylphosphonate (22.5 ml, 210 mmol, 210 mol%) is dissolved in 150 ml of THF and cooled to −78° [internal temperature (IT)]. To this solution add 84.0 ml (2.4 *M*, 205 mmol, 205 mol%) *n*-BuLi in hexanes. The reaction mixture is allowed to warm to −20° for 1 hr and is then cooled back to −78° (IT). To this solution is cannulated 25.9 g (100 mmol, 100 mol%) of **1** dissolved in 50 ml of THF. Cannulation is performed at a speed to maintain the temperature (IT) below −75°. After addition, the reaction is allowed to warm to 0° slowly and is then quenched with 10% citric acid and the pH adjusted to 3. The mixture is extracted with EtOAc. After evaporation, the crude clear oil is purified by flash chromatography using first 10% EtOAc in hexanes to remove the possible residual starting material and finally 75% EtOAc in hexanes to remove the excess DMMP to yield 23 g (65.5 mmol, 65.5%) of **2** as white solid. mp. 86–87°. R_f (EtOAc)= 0.28. $[\alpha]_D^{20} = -56.2°$ (*c* 0.94, CHCl$_3$). ^{1}H NMR (200 MHz, CDCl$_3$, major rotamer) δ 1.30 (s, 3H), 1.40 (s, 3H), 1.55 (s, 3H), 3.13 (dd, 2H, *J* = 3.1 Hz, 5.9 Hz), 3.66 (s, 3H), 3.71, (s, 3H), 3.96 (dd, 1H, *J* =

3.7 Hz, 9.5 Hz), 4.1 (dd, 1H, J = 3.7 Hz, 7.2 Hz), 4.39 (dd, 1H, J = 7.2 Hz, 9.5 Hz). ^{13}C NMR (50 MHz, CDCl$_3$) δ 23.7 (CH3), 25.0 (CH$_3$), 28.0 (CH$_3$), 52.7 (CH3), 65.0 (C), 65.1 (CH2), 80.8 (C), 94.4 (CH2), 95 (C), 150.9 (C=O), 199.0 (C=O). HRMS (CI, NH$_3$) m/z calcd. for C$_{14}$H$_{27}$NO$_7$P (M + H)$^+$ 352.1525, found 352.1551. Anal. calcd. for C$_{14}$H$_{26}$NO$_7$P: C 47.9; H 7.5; N 4.0; found: C 48.1; H 7.7; N 4.3.

(S)-1,1-Dimethylethyl 4-(2-(E)-hexadecenyl-1-oxo)-2,2-dimethyl-3-oxa-zolidinecarboxylate (3). The β-ketophosphonate 2 (1.39 g, 3.95 mmol) is dissolved in dry MeCN (25 ml) and K$_2$CO$_3$ (1.10 g, 7.90 mmol, 200 mol%) is added at room temperature. After 15 min, tetradecanal (0.82 g of 80 w%, 3.10 mmol, 78 mol%) is added to this solution at room temperature. The reaction is stirred at room temperature for 24 hr, after which 5% citric acid (50 ml) is added and the mixture is extracted with CH$_2$Cl$_2$ (2 × 50 ml). The organic phases are combined, dried, and evaporated to yield 1.55 g of white solid containing some oily material. This is purified by flash chromatography using 10% EtOAc in hexanes as the eluent to give pure 3 (1.1 g, 2.51 mmol, 81%) as a nonmelting white wax. R_f (20% EtOAc/C$_6$) = 0.37. $[\alpha]_D^{20}$ = −21° (c 0.85, CHCl$_3$). ^{1}H NMR (200 MHz, CDCl$_3$, major rotamer) δ 0.86 (t, 3H, J = 6.1 Hz), 1.24 (s, 20 H), 1.35 (s, 9H), 1.53 (s, 3H), 1.68 (s 3H), 2.20 (tq, 2H J = 6.1 Hz, 6.9 Hz), 3.88 (dd, 1H, J = −9.2 Hz 3.7 Hz), 4.16 (dd, 1H, J = 7.5 Hz, 9.2 Hz) 4.47 (dd 1H, J = 3.7 Hz, 7.5 Hz), 6.32 (d, 1H, J = 15.7 Hz), 6.95 (dt, 1H J = 6.9 Hz, 15.7 Hz). ^{13}C NMR (50 MHz, CDCl$_3$) δ 14.0 (CH3), 22.6 (CH2), 24.1 (CH3), 25.2 (CH3), 28.0 (CH2), 28.3 (CH2), 29.1 (CH3), 29.3 (CH2) 29.6 (CH2), 31.9 (CH2), 32.7 (CH2), 64.1 (CH), 65.8 (CH2), 80.5 (C), 95.1 (C), 125.3 (CH2), 125.9 (CH2), 149.4 (C=O), 196.4 (C=O). HRMS m/z calcd. for C$_{26}$H$_{48}$NO$_4$ (M + H)$^+$ calcd. 438.3583, found 438.3572. Anal. calcd. for C$_{26}$H$_{47}$NO$_4$: C 71.4; H 10.8; N 3.2; found C 71.0, H 10.9, N 3.1.

Ethyl 6-triphenylphosphine hexanoate bromide (14). Triphenylphosphine (4.46 g, 20 mmol, 100 mol%) is dissolved to dry THF (10 ml). Ethyl 6-bromohexanoate (5.24 g, 20 mmol, 100 mol%) and tetrabutylammonium iodide (37 mg, 1 mmol, 5 mol%) are added. The reaction is brought to reflux and kept at that temperature for 4 days. The reaction mixture is allowed to cool to 50° (bath temperature) and 20 ml of Et$_2$O is added. Stirring is continued until the mixture is at room temperature. Upon cooling the reaction mixture thickens. The ether is decanted, and the product is washed with 2 × 10 ml of Et$_2$O. The phosphonium-salt 14 is dissolved in 30 ml of dry THF and taken directly to the next reaction. An analytical sample is purified by washing with Et$_2$O and triturating from wet THF. ^{1}H NMR (200 MHz, CDCl$_3$) δ 1.13 (t, 3H, J = 7 Hz), 1.56 (m, 6H), 2.14 (t, 2H, J = 7 Hz), 2.41 (H$_2$O), 3.61 (m, 2H), 3.93 (q, 2H, J = 7 Hz), 7.6 (m 15H aromatic). ^{13}C NMR (50 MHz, CDCl$_3$) δ 14.0 (CH3), 22.1 (CH2), 22.2

(CH2), 24.1 (CH2), 29.7 (CH2), 33.5 (CH2), 60.1 (CH2), 117.1, 118.8, 130.2, 130.5, 133.3, 133.34, 133.5, 134.87, 134.93 (all arom.) 173.3 (C=O). Anal. calcd. for $C_{26}H_{30}PO_2Br \cdot \frac{1}{2}H_2O$: C 63.2, H 6.4; found C 63.0, H 6.4.

Ethyl 7-(1-pyrenyl)-hept-6-enoate (16). To the crude phosphonium-salt **15** in dry THF (30 ml) from the previous reaction (9.3 g, 19 mmol, 150 mol%) is added another 30 ml of dry THF. This solution is cooled to −20° (IT), and the salt is solidified. Warming the mixture to 0° redissolves it and *n*-BuLi (19 mmol, 8.3 mmol 2.3 *M*, 150 mol%) is added dropwise. An immediate color change from faint fluorescent green to dark orange brown is observed. Cooling is attempted after *n*-BuLi addition, but at around −5° the starting material starts to solidify again. After stirring this mixture at 0° for 30 min, all the starting material is redissolved and the mixture can be cooled to −20° at which temperature pyrene-1-carboxaldehyde (12.6 mmol, 2.9 g, 100 mol%) in 5 ml of dry THF is added. The reaction is stirred at −20° for 1 hr and then allowed to warm to room temperature over 30 min. The reaction is quenched with 5 w% citric acid and extracted with CH_2Cl_2 (3 × 30 ml), dried, and evaporated to yield 14 g of a brown solid. This is dissolved in 5% EtOAc/C_6 and filtered through a pad of silica topped with celite. The silica is rinsed with CH_2Cl_2. The filtrate is evaporated to yield **16** [3.64 g (10.2 mmol, 81%)] as a fluorescent green oil. R_f (10% EtOAc/C_6) = 0.24. Due to the ease of polymerization, this was taken directly to the next step without further purification. An analytical sample is purified by chromatography with 10% EtOAc/C_6. ^{1}H NMR (200 MHz, CDCl₃) *cis* isomer δ 1.3 (t, 3H, *J* = 7 Hz), 1.5 (broad m 4H), 2.4 (td, 2H, *J* = 1.5 Hz, 7 Hz, 11.5 Hz), 4.1 (q, 2H, *J* = 7 Hz), 6.4 (ddt, 1H, *J* = 2 Hz, 7 Hz, 16.5 Hz), 7.4 (td, 1H, *J* = 1.5 Hz, 16.5 Hz), 8.1 (m, 9H). *trans* isomer δ 1.2 (t, 3H, *J* = 7 Hz), 1.5 (broad m 4H), 2.2 (td, 2H, *J* = 1.5 Hz, 7 Hz 14 Hz), 4.0 (q, 2H, *J* = 7 Hz), 6.0 (dtt, 1H, *J* = 2 Hz, 7 Hz, 11.5), 7.1 (d, 1H, *J* = 11.5 Hz), 8.1 (m, 9H). ^{13}C NMR (50 MHz, CDCl₃) (mixture of isomers) δ 14.0 (CH3), 22.1 (CH2), 23.2 (CH2), 24.6 (CH2), 25.6 (CH2), 28.8 (CH2), 30.15 (CH2), 33.5 (CH2), 60.1 (CH2), 74.3 (CH2), 123–135 (arom.) 173 (C=O). UV {λ(ε)} 206(56.5), 244(37.5), 278(24.4), 343(25.8), 364.5(19.6). HRMS *m/z* calcd. for $C_{25}H_{24}O_2$ (M)$^+$ 356.1776, found 356.1758.

Ethyl 7-(1-pyrenyl)-heptanoate (17). The catalyst, 10% Pd/C (370 mg, 10 w%), is weighed into the Parr shaker reaction flask. Absolute ethanol (150 ml) is added. Ethyl 7-(1-pyrenyl)heptenoate **16** (3.64 g, 10 mmol) is dissolved in 100 ml of abs. EtOH and added to the reaction vessel. The reaction is stirred using a Parr shaker starting with 50 psig (3.33 atm) H_2 pressure. The reaction is stirred until the pressure ceases to drop (1.5 hr) and then the pressure is increased back to 50 psig. Because no further pressure drop is detected, the reaction mixture is brought to ambient pressure and a UV spectrum is recorded. This shows no olefinic absorption ($\lambda = 365$, $\varepsilon = 19.6$). The solution is filtered through a pad of celite and evaporated to yield a light yellow solid. This is purified by flash chromatography using 20% EtOAc in hexanes as the eluent followed by recrystallization from EtOAc/Hexanes to give **17** (3.5 g, 9.6 mmol, 96%) as white crystals. mp. 78.5°. R_f (10% EtOAc/C_6) = 0.24. ^{1}H NMR (200 MHz, CDCl$_3$) δ 1.27 (t, 3H, $J = 7$ Hz), 1.45 (t, 2H, $J = 7.5$ Hz), 1.51 (t, 2H, $J = 7.5$ Hz), 1.67 (t, 2H, $J = 7.5$ Hz) 1.87 (t, 2H, $J = 7.5$ Hz), 2.32 (t, 2H, $J = 7.5$ Hz) 3.34 (t, 2H, $J = 7.5$ Hz) 4.15 (q, 2H, $J = 7$ Hz), 8.1 (m, 9H). ^{13}C NMR (50 MHz, CDCl$_3$) δ 14.2 (CH3), 24.9 (CH2), 29.0 (CH2), 29.4 (CH2), 31.6 (CH2), 33.4 (CH2), 34.3 (CH2), 60.1 (CH2), 123–137 (arom.) 173.7 (C=O). UV $\{\lambda(\varepsilon)\}$ 206(56.5), 244(37.5), 278(24.4), 343(25.8), HRMS m/z calcd for C$_{25}$H$_{26}$O$_2$ (M)$^+$, calcd. 358.1933, found 358.1952. Anal. calcd. for C$_{25}$H$_{26}$O$_2$ C 83.8, H 7.3; found C 84.0, H 7.3.

7-(1-Pyrenyl)-heptanol (18). Ethyl 7-(1-pyrenyl)ethylheptanoate **17** (358 mg, 1.0 mmol, 100 mol%) is dissolved in 5 ml of dry THF at room temperature. LiAlH$_4$ (76 mg, 2.0 mmol, 200 mol%) is added. TLC immediately after addition shows that all the starting material has disappeared. The reaction is quenched with saturated Na$_2$SO$_4$ (15 ml) and extracted with CH$_2$Cl$_2$ (3 × 10 ml) and brine (10 ml). Evaporation of the solvent gave **18** (316 mg, 1.0 mmol, 100%) as a white solid. mp. 86°. R_f (25% EtOAc/C_6) = 0.1. ^{1}H NMR (200 MHz, CDCl$_3$) δ 1.4 (m, 6H), 1.86 (t, 2H, $J = 7.5$ Hz) 3.33 (t, 2H, $J = 7.5$ Hz), 3.45 (broad s, OH), 3.63 (t, 2H, $J = 7.5$ Hz), 8.1 (m, 9H). ^{13}C NMR (50 MHz, CDCl$_3$) δ 25.6 (CH2), 29.2 (CH2), 29.7 (CH2), 31.7 (CH2), 32.6 (CH2), 33.5 (CH2), 62.9 (CH2), 123–137 (arom.). HRMS m/z (M)$^+$, calcd. for 316.1827, found 316.1833. Anal. calcd. for C$_{23}$H$_{24}$O: C 87.3, H 7.6; found C 87.3, H 7.4.

7-(1-Pyrenyl)-heptanal (19). Oxalyl chloride (104 μl, 1.2 mmol, 150 mol%) is dissolved in 5 ml of CH$_2$Cl$_2$ and the reaction mixture is cooled to $-60°$. To this solution is added dropwise DMSO (185 μl, 2.4 mmol, 300 mol%) dissolved in 5 ml of CH$_2$Cl$_2$ and the solution is stirred at $-60°$ for 15 min. 7-(1-pyrenyl)heptanol **18** (250 mg, 0.8 mmol, 100 mol%) dissolved in 5 ml of CH$_2$Cl$_2$ is added at $-60°$. The reaction mixture is then stirred at this temperature for 1 hr after which Et$_3$N (555 μl, 4.0 mmol, 500 mol%) is added at $-78°$. The reaction mixture is allowed to warm to room temperature and is then poured into saturated NaHCO$_3$ solution (30 ml), extracted with CH$_2$Cl$_2$ (3 × 20 ml). The organic phases are combined, washed with water (30 ml), dried, and evaporated to yield **19** (248 mg, 0.8 mol, 100%) as a white solid. mp. 106°. R_f (25% EtOAc/C_6) = 0.46. ^{1}H NMR (200 MHz, CDCl$_3$) δ 1.35–1.71 (m, 6H), 1.86 (t, 2H, $J = 7.5$ Hz), 2.40 (td, 2H, $J = 1.7$ Hz, 7.5 Hz), 3.33 (t, 2H, $J = 7.5$ Hz), 7.83–8.29 (Arom, 9H), 9.75 (t, 1H,

J = 1.7 Hz).^{13}C NMR (50 MHz, CDCl$_3$) δ 22.0 (CH2), 29.1 (CH2), 29.4 (CH2), 31.6 (CH2), 33.4 (CH2), 43.8 (CH2), 123.4–137.0 (9 C-Arom.), 202.4 (C=O). HRMS m/z calcd. for C$_{23}$H$_{22}$O (M)$^+$ 314.1671, found 314.1682.

(S)-1,1-Dimethylethyl 4-{1-oxo-2-(E)-[7-(1-pyrenyl)-nonenyl]}-2,2-di-methyl-3-oxazolidinecarboxylate (12). The β-ketophosphonate **2** (112.3 mg, 0.32 mmol, 110 mol%) is dissolved in dry MeCN (5 ml), and K$_2$CO$_3$ (80 mg, 0.58 mmol, 200 mol%) is added at room temperature. After 15 min, 7-(1-pyrenyl)heptanal **19** (92 mg, 0.29 mmol, 100 mol%) in 3 ml of MeCN is heated to 50° temperature and added to this solution. The reaction mixture is brought to 50° and stirred at this temperature for 16 hr, after which 5% citric acid (10 ml) is added and the mixture is extracted with hexanes (3 × 10 ml) and CH$_2$Cl$_2$ (2 × 5 ml). The organic phases are combined, dried, and evaporated to yield 216 mg of a white solid containing some oily material. This is purified by flash chromatography using 10% EtOAc in hexanes as the eluent to give pure **12** (121 mg, 0.23 mmol, 72%) as fluorescent green oil. R_f (25% EtOAc/C$_6$) = 0.40. [α]$_D^{20}$ = −31° (c 1.0, CHCl$_3$). ^{1}H NMR (200 MHz, CDCl$_3$, major rotamer) δ 1.4–2.0 (23 H), 2.2 (m, 2H), 2.4 (t, 2H, J = 8.0 Hz), 3.3 (dd, 1H, J = 7.0 Hz, 8.0 Hz) 3.9 (dd 1H, J = 3.5 Hz, 8.0 Hz), 4.5 (dd, 1H, J = 3.5 Hz, 7.0 Hz), 6.28 (dd, 1H, J = 5.0 Hz, 15.7 Hz), 7.00 (dt, 1H, J = 7.0 Hz, 15.7 Hz). ^{13}C NMR (50 MHz, CDCl$_3$, δ 23.0 (CH3), 25.2 (CH3), 27.9 (CH2), 29.0 (CH2), 29.3 (CH2), 29.4 (CH2), 31.6 (CH2), 32.5 (CH2), 33.4 (CH2), 64.0 (CH), 65.8 (CH2), 80.3 (C), 95.0 (C), 123.3–138.6 (2 × CH, arom.), 149.2 (C=O), 196.4 (C=O). HRMS m/z calcd. for C$_{35}$H$_{41}$NO$_4$ (M)$^+$ calcd. 539.3035, found 539.3077. Anal. calcd. for C$_{35}$H$_{41}$NO$_4$: C 77.9, H 7.7, N 2.6; found C 77.8, H 7.6, N 3.0.

(S)-1,1-Dimethylethyl 4-(1-(R)-hydroxy-2-(E)-hexadecenyl)-2,2-dimeth-yl-3-oxazolidinecarboxylate (4a). The enone **2** (51 mg, 0.12 mmol) is dissolved in 2 ml of dry toluene and cooled to −74° (IT). DIBAL-H [360 μl (1.0 M), 0.36 mmol, 300 mol%] is added dropwise at −74° (IT). The starting material disappears immediately after the addition of DIBAL-H. The reaction is quenched at −74° (IT) with dry acetone (1 ml). This mixture is allowed to warm to 0° and the aluminum complex is hydrolyzed by stirring with 1 M HCl (1 ml) for 15 min in an ice bath. The reaction is extracted with CH$_2$Cl$_2$ (3 × 5 ml). The organic phases are combined, dried, and evaporated, yielding 51 mg of pale yellow oil. By HPLC this material is pure *anti* (>20:1) diastereomer **4a**, but contains ~3% of 1,4-reduced product. The product is purified by flash chromatography using 10% EtOAc/ hexanes as the eluent to give **4a** (48.5 mg, 0.11 mmol, 92%) as a nonmelt-ing white wax. R_f (20% EtOAc/C$_6$) = 0.23. [α]$_D^{20}$ = −29.9° (c 0.65, CH$_3$Cl). ^{1}H NMR (200 MHz, CDCl$_3$), δ 0.86 (t, 3H, J = 6.5 Hz), 1.23 (s, 22H), 1.47 (s, 3H), 1.48 (s, 9H), 1.55 (s, 3H), 2.05 (q, 2H, J = 6.5 Hz), 3.48

(broad m, 1H), 3.90 (broad m, 3H), 4.14 (t, 1H, J = 7.0 Hz), 5.40 (dd, 1H, J = 6.9 Hz, 15.3 Hz), 5.75 (dt, 1H, J = 6.6 Hz, 15.3, Hz). ^{13}C NMR (50 MHz, CDCl$_3$) δ 14.0 (CH3), 22.6 (CH2), 24.3 (CH3), 27.1 (CH3) 28.3 (CH3), 29.0 (CH2), 29.2 (CH2), 29.3 (CH2), 29.6 (CH2), 31.9 (CH2), 32.3 (CH2), 62.0 (CH), 64.7 (CH2) 76.3 (CH), 81.3 (C), 94.3 (C), 129.4 (CH), 135.3 (CH) 155.0 (C=O). HRMS calcd. for C$_{26}$H$_{49}$NO$_4$ (M)$^+$ 439.36616, found 439.36090. Anal. calcd. for C$_{26}$H$_{49}$NO$_4$: C 71.0, H 11.2, N 3.2; found C 70.7, H 11.4, N 3.1.

(S)-1,1-Dimethylethyl 4-1-(S)-hydroxy-2-(E)-hexadecenyl)-2,2-dimethyl-3-oxazolidinecarboxylate (4b). The enone **3** (500 mg, 1.14 mmol) is dissolved in 5 ml of THF and the mixture is cooled to −74° (IT). L-selectride® (3.42 ml, 3.42 mmol, 300 mol%) is added dropwise. After addition the reaction is allowed to warm slowly to room temperature (ca. 4 hr), upon which TLC shows that all starting material has been used and the reaction is then quenched with 10% citric acid (ca. 10 ml, until no gas is evolving) and extracted with *i*-octane (2 × 20 ml). The organic phases are combined, dried, and evaporated to yield 0.69 g of pale yellow oil. By HPLC this material is pure *syn* (>20:1) diastereomer **4b**. This material is purified by flash chromatography using 10% EtOAc in hexanes as eluent to yield pure **4b** (0.33 g, 0.75 mmol, 54%). R_f (20% EtOAc/C$_6$) = 0.33. $[\alpha]_D^{20}$ = −34.3° (*c* 0.61, CHCl$_3$). ^{1}H NMR (200 MHz, CDCl$_3$) δ 0.86 (t, 3H, J = 6.7 Hz), 1.23 (s, 20H), 1.46 (s, 3H), 1.48 (s, 9H), 1.55 (s, 3H), 2.00 (dt, 2H, J = 6.7 Hz), 3.85 (broad m, 2H), 4.15 (broad m, 1H) 5.40 (td, 1H, J = 6.4 Hz, 15.3 Hz), 5.71 (td, 1H, J = 6.7, 15.3 Hz). ^{13}C NMR (50 MHz, CDCl$_3$) δ 14.0 (CH3), 22.6 (CH2), 24.3 (CH3), 27.1 (CH3), 28.3 (CH3), 29.0 (CH2), 29.2 (CH2), 29.3 (CH2), 29.6 (CH2), 31.9 (CH2), 32.3 (CH2), 62.0 (CH), 64.7 (CH2), 76.4 (CH), 81.3 (C), 94.3 (C), 129.4 (CH), 135.3 (CH) 155.0 (C=O). HRMS *m/z* calcd. for C$_{26}$H$_{47}$NO$_4$ (M)$^+$ 439.3662, found 439.3669.

(S)-1,1-Dimethylethyl 4-{1-(R)-hydroxy-2-(E)-[7-(1-pyrenyl)]-heptenyl}-2,2-dimethyl-3-oxazolidine-carboxylate (13). Procedure as for **4a**. Diastereomeric ratios are presented in Table III. Yield: 40%. R_f (25% EtOAc/C$_6$) = 0.12. $[\alpha]_D^{20}$ = −10.4° (*c* 1.1, CH$_3$Cl). ^{1}H NMR (200 MHz, CDCl$_3$) δ 1.40–1.65 (6s, 21H), 1.81 (m, 2H), 2.05 (m, 2H), 3.31 (m, 2H), 3.85–4.18 (broad m, 5H), 5.45 (d, 1H, J = 16 Hz), 5.73 (td, 1H, J = 5.7 Hz, 16 Hz), 7.84–8.30 (m 9H, Ar).

[42] Radiolabeling of the Sphingolipid Backbone

By ALICIA BIELAWSKA, YUSUF A. HANNUM, and ZDZISLAW SZULC

Introduction

Effective and specific procedures for the preparation of sphingolipids (SLs) radiolabeled at the designated parts of the molecule produce powerful tools that enable the analysis of naturally occurring compounds, trace their biosynthetic and metabolic pathways, and assist in the investigation of the fate of those compounds after being added exogenously to different cells and animals (Fig. 1 and Scheme 1).

Sphingolipids constitute one of the most structurally diversified class of amphipathic lipids abundant in all living organisms. Variation in the nature of the head group attached to the primary hydroxyl group (R_2), *N*-acyl group (R_1), and sphingosine backbone generate the complex sphingolipids where sphingosine **IB** sphinganine **IA** or phytosphingosine **ID** are the major core structural subunits in mammals. SLs perform not only structural and functional roles as components of cell membranes, but also participate in cell recognition, cell regulation, and signal transduction.[1] Ceramide **IIB** is a central metabolic intermediate for the biosynthesis of mammalion phosphosphingolipids (phospho-SLs) **III**, **IV**, **V**, and glycosphingolipids **VI** respectively, and is considered the structural unit that anchors sphingolipids to biological membranes.

Methods that have been reported for labeling Sls in the sphingoid backbone include (i) a biosynthetic approach using radioactive L-serine,[2,3] (ii) catalytic reduction (Fig. 2) in which tritium is added to the 4,5-double bond of sphingosine **IB**[4–12] or to the 4,5-triple bond of dehydrosphingosine

[1] Y. A. Hannun, *Science* **274**, 1855 (1996)

[2] J. N. Kanfer and S. Bates, *Lipids* **5**, 718 (1970).

[3] Y. Igarashi, K. Kitamura, T. Toyokuni, B. Dean, B. Fenderson, T. Ogawa, and S. Hakamori, *J. Biol. Chem.* **265**, 5385 (1990).

[4] S. Gatt, *Biochem. J.* **101**, 687 (1966).

[5] W. Stoffel and K. Bister, *Hoppe-Seyler's Z. Physiol. Chem.* **355**, 911 (1974).

[6] G. Schwarzmann, *Biochim. Biophys. Acta* **529**, 106 (1978).

[7] G. Schwartzmann and K. Sandhoff, *Methods Enzymol.* **138**, 319 (1987).

[8] P. P. Van Veldhoven and G. P. Mannaerts, *J. Biol. Chem.* **266**, 12502 (1991).

[9] J. Rother, G. van Echten, G. Schwartzmann, and K. Sandhoff, *Biochem. Biophys. Res. Commun.* **189**, 14 (1992).

[10] C. Michel, G. van Echten-Deckert, J. Rother, K. Sandhoff, E. Wang, and A. H. Merrill, Jr., *J. Biol. Chem.* **272**, 22432 (1997).

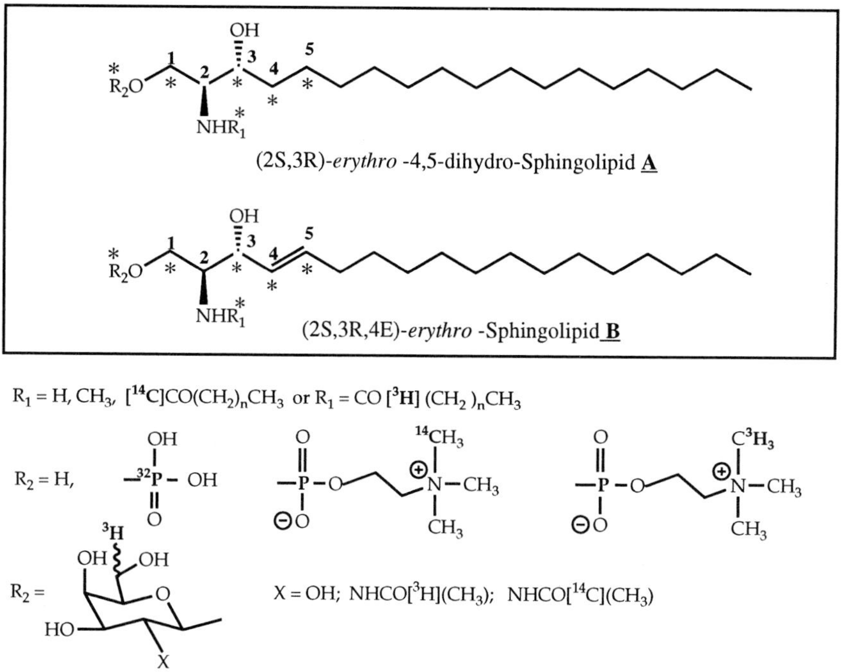

FIG. 1. ^{3}H-, ^{14}C-, and ^{32}P-labeling sites in sphingolipid structures (denoted with asterisks).

IC[13–15] (iii) an oxidation–reduction reaction sequence in which the secondary hydroxyl group is oxidized to the 3-keto derivative (Fig. 3)[16–28] or the

[11] J. W. Kok, M. Nikolova-Karakashian, K. Klappe, C. Alexander, and A. H. Merrill, Jr., *J. Biol. Chem.* **272**, 21128 (1997).

[12] G. van Echten-Deckert, A. Giannis, A. Schwarz, and A. H. Futerman, *J. Biol. Chem.* **273**, 1184 (1998).

[13] C. A. Grob and F. Gadient, *Helv. Chim. Acta* **40**, 1145 (1957).

[14] A. E. Gal, *J. Label. Comp.* **3**, 112 (1967).

[15] A. Schick, G. Schwarzmann, D. Kolter, and K. Sandhoff, *J. Label. Comp. Radiopharm.* **39**, 441 (1997).

[16] R. C. Gaver and C. C. Sweeley, *J. Am. Chem. Soc.* **88**, 3643 (1966).

[17] T. Taketomi and N. Kawamura, *J. Biochem. (Tokyo)* **68**, 475 (1970).

[18] G. Sticht, D. Lekim, and W. Stoffel, *Chem. Phys. Lipids* **8**, 10 (1972).

[19] W. Stoffel and K. Bister, *Hoppe-Seyler's Z. Physiol. Chem.* **355**, 911 (1974).

[20] N. S. Radin, *Lipids* **9**, 358 (1974).

[21] M. Iwamori, H. M. Moser, and Y. Kishimoto, *J. Lipid Res.* **16**, 332 (1975).

[22] M. Iwamori, H. W. Moser, R. H. McCluer, and Y. Kishimoto, *Biochim. Biophys. Acta* **380**, 308 (1975).

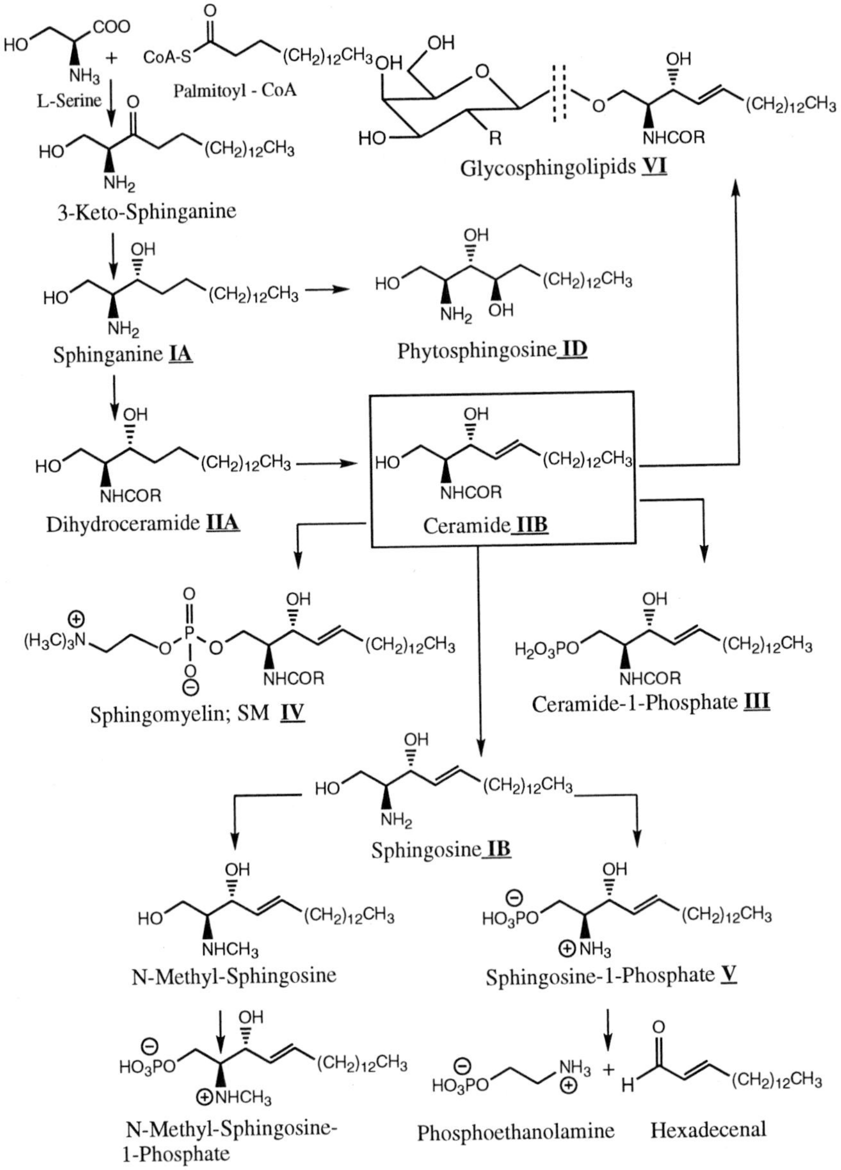

SCHEME 1. Biosynthesis and metabolism of sphingolipids.

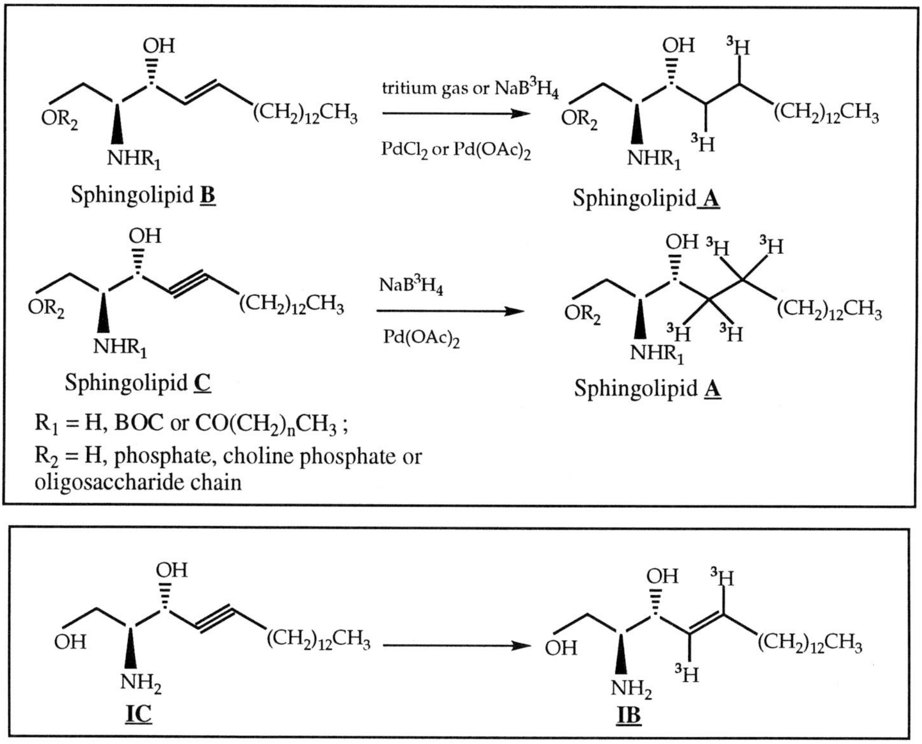

FIG. 2. Approaches to the synthesis of sphingolipids labeled with tritium at the C-4 and C-5 positions in the sphinganine or sphingosine backbone.

primary OH-group is oxidized to the corresponding aldehyde derivative,[29] respectively; these are then reduced back with NaB^3H_4 or KB^3H_4, (iv) reduction of the ester group of sphingoid precursors with NaB^3H_4 to intro-

[23] R. Ghidoni, S. Sonnino, M. Masserini, P. Orlando, and G. Tettamanti, *J. Lipid Res.* **22**, 1286 (1981).

[24] R. Birk, G. Brenner-Weib, A. Giannis, K. Sandhoff, and R. R. Schmidt, *J. Label. Comp. Radiopharm.* **29**, 289 (1990).

[25] R. Wild and R. R. Schmidt, *Tetrahedron Assymetry* **5**, 2195 (1994).

[26] A. Kisic, M. Tsuda, R. J. Kulmacz, W. K. Wilson, and G. J. Schroepfer, *J. Lipid Res.* **36**, 787 (1995).

[27] S. Sonino, M. Nicolini, and V. Chigorno, *Glycobiology* **6**, 479 (1996).

[28] V. Chigorno, E. Negroni, M. Nicolini, and S. Sonnino, *J. Lipid Res.* **38**, 1163 (1997).

[29] T. Toyokuni, M. Nisar, B. Dean, and S.-I. Hakamori, *J. Label. Comp. Radiopharm.* **29**, 567 (1991).

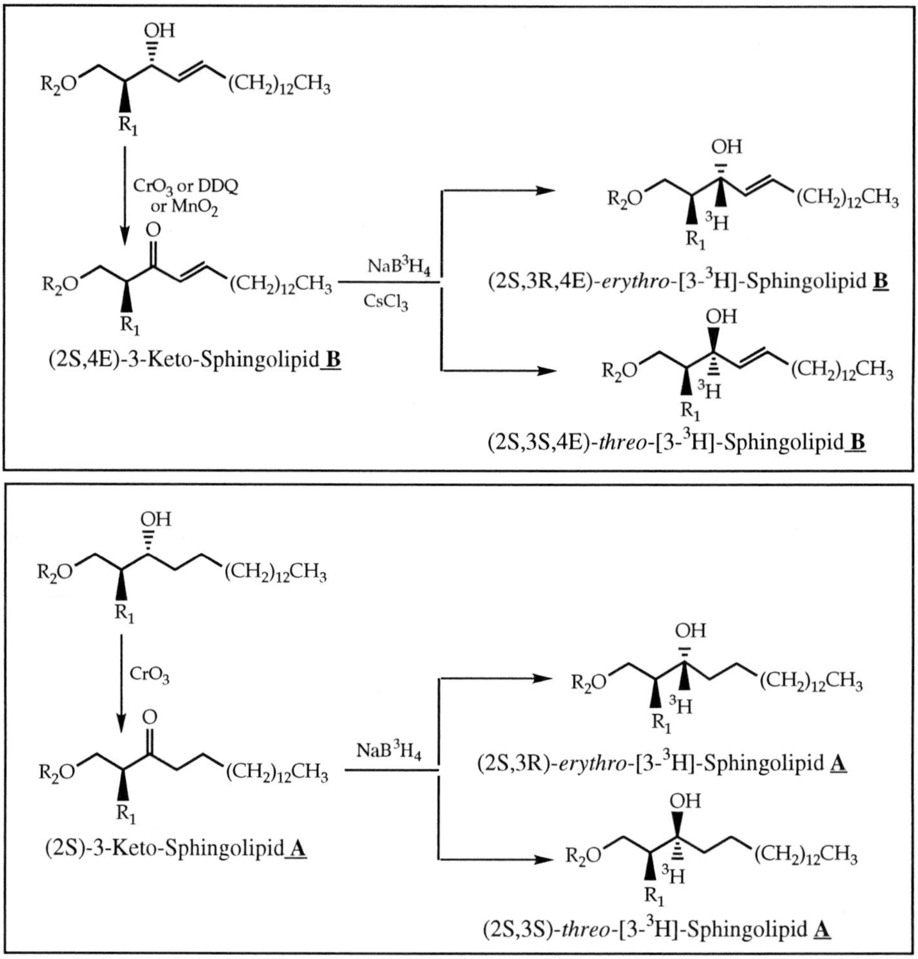

R_2 = H, phosphate, choline phosphate, oligosaccharide chain
R_1 = NH_2, NH_2 *HCl, NHBoc, $NHCO(CH_2)nCH_3$ or N_3

FIG. 3. Approaches to the synthesis of [3-³H]sphingolipid **B** and [3-³H]dihydrosphingo-lipid **A**.

duce tritium at the C-1 position of sphingosine or ceramide,[30,31] and (v) labeling in the course of total synthesis.[32,33]

[30] D. Shapiro, H. Segal, and H. M. Flowers, *J. Am. Chem. Soc.* **80**, 1194 (1958).
[31] Y. Shoyama, H. Okabe, Y. Kishimoto, and C. Costello, *J. Lipid Res.* **19**, 250 (1978).
[32] W. Stoffel and G. Sticht, *Hoppe-Seyler's Z. Physiol. Chem.* **348**, 1561 (1967).
[33] W. Stoffel and K. Bister, *Hoppe-Seyler's Z. Physiol. Chem.* **354**, 169 (1973).

^{3}H-Labeling at the C-1, C-3, and C-4–C-5 positions and ^{14}C-labeling at the C-3 position of the sphingoid base backbone allow investigators to follow metabolism/catabolism of sphingolipids to the final metabolites of sphingosine-1-phosphate i.e., [^{3}H]- or [^{14}C]palmitaldehyde and [^{3}H]phosphoethanolamine. N-Acylation of [^{3}H]- or [^{14}C]sphingoid bases with ^{3}H- or ^{14}C-labeled fatty acids or [^{32}P]phosphorylation of [^{3}H]- or [^{14}C]ceramide or related analogs can be a versatile approach to synthesize double- or even triple-labeled SLs that would provide particularly wider perspective *in vivo* metabolic studies.

This article presents synthetic methods for the preparation of [3-^{3}H]sphingosines **IB**, [4,5-^{3}H]- and [3-^{3}H]sphinganines **IA** and related ceramides/dihydroceramides **IIB/IIA**. To prepare [^{3}H]ceramide/dihydroceramide labeled in the sphingoid backbone, we prefer acylation of the radiolabeled sphingoid bases **IA** and **IB** with nonradioactive fatty acids rather than direct labeling at the ceramide/dihydroceramide level. The last approach causes problems with the separation of ceramide from dihydroceramide and between the erythro and the *threo*-stereoisomers that are formed under this preparation method.

Materials

[^{3}H]Sodium borohydride (450 mCi/mmol) is obtained from American Radiolabeled Chemicals (St. Louis, MO), palladium(II) acetate and trifluoroacetic acid (TFA) are from Sigma (St. Louis, MO), Amberlite IRA 400 (OH-) resin and palmitic chloride are from Fluka Chem. Corp. (Milwaukee, WI), CH$_3$COONa $\times$ 3H$_2$O is from Aldrich Chemical Company, Inc. (Milwaukee, WI), and silica gel 60 (particle size 0.040–0.063 mm) (230–400 Mesh ASTM) for column chromatography, CAS 63231-67-4, is from EM Separation Technology.

Thin-Layer Chromatography Plates

(1) Precoated thin-layer chromatography (TLC) sheets silica gel 60 F 254 (0.2 mm layer thickness) and (2) precoated thin-layer plates Kieselgel 60 (0.25 mm layer thickness) are from EM Separation Technology; (3) precoated thin-layer plates PK5 silica gel 150 A (0.5 mm layer thickness) and (4) Linear-K preadsorbent strip, silica gel 150A (1.0 mm layer thickness) are from Whatman Laboratory Division; (5) TLC plates coated with silica gel G 60/ Na$_2$B$_4$O$^7 \cdot$ 10 H$_2$O are prepared according to the published method.[9]

Silica gel 60 (particle size 0.040–0.063 mm) (230–400 Mesh ASTM) for column chromatography CAS 63231-67-4 is from EM Separation Technology.

All solvents are analytical grade or better and are obtained from Mallinckrodt, Burdick and Jackson, or J. T. Baker.

General Methods Used during Synthesis

Solvents

Tetrahydrofuran (THF) is freed of peroxides by passing over basic alumina, followed by saturation with dry argon.

Key Precursors for Preparation of Radiolabeled Sphingolipids and Analytical Standards

Stereoisomers of sphingosine **IB**, sphinganine **IA**, 4,5-dehydrosphingosine **IC**, and their *N-tert*-butoxycarbonyl-(*N*-Boc-), *N*-acetyl-, and (2*S*)- or (2*R*)-3-keto derivatives are synthesized according to procedures presented elsewhere in this volume.[34]

Thin-Layer Chromatography

Purity of the various compounds, the reaction progress, and chromatographic elution profiles are checked routinely by thin-layer chromatography TLC-1. Purification is performed using preparative TLC plates: TLC-2, -3, -4, or -5.

Flash Column Chromatography

Separation of reaction products and purification of the obtained compounds are achieved by flash chromatography.[35] For purification of the obtained radiolabeled compounds, small glass columns (0.6 mm inner diameter) are used, and 2.0- to 2.5-ml fractions are collected. Fractions containing impure material are rechromatographed.

Solvent Systems

The solvent systems used for TLC and/or flash column chromatography purification are (A) chloroform–methanol–2 N ammonium hydroxide (4:1:0.1, v/v); (B) methylene chloride–methanol (93:7, v/v); (C) chloroform–methanol (9:1, v/v); (D) methanol–water (95:5, v/v); (E) methylene chloride–methanol (9:1, v/v); (F) methylene chloride–methanol–10% ammonium hydroxide (79:20:1, v/v); (G) methylene chloride–methanol (4:1,

[34] A. Bielawska, Z. Szulc, and Y. A. Hannun, *Methods Enzymol.* **311** [44] 1999 (this volume).
[35] W. C. Still, M. Kahn, and A. Mitra, *J. Org. Chem.* **43**, 2923 (1978).

v/v); (H) ethyl acetate–hexane (4:1, v/v); and (I) chloroform–methanol (95:5, v/v).

Analytical Assays

Composition and Diastereoisomeric Purity of Sphingoid Bases. Analysis of sphingosine **IB**, sphinganine **IA**, and dehydrosphingosine **IC** is performed by high-performance liquid chromatography (HPLC) method after their derivatization with *o*-phalaldehyde using eicosasphingosine as an internal standard as described by Merrill *et al.*[36]

Determination of Radioactivity. Radioactivity is measured on a Wallac LKB 1214 Rackbeta liquid scintillation counter with High Flash Point Cocktail Safety-Solve (Res. Product Int. Corp., Mount Prospect, IL) as the scintillation fluid.

Detection of Radioactivity. Following purification by TLC or the flash chromatography method, radioactivity of the tritiated compounds is determined and quantified using TLC/Scanner (System 200 Imaging Scanner, Bioscan, Inc. Washington, DC) or Phosphoimager (Molecular Dynamics, Sunnyvale, CA). Analytical TLC plates containing tritium can be exposed to radiography [after using surface autoradiography enhancer (En³Hance spray, NEN Research Products, Boston, MA].

Lipid Extractions. Extraction from the crude reaction mixtures is performed using the Bligh and Dyer[37] extraction method.

Preparation of [4,5-³H]-Labeled Sphingolipids

Principles

Target compounds may be synthesized by hydrogenation of the double bonds (**B**) present in sphingoid molecules or by hydrogenation of the triple bonds (**C**) present in 4,5-dehydrosphingoid precursors.

Classical methods for tritium incorporation employ potassium boro-[³H]hydride and palladium[II] salts as catalysts.[6] The specific activity of the hydrogenated natural SLs depends on the degree of unsaturation of the ceramide moiety due to the nondiscriminatory hydrogenation process resulting in the saturation of all multiple bonds present. The reduction method produces highly radioactive compounds; however, due to complete satura-

[36] A. H. Merrill, Jr., E. Wang, R. E. Mullins, W. C. L. Jamison, S. Ninkar, and D. Liotta, *Anal. Biochem.* **171,** 373 (1998).
[37] E. A. Bligh and W. J. Dyer, *Can. J. Biochem. Physiol.* **37,** 911 (1959).

tion, the final products may not represent their natural counterparts satisfactorily, which usually include unsaturated species. This may lead to biased biological evaluations. This approach is commonly used for the total ^{3}H-labeling of natural SLs as well as for the preparation of [^{3}H]dihydrosphingolipid standards.[10–12]

The triple bond **C** precursors can also serve as excellent substrates for the preparation of [4,5-^{3}H]SL standards with higher specific activity than obtained by hydrogenation of the double-bond precursors. Our method for the preparation of stereoisomers of [4,5-^{3}H]sphinganine (presented in this article), [4,5-^{3}H]dihydroceramide, and [4,5-^{3}H]sphingosine (project under investigation in our laboratory), as well as published methods for the preparation of [4,5-^{3}H]sphingosine[14] and tritium-labeled phosphonate analogs of sphinganine-1-phosphate[15] using their **C** precursors, suggest that this synthetic approach will be receiving much more attention.

This section present the synthesis of stereoisomers of [4,5-^{3}H]dihydrosphingosine by methods adapted from the general procedure for ^{3}H-labeling of SLs.[6]

Preparation of [4,5-^{3}H]D-erythro-dihydrosphingosine IA

*From D-erythro-Sphingosine **IB**.* D-*erythro*-Sphingosine **IB** (4.2 mg, 14 μmol), dissolved after sonication in 500 μl of THF and 50 mM palladium acetate (240 μl, 12 μmol) in THF, is added via syringes to a screw-capped, Teflon-sealed, air-evacuated viral containing [^{3}H]sodium borohydrate (55.6 μmol, 25 mCi). After flushing with argon, the hydrogenation reaction is started by the addition of 1 M sodium hydroxide (150 μl) via syringe and allowed to proceed at room temperature with vigorous agitation for 2 hr, TLC analysis in solvent A shows formation of **IA** (major product, R_f = 0.15), some unidentified nonpolar by-products, and a small amount of **IB** (R_f = 0.29). Quantitation from HPLC analysis[36] gives ~90% of **IA** and ~10% unreacted **IB**. The reaction mixture is filtered off and tritium gas is removed by a vigorous stream of nitrogen, with the vapors being passed into a 5 M sulfuric acid trap. Chloroform–methanol (3 ml, 2:1, v/v) is added, and the pH of the reaction mixture is adjusted to 5 by dropwise addition of diluted acetic acid. After centrifugation (3000 rpm, 5 min, at room temperature), the supernatant is evaporated under nitrogen and dissolved in methanol–water (2 ml, 1:1 v/v), and the pH is adjusted to 11 with concentrated ammonium hydroxide solution. Bligh and Dyer[37] extraction is then performed by adding chloroform (2 ml), methanol (1 ml), and water (0.8 ml). Phases are separated by centrifugation (3000 rpm, 5 min), and the lower organic phase (containing product) is evaporated to dryness under a nitrogen gas stream. The labeled residue is dissolved in a small volume of chloroform–methanol (~0.1 ml, 2 : 1, v/v) and purified by preparative TLC-2 in solvent A. The area corresponding to the **IA** band (detected by radioscanning) is removed from the plate, and the tritiated product is extracted three times with 3 ml of chloroform–methanol (2 : 1, v/v) and evaluated quantitatively by liquid scintillation counting. Evaporation of the solvent by a nitrogen gas gives [4,5-^{3}H]dihydrosphingosine **IA** with 76% yield (3.2 mg, ~2.3 mCi). The product is characterized by analytical TLC in system A; **IA** (R_f = 0.15), and additionally by HPLC method and is 98% pure.

*From D-erythro-4,5-Dehydrosphingosine **IC***. D-*erythro*-4,5-Dehydro-sphingosine **IC** (4.2 mg, 14 μmol), dissolved in freshly distilled THF (300 μl) and 50 m*M* palladium acetate (240 μl, 12 μmol) in THF, is added via syringes to a screw-capped, Teflon-sealed, air-evacuated vial containing [^{3}H]sodium borohydrate (55.6 μmol, 25 mCi). After flushing with argon, the hydrogenation reaction is started by the addition of 1 *M* sodium hydrox-ide (150 μl) via syringe and allowed to proceed at room temperature with vigorous agitation for 1 hr, TLC-1 analysis in solvent A shows the formation of **IA** (major product, R_f = 0.15) and three by-products (R_f = 0.29, 0.24, and 0.18) with no presence of **IC** (R_f = 0.44). Quantitation from HPLC analysis gives ~65% of **IA**; the rest is a mixture of unidentified products with the major product corresponding to D-*erythro*-sphingosine **IB**. Tritium gas is removed by a vigorous stream of nitrogen, with the vapors being passed into a 5 *M* sulfuric acid trap. Chloroform–methanol (3 ml, 2 : 1, v/v) is added, and the pH of the reaction mixture is adjusted to 5 by the dropwise addition of diluted acetic acid. After centrifugation (3000 rpm, 5 min, room temperature), the supernatant is evaporated under nitrogen gas and dissolved in methanol–water (2 ml, 1 : 1, v/v) and the pH is adjusted to 11 with concentrated ammonium hydroxide solution. Bligh and Dyer[37] extraction is continued by adding 2 ml chloroform, 1 ml methanol, and 0.8 ml water. Phases are separated by centrifugation (3000 rpm, 5 min) and the product is separated in the lower organic phase. Solvents are evaporated to dryness in a nitrogen stream. The labeled residue is dissolved in a small volume of chloroform–methanol (2 : 1, v/v) and purified by preparative TLC-2 in solvent A or D. The area corresponding to the **IA** band (detected by radioscanning) is removed from the plate, and the tritiated product is extracted three times with 3 ml of chloroform–methanol (2 : 1, v/v) and evaluated quantitatively by liquid scintillation counting. Evaporation of the solvents by a nitrogen stream gives [4,5-^{3}H]**IA** with 60% yield (1.7 mg, ~2.3 mCi). The product is characterized by analytical TLC-1 in solvent A; **IA**, R_f = 0.15, or in solvent D[13,14]; **IA**, R_f = 0.22; **IC**, R_f = 0.64 and additionally by the HPLC method and is ca. 95% pure.

Using (2*S*,3*S*), (2*R*,3*S*), and (2*R*,3*R*) stereoisomers of **IB** or **IC** and following the approaches described previously, the remaining stereoisomers of **IA** can be obtained.

Preparation of [4,5-^{3}H]D-erythro-C$_{16}$-Dihydroceramide **IIA**

This section presents the synthesis of stereoisomers of [4,5-^{3}H]dihydroceramides **IIA** by methods that have been adapted from the general procedure for ^{3}H-labeling of SLs[6] and by the acylation of [4,5-^{3}H]sphinganine with fatty acids chlorides following our published method for the preparation of nonradioactive ceramides and their analogs.[37-39]

*From D-erythro-C$_{16}$-Ceramide **IIB**.* D-erythro-C$_{16}$-Ceramide **IIB** (1.5 mg, 2.78 μmol), dissolved after sonication in 4 ml of THF and 50 m*M* palladium acetate (38 μl, 1.85 μmol) in THF, is added via syringes to a screw-capped, Teflon-sealed, air-evacuated vial containing [^{3}H]sodium borohydrate (11.1 μmol, 5 mCi). After flushing with argon, the hydrogenation reaction is started by the addition of 1 *M* sodium hydroxide (15 μl) via a syringe and allowed to proceed at room temperature with vigorous agitation for 2 hr. Analysis on the TLC-5 plate, developed twice in solvent C, shows the formation of **IIA** (R_f = 0.88) and an unidentified radioactive compound with R_f = 0.7 (similar to R_f of **IIB**) in the ratio 3:1. This system is used to separate ceramide from dihydroceramide,[9] as other TLC systems used for the identification or purification of ceramides do not differentiate these two compounds. Tritium gas is removed by a vigorous stream of nitrogen, with the vapors being passed into a 5 *M* sulfuric acid trap. Chloroform–methanol (3 ml, 2:1, v/v) is added, and the pH of the reaction mixture is adjusted to 5 by the dropwise addition of diluted acetic acid. After centrifugation (3000 rpm, 5 min, room temperature). The supernatant is evaporated under nitrogen and dissolved in 2 ml methanol–water (1:1, v/v), and Bligh and Dyer[37] extraction is continued by adding 2 ml chloroform, 1 ml methanol, and 0.8 ml water. Phases are separated by centrifugation (3000 rpm, 5 min), and the product is separated in the lower organic phase. Solvents are evaporated to dryness in a nitrogen stream. The labeled residue is dissolved in a small volume of chloroform–methanol (0.1 ml, 1:1, v/v), purified by TLC-5, and developed twice in solvent C. Radiolabeled dihydroceramide is detected by radioscanning and scraped from the plate. The product is extracted three times, with 5 ml of chloroform–methanol (2:1, v/v) and evaluated quantitatively by scintillation counting. Evaporation of the solvent by a nitrogen stream gives ~0.75 mCi of [4,5-^{3}H]**IIA**

[38] S. Jayadev, B. Liu, A. Bielawska, F. Nazaire, M. Pushkareva, L. M. Obeid, and Y. A. Hannun, *J. Biol. Chem.* **270,** 2047 (1994).

[39] A. Bielawska, C. M. Linardic, and Y. A. Hannun, *J. Biol. Chem.* **267,** 18493 (1992).

(~30% yield, calculation based on the radioactivity of the tritiated sodium borohydrate).

From [4,5-³H]-D-erythro-Sphinganine. Palmitic chloride (3.25 mg, 12 μmol), dissolved in 0.5 ml of THF, is added via a syringe to a solution of [4,5-³H]D-*erythro*-sphinganine (10 μmol, specific activity: 220 mCi/mol) in 1.5 ml of THF followed by the addition of 1.25 ml of 50% CH_3COONa. The reaction mixture is stirred vigorously for 4 hr. [³H]Dihydroceramide is isolated by partitioning with chloroform (6 ml)–methanol (3 ml)–water (2.25 ml). Phases are separated, and the lower phase (8 ml) is washed twice with 2 ml of water. More than 95% of the theoretical radioactivity is recovered in the lower organic phase. TLC-1 analysis of the organic phase in solvent G (separates ceramides from esterified ceramides) and in solvent A (separates dihydroceramides from unreacted dihydrosphingosine) shows that ~90% of the radioactivity is incorporated into [4,5-³H]D-*erythro*-C_{16}-dihydroceramide. Dihydroceramide **IIA** is purified by preparative TLC-3 in solvent A and exposed to phosphoimager reading. The band corresponding to **IIA** is scraped from the plate, extracted three times with 5 ml of chloroform–methanol (2 : 1, v/v), and evaluated quantitatively by scintillation counting. Evaporation of the solvent by a nitrogen stream gives ~1.6 mCi of [4,5-³H]**IIA** (~72% yield, calculation based on the radioactivity of tritiated sphinganine).

Using these procedures, the remaining [4,5-³H]C_{16}-dihydroceramide isomers—L-*threo*-(2S,3S), D-*threo*-(2R,3R), and L-*erythro*-(2R,3S)—can be prepared starting from the related [4,5-³H]sphinganine isomers. Other [4,5-³H]-Cn-dihydroceramides can also be prepared following this procedure.

Note: Because the first method for the preparation of [4,5-³H]dihydroceramides causes problems with the separation of unreacted ceramide from dihydroceramide, we prefer to use direct acylation of sphinganine **IA** to dihydroceramides **IIA**.

Preparation of [3-³H]-Labeled Sphingolipids

Principles

Analogous to the *de novo* biosynthesis of sphingolipids (Scheme 1), tritium labeling at the C-3 position of the sphingoid backbone can be achieved by reduction of 3-keto derivatives. Unfortunately, chemical reduction of the (2S)-3-ketosphingolipids (3-keto-SLs) is not stereospecific and generates two diastereoisomers with 3R and 3S configurations. The ratio of the *erythro* to the *threo* form of this [3-³H]diastereoisomers depends on the type of the 3-keto precursors used and the conditions of the reduction reaction.[20,24,26,40,41] Performed in our laboratory, the reduction of (2S)-N-Boc-3-ketosphingosine 3-keto-**XB**, (2S)-3-ketosphingosine hydrochloride 3-keto-**IB**, and 3-ketosphinganine hydrochloride 3-keto-**IA** gives products with 50–70% (de) of D-*erythro* form (based on HPLC analysis and ³H quantitation) (Fig. 3).

High diastereoselective reduction of specifically double-protected 3-keto-SLs bases leading to 91–95% (de) was reported by Koskinen and Koskinen[42] and by Hoffman and Tao.[43] This approach may find broad applications in the synthesis of radiolabeled stereoisomers of sphingosine and sphinganine.

Preparation of [3-³H]-(2S,3R)- and [3-³H]-(2S,2S)-Sphingosines IB

This section presents the synthesis of stereoisomers of [3-³H]sphingosine **IB** and [3-³H]dihydrosphingosine **IA** by methods using (2S)- or (2R)-3-keto derivatives of sphingosine HCl, N-Boc-sphingosine, or C_2-ceramide as alternative key precursors and 3-ketosphinganine HCl. Preparation methods have been adopted from the general procedure for [3-³H]-labeling of SLs.[20,21,24]

*From (2S)-3-Keto-sphingosine HCl (3-keto-**IB**).* A mixture of (2S)-3-ketosphingosine hydrochloride (6.4 mg, 18.5 μmol) and $CsCl_3 \cdot 7H_2O$ (7 mg, 18.7 μmol) in dry cold methanol (2 ml) is added via syringes to a screw-capped, Teflon-sealed, air-evacuated vial containing [³H]sodium borohydrate (55.6 μmol, 25 mCi) followed by the addition of 0.05 N NaOH in dry methanol (0.1 ml). The reaction mixture is cooled to 0° and stirred at this temperature for 30 min. The progress of the reaction is monitored by TLC-1 in solvent A (3-keto-**IB**: R_f = 0.45, **IB**: R_f = 0.29). The tritium gas is removed by a vigorous stream of nitrogen, with the vapors being passed into a 5 M sulfuric acid trap. The crude product is evaporated to dryness in a nitrogen stream and extracted into the lower phase of the Bligh or Dyer[37] extraction procedure. The upper phase (after the pH adjustment to 11) is reextracted with an additional 2 ml of chloroform. The combined organic phases are washed twice with water (2 ml) and concentrated *in vacuo*. TLC-1 analysis of the crude organic mixture in solvent A shows the formation of D-*erythro*-[3-³H]sphingosine (~50%), L-*threo*-[3-³H]sphingosine (~40%), and unknown nonpolar by-products (~10%), as estimated

[40] A. Bielawska, D. Perry, S. Jayadev, C. McKay, J. Shayman, and Y. A. Hannun, *J. Biol. Chem.* **271**, 12642 (1996).

[41] A. Bielawska, unpublished observations.

[42] A. M. P. Koskinen and P. M. Koskinen, *Tetrahedron Lett.* **34**, 6765 (1993).

[43] R. V. Hoffman and J. Tao, *J. Org. Chem.* **63**, 3979 (1998).

from total radioactivity. HPLC analysis shows a 3:2 ratio of the *erythro/threo* forms. The crude product (5.4 mg, 97.5% yield) is purified by flash chromatography applying a gradient solvent system E–F. Each 2.5-ml fraction is monitored by scintillation counting and by TLC-1 radioscanning in solvent A. Elution with 20 ml of solvent E delivers less polar by-products. [3-^{3}H]D-*erythro*-L-sphingosine—2.5 mg, 1 mCi, 95% (de)—is eluted with 20 ml of solvent G. Further elution with 20 ml of solvent G followed by 20 ml of solvent F gives a mixture of diastereoisomers with an increased amount of the the *threo* isomer. This diastereoisomeric mixture is repurified twice by flash chromatography applying the just-described solvent systems. Under this condition, [3-^{3}H]L-*threo*-sphingosine—0.6 mg, ~0.2 mCi, ~90% (de)—is isolated.

Alternatively, diastereoisomers of [3-^{3}H]sphingosine can be separated on preparative TLC-2 plates using solvent A. Bands corresponding to D-*erythro*-isomer (R_f = 0.225) and to L-*threo*-isomer (R_f = 0.169) are scraped from the plate, extracted three times with 5 ml of chloroform–methanol (1:1, v/v), and evaluated by scintillation counting. Evaporation of the solvents by a nitrogen stream gives [3-^{3}H]D-*erythro*-sphingosine—2.2 mg, 0.85 mCi, 80% (de)—and L-*threo*-sphingosine—1.4 mg, 0.5 mCi, 70% (de). Repurification under this procedure affords [3-^{3}H]D-*erythro*-sphingosine—1.8 mg, ~0.7 mCi, 92% (de)—and L-*threo*-sphingosine—0.8 mg, 0.3 mCi, 90% (de).

Starting from the (2*R*)-3-keto-**IB** precursor, the corresponding L-*erythro*- and D-*threo*-[3-^{3}H]sphingosine can be prepared with similar results. [3-^{3}H]D-*erythro*-Sphingosine can be also isolated from the crude product by the preparative HPLC method on Zorbax Stable Bond C$_{18}$ column using methanol–THF–water–acetic acid (60:20:20:0.5, v/v) as a mobile phase.[44] This compound is ~90% (de) pure (our observation). Direct reduction of (2*S*)-3-keto-sphingosine is unsatisfactory because it producers a low yield and the major product is a nonpolar, unknown radioactive compound).[41]

From (2S)-3-keto-N-tert-butoxycarbonyl-sphingosine: (2S)-3-keto-N-Boc-sphingosine (3-keto-XB). A mixture of (2*S*)-3-keto-*N*-Boc-sphingosine 3-keto-**XB** (7.3 mg, 18.5 μmol) and CsCl$_3$ · 7H$_2$O (7 mg, 18.7 μmol) in dry, cold methanol (2 ml) is added via syringe to a screw-capped, air-evacuated flask containing [^{3}H]sodium borohydrate (55.6 μmol, 25 mCi) in 0.5 ml of cold, dry methanol, followed by the addition of 0.05 *N* NaOH in dry methanol (0.1 ml). The reaction mixture is cooled to 0° and stirred at this temperature for 30 min. Progress is monitored by TLC-1 in solvent A (3-keto-*N*-Boc-sphingosine: R_f = 0.786; *N*-Boc-sphingosine: R_f = 0.685; D-*erythro*-sphingosine: R_f = 0.29). The tritium gas is removed by a vigorous stream

---

[44] American Radiolabeled Chemicals, St. Louis, MO, published information.

of nitrogen, with the vapors being passed into a 5 M sulfuric acid trap. The crude product is evaporated to dryness in a nitrogen stream and extracted into the lower phase using the Bligh and Dyer[37] procedure. The organic phase is washed twice with water (2 ml) and concentrated *in vacuo*. Without further purification the crude product (7.0 mg, 96% yield) is dissolved in 2 ml of dry methylene chloride and treated with dry trifluoroacetic acid (500 μl) under a dry nitrogen atmosphere for 15 min at 0°. Concentration and treatment with Amberlite IRA 400 (OH-) resin in methanol produces D-*erythro*- and L-*threo*-[3-³H]sphingosines at a ratio of 3:1 (estimated from TLC-1 in solvent A). Two stereoisomers of [3-³H]sphingosine are separated via flash chromatography (see earlier procedure). This procedure gives D-*erythro*- [³H]sphingosine—1.5 mg, ~0.65 mCi, 95% (de). TLC-1 analysis of the formed [3-³H]-*N*-Boc-sphingosine in solvent H shows the presence of two closely located spots; R_f = 0.67 and R_f = 0.64 corresponding to D-*erythro*- and L-*threo* isomers and no presence of 3-keto-*N*-Boc-sphingosine (R_f = 0.74). This result may indicate the possibility for the separation of stereoisomers of sphingosine at the level of their *N*-Boc-derivatives.

*From (2S)-3-Keto-C₂-ceramide: 3-keto-**IIB** or 3-Keto-dihydro-C₂-cer-amide (3-keto-**IIA**).* A mixture of 3-keto-C₂-ceramide (6.3 mg, 18.5 μmol) and $CsCl_3 \cdot 7H_2O$ (7 mg, 18.7 μmol) in dry, cold methanol (2 ml) is added via syringe to a screw-capped, air-evacuated flask containing [³H]sodium borohydride (55.6 μmol, 25 mCi) in 0.5 ml of cold, dry methanol, followed by the addition of 0.05 N NaOH in dry methanol (0.1 ml). The reaction mixture is cooled to 0° and stirred at this temperature for 30 min with TLC monitoring in solvent I (3-keto-C₂: R_f = 0.46, D-*erythro*-C₂: R_f = 0.065). The tritium gas is removed by a vigorous stream of nitrogen, with the vapors being passed into a 5 M sulfuric acid trap. The crude product is evaporated to dryness in a nitrogen stream and extracted into the lower phase using the Bligh and Dyer[37] procedure. The organic phase is washed twice with water (2 ml) and concentrated *in vacuo*. Without further purifi-cation, the crude product (5.5 mg, 88% yield) is subjected to acid hydrolysis with 0.5 N methanolic HCl (3 ml) at 60° for 10 hr.[45] After cooling, the solution is extracted three times with 2 volumes of petroleum ether, and methanol is concentrated *in vacuo* to about one-half of its original volume and cooled in an ice bath. The pH of the reaction mixture is adjusted to 12 with a concentrated aqueous ammonium hydroxide solution. The crude product is evaporated to dryness in a nitrogen stream and extracted into the lower phase using the Bligh and Dyer[37] procedure. TLC-1 analysis in solvent A shows the formation of D-*erythro*- and L-*threo*-[3-³H]sphingosines at a ratio of 2:1. Stereoisomers of [3-³H]sphingosine are separated via flash

[45] R. C. Gaver and C. C Sweeley, *J. Am. Oil Chem. Soc.* **42**, 294 (1965).

chromatography (see previous procedure). This procedure gives D-*erythro*-[3-³H]**IB**—1.4 mg, ~0.6 mCi, 95% (de). Alternatively, [3-³H]D-*erythro*- and [3-³H]L-*threo*-sphingosines with high (de) purity can be prepared by the reduction of (2S)-3-ketoazidosphingosine. Diastereoisomers are separated at the level of their [3-³H]azido derivatives.[24]

[3-³H]Sphingosine is also prepared from (2S)-3-ketogalactocerebrosides[20,21] and from [3-³H]psychosines.[17]

Preparation of [3-³H]-(2S,3R)- and [3-³H]-(2S,3S)-Sphinganines **IA**

*From (2S)-3-ketosphinganine HCl: 3-keto-***IA***. A mixture of (2S)-3-ke-tosphinganine hydrochloride (6.4 mg, 18.5 μmol) in dry, cold methanol (2 ml) is added via syringe to a screw-capped, Teflon-sealed, air-evacuated vial containing [³H]sodium borohydrate (55.6 μmol, 25 mCi) followed by the addition of 0.05 N NaOH in dry methanol (0.1 ml). The reaction mixture is cooled to 0° and stirred at this temperature for 30 min. The progress of the reaction is monitored by TLC-1 in solvent A (3-keto-**IA**: R_f = 0.60, **IA**: R_f = 0.15). The tritium gas is removed by a vigorous stream of nitrogen with the vapors being passed into a 5 M sulfuric acid trap. The crude product is evaporated to dryness in a nitrogen stream and extracted into the lower phase using the Bligh and Dyer[37] procedure. The upper phase (after the pH adjustment to 11) is reextracted with a additional 2 ml of chloroform. The combined organic phases are washed twice with water (2 ml) and concentrated *in vacuo*. TLC-1 analysis of the crude organic mixture in solvent A shows the formation of [3-³H]sphinganine (R_f = ~0.15; 90% radioactivity) and unknown nonpolar by-products (10%). HPLC analysis showed a 1 : 1 ratio of the *erythro/threo* forms. The crude product (5.4 mg, 97.5% yield) is purified by flash chromatography applying a gradient solvent system E–F. Each 2.5-ml fraction is monitored by scintillation counting and by TLC-1 radioscanning in system A. Elution with 20 ml of solvent delivers less polar by-products. [3-³H]D-*erythro*-Sphinganine [1.0 mg, 0.4 mCi, 85% (de)] is eluted with 20 ml of methylene chloride–methanol (2 : 1 v/v). Further elution with 20 ml of methylene chloride–methanol (2 : 1, v/v), followed by 20 ml of solvent F, gives a mixture of diastereoisomers with increased amounts of the *threo* isomer. This diastereoisomeric mixture is repurified by flash chromatography applying the just-described solvent system. Under these conditions, [3-³H]L-*threo*-sphinganine [0.8 mg, ~0.3 mCi, ~80% (de)] is isolated.

*From (2S)-3-ketodihydro-C₂-ceramide: 3-keto-***IIA***. Alternatively, [³H]-(2S,3R)- and [3-³H]-(2S,3S)-sphinganines **IA** can be obtained by the procedure presented earlier for the preparation of [3-³H]sphingosine from (2S)-3-keto-C_2-ceramide.

Preparation of [3-³H]Ceramide **IIB** and [3-³H]Dihydroceramide **IIA**

This section presents the synthesis of stereoisomers of [3-³H]ceramides and [3-³H]dihydroceramides by the acylation of [3-³H]sphingosine and [3-³H]sphinganine with fatty acids and their chlorides or anhydrides following our published method for the preparation of nonradioactive ceramides and their analogs.[38–40] An alternative method for the preparation of [3-³H]ceramides, a direct reduction of their 3-ketoanalogs with [³H]sodium borohydride, generates two stereoisomers of [3-³H]ceramides. Because both diastereoisomers run together in the solvent systems recommended for ceramide purification, we prefer the following method. In order to achieve separation of diastereoisomeric mixtures of long chain ceramides, Shoyama et al.[31] used a borate-impregnated TLC-5 plate in solvent C. To separate a diastereoisomeric mixtures of short chain ceramides, Ridgway et al.[46] used TLC plates developed in diethyl ether–methanol (9:1, v/v for C_2-ceramide and 31:1, v/v for C_6-ceramide). These separation methods did not work successfully in our hands.

Preparation of [3-³H]D-erythro-C_{16}-ceramide **IIB**

Palmitoyl chloride (1.65 mg, 6 μmol) dissolved in 0.5 ml of THF is added via syringe to the solution of [3-³H]D-*erythro*-sphingosine [95% (de), [5 μmol, 0.58 mCi] in 0.5 ml of THF followed by the addition of 0.625 ml of 50% CH_3COONa. The reaction mixture is stirred vigorously for 4 hr. [³H]Ceramide is isolated by partitioning with chloroform (3 ml)–methanol (1.5 ml)–water (1.125 ml). The phases are separated, and the lower phase (4 ml) is washed twice with 1 ml of water. More than 95% of the theoretical radioactivity is recovered in the lower organic phase. TLC analysis of the organic phase, performed in solvent B (this solvent system separates ceramides from esterified ceramides) and in solvent A (this solvent system separates ceramides from unreacted sphingosine), shows that ~85% of radioactivity is incorporated into [3-³H]C_{16}-ceramide. Ceramide **IIB** is purified by preparative TLC-3 in solvent A and exposed to phosphoimager reading. The band corresponding to **IIB** is scraped from the plate, and the product is extracted three times with 3 ml of chloroform–methanol (2:1 v/v) and quantified by scintillation counting. Evaporation of the solvent by a nitrogen stream gives ~0.45 mCi of [3-³H] **IIB** (77% yield, based on radioactivity of the tritiated sphingosine).

[46] N. D. Ridgway, *Biochim. Biophys. Acta* **1256,** 39 (1994).

Preparation of [3-³H]-D-erythro-C₁₆-dihydroceramide **IIA**

Following the just-described protocol, this compound can be prepared from [3-³H]D-*erythro*-dihydrosphingosine [85% (de) with 70% yield]. [3-³H]Ceramides and [3-³H]dihydroceramides can be also synthesized from [3-³H]sphingosine and [3-³H]sphinganine by the preparation methods presented elsewhere in this volume.[47]

General Synthetic Methods for Labeling in the Sphingoid Backbone

Labeling in the sphingoid-backbone of glycosphingolipids has been reviewed by Sonnino *et al.*[27] Methods used for the preparation of SLs labeled in the sphingoid backbone (sphingosine, ceramide, and phosphosphingolipids) and published by other research groups-include

1. 4,5-³H labeling: [4,5-³H]dihydrosphingomyelin **IVA**,[6,8] [4,5-³H]dihydroceramide **IIA**,[10–12] [4,5-³H]phosphonate analog of sphinganine-1-phosphate,[15] and [4,5-³H]sphingosine **IB** from 4,5-dehydrosphingosines **IC**.[13,14]
2. 3-³H labeling: [3-³H]D-*erythro*- and [3-³H]-L-*threo*-sphingosines from (2S) 3-ketoazidosphingosine,[24] [3-³H]sphingosine from (2S) 3-keto-galactocerebrosides,[20,21] [3-³H]sphingosine from [3-³H]psychosines,[17] [3-³H]ceramides **IIB**,[31,44] [3-³H]SM **IVB**,[22] and [3-³H]dihydro-SM **IVA**.[19]
3. 1-³H labeling: [1-³H]D-*erythro*-sphingosine,[29] [1-³H]sphingosine stereoisomers,[31] and [1-³H]ceramide sterereoisomers.[31]
4. Radiolabeling in the course of total synthesis[32,33]: [3-¹⁴C]- and [1-³H; 3-¹⁴C]*erythro*-DL-dihydrosphingosine, [7-³H]*erythro*-DL-sphingosine, and [5-³H]*erythro*- and -*threo*-DL-dihydrosphingosine.

Conclusions and Future Directions

Introduction of radiolabeled markers into C-1, C-3, C-4, and C-5 positions of the sphingoid backbone is accomplished most commonly through the reduction of multiple bonds already present in the molecule or by the reduction of specifically modified functional groups with metallic hydrides, e.g., NaB^3H_4, KB^3H4, or $LiAl^3H_4$. The ratio of the reducing agent to substrate plays a critical role in the reaction course and has to be optimized for the specific applications. To avoid double-bond reduction in conjugated ketones, $CsCl_3$ can be used as a suppressing agent.[43]

---

[47] A. Bielawska and Y. A. Hannun, *Methods Enzymol.* **311** [43] 1999 (this volume).

Tritium labeling at the C-4 and C-5 positions of the natural SLs and synthetic standards through reduction of their carbon–carbon double bond is the simplest method to introduce radioactivity into the sphingosine backbone. On the one hand, properties of the saturated molecules may no longer represent the starting SLs. On the other hand, however, introduction of the radioactive markers into the C-1 or C-3 positions of the sphingosine backbone retains their properties.

Reduction of stereo- and enantiomerically pure 4,5-dehydrosphingosines **IC** may be the most effective and simple tritiation procedure developed so far for the preparation of [4,5-^{3}H]sphinganine, [4,5-^{3}H]sphingosine, and their stereoisomers. Triple-bond precursors are already available synthetically for both sphingosine and ceramide stereoisomers. This method needs more investigation.

Introduction of tritium at the C-3 position of SLs by reduction of their 3-ketoprecursor is the most common method applied for the preparation of all classes of SLs. By working out the specific conditions, particularly to improve stereospecificity at the reduction step and the separation techniques, reduction of the 3-ketoprecursors may become the most universal synthetic method for the introduction of tritium into the sphingosine backbone.

Alternatively, incorporation of tritium into the C-1 position, achieved by reduction of the 1-O derivatives, seems to be an appropriate way, leading to optically pure sphingosine and ceramide stereoisomers. Moreover, this operation does not require a tedious separation of stereoisomers because the transformation takes place at the achiral part of the molecule, leaving no possibility for racemization. Unfortunately, 1-^{3}H labeling can be applied only to SLs with a free primary hydroxyl group. Synthetic methods leading to the preparation of the precursors to this method need to be improved.

In conclusion, from the overview of the available methods of incorporation of tritium into the sphingosine backbone at the C-1 and C-3 positions, the oxidation–reduction methods of the secondary hydroxyl group or primary hydroxyl group of the properly protected precursors, as well as the further development of highly effective chiral HPLC techniques, seem to be the most applicable and effective approaches to obtain enantiomerically pure radiolabeled sphingolipids. Incorporation of tritium into the sphingosine backbone at the C-4–C-5 positions, used commonly at the double bond level, should be investigated further for triple-bond precursors. This method gives highly radioactive isomers of sphinganine and may also provide high-purity isomers of sphingosine.

[43] Preparation of Radiolabeled Ceramides and Phosphosphingolipids

By Alicja Bielawska *and* Yusuf A. Hannun

Introduction

Radiolabeled sphingolipids (radio-SLs) with high specific activity, labeled at designated parts of the molecule, allow thorough investigation of their metabolic fate and quantitation of enzyme activities in sphingolipid metabolic pathways. To prepare structurally homogeneous radio-SLs, well-designed synthetic approaches and precursors should be employed, which are available only in the course of stereospecific synthesis.

In general, a few methods can be considered for the preparation of radiolabeled phospho-SLs: (1) biosynthetic approach using radioactive precursors such as L-serine, fatty acids, choline, or ^{32}P. This approach generally results in low activity and leads to mixtures. (2) Labeling in the course of total synthesis following established synthetic methods. (3) Synthetic or enzymatic approaches that are mainly limited to the transformation of existing functional groups present in the structure of natural and synthetic phospho-SLs. This latter approach was commonly used during the early stage of sphingolipid biochemistry and, due to its simplicity and effectiveness, still remains a "method of choice" for many radiolabeled analogs.

Homogeneous simple synthetic radio-SLs are available at the level of sphingosine and ceramide stereoisomers. Unfortunately, to date, full access to structurally homogenic complex radio-SLs is limited because of deficiency of the suitable synthetic precursors. Radioactive complex SLs are mainly reconstructed from their natural forms. We would like to stress two general problems encountered with the utilization of natural SLs as precursors in the radiolabeled synthesis. First, naturally occurring SLs exist as nonhomogeneous mixtures due to variation in the length, saturation, and hydroxylation of the sphingoid backbone and the *N*-acyl groups. Second, the commonly applied procedures for the isolation and preparation of simple SLs from complex precursors (such as lysosphingomyelin from sphingomyelin) can lead to a mixture of *erythro-* and *threo*-diastereoisomers due to epimerization at the C-3 position, a result of the tendency of the allylic alcohol to racemize during acidic hydrolysis.[1]

This article presents synthetic and enzymatic methods for the prepara-

---

[1] K. A. Karlsson, *Chem. Phys. Lipids* **5,** 6 (1970).

METHODS IN ENZYMOLOGY, VOL. 311

tion of ceramide, SM, and ceramide phosphate labeled in the N-acyl group; SM and lyso-SM labeled in the choline group; ceramide phosphate and sphingosine phosphate labeled in the phosphate group; and the combination thereof leading to multiple-labeled analogs.

Materials

[^{3}H]Sodium borohydride (450 mCi/mmol), [^{3}H]acetic anhydride (85 mCi/mmol), [1-^{14}C]hexanoic acid (55 mCi/mmol), N-hydroxysuccinimidyl ester of [1-^{14}C]hexanoic acid (50 mCi/mmol), [9,10-^{3}H]palmitic acid (50 Ci/mmol), [1-^{14}C]palmitic acid (54.2 Ci/mmol), [^{14}C]methyl iodide (55.7 mCi/mmol), and [^{3}H]methyl iodide (82 Ci/mmol) are obtained from American Radiolabeled Chemicals (St. Louis, MO). Trifluoroacetic acid (TFA) and phytosphingosine hydrochloride are from Sigma. Amberlite IRA 400 (OH-) resin and palmitic acid are from Fluka Chem. Corp. (Milwaukee, WI). $CH_3COONa \times 3H_2O$, cyclohexylamine, diisopropylethylamine, $Na_2S_2O_3$, and diethylphosphoryl cyanide (DEPC) are from Aldrich Chemical Company, Inc. (Milwaukee, WI).

TLC plates: The following TLC plates are used as indicated in the specific sections: (1) TLC-1; precoated TLC sheets of silica gel 60 F 254 (0.2 layer thickness) are from EM Separation Technology, (2) TLC-2; precoated thin-layer plates Kieselgel 60 (0.25 mm layer thickness) are from EM Separation Technology, (3) TLC-3; precoated thin-layer plates PK5 silica gel 150 A (0.5 mm layer thickness) are from Whatman Laboratory Division, and (4) TLC-4; Linear-K preadsorbent strip, silica gel 150A (1.0 mm layer thickness) are from Whatman Laboratory Division.

Silica gel 60 (particle size 0.040–0.063 mm) (230–400 Mesh ASTM) for column chromatography CAS 63231-67-4 is from EM Separation Technology. Sep-Pac Plus Silica, Sep-Pac Plus CM, and Sep-Pac Plus C$_{18}$ cartridges are from Millipore Corporation, Waters Chromatography Division (Milford, MA).

All solvents are of analytical grade or better and are obtained from Mallinckrodt, Burdick and Jackson, or J. T. Baker.

Anhydrous N,N-dimethylformamide: DMF (packaged in prescored ampoules) is from Aldrich Chemical Company, Inc.

General Methods Used during Synthesis

Tetrahydrofuran (THF) is freed of peroxides by passing over basic alumina, followed by saturation with dry argon.

Cyclohexylamine, diisopropylethylamine, and diethylphosphoryl cyanide are distilled before using.

Key Precursors for Preparation of Radiolabeled Sphingolipids and Analytical Standards

Stereoisomers of sphingosine, sphinganine, and their N-*tert*-butoxycarbonyl and N-acyl derivatives, N-hydroxysuccinimidyl ester of $[1-{}^{14}C]$hexanoic acid, ceramide-1-phosphoryl-N,N-dimethylethanolamine, and sphingosylphosphorylamine are synthesized according to the procedures presented elsewhere in this volume.[2]

Preparation of stereoisomers of $[3-{}^{3}H]$sphingosine and $[3-{}^{3}H]$ceramide is presented elsewhere in this volume.[3]

Thin-Layer Chromatography

Purity of the compounds, the reaction progress, and chromatographic elution profiles are checked routinely by thin-layer chromatography using analytical TLC plate: TLC-1. Purification is performed using preparative TLC plates: TLC-2, -3, or -4.

Flash Column Chromatography

Separation of reaction products and purification of the obtained compounds are achieved by flash chromatography under positive pressure of nitrogen gas as described previously.[4] Fractions containing impure material are rechromatographed.

Solvent Systems

The solvent systems used for TLC and/or flash column chromatography purification are (A) chloroform–methanol–2 N ammonium hydroxide (4:1:0.1, v/v); (B) methylene chloride–methanol (93:7, v/v); (C) chloroform–methanol–15 mM $CaCl_2$ (60:35:8, v/v); (D) chloroform–methanol (19:1, v/v); (E) chloroform–methanol–water (65:25:4, v/v); (F) chloroform–methanol–ammonium hydroxide–water (160:40:1:3, v/v); (G) chloroform–methanol–concentrated ammonium hydroxide (65:35:8, v/v); (H) chloroform–methanol–ammonium hydroxide–water (65:35:2.5:2.5, v/v); (I) chloroform–methanol–water (3:48:47, v/v); (J) chloroform–methanol–water (86:14:1, v/v); (K) methanol (containing 7% water and 1% concentrated ammonium hydroxide, v/v) and chloroform (1:1, v/v); (L) chloroform–methanol–water (60:35:8, v/v); (M) n-butanol–acetic acid–water (3:1:1, v/v); (N) chloroform–methanol–ammonium hydroxide (60:30:5,

---

[2] A. Bielawska, Z. Szulc, and Y. A. Hannun, *Methods Enzymol.* **311** [44] 1999 (this volume).
[3] A. Bielawska, Y. A. Hannun, and Z. Szulc, *Methods Enzymol.* **311** [42] 1999 (this volume).
[4] W. C. Still, M. Kahn, and A. Mitra, *J. Org. Chem.* **43,** 2923 (1978).

v/v); (P) chloroform–acetone–methanol–acetic acid–water (10:4:3:2:1, v/v); (R) chloroform–methanol (2:1, v/v); and (S) chloroform–methanol (1:2, v/v).

Analytical Assays

Composition and Diastereoisomeric Purity of Sphingoid Bases

The purity of sphingosine, sphinganine, and their radiolabeled analogs is evaluated by high-performance liquid chromatography (HPLC) using eicosasphingosine (C_{20}-sphingosine) as an internal standard, as described by Merrill *et al.*[5]

Radioactivity Measurement

Radioactivity is measured on a Wallac LKB 1214 Rackbeta liquid scintillation counter with High Flash Point Cocktail Safety-Solve (Res. Product Int. Corp. Mount Prospect, IL) as the scintillation fluid.

Radioactivity Detection

Detection and quantitation of the radioactivity of the obtained radiolabeled compounds are performed by exposing TLC plates containing radioactive compounds to TLC/Scanner (System 200 Imaging Scanner, Bioscan, Inc. Washington, DC) or to Phosphoimager (Molecular Dynamics, Sunnyvale, CA) Analytical TLC plates containing tritium can be exposed to radiography [after using surface autoradiography enhancer (En³Hance spray, NEN Research Products, Boston, MA].

Lipid Extractions

Lipid extractions from the crude reaction mixtures are performed using the Bligh and Dyer extraction method[6] or the Folch *et al.*[7] procedure.

Phosphorous measurement is performed using a published method.[8]

DG-kinase phosphorylation is performed using a published method[9]; this method is also presented in volume 312.[10]

[5] A. H. Merrill, Jr., E. Wang, R. E. Mullins, W. C. L. Jamison, S. Ninkar, and D. Liotta, *Anal. Biochem.* **171,** 373 (1998).

[6] E. A. Bligh and W. J. Dyer, *Can. J. Biochem. Physiol.* **37,** 911 (1959).

[7] J. Folch, M. Lees, and G. H. Sloane-Stanley, *J. Biol. Chem.* **226,** 497 (1957).

[8] B. N. Ames, *Methods Enzymol.* **8,** 115 (1966).

[9] P. P. Van Veldhoven, W. R. Bishop, D. A. Yurivich, and R. M. Bell, *Biochem. Mol. Biol. Inter.* **36,** 21 (1995).

[10] D. Perry, A. Bielawska, and Y. A. Hannun, *Methods Enzymol.* **312** (1999).

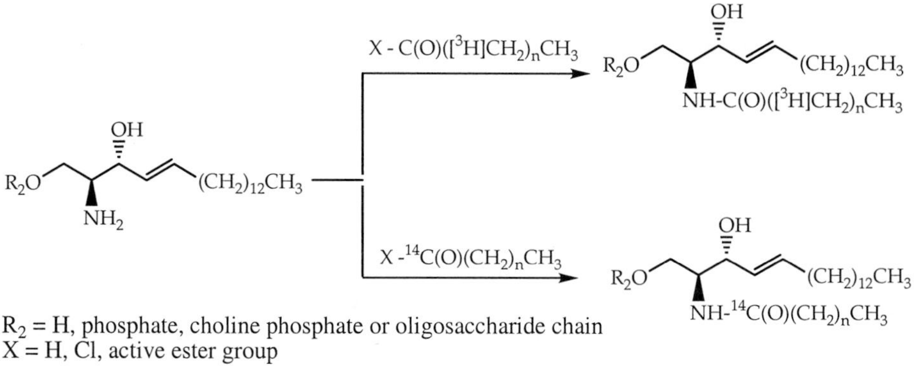

R_2 = H, phosphate, choline phosphate or oligosaccharide chain
X = H, Cl, active ester group

FIG. 1. Approach to the synthesis of [^{3}H]- and [^{14}C]sphingolipids labeled in the N-acyl group.

Preparation of Sphingolipids Labeled in the Amide Moiety

Principles

Ceramide and its more complex derivatives are synthesized from the corresponding sphingoid bases prepared synthetically,[2] naturally,[11-16] or from lyso-SLs by classical methods used in peptide chemistry. Stringent competition in the reactivity of the amino and hydroxyl groups in SLs toward O- and N-acylation favors N-acylation products. The amide bond is usually prepared by the action of acid anhydrides[17,18] or acid halides[19,20] by transacylation of the activated fatty acids as N-hydroxysuccinimide esters,[21,22] or by p-nitrophenyl esters.[23] Alternative methods have been

[11] R. C. Gaver and C. C. Sweeley, *J. Am. Oil Chem. Soc.* **42,** 294 (1965).

[12] R. Cohen, Y. Barenholtz, S. Gatt, and A. Dagan, *Chem. Phys. Lipids* **35,** 371 (1984).

[13] P. P. Van Veldhoven, R. J. Foglesong, and R. M. Bell, *J. Lipid Res.* **30,** 611 (1989).

[14] P. P. Van Veldhoven and G. P. Mannaerts, *J. Biol. Chem.* **266,** 12502 (1991).

[15] M. Buneman, K. Liliom, B. T. Brandts, L. Pott, J. L. Tseng, D. M. Desiderio, G. Sun, D. Miller, and G. Tigyi, *EMBO J.* **15,** 101 (1996).

[16] A. Kisic, M. Tsuda, R. J. Kulmacz, W. K. Wilson, and G. J. Schroepfer, *J. Lipid Res.* **36,** 787 (1995).

[17] R. C. Gaver and C. C. Sweeley, *J. Am. Chem. Soc.* **88,** 3643 (1966).

[18] A. Bielawska, H. M. Crane, D. Liotta, L. M. Obeid, and Y. A. Hannun, *J. Biol. Chem.* **268,** 26226 (1993).

[19] K. C. Kopaczyk and N. S. Radin, *J. Lipid Res.* **6,** 140 (1965).

[20] A. Bielawska, D. Perry, S. Jayadev, C. McKay, J. Shayman, and Y. A. Hannun, *J. Biol. Chem.* **271,** 12642 (1996).

[21] Y. Lapidot, S. Rapport, and Y. Wolman, *J. Lipid Res.* **8,** 142 (1967).

[22] A. Futerman and R. E. Pagano, *Methods Enzymol.* **209,** 437 (1986).

[23] B. Neises, T. Andries, and W. Steglich, *J. Chem. Soc. Chem. Commun.* **1132,** (1982).

developed using coupling agents such as N-ethoxycarbonyl-2-ethoxy-1,2-dihydroquinoline,[24] dicyclohexylcarbodiimide (DCC),[12] diphenyl phosphorazidate,[25] 1-(3-dimethylaminopropyl)-3-ethylcarbodiimide,[26,27] ethylchloroformate,[28] pentafluorophenol,[29] or diethylphosphoryl cyanide (DEPC).[30] A new enzymatic method for the preparation of radioactive ceramides using the reverse hydrolysis reaction of sphingolipid ceramide N-deacylase (SCDase) has been published.[31]

Four approaches for the radio acylation of the amino group of sphingoid bases employ (1) acetic anhydride, (2) acyl chlorides, (3) N-hydroxy succinimidyl esters, and (4) fatty acids and DEPC as a coupling agent. To synthesize C_2-ceramides, use the first approach; to prepare long chain ceramides, use the second approach; and to synthesize C_6- and C_8-ceramides, use the last approach.

The DEPC method employs diethylphosphoryl cyanide as a coupling agent in the presence of triethylamine and is rapid, free of racemization, utilizes acids without prior derivatization, and generates relatively pure products. We have noticed that this reaction needs to be carried out in anhydrous conditions, under nitrogen atmosphere, and at 0°. The use of freshly distilled DEPC and triethylamine is critical. Following these rules we have synthesized all stereoisomers of [^{14}C]C_6- and [^{14}C]dihydro-C_6-ceramides with about 90% yield. The acid chloride method (the second approach), which we have named the "one pot" procedure, employs a simple and effective protocol of in situ generation of the acid chlorides from fatty acids and thionyl chloride or oxalyl chloride[19,20] following the introduction of a THF-aqueous CH_3COONa reaction system and selected sphingosines. Using this method, stereoisomers of various radiolabeled long chain ceramide and dihydroceramides can be synthesized with 70–85% yield.

[24] Z. Dong and J. A. Butcher, Jr., *Chem. Phys. Lipids* **66**, 41 (1993).

[25] S. Yamada, N. Ikota, T. Shioriand, and S. Tachibana, *J. Am. Chem. Soc.* **97**, 7174 (1975).

[26] S. Hammarstrom, *J. Lipid Res.* **12**, 760 (1971).

[27] T. Levade, S. Gatt, A. Maret, and R. Salvayre, *J. Biol. Chem.* **266**, 13519 (1991).

[28] D. Aquotti, S. Sonnino, M. Masserine, L. Casella, G. Fronza, and G. Tettamanti, *Chem. Phys. Lipids* **40**, 71 (1986).

[29] R. Jennemann, C. Gnewuch, S. Bolets, B. L. Bauer, and H. Wiegandt, *J. Biochem.* **115**, 1947 (1994).

[30] J. K. Anand, K. K. Sadozai, and S.-I. Hakamori, *Lipids* **30**, 995 (1996).

[31] S. Mitsutake, K. Kita, N. Okino, and M. Ito, *Anal. Biochem.* **247**, 52 (1997).

Preparation of Radiolabeled Ceramides

Synthesis of N-[1-³H-acetyl]-D-erythro-sphingosine:
N-[1-³H]D-erythro-C₂-ceramide

This section presents the synthesis of stereoisomers of C_2-ceramide and C_2-dihydroceramide and the preparation of N-acetylphytosphingosine, all labeled in the acetyl group, by a method[18] adapted from the procedure for the preparation of nonradioactive C_2-ceramide.[17,18]

[³H]Acetic anhydride (5 mCi, 59 μmol) is cooled in an ice bath and diluted with 100 μmol (5 μl) of cold acetic anhydride in 5 ml of dry methanol. D-*erythro*-Sphingosine (12 mg, 40 μmol) is dissolved in 2 ml of dry methanol and is added to the methanol solution of [³H]acetic anhydride. The reaction mixture is stirred in an ice bath for 4 hr, when the TLC-1 solvent shows that the reaction is complete. Cold water (5 ml) is added dropwise to produce a white precipitate, and the mixture is stirred in an ice bath for an additional 30 min. The precipitate is separated by centrifugation (3000 rpm, 5 min, 5°), separated from the supernatant, dissolved in 5 ml of chloroform, and washed with 1 ml of 0.1 N NaOH. Phases are separated by centrifugation, and the organic phase is washed with water until no radioactivity is present in the upper phase. The organic phase is dried under nitrogen, dissolved in solvent R, and purified by preparative TLC-4 in solvent A. The area corresponding to C_2-ceramide is scraped, and ceramide is extracted three times with 5 ml of solvent R. The extraction efficiency is evaluated quantitatively by scintillation counting. Combined extracts are dried under vacuum to give 12.5 mg (91.5% yield) of [³H]D-*erythro*-C_2-ceramide (specific activity: 15.75 mCi/mmol). The purity of this product (~96% pure) is established by analytical TLC-1 in solvent B ($R_f = 0.21$) and in solvent A ($R_f = 0.64$). This product is dissolved in solvent R at 10 mM and is stored at −20° for several months.

Using the just-described procedures, the remaining C_2-ceramide and C_2-dihydroceramide stereoisomers—L-*threo*-(2S,3S),D-*threo*-(2R,3R), and L-*erythro*-(2R,3S)—can be prepared starting from the corresponding sphingosine and sphinganine isomers. [³H]C_2-ceramides/dihydroceramides labeled in the sphingosine/sphinganine backbone can be prepared by the acylation of [3-³H]sphingosine and [3-³H]- or [4,5-³H]sphinganines with unlabeled acetic anhydride following the just-described procedure.

Synthesis of N-[1-³H-acetyl]phytosphingosine: N-[1-³H]C₂-phytoceramide

[³H]Acetic anhydride (5 mCi, 59 μmol) is cooled down in an ice bath and diluted with 100 μmol (5 μl) of cold acetic anhydride in 5 ml of dry methanol. Phytosphinogosine hydrochloride (14.2 mg, 40 μmol) is dissolved in 2 ml of dry methanol, and the pH is adjusted to 9 with 2 N NaOH in methanol and added to the methanol solution of [^{3}H]acetic anhydride. The reaction mixture is stirred in an ice bath for 4 hr. TLC-1 in solvent A shows that the reaction is complete. The product is isolated and purified following the procedure developed for the C_2-ceramide preparation. After final purification, [^{3}H]C_2-phytoceramide is obtained in 77% yield (12.2 mg, specific activity: 15.75 mCi/mmol), with a purity of ~98% as established by analytical TLC-1 in solvent A (R_f = 0.4).

Synthesis of N-[1-^{14}C-Hexanoyl]-D-erythro-sphingosine: N-[1-^{14}C]-D-erythro-C$_6$-ceramide

This section presents the synthesis of stereoisomers of C_6-ceramide and C_6-dihydroceramide by methods that have been adapted from general procedures for the preparation of radioactive ceramides by the DEPC method[30] and by the NHS ester method.[21,22]

Via the DEPC Method. D-*erythro*-Sphingosine (4.9 mg, 16.4 μmol), in freshly distilled and dry DMF–methylene chloride 1:3 mixture (2 ml), is added via syringe to a screw-capped, Teflon-sealed, air-evacuated vial containing [1-^{14}C]hexanoic acid (0.9 mCi, 16.4 μmol) in 900 μl of dry methylene chloride followed by the addition of 32.8 μmol (11 μl) of freshly distilled DEPC. The volume of the solvents is adjusted to 11 ml, keeping the proportion DMF–methylene chloride as 1:3 (v/v). After flushing with argon, the reaction mixture is cooled to 0°, and freshly distilled triethylamine (16.4 μmol, 9 μl) is added under argon. Stirring under argon is continued for 2 h, solvents are evaporated, and the product is extracted by the Bligh and Dyer[6] procedure. The upper phase is reextracted with 2 ml of chloroform; liquid scintillation counting detects almost no radioactivity in the upper phase. TLC analysis of the radioactive organic phase is performed in solvent B (C_6-ceramide: R_f = 0.34) and in solvent A (C_6-ceramide, R_f = 0.74). Solvent A, which separates well ceramides from unreacted fatty acids and sphingosine, shows that ~80% of radioactivity is incorporated into ceramide. The organic phase is dried under nitrogen, dissolved in solvent R, and purified by preparative TLC-4 in solvent A. The area corresponding to C_6-ceramide is scraped, and ceramide is extracted three times with solvent R. Extraction efficiency is evaluated by liquid scintillation counting. Combined extracts are dried under vacuum to give 5.54 mg (85% theoretical yield) of N-[1-^{14}C]D-*erythro*-C_6-ceramide. The purity of this product was ~95% as established by analytical TLC-1 in solvents C and A. This product is dissolved in solvent R at 10 mM (specific activity: 1.22×10^5 cpm/nmol) and stored at −20° for several months.

Using the just-described procedures, the remaining C_6-ceramide and C_6-dihydroceramide stereoisomers—L-*threo*-(2S,3S), D-*threo*-(2R,3R), and L-*erythro*-(2R,3S) can be prepared starting from the corresponding sphingosine and sphinganine isomers.

Via the NHS Ester Method. D-*erythro*-Sphingosine (6.0 mg, 20 μmol) in 1 ml of dry DMF is added via syringe to a screw-capped, Teflon-sealed, air-evacuated vial containing N-hydroxysuccinimidyl ester of [1-^{14}C]hexanoic acid (1.0 mCi, 20 μmol) in 1 ml of dry DMF followed by the addition of 10 μl of freshly distilled diisopropylethylamine. The reaction is allowed to proceed at room temperature under nitrogen atmosphere in the dark, with stirring. Progress is monitored periodically by TLC-1 in solvent C. After 15 hr, the reaction is almost complete, the reaction mixture is acidified with 20 μl of 3 N HCl, and the mixture is dried completely under a stream of nitrogen. The product is extracted by the Bligh and Dyer[6] procedure. The upper phase is reextracted with 2 ml of chloroform, dried under nitrogen gas, dissolved in solvent R, and purified by preparative TLC-3 in solvent E or F. The area corresponding to C_6-ceramide is scraped, and ceramide is extracted three times with 5 ml of solvent R. The extraction efficiency is evaluated by liquid scintillation counting. The combined extracts are dried under vacuum to give 6.4 mg (80% theoretical yield) of N-[1-^{14}C]D-*erythro*-C_6-ceramide. The purity of this product is ∼95% as was established by analytical TLC-1 in solvents C and A.

Synthesis of N-[^{3}H or ^{14}C-palmitoyl]D-erythro-sphingosine: N-[9,10-^{3}H]D-erythro-C_{16}-ceramide or N-[1-^{14}C]D-erythro-C_{16}-ceramide

This section presents the synthesis of stereoisomers of C_{16}-ceramide and C_{16}-dihydroceramide via an acyl chloride method that has been adapted from the procedures for the preparation of nonradioactive ceramides and their analogs.[19,20]

[9,10-^{3}H]Palmitic acid (4.8 mCi, 0.08 μmol), diluted with 8 μmol of cold palmitic acid, is refluxed with 0.5 ml of oxalyl chloride in dry benzene for 1 hr under nitrogen atmosphere.[16] The solution is then evaporated to dryness under vaccum. D-*erythro*-Sphingosine (2.6 mg, 8 μmol) dissolved in 2.5 ml of THF is added via syringe to the crude [^{3}H]palmitic chloride followed by the addition of 1.25 ml of 50% CH$_3$COONa. The reaction mixture is stirred vigorously for 4 h. Ceramide is isolated by partitioning with chloroform (6 ml)–methanol (3 ml)–water (2.25 ml). The phases are separated, and the lower phase (8 ml) is washed twice with 2 ml of water. More than 90% of the theoretical radioactivity (calculation from the specific activity of the original [^{3}H]palmitic acid) is recovered in the lower organic phase with almost no radioactivity in the upper phase. TLC-1 analysis of the organic phase, performed in solvent B (separates ceramides from esterified ceramides) and in solvent A (separates ceramides from unreacted fatty acids and sphingosine), shows that ~90% of radioactivity is incorporated into N-[9,10-^{3}H]D-*erythro*-C$_{16}$-ceramide. Ceramide is purified by preparative TLC-3 in solvent A, the area corresponding to C$_{16}$-ceramide is scraped, and ceramide is extracted three times with solvent R. The extraction efficiency is monitored by liquid scintillation counting. The combined extracts are dried under vacuum and repurified with solvent B. The final [^{3}H]D-*erythro*-C$_{16}$-ceramide (82% yield) is dissolved in solvent R at 10 mM (specific activity: 1.3×10^5 dpm/nmol). The purity is ~96%, as established by analytical TLC-1 in solvent B ($R_f = 0.21$) and in solvent A ($R_f = 0.64$).

This procedure can be used to prepare the remaining C$_{16}$-ceramide and C$_{16}$-dihydroceramide isomers and other Cn-ceramides.

Preparation of Sphingomyelin Labeled in the N-Acyl Group

Synthesis of N-[1-^{14}C-Hexanoyl]-D-erythro-sphingosylphosphorylcholine: N-[1-^{14}C]-C$_6$-SM

Sphingomyelin labeled in the N-acyl group is obtained by N-acylation of D-*erythro*-sphingosylphosphorylcholine via the DEPC method or via the NHS ester method, following procedures developed for N-[1-^{14}C]ceramide preparation.[26,34]

D-*erythro*-Sphingosylphosphorylcholine (6.0 mg, 13 μmol) in 2 ml of dry DMF is added via syringe to a screw-capped, Teflon-sealed, air-evacuated vial containing N-hydroxysuccinimidyl ester of [1-^{14}C]hexanoic acid (0.65 mCi, 13 μmol) in 1 ml of dry DMF followed by the addition of 10 μl of freshly distilled diisopropylethylamine. The reaction is allowed to proceed at room temperature under nitrogen atmosphere in the dark, with stirring. Progress is monitored periodically by TLC-1 in solvent C. After 25 hr the reaction is almost complete, and the reaction mixture is acidified with 20 μl of 3 N HCl and then dried completely under a stream of nitrogen. The product is extracted by the Bligh and Dyer procedure. The upper phase is reextracted with 2 ml of chloroform, dried under nitrogen, dissolved in solvent R, and purified by preparative TLC-3 in solvent E or F. The area corresponding to C$_6$-SM is scraped, and sphingomyelin is extracted five times with 5 ml of solvent R. The extraction efficiency is evaluated by liquid scintillation counting. The combined extracts are dried under vacuum to give 5.3 mg (75% theoretical yield) of N-[1-^{14}C]D-*erythro*-C$_6$-SM. The purity of this product is ~95% as established by analytical TLC-1 in solvents C and G.

Preparation of Radiolabeled Sphingomyelin

Principles

Sphingomyelin can be radiolabeled (Fig. 2) in various parts of the molecule, depending on the experimental needs. Reduction of 3-keto-SM labels the C-3 position of the sphingosine backbone ([3-^{3}H]-SM). N-acylation of sphingosylphosphorylcholine (lyso-SM) with radioactive fatty acids generates SM labeled in the N-acyl group (N-[1-^{14}C-acyl]-labeled SM, see procedure described earlier). The quaternization procedure of ceramide-1-phosphoryl-N,N-dimethylethanolamine (demethylated-SM) generates [^{14}C-choline]-labeled SM, and this method is presented next.

Chemically homogeneous SMs are available presently in a limited scope as their syntheses were described only for a few selected D-*erythro* series.[32,33] These multistep methods have not been adapted and optimized yet to the needs of SM radiochemistry. A synthetic approach used by Bitmann's group[33] may be considered as a method for the preparation of stereoisomers of radiolabeled SM if radiolabeled ceramide stereoisomers are used as substrates. Also, by employing the [^{3}H]trimethylamine at the last synthetic step, [^{3}H-choline]-SM or double-labeled-SM (in the choline and in the amide part) may be prepared.

Radiolabeling in the Choline Moiety. This method was developed for natural SM[34,35] and involves the demethylation of SM to ceramide 1-phosphoryl-N,N-dimethylethanolamine (de-SM) and subsequent quaternization with radioactive [^{14}C]- or [^{3}H]methyl iodide. Quaternization of de-SM with [^{3}H]methyl iodide allows the preparation of [^{3}H-choline]SM with a high specific activity (~80 Ci/mmol). The demethylation process is presented elsewhere in this volume.[2]

Preparation of [^{14}C-choline]Sphingomyelin

This section presents the synthesis of [^{14}C-choline]SM by a method that has been adapted from the procedure published by Stoffel *et al.*[34] and the synthesis of [^{14}C-choline]lyso-SM, obtained from [^{14}C-choline]SM by a method that has been adapted from previously described procedures.[12-15]

[^{14}C]Methyl iodide (5 mCi, 89.8 μmol) is dissolved in 2 ml of dry methanol and transferred to an air-evacuated flask containing 5 ml of milky

[32] R. R. Schmidt, *in* "Synthesis in Lipid Chemistry" (H. P. Tyman, ed.), p. 93. Brunel University, Uxbridge, UK, 1996.
[33] H. S. Byun, R. K. Erukulla, and R. Bittman, *J. Org. Chem.* **59,** 6495 (1994).
[34] W. Stoffel, D. LeKim, and T. S. Tsung, *Hoppe-Seyler's Z. Physiol. Chem.* **352,** 1058 (1971).
[35] W. Stoffel, *Methods Enzymol.* **35,** 533 (1975).

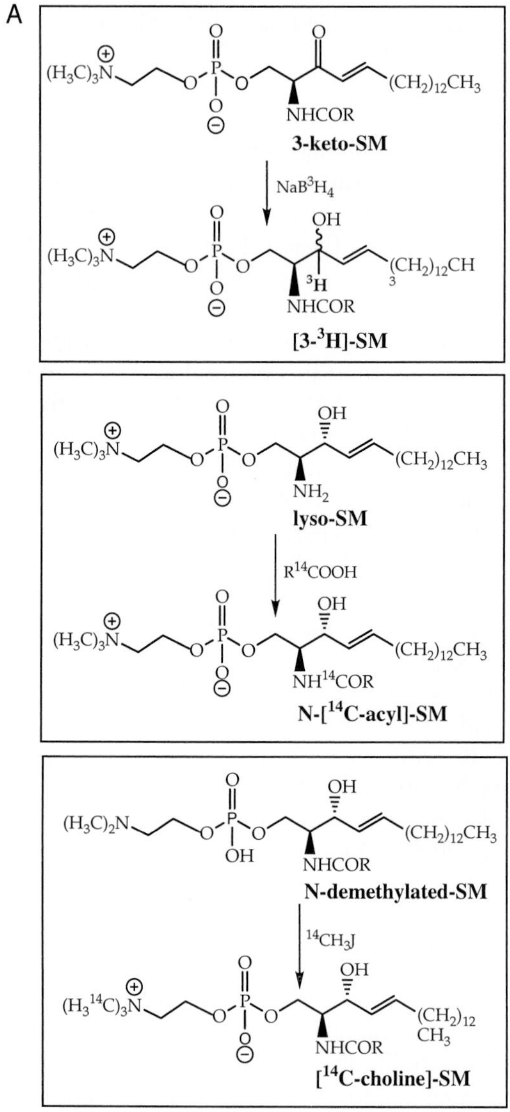

FIG. 2. (a) Approaches to the preparation of sphingomyelin radiolabeled in various parts of the molecule. (b) Approach to the synthesis of [³H-] or [¹⁴C-choline]-labeled sphingomyelin and lysosphingomyelin.

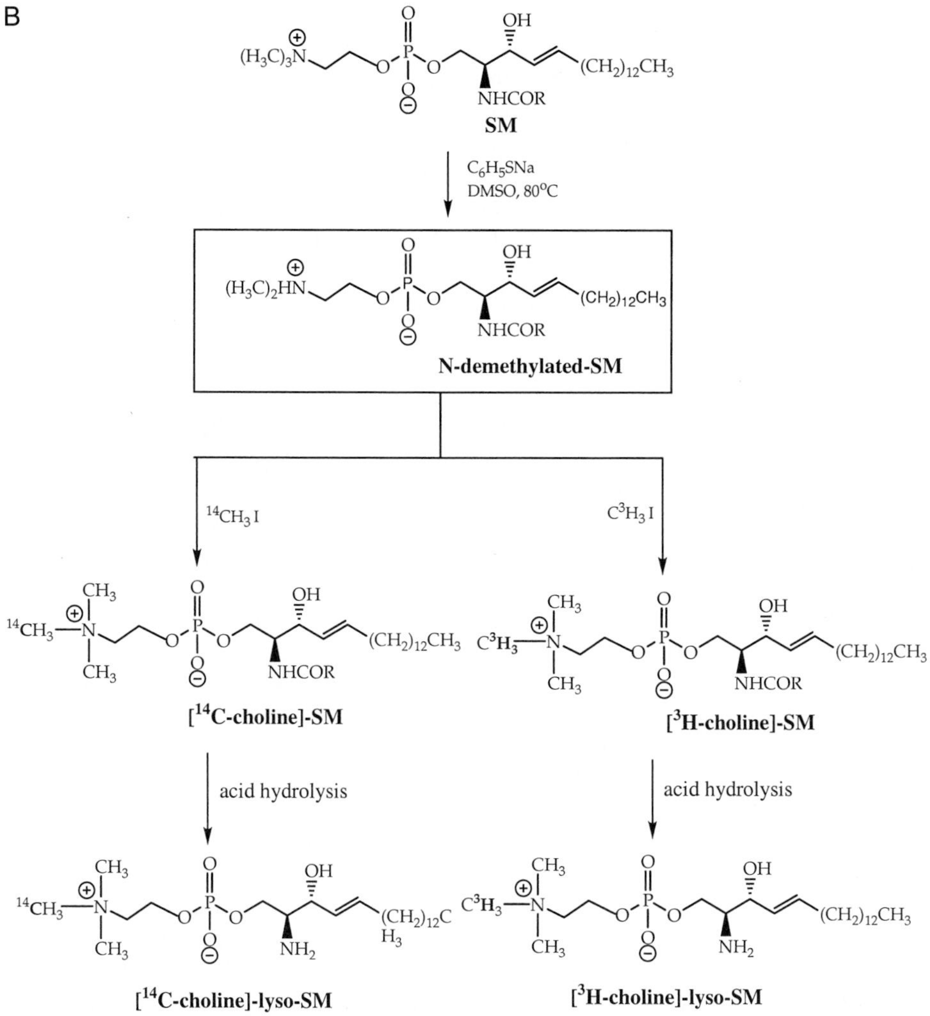

B

FIG. 2. (*continued*)

solution of demethylated sphingomyelin (77.3 mg, 108 μmol) in dry methanol. This process is repeated four times followed by the addition of freshly distilled cyclohexylamine (11.1 mg, 13 μl, 112 μmol) in 200 μl of dry methanol. This mixture contains more than 80% of the theoretical radioactivity of [^{14}C]methyl iodide. The milky solution becomes progressively clear after ~1 hr and is stirred at room temperature in the dark for 24 hr. TLC-1 analysis of the reaction mixture in solvent G shows formation of [^{14}C]SM

(~90% of the total radioactivity), some unidentified minor radioactive by-products, and the ureacted substrate. Methanol is removed under a stream of nitrogen, and the reaction mixture is treated twice with 3 ml of chloroform (and redried) and dissolved in 10 ml of chloroform. The chloroform solution is washed twice with 5 ml of freshly prepared 5% $Na_2S_2O_3$ followed by 2 N HCl (2×4 ml) and water (2×4 ml). After centrifugation (3000 rpm, 15 min, 5°), the upper phase is discarded and the lower phase, with the milky interphase, is diluted with chloroform to get a clear solution (15 ml total) (during the washing process, ~15% of radioactivity is lost). The crude product is purified by gradient flash chromatography (11-mm-diameter column) using chloroform–methanol (65:35–5:95, v/v) as an eluting system. Each of the 5-ml fractions is monitored by counting, and elution progress is followed by TLC-1 in solvent G. Unreacted de-SM is eluted with chloroform–methanol (3:7, v/v); pure N-[14C-choline]SM elutes with chloroform–methanol (5:95, v/v). Fractions containing pure sphingomyelin are combined and evaporated under nitrogen gas to give a white solid product (45.9 mg, 3.5 mCi, 88% yield) (calculated from the specific activity of [14C]methyl iodide). Fractions containing pure de-SM are also combined and evaporated under nitrogen to give a white solid product (22 mg).

[14C]Sphingomyelin is dissolved in a mixture of methanol–toluene (1:1, v/v) to yield an activity of 1×10^6 cpm/μl (8.2 mM as determined by phosphorus assay).[12] [14C]SM can be kept at $-20°$ for several months without decomposition.

Preparation of [14C-choline]Lysosphingomyelin

N-[14C-choline]Sphingomyelin (0.56 mCi, 6.3 mg, 8.55 μmol) is hydrolyzed under reflux in 3.5 ml of 1 N HCl in methanol for several hours with monitoring by TLC-1 in solvent H. After 12 hr, the reaction is complete. Solvent is evaporated and the product is partitioned into the upper phase of the Folch procedure. Briefly, the reaction mixture is redissolved in 2 ml of methanol followed by the addition of 4 ml of chloroform and 1.5 ml of water. Phases are separated and the lower phase (4 ml) is reextracted twice with 3.5 ml of solvent I. The combined upper phases contain more than 90% of [14C-choline]lyso-SM, and sphingosine and N-[14C]phosphocholine. The pH of the combined upper phases is adjusted to 12–13 with 5 N NaOH and 12 ml of solvent J is added to transfer lyso-SM and sphingosine to the lower phase. This process is repeated three times, with the extraction efficiency evaluated by liquid scintillation counting. The lower phases are combined and evaporated *in vacuo.* The crude product is purified by preparative TLC-3 in solvent G. The area corresponding to N-[14C-choline]lyso-SM is scraped from the plate, and the product is extracted several times with 3 ml of solvent K.[15] Extracts containing lyso-SM are combined and evaporated under nitrogen to give a white solid product (2.5 mg, 65% yield, 0.36 mCi). Purity is assessed by TLC-1 in solvents H, L, or N.[14]

Preparation of Radiolabeled Ceramide-1-Phosphate and Sphingosine-1-phosphate

Principles

Synthesis of this class of compounds is difficult to perform, particularly for sphingosine-1-phosphate. The only practical approaches described so far for the preparation of these radiolabeled compounds are enzymatic methods.

Enzymatic Methods for the Preparation of [^{32}P, ^{3}H or ^{14}C]Ceramide-1-phosphate and [^{32}P or ^{3}H]Sphingosine-1-phosphate

An enzymatic method employing *E. coli* diacylglycerol kinase (DG-kinase) can be applied for the preparation of radioactive ceramide-1-phosphate (Fig. 3a) and sphingosine-1-phosphate (Fig. 3b). DG-Kinase phosphorylates diacyl and monoacyl glycerols using adenosine triphosphate (ATP) as a phosphate source.[36] This enzyme can also phosphorylate the primary hydroxyl group of ceramide[9,13,20,37] and their analogs.[38] This method was successfully adopted for the determination and quantitation of the endogenous ceramide in cells and tissue as well as for the fast preparation of a variety of [^{32}P]-, [^{3}H], or [^{14}C]-labeled ceramide-1-phosphates in moderate yield (~70%). DG-kinase phosphorylates only ceramides posessing the 2*S* configuration at the C-2 position of sphingosine backbone and generates only (2*S*,3*R*) and (2*S*,3*S*) stereoisomers.[38]

We find that DG-kinase also phosphorylates *N*-Boc-protected D-*erythro*-and L-*threo*-sphingosines and sphinganines. Based on that observation and a simple protocol for *N*-Boc-deprotection [mild acidic, free aminolipid liberation protocol: TFA, ~0°, Amberlite IRA 400 (OH-), MeOH], we have successfully prepared [^{32}P]- or [^{3}H]-labeled sphingosine-1-phosphate and sphinganine-1-phosphate without contamination of one stereoisomer by another.[38] The acidic hydrolysis approach, applied pre-

[36] J. Preiss, C. R. Loomis, W. R. Bishop, R. Stein, J. E. Niedel, and R. M. Bell, *J. Biol. Chem.* **261**, 8597 (1986).

[37] K. A. Dressler and R. N. Kolesnik, *J. Biol. Chem.* **265**, 14917 (1990).

[38] A. Bielawska, unpublished observations.

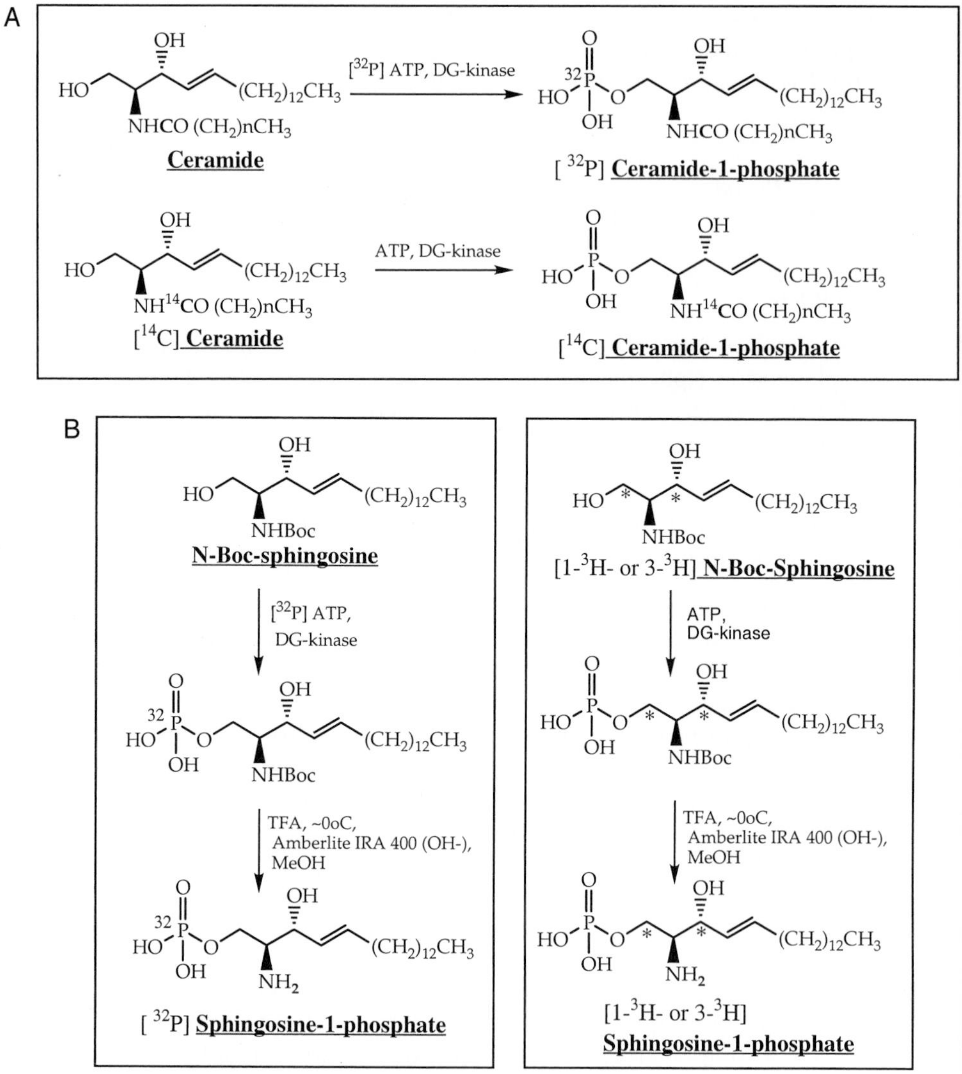

FIG. 3. (a) Enzymatic method for the preparation of ceramide-1-phosphate **III** labeled in the phosphate group or in the ceramide portion. (b) Enzymatic method for the preparation of sphingosine-1-phosphate **VB** labeled in the phosphate group or in the sphingosine portion. Asterisk indicates tritium labeling. (c) Enzymatic method for the preparation of ceramide-1-phosphate **III B** and sphingosine-1-phosphate **VB**.

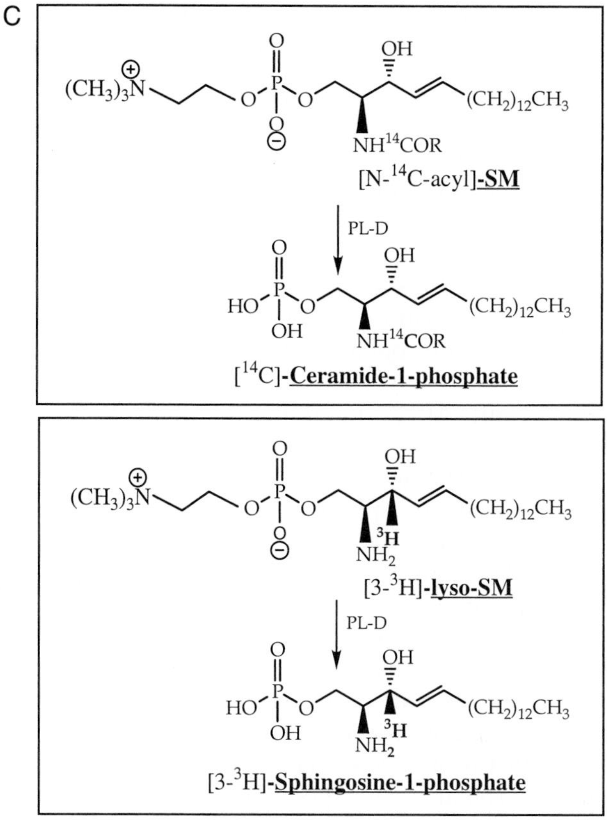

Fig. 3. (*continued*)

viously for the generation of sphingosine-1-phosphate from ceramide phosphates, afforded sphingosine-1-phosphate as a diastereoisomeric mixture.[13,37,38]

Other enzymatic methods employ bacterial phospholipase D treatment of SM or lyso-SM[13,14] (Fig. 3c) or sphingosine kinase[39,40] for the preparation of radioladeled ceramide- and sphingosine-1-phosphates.

[39] W. Stoffel, G. Assmann, and E. Binczek, *Hoppe-Seyler's Z. Physiol. Chem.* **351,** 635 (1970).
[40] N. Mazurek, T. Megidish, S.-I Hakamori, and Y. Igarashi, *Biochem. Biophys. Res. Commun.* **198,** 1 (1994).

Preparation of N-[³H-palmitoyl]D-erythro-Sphingosine-1-phosphate:
[³H]-D-erythro-C₁₆-ceramide-1-phosphate

This section presents an enzymatic method for the preparation of D-*erythro*-C₁₆-ceramide phosphate and D-*erythro*-sphingosine-1-phosphate (labeled with ³²P, ¹⁴C, or ³H) by DG-kinase action. This method has been adopted from the general method for ATP-phosphorylation of diacylglycerols and ceramides by DG-kinase.[9,36]

N-[³H]Palmitoyl-D-*erythro*-sphingosine (0.2 mCi, 200 nmol) is sonicated for 5 min with 100 μl of 3.75% (w/v) octylglucoside–12.5 mM dioleoylphosphatidylglycerol, made up in 1 mM DTPA, followed by the addition of 350 μl of the reaction mixture containing 120 mM HEPES buffer, pH 7.0–100 mM LiCl–25 mM MgCl₂–2 mM EGTA–2 mM dithiothreitol, and 25 μl of DG-kinase (7 mg/ml). After 10 min, the reaction is started by adding 50 μl of 10 mM ATP in 20 mM imidazole buffer, pH 6.6–1 mM DTPA. After vigorous mixing, the reaction mixture is left at room temperature for 1 hr, 3 ml of solvent S is added, and the reaction mixture is mixed vigorously for 1 min. Bligh and Dyer extraction is continued by adding 350 μl of 1% perchloric acid followed by 1 ml of chloroform and 1 ml of 1% perchloric acid. Phases are separated by centrifugation (3000 rpm, 5 min), and the lower phase is transferred to a new vial. This organic phase is dried under nitrogen and subjected to the base cleavage in 2 ml of 0.1 N NaOH in methanol for 30 min, at room temperature. The product is extracted by the Bligh and Dyer method. The lower phase is concentrated under nitrogen gas, and the product is purified by TLC-3 in solvent P and detected using a phosphoimager. The area corresponding to [³H]D-*erythro*-C₁₆-ceramide-1-phosphate (R_f = 0.53) is scraped and extracted five times with 3 ml of chloroform–methanol (1:1, v/v) and is evaluated quantitatively by liquid scintillation counting. The combined exstracts are dried under vacuum to give the final product with ~55% yield (calculation is made from the specific activity of the original [³H]D-*erythro*-C₁₆ ceramide). The product is characterized by TLC-1 in solvents P and M and is found to be 95% pure.

Following this procedure, other D-*erythro*- and L-threo-ceramide phosphates and dihydroceramide phosphates can be prepared. However, the phosphorylation efficiency for short chain ceramides is much lower (~35% for C₆-ceramide and only ~15% for C₂-ceramide).[40]

Preparation of D-erythro-[³²P]Sphingosine-1-phosphate

D-*erythro*-N-Boc-sphingosine (200 nmol) is sonicated for 5 min with 100 μl of 3.75 (w/v) % octylglucoside–12.5 mM dioleoylphosphatidylglycerol, made up in 1 mM DTPa, followed by the addition of 350 μl of the reaction mixture containing 120 mM HEPES buffer, pH 7.0–100 mM LiCl–25 mM MgCl₂–2 mM EGTA–2 mM dithiothreitol, and 25 μl of DG-kinase (7 mg/

ml). After 10 min, the reaction is started by adding 50 μl of 10 mM [^{32}P]ATP (specific activity 1.6×10^5 cpm/nmol) in 20 mM imidazole buffer, pH 6.6–1 mM DTPA. After vigorous mixing, the reaction mixture is left at room temperature for 1 hr. Three milliliters of solvent S is added, and the reaction mixture is mixed vigorously for 1 min. The Bligh and Dyer extraction is continued by adding 350 μl of 1% perchloric acid followed by the addition of 1 ml of chloroform and 1 ml of 1% perchloric acid. Phases are separated by centrifugation (3000 rpm, 5 min) and the lower phase is transferred to a new vial. Analytical TLC-1 in solvent P shows formation of [^{32}P]N-Boc-sphingosine-1-phosphate ($R_f = 0.42$) in ~40% yield. The organic phase is dried under nitrogen, 0.5 ml of 50% trifluoroacetic acid in methylene chloride is added with vigorous stirring, and reaction mixture is left at room temperature for 15 min. Solvents are evaporated under nitrogen and 3 ml of solvent S is added. The reaction mixture is sonicated for 15 min, and chloroform (2 ml) and 1 N KCl (2 ml) are added. The pH of the reaction mixture is adjusted to 11 by adding 50 μl of concentrated ammonium hydroxide. Phases are separated by centrifugation (3000 rpm, 5 min) and analyzed by TLC-1 in solvent P. The lower phase (~90% of the total radioactivity) contains almost pure [^{32}P] sphingosine-1-phosphate ($R_f = 0.34$). The upper phase contains [^{32}P]sphingosine-1-phosphate ($R_f = 0.34$) and unhydrolyzed [32^P]N-Boc-sphingosine-1-phosphate ($R_f = 0.46$) in 1:1 proportion. The upper phase is acidified to pH 1 and extraction is repeated. After this step, the upper acidic phase contains [^{32}P]N-Boc-sphingosine-1-phosphate and the lower acidic phase [^{32}P]sphingosine-1-phosphate. Lower phases from both extractions are combined and evaporated *in vacuo* to provide [^{32}P]sphingosine-1-phosphate in ~30% of the theoretical yield. The product is characterized by TLC-1 in solvents P and M and is 93% pure.

Following this procedure, but using L-*threo*-N-Boc-sphingosine as a substrate, the corresponding L-*threo*-[^{32}P]sphingosine-1-phosphate can be prepared in 35% yield.

Preparation of [4,5-^{3}H]Sphinganine-1-phosphate

This section presents the enzymatic method for preparation of [4,5-^{3}H]sphinganine-1-phosphate by phospholipase D action by the method described by Van Veldhoven and Mannaerts.[14]

[^{3}H]lysodihydro-SM (40 μmol) is treated with 750 U phospholipase D from *Streptomyces chromofuscus* in 8 ml of 50 mM ammonium acetate buffer, pH 8.0, at 30°. After 1 hr, the insoluble reaction product is removed by centrifugation. The supernatant is allowed to react for another hour, cooled to 4°, and centrifuged. The pellets are combined, dispersed in water (4 ml) by sonication, and cooled to 4°, and this precipitate is sedimented by centrifugation. These steps are repeated, and the final pellet is dried under vacuum and then dispersed in acetone by sonication. After cooling to 4°, the solution is centrifuged and the supernatant is removed. This step is repeated, and the acetone-insoluble material is dried under nitrogen, dissolved in chloroform–methanol (1 : 1, v/v) by sonication, cooled to 4°, centrifuged, and the supernatant removed. This step is repeated and the final product is dissolved in methanol (approximately 0.15 mM) and is characterized by TLC-1 in solvents L (R_f = 0.26) and M (R_f = 0.34) to be ~80% pure.

Acknowledgment

Supported in part by NIH Grant GM-43825.

[44] Synthesis of Key Precursors of Radiolabeled Sphingolipids

By ALICJA BIELAWSKA, ZDZISLAW SZULC, and YUSUF A. HANNUN

Introduction*

This article presents synthetic methods used in the preparation of intermediates for sphingolipids (SLs) discussed elsewhere in this volume.[1,2] We

---

* Sphingolipid nomenclature: Sphingosine: 4E-octadecene-1,3-diol-2-amino [CA: 123-78-4] contains two chiral centers on carbon atoms C-2 and C-3 that create four stereoisomers: an enantiomeric pair of the *erythro* and *threo* diastereoisomers. Extending geometrical (4E) "*trans*" isomer to its "*cis*" (4Z) isomer, sphingosine can create eight stereoisomers. There is a great deal of confusion regarding the DL nomenclature of optical isomers of sphingolipids, particularly the *threo* isomers, as both chiral centers have been chosen as a point of reference of sphingolipid chirality.[2a,2b] Using DL nomenclature, the natural *erythro*-sphingosine is (2D,3D) and its enantiomer is (2L,3L). The *threo* isomers are (2D,3L) and (2L,3D). Using chiral carbon C-2 as a reference, the (2D,3L) isomer is D-*threo*, but using chiral carbon C-3 as a reference this isomer is L-*threo*. To avoid these confusions, we have decided to name the absolute configuration of sphingolipids using the (CIP) system for stereochemical assignment of the chiral compounds[2c] and keep the 3-D and 3-L nomenclature.[2b] Following these rules, D-*erythro*-sphingosine (2D,3D) constitutes (2S,3R) configuration. The remaining sphingosine stereoisomers are L-*erythro*-, (2L,3L), (2R,3S); L-*threo*-, (2D,3L), (2S,3S); and D-*threo*-, (2L,3D), (2R,3R).

[1] A. Bielawska, Y. A. Hannun, and Z. Szulc, *Methods Enzymol.* **311** [42] 1999 (this volume).
[2] A. Bielawska and Y. A. Hannun, *Methods Enzymol.* **311** [43] 1999 (this volume).
[2a] H. E. Carter, D. S. Galanos, and J. Fujino, *Can. J. Biochem. Physiol.* **34**, 320 (1956).
[2b] W. Stoffel and G. Sticht, *Hoppe-Seyler's Z. Physiol. Chem.* **348**, 1561 (1967).
[2c] E. L. Eliel, S. H. Wilen, and L. N. Mander, "Stereochemistry of Organic Compounds." Wiley, New York, 1994.

have selected the compounds listed in this article based on availability of the starting materials, simplicity, and effectiveness of the synthesis, as well as chemical stability and versatility of the prepared synthon. It should be stressed at the outset that access to all four stereoisomers of sphingosine or sphinganine and their derivatives, as well as differentially N- and O-protected forms, requires a multistep synthetic approach such as depicted in Schemes 1–3. This article specifically presents synthetic techniques for the preparation of (i) stereoisomers of the dehydro precursor **IC** used for the preparation of [4,5-^{3}H]-labeled sphinganines **IA**; (ii) stereoisomers of ceramide **IIB** used for the preparation of stereoisomers of [4,5-^{3}H]dihydro-ceramide **IIA**; (iii) 3-keto precursors used for the preparation of [3-^{3}H]-labeled sphingosines, dihydrosphingosines, ceramides, and higher analogs; (iv) active forms of fatty acids used for the preparation of SLs labeled in the acyl chain; (v) N-Boc-sphingosine **XB** and N-Boc-sphinganine **XA**, which are precursors for the preparation of radiolabeled sphingosine-1-phosphate **VB** and sphinganine-1-phosphate **VA**; (vi) ceramide **IIB** and dihydroceramide **IIA**, which are precursors for the preparation of radiola-beled ceramide-1-phosphate **IIIB** and dihydroceramide-1-phosphate **IIIA**; (vii) methods for the preparation of sphingosylphosphorylcholine (lyso-SM) used in the synthesis of the radiolabeled sphingosine-1-phosphate **VB/VA** or N-acyl-labeled sphingomyelin; and (viii) ceramide-1-phosphoryl-Np,N-dimethylethanolamine **de-SM**, used to synthesize [^{14}C- or ^{3}H-cho-line]**SM**.

Materials

Sphingomyelin (from bovine brain) is obtained from Avanti Polar Lipids. Sodium acetate trihydrate, acetic anhydride, chromium anhydride, 2,3-dichloro-5,6-dicyanobenzoquinone, phosphorous pentoxide, manga-nese(IV) oxide (activated 85%), potassium sodium tartrate tetrahydrate, D-serine, methyl-(S)-tert-butoxycarbonyl-2,2-dimethyl-4-oxazolidinecar-boxylate, DIBAL-H, sodium bis(2-methoxyethoxy)aluminum hydride (Red-Al), di-tert-butyldicarbonate, and diisopropylethylamine (DIPEA) are obtained from Aldrich. Amberlite IRA-400, Celite 535, 10% palladium on activated charcoal, palmitoyl chloride, sodium thiophenolate are ob-tained from Fluka. Trifluoroacetic acid is obtained from Sigma. Concen-trated hydrochloric acid (36.5–38.5%), sodium hydroxide (pellets), and con-centrated ammonium hydroxide (26–30%) are from J. T. Baker.

Sodium salt of [^{14}C]hexanoic acid (specific activity 55 mCi/mmol) is obtained from American Radiolabeled Chemicals, Inc. (St. Louis, MO).

TLC Plates

TLC-1: Precoated TLC sheets (on aluminium), silica gel 60 F 254 (0.2 mm layer thickness) are from Merck, TLC-2: Precoated TLC sheets (on foil), silica gel 60 F 254 (0.2 mm layer thickness) and TLC-3: precoated thin-layer plates, Kieselgel 60 (0.25 mm layer thickness) are from EM Separation Technology. TLC-4: Precoated thin-layer plates, PK5 silica gel 150 A (0.5 mm layer thickness) and TLC-5: Linear-K preadsorbent strip, silica gel 150A (1.0 mm layer thickness) are from Whatman Laboratory Division.

Silica gel 60 (particle size 0.040–0.063 mm) (230–400 Mesh ASTM), CAS 63231-67-4, is from EM Separation Technology.

All solvents are analytical grade and are obtained from Aldrich, Mallinckrodt, Burdick and Jackson, or EM Science.

Anhydrous solvents such as methylene chloride, pyridine, dimethyl sulfoxide, dioxane, benzene, toluene, and methanol are obtained from Aldrich.

General Methods Used during Syntheses

Stereoisomers of sphingosine **IB**, sphinganine **IA**, and their derivatives are prepared from the optically pure precursors **VIII C** and **IX C** by the addition of pentadecyne to the Garner's aldehyde **VII** (Scheme 1) according

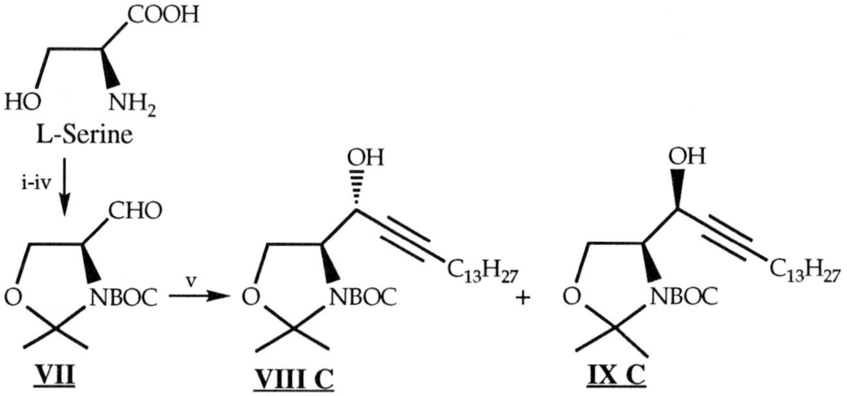

i : MeOH, CH_3COCl ; ii : $(BOC)_2O$, TEA, CH_2Cl_2 ;

iii : DMP , acetone, $BF_3 \cdot Et_2O$; iv : DIBALH, toluene, $-75°$;

v: pentadecyne, n-BuLi,,THF, $-78°$;

SCHEME 1. Approach to the synthesis of precursors **VIII C** and **IX C** used for the preparation of optically pure (2*S*,3*R*) and (2*S*,3*S*) sphingosines.

to published procedures.[3–5] Stereoisomers of **IB/IA** have become available commercially (Matreya, Inc., Pleasant Gap, PA Biomol Research Laboratories, Inc., Plymouth Meeting, PA, and Universal Biologicals Limited).

Lysosphingomyelin is prepared by the acid hydrolysis of bovine brain sphingomyelin according to published procedures.[6–9] This compound is also available commercially (Matreya, Inc.).

Ceramide-1-phosphoryl-*N*,*N*-dimethylethanolamine: (demethylated sphingomyelin; de-SM) is prepared by demethylation of bovine brain sphingomyelin according to published procedures.[10–15]

N-Hydroxysuccinimide (NHS) ester of the radiolabeled fatty acids is prepared according to published procedures.[12–14] Some radioactive *N*-hydroxysuccinimide esters are also available commercially (American Radiolabeled Chemicals, Inc., St. Louis, MO).

Thin-Layer Chromatography

Purity of the synthesized compounds, the reaction progress, and chromatographic elution profiles are checked routinely by thin-layer chromatography using TLC-1 or TLC-2 plates. Purification is performed by preparative thin-layer chromatography using TLC-3, TLC-4, or TLC-5 plates.

Column Chromatography

Separation of the reaction products and purification of the desired target sphingolipids are achieved by medium pressure flash chromatography performed on silica gel 60 (EM Science) according to a published method.[15] Fractions containing impure material are rechromatographed.

[3] P. Garner, J. M. Park, and E. Malecki, *J. Org. Chem.* **53**, 4395 (1988).
[4] S. Ninkar, D. Menaldino, A. H. Merrill, and D. Liotta, *Tetrahedron Lett.* **29**, 3037 (1988).
[5] P. E. Herold, *Helv. Chim. Acta* **71**, 354 (1988).
[6] R. C. Gaver and C. C. Sweeley, *J. Am. Oil Chem. Soc.* **42**, 294 (1965).
[7] R. Cohen, Y. Barenholtz, S. Gatt, and A. Dagan, *Chem. Phys. Lipids* **35**, 371 (1984).
[8] P. P. Van Veldhoven and G. P. Mannaerts, *J. Biol. Chem.* **266**, 12502 (1991).
[9] M. Buneman, K. Liliom, B. T. Brandts, L. Pott, J. L. Tseng, D. M. Desiderio, G. Sun, D. Miller, and G. Tigyi, *EMBO J.* **15**, 101 (1996).
[10] W. Stoffel, *Methods Enzymol.* **35**, 533 (1975).
[11] W. Stoffel, D. LeKim, and T. S. Tsung, *Hoppe-Seyler's Z. Physiol. Chem.* **352**, 1058 (1971).
[12] A. Futerman and R. E. Pagano, *Methods Enzymol.* **209**, 437 (1986).
[13] Y. Lapidot, S. Rapport, and Y. Wolman, *J. Lipid Res.* **8**, 142 (1967).
[14] H. Ogura, T. Kobayashi, K. Shimizu, K. Kawabe, and K. Takeda, *Tetrahedron Lett.* **49**, 4745 (1979).
[15] W. C. Still, M. Kahn, and A. Mitra, *J. Org. Chem.* **43**, 2923 (1978).

Solvent Systems

The solvent systems used for TLC and/or column chromatography are (A) chloroform–methanol–2 N ammonium hydroxide (4:1:0.1, v/v); (B) chloroform–methanol–concentrated ammonium hydroxide (4:1:0.1, v/v); (C) ethyl acetate–*n*-hexane (10:1, v/v); (D) ethyl acetate–*n*-hexane (1:1, v/v); (E) ethyl acetate–*n*-hexane (1:5, v/v); (F) chloroform–methanol (9:1, v/v); (G) ethyl acetate–acetone (95:5, v/v); (H) chloroform–methanol (95:5, v/v); (I) chloroform–methanol–water (60:35:5, v/v); (J) chloroform–methanol (4:1, v/v); (K) methylene chloride–methanol (97:3, v/v); (L) ethyl acetate–*n*-hexane (4:1, v/v); (M) chloroform–methanol–concentrated ammonium hydroxide–water (60:35:2.5:2.5, v/v); (N) chloroform–methanol–concentrated ammonium hydroxide (65:35:8, v/v); (P) ethyl acetate–*n*-hexane (1:8, v/v); (R) chloroform–methanol–water (3:48:47, v/v); (S) chloroform–methanol–water (86:14:1, v/v); (T) chloroform–methanol–water (65:35:8, v/v); and (W) chloroform–methanol–concentrated ammonium hydroxide (60:30:5, v/v).

Compounds are visualized by exposing to iodine vapor and by spraying with 5% potassium permanganate in *N*-potassium hydroxide. Compounds with the free amino group are visualized with ninhydrin spray.

Lipid extractions from the crude reaction mixtures are performed using the Bligh and Dyer[16] extraction method or the procedure of Folch *et al.*[17]

Analytical Assays

Composition and Diastereoisomeric Purity of Sphingoid Bases

Analysis of sphingosine **IB**, sphinganine **IA**, and dehydrosphingosine **IC** is performed by the high-performance liquid chromatography (HPLC) method after derivatization with *o*-phthlalaldehyde using eicosasphingosine as an internal standard as described by Merrill *et al.*[18]

^{1}H nuclear magnetic resonance (NMR) spectroscopy is performed at ambient temperature with a G. E. Omega or Bruker XX spectrometers in 5-mm tubes at 500 MHz using trimethylsilane (TMS) as an external reference.

FAB mass spectrometry is carried out with a JEOL HX110/110 high-resolution mass spectrometer using FAB ionization method, glycerol as a matrix, and PEG 400 as an internal standard.

[16] E. A. Bligh and W. J. Dyer, *Can. J. Biochem. Physiol.* **37,** 911 (1959).

[17] J. Folch, M. Lees, and G. H. Sloane-Stanley, *J. Biol. Chem.* **226,** 497 (1957).

[18] A. H. Merrill, Jr., E. Wang, R. E. Mullins, W. C. L. Jamison, S. Ninkar, and D. Liotta, *Anal. Biochem.* **171,** 373 (1998).

Precursors for Preparation of Sphingoid Bases Labeled with Tritium

Basic substrates in the synthesis of the desired radiolabeled-SLs include sphingosines (**IB**), sphinganines (**IA**), and (4,5)-dehydrosphingosines (**IC**) of specific absolute configurations: (2S,3R)-D-*erythro*-, (2S,3S)-L-*threo*-, (2R,3S)-L-*erythro*-, and (2R,3R)-D-*threo*-, their N-Boc derivatives **X A-C** and N-acyl derivatives **IIA, IIB**, 3-keto analogs: 3-keto-**I A-C** and 3-keto-**X A-C** as well as double protected analogs: **VIII, IX** and keto-**VIII**, keto-**IX**. Schemes 1–3 outline our general synthetic strategy for the preparation of these key precursors.

Stereocontrolled Synthesis of Sphingoid Bases: IA-C and Their N-Boc- and N-acyl derivatives

Principles

Synthesis of (2S,3R)-D-*erythro*-sphingosine **IB** has been performed using configurationally stable chiral auxiliary **VII** (Garner's aldehyde) prepared from L-serine.[19,20]

The stereocontrolled addition of lithium 1-pentadecyne and sequential reduction–deprotection hybrid methods applied to D-*erythro*- and L-*threo*-alkynols **VII C** and **IX C** lead to the synthesis of all regio- and diastereoisomeric 2S-sphingosines.[3–5] Starting with chiral auxiliary **VII** derived from D-serine[4,5] and using the reaction sequence described previously will access all remaining 2R-sphingosines.

Methods presented in Scheme 1 are described briefly: (1) Chiral L-serine adehyde **VII** is prepared directly by reduction of methyl(S)-*tert*-butoxycarbonyl-2,2-dimethyl-4-oxazolidinecarboxylate (Aldrich) with DIBAL-H or synthesized in a five-step synthesis from L-serine. Chiral D-serine adehyde **VII** is prepared in a five-step synthesis from D-serine. (2) Aldehyde **VII** is transformed to alkynols **VIII C, IX C** by the addition of cation-complexing agents: HMPT to obtain the *erythro* form: **VIII C**, ~90% (de) and ZnBr$_2$ to obtain the *threo* form: **IX C**, ~90% (de). (3) Alkynols **VIII C, IX C** are separated by flash chromatography.[3–5] Separated stereoisomers—**VIIIC** and **IXC**, after sequential deprotection and/or reduction reactions—are used as shown in Scheme 2. (4) Alkynols **VIII C + IX C** are oxidized directly (without separation of stereoisomers) to the 3-keto derivative 3-keto-**XI C** and used as shown in Scheme 3.

[19] P. Garner and J. M. Park, *J. Org. Chem.* **52**, 2361 (1987).
[20] A. McKillop, R. J. K. Taylor, R. J. Watson, and N. Levis, *Synthesis* **1**, 31 (1994).

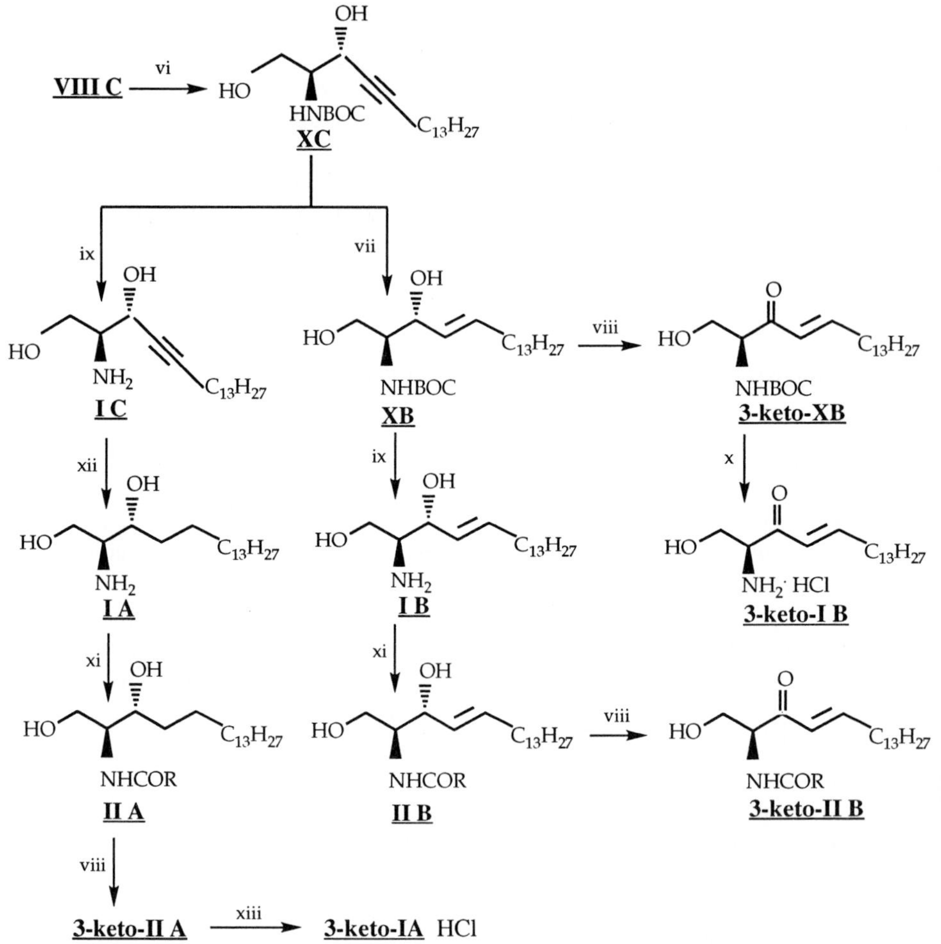

vi: Amberlyst 15, MeOH, r.t.; vii: Red-Al , Et$_2$O, 0° ; viii: CrO$_3$, pyridine;
ix: TFA, CH$_2$Cl$_2$, r.t.; x : 3N HCl, EtOAc, 0° ; xi : (CH$_3$CO)$_2$O or CH$_3$(CH$_2$)$_n$COCl
xii: H$_2$, 10% Pd/C, MeOH; xiii: 5% HCl in Et$_2$O

SCHEME 2. Approaches to the synthesis of precursors used in the preparation of radiola-beled sphingolipids.

Procedures

This section presents the synthesis of (i) stereoisomers of dehydrosphin-gosine **IC**, sphingosine **IB**, and sphinganine **IA** by methods that have been adapted from the stereocontrolled synthesis of sphingosine,[3–5] (ii) *N*-Boc

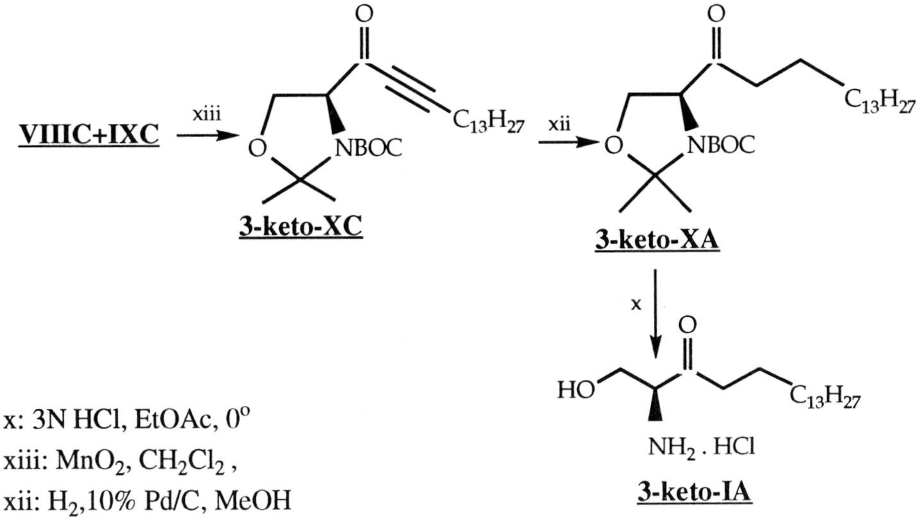

x: 3N HCl, EtOAc, 0°
xiii: MnO$_2$, CH$_2$Cl$_2$,
xii: H$_2$,10% Pd/C, MeOH

SCHEME 3. Approach to the synthesis of 3-ketosphinganine (3-keto-**IA**)

derivatives of the previous compounds by methods that have been adapted from the stereocontrolled synthesis of sphingosine or by derivatization of sphingoid bases with di-*tert*-butyldicarbonate, following published methods,[21,22] and (iii) the corresponding ceramides by the acylation of sphingoid bases with fatty acid chlorides or anhydrides, following published methods.[23,24] Synthetic procedures presented in this section begin at the level of *N*-Boc-dehydrosphingosine **X C**.[4,5]

Preparation of (2S,3R)-2-Amino-4-octadecyne-1,3-diol; (2S,3R)-
*D-erythro-4,5-dehydrosphingosine; **IC***

Anhydrous trifluoroacetic acid (2 ml) is added dropwise over 10 min to a well-stirred and cooled (in an ice-water bath) solution of (2S,3R)-D-*erythro*-*N*-Boc-4,5-dehydrosphingosine **X C**[4,5] (191 mg, 0.4 mmol) in dry methylene chloride (5.0 ml). After stirring for 30 min at room temperature, the solution is evaporated under reduced pressure and the residue is dis-

[21] T. Toyokuni, M. Nisar, B. Dean, and S.-I. Hakamori, *J. Label. Comp. Radiopharm.* **39**, 567 (1991).

[22] V. Chigorno, E. Negroni, M. Nicolini, and S. Sonnino, *J. Lipid Res.* **38**, 1163 (1997).

[23] A. Bielawska, H. M. Crane, D. Liotta, L. M. Obeid, and Y. A. Hannun, *J. Biol. Chem.* **268**, 26226 (1993).

[24] A. Bielawska, D. Perry, S. Jayadev, C. McKay, J. Shayman, and Y. A. Hannun, *J. Biol. Chem.* **271**, 12642 (1996).

solved in anhydrous methanol (4 ml). The solvent is evaporated, and the residue is dissolved in methanol and recovered by evaporation. The crude product is dissolved in chloroform (25 ml), and the mixture is treated with cold 2 N ammonium hydroxide (~4 ml). The organic layer is separated and washed with water (20 ml) and a NaCl solution (20 ml). The collected aqueous extracts are extracted back with chloroform (2 × 25 ml). The collected organic extracts are dried over anhydrous sodium sulfate, filtered, and evaporated. The obtained crude product is purified by flash column chromatography, and fractions are eluted with solvent J and then solvent A. Evaporation of the solvents gives 87 mg of **IC** (60% yield) as a white solid. The product is characterized by analytical TLC-1 in solvent A; **IC** (R_f = 0.44) and by HPLC and is more than 95% pure.

Using (2S,3S), (2R,3S), and (2R,3R) stereoisomers of **X C**, the remaining stereoisomers of *IC* can be obtained. Also, to liberate the free amino group from *N*-Boc-protected derivatives, a protocol using trifluoroacetic acid and treatment with Amberlite IRA 400 (OH-) resin in methanol can be used.[21]

Preparation of (2S,3R,4E)-2-N-(1,1-Dimethylethylcarboxy)-amino-4-octadecene-1,3-diol; D-erythro-N-Boc-sphingosine; **XB**

By a Direct Reduction of N-Boc-4,5-dehydro-sphingosine **X C** *as shown in Scheme 2.* To a well-stirred and cooled (in an ice-water bath) solution of Red-Al (65% solution in toluene, 23 ml, 76.6 mmol) in anhydrous diethyl ether (30 ml), a solution of D-*erythro*-N-Boc-dehydrosphingosine (6.2 g, 15.6 mmol) in anhydrous diethyl ether (35 ml) is added dropwise over 15 min under a dry nitrogen atmosphere. After the addition is complete, the mixture is stirred at room temperature under a dry nitrogen atmosphere for 24 hr. The reaction mixture is cooled to ~0° and quenched by a very slow dropwise addition of anhydrous methanol (11 ml) with vigorous stirring. Upon the addition of the methanol, the ice bath is removed and diethyl ether is added (140 ml), followed by the addition of the solution of saturated potassium sodium tartrate (140 ml), and the reaction mixture is stirred at room temperature for 2 hr. The reaction mixture is transferred to a separatory funnel and the organic layer is separated. Any remaining traces of product from the aqueous layer are extracted back into the organic phase using diethyl ether (2 × 30 ml). The combined organic extracts are washed with aqueous NaCl (100 ml), dried over anhydrous sodium sulfate, and evaporated under reduced pressure to dryness. Crude product is purified by flash column chromatography in solvent C to give 5.1 g (80% yield) of D-*erythro*-N-Boc-sphingosine **XB** as a white solid. The product is characterized by analytical TLC-1 in solvent D; **XB** (R_f = 0.21) and in solvent F; (R_f = 0.53) and is more than 95% pure.

Using (2*S*,3*S*), (2*R*,3*S*), and (2*R*,3*R*) stereoisomers of **XC**, the remaining stereoisomers of **XB** can be obtained.

By Derivatization of the Amino Group of D-*erythro-Sphingosine with di-tert-Butyl-dicarbonate.* To a stirred solution of D-*erythro*-sphingosine (100 mg, 0.33 mmol) in dry methylene chloride (4 ml), di-*tert*-butyl dicarbonate (144 mg, 0.66 mmol) is added. The solution is cooled to 0° and DIPEA (85 mg, 0.66 mmol) is added dropwise. The reaction mixture is stirred at room temperature for 5 hr, later diluted with methylene chloride (30 ml), and washed with water (15 ml) and aqueous NaCl (20 ml). The organic phase is dried over anhydrous sodium sulfate and evaporated to provide a crude material (177 mg). Purification by flash chromatography in solvent D gives 112.7 mg (85% yield) of *N*-Boc-sphingosine **XB**, with a purity of (95%, established by TLC-1 in two solvent systems: D (R_f = 0.21) and F (R_f = 0.53).

Preparation of (2S,3R,4)-2-Amino-4-octadecene-1,3-diol; (2S,3R)-
   D-*erythro-sphingosine;* **IB**

Anhydrous trifluoroacetic acid (2 ml) is added dropwise over 10 min to a well-stirred and cooled (in an ice-water bath) solution of (2*S*,3*R*)-D-*erythro-N*-Boc-sphingosine **XB** (119.8 mg, 0.4 mmol) in dry methylene chloride (5.0 ml). After stirring for 30 min at room temperature, the solution is evaporated under reduced pressure, and the residue is dissolved in anhydrous methanol (4 ml). The solvent is evaporated, and the residue is again dissolved in methanol and dried by evaporation. The obtained crude product is dissolved in chloroform (25 ml) and the mixture is treated with cold 2 *N* ammonium hydroxide (~4 ml). The organic layer is separated and washed with water (20 ml) and aqueous NaCl (20 ml). The collected aqueous extracts are extracted back with chloroform (2 × 25 ml), the collected organic extracts are dried over anhydrous sodium sulfate, and the solvent is evaporated. The obtained crude product is purified by flash column chromatography, and fractions are eluted with solvent J and then with solvent A. Evaporation of the solvents gives 58.7 mg of **IB** (65% yield) as a white solid. The product is characterized by analytical TLC-1 in solvent A; **IB** (R_f = 0.29) and by HPLC, and is 95% pure.

Using (2*S*,3*S*), (2*R*,3*S*), and (2*R*,3*R*) stereoisomers of **XB** and the just-described procedure, the remaining stereoisomers of **IB** can be obtained.

Preparation of (2S,3R,4E)-2-N-Acetylamino-4-octadecene-1,3-diol;
   D-*erythro-C₂-ceramide;* **IB**

D-*erythro*-Sphingosine (60 mg, 0.2 mmol) is dissolved in 8 ml of methanol and acetic anhydride (2 ml, 20 mmol) in 2 ml of methanol is added dropwise. The reaction mixture is stirred at room temperature for 4 hr when TLC-1 analysis in solvent A shows that the reaction is complete. The reaction mixture is cooled to 4°, and 10 ml of cold water is added dropwise to get a white precipitate. The mixture is stirred in an ice bath for an additional 45 min, and then the precipitate is recovered by centrifugation (3000 rpm, 5 min, 4°), separated from the supernatant, washed with 0.1 N NaOH (2 ml) and water (8 ml) to neutral pH, and dried *in vacuo* over P_2O_5 to give 64.8 mg (95% yield) of D-*erythro*-C_2-ceramide. The purity of this product is (~100%) as established by analytical TLC in solvents: K ($R_f = 0.21$) and A ($R_f = 0.64$).

Using these procedures, the remaining stereoisomers of C_2-ceramide and C_2-dihydroceramide—L-*threo*-(2S,3S), D-*threo*-(2R,3R), and L-*erythro*-(2R,3S)—were prepared starting from the related sphingosine and sphinganine isomers.

Preparation of (2S,3R,4E)-2-N-Palmitoyl-amino-4-octadecene-1,3-diol; D-erythro-C₁₆-ceramide; *IB*

D-*erythro*-sphingosine **IB** (58 mg, 0.19 mmol) is dissolved in 3 ml of freshly distilled THF and added to a well-stirred and cold solution of 50% CH_3COONa (2.5 ml), followed by the addition of palmitoyl chloride (65 mg, 0.24 mmol) in 1.0 ml of THF. The reaction mixture is stirred magnetically at room temperature for 18 hr, and then ceramide is isolated by partitioning with chloroform (12 ml)–methanol (6 ml)–water (4.5 ml). The lower phase (16 ml) is washed twice with 4 ml of water and is evaporated under reduced pressure. This material is redissolved in chloroform (15 ml), washed with 0.1 N ammonium hydroxide solution (2 × 4.0 ml), washed with aqueous NaCl, water, and dried over anhydrous sodium sulfate. After evaporating to dryness, the crude product is purified by recrystallization from acetone–methanol (3 : 1, v/v) to give 92 mg (88% yield) product as a white powder. The purity of this compound (96%) is established by TLC-1 in solvent G ($R_f = 0.38$).

Synthesis of 3-Keto-sphingoid Bases: **IA, IB, IC,** and Their 3-Keto-N-Boc- and 3-Keto-N-acyl Derivatives

Principles

Tritium labeling at the C-3 position of the sphingoid backbone is a widely applied method for the preparation of radiolabeled sphingolipids. Synthetic 3-keto precursors can be quantitatively reduced back to the original alcohols as has been demonstrated for 3-keto-C_2-ceramide; 3-keto-C_2-

dihydroceramide,[25] 3-ketodihydrosphingomyelin,[26] and 3-ketocerebrosides.[27]

3-Ketosphingolipid 3-keto-SLs are prepared from SLs by oxidation of the secondary hydroxyl group[25–31] or in the course of the total synthesis of 3-ketosphingosine and 3-ketosphinganine.[32] The oxidizing agents are chromium(VI) oxide and 2,3-dichloro-5,6-dicyanobenzoquinone (DDQ). The chromium oxide–pyridine complex in benzene solution[25,26,28] or chromium oxide in dimethylformamide (DMF) catalyzed by concentrated sulfuric acid[29] is used to oxidize selectively the secondary hydroxyl group of either saturated or unsaturated SLs. DDQ specifically oxidizes only the allylic hydroxyl group of SLs, leaving dihydro analogs untouched.[27,30,31] MnO_2 in chloroform solution has been used for the oxidation of N-Boc-D-erythro-sphingosine,[33] and dehydro-derivatives type **C**.[34]

Procedures

This section presents the synthesis of (2S) and (2R) isomers of 3-ketosphingosine; 3-keto-**IB**, 3-keto-N-Boc-sphingosine; 3-keto-**XB** and 3-keto-ceramides; 3-keto-**IIB** as shown in Scheme 2, and 3-keto-sphinganine; 3-keto-**IA** as shown in Schemes 2 and 3.[35] These compounds are prepared by oxidation of the secondary hydroxyl group by methods that have been adapted from published procedures.[25,26,28,33]

Preparation of (2S)-3-Keto-C_2-ceramide; 3-keto IIB

A solution of D-e-C_2-ceramide (24.2 mg, 0.71 mmol) in 1.5 ml of dry benzene is mixed with 1.5 ml of a solution containing 65 mg of dry chromium(VI) oxide in 10 ml of pyridine (freshly distilled over NaOH) under a nitrogen atmosphere. The reaction mixture is shaken thoroughly and left at room temperature for 10 hr during which the mixture becomes progressively more turbid with a brown precipitate. The reaction mixture

[25] R. C. Gaver and C. C. Sweeley, *J. Am. Chem. Soc.* **88,** 3643 (1966).
[26] W. Stoffel and K. Bister, *Hoppe-Seyler's Z. Physiol. Chem.* **355,** 911 (1974).
[27] M. Iwamori, H. M. Moser, and Y. Kishimoto, *J. Lipid Res.* **16,** 332 (1975).
[28] M. Iwamori, H. W. Moser, R. H. McCluer, and Y. Kishimoto, *Biochim. Biophys. Acta* **380,** 308 (1975).
[29] G. Sticht, D. Lekim, and W. Stoffel, *Chem. Phys. Lipids.* **8,** 10 (1972).
[30] Y. Kishimoto and M. T. Mitry, *Arch. Biochem. Biophys.* **161,** 426 (1974).
[31] R. Birk, G. Brenner-Weib, A. Giannis, K. Sandhoff, and R. R. Schmidt, *J. Label. Comp. Radiopharm.* **39,** 289 (1990).
[32] R. V. Hoffman and J. Tao, *J. Org. Chem.* **63,** 3979 (1998).
[33] V. Chigorno, E. Negroni, M. Nicolini, and S. Sonnino, *J. Lipid Res.* **38,** 1163 (1997).
[34] A. Bielawska, unpublished observation.
[35] Z. Szulc, Y. A. Hannun, and A. Bielawska, manuscript in preparation.

is added to a mixture of 17.4 ml of cold 5 N hydrochloric acid, 17.4 ml of cold distilled water, and 67 ml of diethyl ether. After shaking, the ether layer is removed, the lower layer is extracted four times with 10 ml of ether, and the combined extracts are washed to neutrality with water (4 × 5 ml), dried over anhydrous sodium sulfate, and evaporated *in vacuo* to give 17 mg of a white-yellow solid. The crude product is purified by preparative TLC in solvent H. Areas corresponding to (2S)-3-keto-C_2-ceramide and unreacted D-e-C_2-ceramide are scraped, extracted with 3 × 5 ml of chloroform–methanol (1 : 1, v/v), and evaporated under a nitrogen stream to give 12.5 mg (52% yield) of (2S)-3-keto-C_2-ceramide. The purity of this compound (95%) is established by TLC-1 in solvent H (3-keto-**IIB**: R_f = 0.46, **IIB**: R_f = 0.065). This compound can be crystallized from pentane.[25]

Preparation of (2S)-3-Ketodihydro-C_2-ceramide; 3-keto-**II A**

A solution of D-e-*dihydro*-C_2-ceramide (18.8 mg, 0.55 mmol) in 6 ml of dry benzene is mixed with 2 ml of a solution containing 65 mg of dry chromium(VI) oxide in 10 ml of pyridine (freshly distilled over NaOH) under a nitrogen atmosphere. The reaction mixture is shaken occasionally and left at room temperature for 8 hr. During this time the reaction mixture becomes progressively more turbid with a brown precipitate. The reaction mixture is added to a mixture of 17.4 ml of cold 5 N hydrochloric acid, 17.4 ml of cold distilled water, and 67 ml of diethyl ether. After shaking, the ether layer is removed and the lower layer is extracted four times with 10 ml of ether. The combined extracts are washed to neutrality with water (4 × 5 ml), dried over anhydrous sodium sulfate, and evaporated *in vacuo* to give 17 mg of a white-yellow solid. The crude product is purified by preparative TLC in solvent system H. Areas corresponding to (2S)-3-keto-dihydro-C_2-ceramide and unreacted D-e-dihydro-C_2-ceramide are scraped and extracted with 3 × 5 ml of chloroform–methanol (1 : 1, v/v) and evaporated under a nitrogen stream to give 11 mg (58.5% yield) of (2S)-3-keto-C_2-dihydroceramide. The purity of this compound (95%) is established by TLC-1 in solvent H (3-keto-**IIA**: R_f = 0.44, **IIA**: R_f = 0.85).

Preparation of (2S)-2-N-(1,1-Dimethylethylcarboxy)-amino-3-ketooctadecene-1,3-diol; (2S)-3-Keto-N-Boc-sphingosine: 3-keto-**XB**

A solution of N-Boc-D-*erythro*-sphingosine **XB** (1.19 g, 3 mmol) in dry benzene (75 ml) is mixed with a solution containing dry CrO_3 (500 mg, mmol) in dry pyridine (75 ml) under a nitrogen atmosphere. The reaction mixture is stirred at room temperature for 6 hr. During this time, the reaction mixture becomes progressively more turbid with a brown precipitate. The reaction mixture is added to a solution of 200 ml of cold 5 N hydrochloric acid, 150 ml of cold distilled water, and 250 ml of diethyl ether. After shaking, the ether layer is removed and the lower layer is extracted several times with small volumes of ether. The combined extracts are washed to neutrality with 10% sodium bicarbonate solution, dried over anhydrous sodium sulfate, and evaporated *in vacuo* to give crude product. The crude product is purified by flash column chromatography: elution with 100 ml of chloroform removes a mixture of unidentified products (110 mg), and the desired product (532 mg, 62% yield) is recovered by elution with 150 ml of 1% methanol in chloroform. Unreacted starting material (325 mg) is eluted with solvent H. The purity of 3-keto-**XB** is established by TLC-1 in solvent L (R_f = 0.42).

Preparation of (2S)-3-Ketosphingosine hydrochloride; 3-keto-**IB**

A solution of (2S)-3-keto-N-Boc-sphingosine **XB** (1.08 g, 2.7 mmol) in ethyl acetate (35 ml) is added dropwise over 5 min to a well-stirred and cooled (in an ice-water bath) mixture of 3.5 N HCl (22 ml) in ethyl acetate. The mixture is stirred with cooling in an ice-water bath for 4 hr. The resulting white precipitate is separated by filtration, washed with cold anhydrous diethyl ether, and dried in vacuum to give pure (2S)-3-ketosphingosine hydrochloride (375 mg, 45% yield) as a white microcrystalline powder. The purity of this compound is established by TLC-1 in solvent B (R_f = 0.72).

Preparation of (2S)-3-Ketosphinganine-hydrochloride; 3-keto-**IA**

Following Scheme 3.[35]

PREPARATION OF 3-KETO-**XIC**: A mixture of 1,1-dimethyl-ethyl-2,2-di-methyl-4-(1-hydroxy-2-hexadecynyl)-3-oxazolidinecarboxylate **VIIIC** + **IXC** (6.25 g, 14.3 mmol) and activated MnO_2 (29 g, 280 mmol) in anhydrous methylene chloride (250 ml) is stirred vigorously with exclusion of moisture at room temperature for 3 hr. The reaction mixture is filtered through the Celite bed, and the filtrate is evaporated under reduced pressure to dryness to give a crude product. This material is purified by flash colum chromatography and eluted with solvent P to give 3-keto-**XC** (4.80 g, 77% yield) as a colorless semisolid, which slowly solidifies (on standing in the refrigerator) to a white solid. The purity of this compound is established by TLC-2 in solvent E (R_f = 0.56).

PREPARATION OF 3-KETO-**XA**. A mixture of 3-keto-**XI C** (2.6 g, 5.9 mmol) and 10% Pd/C (300 mg) in anhydrous methyl alcohol (150 ml) is stirred vigorously under H_2 atmosphere at room temperature for 24 hr. The reaction mixture is filtered through Celite, evaporated under reduced pressure to dryness, and purified by flash column chromatography. It is eluted with solvent P to give 3-keto-**XIA** (2.21 g, 84.7% yield) as a colorless semisolid, which slowly solidifies on standing in the refrigerator to a white solid. The purity of this compound is established by TLC-2 in solvent E (R_f = 0.52).

PREPARATION OF 3-KETO-**IA**. To a well-stirred and cooled (in an ice-water bath) mixture of 3.5 N HCl (12 ml) in ethyl acetate, a solution of 3-keto-**XIA** (0.535 g, 1.23 mmol) in ethyl acetate (25 ml) is added dropwise over 5 min. The mixture is stirred for 3.5 hr with cooling in an ice-water bath. The reaction mixture is concentrated under reduced pressure at room temperature to one-eighth the original volume, diethyl ether (25 ml) is added, the white precipitate is separated by filtration, washed with cold anhydrous diethyl ether, and dried in vacuum to give pure (2S)-3-ketosphinganine hydrochloride (375 mg, 91% yield) as a white microcrystalline powder. The purity of this compound is established by TLC-2 in solvent B (R_f = 0.70).

From 3-Ketodihydro-C_2-ceramide. 3-Ketodihydro-C_2-ceramide (1 g, 2.93 mmol) is added to a mixture of 50 ml of 5% ethereal hydrochloric acid and 60 μl water and is refluxed with vigorous stirring, with exclusion of moisture. The progress of the cleavage is followed by TLC-2 in solvent H: 3-keto-**IA**, (R_f = 0.15), 3-keto-C_2-**IIA**, (R_f = 0.44), unknown by-product, (R_f = 0.735). The reaction mixture is allowed to stand overnight and the resulting white precipitate is centrifuged with the exclusion of moisture. The supernatant is decanted and the precipitate is washed three times with ether and dried under vacuum to give 530 mg (67% yield) of a white microcrystalline powder.

Synthesis of Precursors for Preparation of Labeled Sphingomyelin

Principles

Methods for the preparation of radiolabeled SM via the methylation of de-SM and via the acylation of lyso-SM are presented elsewhere in this volume.[2] Demethylation of SM can be performed in a dimethyl sulfoxide solution using sodium thiophenolate or in a dimethylformamide solution using 1,4-diazabicyclo-(2,2,2)-octane.[10,11]

Because deacylation of SM is performed under acid hydrolysis conditions that are known to cause epimerization at the C-3 position and to generate the *threo* isomer,[6–8] separation of the resulting diastereoisomeric mixture is necessary.

Procedures

This section presents methods for the preparation of the precursors necessary to synthesize SM radiolabeled in acyl or choline groups. They include sphingosylphosphorylcholine (lyso-SM) and ceramide-1-phosphoryl-N,N-dimethylethanolamine (demethylated-SM, de-SM) which are prepared from bovine brain sphingomyelin using published procedures.[6,7,10,11] 3-Ketosphingomyelin can be prepared by the oxidation of SM by general published methods.[26,28]

Preparation of Sphingosylphosphorylcholine; Lyso-SM

Sphingomyelin (65 mg) is hydrolyzed under reflux in 7 ml of 1.5 N HCl in methanol for several hours by monitoring the reaction progress by TLC-1 in solvent M. After 12 hr, the reaction is complete and lyso-SM, as well as sphingosine, is generated. The solvent is evaporated, and the product is partitioned into the upper phase of the Folch *et al.*[17] procedure. Briefly, the reaction mixture is redissolved in 4 ml of methanol followed by the addition of 8 ml of chloroform and 3 ml of water. The phases are separated and the lower phase (8 ml) is reextracted twice with 7 ml of solvent R. The combined upper phases contain mainly lyso-SM with sphingosine and choline phosphate as minor by-products. The pH of the combined upper phases is adjusted to 12–13 with 5 N NaOH, and 24 ml of solvent R is added to force lyso-SM and sphingosine to the lower phase. This process is repeated three times. The lower phases are combined and evaporated *in vacuo*. The crude product is purified by preparative TLC-3 in solvent M. The area corresponding to lyso-SM is scraped from the plate, and the product is extracted four times with 6 ml of the solvent mixtures: methanol (containing 7% H_2O and 1% concentrated NH_4OH, v/v) in chloroform (1:1, v/v). Extracts are combined and evaporated under nitrogen gas to give a solid white product (25 mg, 65% yield). The purity of this product is assessed by TLC-2 in solvents M, T, and W. TLC analysis in solvent T of the purified lyso-SM shows two spots corresponding to the *erythro*- (R_f = 0.2) and to the *threo*-lyso-SM (R_f = 0.12). Both isomers are partially preparatively recovered from the plate by scraping off the appropriate areas to provide 13 mg of the *erythro* isomer and 3.5 mg of the *threo* isomer. Analysis using TLC-1 in solvent T shows single spots for both diastereoisomers. Alternatively, *erythro*- and *threo*-diastereoisomers can be separated from the crude lyso-SM by HPLC.[9]

Preparation of Ceramide-1-phosphoryl-N,N-dimethylethanolamine (Demethylated-sphingomyelin; de-SM)

Sodium thiophenolate (3.5 g, 28 mmol) is dissolved in 30 ml of freshly distilled, dry dimethyl sulfoxide and added dropwise under a nitrogen atmosphere to a solution of 3 g (~4 mmol) of bovine brain sphingomyelin in 100 ml of freshly distilled, dry dimethyl sulfoxide (after warming to 100°). The reaction mixture is stirred at 100° for 4 hr under a nitrogen atmosphere.

Analysis of the reaction mixture using TLC-1 in solvent N shows no presence of SM and formation of the demethylated product. The reaction mixture is cooled to room temperature, hydrochloric acid in cold water (80 ml of 1 N HCl in 700 ml of water) is added slowly with vigorous stirring, and the pH of the reaction mixture is changed to ~2.5, which causes formation of a white suspension. The product is extracted from the reaction mixture with chloroform (3 $\times$ 250 ml), and the phases are separated by centrifugation (3000 rpm, 10 min, 5°). The combined lower phases (including milky interphase) are dried over anhydrous sodium sulfate, evaporated under vacuum, and a precipitate forms. The small volume of DMSO is removed and the precipitate is washed with petroleum ether and dried *in vacuo* over P_2O_5 to give a sticky white powder (2.8 g, 95% yield). This product is purified further by column flash chromatography applying a gradient solvent system: chloroform–methanol (1 : 0–3 : 7, v/v). Each 25-ml fraction is monitored by TLC-1 in solvent N. The product elutes with chloroform–methanol (3 : 7, v/v). Fractions containing pure product are combined and evaporated under vacuum to give 2.5 mg (85% yield) of a white solid. The purity of this compound is established by TLC-1 in solvent N: R_f = 0.58 (SM: R_f = 0.33).

Preparation of N-Succinimidyl Esters of Labeled Fatty Acids

Principles

Transacylation reactions using activated fatty acid as their N-hydroxysuccinimide esters are commonly used for the preparation of N-acylsphingoid bases and their radiolabeled analogs. N-Succinimidyl esters of labeled fatty acids can be prepared by the method of Lapidot et al.[13] (reaction of an equimolar amount of the fatty acid and N-dihydroxysuccinimide in dry ethyl acetate) or, alternatively, by the procedure of Ogura et al.[14] (reaction of an equimolar amount of the fatty acid, N,N-disuccinimidyl carbonate, and pyridine in acetonitrile).

Procedures

This section presents a method for the preparation of the N-hydroxysuccinimide ester of hexanoic acid.[12,13]

Preparation of N-Hydroxysuccinimide Ester (NHS Ester) of [14C]Hexanoic acid

The sodium salt of [14C]hexanoic acid (2 mCi, 36.4 μmol) is converted to free acid by the addition of 100 μl of 0.1 N HCl and partitioning into diethyl ether (3 × 5 ml). The diethyl ether extracts are combined and dried over anhydrous sodium sulfate, the solvent is evaporated under a stream of dry nitrogen, and the dry material is dissolved in 500 μl of dry DMF followed by the addition of dicyclohexylcarbodiimide (37 μmol in 100 μl DMF) and N-didroxysuccinimide (37 μmol in 100 μl DMF). The reaction is allowed to proceed for 2 days at room temperature in the dark with continual stirring. The insoluble N,N-dicyclohexylurea is sedimented by centrifugation, and analysis of the supernatant using TLC-1 in solvent F shows formation of the NHS ester and the presence of the unreacted hexanoic acid in the proportion ~9:1. The NHS ester of hexanoic acid is purified by TLC-2 in solvent F (NHS ester: R_f = 0.8, hexanoic acid: R_f = 0.4). The area corresponding to the product is scraped from the plate and eluted quantitatively with chloroform–methanol (1:1, v/v) to provide the NHS ester of [^{14}C]hexanoic acid with an 80% yield.

[45] Practical Synthesis of N-Palmitoylsphingomyelin and N-Palmitoyldihydrosphingomyelin

By ANATOLIY S. BUSHNEV and DENNIS C. LIOTTA

Introduction

Sphingolipids play a significant role in the structural organization of biological membranes, and numerous intra- and intercellular processes are dependent on their metabolism and catabolism. Over the last decade, sphingolipids and their metabolites have generated a great deal of interest due to the discovery of the sphingomyelin cycle[1] in which ceramide plays an important role in the regulation of cell growth, differentiation, and cell death. Sphingolipids have also been implicated in the development of general types of tumors, and there are several reports that the dietary sphingolipids can affect and even reverse the development of carcinogenesis.[2]

The success of these and many other investigations depends on the accessibility of sphingolipids, which are used as substrates. Many kinds of sphingolipids can be isolated easily from natural sources; however, these compounds are not chemically homogeneous with respect to sphingosine bases or fatty acids. Furthermore, it is clear that the heterogeneity of

[1] Y. A. Hannun and L. M. Obeid, *Biochem. Soc. Trans.* **25**, 1171 (1997).

[2] A. H. Merrill, Jr., E.-M. Schmelz, D. L. Dillehay, S. Spiegel, J. A. Shayman, J. J. Schroeder, R. T. Riley, K. A. Voss, and E. Wang, *Toxicol Appl. Pharmacol.* **142**, 208 (1997).

these moieties affects both physicochemical characteristics and biological functions of the sphingolipids. The semisynthetic preparation of sphingomyelins and other sphingolipids through their lyso derivatives provides a way to obtain compounds in which the fatty acid is homogeneous, while the sphingosine base composition remains nonhomogeneous. Prepared by this method, substrates may be contaminated with the *threo* isomers due to epimerization at the C-3-atom in the process of acidic deacylation of the natural materials.[3] Moreover, commercial sphingolipids can be contaminated with other biologically active compounds, which can exhibit the same or different activity.[4]

Total chemical synthesis allows the preparation of individual homogeneous natural compounds with a completely defined structure in any desirable quantities for biophysical and biochemical studies. Furthermore, chemically prepared substrates with variations in a core structure may be used for elucidating variations in the structure-dependent biochemical and biophysical properties. With this in mind, we undertook the preparative total syntheses of N-palmitoylsphingomyelin (**1a**) and N-palmitoyldihydrosphingomyelin (**1b**):

1a: X-X = *trans*-CH$=$CH
1b: X-X = CH$_2$CH$_2$

Problems Related to Sphingomyelin Synthesis

The total chemical synthesis of sphingomyelins consists of two main stages: (1) the synthesis of 3-O-protected ceramides and (2) the introduction of the phosphocholine moiety into 3-O-protected ceramides, followed if necessary by removal of the protecting group(s) from the ceramide counterpart and phosphocholine residue. Buyn *et al.*[5] utilized unprotected N-octanoylceramide in the synthesis of the short chain N-octanoylsphingomyelin, but this approach could not be employed in the syntheses of long chain sphingomyelins due to limited solubility of the corresponding ceramides.

[3] P. K. Sripada, P. R. Maulik, J. A. Hamilton, and G. G. Shipley, *J. Lipid Res.* **28,** 710 (1987).
[4] K. Liliom, D. J. Fischer, T. Virág, G. Sun, D. D. Miller, J.-L. Tseng, D. M. Desiderio, M. C. Seidel, J. R. Erickson, and G. Tigyi, *J. Biol. Chem.* **273,** 13461 (1998).
[5] H. S. Byun, R. K. Erukulla, and R. Bittman, *J. Org. Chem.* **59,** 6495 (1994).

To date, numerous sphingomyelin syntheses have been reported, and modern approaches employ enantioselective methods for the preparation of sphingoid bases (3-O-protected ceramides)[6-11] and the highly efficient phosphoramidite (or phosphitetriester) technique of phosphorylation.[5-7,9-12] However, these methods involve many steps[8-10] or lack *erythro/threo* diastereselectivity[6,7] in the preparation of 3-O-protected ceramides.

For these reasons, we developed another synthetic approach[13-16] based on the attachment of long chain alkynes to an N,O-diprotected L-serine aldehyde, or the (S)-Garner aldehyde [(S)-4-formyl-2,2-dimethyl-3-oxazolidinecarbonic acid *tert*-butyl ester], to form the C-3–C-4 bond of (2S,3R)-sphingosine bases. These methods have a minimal number of steps and are highly efficient. Moreover, the synthesis of our target compounds could be started with L-serine or the commercially available N-Boc-L-serine methyl ester or (S)-Garner aldehyde. It is thus possible to obtain sphingosine bases in either five or three steps, respectively. High yields (70–93%)[13,15] and very good *erythro/threo* selectivity (ca. 20:1) of the key addition reaction are ensured by the addition of hexamethylphosphoramide (HMPA).[13] Furthermore, reduction of the intermediate propargyl alcohols with lithium in liquid ammonia leads to the protected sphingosine,[15] whereas the same reaction in ethylamine immediately yields sphingosine bases,[14,17,18] thus reducing the length of the synthesis by one more step. This method appears to be universal, as shown by the conversion of (S)-Garner aldehyde to unnatural (2S,3S)-*threo*-sphingosine by the Mitsunobu isomerization of the intermediate propargylic alcohol[16] or by direct conversion of (S)-Garner aldehyde into (2S,3S)-*threo*-sphingosines through the reactions with corresponding alkynes in the presence of zinc bromide.[13] If D-serine or the

[6] K. S. Bruzik, *J. Chem. Soc. Chem. Commun.* 329 (1986).

[7] K. S. Bruzik, *J. Chem. Soc. Perkin Trans. 1* 423 (1988).

[8] Z. Dong and J. A. Butcher, Jr., *Chem. Phys. Lipids* **66,** 41 (1993).

[9] B. Kratzer, T. G. Mayer, and R. R. Schmidt, *Tetrahedron Lett.* **34,** 6881 (1993).

[10] B. Kratzer and R. R. Schmidt, *Liebigs Ann.* 957 (1995

[11] J. J. Gaudino, K. Bjergarde, P.-Y. Chan-Hui, C. D. Wright, and D. S. Thomson, *Bioorg. Med. Chem. Lett.* **7,** 1127 (1997).

[12] A. Yu. Frantova, A. S. Bushnev, E. N. Zvonkova, and V. I. Shvets, *Bioorg. Khim.* (*Russian*) **17,** 1562 (1991).

[13] P. Herold, *Helv. Chim. Acta* **71,** 354 (1988).

[14] P. Garner, J. Park, and E. Malecki, *J. Org. Chem.* **53,** 4395 (1988).

[15] H.-E. Radunz, R. Devant, and V. Eiermann, *Liebigs Ann. Chem.* 1103 (1988).

[16] S. Nimkar, D. Menaldino, A. Merrill, and D. Liotta, *Tetrahedron Lett.* **29,** 3037 (1988).

[17] K. Mori and H. Matsuda, *Liebigs Ann. Chem.* 529 (1991).

[18] T. Abe and K. Mori, *Biosci. Biotechnol. Biochem.* **58,** 1671 (1994).

corresponding (R)-Garner aldehyde[19–23] was chosen as the starting material, the unnatural (2R,3S)-erythro-sphingosine bases could be prepared.[16]

The problem of the synthesis of dihydrosphingosine and derivatives may be solved rather simply by employing the reduction of the triple bond in the propargylic intermediate(s) via the hydrogenation over platinum oxide, as exemplified in syntheses of the cerebroside symbioramide.[24–26]

There were several possibilities for introducing the phosphocholine moiety. A thorough review of the chemical literature on the syntheses of sphingomyelins[5–7,9–12] and other sphingophospholipids—ceramide-1-phosphonucleosides,[27,28] ceramide-1-phosphoinositols,[9,29–32] ceramide-1-phosphates,[10] sphingosine-1-phosphocholine,[10] N,N-dimethylsphingoethanolamine,[12] and sphingosine-1-phosphates and analogs[33–35]—suggests that the application of phosphoramidite (phosphitetriester) methodology to the phosphorylation of 3-O-protected ceramides should proceed in a similar fashion to syntheses of other natural phosphoesters.[36–40] In this approach, building of the phosphodiester structures includes four steps: (1) phosphitylation of alcohols with the appropriate phosphitylation reagents; (2) condensation of the resulting phosphoramidites with the second alcohol

[19] P. Garner and J. M. Park, J. Org. Chem. 52, 2361 (1987).

[20] P. Garner and J. M. Park, Org. Synth. 70, 18 (1991).

[21] A. Dondoni and D. Perrone, Synthesis 527 (1997).

[22] J. S. R. Kumar and A. Datta, Tetrahedron Lett. 38, 6779 (1997).

[23] A. Avenoza, C. Cativiela, F. Corzana, J. M. Peregrina, and M. M. Zurbano, Synthesis 1146 (1997).

[24] M. Nakagawa, J. Yoshida, and T. Hino, Chem. Lett. 1407 (1990).

[25] J. Yoshida, M. Nakagawa, H. Seki, and T. Hino, J. Chem. Soc. Perkin Trans. 1 343 (1992).

[26] K. Mori and K. Uenishi, Liebigs Ann. Chem. 41 (1994).

[27] O. V. Oskolkova, A. Y. Zamyatina, V. I. Shvets, D. S. Esipov, and V. G. Korobko, Bioorg. Khim. (Russian) 22, 307 (1996).

[28] O. V. Oskolkova, A. V. Perepelov, D. S. Esipov, A. Y. Zamyatina, S. G. Alekseeva, and V. I. Shvets, Bioorg. Khim. (Russian) 23, 591 (1997).

[29] A. Yu. Frantova, A. E. Stepanov, A. S. Bushnev, E. N. Zvonkova, and V. I. Shvets, Tetrahedron Lett. 33, 3539 (1992).

[30] A. Y. Zamyatina, A. E. Stepanov, A. S. Bushnev, E. N. Zvonkova, and V. I. Shvets, Bioorg. Khim. (Russian) 19, 347 (1993).

[31] A. Y. Zamyatina and V. I. Shvets, Chem. Phys. Lipids 76, 225 (1995).

[32] B. Kratzer, T. G. Mayer, and R. R. Schmidt, Eur. J. Org. Chem. 291 (1998).

[33] A. Boumendjel and S. P. F. Miller, J. Labeled Comp. Radiopharm. 36, 243 (1994).

[34] B. Kratzer and R. R. Schmidt, Tetrahedron Lett. 34, 1761 (1993).

[35] A. Boumendjel and S. P. F. Miller, J. Lipid Res. 35, 2305 (1994).

[36] A. Yu. Zamyatina, A. S. Bushnev and V. I. Shvets, Bioorg. Khim. (Russian) 20, 1253 (1994).

[37] S. L. Beaucage and R. P. Iyer, Tetrahedron 49, 10441 (1993).

[38] E. E. Nifant'ev and D. A. Predvoditelev, Russian Chem. Rev. 63, 71 (1994).

[39] E. E. Nifant'ev and M. K. Grachev, Russian Chem. Rev. 63, 575 (1994).

[40] F. Paltauf and A. Hermetter, Prog. Lipid Res. 33, 239 (1994).

component in the presence of 1*H*-tetrazole; (3) oxidation of the forming phosphitetriesters into the phosphatetriesters (i.e., the completely protected target compounds); and (4) deprotection of the phosphoric acid residue. Despite the apparent complexity and number of steps, these processes are easily conducted due to high reactivity of both phosphitylation reagents and intermediate phosphite derivatives. Under careful thin-layer chromatography (TLC) and nuclear magnetic resonance (NMR)[12,30,31] monitoring, these conversions could be carried out without isolation of the intermediates in pure form. As a whole, this approach leads to desired phosphodiesters with good efficiency.

In summary, we have designed a sphingomyelin synthesis [D-*erythro-N*-palmitoylsphingomyelin and D-*erythro-N*-palmitoyl dihydrosphingomyelin (**1a,b**)] based on the enantioselective synthesis of 3-*O*-benzoylceramides (**9a,b**) from *N*-Boc-L-serine methyl ester (**2**) (Scheme 1) and the introduction of phosphorylcholine moiety using phosphitetriester methodology (Scheme 2).

Preparation of 3-*O*-Benzoylceramides (9a,b)

Scheme 1 outlines a seven-step synthesis of these compounds. Preparation of the (*S*)-Garner aldehyde (**4**) and condensation of this compound with lithium pentadecyne are the key steps in this process. The (*S*)-Garner aldehyde synthesis was performed according to published procedures[19,20]; however, commercially available *N*-Boc-L-serine methyl ester (**2**) was used as the starting material.

The synthesis of the propargylic intermediate **5**, common for the syntheses of both unsaturated **9a** and saturated **9b** 3-*O*-benzoylceramides, was accomplished by known methods. This stage is very critical for the overall success of the synthesis. Addition of HMPA[13] increases the *erythro/threo* ratio in favor of the *erythro* isomer of the propargylic alcohol **5** (ca. 20 : 1). As result, isolation and purification of *erythro*-**5** by column chromatography are facilitated. The yields of this reaction are reported to be 70–93%.[13,15] Our average yields of this product were approximately 60%. To convert propargylic alcohol **5** to olefinic derivative **6a**, reduction with lithium in liquid ammonia was employed.[15] This method has the advantage of being simple to perform and work up. The crude product **6a** was adequate for use in the next step without additional purification. Transformation of **5** into saturated **6b** was conducted by hydrogenation over platinum oxide in ethyl acetate using known procedures.[25,26] The last compound was also used in the next step without purification. The presence of single unprotected hydroxy groups in **6a** and **6b** allowed facile conversion to 3-*O*-benzoylceramides (**9a**) and (**9b**). By direct benzoylation with three equivalents of

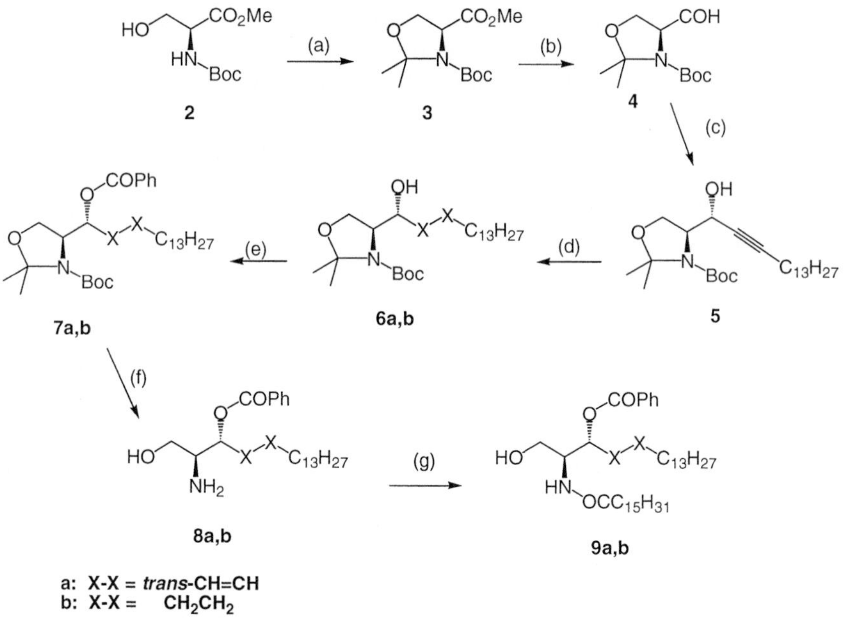

a: X-X = *trans*-CH=CH
b: X-X = CH₂CH₂

SCHEME 1. Synthetic route to 3-*O*-benzoylceramides (**9a,b**) through (*S*)-Garner aldehyde (**4**): (a) Me₂C(OMe)₂TsOH; (b) DIBAL, toluene; (c) pentadecyne/HMPA, THF; (d) Li/NH₃ or H₂/PtO₂; (e) PhCOCl/PyCH₂Cl₂; (f) Me₃Sil/CH₂Cl₂; and (g) PalmCl/NaOAc/THF.

benzoylchloride in the presence of pyridine, a benzoyl group was introduced at C-3 to give the fully protected sphingosine bases **7a** and **7b**, which were purified by column chromatography. From this point, the synthesis of 3-*O*-benzoylceramides (**9a**) and (**9b**) follows Shapiro's synthesis of these compounds[41,42]: transformation of **7a** and **7b** into the corresponding 3-*O*-benzoylsphingosine (**8a**) and 3-*O*-benzoyldihydrosphingosine (**8b**), followed by acylation with palmitoylchloride in the presence of sodium acetate. To obtain **8a** and **8b** from the fully protected bases **7a** and **7b**, *tert*-Boc and isopropylidene protecting groups were removed simultaneously by treatment with trimethylsilyliodide in methylene chloride, followed by quenching with a 4:1 mixture of methanol and water.[43] This reaction was carried out under ambient conditions, and no by-products were detected by TLC. In the synthesis of short chain C₉-dihydrosphingosines[44] (published by Vil-

[41] D. Shapiro, "Chemistry of Sphingolipids." Herman, Paris 1969.

[42] D. Shapiro and H. M. Flowers, *J. Am. Chem. Soc.* **84,** 1047 (1962).

[43] In: T. W. Greene and P. G. M. Wuts, "Protective Groups in Organic Synthesis," 2nd ed., p. 309. Wiley, New York, 1984.

[44] R. Villard, F. Fotiadu, and G. Buono, *Tetrahedron: Asymmetry* **9,** 607 (1998).

lard *et al.* after our work was completed), 5 *N* hydrochloric acid in refluxing dioxane was used to perform this operation. After the conversion of **7a** and **7b** was completed, the reaction mixture was evaporated to dryness, and crude **8a** and **8b** were used directly in the next step. To acylate **8a** and **8b**, a solution of palmitoyl chloride in THF and saturated aqueous sodium acetate were added simultaneously to a solution of these crude compounds in a mixture of 1 *N* AcOH and THF under vigorous stirring. TLC revealed complete conversion to the desired product with no by-products. 3-*O*-Benzoylceramides (**9a**) and (**9b**) were isolated by pouring the reaction mixture into water, filtering off the reaction mixture, drying the resulting precipitate, and recrystallizing from ethanol. The yields, over seven steps, of the 3-*O*-benzoylceramides (**9a**) and (**9b**) from **2** were 26 and 30%, respectively. This approach could also be employed for the preparation of other 3-*O*-protected ceramides (Ac, Piv, TBDMS, and TBDPS) used commonly in sphingophospholipid and glycosphingolipids syntheses.

Preparation of Sphingomyelins (1a,b) (Scheme 2)

In the section "Problems Related to Sphingomyelin Synthesis," we discussed the basic principles of creating phosphodiester structures using phosphoramidite methodology. Commonly, two phosphitylating reagents are used for performing the first stages of phosphorylation of 3-*O*-protected ceramides: (1) bifunctional chloro(*N*,*N*-diisopropylamino)methoxyphosphine (**14**)[5–7,11,12] and monofunctional bis(diisopropylamino)cyanoethoxyphosphine (**15**).[9,10,12,27–32] Phosphitylation with reagent **14** is conducted in the presence of tertiary amines, usually triethylamine, to give corresponding phosphoramidites in yields greater than 90%. The resulting phosphoramidites are stable and can be isolated and purified by column chromatography. Reactions with reagent **15** were carried out in the presence of 1*H*-tetrazole or its diisopropylammonium salt and also resulted in very good yields. Both processes are monitored easily by TLC and are very rapid. However, when chloroamidite **14** and tertiary amines are used in the phosphitylation reactions, excess reagents can be removed easily by gentle distillation and the resulting crude products could be used without further purification.[5–7]

In the present work, optimal results were obtained by the combination of the following procedures: phosphitylation of 3-*O*-benzoylceramides (**9a**) and (**9b**) with phosphitylating reagent (**14**) in the presence of triethylamine, condensation of the amidites **10a,b** with an excess choline tosylate in the presence of 1*H*-tetrazol, further oxidation of the corresponding phosphitetriesters **11a** and **11b** with *tert*-butyl hydroperoxide, and removal of the blocking groups in the resulting phosphoric acid residues of **12a** and **12b** with *tert*-butylamine and then the hydroxyl group in C-3 with methanolic

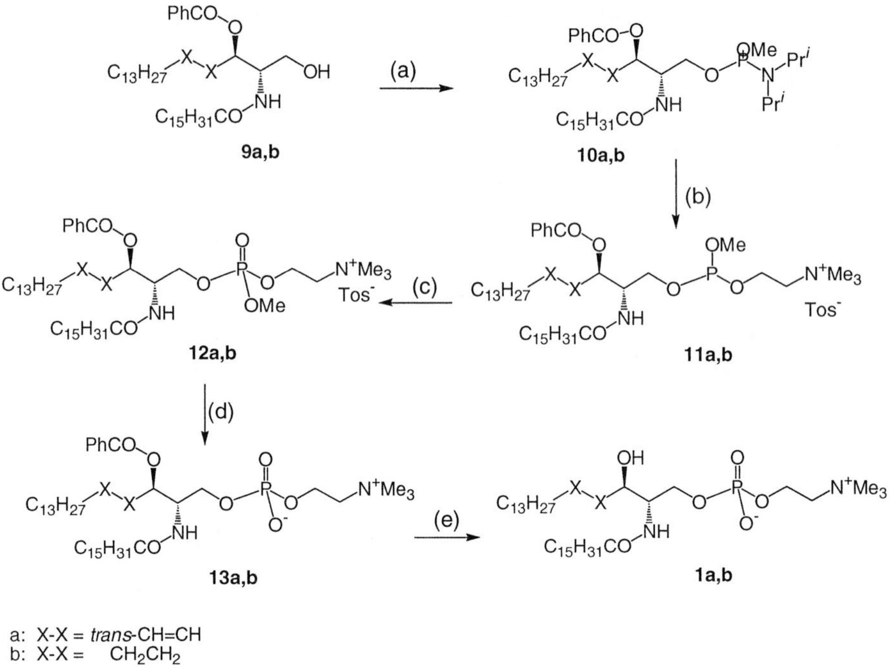

a: X-X = *trans*-CH=CH
b: X-X =   CH₂CH₂

SCHEME 2. Synthetic transformation of 3-*O*-benzoylceramides (**9a,b**) to the target *N*-palmitoylsphingomyelins (**1a,b**) using the phosphoramidite approach: (a) **14**, Et_3N, CH_2Cl_2, rt, 1 hr; (b) choline tosylate/1*H*-tetrazole, MeCN-THF, rt, 4.5 hr; (c) *t*BuOOH, CH_2Cl_2, rt, overnight; (d) *t*BuNH₂, CH_2Cl_2, rt, overnight; and (e) MeONa, MeOH, rt, 2 hr.

sodium methoxide. As observed in the literature,[7,12] the products of the phosphitylation—amidites (**10a,b**)—were the only reaction products and were taken into the condensation with choline tosylate immediately after the excess reagents and solvents were removed. After TLC showed that the transformation of **10a,b** to **11a,b** was complete, excess *tert*-butyl hydroperoxide (5.0–6.0 *M* solution in 2,2,3-trimethylpentane) was added to the same reaction. After several hours, the reaction mixtures containing phosphotriesters **12a** or **12b** were concentrated *in vacuo,* dissolved in methylene chloride, and partially deprotected with excess *tert*-butylamine. The resulting 3-*O*-benzoylsphingomyelins (**13a,b**) were isolated as pure compounds after column chromatography on silica gel and were deionized with Amberlyte MB-3. The yields of desired products were 67–72% and were comparable to the yields described in the literature. These procedures allowed us to prepare 3-*O*-benzoylsphingomyelins (**13a**) and (**13b**) on a gram scale. The final compounds sphingomyelins (**1a**) and (**1b**) were obtained in yields 92–97% after deprotection of the ceramide portions with sodium methoxide in methanol, column chromatography (for **1a**), and treatment with Amberlyte MB-3. The purity of intermediate and final products was established by the specific optical rotation, IR, NMR spectroscopy, and mass spectrometry, and the overall yields of **1a** and **1b** from N-Boc-L-ser-OMe (**2**) were 17 and ca. 20%, respectively.

Materials

Palmitoyl chloride, benzoyl chloride, N-Boc-L-serine methylate (**2**), chloro(N,N-diisopropylamino)methoxyphosphine (**14**), $1H$-tetrazole (sublimated), triethylamine, were purchased from Aldrich and used without further purification. *tert*-Butyl hydroperoxide (5.0–6.0 M solution in 2,2,3-trimethylpentane, purity 85%), diisobutylaluminium hydride (DIBAL) (1.5 M solution in toluene), and n-BuLi (1.6 M solution in hexanes) were purchased from Aldrich. 1-Pentadecyne was purchased from Lancaster. Choline tosylate was purchased from Sigma or was prepared from choline hydroxide (from Aldrich) and p-toluenesulfonic acid.[45] THF, methylene chloride, and acetonitrile were stored over molecular sieves (5 Å) and activated in a microwave oven (high power) for 4–5 min.

General Methods

All reactions that require dry conditions are run under an argon atmosphere using anhydrous solvents. The melting points are uncorrected. Infrared spectra are recorded on sodium chloride disks with an Impact Model 400 spectrophotometer, and only the structurally important peaks are listed. Positive and negative ion FAB mass spectra are taken on a JEOL spectrometer. Merck silica gel 60 TLC plates (20 × 50 mm of 0.25 mm thickness) are used to monitor the reactions, with visualization by heating with orcinol spray reagent (0.5% solution of orcinol in diluted sulfuric acid, 1 : 9) or with Sigma phosphate-specific spray reagent Molybdenum blue. Flash column chromatography is performed using silica gel 60 (Merck, 230–400 mesh ASTM). Solvent systems for TLC and preparative flash column chromatography are: **I**, **II**, and **III** hexane–ethyl acetate, 6 : 1, 9 : 2, and 9 : 1, respectively; **IV**, hexane–triethylamine, 20 : 1; **V**, chloroform–methanol–acetone, 10 : 0.5 : 0.5; **VI** chloroform–methanol–25% ammonium hydroxide, 6 : 1 : 0.2; **VII**, chloroform–methanol–25% ammonium hydroxyde, 7 : 1 : 0.1; **VIII**, chloroform–methanol–water, 165 : 35 : 4, **IX** chloroform–methanol–water, 65 : 35 : 4; and **X**, chloroform–methanol–acetic acid–water, 100 : 50 : 16 : 8.

[45] L. Clary, C. Santaella, and P. Vierling, *Tetrahedron* **51**, 13073 (1995).

All proportions of solvents are v/v. Optical rotations are measured in 1 dm tube with a Perkin-Elmer Model 241C polarimeter. Concentrations of the solutions are conducted by rotary evaporation with bath temperature kept below 30°; when concentrations are for drying purposes, an oil pump with a CO_2–acetone trap is employed.

Procedures

Preparation of 3-O-Benzoylceramides (9a,b)

According to the method of Garner and Park,[19,20] the synthesis of all-protected L-serine methyl ester (**3**) from *N*-Boc-L-ser-OMe (**2**) is carried out with a yield of 76–78%, $[\alpha]_D^{25}$ −57.63° (*c* 1.30, $CHCl_3$). Lit. $[\alpha]_D$ −46.7° (*c* 1.30, $CHCl_3$).[19,20] The reduction of **3** with DIBAL[19,20] leads to Garner aldehyde (**4**) in 84–85% yield, $[\alpha]_D^{23}$ −96.40° (*c* 1.34, $CHCl_3$). Lit. $[\alpha]_D$ −91.7° (*c* 1.34, $CHCl_3$).[19,20] The condensation of pentadecyne to aldehyde **4** and characterization of the resulting (4*S*,1′*R*)-4-(1′-hydroxy-2′-hexadecinyl)-2,2-dimethyloxazolidine-3-carboxylic acid *tert*-butyl ester (**5**) are performed as described in the literature.[13,14] Yield 58.0%, $[\alpha]_D^{25}$ −39.0° (*c* 1.43, $CHCl_3$). Lit.: $[\alpha]_D$ −39.7° (*c* 1.41, $CHCl_3$).[14]

(4S,1′R)-4-(1′-Benzoxy-2′-hexadecenyl)-2,2-dimethyloxazolidine-3-carboxylic acid tert-butyl ester (7a). According to the protocol of Radunz *et al.*,[15] a solution of derivative **5** (6.00 g, 13.71 mmol) in THF (400 ml) is added dropwise to a stirring solution of Li (5.0 g) in liquid ammonia (500 ml) at −75° over a period of 1.5 hr, and the reaction mixture is allowed to stir at −75–78° for an additional 1.5 hr. Ammonium chloride (120 g), ethyl acetate (600 ml), and water (400 ml) are then added sequentially. The reaction mixture is transferred into a separatory funnel, the organic phase is separated, and the aqueous phase is extracted with ethyl acetate (2 × 500 ml). The organic solutions are combined, dried over magnesium sulfate, and concentrated *in vacuo*. The crude (4*S*,1′*R*)-4-(1′-hydroxy-2′-hexadecenyl)-2,2-dimethyloxazolidine-3-carboxylic acid *tert*-butyl ester (**6a**) [6.00 g, ~100%, R_f 0.43 (**II**), MS: *m/z*: 440.3741. (M + H)$^+$ $C_{26}H_{50}NO_4$ requires 440.3740] is dissolved in a mixture of methylene chloride (100 ml) and pyridine (10 ml). To this is added benzoyl chloride (5.0 ml, 6.1 g, 43 mmol) at 0° over a period of 15–20 min. The reaction mixture is allowed to warm to room temperature over a period of 6–7 hr. Saturated sodium bicarbonate solution (3 ml) is then added and stirring is continued overnight. The reaction mixture is diluted with ether (1 liter), and the ether solution is washed sequentially with 10% sodium bicarbonate solution (2 × 100 ml), brine (2 × 100 ml), dried over magnesium sulfate, and concentrated *in*

vacuo. The residue is chromatographed on silica gel (300 g) using solvent mixture **III** as eluent. Yield 7.20 g (96.6%). R_f 0.56 (**I**), $[\alpha]_D^{23}$ −32.9° (*c* 1.00, CHCl₃). IR (neat): 1723 (C=O of benzoate), 1702 (C=O of carbamate), 1268 (benzoate) and 968 (*trans* double bond) (cm⁻¹). MS: *m/z* 550.4080 (M + Li)⁺ C₃₃H₅₃NO₅Li requires 550.4084.

(4S,1'R)-4-(1'-Benzoxyhexadecyl)-2,2-dimethyloxazolidine-3-carboxylic acid tert-butyl ester (7b). According to the literature,[25,26] the propargyl alcohol **5** (16.80 g, 38.39 mmol) is hydrogenated at room temperature in ethyl acetate (200 ml) over platinum oxide with hydrogen at atmospheric pressure for 2 hr to yield ~100% of the compound **6b**. R_f 0.46 (**II**), $[\alpha]_D^{25}$ −13.74° (*c* 0.84, CHCl₃), IR (neat): 3434 (OH), 1697 and 1675 (C=O of carbamate) (cm⁻¹), MS: *m/z*: 442.3889. (M + H)⁺ C₂₆H₅₂NO₄ requires 442.3896. Lit.: $[\alpha]_D^{21}$ −13.0° (*c* 1.03, CHCl₃),[26] or $[\alpha]_D^{20}$ −12.7° (*c* 1.090, CHCl₃).[25] This product in two equal portions is transformed into **7b** using the procedure for the conversion of **6a** to **7a**. Yield 10.00–10.30 g (95–98%), R_f 0.55 (**I**), $[\alpha]_D^{23}$ −48.91° (*c* 1.19, CHCl₃), IR (neat): 1722 (C=O of benzoate), 1700 (C=O of carbamate), 1270 (benzoate) (cm⁻¹). MS: *m/z*: 546.4141. (M + H)⁺ C₃₃H₅₆NO₅ requires 546.4158.

D-*erythro-3-O-Benzoyl-2-N-palmitoylsphingosine   (3-O-benzoylceramide) (9a).* To a solution of the benzoyl derivative **7a** (6.75 g, 12.4 mmol) in methylene chloride (150 ml) is added via syringe trimethylsilyl iodide (2.2 ml, 3.1 g, 15.5 mmol). The reaction mixture is allowed to stir at 20–25° for 2–3 hr. After this time, TLC shows the disappearance of the starting material, a mixture of water–methanol (1:4, 10 ml) is added, and the mixture is stirred for 15 min. The reaction mixture is concentrated *in vacuo,* and then the residue, crude 3-*O*-benzoylsphingosine (**8a**) [R_f 0.58 (**VII**), IR (mineral oil): 3429 (OH, NH₂), 1720 (C=O of benzoate), 1264 (benzoate) and 971 (*trans* double bond) (cm⁻¹)] is taken up into a mixture of THF (160 ml) and 1 *M* acetic acid (60 ml). A solution of palmitoyl chloride (3.70 g, 13.5 mmol) in THF (30 ml) and a saturated solution of sodium acetate (160 ml) are added simultaneously over a period of 5–10 min, and the reaction mixture is stirred at 20–25° for 2 hr. After this reaction is complete, the reaction mixture is diluted with water (250–300 ml), the precipitate is separated by filtration, washed sequentially with 5% sodium bicarbonate and water, dried carefully in air, and crystallized from ethanol (150 ml). The yield is 5.71 g (71.7%). mp 88.0–88.5°, $[\alpha]_D^{23}$ +21.0° (*c* 0.97, CHCl₃). R_f 0.52 (**V**). IR (mineral oil): 3414 (OH), 3328 (NH), 1720 (C=O of benzoate), 1630 (amide I), 1535 (amide II), 1266 (benzoate), 970 (*trans* double bond) (cm⁻¹). Lit.: mp 88–90°[41,42] or 90–92°[46], $[\alpha]_D^{25}$ +13.0° (*c* = 1.05, CHCl₃)[41,42] or $[\alpha]_D^{22}$ +17.8° (*c* = 0.56, CHCl₃).[46]

[46] T. Murakami, H. Minamikawa, and M. Hato, *J. Chem. Soc. Perkin Trans. 1* 1875 (1992).

D-*erythro-3-O-Benzoyl-2-N-palmitoyldihydrosphingosine (3-O-benzo-yldihydroceramide)* (**9b**). Starting from the benzoyl derivative **7b** (4.63 g, 8.48 mmol), 3-*O*-benzoylceramide (**9b**) is prepared in 82–83% yield using the previous method. mp 75.0–75.5° (MeOH), $[\alpha]_D^{25}$ +32.56° (*c* 1.26, CHCl$_3$). R_f 0.59 (**V**). IR (mineral oil): 3427 (OH), 3300 (NH), 1717 and 1693 (C=O of benzoate), 1638 (amide I), 1537 (amide II), 1276 (C benzo-ate) (cm^{-1}). MS: *m/z*: 644.5638. (M + H)$^+$ C$_{41}$H$_{74}$NO$_4$ requires 644.5618. Lit.:[41,42] mp 78–80° and $[\alpha]_D^{28}$ +26.4° (*c* = 1.25, CHCl$_3$).

Synthesis of Sphingomyelins (*1a*) and (*1b*)

D-*erythro-3-O-Benzoyl-2-N-palmitoylsphingosine-1-phosphorylcholine (3-O-Benzoylsphingomyelin)* (**13a**). To a solution of the carefully dried *in vacuo* 3-*O*-benzoylceramide (**9a**) (2.79 g, 4.34 mmol) in methylene chloride (25 ml) under argon is added via syringe triethylamine (1.40 ml, 1.02 g, 10.08 mmol) and phosphitylating reagent (**14**) (1.20 g, 1.18 ml, 6.1 mmol), and the resulting mixture is allowed to stir at 20–25° for 1 hr. After this time TLC shows the disappearance of **9a**, R_f ~0.5 in solvent system **V**, and the formation of a new product **10a** with R_f ~0.45 in solvent system **IV**. The reaction mixture is concentrated *in vacuo* and dried by reevaporation with dry toluene and then *in vacuo*. This crude amidite **10a** is dissolved in THF (20 ml), and to this solution under argon is added choline tosylate (4.60 g, 16.7 mmol) in acetonitrile (35 ml) and 1*H*-tetrazole (1.20 g, 16.9 mmol) in acetonitrile (25 ml). The reaction mixture is allowed to stir at 20–25° for 2.5 hr. After this reaction appears to be complete (TLC shows the disappearance of the compound **10a** with R_f ~0.45, solvent system **IV**, and the formation of the phosphitetriester **11a** with R_f ~0.5, solvent system **VI**), *tert*-butyl hydroperoxide (5.0–6.0 *M* solution in 2,2,3-trimethylpentane) (2.5 ml) is added and the reaction mixture is allowed to stir at 20–25 ° for 2 hr. After this time, TLC shows the disappearence of compound **11a** with R_f ~0.5 (solvent system **VI**) and the formation of the phosphatetriester **12a** with R_f ~0.4 (solvent system **VI**). The reaction mixture is concentrated *in vacuo,* and the residue is dissolved in methylene chloride (20 ml). *tert*-Butylamine (10 ml) is then added to this solution, and the reaction mixture is allowed to stir at 20–25° overnight. The mixture is then concentrated *in vacuo,* and the residue is chromatographed on silica gel (175 g) using solvent mixture **IX**. The resulting product is purified further using an Amberlyte MB-3 column with a mixture of chloroform–methanol–water (7:7:1) as eluent to yield 2.35 g (65.6%) of amorphous 3-*O*-benzoylsphingomyelin (**13a**) (monohydrate), R_f 0.50 (**X**), $[\alpha]_D^{23}$ +6.08° (*c* 1.15, CHCl$_3$) IR (film): 3500, 3300, 1715 (C=O of benzoate), 1645 (amide I), 1540 (amide II), 1262 (benzoate), 1119–1034 (PO_3^-), 966 (*trans* double bond) (cm^{-1}). MS: m/z: 807.6023. $(M + H)^+$ $C_{46}H_{84}N_2O_7P$ requires 807.6016.

D-erythro-2-N-Palmitoylsphingosine-1-phosphorylcholine   (N-Palmitoylsphingomyelin) (1a). To a solution of 3-*O*-benzoylsphingomyelin (**13a**) (2.06 g, 2.55 mmol) in methanol (25 ml) is added 2 M sodium methoxide solution in methanol (2.5 ml), and the reaction mixture is allowed to stir at 20–25° for 2 hr. The reaction mixture is neutralized with acetic acid, concentrated *in vacuo,* and the residue is chromatographed on silica gel using solvent mixture **IX** to yield 1.74 g (97.2%) of sphingomyelin (**1a**). mp ~ 190°, R_f 0.36 (**X**). $[\alpha]_D^{22}$ +5.6° (c 1.02, $CHCl_3$–MeOH, 1:1) or +4.3° (c 1.07, $CHCl_3$–MeOH–H_2O, 65:35:4). MS: m/z: 703.5767 (M + H)$^+$ $C_{39}H_{80}N_2O_6P$ requires 703.5754. Lit.: mp 215–217° [41], $[\alpha]_D^{25}$ +6.1° [41] or $[\alpha]_D$ +6° [c 1, $CHCl_3$–MeOH, 1:1).[10] IR (film): 3285, 1645 (amide I), 1575 (amide II), 1009–1058 (PO_3^-), 969 (*trans* double bond) (cm^{-1}).

D-erythro-3-O-Benzoyl-2-N-palmitoyldihydrosphingosine-1-phosphorylcholine (3-O-Benzoyldihydrosphingomyelin) (13b). Using the same procedure employed for converting **9a** into **13a**, **13b** (amorphous monohydrate) is prepared from 3-*O*-benzoyldihydroceramide (**9b**) (4.84 g, 7.51 mmol). Yield 4.39 g (71.9%), R_f 0.50 (**X**). IR (mineral oil): 3343, 1704 (C=O of benzoate), 1651 (amide I), 1531 (amide II), 1277 (benzoate), 1180 and 1117 (PO_3^-) (cm^{-1}). MS: m/z: 809.6171 $(M + H)^+$ $C_{46}H_{86}N_2O_7P$ requires 807.6173.

D-erythro-2-N-Palmitoyldihydrosphingosine-1-phosphorylcholine   (N-Palmitoyldihydrosphingomyelin) (1b). 3-*O*-Benzoyldihydrosphingomyelin (**13b**) (1.97 g, 2.38 mmol) is deprotected as described earlier to yield **1b**, 1.58 g (91.9%), monohydrate. mp 220–221°, R_f 0.36 (**X**). $[\alpha]_D^{25}$ +24.04° (c 0.92, $CHCl_3$–MeOH, 1:1). IR (mineral oil): 3286 (NH and OH), 1641 (amide I), 1549 (amide II), 1087 and 1059 (PO_3^-) (cm^{-1}). MS: m/z: 705.5928. $(M + H)^+$ $C_{39}H_{82}N_2O_6P$ requires 705.5911. Lit.:[41] mp 222–223° and $[\alpha]_D^{25}$ +22.5°.

[46] Synthesis and Biological Activity of Glycolipids, with a Focus on Gangliosides and Sulfatide Analogs

By Takao Ikami, Hideharu Ishida, and Makoto Kiso

Introduction

Glycosphingolipids, which are composed of a complex carbohydrate chain and a hydrophobic ceramide, are localized primarily on outer cell surfaces and are particularly abundant in nervous tissues.[1,2] Sialic acid and sulfate-containing glycoconjugates have been recognized to play important roles in many biological processes, such as cell adhesion, cell differentiation and proliferation, immune responses, receptor functions for viruses and bacterial toxins, and tumor progression. In particular, studies of the selectin family[3,4] and sialoadhesin family (I-type lectins)[5,6] have demonstrated the critical importance of protein–carbohydrate binding in cell–cell interactions, providing new opportunities for the development of therapeutic agents based on carbohydrates as binding inhibitors. Furthermore, studies using chemically synthesized gangliosides have shown that gangliosides are potent inhibitors of cellular immune responses and that the structures of gangliosides influence their activity greatly.[7]

This article describes mainly the synthesis and biological evaluation of glycolipids, with a focus on gangliosides and sulfatide analogs.

Selectin Ligands

Cell adhesion molecules (CAMs) play a crucial role during inflammatory conditions and in several immune system disorders by recruiting leukocytes to the injured area. The selectins are a family of CAMs composed of three structurally related carbohydrate-binding proteins [E-selectin (ELAM-1), L-selectin (LECAM-1) and P-selectin (GMP-140, PADGEM)].[8–11] E- and P-selectins are induced on the endothelial surface in response to inflammatory signals, whereas L-selectin is constitutively expressed on all classes of circulating leukocytes and interacts with cognate ligands on endothelial cells. They share a common mosaic structure consisting of an N-terminal C-type lectin domain (carbohydrate-binding site), a single epidermal growth

[1] S. Hakomori, S. Yamamura, and K. Handa, *Ann. N.Y. Acad. Sci.* **845,** 1 (1998).
[2] A. Varki, *Glycobiology* **2,** 97 (1993).
[3] L. A. Lasky, *Science* **258,** 964 (1992).
[4] M. P. Bevilacqua and R. M. Nelson, *J. Clin. Invest.* **91,** 379 (1993).
[5] S. Kelm, A. Pelz, R. Schauer, M. T. Filbin, T. Song, M. E. de Bellard, R. L. Schnaar, J. A. Mahoney, P. Bradfield, and P. R. Crocker, *Curr. Biol.* **4,** 965 (1994).
[6] L. D. Powell, and A. Varki, *J. Biol. Chem.* **270,** 14243 (1995).
[7] S. Ladisch, A. Hasegawa, R. Li, and M. Kiso, *Biochem. Biophys. Res. Commun.* **203,** 1102 (1994).
[8] M. P. Bevilacqua, S. Stengelin, M. A. Gimbrone, Jr., and B. Seed, *Science* **243,** 1160 (1989).
[9] G. I. Johnston, R. G. Cook, and R. P. McEver, *Cell* **56,** 1033 (1989).
[10] L. A. Lasky, M. S. Singer, T. A. Yednock, D. Dowbenko, C. Fennie, H. Rodriguez, T. Nguyen, S. Stachel, and S. D. Rosen, *Cell* **56,** 1045 (1989).
[11] T. F. Tedder, C. M. Isaacs, E. J. Ernst, G. D. Demetri, D. A. Adler, and C. M. Disteche, *J. Exp. Med.* **170,** 123 (1989).

factor domain, a discrete number of sequence modules similar to those found in regulatory proteins that bind the complement, a transmembrane domain, and a C-terminal cytoplasmic domain. The selectin family appears to be involved in the earliest events of the acute inflammatory response, and the selectin-dependent adhesion-promoting process is thought to be responsible for the transient "rolling" phenomenon of leukocytes along the endothelial surfaces.[12] A number of reports have focused on the identification of the carbohydrate ligand for the selectin family.[13]

There is now general agreement that all three selectins can efficiently recognize sialyl Lewis X [Neu5Acα2-3Galβ1-4(Fucα1-3)GlcNAc] and, therefore, the attention of many laboratories has focused on sialyl Lewis X.[13–15] However, sulfatide, one of the major acidic glycosphingolipids in mammalian tissues, is found to be a good ligand for L- and P-selectin and shows highly protective effects against selectin-dependent inflammatory lung injury.[16,17] These results suggest that selectin adhesion inhibitors may be employed as novel curing or preventive drugs for clinical use.

Sialyl Lewis X Gangliosides and Analogs

The first total syntheses of pentasaccharidic sialyl Lewis X (1),[18] hexasaccharidic sialyl Lewis X (2),[19] and the sialyl α(2-6)-Lewis X (3)[20] as the positional isomer (Fig. 1) can be achieved according to established methods.[21] The activity of these synthetic gangliosides and analogs in adhesion assays revealed that both sialyl Lewis X (1) and (2) were effectively recognized by E-selectin in dose-dependent manners, whereas the sialyl α(2-6) isomer (3) was not recognized.[15] Furthermore, the use of ELISA assay developed with immobilized sialyl Lewis X gangliosides and selectin–IgG chimeras demonstrated that three members of the selectin family (E-, P-,

[12] E. B. Finger, K. D. Puri, R. Alon, M. B. Lawrence, U. H. von Andrian, and T. A. Springer, *Nature* **379**, 266 (1996).

[13] A. Varki, *Proc. Natl. Acad. Sci. U.S.A.* **91**, 7390 (1994).

[14] B. K. Brandley, S. J. Swiedler, and P. W. Robbins, *Cell* **63**, 861 (1990).

[15] D. Tyrrell, P. James, N. Rao, C. Foxall, S. Abbas, F. Dasgupta, M. Nashed, A. Hasegawa, M. Kiso, D. Asa, J. Kidd, and B. K. Brandley, *Proc. Natl. Acad. Sci. U.S.A.* **88**, 10372 (1991).

[16] M. S. Mulligan, M. Miyasaka, Y. Suzuki, H. Kawashima, M. Iizuka, A. Hasegawa, M. Kiso, R. L. Warner, and P. A. Ward, *Int. Immunol.* **7**, 1107 (1995).

[17] M. S. Mulligan, R. L. Warner, J. B. Lowe, P. L. Smith, Y. Suzuki, M. Miyasaka, S. Yamaguchi, Y. Ohta, Y. Tsukada, M. Kiso, A. Hasegawa, and P. A. Ward, *Int. Immunol.* **10**, 569 (1998).

[18] A. Hasegawa, T. Ando, A. Kameyama, and M. Kiso, *J. Carbohydr. Chem.* **11**, 645 (1992).

[19] A. Kameyama, H. Ishida, M. Kiso, and A. Hasegawa, *J. Carbohydr. Chem.* **10**, 549 (1991).

[20] A. Kameyama, H. Ishida, M. Kiso, and A. Hasegawa, *J. Carbohydr. Chem.* **10**, 729 (1991).

[21] A. Hasegawa and M. Kiso, *Methods Enzymol.* **242**, 158 (1994).

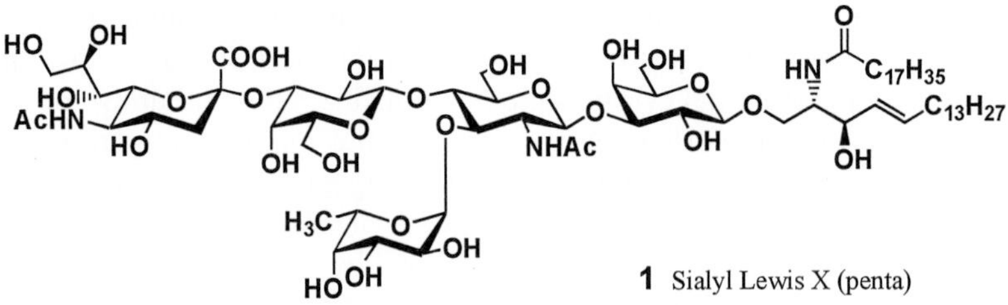

1 Sialyl Lewis X (penta)

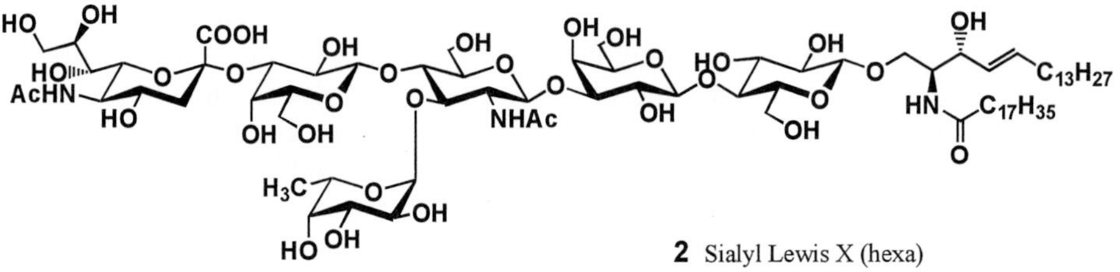

2 Sialyl Lewis X (hexa)

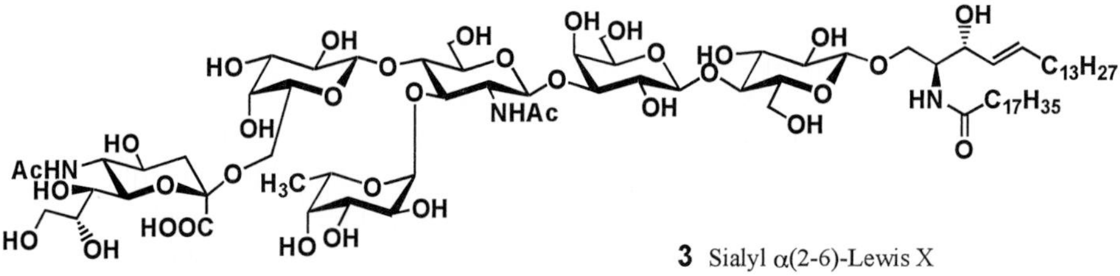

3 Sialyl α(2-6)-Lewis X

FIG. 1. Sialyl Lewis X gangliosides (**1**, **2**) and sialyl α(2-6) isomer (**3**).

and L-selectin) recognized a common carbohydrate epitope sialyl Lewis X in a calcium-dependent manner.[22]

We have also systematically synthesized three analogs of sialyl Lewis X gangliosides (**7**, **8**, and **9**) with altered fucose residue to examine in finer detail the requirement for binding of the three selectins (Scheme 1).[23] In the ELISA assay using modified sialyl Lewis X gangliosides with a variety of deoxyfucose residues, these deoxyfucose-containing compounds (**7**, **8**, and **9**) did not bind to E- and L-selectin, demonstrating that the hydroxyl groups at the 2, 3, and 4 positions of the fucose residue are required for

[22] C. Foxall, S. R. Watson, D. Dowbenko, C. Fennie, L. A. Lasky, M. Kiso, A. Hasegawa, D. Asa, and B. Brandley, *J. Cell Biol.* **117,** 895 (1992).

[23] A. Hasegawa, T. Ando, M. Kato, H. Ishida, and M. Kiso, *Carbohydr. Res.* **257,** 67 (1994).

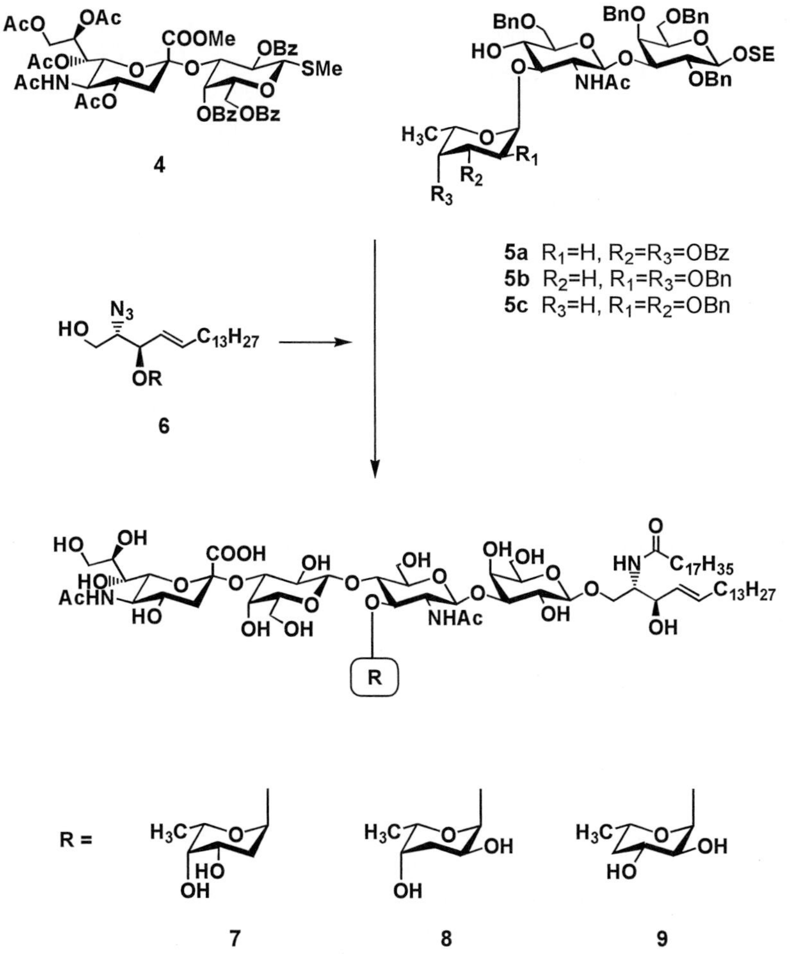

SCHEME 1. Synthesis of sialyl Lewis X ganglioside analogs containing modified L-fucose residues.

binding. Removal of any one of these hydroxyl groups completely eliminated recognition. P-selectin, however, required only 3-hydroxyl substitution and removal of the 2- or 4-hydroxyl groups did not markedly affect binding.[24]

[24] B. K. Brandley, M. Kiso, S. Abbas, P. Nikrad, O Srivasatava, C. Foxall, Y. Oda, and A. Hasegawa, *Glycobiology* **3,** 633 (1993).

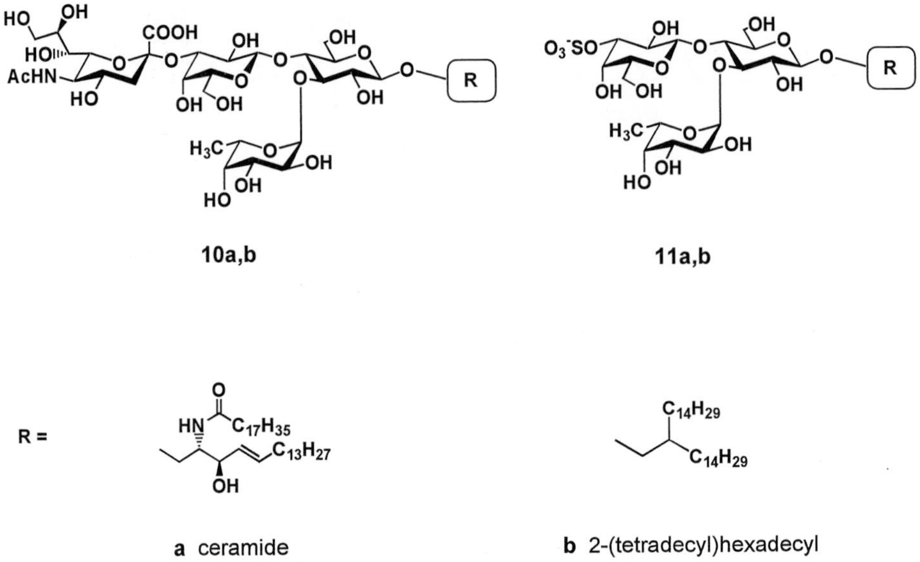

FIG. 2. Sialyl and sulfo-Lewis X analogs.

Sulfo-Lewis X Gangliosides and Analogs (Fig. 2)

Sulfated carbohydrates appear to have binding activity for selectins. Feizi's group[25] reported that E- and L-selectin can bind to sulfo-Lewis X-like structures isolated from an ovarian cystadenoma, and Imai et al.[26] described a requirement for sulfate in the native ligand for L-selectin. We have synthesized sialo (10a,b)- and sulfo-Lewis X analogs (11a,b) containing a ceramide and also a branched alkyl residue in order to clarify the role of the ceramide moiety for selectin recognition.[27] Even though the synthetic sulfo-Lewis X analogs (11a,b) were significantly shorter (three sugar residues) than the hexasaccharidic sialyl Lewis X (2), they were bound with equal (E-selectin) or greater (L- and P-selectin) potency than 2.[24] The recognition of sulfo-Lewis X analogs (11a,b) by E-selectin was calcium dependent, whereas that by L- and P-selectin was less dependent on calcium, maintaining ≥50% of control binding with the addition of EDTA. This suggests two modes of interaction: one that is calcium dependent and the other calcium independent. In this respect, sulfo-Lewis X analogs are

[25] P. J. Green, T. Tamatani, T. Watanabe, M. Miyasaka, A. Hasegawa, M. Kiso, C.-T. Yuen, M. S. Stoll, and T. Feizi, Biochem. Biophys. Res. Commun. 188, 244 (1992).
[26] Y. Imai, L. A. Lasky, and S. D. Rosen, Nature 361, 555 (1993).
[27] A. Hasegawa, K. Ito, H. Ishida, and M. Kiso, J. Carbohydr. Chem. 14, 353 (1995).

recognized by L- and P-selectin in a manner similar to sulfatides, as described in the next section.[28] Furthermore, selectin binding to sulfo-Lewis X lipid (**11b**) containing a branched alkyl residue was effectively distinct from that of the sulfo-Lewis X ceramide (**11a**), indicating that lipid aglycon can affect selectin recognition.

Synthesis of Sulfo-Lewis X Derivatives Containing a Ceramide or the 2-(Tetradecyl)hexadecyl Residue (11a,b) (Scheme 2)

O-(4-*O*-Acetyl-2,6-di-*O*-benzoyl-3-*O*-levulinyl-β-D-galactopyranosyl)-(1-4)-[*O*-(2,3,4-tri-*O*-acetyl-α-L-fucopyranosyl)-(1-3)]-2,6-di-*O*-benzoyl-α-D-glucopyranosyl trichloroacetimidate (**12**) was selected as the glycosyl donor, whereas (2*S*,3*R*,4*E*)-2-azido-3-*O*-benzoyl-4-octadecene-1,3-diol (**13**)[29] and 2-(tetradecyl)hexadecan-1-ol (**14**)[7] serve as the acceptors in the synthesis of target sulfo-Lewis X analogs. Glycosylation of **13** or **14** with **12**, in dichloromethane in the presence of boron trifluoride etherate and molecular sieves 4 Å (MS-4A), gives the desired β-glycosides **15** and **18** in 85 and 69% yields, respectively. Selective reduction of the azido group in **15** with hydrogen sulfide in aqueous 83% pyridine and subsequent condensation with octadecanoic acid using 1-(3-dimethylaminopropyl)-3-ethylcarbodiimide (WSC) in dichloromethane furnish a good yield (75%) of the per-*O*-acylated Lewis X sphingolipid **16**. Treatment of **16** or **18** in ethanol-tetrahydrofuran with hydrazine monoacetate at room temperature gives the 3-hydroxy derivatives **17** and **19** in high yields, respectively. Treatment compounds of **17** or **19** with the sulfur trioxide–pyridine complex in *N,N*-dimethylformamide for 1 hr at room temperature and subsequent *O*-deacylation with sodium methoxide in methanol afford the desired sulfo-Lewis X derivatives **11a** and **11b** as their sodium salts in good yields.

Sulfatide and Analogs

It has been found that sulfated carbohydrates such as sulfatides, fucoidan, a sulfated glucuronic acid (HNK-1) epitope, and heparin bind strongly to L- and P-selectins.[30–32] In particular, sulfatide, one of the major acidic glycosphingolipids in mammalian tissues, is found to be a good ligand

[28] D. Asa, T. Gant, Y. Oda, and B. K. Brandley, *Glycobiology* **2,** 395 (1992).
[29] Y. Ito, M. Kiso, and A. Hasegawa, *J. Carbohydr. Chem.* **8,** 285 (1989).
[30] K. Ley, G. Linnemann, M. Meinen, L. M. Stoolman, and P. Gaehtgens, *Blood* **81,** 177 (1993).
[31] R. M. Nelson, *Blood* **82,** 3253 (1993).
[32] Y. Suzuki, Y. Toda, T. Tamatani, T. Watanabe, T. Suzuki, T. Nakao, K. Murase, M. Kiso, A. Hasegawa, K. T. Aritomi, I. Ishizuka, and M. Miyasaka, *Biochem. Biophys. Res. Commun.* **190,** 426 (1993).

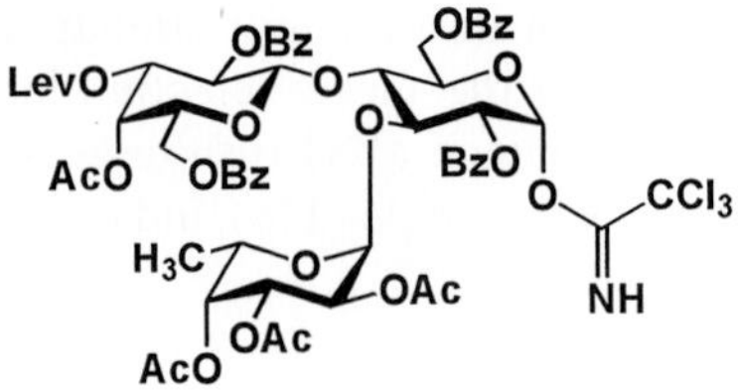

12

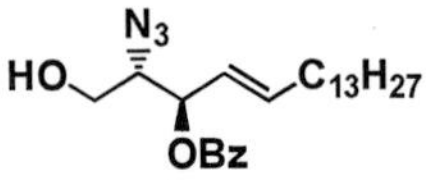

13

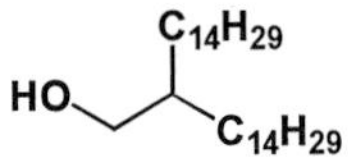

14

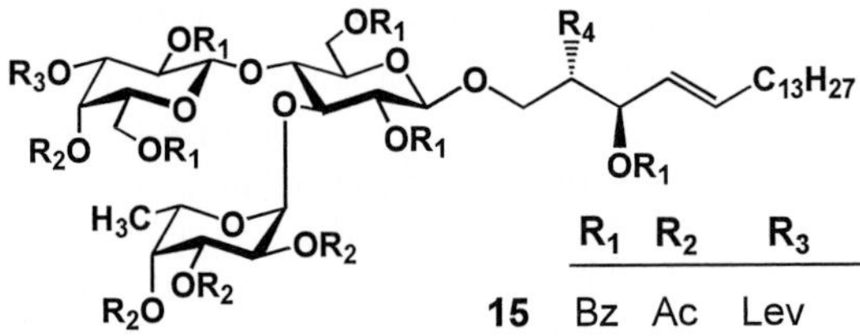

	R₁	R₂	R₃	R₄
15	Bz	Ac	Lev	N₃
16	Bz	Ac	Lev	NHCO(CH₂)₁₆CH₃
17	Bz	Ac	H	NHCO(CH₂)₁₆CH₃
11a	H	H	SO₃Na	NHCO(CH₂)₁₆CH₃

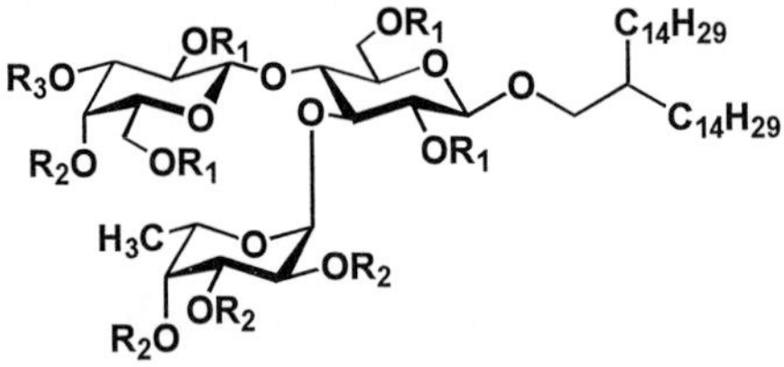

	R₁	R₂	R₃
18	Bz	Ac	Lev
19	Bz	Ac	H
11b	H	H	SO₃Na

Lev = CH₃CCH₂CH₂C —

SCHEME 2. Synthesis of sialyl- or sulfo-Lewis X analogs containing a ceramide or 2-(tetradecyl)hexadecyl residue.

for L- and P-selectin and shows highly protective effects against selectin-dependent inflammatory lung injury.[16,17] In view of these facts, we have synthesized novel, sulfated or phosphorylated β-D-galactopyranosides and β-D-lactopyranosides anchored with branched fatty alkyl residues in place of ceramide as the sulfatide mimetics.[33-35] The regioselective, one-pot sulfations of the parent glycolipids through the dibutylstannylene acetal with a certain amount of the sulfur trioxide–trimethylamine complex produced target-sulfated glycolipids, whereas stepwise phosphorylation by treatment of the properly protected diol with dibenzyloxy (diisopropylamino)phosphine gave phosphorylated glycolipids (Fig. 3). Structures of the sulfated products were determined by NMR and MS analyses.^{1}H and ^{13}C NMR have also been used to locate the positions of sulfate groups. Comparison of ^{1}H NMR data (Table I) of the sulfated galactosides **21c**, **22c**, **23c**, and **24c** with those of the unsulfated precursor glycolipid **20c** demonstrated that the sulfate groups deshield the geminal and vicinal protons. The secondary sulfate groups in sulfated derivatives cause α effects of 0.7–0.8 ppm, whereas the primary sulfate groups show 0.4–0.5 ppm for this effect. The β effects were 0.1–0.4 ppm depending on the axial or equatorial orientation of the vicinal proton. Comparison of ^{13}C NMR data (Table II) of the sulfated galactosides **21c**, **22c**, **23c**, and **24c** with those of the parent glycolipid **20c** demonstrate that the specific downfield shifts of 4–7 ppm for the signals of the α-carbon atoms bearing the sulfate groups were observed, whereas β-carbon atoms were upfield 2–4 ppm. Furthermore, negative FAB mass spectrometry gave (M-Na)⁻ ions as base peaks, confirming the number of sulfate groups in the molecules.

The activity of target glycolipids *in vitro* was measured in adhesion assays in terms of inhibition of the binding of HL-60 cells (sLe[x] expressing) to recombinant human selectin–IgG fusion proteins on plates.[35] Several of the synthesized glycolipids inhibit HL-60 cell binding to the selectin fusion proteins with greater potency than the sLe[x] tetrasaccharide. In particular, compounds **22c**, **23c**, and **24c** were significantly more potent than sLe[x] and phosphorylated β-D-galactopyranosides in blocking adhesion to P- and L-selectins. In addition, when the branched fatty alkyl residue was long, there was greater potency of the blocking adhesion to P- and L-selectins. These data indicate that the sulfate groups on positions 3 and 6 of the galactopy-

[33] E. Tanahashi, K. Murase, M. Shibuya, Y. Igarashi, H. Ishida, A. Hasegawa and M. Kiso, *J. Carbohydr. Chem.* **16**, 831 (1997).

[34] T. Ikami, H. Hamajima, T. Usui, T. Mitani, H. Ishida, M. Kiso, and A. Hasegawa, *J. Carbohydr. Chem.* **16**, 859 (1997).

[35] T. Ikami, N. Tsuruta, H. Inagaki, T. Kakigami, Y. Matsumoto, N. Tomiya, T. Jomori, T. Usui, Y. Suzuki, H. Tanaka, D. Miyamoto, H. Ishida, A. Hasegawa, and M. Kiso, *Chem. Pharm. Bull.* **46**, 797 (1998).

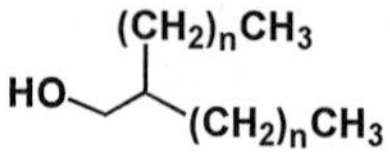

	n
19a	0
19b	2
19c	13

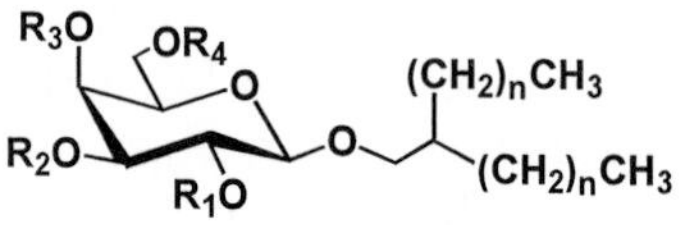

	R_1	R_2	R_3	R_4
20a-c	H	H	H	H
21a-c	H	SO_3Na	H	H
22a-c	H	SO_3Na	H	SO_3Na
23a-c	H	SO_3Na	SO_3Na	SO_3Na
24a-c	SO_3Na	SO_3Na	SO_3Na	SO_3Na
25a-c	H	$PO(ONa)_2$	$PO(ONa)_2$	H
26a-c	$PO(ONa)_2$	$PO(ONa)_2$	$PO(ONa)_2$	$PO(ONa)_2$

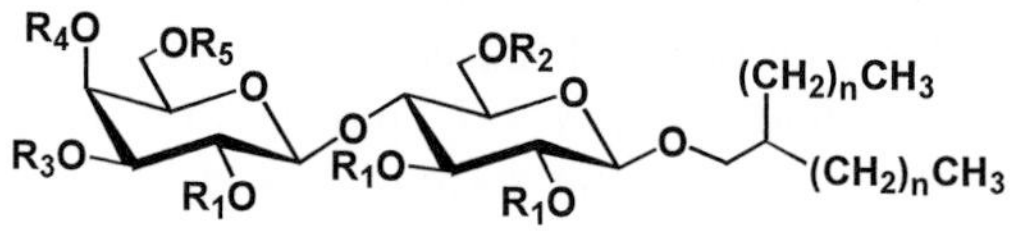

	R_1	R_2	R_3	R_4	R_5
27a-c	H	H	H	H	H
28a-c	H	H	SO_3Na	H	H
29c	H	H	SO_3Na	H	SO_3Na
30c	H	SO_3Na	SO_3Na	H	SO_3Na
31b,c	H	H	$PO(ONa)_2$	$PO(ONa)_2$	H

FIG. 3. Sulfated or phosphorylated β-D-galacto- and lactopyranosides containing fatty alkyl residues of different carbon chain lengths.

ranoside and the attachment of a long branched fatty alkyl residue to β-D-galactopyranoside were important for binding to P- and L-selectins. However, 3,4-bisphosphorylated β-D-galactopyranosides (**25a–c**) and 3′,4′-bisphosphorylated β-D-lactopyranosides (**31b,c**) were less potent than the sulfated glycolipids toward P- and L-selectins, but more potent toward E-selectin.

In addition, the multivalent derivatives of the sulfated or phosphorylated

TABLE I
^{1}H NMR Chemical Shift[a] of Unsulfated and Sulfated Galactopyranosides **20c–24c**

Compound	Chemical shift (and shift relative to **20c**)						
	H-1	H-2	H-3	H-4	H-5	H-6	H-6′
20c (unsulfated)	4.18	3.52	3.45	3.85	3.49	3.75	3.75
21c (3-sulfate)	4.28	3.70	4.21	4.25	3.53	3.74	3.74
	(+0.10)	(+0.18)	(+0.76)	(+0.40)	(+0.04)	(−0.01)	(−0.01)
22c (3,6-disulfate)	4.30	3.70	4.25	4.27	3.82	4.16	4.22
	(+0.12)	(+0.18)	(+0.80)	(+0.42)	(+0.33)	(+0.41)	(+0.47)
23c (3,4,6-trisulfate)	4.45	3.57	4.40	5.00	4.00	4.18	4.23
	(+0.27)	(+0.05)	(+0.95)	(+1.15)	(+0.51)	(+0.43)	(+0.48)
24c (2,3,4,6-terasulfate)	4.60	4.40	4.55	5.10	4.10	4.20	4.25
	(+0.42)	(+0.88)	(+1.10)	(+1.25)	(+0.61)	(+0.45)	(+0.50)

[a] Recorded in CD_3OD with TMS as an internal standard.

β-D-galactopyranosides[36,37] and 2-*O*-fucosyl sulfatides[38,39] containing 2-branched fatty alkyl residues in place of the ceramide have been designed

[36] T. Ikami, N. Tomiya, T. Morimoto, N. Iwata, T. Jomori, T. Usui, Y. Suzuki, H. Tanaka, D. Miyamoto, H. Ishida, A. Hasegawa, and M. Kiso, *Chem. Pharm. Bull.* **45**, 1726 (1997).
[37] T. Ikami, N. Tomiya, T. Morimoto, N. Iwata, R. Yamashita, T. Jomori, T. Usui, Y. Suzuki, H. Tanaka, D. Miyamoto, H. Ishida, A. Hasegawa, and M. Kiso, *J. Carbohydr. Chem.* **17**, 499 (1998).
[38] H. Ishida, S. Sago, T. Ikami, M. Kiso, and A. Hasegawa, *Biosci. Biotechnol. Biochem.* **61**, 1615 (1997).
[39] T. Ikami, T. Kakigami, K. Baba, H. Hamajima, T. Jomori, T. Usui, Y. Suzuki, H. Tanaka, H. Ishida, A. Hasegawa, and M. Kiso, *J. Carbohydr. Chem.* **17**, 453 (1998).

TABLE II
[1]C NMR Chemical Shift[a] of Unsulfated and Sulfated Galactopyranosides **20c–24c**

Compound	Chemical shift (and shift relative to **20c**)					
	C-1	C-2	C-3	C-4	C-5	C-6
20c (unsulfated)	105.4	72.8	75.2	70.4	76.6	62.5
21c (3-sulfate)	105.0	70.7	82.2	68.5	76.0	62.2
	(−0.4)	(−2.1)	(+7.0)	(−1.9)	(−0.6)	(−0.3)
22c (3,6-disulfate)	104.3	69.9	81.3	67.2	72.6	66.3
	(−1.1)	(−2.9)	(+6.1)	(−3.2)	(−4.0)	(+3.8)
23c (3,4,6-trisulfate)	104.1	70.0	78.6	75.8	72.5	67.3
	(−1.3)	(−2.8)	(+3.4)	(+5.4)	(−4.1)	(+4.8)
24c (2,3,4,6-terasulfate)	102.6	76.3	76.9	76.1	72.6	67.4
	(−2.8)	(+3.5)	(+1.7)	(+5.7)	(−4.0)	(+4.9)

[a] Recorded in CD_3OD with TMS as an internal standard.

and synthesized. In particular, 2-O-fucosyl sulfatide containing 2-(tetrade-cyl)hexadecyl residue (**32**), which was designed with a view to obtaining more potent E-selectin, was significantly more potent than sialyl Lewis X and the corresponding nonfucosylated sulfatide (**33**) during the blocking adhesion to the P- and L-selectins and was most potent of all the synthesized sulfatide mimetics. Contrary to our expectations, the synthesized 2-O-fuco-syl sulfatide was less potent toward E-selectin.

Synthesis of 2-O-Fucosyl Sulfatide (**32**) (Scheme 3)

2-(Tetradecyl)hexadecyl β-D-galactopyranoside (**34**)[33,34] was selected as the starting material. Treatment of **34** with benzaldehyde dimethyl acetal in THF containing H_2SO_4 affords the 4,6-O-benzylidene derivative **35** in 77% yield. Regioselective 3-O-4-methoxybenzylation of **35** is achieved by treatment of the corresponding stannylene intermediates with 4-meth-oxybenzyl chloride to give **36**, exclusively in 74% yield. The H-3 proton in the ^{1}H NMR spectra of **36** appears at δ 3.47 ($J_{2,3}$ = 9.9, $J_{3,4}$ = 3.5 Hz), indicating that the 4-methoxybenzylated position is O-3. Glycosylation of **36** with **37**[40] is performed in the presence of N-iodosuccinimide (NIS), trifluoromethanesulfonic acid (TfOH), and molecular sieves 4 Å (MS-4A) in toluene for 1 hr at $-20°$ to afford the corresponding disaccharide **38** in 82% yield. In the ^{1}H NMR spectrum of **38**, a significant one-proton doublet appears at δ 5.68 ($J_{1,2}$ = 2.0 Hz, H-1 of Fuc), showing the characteristic of the α-L-fucopyranosyl unit. Selective removal of the 4-methoxybenzyl groups at O-3 in **38** in a dichloromethane–water solution in the presence of 2,3-dichloro-5,6-dicyanobenzoquinone (DDQ) for 1–3 hr at room temperature gives **39**. The sulfation of compound **39** with a sulfur trioxide–pyridine complex in DMF for 2–4 hr at room temperature affords the corresponding sulfate **40** in 93% yield. Reductive removal of the benzylidene and benzyl groups in **40** and sequential treatment by a cation-exchange resin affords the desired, sulfated 2-O-α-L-fucopyranosyl β-D-galactopyranoside **32** in 67% yield. FABMS spectra (negative ion mode) show the base peaks at m/z (M-Na)$^-$ 825.6.

Current approaches for the *in vivo* blocking of selectin-dependent in-flammatory reactions have shown that sialyl Lewis X derivatives are highly protective against reperfusion injuries,[41] traumatic shock,[42] and acute lung

[40] S. Komba, H. Ishida, M. Kiso, and A. Hasegawa, *Bioorg. Med. Chem.* **4,** 1833 (1996).
[41] B. Michael, S. W. Andrew, Z. Zhongli, C. G. Federico, J. F. Michael, and M. L. Allan, *J. Clin. Invest.* **93,** 379 (1994).
[42] C. Skurk, M. Buerke, J.-P. Guo, J. Paulson, and A. M. Lefer, *Am. J. Physiol.* **267,** H2124 (1994).

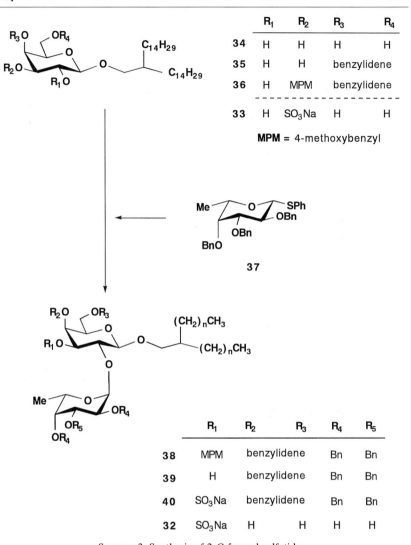

	R$_1$	R$_2$	R$_3$	R$_4$
34	H	H	H	H
35	H	H	benzylidene	
36	H	MPM	benzylidene	
33	H	SO$_3$Na	H	H

MPM = 4-methoxybenzyl

	R$_1$	R$_2$	R$_3$	R$_4$	R$_5$
38	MPM	benzylidene		Bn	Bn
39	H	benzylidene		Bn	Bn
40	SO$_3$Na	benzylidene		Bn	Bn
32	SO$_3$Na	H	H	H	H

Scheme 3. Synthesis of 2-O-fucosyl sulfatide.

injury[43] in many animal models. Similarly, a sulfo-Lewis X analog has been found to have potent anti-inflammatory activity against an IgE-mediated

[43] M. S. Mulligan, J. B. Lowe, R. D. Larsen, J. Paulson, Z.-L. Zheng, S. DeFrees, K. Maemura, M. Fukuda, and P. A. Ward, *J. Exp. Med.* **178,** 623 (1993).

skin reaction in mouse.[44] It was also demonstrated that sulfatide and its synthetic analogs are highly protective against immune complex-induced and neutrophil-mediated lung injury in rats.[16,17]

Although there are many questions regarding selectin ligands and their binding modes, a variety of approaches, including the molecular basis for the ligand–selectin complex formation using computer modeling techniques,[35,45] for the development of therapeutic agents based on carbohydrates as binding blockers are now in progress.

Sialic Acid-Dependent Immunoglobulin Family Lectin (SIGLEC)

Sialoadhesins are a structurally and functionally related family consisting of five immunoglobulin superfamily lectins (I-type lectins), including the myelin-associated glycoprotein (MAG), Schwann cell myelin protein (SMP), the B-cell antigen CD22, the marker of myelomonocytic cells CD33, and sialoadhesin.[5,6,46,47] Each sialoadhesin recognizes the carbohydrate chain containing the sialic acid at the nonreducing terminal. Among them, MAG, a quantitatively minor protein constituent of the myelin of both the central (1%) and peripheral nervous systems (0.1%), mediates certain myelin–neuron cell–cell interactions.[5,6] MAG is also thought to be a major inhibitor of neurite outgrowth (axon regeneration) in the adult central nervous system. Gangliosides are the most abundant sialylated glycoconjugates in nerve tissue, representing about 80% of total conjugated sialic acid in brains of various vertebrates from chickens to humans.[48] In the brain, O-linked glycoproteins are quantitatively minor compared to N-linked glycoproteins.[49] Therefore, for MAG, a brain lectin that prefers "3-O" type sialic acid linkages, gangliosides are the most abundant potential targets.

The ability of gangliosides to support MAG binding was tested using a model system in which full-length MAG was expressed transiently on the surface of an otherwise nonadhesive cell line. It has been found that MAG binds the best to α2,3-linked sialic acid on a β-D-Gal-(1-3)-β-D-GalNAc core structure, which is often carried on gangliosides in the nervous

[44] Y. Wada, T. Saito, N. Matsuda, H. Ohmoto, K. Yoshino, M. Ohashi, H. Kondo, H. Ishida, M. Kiso, and A. Hasegawa, *J. Med. Chem.* **39**, 2055 (1996).

[45] H. Tsujishita, Y. Hiramatsu, N. Kondo, H. Ohmoto, H. Kondo, M. Kiso, and A. Hasegawa, *J. Med. Chem.* **40**, 362 (1997).

[46] P. R. Crocker, S. D. Freeman, S. Gordon, and S. Kerm, *J. Clin. Invest.* **95**, 635 (1995).

[47] S. D. Freeman, S. Kerm, E. K. Barker, and P. R. Crocker, *Blood* **85**, 2005 (1995).

[48] G. Tettamanti, F. Bonali, S. Marchesini, and V. Zambotti, *Biochim. Biophys. Acta* **296**, 160 (1973).

[49] R. K. Margolis and R. U. Margolis, *Biochim. Biophys. Acta* **304**, 421 (1973).

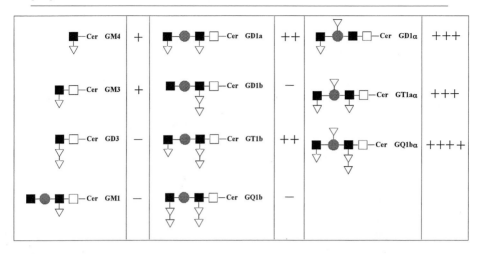

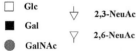

FIG. 4. Structure–function studies of MAG-mediated cell adhesion to gangliosides.[53]

system (Fig. 4).[50–53] In our continuing efforts on the chemical synthesis and structural determination of ganglioside ligands of cell-adhesion molecules, we have reported the first total synthesis of α series gangliosides, GT1aα **(41)**[54] and GQ1bα **(42)**[55] (Fig. 5). Structure–function studies of MAG-mediated cell adhesion to the synthetic ganglioside demonstrated that the extremely high potency of GQ1bα as ligands of MAG is nearly 10-fold higher than that of GT1b.[50]

Synthesis of Ganglioside GT1aα **(41)** (Scheme 4)

For the synthesis of GT1aα **(41)**, we selected the suitably protected sialyl α-(2-6)-gangliotriose **43** as a key glycosyl acceptor, which had served

[50] L. J.-S. Yang, C. B. Zeller, N. L. Shaper, M. Kiso, A. Hasegawa, R. E. Shapiro, and R. L. Schnaar, *Proc. Natl. Acad. Sci. U.S.A.* **93**, 814 (1996).
[51] B. E. Collins, L. J.-S. Yang, G. Mukhopadhyay, M. T. Filbin, M. Kiso, A. Hasegawa, and R. L. Schnaar, *J. Biol. Chem.* **272**, 1248 (1997).
[52] B. E. Collins, M. Kiso, A. Hasegawa, M. B. Tropak, J. C. Roder, P. R. Crocker, and R. L. Schnaar, *J. Biol. Chem.* **272**, 16889 (1997).
[53] B. E. Collins, H. Ito, H. Ishida, M. Kiso, and R. L. Schnaar, 19th Int. Carbohydr. Sympo. Abstract Book CK005, San Diego, 1998.
[54] H. Ito, H. Ishida, M. Kiso, and A. Hasegawa, *Carbohydr. Res.* **304**, 187 (1997).
[55] K. Hotta, H. Ishida, M. Kiso, and A. Hasegawa, *J. Carbohydr. Chem.* **14**, 491 (1995).

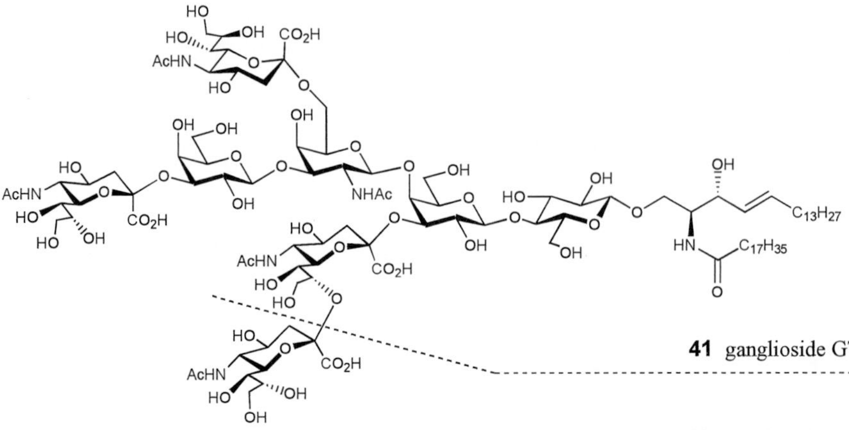

41 ganglioside GT1aα

42 ganglioside GQ1bα

FIG. 5. Ganglioside GT1aα and GQ1bα.

as the intermediate for GQ1bα (**42**).[55] The glycosylation of 2-(trimethylsilyl)ethyl *O*-(methyl 5-acetamido-4,7,8,9-tetra-*O*-acetyl-3,5-dideoxy-D-*glycero*-α-D-galacto-2-nonulopyranosylonate)-(2-6)-*O*-(2-acetamido-2-deoxy-3,4-*O*-isopropylidene-β-D-galactopyranosyl)-(1-4)-*O*-(2,6-di-*O*-benzyl-β-D-galactopyranosyl)-(1-4)-*O*-2,3,6-tri-*O*-benzyl-β-D-glucopyranoside (**43**)[55] with methyl (phenyl 5-acetamide-4,7,8,9-tetra-*O*-acetyl-3,5-dideoxy-2-thio-D-*glycero*-D-*galacto*-2-nonulopyranosid)onate (**44**)[56] in acetonitrile for 72 hr at 15° in the presence of *N*-iodosuccinimide (NIS)-trimethylsilyl trifluoromethanesulfonate (TMSOTf) and powdered molecular sieves 3 Å (MS-3A) give the desired α-glycoside **45** {$[\alpha]_D$ +2.8° (*c* 0.7, $CHCl_3$)} in 40% yield. Removal of the isopropyridene group from **45** with aqueous 80% acetic acid gives the glycosyl acceptor **46** in 85% yield. Glycosylation of **46** with methyl(methyl-5-acetamide-4,7,8,9-tetra-*O*-acetyl-3,5-dideoxy-D-*glycero*-α-D-*galacto*-2-nonulopyranosylonate)-(2-3)-2,4,6-tri-*O*-benzoyl-1-thio-β-D-galactopyranoside (**47**)[57] in dichloromethane for 24 hr at 0° in the presence of dimethyl(methylthio)sulfonium triflate (DMTST) and MS-4A gives the protected GT1aα oligosaccharide **48** {amorphous mass, $[\alpha]_D$ +0.4° (*c* 0.6, $CHCl_3$)} in 95% yield. Hydrogenolytic removal of the benzyl groups in **48** over palladium hydroxide in 9:1 ethanol–acetic acid for 72 hr at 30°, followed by acetylation of the free hydroxyls with acetic anhydride and pyridine for 24 hr at 40°, affords the fully acylated oligosaccharide **49**

[56] A. Hasegawa, H. Ohki, T. Nagahama, H. Ishida, and M. Kiso, *Carbohydr. Res.* **212,** 277 (1991).

[57] A. Kameyama, H. Ishida, M. Kiso, and A. Hasegawa, *Carbohydr. Res.* **200,** 269 (1990).

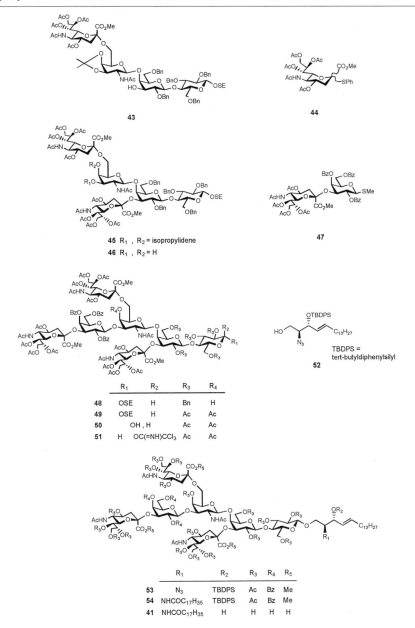

	R_1	R_2	R_3	R_4
48	OSE	H	Bn	H
49	OSE	H	Ac	Ac
50	OH , H		Ac	Ac
51	H	OC(=NH)CCl$_3$	Ac	Ac

	R_1	R_2	R_3	R_4	R_5
53	N$_3$	TBDPS	Ac	Bz	Me
54	NHCOC$_{17}$H$_{35}$	TBDPS	Ac	Bz	Me
41	NHCOC$_{17}$H$_{35}$	H	H	H	H

45 R_1 , R_2 = isopropylidene
46 R_1 , R_2 = H

TBDPS = tert-butyldiphenylsilyl

SCHEME 4. Synthesis of ganglioside GT1aα.

in 81% yield. For the selective removal of the 2-(trimethylsilyl)ethyl group, the fully acylated oligosaccharide **49** is treated with trifluoroacetic acid in dichloromethane for 3 hr at room temperature to give the 1-hydroxy compound **50** in 92% yield, which on further treatment with trichloroaceto-nitrile in the presence of 1,8-diazabicyclo[5.4.0]undec-7-ene (DBU) in di-chloromethane for 0.5 hr at 0° gives the trichloroacetimidate **51** in 65% yield. The ^{1}H NMR spectrum of the trichloroacetimidate contains a one-proton doublet at δ 6.51 ($J_{1,2}$ 3.67 Hz, H-1) and a one-proton singlet at δ 8.7 (C=NH), showing the imidate to be an α anomer. Glycosylation of (2S,3R,4E)-2-azido-3-O-(*tert*-butyldiphenylsilyl)-4-octadecene-1,3-diol (**52**)[58] is carried out in the presence of TMSOTf and MS-4A (AW300) for 45 hr at 0° to give the desired β-glycoside **53** {[α]$_D$ −14.3° (c 0.8, CHCl$_3$)} in 62% yield. Selective reduction of the azido group with triphenylphosphine in 5:1 benzene–water gives the amine, which on condensation with stearic acid using 1-(3-dimethylaminopropyl)-3-ethylcarbodiimide hydrochloride (WSC) in dichloromethane gives the fully protected ganglioside GT1aα **54** {[α]$_D$ −2.2° (c 1.2, CHCl$_3$)} in 76% yield. Finally, removal of the *tert*-butyldiphenylsilyl group in **54** with 1.0 M tetrabutylammonium fluoride in acetonitrile, O-deacetylation of **54** with sodium methoxide in methanol for 48 hr at 40° and subsequent saponification of the methyl ester groups afford the ganglioside GT1aα **41** as an amorphous mass in quantitative yield, after chromatography on a column of Sephadex LH-20 with 5:5:1 CHCl$_3$–MeOH–H$_2$O. ^{1}H NMR data of the product thus obtained are consistent with the structure assigned.

Immunosuppressive Activity

Ganglioside GM3 **55** (Fig. 6) was first isolated[59] from horse erythrocytes in 1952 and is the major ganglioside component in erythrocytes of many animal species.[60] Ganglioside GM3 has various types of important biological functions,[61–63] and these activities are strictly related to the structure of sialic acid, oligosaccharide chain, and fatty acyl residue at the ceramide moiety.[7] In particular, gangliosides are potent inhibitors of cellular immune responses, and the structures of gangliosides influence their activity greatly. Studies using chemically synthesized gangliosides GM3, GM4, and their

[58] M. Mori, Y. Ito, and T. Ogawa, *Carbohydr. Res.* **195,** 199 (1990).
[59] T. Yamakawa and S. Suzuki, *J. Biochem.* **39,** 383 (1952).
[60] G. J. M. Hooghwinkel, P. F. Barri, and G. W. Bruyn, *Neurology* **16,** 934 (1966).
[61] Y. Suzuki, M. Matsunaga, and M. Matsumoto, *J. Biol. Chem.* **260,** 1362 (1985).
[62] H. Nojiri, F. Takaku, Y. Terui, Y. Miura, and Y. Saito, *Proc. Natl. Acad. Sci. U.S.A.* **83,** 782 (1986).
[63] S. Ladisch, A. Hasegawa, R. Li, and M. Kiso, *Biochemistry* **34,** 1197 (1995).

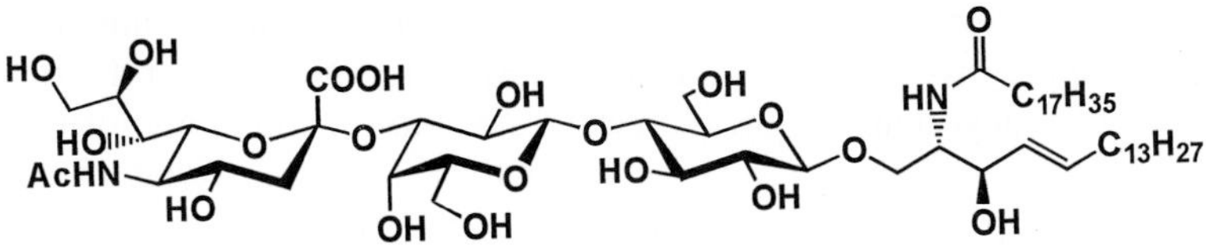

55 GM3

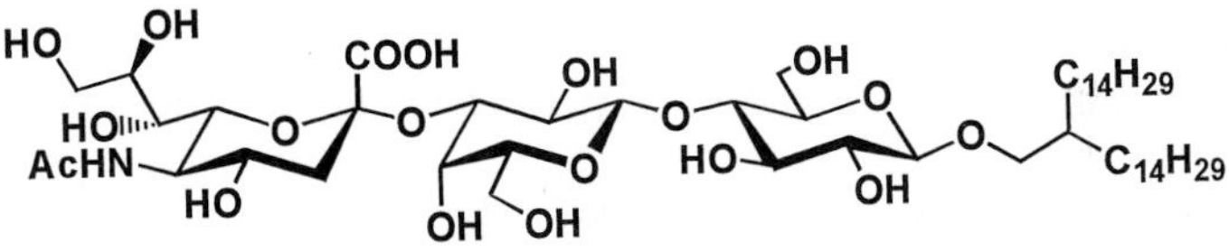

56

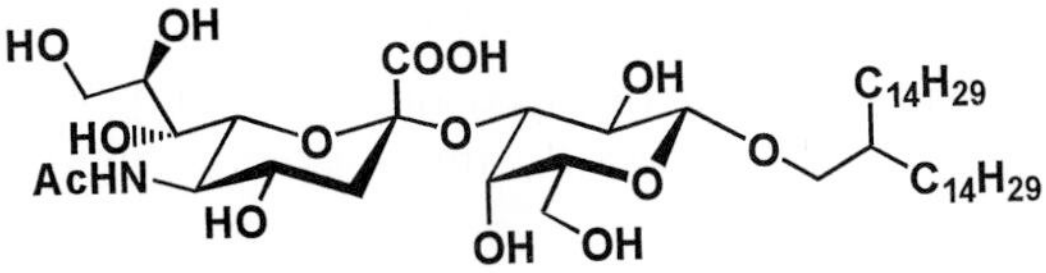

57

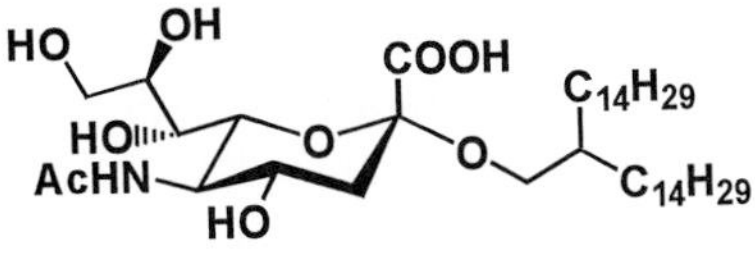

58

FIG. 6. Ganglioside GM3 and synthetic ganglioside analogs.

analogs have shown that the influence of certain structure details of their immunosuppressive activities are as follows: (1) an inverse relationship between fatty acyl chain length at the ceramide moiety and immunosuppressive activity is an important factor for activity, (2) hydroxylation of the fatty acyl group decreases immunosuppressive activity, (3) substitution of an *S*-glycosidic linkage for an *O*-glycosidic linkage in sialic acid linkage does not alter its activity, and (4) modification of the sialic acid structure variably influences activity, as KDN (3-deoxy-D-*glycero*-D-*galacto*-2-nonulopyranosonic acid)-GM3 and -GM4 analogs retain potent activity, whereas other modifications such as 8-epi-GM3 and 9-deoxy-GM3 reduce immunosuppressive activity.[7,63] In view of these facts, we have systematically synthesized gangliosides GM3 and GM4 analogs and 1-*O*-sialyl ceramide derivatives containing the 2-(tetradecyl)hexadecyl residue in place of a ceramide (**56**, **57**, and **58**)[64–66] to examine in finer detail the intact core structure (Fig. 6).

Immunosuppressive activities of the chemically synthesized gangliosides GM3 **56** and GM4 **57** and the minimal structural unit of ganglioside **58** were assessed in both an antigen-induced human lymphocyte proliferation assay and an established murine model of the cellular allogeneic immune response.[7,63] Results have shown that all three compounds have significant and comparable immunosuppressive activity *in vitro* and *in vivo*. Importantly, these studies suggest that (1) the intact ceramide structure is not a requirement and (2) synthetic glycoconjugates may be potent immunosuppressive agents (S. Ladisch *et al.*, unpublished results).

Synthesis of Ganglioside GM3 and GM4 Analogs Containing 2-(Tetradecyl)hexadecyl Residue in Place of a Ceramide (56, 57) (Scheme 5)

For the synthesis of desired ganglioside GM3 and GM4 analogs, we employed *O*-(methyl 5-acetamido-4,7,8,9-tetra-*O*-acetyl-3,5-dideoxy-D-*glycero*-α-D-*galacto*-2-nonulopyranosylonate)-(2-3)-*O*-(2,4-di-*O*-acetyl-6-*O*-benzoyl-β-D-galactopyranosyl)-(1-4)-3-*O*-acetyl-2,6-di-*O*-benzoyl-α-D-glucopyranosyl trichloroacetimidate (**59**)[67] and *O*-(methyl 5-acetamido-4,7,8,9-tetra-*O*-acetyl-3,5-dideoxy-D-*glycero*-α-D-*galacto*-2-nonulopyranosylonate)-(2-3)-2,4,6-tri-*O*-benzoyl-α-D-galactopyranosyl trichloroacetimi-

[64] M. Kiso, A. Nakamura, and A. Hasegawa, *J. Carbohydr. Chem.* **6**, 411 (1987).
[65] A. Hasegawa, N. Suzuki, H. Ishida, and M. Kiso, *J. Carbohydr. Chem.* **15**, 623 (1996).
[66] A. Hasegawa, N. Suzuki, F. Kozawa, H. Ishida, and M. Kiso, *J. Carbohydr. Chem.* **15**, 639 (1996).
[67] T. Murase, H. Ishida, M. Kiso, and A. Hasegawa, *Carbohydr. Res.* **188**, 71 (1989).

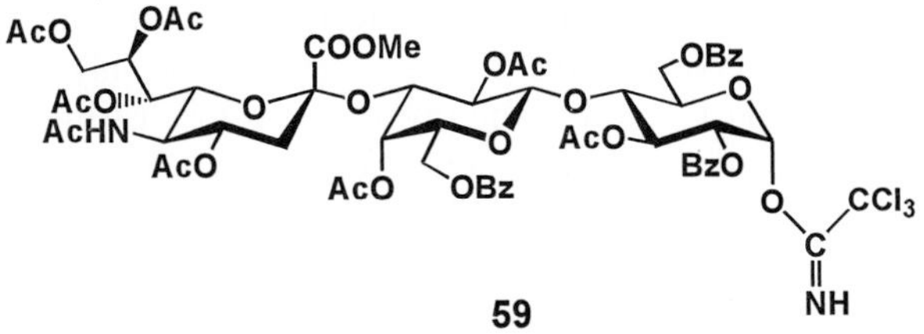

59

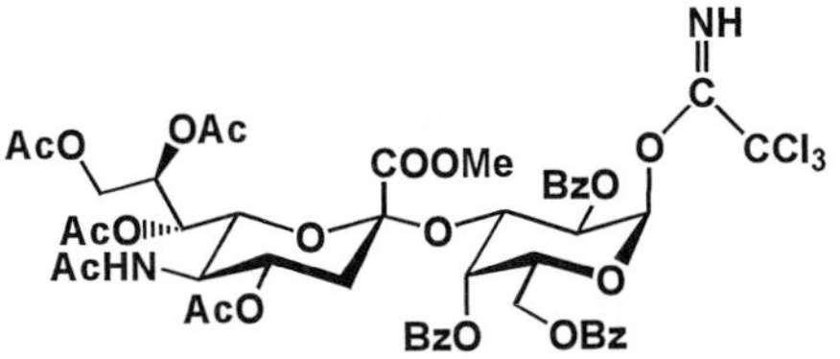

60

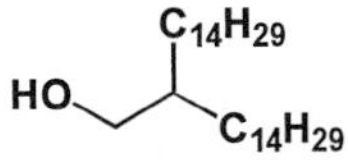

61

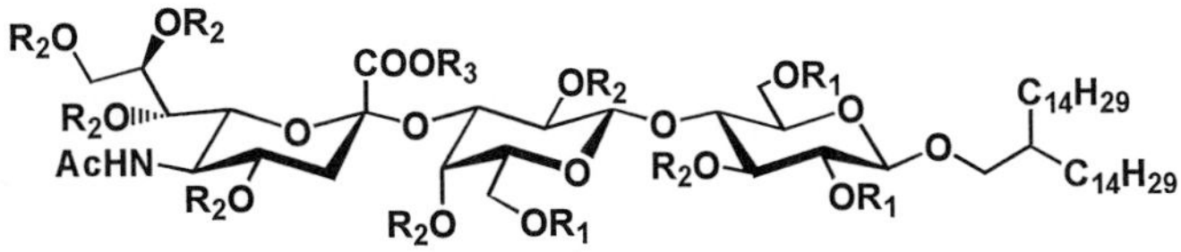

62 R_1 = Bz, R_2 = Ac, R_3 = Me

56 $R_1 = R_2 = R_3$ = H

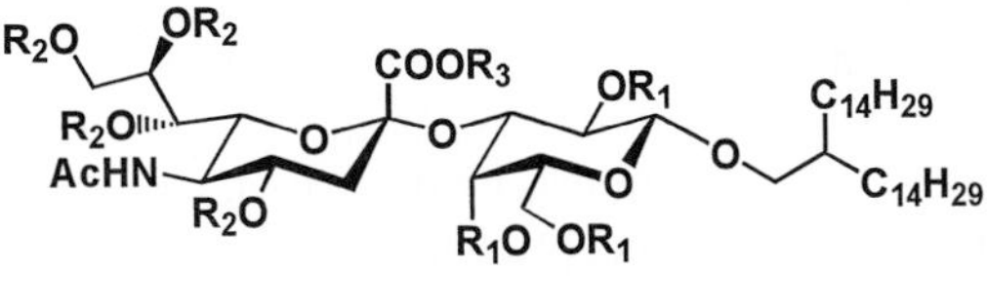

63 R_1 = Bz, R_2 = Ac, R_3 = Me

57 $R_1 = R_2 = R_3$ = H

SCHEME 5. Synthesis of ganglioside GM3 and GM4 analogs containing 2-(tetradecyl)hexadecyl residue in place of ceramide.

date (**60**)[68] as the glycosyl donors and 2-(tetradecyl)hexadecan-1-ol prepared from corresponding fatty acid as the glycosyl acceptor. The glycosylation of **59** and **60** with **61** is performed in the presence of boron trifluoride etherate as a promoter and powered molecular sieves 4 Å [MS-4A(AW-300)] in dry dichloromethane overnight at 0° to afford exclusively the β-glycosides **62** (89%) and **63** (84%), respectively. Significant signals in ^{1}H NMR spectra of **62** and **63** are one-proton doublets at δ 4.59 ($J_{1,2}$ = 8.1 Hz, H-1 of Glc in **62**) and 4.88 ($J_{1,2}$ = 8.1 Hz, H-1 of Gal in **63**), showing characteristics of the β-glycosidic linkage. *O*-Deacylation of **62** and **63** with sodium methoxide in methanol, with subsequent saponification of the sialate methyl ester group, yields the desired products **56** and **57**, both in quantitative yields.

[68] T. Murase, A. Kameyama, K. P. R. Kartha, H. Ishida, M. Kiso, and A. Hasegawa, *J. Carbohydr. Chem.* **8,** 265 (1989).

[47] Sphingolipid Photoaffinity Labels

By Friederike Knoll, Thomas Kolter, and Konrad Sandhoff

I. Glycosphingolipids and Their Metabolic Products

Glycosphingolipids (GSLs) are components of eukaryotic plasma membranes.[1] They are anchored in the lipid bilayer by a hydrophobic ceramide moiety and expose a hydrophilic oligosaccharide chain into the extracellular space. Variations in the type, number, and linkage of sugar residues in the oligosaccharide chain give rise to the wide range of naturally occurring GSLs. These form cell-type-specific patterns at the cell surface that change with cell growth, differentiation, viral transformation, and oncogenesis.[2] At the cell surface, GSLs can interact with toxins, viruses, and bacteria, as well as with membrane-bound receptors and enzymes.[3] They are also involved in cell-type-specific adhesion processes. In addition, lipophilic products of GSL metabolism, such as sphingosine-1-phosphate and ceramide, play a role in signal transduction events.[4] Finally, GSLs form a protective layer on biological membranes, protecting them from inappropriate degradation.

[1] H. Wiegandt, *in* "New Comprehensive Biochemistry" (A. Neuberger and L. L. M. van Deenen, eds.), p. 199. Elsevier, 1985.
[2] S. Hakomori, *Annu. Rev. Biochem.* **50,** 733 (1981).
[3] K.-A. Karlsson, *Annu. Rev. Biochem.* **58,** 309 (1989).
[4] L. Riboni, P. Viani, R. Bassi, A. Prinetti, and G. Tettamanti, *Prog. Lipid Res.* **36,** 153 (1997).

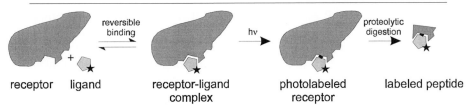

SCHEME 1. Formation of a covalent bond between a receptor and its ligand using labeling techniques.

Sphingolipid biosynthesis[5] starts with the formation of ceramide at the membranes of the endoplasmic reticulum (ER). In the Golgi apparatus, ceramide is modified by the addition of a phosphorylcholine moiety to form sphingomyelin or by the stepwise addition of sugar residues and sialic acid to form glycosphingolipids. Products of these reactions are transported to the plasma membrane by exocytosis.

GSL degradation occurs in the acidic compartments of the cell, the endosomes, and the lysosomes.[6] Fragments of the plasma membrane containing GSLs destined for degradation are endocytosed and reach the lysosomes, presumably on the surface of intralysosomal vesicles or related membrane structures. Hydrolyzing enzymes sequentially cleave off the sugar residues to produce ceramide, which is deacylated to sphingosine. This can leave the lysosome and reenter the biosynthetic pathway or can be degraded further.

II. Labeling of (Glyco)sphingolipid-Binding Proteins

Different experimental approaches can be used for the detailed analysis of sphingolipid metabolism and of recognition phenomena on the molecular level. For the determination of active sites of enzymes or the ligand-binding sites of receptors, several methods are suitable. Among them is the crystallographic analysis of receptor–ligand complexes, spectroscopic analysis,[7] chemical cross-linking, affinity labeling, and photoaffinity labeling. The aim of these labeling techniques is the formation of a covalent bond between a receptor and its ligand (Scheme 1). The covalently labeled receptor obtained from the labeling experiment can be detected if an appropriate label was introduced into the ligand prior to the labeling event. Most conveniently, this can be achieved by the incorporation of a radioisotope

[5] G. van Echten and K. Sandhoff, *J. Biol. Chem.* **268,** 5341 (1993).
[6] K. Sandhoff and T. Kolter, *Trends Cell Biol.* **6,** 98 (1996).
[7] S. B. Shuker, P. J. Hajduk, R. P. Meadows, and S. W. Fesik, *Science* **274,** 1531 (1996).

such as ^{14}C, ^{3}H, ^{35}S, ^{32}P, or ^{125}I into the ligand. The choice of the radioisotope is dependent on the required specific radioactivity and of the availability of a suitable labeling procedure. Due to minimal disturbance of the native structure, radiolabeling is superior to most other labeling techniques, e.g., fluorescence.

This article gives an overview of photoaffinity ligands that have been used for the identification of (glyco)sphingolipid receptors or for the analysis of (glyco)sphingolipid metabolism. This includes the determination of binding sites within sphingolipid-binding proteins. Applications and some basic problems associated with this approach are discussed.

III. Labeling Techniques

Labeling techniques replace weak and/or reversible interactions between two components by a stable covalent bond. This approach can be used to identify an unknown receptor for a distinct ligand. In cases where the receptors are already known and available in pure or enriched form, the binding site for the ligand can also be analyzed on the molecular level.

A. Cross-Linking

Assemblies or complexes of protein subunits, and also receptor–ligand complexes, can be analyzed by chemical cross-linking.[8] Covalent bridges, e.g., between neighboring subunits, are formed. They are not necessarily a result of affinity between domains but can be a product of immediate vicinity. Cross-linking employs bifunctional reagents that usually modify amino acid side chains of proteins. The bifunctional reagents are classified according to their chemical specificity, the length of the cross-bridge formed; whether both reactive cross-linking groups are the same (homobifunctional and heterobifunctional); whether the groups react chemically or are photochemically activated; and whether the reagents contain a cleavable bond. Heterobifunctional reagents have one functional group that is first allowed to react chemically with lysine or cysteine residues on a peptide or protein. Subsequently, the modified component is bound to the complex and a photolabile group is irradiated to trigger a reaction with a neighboring component, usually a protein. Homobifunctional and heterobifunctional reagents can also be used to achieve stable linkages between a ligand and its receptor. This requires a chemical reaction between suitable functional groups on receptor and ligand with complementary reactive sites present on

---

[8] R. R. Traut, C. Casiano, and N. Zecherle, in "Protein Function: A Practical Approach" (T. E. Creighton, ed.), p. 101. IRL Press at Oxford Univ. Press, Oxford, 1991.

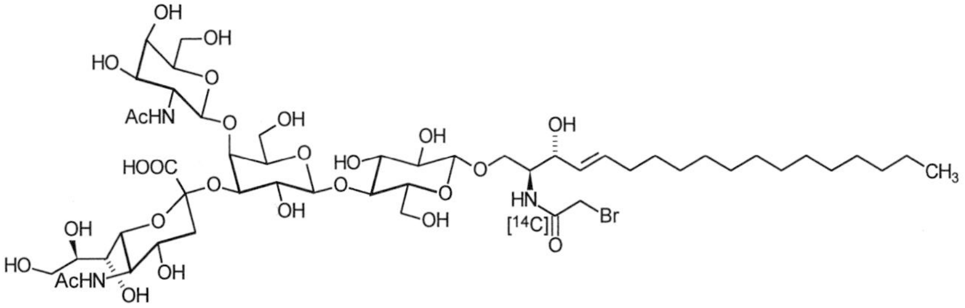

(1)

Fig. 1. N-(Bromo-[1-^{14}C]acetyl)lyso-GM2 (**1**).

the cross-linker. Stepwise cross-linking by heterobifunctional cross-linkers reduces the occurrence of unwanted side reactions such as homoprotein polymers. Usually, heterobifunctional cross-linkers contain a primary amine reactive group, e.g., N-hydroxysuccinimide (NHS), or its water-soluble analog N-hydroxysulfosuccinimide (sulfo-NHS).

B. Affinity Labeling

Affinity labeling refers to a variety of methods developed to analyze both structural and functional properties of biological targets. The method is based on the fact that the binding of most biological ligands to their specific receptor site involves a number of favorable interactions that together make up the ligand–receptor recognition process. The receptor-binding site may be specifically occupied by an analog of its natural ligand into which a chemically reactive group is incorporated. The modified ligand interacts with the binding site of the receptor and forms a stable covalent bond with the receptor, usually by electrophilic attack at or near the binding site. This approach has also been applied to the identification of binding proteins or binding regions of proteins that interact with sphingolipids. For example N-(bromo-[1-^{14}C]acetyl)lyso-GM2 (**1**) (Fig. 1) was synthesized as an affinity ligand that inactivated the purified GM2 activator protein by a covalent modification.[9] The GM2 activator protein is a ganglioside-binding protein and stimulates the hydrolysis of ganglioside GM2 by the enzyme β-hexosaminidase A.[10] Amino acid residues near or within the lipid-binding site might be identified by this method after enzymatic degradation of the protein and the analysis of labeled fragments.

[9] S. Neuenhofer and K. Sandhoff, FEBS Lett. **185,** 112 (1985).
[10] E. Conzelmann and K. Sandhoff, in "Methods of Biochemical Analysis" (V. Ginsburg, ed.), p. 792. Academic Press, Orlando, 1987.

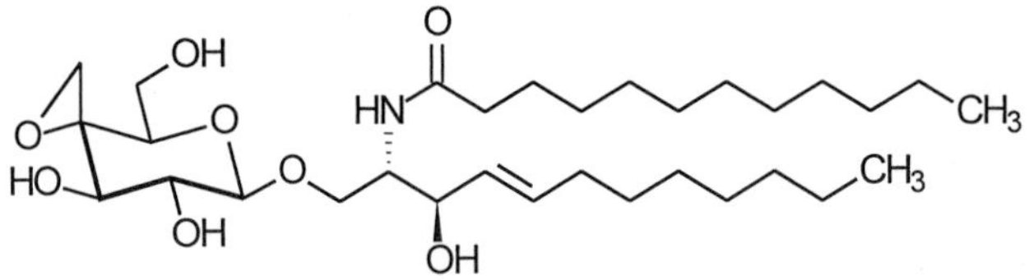

(2)

Fig. 2. A synthetic GlcCer derivative with an epoxide group in the 4-position of the glucose residue (**2**).

Affinity labeling requires the introduction of a functionality of intermediate reactivity. Several examples indicate an epoxide function as an appropriate structural element. Here, the reactivity of the oxirane decreases with the degree of substitution by electron-withdrawing groups.[11] A synthetic GlcCer derivative with an epoxide group in the 4-position of the glucose residue (**2**) (Fig. 2)[12] causes an irreversible and concentration-dependent decrease of LacCer synthase activity in cultured cells.[13] In primary cultured neurons from chick embryos, biosynthetic labeling of GSLs downstream of GlcCer is reduced with label accumulating in GlcCer. The corresponding galacto derivative shows no effect. Inhibition of LacCer formation is much less pronounced *in vitro* (30% inhibition at 250 μM concentration). Therefore, it cannot be excluded that the effects observed *in vivo* are due to inhibition of a GlcCer translocator or a transcriptional factor.

Also, affinity labeling with monosaccharide derivatives can be useful in the determination of GSL recognition sites. Conduritol B-epoxide (**3**) (Fig. 3), an irreversible inhibitor of glucocerebrosidase,[14,15] has been used in the identification of the enzyme's active site. The amino acid residue Asp-443 has been labeled by this approach.[16] A contribution of this amino acid side chain to the active site, however, was excluded by site-directed mutagenesis.[17] However, Glu-340, which has been labeled by 2-deoxy-2-fluorogluco-

---

[11] Z. Yu, P. Caldera, F. McPhee, J. J. DeVoss, P. R. Jones, A. L. Burlingame, I. D. Kuntz, C. S. Craig, and P. R. O. de Montanello, *J. Am. Chem. Soc.* **118**, 5846 (1996).

[12] M. Plewe, K. Sandhoff, and R. R. Schmidt, *Liebigs Ann. Chem.* 699 (1992).

[13] C. Zacharias, G. van Echten, M. Plewe, R. R. Schmidt, and K. Sandhoff, *J. Biol. Chem.* **269**, 13313 (1994).

[14] G. Legler, *Biochim. Biophys. Acta* **151**, 728 (1968).

[15] G. Legler, *in* "Carbohydrate Mimics" (Y. Chapleur, ed.), p. 463. Wiley-VCH, Weinheim, 1998.

[16] T. Dinur, K. M. Osiecki, G. Legler, S. Gatt, R. J. Desnick, and G. A. Grabowski, *Proc. Natl. Acad. Sci. U.S.A.* **83**, 1660 (1986).

[17] M. E. Grace, K. M. Newman, V. Cheinker, A. Berg-Fussmann, and G. A. Grabowski, *J. Biol. Chem.* **269**, 2291 (1994).

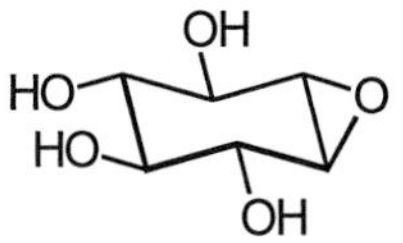

(3)

Fig. 3. Conduritol B-epoxide (**3**).

pyranose, obviously represents one of the active site amino acids.[18] This indicates that (photo)affinity reagents derived from GSL substructures can be employed and that the results obtained should be confirmed by an independent method.

Although the three examples just described illustrate the application of classical electrophilic labeling reagents, two major limitations have to be considered regarding the affinity labeling concept.[19,20] First, the ligand-binding sites do not necessarily contain reactive nucleophilic centers. Second, the electrophilic labels can react with other cellular components or with nucleophiles not located in the binding site, which leads to unspecific labeling. In principle, these limitations of classical chemical affinity labeling can be circumvented to some extent by the use of photoaffinity labeling.

C. Photoaffinity Labeling

Several review articles have appeared on this issue.[19–31] Photoaffinity labeling represents a special category of general affinity labeling that has been developed to overcome these two major limitations of classical electrophilic labeling. It requires the modification of the natural ligand by the

[18] S. Miao, J. D. McCarter, M. E. Grace, G. A. Grabowski, R. Aebersold, and S. G. Withers, *J. Biol. Chem.* **260,** 10975 (1994).

[19] R. J. Guillory, *Pharmacol. Ther.* **41,** 1 (1989).

[20] H. Bayley and J. R. Knowles, *Methods Enzymol.* **46,** 69 (1977).

[21] J. R. Knowles, *Acc. Chem. Res.* **5,** 155 (1972).

[22] K. Peters and F. M. Richards, *Annu. Rev. Biochem.* **46,** 523 (1977).

[23] T. H. Ji, *Biochim. Biophys. Acta* **559,** 39 (1977).

[24] V. Chowdhry and F. H. Westheimer, *Annu. Rev. Biochem.* **48,** 292 (1979).

[25] H. Bayley, *in* "Laboratory Techniques in Biochemistry and Molecular Biology" (T. S. Work and R. H. Burdon, eds.). Elsevier, Amsterdam, 1983.

[26] J. Brunner, *Methods Enzymol.* **172,** 628 (1989).

[27] B. J. Gaffney, *Biochim. Biophys. Acta* **822,** 289 (1985).

[28] J. Brunner, *Annu. Rev. Biochem.* **62,** 483 (1993).

[29] F. Kotzyba-Hibert, I. Kapfer, and M. Goeldner, *Angew. Chem. Int. Ed.* **107,** 1391 (1995).

[30] S. A. Fleming, *Tetrahedron* **51,** 12479 (1995).

[31] J. Brunner, *Trends Cell Biol.* **6,** 154 (1996).

incorporation of a chemically inert but photochemically labile functionality. On irradiation, the photolabile group is converted into a species of very high chemical reactivity that reacts with hydrophobic or polar residues of the receptor with concomitant formation of a covalent bond.

IV. Principles of Photoaffinity Labeling

Criteria of an ideal photoaffinity reagent are as follows:

it should be synthesized easily

it should be chemically inert; stable in the dark and in biological systems under the conditions used for the labeling procedure

the photochemically derived species should be highly reactive and of very short half-life

no intramolecular rearrangements should occur after photolysis

it should be recognized by the receptor or the enzyme of interest and bind to it with sufficient affinity. Photolabels with bulky functional groups might change binding specificity of the ligand and should therefore be avoided. Analogs of enzyme substrates should not be cleaved by enzymatic reaction

activation of the photoprobe should be accomplished at wavelengths remote from protein-damaging regions

reactive species should form a stable covalent bond with the receptor. Stability should be maintained during subsequent chemical workup

photoreaction and covalent insertion should be independent of a particular chemical function at the binding site

a suitable label should be incorporated at an appropriate position of the reagent

Most of the sphingolipid-based photoaffinity reagents are precursors of nitrenes or carbenes. These are isoelectronic species that can occur in the singlet or triplet state. They are highly electrophilic due to the presence of only six valence electrons. Singlet species add to carbon–carbon double bonds, insert into carbon–hydrogen bonds, or react with free electron pairs on heteroatoms, also with the solvent. Dependent on their structure, they can undergo rearrangement reactions, e.g., Wolff rearrangement. Triplet species are formed from electronically excited precursors and undergo radical attack on π, σ electrons, or free electron pairs. After abstraction of a hydrogen atom from the receptor, recombination of the thereby generated radicals leads to the formation of a covalent bond. Triplet species are usually inert to water and rearrange only after intersystem crossing, but oxygen should be excluded in the labeling experiment.

Nitrene precursors most often used in the synthesis of sphingolipid photoaffinity labels are azides, especially arylazides, as alkyl nitrenes generated from alkylazides undergo Wolff rearrangement to imines. Wavelengths of about 250 nm are used for the photolysis of arylazides; the molar extinction coefficient is in the range of $10^4 M^{-1}$ cm^{-1}.

Reactivity of the reactive species generated from arylazides can vary considerably (up to 10^6-fold) between amino acid side chains, with cysteine residues being the most reactive.[29] It can be assumed that intramolecular rearrangement of the initially formed nitrene to electrophilic species such as didehydroazepines or benzazirines occurs prior to labeling. The formation of didehydroazepines is suppressed by arylazides substituted with electron-withdrawing groups. Especially iodo or nitro substituents significantly reduce the yield of didehydroazepines that are formed on irradiation.[32] The substitution also causes a shift of wavelength needed for photoactivation (about 360 nm). Arylazides are not suitable in the presence of dithiothreitol or mercaptoethanol due to azide reduction. In the presence of oxygen, oxidation of the azide to a nitro group might occur. Carbene precursors used most commonly for sphingolipid-based photoaffinity labeling are diazirines,[33] especially trifluoromethyl diazirines.[28,31] The latter share favorable photochemical properties (excitation wavelength of about 360 nm) and sufficient high chemical stability. A side reaction that occurs on irradiation is the rearrangement to the corresponding diazo compounds that can give rise to undesired alkylation reactions in the absence of light. This side reaction is suppressed by the trifluoromethyl group. The diazo compounds themselves can also be used as carbene precursors in photoaffinity labeling experiments. They are photolyzed with light wavelengths of 240 to 280 nm ($\varepsilon = 10^4 M^{-1}$ cm^{-1}). A side reaction is the Wolff rearrangement to ketenes, which might subsequently react with nucleophiles.

V. Applications

Ganglioside nomenclature used throughout the text is according to Svennerholm.[34,35] To investigate details of sphingolipid function and metabolism, photoaffinity ligands have been used for the identification of sphingolipid-binding proteins and for the analysis of the binding sites of various proteins (toxins, glycosidases). According to the position of photolabile functionality, the ligands can be divided into those that aim at the carbohydrate-binding site and those directed against the lipid-binding site (Scheme

[32] Y.-Z. Li, J. P. Kirby, M. W. George, M. Poliakoff, and G. B. Schuster, *J. Am. Chem. Soc.* **110**, 8092 (1988).
[33] R. A. G. Smith and J. R. Knowles, *J. Chem. Soc. Perkin II* 686 (1975).
[34] L. Svennerholm, *Prog. Brain Res.* **101**, 391 (1994).
[35] IUPAC Recommendations, *Pure Appl. Chem.* **69**, 2475 (1997).

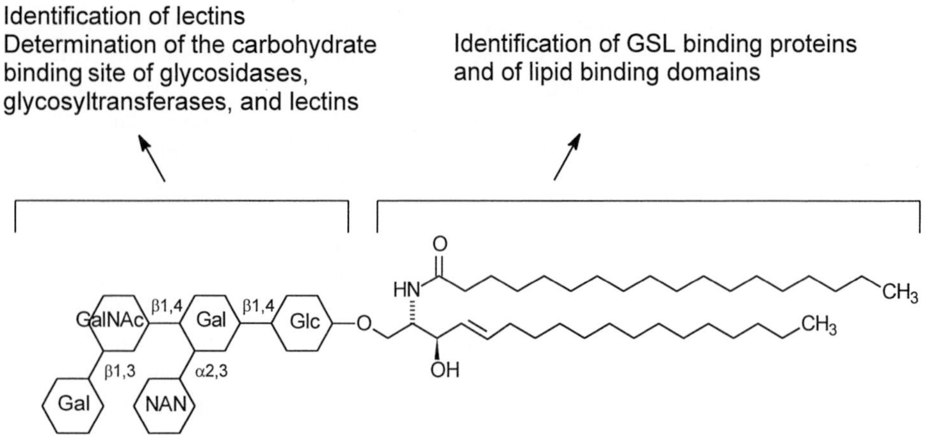

Identification of lectins
Determination of the carbohydrate
binding site of glycosidases,
glycosyltransferases, and lectins

Identification of GSL binding proteins
and of lipid binding domains

SCHEME 2. Use of photoaffinity ligands for the identification of sphingolipid-binding proteins and for the analysis of binding sites of various proteins. According to the position of the photolabile functionality, the ligands can be divided into those that aim at the carbohydrate site and those directed against the lipid-binding site.

2). The latter include chemically modified sphingolipids such as ceramide or sphingomyelin, which contain no carbohydrates.

A. Photoaffinity Labeling of Carbohydrate-Binding Sites

    1. Photoaffinity Labeling of GSL–Carbohydrate-Binding Domains. An example of a photoaffinity reagent containing a nitrene precursor positioned in the carbohydrate moiety of a GSL is [^{125}I]GD1b-derivative (**4**).[36] It has been synthesized to identify the ganglioside-binding domain of tetanus toxin. Tetanus toxins belong to a family of homologous proteins with selective toxicity for vertebrate motor neurons[37] and are believed to gain neural entry by recognition of cell surface glycolipids. A carbohydrate recognition site is suggested to be responsible for toxin–ganglioside binding. Gangliosides containing the "1b" structure (a NeuAc-α2,8-NeuAc unit on the internal galactose residue) were thought to mediate toxin binding.[37–39] The

---

[36] R. E. Shapiro, C. D. Specht, B. E. Collins, A. S. Woods, R. J. Cotter, and R. L. Schnaar, *J. Biol. Chem.* **272,** 30380 (1997).

[37] H. Niemann, *in* "Sourcebook of Bacterial Protein Toxins" (J. E. Alouf and J. Freer, eds.), p. 303. Academic Press, London, 1991.

[38] T. B. Rogers and S. H. Snyder, *J. Biol. Chem.* **256,** 2402 (1981).

[39] N. P. Morris, E. Consiglio, L. D. Kohn, W. H. Habig, M. C. Hardegree, and T. B. Helting, *J. Biol. Chem.* **255,** 6071 (1980).

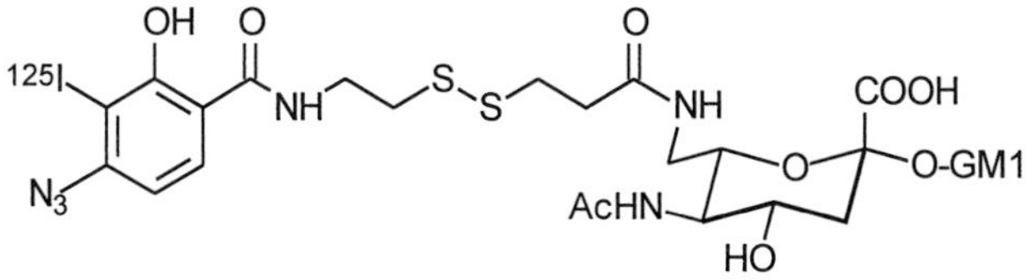

(4)

FIG. 4. [^{125}I]GD1b-derivative (**4**).

toxin photolabeled with (**4**) (Fig. 4) was proteolyzed enzymatically and the radiolabeled peptides purified. A modification at a single residue (His-1293) in the ganglioside-binding domain could be identified by peptide sequencing and mass spectrometry.

A GSL derivative bearing a carbene precursor in the carbohydrate moiety has been developed for photoaffinity labeling of enzymes involved in ganglioside biosynthesis. In order to identify the glycosyl acceptor-binding domains of two glycosyltransferases, (**5**) has been synthesized. It turned out to be an inhibitor of GM2 synthase ($K_i = 0.8 \ \mu M$, $K_m = 62 \ \mu M$) and of GD3 synthase ($K_i = 70 \ \mu M$, $K_m = 270 \ \mu M$).[40] In labeling experiments, only the GD3 synthase was irreversibly inactivated by this reagent. In GM2 synthase, the reactive group liberated on irradiation is presumably exposed to the aqueous environment and not to the active site of the transferase. Because the compound has not been synthesized in radiolabeled form, the analysis of the substrate-binding site of GD3 synthase has not been completed. The synthesis of (**5**) (Fig. 5) is given in the experimental part.

2. *Modified GSL Oligosaccharides in Photoaffinity Labeling.* Regarding the photo probe criteria mentioned earlier, the reagent should contain the recognition element required for binding at the active site. In animal glycolipids, the reducing end of the oligosaccharide moiety is usually linked β-glycosidically to the 1-hydroxyl group of ceramide. Experiments have indicated that the ceramide portion of gangliosides is not essential for all recognition processes with GSL receptors. For instance, the addition of sialylgangliotetraose, the oligosaccharide moiety of ganglioside GM1 (oligo-GM1), to neuroblastoma cells was found to evoke a similar neuritogenic response such as the corresponding GSL.[41,42] To identify proteins with affinity to the oligo-GM1, a radiolabeled (^{125}I), photoactivatable derivative

[40] R. Kaufmann, thesis, University of Bonn, 1992.
[41] J. Nakajima, S. Tsuji, and Y. Nagai, *Biochim. Biophys. Acta* **876,** 65 (1986).
[42] C.-L. Schengrund and C. Prouty, *J. Neurochem.* **51,** 277 (1988).

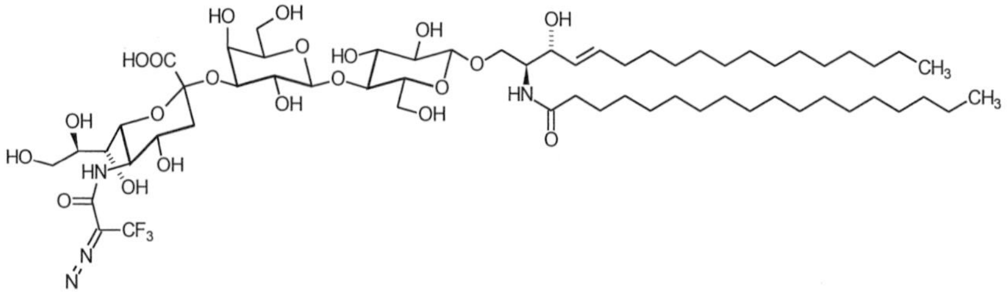

(5)

FIG. 5. A GSL derivative bearing a carbene precursor in the carbohydrate moiety (5).

of oligo-GM1 (6) was prepared.[43] This was accomplished by reductive amination of the glucosyl moiety of oligo-GM1 to the corresponding 1-deoxy-1-aminoglucitol derivative, followed by reaction of the amine with sulfosuccinimidyl 2-(p-azidosalicylamido)ethyl-1′,3′-dithiopropionate (SASD) to (6) (Fig. 6). On photolysis, the precursor labeled a protein with an apparent molecular mass of approximately 71 kDa. In competition experiments, as little as a 10-fold molar excess of oligo-GM1 resulted in a selective reduction in labeling of this protein; preincubation with a 200-fold molar excess of sialyllactose was required to observe the same change in the labeling pattern. These results show that the protein specifically associates with oligo-GM1. Cell surface location of the oligo-GM1-binding protein was confirmed using subcellular fractionation and morphological analyses.

A photoreactive, radioiodinated derivative of the GM1 oligosaccharide (GM1OS) (7) (Fig. 6) was generated from GM1 by ozonolysis and subsequent alkaline fragmentation, followed by reductive amination to the corresponding glycosyl amine.[44] The latter compound was then reacted with N-hydroxysuccinimidyl-4-azidosalicylic acid (NHS-ASA) to form (7), which was radioiodinated and purified further. To test the photoaffinity label as a probe for ganglioside-binding proteins, the derivative was incubated with cholera toxin, which specifically binds GM1, followed by photolysis and SDS–PAGE. The probe only labeled the B subunit of cholera toxin, but not the A subunit. Labeling was inhibited by excess GM1 oligosaccharide, but not by the oligosaccharides derived from gangliosides GD1a and GD1b. GSL structures can be simplified further for labeling purposes. In principle, the carbohydrate structure can be formally reduced via spacer-modified

[43] S. M. Fueshko and C. L. Schengrund, *J. Neurochem.* **59,** 527 (1992).
[44] T. Pacuszka and P. H. Fishman, *Glycobiology* **2,** 251 (1992).

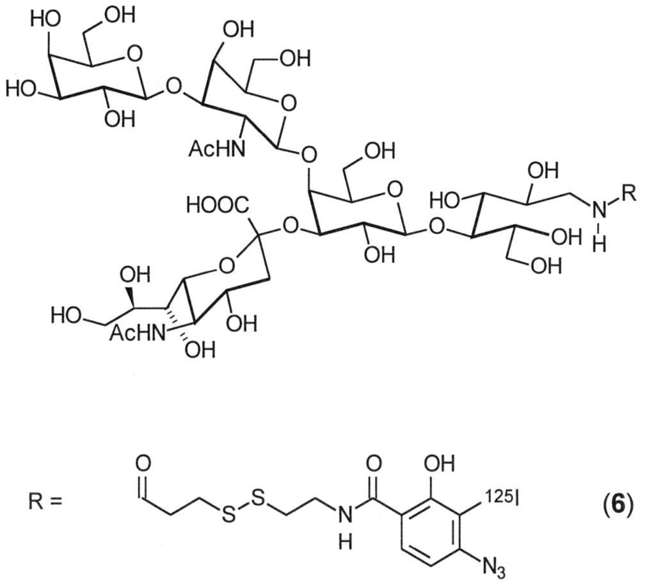

FIG. 6. A radiolabeled (^{125}I) photoactivatable derivative of oligo-GM1 (**6**) and a photoreactive radioiodinated derivative of the GM1 oligosaccharide (GM1OS) (**7**).

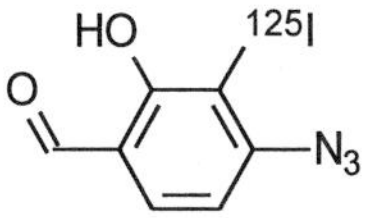

oligosaccharides (e.g., Ref. 45) to monosaccharide-based photoaffinity reagents.

3. *Photoaffinity Labeling of Monosaccharide-Binding Domains.* Enzymatic glycoside hydrolysis often requires only the terminal carbohydrate residue in the correct anomeric linkage as the recognition element. Therefore, photoaffinity ligands on the basis of monosaccharides allow the identification and analysis of glycosidases and, eventually, glycosyltransferases. The carbene precursor (**8**) (Fig. 7)[46] turns out to be a competitive inhibitor of β-hexosaminidase B. The essential structural element for recognition by human β-hexosaminidase B is the terminal *N*-acetylhexosamine residue. The active site of β-hexosaminidase B was analyzed by photoaffinity label-

[45] J. Lehmann and M. Schmidt-Schuchardt, *Methods Enzymol.* **247,** 265 (1994).
[46] C.-S. Kuhn, J. Lehmann, and K. Sandhoff, *Bioconj. Chem.* **3,** 230 (1992).

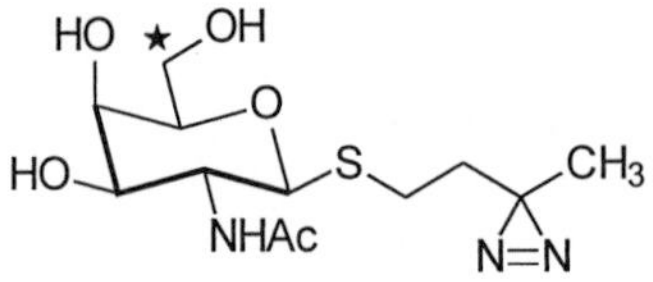

(8)

FIG. 7. The active site of β-hexosaminidase B was analyzed by photoaffinity labeling using thioglycoside (**8**).

ing using thioglycoside (**8**).[47] A single amino acid residue (Glu-355) was modified, which is located within a region of β-hexosaminidase B that shows considerable homology to the α subunit of human β-hexosaminidase A and other hexosaminidases from various species. This result was confirmed later by X-ray crystallography of a closely related chitobiase.[48] Identification of the active site residue by labeling techniques is one approach toward understanding the mechanisms of enzymatic glycoside hydrolysis,[49,50] e.g., in the hexosaminidases.[51]

The same concept was employed to investigate the active site of β-D-galactosidase (*Escherichia coli*).[52,53] Segments detected as part of the active site of the enzyme have been identified by the photoaffinity labels (**9**)–(**11**) (Fig. 8). The heterobifunctional substrate analog (**11**) offers the possibility to investigate the topology of the substrate-binding site and to gain insight into the mechanism of action of such enzymes.

The contribution of particular functional groups to recognition and binding may be evaluated by the use of photoaffinity labeling. For example, (**12**) (Fig. 9) was established as a photoreactive competitive inhibitor of human lysosomal neuraminidase in cultured skin fibroblasts.[54] Its synthesis started from NeuAc-2-en methyl ester. Key steps are the activation of the 9-hydroxyl group as *p*-toluenesulfonate, nucleophilic substitution with KSAc, and introduction of the photolabel with 4-fluoro-3-nitrophenylazide. Results of competitive inhibition kinetics suggested that the terminal hy-

[47] B. Liessem, G. J. Glomitza, F. Knoll, J. Lehmann, J. Kellermann, F. Lottspeich, and K. Sandhoff, *J. Biol. Chem.* **270,** 23693 (1995).

[48] I. Tews, A. Perrakis, A. Oppenheim, Z. Dauter, K. S. Wilson, and C. E. Vorgias, *Nature Struct. Biol.* **3,** 638 (1996).

[49] J. D. McCarter and S. G. Withers, *Curr. Opin. Struct. Biol.* **4,** 885 (1994).

[50] S. G. Withers and R. Aebersold, *Protein Sci.* **4,** 361 (1995).

[51] E. C. K. Lai and S. G. Withers, *Biochemistry* **33,** 14743 (1994).

[52] C.-S. Kuhn, J. Lehmann, G. Jung, and S. Stevanovic, *Carbohydr. Res.* **232,** 227 (1992).

[53] R. E. Huber, J. Lehmann, and L. Ziser, *Carbohydr. Res.* **214,** 35 (1991).

[54] T. G. Warner, *Biochem. Biophys. Res. Commun.* **148,** 1323 (1987).

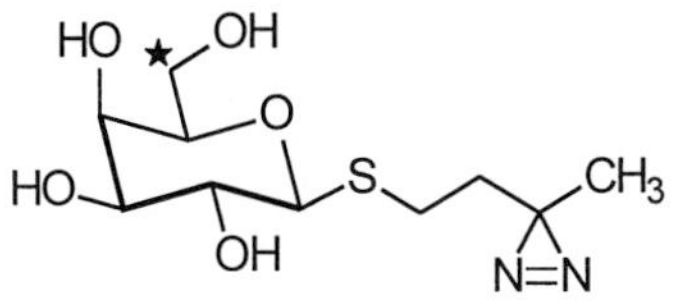

(9)

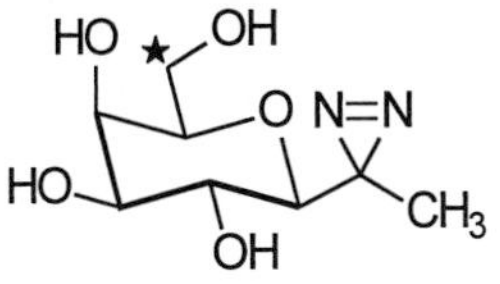

(10)

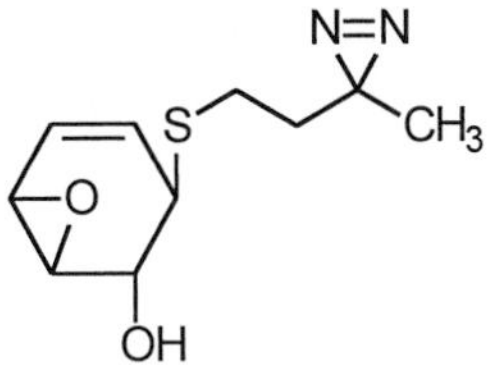

(11)

FIG. 8. Segments detected as part of the active site of β-D-galactosidase (*E. coli*) identified by photoaffinity labels (**9–11**).

droxyl group at C-9 is not important in the recognition and binding of the substrate by the enzyme. The compound is of limited use because a suitable radioactive label cannot be introduced easily into the target structure.

Photosensitive analogs of sialic acid such as (**13**) (Fig. 10) are of interest

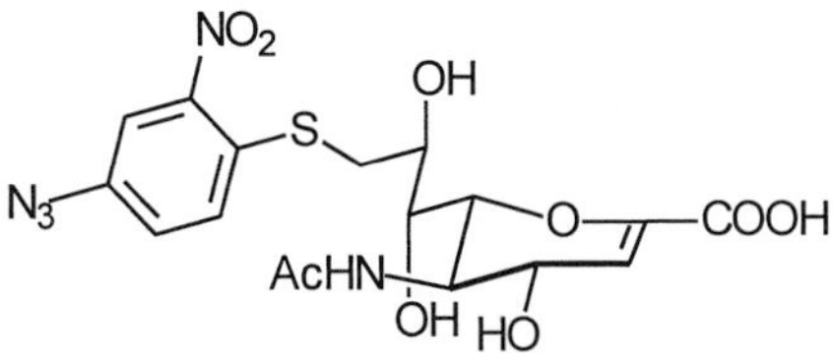

(12)

FIG. 9. A photoreactive competitive inhibitor of human lysosomal neuraminidase in cultured skin fibroblasts (**12**).

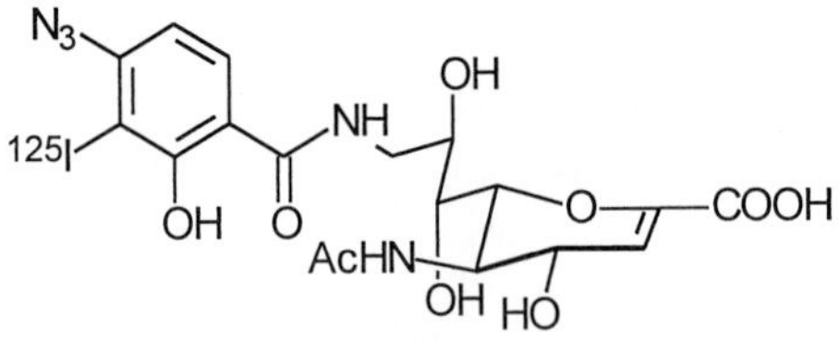

(13)

FIG. 10. A photosensitive analog of sialic acid (**13**).

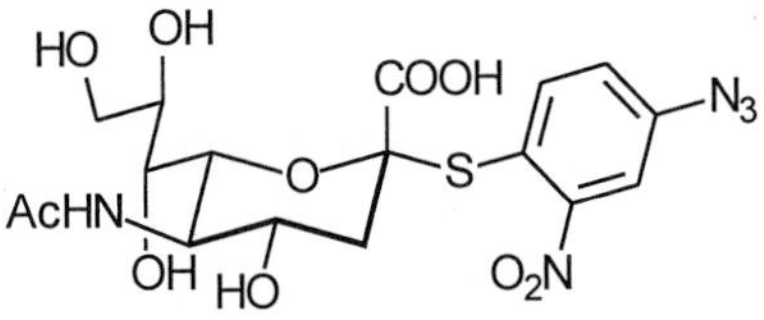

(14)

FIG. 11. An azidoaryl thioglycoside of sialic acid (**14**), another potential photoaffinity probe of sialic acid.

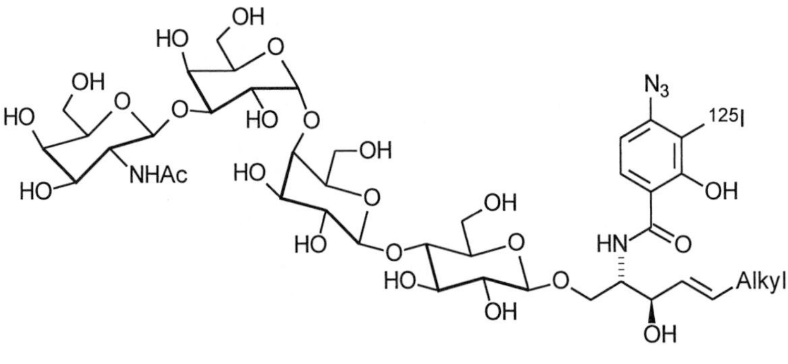

(15)

FIG. 12. A radioiodinated photoactivatable derivative of globoside (**15**).

for isolation and characterization of a number of proteins involved in the catabolism of sialic acid-containing glycoconjugates.[55-57] The photoreactive arylazide derivative of sialic acid specifically labeled a polypeptide in bovine testis β-galactosidase/neuraminidase/protective protein complex containing the catalytic site of neuraminidase.[55]

Another potential photoaffinity probe of sialidase is an azidoaryl thioglycoside of sialic acid (14) (Fig. 11).[58]

B. Photoaffinity Labeling of Lipid-Binding Sites

1. Photoaffinity Labeling of GSL–Lipid-Binding Domains. Photoaffinity labeling is used to study membrane topology, including spatial relations between glycosphingolipids and proteins. A radioiodinated photoactivatable derivative of globoside (15) (Fig. 12) was employed to photolabel human erythrocyte membrane proteins and lipids, which were analyzed by SDS–PAGE and immunostaining.[59]

Gangliosides added to cultured cells bind to the cell membranes and are in part integrated into the lipid bilayer. Eventually, the insertion is mediated by surface-located membrane proteins.[60,61] Proteins of the plasma membrane and other subcellular compartments have been identified by photoaffinity labeling.[62-64] For this purpose, a derivative of ganglioside GM1 (16) (Fig. 13) containing a photoreactive nitrophenyl azide group at the terminus of the fatty acid moiety and a tritium label at the acetyl group of *N*-acetylneuraminic acid has been synthesized. Cultured rat cerebellar granule cells were photolabeled with (16).[62] After a 30-min incubation, only few proteins became radiolabeled. Increasing the incubation time to 24 hr resulted in a larger number of radiolabeled proteins, probably as a consequence of the internalization and metabolic processing of adminis-

[55] G. T. J. van der Horst, U. Rose, R. Brossmer, and F. W. Verheijen, *FEBS Lett.* **277,** 42 (1990).
[56] G. T. J. van der Horst, G. M. S. Mancini, R. Brossmer, U. Rose, and F. W. Verheijen, *J. Biol. Chem.* **265,** 10801 (1990).
[57] J. Kopitz, K. Sinz, R. Brossmer, and M. Cantz, *Eur. J. Biochem.* **248,** 527 (1997).
[58] T. G. Warner and L. A. Lee, *Carbohydr. Res.* 211 (1988).
[59] T. Pacuszka and M. Panasiewicz, *Biochim. Biophys. Acta* **1257,** 265 (1995).
[60] K. C. Leskawa, R. E. Erwin, A. Leon, G. Toffano, and E. L. Hogan, *Neurochem. Res.* **14,** 547 (1989).
[61] S. Sonnino, V. Chigorno, M. Valsecchi, R. Bassi, D. Acquotti, L. Cantu, M. Corti, and G. Tettamanti, *Indian J. Biochem. Biophys.* **27,** 353 (1990).
[62] S. Sonnino, V. Chigorno, M. Valsecchi, M. Pitto, and G. Tettamanti, *Neurochem. Int.* **20,** 315 (1992).
[63] S. Sonnino, V. Chigorno, D. Acquotti, M. Pitto, G. Kirschner, and G. Tettamanti, *Biochemistry* **28,** 77 (1989).
[64] V. Chigorno, M. Valsecchi, D. Acquotti, S. Sonnino, and G. Tettamanti, *FEBS Lett.* **263,** 329 (1990).

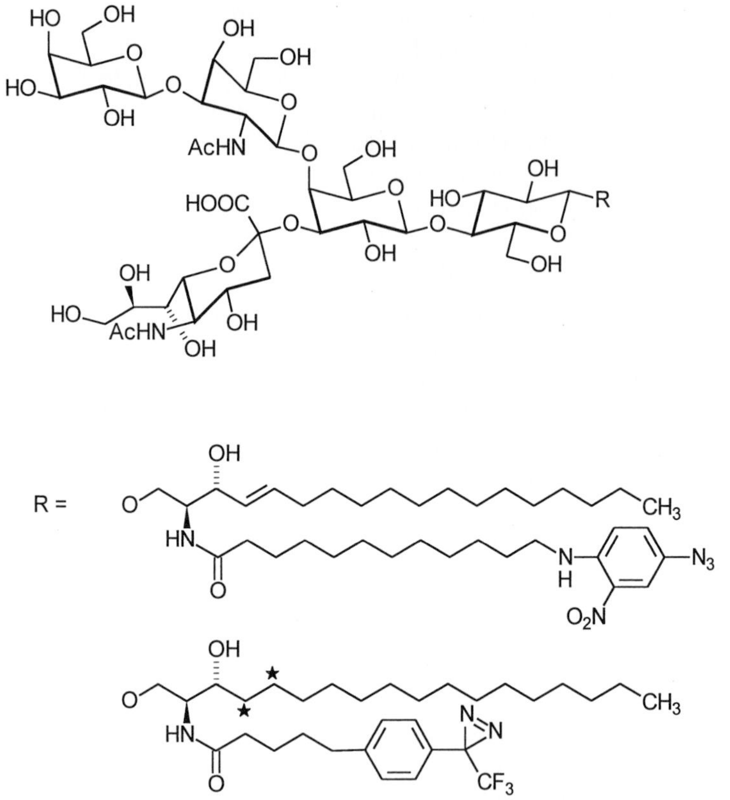

FIG. 13. A derivative of ganglioside GM1 (**16**) containing a photoreactive nitrophenyl azide group at the terminus of the fatty acid moiety and a tritium label at the acetyl group of *N*-acetylneuraminic acid and a photoreactive ganglioside derivative *N*-diazirinyl-lyso-GM1 with a carbene precursor in its hydrophobic moiety (**17**).

tered (**16**). After a 24-hr incubation, some radioactivity was also associated with cytosolic proteins.

In previous experiments, cultured human fibroblasts were exposed to mixtures of (**16**) and native GM1 (1 : 10 by mol) for different times and then irradiated and the radioactive protein pattern studied by SDS–PAGE.[63,64] After 2 hr of exposure, the label was stably associated to the cells and underwent almost no metabolic processing, behaving like the underivatized native GM1. Under these conditions, only few proteins became labeled. When the incubation was prolonged to 24 hr, the derivative of ganglioside GM1 underwent extensive metabolic processing. Under these conditions, many proteins became labeled radioactively.

Previous studies have shown that sphingolipids may be enriched in caveolae, plasma membrane invaginations implicated in endocytosis, and signal transduction.[65] Evidence for ganglioside segregation in the biogenesis of caveolae was implicated by the use of a modified derivative of (16). A radioactive derivative of ganglioside GM1 with a tritium label in the galactose residue was inserted into the plasma membrane of cultured A431 or MDCK cells.[66] On photoactivation, the main protein labeled by the GM1 derivative was VIP21-caveolin, an essential structural component of caveolae.

The core of many eukaryotic membranes is formed by alkyl chains present in phospholipids and sphingolipids and by cholesterol. In contrast to the hydrated region of the lipid head groups, the core of the membranes is generally assumed to be highly apolar. Based on the extraordinary high permeability of membranes for water[67,68] and protons[69-74] and on measurements of local microviscosity,[75] it has been suggested that water might be present within the hydrophobic core of the membranes.

Temperature dependence of photoaffinity labeling in liposomes consisting of 1,2-dipalmitoyl-*sn*-glycero-3-phosphoryl-choline or 1,2-distearoyl-*sn*-glycero-3-phosphorylcholine and on photoaffinity labeling experiments with cholera toxin support the hypothesis that water is present in the hydrophobic core of artificial and biological membranes.[76] The photoreactive ganglioside derivative *N*-diazirinyl-lyso-GM1 with a carbene precursor in its hydrophobic moiety (17) (Fig. 13) was incorporated into liposomes and calf brain microsomes. After photoactivation at 350 nm it was found to be linked to phospholipids such as phosphatidylcholine and phoshatidylserine and to cholesterol. The predominant covalent reaction product, however, was the alcohol, resulting from the reaction with water. It amounted to about 45% of the covalent reaction products in calf brain microsomes and to about 58% in pure phosphatidylcholine liposomes.

A lipid-binding site is expected on a GSL-binding protein, the GM2

[65] K. Simons and E. Ikonen, *Nature* **387,** 569 (1997).

[66] A. M. Fra, M. Masserini, P. Palestini, S. Sonnino, and K. Simons, *FEBS Lett.* **375,** 11 (1995).

[67] A. D. Bangham, J. DeGier, and G. D. Greville, *Chem. Phys. Lipids* **1,** 225 (1967).

[68] B. M. Fung, *Methods Enzymol.* **127,** 151 (1986).

[69] J. W. Nichols and D. W. Deamer, *Proc. Natl. Acad. Sci. U.S.A.* **77,** 2038 (1980).

[70] J. W. Nichols, M. W. Hill, A. D. Bangham, and D. W. Deamer, *Biochim. Biophys. Acta* **596,** 393 (1980).

[71] H. Hauser, D. Oldani, and M. C. Phillips, *Biochemistry* **12,** 4507 (1986).

[72] A. Blume, *Methods Enzymol.* **127,** 480 (1986).

[73] N. A. Dencher, P. A. Burghaus, and S. Grzesiek, *Methods Enzymol.* **127,** 746 (1986).

[74] K. Elamrani and A. Blume, *Biochim. Biophys. Acta* **727,** 22 (1983).

[75] I. R. Miller, *Biophys. J.* **52,** 497 (1987).

[76] E. M. Meier, D. Schummer, and K. Sandhoff, *Chem. Phys. Lipids* **55,** 103 (1990).

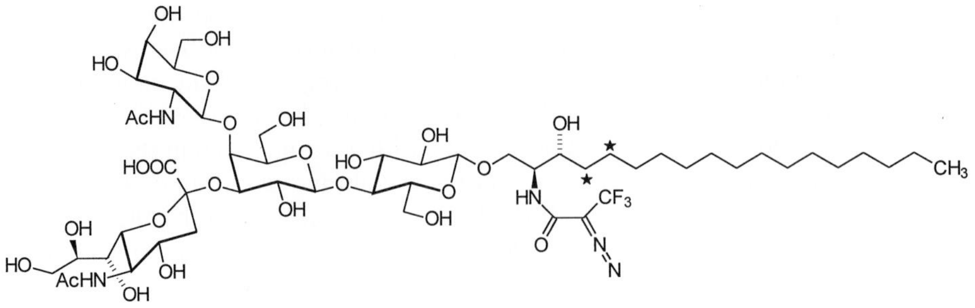

(18)

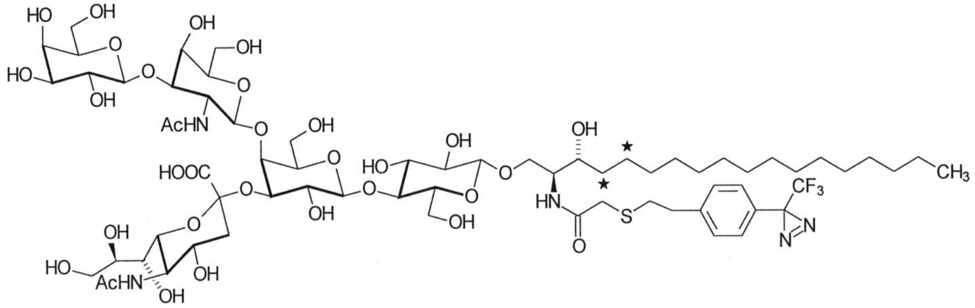

(19)

Fig. 14. Two photoaffinity ligands bearing a carbene precursor (18 and 19).

activator (Section III,A).[10] In addition to affinity labeling using derivative (1), two photoaffinity ligands (18 and 19) (Fig. 14) bearing a carbene precursor have been synthesized to examine different regions within the putative lipid-binding site.[77] In liposomal transfer studies[78] and ultracentrifugation experiments[78] it was shown that ganglioside GM1 is also recognized and bound by the GM2 activator. Therefore, the GM1 derivative (19) was also employed in photoaffinity labeling of the GM2 activator protein. Two tryptic peptides (amino acids 52–81 and 97–130) of the GM2 activator were labeled predominantly by derivatives (1), (18), and (19). Identification of the modified amino acid residues within the peptide sequence failed. The label was lost during isolation and purification (see Section VI,D). There-

[77] F. Knoll, thesis, University of Bonn, 1998.
[78] E. Conzelmann, J. Burg, G. Stephan, and K. Sandhoff, *Eur. J. Biochem.* **123,** 455 (1982).

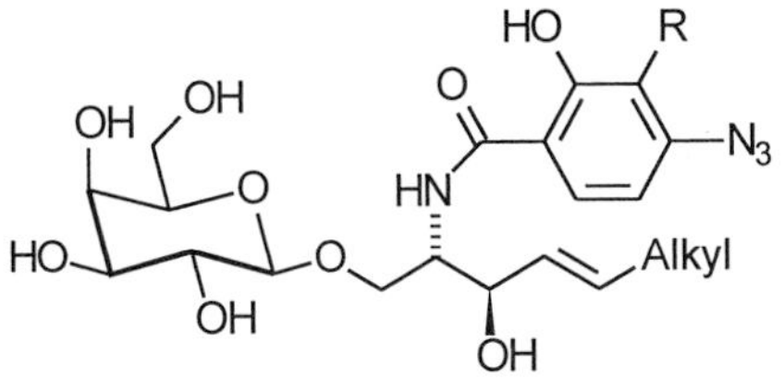

FIG. 15. An azido derivative of galactosylceramide (**20**) and a photoreactive and iodinated derivative of galactosylceramide (**21**).

fore, the labeling approach revealed only an approximate location of the lipid-binding site within the protein.

Rat testicular galactolipid sulfotransferase has been purified by affinity chromatography using a nucleotide matrix.[79] Subsequent photoaffinity labeling using an azido derivative of galactosylceramide (**20**) (Fig. 15)[80] was used to identify this protein, both in crude extracts and when purified.

A photoreactive and iodinated derivative of galactosylceramide (**21**) (Fig. 15) was also used to examine the galactosylceramide-binding protein in normal human cerebrospinal fluid.[81] Affinity-bound peptides appeared at 66, 36 and 28 kDa as radiolabeled bands. The latter two peptides, 36 and 28 kDa, were identified independently by immunostaining of the isolated peptides using biotinylated GalSph and immobilized avidin and by immunoprecipitation of the photolabeled peptide, as apolipoproteins (ALPs) E and A-I, respectively.

GM3 synthase catalyzes the common step in the biosynthesis of most gangliosides.[5] A biotinylated lactosylceramide bearing a ^{14}C-labeled phenyldiazirine (**22**) (Fig. 16) was synthesized as a photoaffinity ligand to examine GM3 synthase.[82] The K_m value of this biotinylated photoprobe was determined as 180 μM using rat liver Golgi as the enzyme source. The biotinylated photoprobe might become helpful in the molecular analysis of the glycosyl-acceptor region of this transferase.

The biotinylated lactose derivative bearing a diazirine and a biotinyl moiety enabled a convenient assay of GM3 synthase; the K_m values of this biotinylated photoprobe were determined as 40 and 47 μM using bovine

[79] D. Sakac, M. Zachos, and C. A. Lingwood, *J. Biol. Chem.* **267,** 1655 (1992).
[80] C. Lingwood and T. Taylor, *Biochem. Cell Biol.* **64,** 631 (1986).
[81] M. Chiba, K. Tsuchihashi, K. Suetake, Y. Ibayashi, S. Gasa, and K. Hashi, *Biochem. Mol. Biol. Int.* **32,** 961 (1994).
[82] Y. Hatanaka, M. Hashimoto, K. Hidari, Y. Sanai, Y. Nagai, and Y. Kanaoka, *Bioorg. Med. Chem. Lett.* **5,** 2859 (1995).

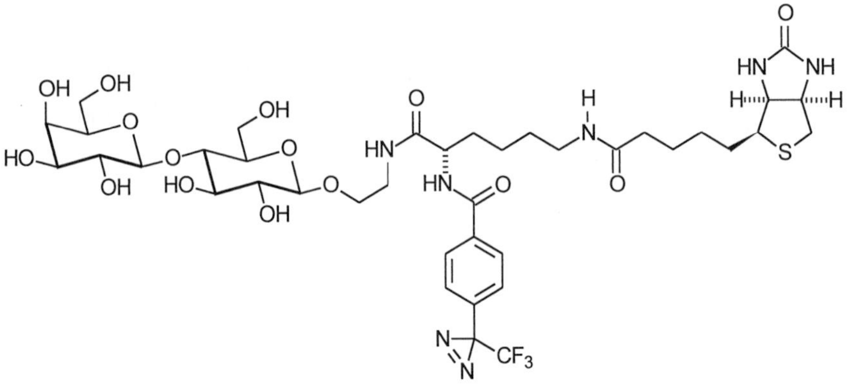

(22)

FIG. 16. A biotinylated lactosylceramide bearing a ^{14}C-labeled phenyldiazirine (22).

brain and rat liver Golgi as enzyme sources, respectively.[83] Sialylation of the natural acceptor substrate for GM3 synthase, lactosylceramide, was inhibited competitively by this synthetic analog. The reagent was introduced as a potentially useful photoprobe for GM3 synthase.

2. Photoaffinity Labeling of Sphingolipid-Binding Proteins. In addition to the physiological function of glycosphingolipids per se, they serve as metabolic precursors for sphingolipids that are involved in the transduction of extracellular signals into the interior of cells.[84,85] The observation that ceramide liberation by sphingomyelin hydrolysis can be induced by extracellular agents in various cell types led to the discovery of the so-called sphingomyelin cycle.[86] In general, ceramide appears to mediate antimitogenic effects such as cell differentiation, cell cycle arrest, and cell senescence. Several lines of evidence indicate ceramide to be a physiological mediator of apoptosis. The identity of the cellular targets of ceramide and other molecules downstream within the signal flow is not known unambiguously. A ceramide-dependent kinase,[87] a phosphatase,[88] and a protein kinase C

[83] Y. Hatanaka, M. Hashimoto, K. Hidari, Y. Sanai, Y. Tezuka, Y. Nagai, and Y. Kanaoka, *Chem. Pharm. Bull. (Tokyo)* **44**, 1111 (1996).
[84] R. Kolesnick and D. W. Golde, *Cell* **77**, 325 (1994).
[85] Y. A. Hannun, *J. Biol. Chem.* **269**, 3125 (1994).
[86] T. Okazaki, R. M. Bell, and Y. A. Hannun, *J. Biol. Chem.* **264**, 19076 (1989).
[87] S. Mathias, K. A. Dressler, and R. N. Kolesnick, *Proc. Natl. Acad. Sci. U.S.A.* **88**, 10009 (1991).
[88] R. T. Dobrowsky and Y. A. Hannun, *J. Biol. Chem.* **267**, 5048 (1992).

subtype[89] are currently under investigation. Protein kinase c-Raf has been identified by photoaffinity labeling within this signaling pathway.[90] Other ceramide-binding proteins identified by this approach in rat mesangial cells are the protein kinase isoenzymes α and δ.[91] A large number of events are reported to be influenced by other intermediates of glycosphingolipid metabolism, e.g., inhibition of protein kinase C by sphingosine and lysoGSLs lacking the amide-bound fatty acid.[92] Sphingosine-1-phosphate operates as a mitogenic regulator in certain cell types.[93] An extracellular receptor for sphingosine-1-phosphate has been cloned,[94,95,96] but intracellular functions of this lipid messenger are not mediated by this receptor.[96a] Interpretation of experimental results obtained in cultured cells is complicated due to the fact that ceramide, sphingosine, and sphingosine-1-phosphate are coupled metabolically. With the exception of an extracellular receptor for sphingosine-1-phosphate (see earlier discussion), a sphingolipid receptor involved in signal transduction has not been characterized unambiguously to date.

To identify downstream signaling targets of ceramide, a radioiodinated photoaffinity labeled analog of ceramide (23) (Fig. 17) has been employed.[97] It could be shown that this ceramide analog specifically binds to and activates protein kinase c-Raf, leading to a subsequent activation of the MAPK cascade. Binding of this derivative to any other member of the MAPK module or to protein kinase C-j was not detected. Using this approach, protein kinase c-Raf has been identified as a specific molecular target for interleukin 1b-stimulated ceramide formation. The photoaffinity ligand was radioiodinated by halogen-metal exchange.[97]

Photoactivatable derivatives of ceramide, glucosylceramide, and sphin-

[89] M. T. Diaz-Meco, E. Berra, M. M. Municio, L. Sanz, J. Lozano, I. Dominguez, V. Diaz-Golpe, M. T. Lain de Lera, J. Alcami, C. V. Paya, F. Arenzana-Seisdedos, J. -L. Virelizier, and J. Moscat, *Mol. Cell Biol.* **13,** 4770 (1993).
[90] A. Huwiler, J. Brunner, R. Hummel, M. Ver Voordeldonk, S. Stabel, H. Van Den Bosch, and J. Pfeilschifter, *Proc. Natl. Acad. Sci. U.S.A.* **93,** 6959 (1996).
[91] A. Huwiler, D. Fabbro, and J. Pfeilschifter, *Biochemistry* **37,** 14556 (1998).
[92] Y. Igarashi, *J. Biochem.* **122,** 1080 (1997).
[93] S. M. Mandala, R. Thornton, Z. Tu, M. B. Kurtz, J. Nickels, J. Broach, R. Menzeleev, and S. Spiegel, *Proc. Natl. Acad. Sci. U.S.A.* **95,** 150 (1998).
[94] M. Lee, J. R. Van Brocklyn, S. Thangada, C. H. Liu, A. R. Hand, R. Menzeleev, S. Spiegel, and T. Hla, *Science* **279,** 1552 (1998).
[95] G. C. M. Zondag, F. R. Postma, I. V. Etten, I. Verlaan, and W. H. Moolenaar, *Biochem. J.* **330,** 605 (1998).
[96] J. R. Van Brocklyn, M. J. Lee, R. Menzeleev, A. Olivera, L. Edsall, O. Cuvillier, D. M. Thomas, P. J. P. Coopman, S. Thangada, C. H. Liu, T. Hla, and S. Spiegel, *J. Cell Biol.* **142,** 229 (1998).
[96a] G. Cavallo, C. Iavarone, and E. Tubaro, *Carbohydr. Res.* **248,** 251 (1993).
[97] T. Weber and J. Brunner, *J. Am. Chem. Soc.* **117,** 3084 (1995).

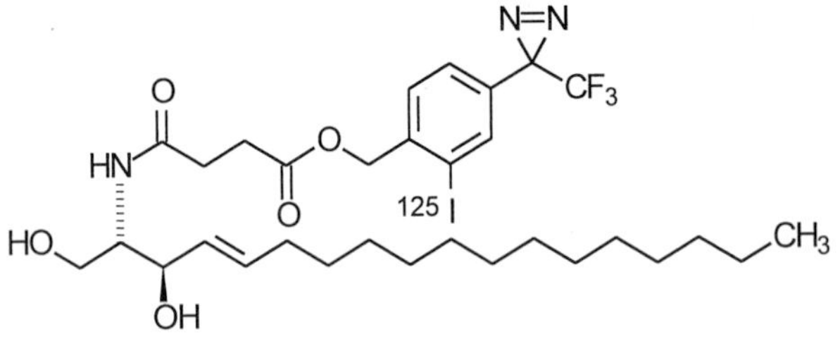

(23)

Fig. 17. A radioiodinated photoaffinity-labeled analog of ceramide (**23**).

gomyelin were synthesized in an attempt to identify compartment-specific proteins involved in sphingolipid sorting or metabolism.[98] In HT29 and BHK cells, the ceramide analog (**24**) (Fig. 18) was internalized efficiently at low temperature (4°). In contrast to this, the photoactivatable glucosylceramide derivative (**26**) (Fig. 18) was internalized only at elevated temperatures (37°), presumably reflecting an endocytic mechanism of uptake. The photoactivatable ceramide was metabolized predominantly to the corresponding sphingomyelin analog (**25**) (Fig. 18), but small amounts of glucosylceramide and galactosylceramide have also been formed. The newly synthesized photoreactive sphingomyelin was subsequently transported to the cell surface, a process that was effectively inhibited by the presence of brefeldin A. Incubation of cells with the photoactivatable analogs at 4°, followed by irradiation, led to the association of the sphingolipid analogs with a specific subset of proteins. The protein labeling pattern of ceramide differed from that of glucosylceramide. A further shift in the labeling pattern was apparent when the cells were incubated with the lipid analogs at 37°. Moreover, most of the proteins labeled by photoreactive sphingomyelin seemed to be detergent insoluble, which indicates a location in sphingolipid-rich microdomains at the plasma membrane.

These results emphasize the role of metabolism in *in vivo* labeling of cellular proteins and the need for suitable sphingolipid analogs of enhanced metabolic stability (compare with Schwarzmann,[98a] especially Section VII).

VI. Synthesis and Practical Approaches

A. Design of Sphingolipid Photoaffinity Labels

In general, native glycosphingolipids and sphingolipids serve as preparative precursors in the chemical synthesis of photoaffinity ligands for the

[98] M. M. P. Zegers, J. W. Kok, and D. Hoekstra, *Biochem. J.* **328,** 489 (1997).
[98a] G. Schwarzmann, *Methods Enzymol.* **311** [48] 1999 (this volume).

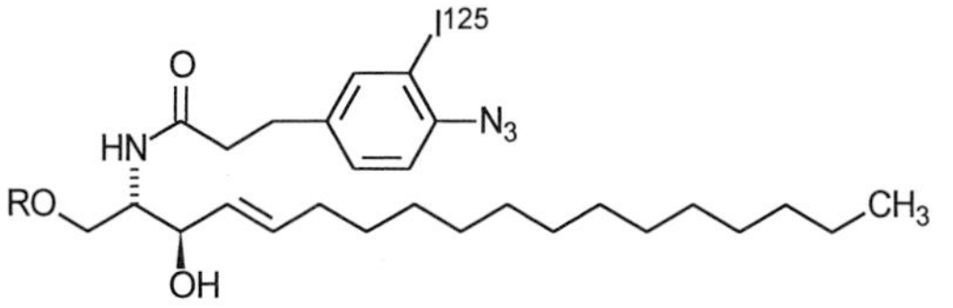

R =	H	(24)
	Choline	(25)
	β-1-D-glucose	(26)

FIG. 18. A ceramide analog (24), a sphingomyelin analog (25), and a photoactivatable glucosylceramide derivative (26).

investigation of GSL metabolism and function. Most methods introduce a reactive functionality into these structures, either an amino group or an aldehyde function, that can be generated by oxidative diol cleavage. Most applications require the additional incorporation of a radioisotope into a defined position within the (glyco)lipid structure.

The majority of glycosphingolipid-based photoaffinity ligands have been synthesized by acylation of one of the nitrogen groups present in these structures. This requires initial deacylation of amide-bound amino groups, either in the sphingosine moiety or in a sialic acid moiety, or an N-acetylga-lactosamine residue. Selectively deacylated glycosphingolipids are accessible by the method of Schwarzmann and Sandhoff,[99,100] although other approaches have been reported, e.g., Ref. 96. Lysogangliosides with a radio-labeled sialic acid residue(s) or with tritium-labeled sphinganine moieties can also be obtained as described by Schwarzmann and Sandhoff.[99] Periodate cleavage of geminal diol groups with a *syn* relationship in rings as the 3′ and 4′ hydroxyl groups in galactose, gives rise to aldehyde functions that can be modified easily by reductive amination. By this way, an additional amino group is incorporated into the GSL structure that can be modified as described earlier. Alternatively, if only the oligosaccharide portion of the GSL is sufficient for the desired study, the oligosaccharide can be released from the native glycolipid by ozonolysis/alkaline fragmentation and its reducing end can be reductively aminated.

Radioactive labels can be introduced into various positions of glyosphin-golipids,[101] e.g., into the lipid backbone by the method of Schwarz-mann.[102,103] Very high specific radioactivity has been introduced into ceramide by exchange of a tributylstannyl residue by iodine, at the expense of a larger preparative effort.[97]

[99] G. Schwarzmann and K. Sandhoff, *Methods Enzymol.* **138**, 319 (1987).
[100] S. Neuenhofer, G. Schwarzmann, H. Egge, and K. Sandhoff, *Biochemistry* **24**, 525 (1985).
[101] S. Sonnino, M. Nicolini, and V. Chigorno, *Glycobiology* **6**, 479 (1996).
[102] G. Schwarzmann, *Biochim. Biophys. Acta* **529**, 106 (1978).
[103] J. K. Anand, K. K. Sadozai, and S. Hakomori, *Lipids* **31**, 995 (1996).

B. Preparative Approaches

1. Synthesis of N-{2-[4-(3-Trifluoromethyl-3H-diazirine-3yl)phenyl] ethylsulfanyl-acetyl}[³H]lyso-GM1 (19). This procedure is an example of the introduction of a photoaffinity probe via a nucleophilic substitution, here, via a thiol group in the photoaffinity label acting as a nucleophile. It offers the possibility of introducing a ^{35}S isotope into the GSL structure. This method requires the availability of radiolabeled lyso-GM1, which is accessible by protocols already published in this series[99] as well as the photoaffinity label, which is accessible in a multistep synthesis. For an alternative approach to sulfur-containing photoaffinity ligands, compare compound **12**.

a. 3-Trifluoromethyl-3-[4-(2-p-toluenesulfonyloxyethyl)phenyl]diazirine. The synthesis of 3-trifluoromethyl-3-[4-(2-p-toluenesulfonyloxyethyl)phenyl]diazirine is described by Brunner and co-workers.[97,105,106]

b. N-Bromoacetyl-lyso-GM1. Triethylamine (94 μmol) is added to a solution of tritium-labeled lyso-GM1[99] (10 mg, 7.8 μmol) in methanol (6 ml). The mixture is cooled to −50° and bromoacetylchloride (70 μmol) is added. The reaction is stirred at −35 to −50° for 2 hr and is monitored by thin-layer chromatography (TLC) (chloroform/methanol/15 mM calcium chloride, 60 : 35 : 8, v/v/v). The product is visible after spraying with anisaldehyde.[107] If significant amounts of lysoganglioside should be present in the reaction mixture, the reaction time should be prolonged and more triethylamine and bromoacetylchloride added. When TLC indicates complete conversion of the starting material, the solvent is removed in a stream of nitrogen. The residue is purified by reversed-phase chromatography (methanol).

c. N-{2-[4-(3-Trifluoromethyl-3H-diazirine-3-yl)phenyl]ethylsulfanyl-acetyl}[³H]lyso-GM1 (19). A solution of 3-trifluoromethyl-3-[4-(2-p-toluenesulfonyloxyethyl)phenyl]diazirine (10.9 mg/28 μmol) and thiourea (2.16 mg, 27.6 μmol) in methanol (0.4 ml) is concentrated in a steady flow of argon. The resulting mixture (0.2 ml) is heated (67–70°, 4 hr) in the dark. The reaction is monitored by TLC (TLC aluminium sheets, silica gel 60 F_{254}, Merck, Darmstadt, Germany; chloroform/methanol, 80 : 20, v/v). Excess thiourea is detected by UV light. Bromoacetyllyso-GM1 and potassium carbonate (3.7 mg/27 μmol in 0.2 ml methanol) are added and the solution

[104] Deleted in proof.
[105] J. Brunner, H. Senn, and F. M. Richards, *J. Biol. Chem.* **255**, 3313 (1980).
[106] J. Brunner and G. Semenza, *Biochemistry* **20**, 7174 (1981).
[107] E. Stahl and U. Kaltenbach, *J. Chromatogr.* **5**, 351 (1961).

is kept under argon atmosphere at 45–50° in the dark for 24 hr. The reaction is monitored by TLC (chloroform/methanol/15 mM calcium chloride, 60:35:8, v/v/v) and the product is visualized by anisaldehyde spray. Separation of reaction products and purification are achieved by medium-pressure column chromatography on LiChroprep Si 60 (chloroform/methanol/15 mM calcium chloride, 60:35:8, v/v/v). The yield is 8.23 mg (80%).

2. *Synthesis of (2-Diazo-3,3,3-trifluoropropionyl)[^{3}H]lyso-GM2* (**18**). A solution of tritium-labeled lyso-GM1[99] (1.0 mg, 0.9 μmol) in DMF (0.1 ml) and diisopropylethylamine (5 μl, 30 μmol) is mixed with 2-diazo-3,3,3-trifluoropropionic acid *p*-nitrophenylester (0.3 mg, 1.1 μmol) in DMF (35 μl). The active ester is available from Sigma, Deisenhoven, Germany. The mixture is kept under argon at room temperature in the dark. After 24 hr, the reaction progress is monitored by TLC (chloroform/methanol, 1:1, v/v). The labeled lysoganglioside derivative is visible in UV light (TLC aluminium sheets, silica gel 60 F$_{254}$, Merck, Darmstadt, Germany) as well as after staining with anisaldehyde spray.[107] The solvents are removed in a nitrogen stream and the residue is purified by column chromatography on LiChroprep Si 60. Excess active ester and 4-nitrophenol are removed by chloroform/methanol (5:1, v/v) and the labeled product is eluted with chloroform/methanol (1:1, v/v). Fractions containing the product are pooled and lyophilized. The yield is 60% of pure (2-diazo-3,3,3-trifluoropropionyl)[^{3}H]lyso-GM2 (**19**).

3. *Synthesis of N-(2-Diazo-3,3,3-trifluoropropionyl)neuraminyllactosylceramide* (**5**). *N*-(2-Diazo-3,3,3-trifluoropropionyl)neuraminyllactosylceramide (**5**) (Fig. 19) is a photoaffinity ligand for GD3-synthase, which transfers sialic acid from CMP-NeuAc to ganglioside GM3.

a. *Neuraminyllactosylceramide (NeuLacCer)*. The amino group in the sialic acid portion of ganglioside GM3 is liberated selectively by base treatment. GM3 (15 mg, 12.1 μmol) is suspended in freshly prepared 0.5 M potassium hydroxide (4 ml) by sonication. The suspension is stirred at 100° for 24 hr. The reaction is monitored by TLC with chloroform/methanol/2 M ammonia (60:35:8, v/v/v) and the product is visualized by fluorescence-quenching TLC (TLC aluminium sheets, silica gel 60 F$_{254}$, Merck, Darm-

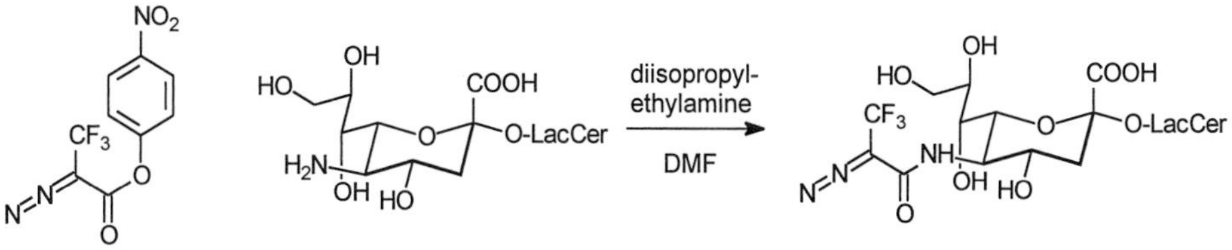

FIG. 19. Synthesis of *N*-(2-diazo-3,3,3-trifluoropropionyl)+ neuraminyllactosylceramide.

stadt, Germany), ninhydrin, or anisaldehyde spray.[107] Salts are removed by reversed-phase chromatography. Fractions containing the product are pooled, freeze-dried, and purified further by column chromatography on LiChroprep Si 60 (chloroform/methanol/2 M ammonia, 60:35:8, v/v/v). Fractions containing the pure product are collected and lyophilized. The yield of NeuLacCer is 8.2 mg (56%).

N-(2-Diazo-3,3,3-trifluoropropionyl)neuraminyllactosylceramide (5). A solution of NeuLacCer (8.1 mg, 6.83 μmol) in DMF (0.35 ml) and diisopropylethylamine (5 μl, 30 μmol) is mixed with 2-diazo-3,3,3-trifluoropropionic acid p-nitrophenylester (4 mg, 14.4 μmol, Sigma, Deisenhoven, Germany) in DMF (50 μl). The mixture is kept under argon at room temperature in the dark. After 10 days the reaction progress is monitored by TLC (chloroform/methanol/2 M ammonia, 60:35:8, v/v/v). The labeled lactosylceramide derivative is detectable in UV light (TLC aluminium sheets, silica gel 60 F_{254}, Merck, Darmstadt, Germany) as well as after spraying with anisaldehyde reagent.[107] Solvents are removed in a stream of nitrogen and the residue is purified by column chromatography on LiChroprep Si 60. Excess active ester and 4-nitrophenol are removed by chloroform/ methanol (5:1, v/v) and the labeled product is eluted by chloroform/ methanol/2 M ammonia (60:35:8, v/v/v). Fractions containing product are pooled and lyophilized. The yield is 1.7 μmol (25%).

4. Synthesis of N-(p-Trifluoromethyldiazirinyl)phenylethyl-2-[^{35}S]-2-thioacetyl-D-erythro-C$_{18}$-sphingosine

a. N-(Bromoacetyl)-D-erythro-C$_{18}$-sphingosine. Triethlyamine (4.2 mg, 42 μmol) is added to a solution of D-*erythro*-C$_{18}$-sphingosine (4.7 mg, 14 μmol) in methanol (1 ml). The mixture is cooled to $-40°$. Bromoacetylchloride (4.4 mg, 28 μmol) is added, and the reaction mixture is stirred at $-40°$ for 1 hr. The solvent is removed in a stream of nitrogen. The product is purified by column chromatography on LiChroprep Si 60 (chloroform/ methanol, 95:5, v/v). The yield is 5.4 mg (12.7 μmol, 91%).

b.   [^{35}S]N-(p-trifluoromethyldiazirinyl)phenylethyl-2-thioacetyl-D-erythro-C$_{18}$-sphingosine. [^{35}S]Thiourea (167 nmol, 5 mCi, specific activity: 30 000 Ci/mol) and 1-(1'-trifluoromethyldiaziminyl)-4-(2'-tosylethyl)benzene (0.19 mg/500 nmol)[97,105,106] are dissolved in methanol (1 ml). The reaction volume is reduced in an argon stream to 0.2 ml. The mixture is kept under argon at 70° in the dark for 4 hr. The reaction is monitored by TLC (chloroform/methanol/H$_2$O, 80:20:1, v/v/v), and excess thiourea is detected by UV light. N-Bromoacetyllyso-GM1 (0.21 mg, 500 nmol) and potassium carbonate (167 nmol in 0.2 ml methanol) are added and the solution is shortly heated at 55–60° and kept in the dark at room temperature under argon overnight. The product is purified by TLC (chloroform/ methanol, 9:1, v/v). The product is identified by autoradiography and scraped off. The silica gel is placed in a column and the product is eluted by methanol. The yield is 65.3 nmol (1.96 mCi, 39.2%) as determined by radioactivity. The photoaffinity label is stable in the dark, but the thioether is oxized easily by air.

C. Photoaffinity Experiment and Workup Procedures

Based on Bayley and Knowles[20] and Brunner[26] and laboratory experiences, the following suggestions are made.

Photolysis. The type of light source and filter combination depends primarily on the spectral properties of the photoaffinity label. For irradiation at 250 or 360 nm, a Rayonet minireactor can be used (e.g., Rayonet 100 reactor, Southern New England Ultraviolet Company, Middletown, CT). Photoactivation is carried out by irradiating the cuvette containing the biological system, including the photoaffinity label. Depending on the irradiation time, the probe may need to be cooled during photoactivation. Usually, an excess of photoaffinity ligand of sufficient specific radioactivity (or other appropriate label) is employed. Ligand uptake into binding sites can be slow, and proper equilibration should be ensured prior to photolysis.

Irradiation times with respect to ligand and to the biological system have to be optimized in control experiments. The stability of the biological system under irradiation conditions can put limits on the photoaffinity experiment, e.g., by denaturation of the protein. Photolysis should be carried out in an atmosphere of nitrogen or argon to prevent chromophore-sensitized photooxidation of the biological system.[108,109] The half-life of the photoaffinity reagent also has to be determined. Photolysis is usually monitored by UV-Vis or IR spectroscopy. It should be considered that the half-life of the ligand alone is only an approximate guide to the behavior of the ligand–receptor complex, as the half-life of the reagent may be increased on binding. In complex systems, photolysis can be monitored by UV-Vis difference spectroscopy or by quantitative thin-layer chromatography of the remaining radioactive ligand.

Photolysis can be controlled as follows: it should be checked whether the receptor still binds to the ligand after photolysis of the biological system in the absence of a photoaffinity reagent. Furthermore, the photoaffinity label can be prephotolyzed. The ligand should not react covalently with the receptor any longer. In another control experiment, the labeling site can be protected by the addition of the natural ligand.

[108] G. Jori, *Photochem. Photobiol.* **21,** 463 (1975).
[109] G. Laustriat and G. Hasselmann, *Photochem. Photobiol.* **22,** 295 (1975).

Workup Procedures. The covalently labeled receptor obtained from photoaffinity labeling can be examined by a variety of methods, such as spectroscopic analysis, fluorophore procedures, visualization by staining, partial and complete hydrolysis, and fragmentation analysis.[110]

First, excess ligand should be removed after photolysis. Removal may be performed by dialysis, size-exclusion chromatography, ion-exchange chromatography, or HPLC. Electrophoresis,[111,112] mass spectrometry, e.g., matrix-assisted laser desorption (MALDI), mass spectrometry,[113,114] or electrospray mass spectrometry may be used to analyze the purified labeled receptor.[115,116] Electroblotting is another method to detect the labeled proteins with antibodies, lectins, and substrates directly on membranes.[117] With the development and introduction of chemically inert membranes, proteins may be amenable to protein-chemical analysis.[118,119] Proteins electroblotted after SDS–PAGE[120,121] can be analyzed by MALDI mass spectrometry.

Further analysis may be achieved by chemical and enzymatic fragmentation of the photolabeled receptor/protein followed by HPLC separation of the labeled peptides. The modified labels may be identified by Edman degradation[122,123] or by sequence analysis by mass spectrometry.[124,125] Unknown labeled proteins and peptides may also be identified by sequence database searching based on mass spectrometric data. Parameters of research may be the total molecular weight or the molecular weight of a set

[110] R. Kellner, F. Lottspeich, and H. E. Meyer, "Microcharacterization of Proteins." VCH, Weinheim, 1994.

[111] U. K. Laemmli, *Nature* **227,** 680 (1970).

[112] H. Schägger and v. G. Jagow, *Anal. Biochem.* **166,** 368 (1987).

[113] M. Karas and F. Hillenkamp, *Anal. Chem.* **60,** 2299 (1988).

[114] M. Karas, U. Bahr, A. Ingendoh, and F. Hillenkamp, *Angew. Chem.* **101,** 805 (1989).

[115] M. Yamashita and J. B. Fenn, *J. Phys. Chem.* **88,** 4451 (1984).

[116] M. Yamashita and J. B. Fenn, *J. Phys. Chem.* **88,** 4671 (1984).

[117] U. Beisiegel, *Electrophoresis* **7,** 1 (1986).

[118] R. Aebersold, *in* "Advances in Electrophoresis" (A. Chrambach, M. J. Dunn, and B. J. Radola, eds.), p. 81. VCH Verlag, Weinheim, 1991.

[119] C. Eckerskorn and F. Lottspeich, *J. Prot. Chem.* **9,** 272 (1990).

[120] C. Eckerskorn, K. Strupat, M. Karas, F. Hillenkamp, and F. Lottspeich, *Electrophoresis* **13,** 664 (1992).

[121] K. Strupat, M. Karas, F. Hillenkamp, C. Eckerskorn, and F. Lottspeich, *Anal. Chem.* **66,** 470 (1994).

[122] P. Edman, *Arch. Biochem.* **22,** 475 (1949).

[123] P. Edman and G. Begg, *Eur. J. Biochem.* **1,** 80 (1967).

[124] K. Biemann, *in* "Methods in Protein Sequence Analysis" (K. Imahori and F. Sakiyama, eds.), p. 119. Plenum Press, New York, 1993.

[125] A. L. Burlingame, *in* "Techniques in Protein Chemistry IV" (R. H. Angeletti, ed.), p. 3. Academic Press, New York, 1993.

of peptides generated by sequence-specific cleavage of the labeled protein. Furthermore, results of database searching may be achieved by the molecular weight of the labeled peptide and its partial sequence.

Labeled peptides can also be separated by capillary electrophoresis (CE), SDS–PAGE, and two-dimensional electrophoresis followed by electroblotting. These techniques enable the direct analysis of separated labeled peptides by mass spectrometry or automated sequence analysis.[126–130]

D. Problems

The photolabile group is required to form a covalent bond between the photoaffinity ligand and the receptor after photoactivation, which has to be stable during the subsequent workup. It has to be considered that affinity labeling may also lead to the formation of covalent, but reversible (labile) bonds, e.g., a labile ester bond[47,131–133] and formation of a labile imidate.[132,134–138]

Indication for the formation of a labile ester bond is the (partial) loss of the label (radioactivity) during isolation and purification procedures of labeled protein/peptide. An ester bond should be stable in neutral solutions and be hydrolyzed under strongly acidic or basic conditions.[47,133] Identification of such a modified amino acid by Edman degradation cannot be expected, as the protocols used for automated sequencing include treatment with 25% aqueous trifluoroacetic acid. Loss of label under these conditions may indicate that the photolabel has bound to the peptide via an ester bond (glutamate or aspartate by carbon insertion into an O-H bond). Nevertheless, identification of the modified amino acid can be achieved by

[126] R. Aebersold, D. B. Teplow, L. E. Hood, and S. B. Kent, *J. Biol. Chem.* **261,** 4229 (1986).
[127] R. H. Aebersold, J. Leavitt, L. E. Hood, and S. H. Kent, *Proc. Natl. Acad. Sci. U.S.A.* **84,** 6970 (1987).
[128] C. Eckerskorn, P. Jungblut, P. Mewes, J. Klose, and F. Lottspeich, *Electrophoresis* **9,** 830 (1988).
[129] C. Eckerskorn and F. Lottspeich, *Chromatographia* **28,** 92 (1989).
[130] C. Eckerskorn and F. Lottspeich, *Electrophoresis* **11,** 554 (1990).
[131] A. H. Ross, R. Radhakrishnan, R. J. Robson, and H. G. Khorana, *J. Biol. Chem.* **257,** 4152 (1982).
[132] Y. Takagaki, R. Radhakrishnan, C. M. Gupta, and H. G. Khorana, *J. Biol. Chem.* **258,** 9128 (1983).
[133] J. C. Gebler, R. Aebersold, and S. G. Withers, *J. Biol. Chem.* **267,** 11126 (1992).
[134] E. H. White, H. M. Perks, and D. F. Roswell, *J. Am. Chem. Soc.* **100,** 7421 (1978).
[135] S. Donadio, H. M. Perks, K. Tsuchiya, and E. H. White, *Biochemistry* **24,** 2447 (1985).
[136] R. K. Chaturvedi and G. L. Schmir, *J. Am. Chem. Soc.* **90,** 4413 (1968).
[137] T. C. Pletcher, S. Koehler, and E. H. Cordes, *J. Am. Chem. Soc.* **90,** 7072 (1968).
[138] D. G. Neilson, *in* "The Chemistry of Amidines and Imidates" (S. Patai and Z. Rappoport, eds.), p. 425. Wiley, Chichester, 1991.

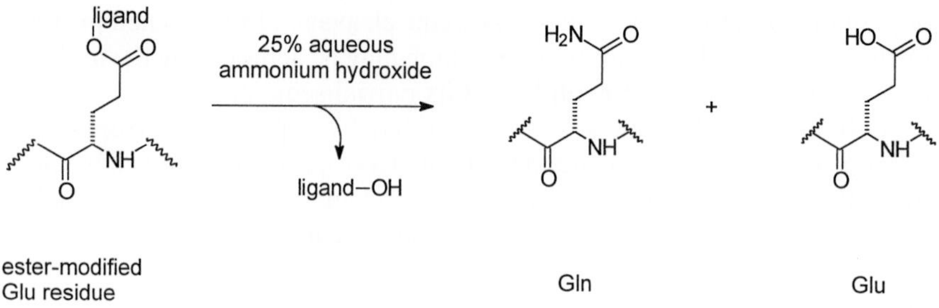

SCHEME 3. Identification of the modified amino acid achieved by treatment with 25% aqueous ammonium hydroxide.

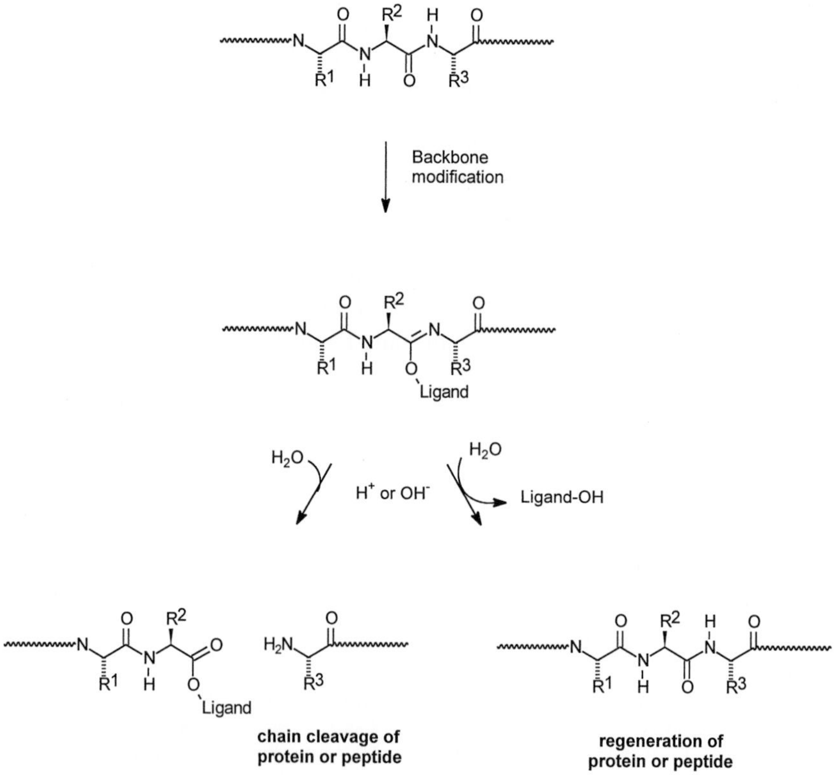

SCHEME 4. Labile substituted imidates may be formed by electrophilic attack of a ligand on the carboxyl oxygen on the amide bond in the peptide backbone. The hydrolysis of these imidates influenced by the presences of general acid–base catalysis is shown.

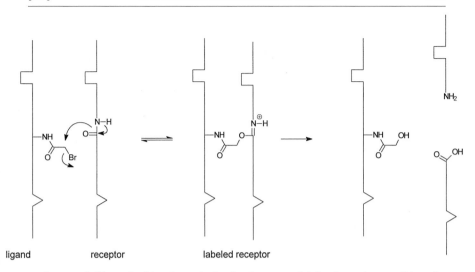

ligand   receptor   labeled receptor

SCHEME 5. Shown in this scheme is the development of defined reaction conditions for selective peptide cleavage which allows photoaffinity cleavage to become a tool in the determination of binding sites with the contribution of the peptide backbone.

treatment with 25% aqueous ammonium hydroxide,[133] which results in the quantitative release of the label from the peptide and concomitant conversion of the ester to an amide and, in part, to the carboxylic acid (Scheme 3). The transformation of Glu to Gln and Asp to Asn can be detected by Edman degradation. This method was used in the identification of the substrate-binding site of β-hexosaminidase B,[47] which was labeled by a radioactive thioglycoside bearing a carbene precursor (see Section V,C). The purified ammonia-treated peptide was then subjected to sequence analysis, which revealed that Glu-355 has been in part converted to Gln-355.

Receptor–ligand interactions are not necessarily mediated by interactions with amino acid side chains of a peptide or a protein, but also with the peptide backbone. This is one explanation for the limited success of the *retro–inverso* peptide modification in the development of peptidomimetics.[139–141] Labile-substituted imidates may be formed by electrophilic attack of a ligand on the carboxyl oxygen on the amide bond in the peptide backbone.[134,135] Hydrolysis of these imidates (Scheme 4, according to Ref. 136) is influenced by pH and by the presence of general acid–base cata-

[139] M. D. Fletcher and M. M. Campbell, *Chem. Rev.* **98**, 763 (1998).
[140] A. Giannis and T. Kolter, *Angew. Chem.* **105**, 1303 (1993).
[141] A. Giannis and T. Kolter, *Angew. Chem. Int. Ed. Engl.* **32**, 1244 (1993).

lysts.[136,137] This can, in principle, give rise to chain cleavage or to regeneration of the protein/peptide.

A hypothetical application of this reaction is the determination of contact sites between the electrophilic ligand and the peptide backbone by affinity cleavage. Hydrolytic conditions that lead to selective chain cleavage would allow the direct determination of the ligand-binding site. Affinity labeling of the GM2 activator protein with (1), (18), or (19) to identify the lipid-binding site led to a continuous loss of radioactive label.[77] This loss can be understood by the formation of labile imidates. Radiolabel was released during workup under regeneration of the protein, no modified amino acids were detected, and the formation of ester bonds was excluded. The development of defined reaction conditions for selective peptide cleavage should allow photoaffinity cleavage to become a novel tool in the determination of binding sites with contribution of the peptide backbone (Scheme 5).

VII. Future Directions

Considerable efforts have been made toward the identification of unknown sphingolipid-binding proteins involved in signal transduction or sphingolipid metabolism. Photoaffinity labeling is a promising tool for the identification of putative sphingolipid receptors. A major problem to be overcome is the demonstration of labeling specificity, as hydrophobic ligands may unspecifically bind to hydrophobic or partial denatured cellular components. Denaturation and, subsequently, nonspecific labeling should be minimized by the application of *in vivo* labeling techniques. However, *in vivo* labeling is of limited suitability for receptor identification as the photoaffinity reagent is subject to metabolism.[98] Ligands of enhanced metabolic stability (compare with Schwarzmann[98a]) are required for this purpose. 1-Deoxy-1-ethyldihydroceramide is such a synthetic derivative that mimics the action of ceramide in cultured cells.[142] A photoaffinity ligand on the basis of this structure is currently used for the identification of ceramide-binding proteins.[143]

Another technique that might facilitate the identification of binding regions within a protein is the affinity cleavage discussed earlier. Affinity[144] and photoaffinity cleavage may become an alternative to affinity and photoaffinity labeling techniques in certain areas of application.

[142] T. Wieder, C. C. Geilen, T. Kolter, F. Sadeghlar, K. Sandhoff, R. Brossmer, P. Ihrig, D. Perry, C. E. Orfanos, and Y. A. Hannun, *FEBS Lett.* **411,** 260 (1997).
[143] R. Betz and T. Kolter, unpublished results (1998).
[144] R. F. Coleman, *FASEB J.* **11,** 217 (1997).

[48] Synthesis and Characterization of Metabolically Stable Sphingolipids

By GÜNTER SCHWARZMANN

Introduction

Over the past decade, sphingolipids have become candidates of prime interest in phenomena such as cell adhesion and signal transduction,[1,2] as well as programmed cell death.[3] Studies on the importance of certain sphingolipids as second messenger molecules[4] and receptor modulators have added ceramides,[5] sphingomyelin,[6] and some glycosphingolipids[7] exogenously to cells in culture or laboratory animals. As these compounds are subject to (sometimes rapid) metabolism, it is difficult to know whether the observed effect on cell behavior is elicited by the applied sphingolipid itself or by its metabolic product(s). In addition, for a complete understanding of the role of sphingolipids, information on their intracellular transport would be essential.[8] Thus, it would be desirable to employ sphingolipids that are metabolically stable or whose metabolism is restricted in either the catabolic or the anabolic pathway of metabolism.

The first part of this article describes the synthesis of a metabolically stable ceramide analog that proves useful in the study by fluorescence microscopy of its intracellular distribution in cultured fibroblasts and perhaps other cells.[9] This analog carries a stable O-methyl group in place of a primary hydroxyl group and, hence, can neither be glycosylated to glycosphingolipids nor acquire a phosphorylcholine residue to become sphingomyelin. This compound or its radioactive and nonfluorescent homo-

[1] S. Hakomori, *J. Biol. Chem.* **265,** 18713 (1990).
[2] K. Iwabuchi, S. Yamamura, A. Prinetti, K. Handa, and S. Hakomori, *J. Biol. Chem.* **273,** 9130 (1998).
[3] I. Flores, A. C. Martinez, Y. A. Hannun, and I. Merida, *J. Immunol.* **160,** 3528 (1998).
[4] Y. A. Hannun, *Adv. Exp. Med. Biol.* **400A,** 305 (1997).
[5] R. T. Dobrowsky and Y. A. Hannun, *J. Biol. Chem.* **267,** 5048 (1992).
[6] A. H. Merrill, Jr., E. M. Schmelz, E. Wang, D. L. Dillehay, L. G. Rice, F. Meredith, and R. T. Riley, *J. Nutr.* **127,** 830S (1997).
[7] A. Rebbaa, J. Hurh, H. Yamamoto, D. S. Kersey, and E. G. Bremer, *Glycobiology* **6,** 399 (1996).
[8] G. Schwarzmann and K. Sandhoff, *Biochemistry* **29,** 10865 (1990).
[9] U. Pütz and G. Schwarzmann, *Eur. J. Cell Biol.* **68,** 113 (1995).

log may be valuable for further investigation on the role of ceramide in apoptosis.[10]

The next section deals with the synthesis of glycosphingolipid analogs (glucosylceramide and lactosylceramide) that are totally resistant to the action of glycohydrolases. These compounds feature a thioglycosidic bond that cannot be split in detectable amounts by mammalian glycohydrolases. It seems that these analogs of glycosphingolipids are becoming valuable tools in studies on their intracellular transport.[11,12]

Materials and General Methods

Materials

Benzoylchloride, *tert*-butyldiphenylchlorosilane, N,N-diisopropylethylamine, 2,2-dimethoxypropane, 4-(dimethylamino)pyridine, tetrabutylammonium iodide, p-toluenesulfonic acid, and trifluoromethanesulfonic anhydride are obtained from Fluka (Buchs, Switzerland). Dimethyl sulfate, LiChroprep RP-18, silica gel Si 60 (15–40 μm), prepacked silica gel chromatography columns, Lobar, and thin-layer chromatography (TLC) plates (silica gel Si 60) are from E. Merck (Darmstadt, Germany). Di-*tert*-butyldicarbonate is from ACROS Organics (a Fisher scientific worldwide company). Methanesulfonylchloride and tetrabutylammonium fluoride are purchased from Sigma-Aldrich Chemie (Steinheim, Germany). Succinimidyl 6-(7-nitrobenz-2-oxa-1,3-diazol-4-yl)-aminohexanoate is obtained from Molecular Probes (Eugene, OR). Sodium-[1-^{14}C]octanoate (2109 MBq/mmol), neutral, acidic, and basic alumina W 200 are from ICN Biomedicals (Eschwege, Germany).

All solvents are analytical grade and are obtained from either E. Merck (Darmstadt, Germany) or Riedel-de Haën (Seelze, Germany). All other chemicals are of the highest purity available.

Methods

Solvents. Some solvents are prepared dry and/or argon saturated prior to use. Anhydrous toluene and ethyl acetate are obtained by purging the analytical grade solvents over neutral alumina. N,N-Dimethylformamide (DMF) is purified and freed from basic decomposition products by passing,

[10] S. Bourteele, A. Hausser, H. Döppler, J. Horn-Müller, G. Schwarzmann, K. Pfizenmaier, and G. Müller, *J. Biol. Chem.* **273**, 31245 (1998).

[11] G. Schwarzmann, P. Hofmann, U. Pütz, and B. Albrecht, *J. Biol. Chem.* **270**, 21271 (1995).

[12] B. Gillard, R. Clement, E. Colucci-Guyon, C. Babinet, G. Schwarzmann, T. Taki, T. Kasama, and D. Marcus, *Exp. Cell Res.* **242**, 561 (1998).

under argon, over acidic alumina. Tetrahydrofurane (THF) is freed of peroxides by passing over basic alumina, followed by saturation with dry argon.

Thin-Layer Chromatography. Column chromatographic elution profiles and the progress of reactions are routinely followed by TLC in tanks (Camag, Muttenz, Switzerland) with vapor saturation. R_f values are determined from 20-cm plates that are developed for 18 cm. Detection on TLC plates of compounds is accomplished by spraying the plates with a mixture of acetic acid/sulfuric acid/anisaldehyde (500:10:2, v/v/v),[13] followed by heating for 10 min at 120° or, alternatively, by dipping the plates into a solution of ceric ammonium nitrate in 20% sulfuric acid followed by heating with a heat gun. This procedure yields blue spots on a slightly yellowish background after cooling. Lysosphingolipids are also detected by spraying with 1% ninhydrin in *n*-butanol prior to heating for 20 min at 120°. Fluorescent bands on TLC plates are localized under UV light (excitation at 366 nm). Radioactive products are localized and quantified by the use of a Fuji BAS 1000 Bio Imaging analyzer (Raytest, Pforzheim, Germany) or by radioscanning using an automatic TLC linear analyzer (Tracemaster 40, Berthold, Wildbad, Germany).

Removal of Salts and Other Polar Compounds from Reaction Products. Sphingosine-containing reaction products are freed of salts and other polar and water-soluble materials by reversed-phase chromatography similar to the procedure described.[14] Briefly, Pasteur pipettes containing a small cotton plug are filled with a 1:1 slurry of LiChroprep RP-18 in methanol and washed successively with methanol, chloroform/methanol (1:1, v/v), methanol, and water prior to use. Lipophilic reaction products (less than 2 mg) are adsorbed onto these minicolumns from a solution in methanol/water (1:3 to 2:3, v/v) and then eluted with methanol after the nonadsorbed salts and other polar materials are washed out. Instead of self-prepared minicolumns, Sep-Pak cartridges (Waters Division of Millipore Corporation, Milford, MA) can be used. If more material (up to 100 mg) has to be desalted, LiChroprep RP-18 can be filled in larger columns; alternatively, Lobar Fertigsäulen LiChroprep RP-18 size A can be used.

Column Chromatography. Separation of reaction products and purification of the desired compounds are achieved by medium-pressure column chromatography using heavy-walled glass columns of up to 100 cm length of various inner diameters (0.6–1.2 cm) and having adjustable column adapters (Latek GmbH, Eppelheim, Germany) and a high-pressure pump (Latek P 400) or any other suitable columns and pumps. The flow rate of

---

[13] E. Stahl and U. Kaltenbach, *J. Chromatogr.* **5,** 351 (1961).
[14] M. Williams and R. McCluer, *J. Neurochem.* **35,** 266 (1980).

solvents is usually equivalent to 2 bed volumes per hour. Fractions of 2–8 ml, depending on the column size, are collected. For chromatography, silica gel Si 60 in glass columns or Lobar Fertigsäulen of size A and B with LiChroprep Si 60 is used. The latter are more expensive than self-packed columns but have better performance.

High Performance Liquid Chromatography (HPLC). For final purification, end products are subjected to HPLC on columns of various sizes (0.4 × 15, 1 × 25, and 2 × 25 cm) containing reversed-phase material (ProSep C18, with 5-μm mean particle diameter; Latek, Eppelheim, Germany).

Fast Atom Bombardment Mass Spectra (FABMS). FABMS spectra are recorded in the positive-ion mode on a ZAB HF instrument (VG Analytical, Manchester, UK).[15,16]

500-MHz ^1H Nuclear Magnetic Resonance (NMR) and 125-MHz ^{13}C NMR Spectroscopy. A Bruker AMX 500 NMR spectrometer (Bruker, Karlsruhe, Germany) is used for ^1H NMR and ^{13}C NMR measurements. Chemical shifts (δ) are indicated in ppm relative to internal tetramethylsilane.

Elemental Analysis. Elemental analysis may be carried out with a CHN-O-Rapid analyzer (Heraeus, Osterode, Germany).

Metabolically Stable and Fluorescent Ceramide

Principle

This section describes the synthesis of 1-*O*-methyl-NBD-C_6-ceramide from 1-*O*-methyl-D-*erythro*-sphingosine as outlined in Figs. 1 and 2. The intermediate 1-*O*-methyl-D-*erythro*-sphingosine (**7**) is obtained in six reaction steps in about 20% overall yield starting from D-*erythro*-sphingosine.[9] For selective methylation of the primary hydroxy group, the secondary hydroxy and the amino group have to be efficiently protected. The amino group is protected by forming a carbamate. Prior to protection of the secondary group, the primary hydroxy group has to be protected by selective silylation.[17] Thereafter, protection of the remaining 3-OH and 2-NH functionalities is achieved by forming a 2,2-dimethyloxazolidine ring using 2,2-dimethoxypropane as reagent and solvent.[18] Selective deprotection of the primary hydroxy residue makes use of tetrabutylammonium fluoride. After methylation is accomplished with dimethyl sulfate in a two-phase system[19]

[15] H. Egge and J. Peter-Katalinic, *Mass Spectrom. Rev.* **6,** 331 (1987).
[16] A. Dell, *Adv. Carbohydr. Chem. Biochem.* **45,** 19 (1987).
[17] S. Hanessian and P. Lavalee, *Can. J. Chem.* **53,** 2975 (1975).
[18] P. Garner and J. M. Park, *J. Org. Chem.* **52,** 2361 (1987).
[19] A. Merz, *Angew. Chem.* **19,** 868 (1973).

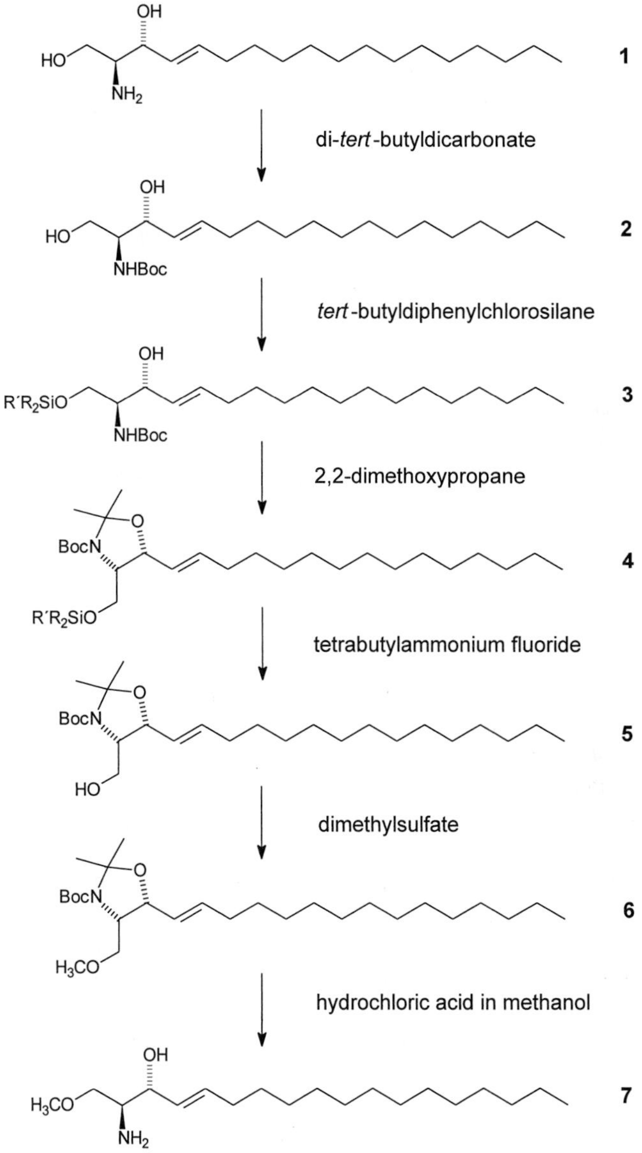

R´ = *tert*-butyl R = phenyl Boc = *tert*-butyloxycarbonyl

FIG. 1. Synthetic scheme for the preparation of 1-*O*-methyl-D-*erythro*-sphingosine.

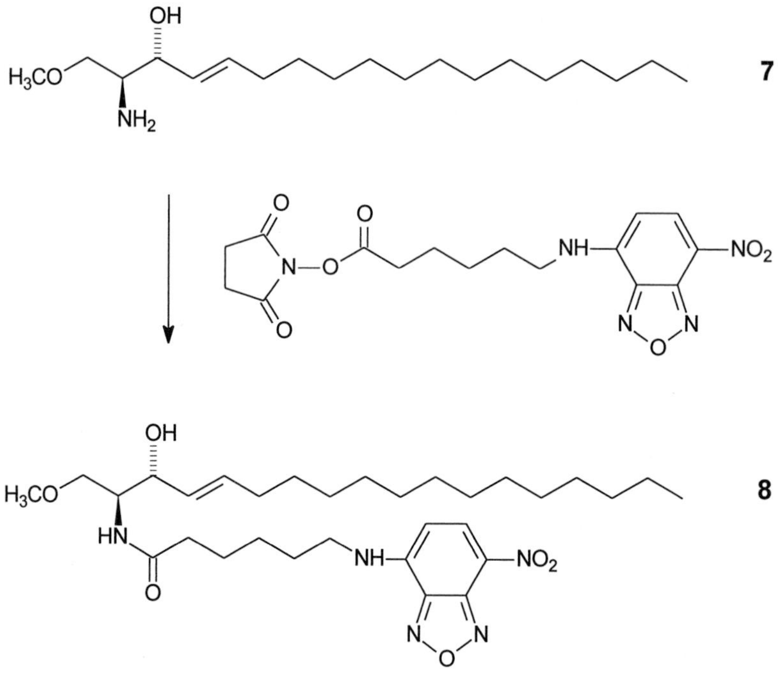

FIG. 2. Synthetic scheme for 1-O-methyl-NBD-C_6-ceramide.

using tetrabutylammonium iodide as a phase-transfer catalyst, both remaining protective groups are removed by acid-catalyzed hydrolysis.[20] The resulting 1-O-methyl-D-*erythro*-sphingosine is converted into a metabolically stable and fluorescent ceramide analog, 1-O-methyl-NBD-C_6-ceramide, with succinimidyl 6-(7-nitrobenz-2-oxa-1,3-diazol-4-yl)aminohexanoate as described for labeled gangliosides.[21]

It should be noted that another strategy for the introduction of a 1-O-methyl group into sphingosine using sodium methanolate as a nucleophilic agent and sphingosine derivatives with a good leaving group, such as mesyl or trifluoromethanesulfonyl, at position 1 failed completely. This approach gives rise mainly to elimination products.

Procedures

D-*erythro*-Sphingosine (*1*), i.e. (2S,3R,4E)-2-amino-octadecene-1,3-diol, is synthesized[20,22] or is obtained by acid hydrolysis of bovine brain sphingo-

---

[20] H.-E. Radunz, R. M. Devant, and V. Eiermann, *Liebigs Ann. Chem.* 1103 (1988).

[21] G. Schwarzmann and K. Sandhoff, *Methods Enzymol.* **138,** 319 (1987).

[22] P. Zimmermann and R. R. Schmidt, *Liebigs Ann. Chem.* 663 (1988).

lipids followed by chromatographic purification essentially as described.[23]
Synthetic sphingosine is also available from Sigma and Matreya.

2-N-tert-Butyloxycarbonyl-D-erythro-sphingosine (2). To an ice-cooled
mixture of 5 μl triethylamine and 130 mg (0.435 mmol) D-*erythro*-sphingo-
sine (1) in 2 ml THF, 105 mg (0.48 mmol) solid di-*tert*-butylcarbonate is
added with stirring. After 30 min at 0°, the solution is slowly warmed to
about 25° over 2 hr while stirring is continued. TLC in chloroform/
methanol/2 *N* NH$_3$ (90:10:1, v/v/v) should demonstrate the clean formation of the desired product 2 (R_f 0.54) at the expense of D-*erythro*-sphingo-
sine (1) (R_f 0.06) near the origin. The reaction is usually nearly complete.
The mixture is then evaporated to dryness under reduced pressure, yielding
about 170 mg (better than 95%) of product that is used in the next step
without prior purification.

*2-N-tert-Butyloxycarbonyl-1-O-tert-butyldiphenylsilyl-D-erythro-sphin-
gosine (3)*. For selective silylation of the primary hydroxyl group of 2-*N*-
tert-butyloxycarbonyl-D-*erythro*-sphingosine (2), the total amount obtained
in the first step (about 170 mg) is dissolved in 1.5 ml CH$_2$Cl$_2$. Triethylamine
(0.1 ml, 0.72 mmol) and 3 mg (20 μmol) 4-(dimethylamino)pyridine are
added to the solution. A slight excess of *tert*-butyldiphenylchlorosilane (0.12
ml, 0.479 mmol) is added quickly and the mixture is stirred for 2 hr in an
atmosphere of argon. After evaporation of the solvent in a stream of
nitrogen, the residue is chromatographed on a column of silica gel Si 60
(1.2 × 90 cm) with toluene/ethylacetate (93:7, v/v). The elution profile is
monitored by TLC in toluene/ethylacetate (93:7, v/v). Fractions containing
the desired pure condensation product (R_f 0.47) are pooled, dried, and
lyophilized from benzene to afford 210 mg (0.33 mmol, 79.7 %) of 3.

*2-N-tert-Butyloxycarbonyl-1-O-tert-butyldiphenylsilyl-2-N,3-O-isopro-
pylidene-D-erythro-sphingosine (4)*. A solution of 210 mg of 3 (0.33 mmol)
and 2 mg *p*-toluenesulfonic acid (as catalyst) in 2 ml 2,2-dimethoxypropane
(the reagent) is heated under reflux for 3 hr. TLC of the cooled reaction
mixture in toluene/ethylacetate (93:7, v/v) should reveal the clean formation of the product 4 (R_f 0.80). Following the addition of 0.2 ml triethylamine, most of the solvent is evaporated in a nitrogen stream, leaving a
residue that is dissolved in 2 ml *n*-hexane, washed with water, dried with
MgSO$_4$, filtered, and concentrated under a stream of nitrogen. The entire
product is used for the next reaction step.

It warrants mention that the addition of a volatile base such as triethylamine or *N*,*N*-diisopropylethylamine prior to evaporation of most of 2,2-
dimethoxypropane prevents the partial hydrolysis of the isopropylidene
compound 4. Without a base, the reaction can be reversed, and a significant

[23] F. Sarmientos, G. Schwarzmann, and K. Sandhoff, *Eur. J. Biochem.* **146,** 59 (1985).

amount of **4** can be destroyed due to *p*-toluenesulfonic acid being concentrated on evaporation of the solvent.

2-N-tert-Butyloxycarbonyl-2-N,3-O-isopropylidene-D-erythro-sphingosine (5). A 1 *M* solution of tetrabutylammmonium fluoride in 0.5 ml THF is added to a solution of **4** in 2 ml THF. The mixture is stirred for 2 hr at 35°. After evaporation of the solvent in a stream of nitrogen, the residue is chromatographed on a column of silica gel Si 60 (1.2 × 90 cm) with toluene/ethylacetate (93 : 7, v/v). The elution profile is monitored by TLC in toluene/ethylacetate (93 : 7, v/v). Fractions containing the desired pure condensation product (R_f 0.16) are pooled, solvents are evaporated in a stream of nitrogen, and the residue is lyophilized from benzene to afford 83 mg of **5** (0.193 mmol, 57.3% with respect to **3**).

1-O-Methyl-2-N-tert-butyloxycarbonyl-2-N,3-O-isopropylidene-D-erythro-sphingosine (6). A solution of 83 mg (0.193 mmol) of **5** and 3 mg of the phase-transfer catalyst tetrabutylammmonium iodide in 2 ml hexane/50% NaOH (1 : 1, v/v) is stirred vigorously at room temperature. After 30 min, the solution is cooled to 0° and, while stirring, dimethyl sulfate (0.1 ml) is slowly added prior to removing the cooling bath. Now the temperature is allowed to rise gradually to room temperature. Stirring is continued for 20 hr at room temperature. The reaction mixture is cooled to 0° and most of NaOH is neutralized by the careful addition of dilute sulfuric acid. After the addition of 5 ml ethyl acetate, the reaction product is extracted into the organic upper layer. The upper phase is washed with water, dried over anhydrous sodium sulfate, and evaporated in a nitrogen stream. The residue is dissolved in 2 ml toluene/ethylacetate (8 : 2, v/v) and chromatographed on a column of silica gel Si 60 (1.2 × 90 cm) with toluene/ethylacetate (93 : 7, v/v). The elution profile is monitored by TLC in toluene/ethylacetate (93 : 7, v/v). Fractions containing the desired pure product (R_f 0.39) are pooled, solvents are evaporated in a nitrogen stream, and the residue is lyophilized from benzene to afford 71 mg (0.156 mmol, 80.8%) of **6**.

1-O-Methyl-D-erythro-sphingosine (7). Both the 2-*N-tert*-butyloxycarbonyl and the 2-*N*,3-*O*-isopropylidene protecting groups are removed at once by acid-catalyzed hydrolysis. Thus, a mixture of 7 *N* HCl and methanol (2 : 5, v/v) (0.5 ml) is added to a solution of 71 mg of **6** (0.156 mmol) in 0.5 ml methanol. The reaction mixture is stirred for 20 min at 40°. After evaporation of the solvent in a nitrogen jet, the residue is dissolved in 1 ml chloroform/methanol/2 *N* NH₃ (40 : 10 : 1, v/v/v) and chromatographed on a column of silica gel Si 60 (0.9 × 90 cm) with chloroform/methanol/2 *N* NH₃ (40 : 10 : 1, v/v/v). The elution profile is checked by TLC in the same mobile phase. Fractions containing the pure product (R_f 0.66) are pooled. After removal of the solvents under reduced pressure, the residue is dissolved in water and freeze-dried to yield 31 mg (99 μmol, 63%) of pure **7**.

1-O-Methyl-C₆-NBD-ceramide (8). For synthesis of 1-*O*-methyl-C₆-NBD-ceramide, a solution of 5.9 mg (15 μmol) *N*-succinimidyl-6-(7-nitrobenz-2-oxa-1,3-diazol-4-yl)aminohexanoate in 0.2 ml dry DMF is added to a solution of 4.1 mg (13 μmol) of 1-*O*-methyl-D-*erythro*-sphingosine (**7**) and 20 μl *N,N*-diisopropylethylamine in 0.3 ml dry DMF. The mixture is kept for 20 hr at 40°. After evaporation of the solvent, the residue is dissolved in a small volume of chloroform/methanol (1 : 1, v/v) and applied as a streak onto thin-layer plates for purification with chloroform/methanol/water (80 : 10 : 1, v/v/v) as the mobile phase. Compound **8** (R_f 0.69) can be detected easily by its fluorescence when exited by UV light and discriminated from remaining *N*-succinimidyl-6-(7-nitrobenz-2-oxa-1,3-diazol-4-yl)aminohexanoate, if any, or free 6-(7-nitrobenz-2-oxa-1,3-diazol-4-yl)aminohexanoic acid. Only the sphingoid-containing material will yield a blue greenish color on spraying with anisaldehyde reagent and heating on a separate plate. The corresponding band is scraped from the thin-layer plate, and the fluorescent ceramide analog is eluted with 1-ml portions of methanol until no more fluorescent material remains with the silica gel. The methanolic solution is diluted with an equal volume of water and applied onto a small column of reversed-phase material (LiChroprep RP-18). Whereas the lipid is adsorbed to the column material, free 6-(7-nitrobenz-2-oxa-1,3-diazol-4-yl)aminohexanoic acid is not. Coeluted silicic acid as well as binder are washed out with water. Compound **8** is then eluted with methanol. After evaporation and drying, 5.3 mg product (9 μmol, 69%) is obtained.

It should be mentioned that in this microscale preparation small amounts of impurities such as ammonia or amines may quench the yield by reacting with the *N*-succinimidyl ester. This unwanted reaction can be minimized by drying in a nitrogen stream the solution of 1-*O*-methyl-D-*erythro*-sphingosine (**7**) in DMF in the presence of 20 μl *N,N*-diisopropylethylamine before dissolving again in 0.3 ml DMF and 20 μl *N,N*-diisopropylethylamine. This less volatile tertiary amine ousts the reactive ones during evaporation. Moisturizing TLC plates with a fine spray of methanol/water (1 : 1, v/v) facilitates scraping of bands. In addition to 1-*O*-methyl-C₆-NBD-ceramide, other metabolically stable ceramides may be prepared using other fluorescent or radioactive and activated fatty acids of various chain lengths (see next section) such as *N*-succinimidyl-8-(7-nitrobenz-2-oxa-1,3-diazol-4-yl)aminooctanoate and *N*-succinimidyl-[1-¹⁴C]octanoate.[24]

---

[24] B. Albrecht, U. Pütz, and G. Schwarzmann, *Carbohydr. Res.* **276**, 289 (1995).

Characterization of 1-O-Methyl-D-erythro-sphingosine (7). The methyl ether **7** is best characterized by NMR spectroscopy and mass spectrometry. The following data are obtained: ^{1}H NMR (500 MHz, CDCl$_3$): 5.72 (H-5), 5.44 (H-4), 4.06 (H-3), 3.47 (H-1), 3.35 (O-CH$_3$), 3.00 (H-2), 2.35 (NH$_2$, OH), 2.04 (H-6), 1.40–1.20 (H-7-H-17), 0.87 (H-18). ^{13}C NMR (125 MHz, CDCl$_3$): 134.25 (C-5), 128.98 (C-4), 74.51 (C-1), 74.46 (C-3), 59.14 (C-19), 54.86 (C-2), 32.42 (C-6), 31.99 (C-16), 29.74-29.26 (C-7-C-15), 22.74 (C-17), 14.15 (C-18). FABMS (C$_{19}$H$_{39}$NO$_2$; 313.53 g/mol) MH$^+$ at *m/z* 314.

Catabolically Stable Fluorescent and Radioactive Glucosylceramides

Principle

To render glucosylceramide resistant to degradation, the replacement of the *O*-glycosidic by an S-glycosidic bond between the glucose to the ceramide moiety is of particular advantage. The formal similarity of oxygen and sulfur is given by their possessing analogous outer valence shells. However, sulfur is less basic than oxygen and thus has the lesser affinity for protons. Hence, as a result the thioglycosidic linkage is relatively resistant to the action of glycohydrolases[25] and is not cleavable by human glucocerebrosidase.[24] However, sulfur has a much greater nucleophilicity than oxygen and this feature can be exploited for the synthesis of 1-thioglycosides.

The synthesis of glucosylceramide analogs bearing a β-thioglycosidic bond between the sugar residue and the sphingosine moiety can be achieved by two different routes. In the first route, the sulfur atom is introduced into the 1 position of a protected sphingosine precursor before condensation with the glycosyl donor.[26] In the second route the sugar is first converted to a derivative containing the sulfur atom in a β-anomeric linkage prior to condensation with a sphingosine or sphingoid derivative bearing a good leaving group.[27] If sphingosine has to be synthesized because of an anomalous chain length, both synthetic routes are appropriate. This section describes the synthesis of glucosylthioceramide by the second route, which makes use of an appropriately protected sphingosine containing iodine as a good leaving group (see Figs. 3 and 4 for synthetic steps).

Procedures

D-erythro-Sphingosine is obtained as described in the first section.

(2S,3R,4E)-2-Dichloroacetamido-4-octadecen-1,3-diol (9). To a solution of D-*erythro*-sphingosine (750 mg, 2.5 mmol) in dry methanol (40 ml) and

[25] D. Horton and D. H. Hutson, *Adv. Carbohydr. Chem. Biochem.* **18**, 123 (1963).
[26] T. Bär and R. R. Schmidt, *Liebigs Ann. Chem.* 185 (1991).
[27] A. Hasegawa, M. Morita, Y. Kojima, H. Ishida, and M. Kiso, *Carbohydr. Res.* **214**, 43 (1991).

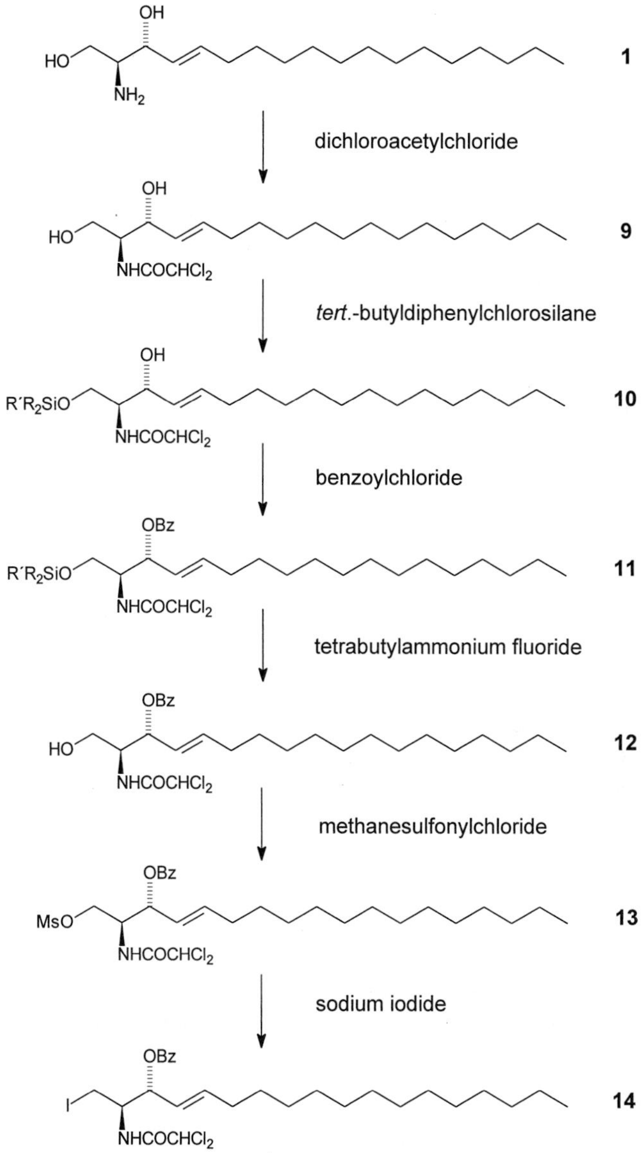

R´ = *tert*-butyl   R = phenyl   Bz = benzoyl   Ms = methanesulfonyl

FIG. 3. Synthetic scheme for glucosylthiosphingosine.

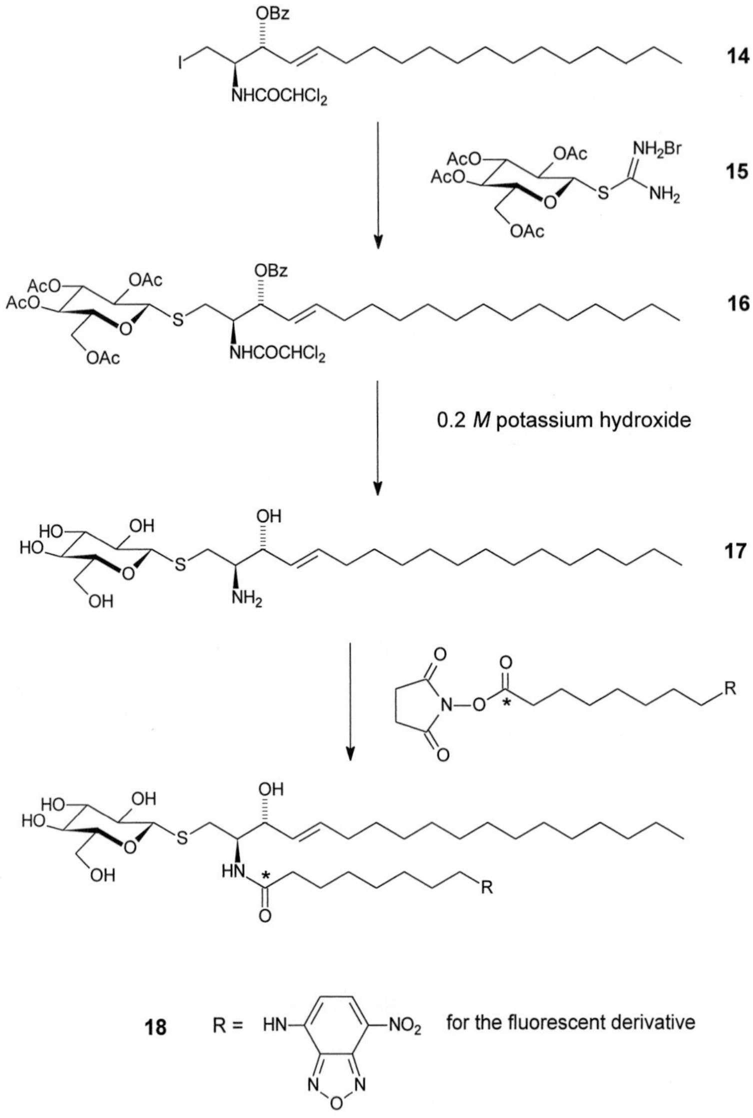

18   R = HN—⬡—NO₂   for the fluorescent derivative

19   R = H for the radioactive derivative

*   position of radiocarbon for the nonfluorescent derivative

Ac = acetyl   Bz = benzoyl

FIG. 4. Synthetic scheme for labeled glucosylthioceramides.

N,N-diisopropylethylamine (0.6 ml), cooled to $-20°$, aliquots (0.2 ml) of a solution of dichloroacetyl chloride (0.77 ml, 8.0 mmol) in dry dichloromethane (4.23 ml) are added over a period of 1 hr with vigorous stirring. The addition of dichloroacetyl chloride is continued until no more ninhydrin-positive material can be detected. The N-acylation is usually complete after a total of 4.5–5 ml of dichloroacetyl chloride in dichoromethane has been added. Following warm up and the addition of water (40 ml), the reaction product is isolated by extraction into ethyl acetate (75 ml). After evaporation of the solvent, the crude product (usually one single spot on TLC in ethyl acetate; R_f 0.42) is freeze-dried from benzene to give **9** (1000 mg, 2.43 mmol, 97%). The product is sufficiently pure for the next reaction step. FABMS: ($C_{20}H_{37}Cl_2NO_3$, 410.38 g/mol) MNa$^+$ at m/z 432, 434, 436, and MH$^+$-water at m/z 392, 394, 396.

*(2S,3R,4E)-1-O-tert-Butyldiphenylsilyl-2-dichloroacetamido-4-octa-decen-1,3-diol (**10**).* To a solution of 1000 mg (2.43 mmol) (2S,3R,4E)-2-dichloroacetamido-4-octadecen-1,3-diol (**9**) in dry dichloromethane (20 ml) and tetrahydrofuran (2 ml), triethylamine (0.6 ml, 4.33 mmol) and N,N-dimethylaminopyridine (20 mg, 0.163 mmol) are added. Then *tert*-butyldi-phenylsilylchloride (0.37 ml, 2.7 mmol) is added and the solution is stirred for 7 hr at room temperature. The reaction is stopped by adding methanol (0.5 ml) and the solvent is evaporated under nitrogen. The residue is dissolved in 5 ml of *n*-hexane/ethyl acetate (8:1, v/v) and chromatographed on silica gel Si 60 (200 g) using *n*-hexane/ethyl acetate (8:1, v/v) as eluent to afford 1505 mg (2.31 mmol, 95%) of **10**; R_f 0.67 (*n*-hexane/ethyl acetate, 1:1, v/v). FABMS: ($C_{36}H_{55}Cl_2NO_3Si$, 648.83 g/mol) MNa$^+$ at m/z 670, 672, 674, and MH$^+$-water at m/z 630, 632, 634.

*(2S,3R,4E)-3-O-Benzoyl-1-O-tert-butyldiphenylsilyl-2-dichloroacet-amido-4-octadecen-1,3-diol (**11**).* To a cold (0°) solution of 1500 mg (2.3 mmol) (2S,3R,4E)-1-O-*tert*-butyldiphenylsilyl-2-dichloroacetamido-4-octa-decen-1,3-diol (**10**) in pyridine (5 ml), benzoyl chloride (0.53 ml, 4.6 mmol) is added and the suspension is stirred for 1 hr at 0° and then overnight at room temperature. After the addition of 0.1 M hydrogen chloride (120 ml), the product is extracted into ethyl acetate (60 ml). The ethyl acetate layer is washed with water, 0.1 M sodium hydrogen carbonate. After ethyl acetate is evaporated, chromatography of the crude product on silica gel Si 60 (200 g) with *n*-hexane/ethyl acetate (8:1, v/v) gives 1686 mg of **11** (2.23 mmol, 96%); R_f 0.58 (*n*-hexane/ethyl acetate, 4:1, v/v). FABMS: ($C_{43}H_{59}Cl_2NO_4Si$, 752.94 g/mol) MNa$^+$ at m/z 774, 776, 778, and MH$^+$ at m/z 752, 754, 756, and MH$^+$-benzoic acid at m/z 630, 632, 634.

*(2S,3R,4E)-3-O-Benzoyl-2-dichloroacetamido-4-octadecen-1,3-diol (**12**).* The solution of 1650 mg (2.18 mmol) (2S,3R,4E)-3-O-benzoyl-1-O-*tert*-butyldiphenylsilyl-2-dichloroacetamido-4-octadecen-1,3-diol (**11**) in

0.17 M tetrabutylammonium fluoride in dry tetrahydrofuran (18 ml) is stirred overnight at 35°. Following the addition of 1 M hydrogen chloride (10 ml), the product is extracted into ethyl acetate (60 ml). Evaporation of the organic solvent furnishes a crude product that is purified by silica gel chromatography (200 g) with n-hexane/ethyl acetate (7 : 3, v/v) to deliver **12** (785 mg, 1.53 mmol, 70%) (R_f 0.29, n-hexane/ethyl acetate, 7 : 3, v/v). ^{1}H NMR (500 MHz, CDCl$_3$): Sphingosine unit δ 0.88 (t, 3 H, CH$_3$), 1.22–1.29 (m, 20 H, 10 CH$_2$), 1.36 (m, 2 H, H-7), 2.05 (m, 2 H, H-6), 2.65 (m, 1 H, OH), 3.71 (m, 1 H, H-1a), 3.80 (m, 1 H, H-1b), 4.21 (m, 1 H, H-2), 5.59 (dd, 1 H, $J_{2,3}$ = 8 Hz, $J_{3,4}$ = 8 Hz, H-3), 5.61 (dd, 1 H, $J_{3,4}$ = 8 Hz, $J_{4,5}$ = 15 Hz, H-4), 5.91 (dd, 1 H, $J_{5,6}$ = 7 Hz, $J_{4,5}$ = 15 Hz, H-5), 7.05 (d, 1 H, $J_{NH,2}$ = 9 Hz, NH); protecting group unit δ 5.92 (s, 1 H, CHCl$_2$), 7.47 (m, 2 H, H-Bz$_{3,5}$), 7.60 (m, 1 H, H-Bz$_4$), 8.05 (m, 2 H, H-Bz$_{2,6}$); FABMS: (C$_{27}$H$_{41}$Cl$_2$NO$_4$, 514.53 g/mol) MNa$^+$ at m/z 536, 538, 540 and MH$^+$-benzoic acid at m/z 392, 394, 396; anal. calculated for C$_{27}$H$_{41}$Cl$_2$NO$_4$ (514.53): C, 63.03; H, 8.03; N, 2.72. Found: C, 63.19; H, 8.36; N, 2.75.

(2S,3R,4E)-3-O-Benzoyl-2-dichloroacetamido-1-O-methanesulfonyl-4-octadecen-1,3-diol (13). To a cooled ($-10°$) solution of (2S,3R,4E)-3-O-benzoyl-2-dichloroacetamido-4-octadecen-1,3-diol (**12**, 410 mg, 0.8 mmol) in pyridine (1.3 ml), methanesulfonylchloride (93 μl, 1.2 mmol) is added and the mixture is kept overnight at 0°. After the mixture is diluted with cold 1 M hydrogen chloride (40 ml), the product is extracted into dichloromethane (40 ml). The organic layer is washed with cold water (20 ml) and 1 M sodium hydrogen carbonate (20 ml). After evaporation of the solvent, the product is purified by silica gel chromatography (200 g) with n-hexane/ethyl acetate (7 : 3, v/v). Fractions are monitored by TLC in n-hexane/ethyl acetate (7 : 3, v/v) as the mobile phase, and those containing the pure product (R_f 0.30) are collected. After evaporation of the solvent, the residue is freeze-dried from benzene to yield 427 mg (0.72 mmol, 90%) of **13**. FABMS: (C$_{28}$H$_{43}$Cl$_2$NO$_6$S, 592.61 g/mol) MH$^+$ at m/z 592, 594, 596 and MH$^+$-benzoic acid at m/z 470, 472, 474; anal. calculated for C$_{28}$H$_{43}$Cl$_2$NO$_6$S (592.61): C, 56.75; H, 7.31; N, 2.36. Found: C, 56.95; H, 7.45; N, 2.30.

(2R,3R,4E)-3-O-Benzoyl-2-dichloroacetamido-1-iodo-4-octadecen-3-ol (14). A solution of 1.5 M sodium iodide (2 ml, 3 mmol) in acetone is added to a solution of 427 mg (0.72 mmol) (2S,3R,4E)-3-O-benzoyl-2-dichloroacetamido-1-O-methanesulfonyl-4-octadecen-1,3-diol (**13**) in acetone (3 ml) and stirred for 7 hr at 60°. After the addition of cold water (20 ml), the product is extracted into dichloromethane (20 ml). Iodine formed during the reaction is reduced by a drop of saturated sodium hydrogen sulfite, and the organic phase is washed once with cold water. After evaporation of dichloromethane, the crude product is freeze-dried from benzene to yield **14** (425 mg, 0.68 mmol, 94%). As this compound usually gives one

single main spot on TLC (R_f 0.52, *n*-hexane/ethyl acetate, 4:1, v/v), it is used without further purification. FABMS: ($C_{27}H_{40}Cl_2INO_3$, 624.43 g/mol) MH^+ at *m/z* 624, 626, 628 and MH^+-benzoic acid at *m/z* 502, 504, 506; anal. calcd for $C_{27}H_{40}Cl_2INO_3$ (624.43): C, 51.94; H, 6.46; N, 2.24. Found: C, 51.35; H, 6.57; N, 2.15.

Note that in the following coupling reaction, 2-*S*-(2,3,4,6-tetra-*O*-acetyl-β-D-glucopyranosyl)-2-thiopseudourea hydrobromide (**15**) is used instead of the commercially available 1-thio-β-D-glucose tetraacetate. This compound and 2-*S*-(2,3,4,6-tetra-*O*-acetyl-β-D-galactopyranosyl)-2-thiopseudourea hydrobromide (**27**) used in the next section are prepared readily from the corresponding 2,3,4,6-tetra-*O*-acetyl-α-D-hexopyranosyl bromides essentially as described.[28]

Protected Glucosylthiosphingosine, i.e., S-(2,3,4,6-Tetra-O-acetyl-β-D-glucopyranosyl)-(1 → 1)-(2R,3R,4E)-3-benzoyloxy-2-dichloroacetamido-4-octadecen-1-thiol (16). Under an argon atmosphere, 2-*S*-(2,3,4,6-tetra-*O*-acetyl-β-D-glucopyranosyl)-2-thiopseudourea hydrobromide (**15**, 245 mg, 0.50 mmol) is dissolved in 1 *M* methanolic sodium acetate (0.6 ml). This solution is added to a solution of (2*R*,3*R*,4*E*)-3-*O*-benzoyl-2-dichloroacetamido-1-iodo-4-octadecen-3-ol (**14**, 125 mg, 0.20 mmol) in 1.4 ml acetone/methanol (1:1, v/v). After the addition of solid potassium carbonate (42 mg, 0.30 mmol), the mixture is stirred for 6 hr at 20°. Potassium carbonate is removed by centrifugation, washed once with acetone (0.6 ml), and again centrifuged off. To buffer the remaining salts, acetic acid (10 μl) is added before drying the clear supernatant in a nitrogen jet. The resulting syrup is chromatographed on a column of silica gel Si 60 (50 g) using a linear gradient from *n*-hexane (300 ml) to ethyl acetate (300 ml). Fractions containing the pure product (R_f 0.23, *n*-hexane/ethyl acetate, 7:3, v/v) are freeze-dried from benzene to afford **16** (121 mg, 0.14 mmol, 70% with respect to **14**) as an amorphous mass. ^{1}H NMR (500 MHz, CDCl$_3$): Glucose unit δ 3.75 (ddd, 1 H, $J_{5,6a}$ = 2 Hz, $J_{5,6b}$ = 5 Hz, $J_{4,5}$ = 9.5 Hz, H-5), 4.14 (dd, 1 H, $J_{5,6a}$ = 2 Hz, $J_{6a,6b}$ = 13 Hz, H-6a), 4.23 (dd, 1 H, $J_{5,6b}$ = 5 Hz, $J_{6a,6b}$ = 13 Hz, H-6b), 4.51 (d, 1 H, $J_{1,2}$ = 10 Hz, H-1), 5.03 (dd, 1 H, $J_{2,3}$ = 9.5 Hz, $J_{1,2}$ = 10 Hz, H-2), 5.08 (dd, 1 H, $J_{4,5}$ = 9.5 Hz, $J_{3,4}$ = 10 Hz, H-4), 5.22 (dd, 1 H, $J_{2,3}$ = 9.5 Hz, $J_{3,4}$ = 10 Hz, H-3); sphingosine unit δ 0.87 (t, 3 H, CH$_3$), 1.2-1.3 (m, 20 H, 10 CH$_2$), 1.36 (m, 1 H, H-7), 2.06 (m, 2 H, H-6), 2.93 (dd, 1 H, $J_{1a,2}$ = 7.5 Hz, $J_{1a,1b}$ = 14 Hz, H-1a), 3.06 (dd, 1 H, $J_{1b,2}$ = 5 Hz, $J_{1a,1b}$ = 14 Hz, H-1b), 4.47 (m, 1 H, H-2), 5.50 (dd, 1 H, $J_{3,4}$ = 6.5 Hz, $J_{4,5}$ = 15 Hz, H-4), 5.67 (dd, 1 H, $J_{2,3}$ = 6.5 Hz, $J_{3,4}$ = 6.5 Hz, H-3), 5.95 (dd, 1 H, $J_{5,6}$ = 7 Hz, $J_{4,5}$ = 15 Hz, H-5), 6.86 (d, 1 H, $J_{NH,2}$ = 7 Hz, NH); protecting group unit δ 2.00, 2.02, 2.02, 2.03 (4s, 12 H,

---

[28] C. Stowell and Y. Lee, *Methods Enzymol.* **83**, 281 (1982).

4 COCH$_3$), 5.94 (s, 1 H, CHCl$_2$), 7.46 (m, 2 H, H-Bz$_{3,5}$), 7.58 (m, 1 H, H-Bz$_4$), 8.04 (m, 2 H, H-Bz$_{2,6}$); ^{13}C NMR (125 MHz, CDCl$_3$): glucose unit δ 62 (1 C, C-6), 68 (1 C, C-4), 70 (1 C, C-2), 74 (1 C, C-3), 76 (1 C, C-5), 84 (1 C, C-1); sphingosine unit δ 14 (1 C, CH$_3$), 23 (1 C, C-17), 29-33 (11 C, 11 CH$_2$), 52 (1 C, C-2), 32 (1 C, C-1), 75 (1 C, C-3), 123 (1 C, C-4), 138 (1 C, C-5); protecting group unit δ 21 (4 C, 4 COCH$_3$), 66 (1 C, CHCl$_2$), 129 (2 C, C-Bz$_{2,6}$), 130 (3 C, C-Bz$_1$, C-Bz$_{3,5}$), 133 (1 C, C-Bz$_4$), 164 (1 C, NHCO), 165 (1 C, PhCO); 169–170 (4 C, 4 COCH$_3$). FABMS: (C$_{41}$H$_{59}$Cl$_2$NO$_{12}$S, 860.88 g/mol) MH$^+$ at m/z 860, 862, and 864, MH$^+$ - benzoic acid at m/z 738, 740, and 742, MNa$^+$ at m/z 882, 884, and 886; anal. calcd. for C$_{41}$H$_{59}$Cl$_2$NO$_{12}$S (860.88): C, 57.20; H, 6.91; N, 1.63. Found: C, 56.67; H, 6.61; N, 1.51.

   Glucosylthiosphingosine, i.e., S-(β-D-Glucopyranosyl)-(1 → 1)-(2R,3R,4E)-2-amino-3-hydroxy-4-octadecen-1-thiol (17). S-(2,3,4,6-Tetra-O-acetyl-β-D-glucopyranosyl)-(1 → 1)-(2R,3R,4E)-3-O-benzoyloxy-2-dichloroacetamido-4-octadecen-1-thiol (**16**, 50 mg, 58 μmol) is treated, under argon, for 3 hr at 65° with 0.2 M potassium hydroxide in methanol (12 ml). After cooling to room temperature, the solution is buffered with acetic acid (0.16 ml) and diluted with water (36 ml). The mixture thus obtained is passed over a column of LiChroprep RP 18 (10 ml) to adsorb the saponified product. Salts and other water-soluble materials are washed out with water and the retained lipid is subsequently eluted with methanol. After evaporation of methanol, the glucosylthiosphingosine **17** is purified on silica gel Si 60 (50 g), using chloroform/methanol/2.5 M ammonia (65 : 25 : 4, v/v/v) as eluent, to afford 24.5 mg **17** (51 μmol, 88%) (R_f 0.26, chloroform/methanol/2.5 M ammonia, (65 : 25 : 4, v/v/v). FABMS: (C$_{24}$H$_{47}$NO$_6$S, 477.70 g/mol) MH$^+$ at m/z 478.

   Fluorescent Glucosylthioceramide, i.e., S-(β-D-Glucopyranosyl)-(1 → 1)-(2R,3R,4E)-3-hydroxy-2-(8-N-(7-nitrobenz-1,3-diazol-2-oxa-4-yl)-amino)-octanamido-4-octadecen-1-thiol (18).[24] S-(β-D-Glucopyranosyl)-(1 → 1)-(2R,3R,4E)-2-amino-3-hydroxy-4-octadecen-1-thiol (**17**) is N-acylated with N-succinimidyl-8-N-(7-nitrobenz-2-oxa-1,3-diazol-4-yl)aminooctanoate as described for the preparation of 1-O-methyl-C$_6$-NBD-ceramide (**8**). Briefly, glycosylthiosphingosine **17** (10 μmol) is dissolved in DMF (0.150 ml) and N,N-diisopropylethylamin (10 μl). Solvents are removed in a nitrogen jet to ensure elimination of volatile amino and imino compounds that will interfere with the N-acylation reaction. The residue, redissolved in DMF (0.150 ml) and N,N-diisopropylethylamine (10 μl), is mixed with N-succinimidyl-[8-N-(7-nitrobenz-2-oxa-1,3-diazol-4-yl]amino)octanoate[24] (15 μmol in 0.150 ml DMF) and stirred, under argon, for 48 hr at 40°. After evaporation of the solvents in a nitrogen jet followed by drying under reduced pressure, the crude product is dissolved in 1 ml methanol/water

(85:15, v/v) and purified by HPLC on ProSep C18 (1 × 25 cm) with methanol/water (85:15, v/v) to yield **18** (9 μmol, 90%). This compound should give one single fluorescent spot on TLC plates (R_f 0.75, chloroform/methanol/15 mM calcium chloride, 60:35:8, v/v/v). FABMS: ($C_{38}H_{63}N_5O_{10}S$, 782.00 g/mol) MH$^+$ at m/z 782, MH$^+$-water at m/z 764, MNa$^+$ at m/z 804.

Radioactive Glucosylthioceramide, i.e., S-(β-D-Glucopyranosyl)-(1 → 1)-(2R,3R,4E)-2-[1-^{14}C]-octanamido-3-hydroxy-4-octadecen-1-thiol (**19**). As radioactive compounds may decompose on prolonged storage, only small amounts are prepared at one time. Thus, 5 μmol of glucosylthiosphingosine **17** is N-acylated with about 7 to 8 μmol of N-succinimidyl-[1-^{14}C]octanoate[24] as described for the fluorescent derivative **18**. The product is purified by HPLC on ProSep C18 (1 × 25 cm) with methanol/water (85:15, v/v) to afford **19** (4.5 μmol, 90%), which should yield one radioactive spot on TLC plates (R_f 0.70, chloroform/methanol/15 mM calcium chloride, 60:35:8, v/v/v). FABMS: ($C_{32}H_{61}NO_7S$, 603.89 g/mol) MH$^+$ at m/z 604, MH$^+$-water at m/z 586, and MNa$^+$ at m/z 626.

For the preparation of N-succinimidyl-[1-^{14}C]octanoate, commercially available sodium [1-^{14}C]octanoate is used. We have found that it is best to place this salt (15–20 μmol) in 0.1 M hydrochloric acid (1 ml) and to extract [1-^{14}C]octanoic acid into *n*-hexane (2 ml). The extract is dried carefully over anhydrous sodium sulfate and the *n*-hexane phase is evaporated in a stream of nitrogen but for a small residue. This ensures that less than 1% of radioactive octanoic acid is evaporated, if at all. The residue is then taken up in dry DMF (0.15–0.20 ml) for conversion into N-succinimidyl-[1-^{14}C]octanoate. Alternatively, sodium [1-^{14}C]octanoate can be dissolved directly in an appropriate amount of dry DMF and titrated with an equimolar amount of dry hydrochloric acid in THF to release the free octanoic acid with formation of sodium chloride.

N-Succinimidyl-(8-N-(7-nitrobenz-2-oxa-1,3-diazol-4-yl)-amino)octanoate and N-Succinimidyl-[1-^{14}C]octanoate. Briefly, these N-succinimidyl esters are prepared as follows: 8-N-(7-nitrobenz-2-oxa-1,3-diazol-4-yl)-aminooctanoic acid[24] (34 mg, 105 μmol) and [1-^{14}C]octanoic acid (2.9 mg, 20 μmol), respectively, are dissolved in dry DMF (1 and 0.20 ml, respectively). Following the addition of N,N'-dicyclohexylcarbodiimide (22 mg, 106 μmol and 4.33 mg, 21 μmol, respectively) and N-hydroxysuccinimide (13 mg, 113 μmol and 2.76 mg, 24 μmol, respectively) the reaction mixtures are stirred, under argon, for 2 days at 25°. N,N'-Dicyclohexylurea formed during the reactions is sedimented by centrifugation for 5 min at 2000g and the clear supernatants are stored at −20°, under argon, until use of the ester(s) without further purification. The yield for both esters is usually in the range of 80 to 88%, as determined by fluorescence and radio scanning of the appropriate bands from TLC plates; R_f 0.60 and 0.58 (ethyl acetate) for N-succinimidyl-[8-N-(7-nitrobenz-2-oxa-1,3-diazol-4-yl)-amino]octanoate and N-succinimidyl-[1-^{14}C]octanoate, respectively.

Catabolically Stable Fluorescent and Radioactive Lactosylceramides

Principle

This section describes the synthesis of lactosylceramide analogs (see Figs. 5 and 6) that contain sulfur in the interglycosidic bond and that are thus resistant to the action of β-galactosidases. To simplify their preparation, use can be made of readily available galactosylceramide from bovine brain.[29] This approach takes advantage of an already existing correct glycosidic bond between the sugar and the lipid part. Galactosylsphingosine is prepared from galactosylceramide by simple alkaline hydrolysis.[29] The attack by the sulfur-containing nucleophile, generated *in situ* from 2-S-(2,3,4,6-tetra-O-acetyl-β-D-galactopyranosyl)-2-thiopseudourea hydrobromide,[28] in an S_N2 reaction of C-4 of an appropriate galactosylsphingosine derivative affords the β-thioglycosidic linkage, with subsequent conversion into the gluco configuration of the sphingosine-bound galactose residue (Fig. 6). The necessary galactosylsphingosine derivative can be obtained via simple chemistry in an overall good yield as outlined in Fig. 5. An intermediate p-methoxybenzylidene derivative is formed that can be selectively benzoylated at position 6 after perbenzoylation and subsequent removal of the benzylidene group due to the known order of reactivity of OH-6 and OH-4 toward benzoylation. The remaining free hydroxy group is converted into the triflate prior to formation of the thioglycosidic bond between the galactose and the glucose moieties.

Procedures

Galactosylsphingosine, i.e., O-(β-D-galactopyranosyl)-(1 → 1)-(2S,3R,4E)-2-amino-4-octadecen-1,3-diol, is prepared from galactosylceramide by alkaline hydrolysis.[29]

O-(β-D-*Galactopyranosyl*)-(1 → 1)-(2S,3R,4E)-2-dichloroacetamido-4-ooctadecen-1,3-diol (**20**). A solution of galactosylsphingosine (420 mg, 0.910 mmol) in dry methanol (20 ml) and N,N-diisopropylethylamine (0.6 ml) is cooled to $-20°$. Aliquots (0.20 ml) of a solution of dichloroacetyl chloride (0.77 ml, 8.0 mmol) in dry dichloromethane (4.23 ml) are then added over a period of 1 to 2 hr with vigorous stirring. The addition of dichloroacetyl

[29] P. Kelly, C. Flanagan, M. Basu, M. Miguel, U. Bradley, E. Ahmad, and S. Basu, *Proc. Int. Glyco Meeting,* Aug 22–26, Tokyo in press, 1991.

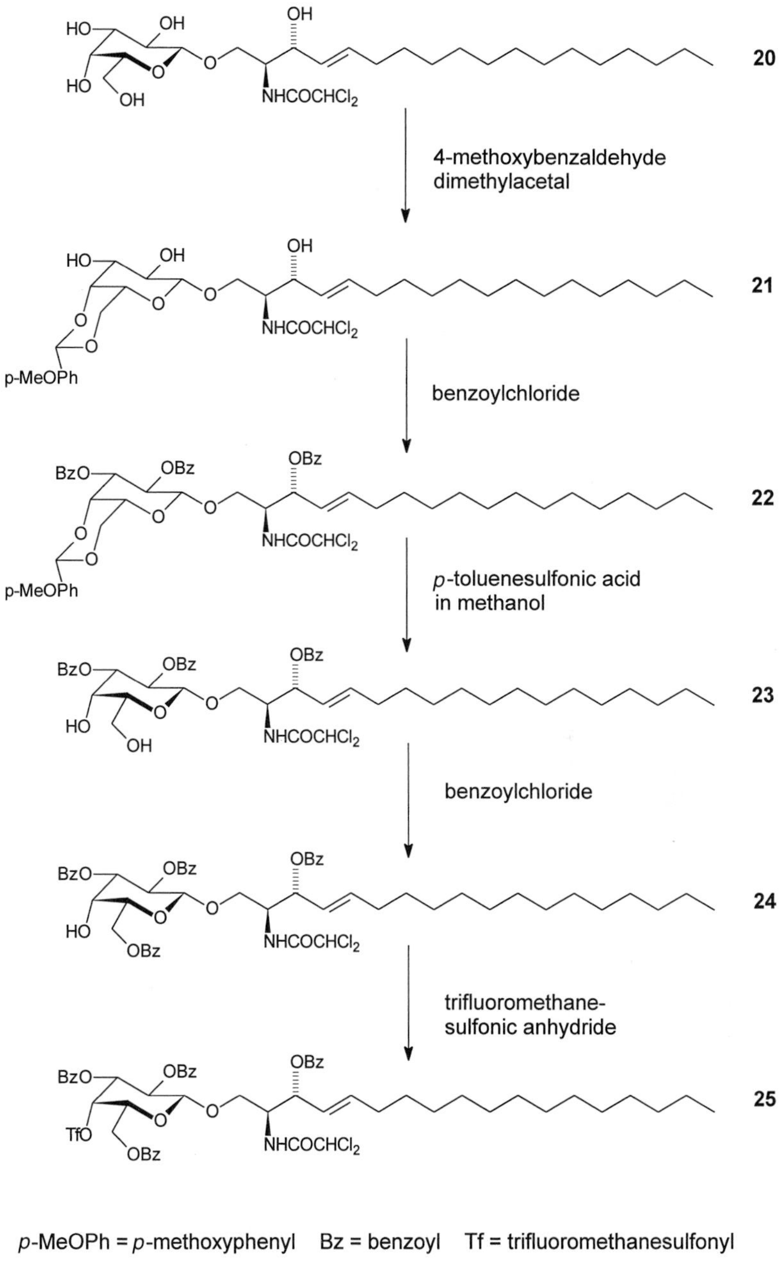

20
4-methoxybenzaldehyde dimethylacetal
21
benzoylchloride
22
p-toluenesulfonic acid in methanol
23
benzoylchloride
24
trifluoromethane-sulfonic anhydride
25
p-MeOPh = p-methoxyphenyl Bz = benzoyl Tf = trifluoromethanesulfonyl

FIG. 5. Synthetic scheme for galactosylthioglucosylsphingosine.

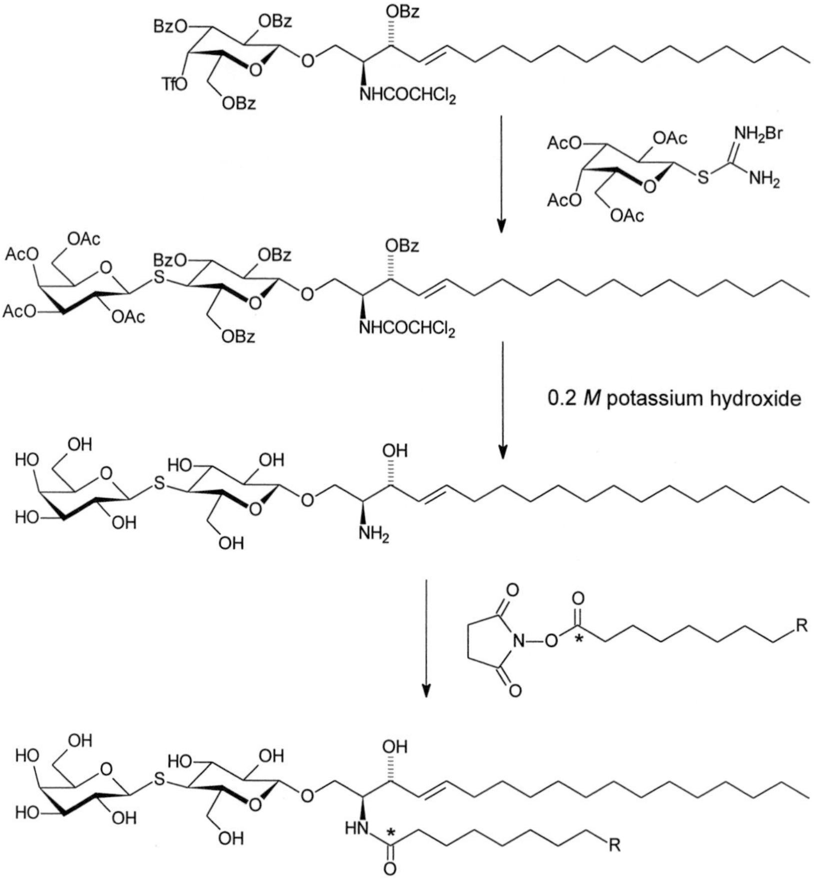

25

27

26

28

29 R = 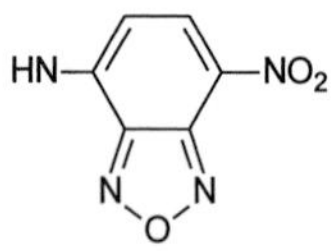 for the fluorescent derivative

30 R = H for the radioactive derivative

* position of radiocarbon for the nonfluorescent derivative

Ac = acetyl Bz = benzoyl Tf = trifluoromethanesulfonyl

FIG. 6. Synthetic scheme for labeled galactosylthioglucosylceramides.

chloride is continued until no more ninhydrin-positive material can be detected. Completion of the reaction should be accomplished after a total of 2 to 2.5 ml of the dilute reagent has been added. Then 50 ml water and a few drops of dilute hydrochloric acid are added until the mixture is slightly acidic. The solution is then extracted six times with ethyl acetate (75 ml). The organic phase is evaporated to a syrup that is chromatographed on a column of silica gel Si 60 (50 g) with chloroform/methanol (9:1, v/v) and the eluent is monitored by TLC. Fractions with the pure product (R_f 0.61, chloroform/methanol/2 M ammonia, 60:40:9, v/v/v) are collected to give compound **20** (460 mg; 0.80 mmol; 88%) as an amorphous mass after evaporation of the solvent. ^{1}H NMR (500 MHz; CDCl$_3$): Galactose unit δ 3.38 (dd, 1 H, $J_{5,6a}$ = 4 Hz, $J_{6a,6b}$ = 10 Hz, H-6a), 3.39 (m, 1 H, H-5), 3.44 (dd, 1 H, $J_{5,6a}$ = 7.5 Hz, $J_{6a,6b}$ = 10 Hz, H-6b), 3.51 (dd, 1 H, $J_{3,4}$ = 3.25 Hz, $J_{2,3}$ = 10 Hz, H-3), 3.77 (d, 1 H, $J_{3,4}$ = 3.25 Hz, H-4), 4.11 (d, 1 H, $J_{1,2}$ = 7.5 Hz, H-1), 4.13 (dd, 1 H, $J_{1,2}$ = 7.5 Hz, $J_{2,3}$ = 10 Hz, H-2); sphingosine unit δ 0.75 (t, 3 H, CH$_3$), 1.1-1.27 (m, 22 H, 11 CH$_2$), 1.89 (m, 2 H, H-6), 3.62 (dd, 1 H, $J_{1a,2}$ = 5 Hz, $J_{1a,1b}$ = 11.5 Hz, H-1a), 3.71 (dd, 1 H, $J_{1b,2}$ = 6.5 Hz, $J_{1a,1b}$ = 11.5 Hz, H-1b), 3.83 (s, 5 H, OH), 3.85 (m, 1 H, H-2), 4.06 (m, 1 H, H-3), 5.34 (dd, 1 H, $J_{3,4}$ = 7.5 Hz, $J_{4,5}$ = 15 Hz, H-4), 5.60 (td, 1 H, $J_{5,6}$ = 6.5 Hz, $J_{4,5}$ = 15 Hz, H-5) 7.02 (d, 1 H, NH); protecting group unit δ 5.93 (s, 1 H, CHCl$_2$); FABMS: (C$_{26}$H$_{47}$Cl$_2$NO$_8$; 572.56 g/mol) MNa$^+$ at m/z 594, 596, 598; anal. calcd. for C$_{26}$H$_{47}$Cl$_2$NO$_8$ $\times$ 1/2 H$_2$O (581,58): C, 53.70; H, 8.32; N, 2.41. Found: C, 53.16; H, 8.56; N, 2.37.

*O-(4,6-O-p-Methoxybenzylidene-β-D-galactopyranosyl)-(1 $\rightarrow$ 1)-(2S,3R,4E)-2-dichloroacetamido-4-octadecen-1,3-diol (**21**)*. A solution of *O*-(β-D-galactopyranosyl)-(1 $\rightarrow$ 1)-(2S,3R,4E)-2-dichloroacetamido-4-octadecen-1,3-diol (**20**, 375 mg, 0.656 mmol), 4-methoxybenzaldehyde dimethylacetal (170 μl; 0.98 mmol), and *p*-toluenesulfonic acid monohydrate (2 mg) in DMF (1 ml) is stirred under argon at 50° for 2 days. The mixture is then poured into a stirred solution of cold 2.5 m*M* sodium hydroxide (4 ml). The mixture is then extracted with diethyl ether (5 $\times$ 2 ml), and the combined organic phases are dried under a nitrogen jet. The residue is purified by column chromatography on silica gel Si 60 (200 g) using *n*-hexane/ethyl acetate (2:1, v/v). Fractions with the pure product (R_f 0.59, chloroform/methanol, 9:1, v/v) are dried to yield compound **21** (215.5 mg, 0.312 mmol, 47.5%). ^{1}H NMR (500 MHz; CDCl$_3$): Galactose unit δ 3.75 (t, 1 H, $J_{5,6}$ = 8.5 Hz, H-5), 4.03 (m, 2 H, H-6), 4.12 (d, 1 H, $J_{3,4}$ = 3 Hz, H-4), 4.23 (dd, 1 H, $J_{3,4}$ = 3 Hz, $J_{2,3}$ = 10 Hz, H-3), 4.27 (dd, 1 H, $J_{1,2}$ = 7.5 Hz, $J_{2,3}$ = 10 Hz, H-2), 4.33 (d, 1 H, $J_{1,2}$ = 7.5 Hz, H-1); sphingosine unit δ 0.88 (t, 3 H, CH$_3$), 1.2-1.4 (m, 22 H, 11 CH$_2$), 2.03 (m, 2 H, H-6), 3.43 (s, 1 H, OH), 3.63 (dd, 1 H, $J_{1a,2}$ = 5 Hz, $J_{1a,1b}$ = 10 Hz, H-1a), 3.82 (dd, 1 H, $J_{1b,2}$ = 6.5 Hz, $J_{1a,1b}$ = 10 Hz, H-1b), 4.26 (m, 1 H, H-3), 4.27 (m,

1 H, H-2), 5.51 (dd, 1 H, $J_{3,4}$ = 6 Hz, $J_{4,5}$ = 15 Hz, H-4), 5.77 (td, 1 H, $J_{5,6}$ = 6 Hz, $J_{4,5}$ = 15 Hz, H-5), 7.47 (d, 1 H, NH); protecting group unit δ 3.80 (s, 3 H, OCH$_3$), 5.49 (s, 1 H, CHCl$_2$), 6.00 (s, 1 H, p-CH$_3$OC$_6$H$_4$CH), 6.89 (d, 2 H, H-Ph$_{3,5}$), 7.41 (d, 2 H, H-Ph$_{2,6}$); FABMS: (C$_{34}$H$_{53}$Cl$_2$NO$_9$; 690.70 g/mol) MNa$^+$ at m/z 712, 714, 716; anal. calcd. for C$_{34}$H$_{53}$Cl$_2$NO$_9$ × 1/2 H$_2$O (699.71): C, 58.36; H, 7.78; N, 2.00. Found: C, 58.17; H, 7.80; N, 1.77.

O-(2,3-Di-O-benzoyl-4,6-O-p-methoxybenzylidene-β-D-galactopyrano-syl)-(1 → 1)-(2S,3R,4E)-3-O-benzoyl-2-dichloroacetamido-4-octadecen-1,3-diol (22). To a solution of *O*-(4,6-*O*-p-methoxybenzylidene-β-D-galactopy-ranosyl)-(1 → 1)-(2S,3R,4E)-2-dichloroacetamido-4-octadecen-1,3-diol (**21**, 205 mg, 0.296 mmol) in anhydrous pyridine (2 ml), cooled to −20°, benzoyl-chloride (104 μl, 0.89 mmol) is added in small portions over a period of 30 min with stirring. The bath temperature is kept below −10° for 4 hr and is then allowed to increase slowly to room temperature. After the addition of ethyl acetate (20 ml), the solution is washed twice successively with 5 ml each of cold 1 *M* hydrogen chloride, 1 *M* sodium hydrogen carbonate, and cold water. The organic phase is evaporated to dryness and the residue is freeze-dried from benzene (4 ml). The crude product (271 mg, 0.270 mmol; 91%) is used without purification. For analysis, a small portion may be purified by HPLC on ProSep C18 in a stainless steel column (0.4 × 15 cm) using methanol/water (85 : 15, v/v). The column eluent is monitored by TLC in *n*-hexane/ethyl acetate (1 : 1, v/v). Fractions containing the pure product (R_f 0.53) are collected. ^{1}H NMR (500 MHz; CDCl$_3$): Galactose unit 3.66 (m, 1 H, H-5), 4.07 (dd, 1 H, $J_{5,6a}$ = 2 Hz, $J_{6a,6b}$ = 12.5 Hz, H-6a), 4.25 (dd, 1 H, $J_{5,6b}$ = 1.5 Hz, $J_{6a,6b}$ = 12.5 Hz, H-6b), 4.55 (d, 1 H, $J_{3,4}$ = 3.5 Hz, H-4), 4.78 (d, 1 H, $J_{1,2}$ = 8 Hz, H-1), 5.35 (dd, 1 H, $J_{3,4}$ = 3.5 Hz, $J_{2,3}$ = 10.5 Hz, H-3), 5.85 (dd, 1 H, $J_{1,2}$ = 8 Hz, $J_{2,3}$ = 10.5 Hz, H-2); sphingosine unit δ 0.88 (t, 3 H, CH$_3$), 1.2-1.35 (m, 22 H, 11 CH$_2$), 1.98 (m, 2 H, H-6), 3.88 (dd, 1 H, $J_{1a,2}$ = 4 Hz, $J_{1a,1b}$ = 10.5 Hz, H-1a), 4.20 (dd, 1 H, $J_{1b,2}$ = 4 Hz, $J_{1a,1b}$ = 10.5 Hz, H-1b), 4.41 (m, 1 H, H-2), 5.52 (dd, 1 H, $J_{3,4}$ = 7 Hz, $J_{4,5}$ = 15 Hz, H-4), 5.65 (m, 1 H, H-3), 5.84 (td, 1 H, $J_{5,6}$ = 6.5 Hz, $J_{4,5}$ = 15 Hz, H-5), 6.82 (d, 1 H, NH); protecting group unit δ 3.81 (s, 3 H, OCH$_3$), 5.69 (s, 1 H, CHCl$_2$), 6.88 (d, 2 H, H-Ph$_{3,5}$), 5.48 (s, 1 H, p-CH$_3$OC$_6$H$_4$CH), 7.35–7.93 (m, 15 H, H-Bz), 8.02 (dd, 2 H, H-Ph$_{2,6}$); FABMS: (C$_{55}$H$_{65}$Cl$_2$NO$_{12}$; 1003.02 g/mol) MNa$^+$ at m/z 1024, 1026, 1028.

O-(2,3-Di-O-benzoyl-β-D-galactopyranosyl)-(1 → 1)-(2S,3R,4E)-3-O-benzoyl-2-dichloroacetamido-4-octadecen-1,3-diol (23). *O*-(2,3-Di-*O*-benzoyl-4,6-*O*-p-methoxybenzylidene-β-D-galactopyranosyl)-(1 → 1)-(2S,3R,4E)-3-O-benzoyl-2-dichloroacetamido-4-octadecen-1,3-diol (**22**, 260 mg, 0.259 mmol) is dissolved in 5 ml of chloroform/methanol (1 : 1, v/v). After the addition of *p*-toluenesulfonic acid monohydrate (2 mg), the solution is stirred under argon for 4 hr at room temperature. After the addition of N,N-diisopropylethylamine (10 μl), the mixture is dried under a stream of nitrogen. The residue is chromatographed on silica gel Si 60 (200 g) using n-hexane/ethyl acetate (1 : 1, v/v). Fractions with the pure compound (R_f 0.26, n-hexane/ethyl acetate, 1 : 1, v/v) are dried to afford **23** as an amorphous white product (158.5 mg, 0.179 mmol, 69%). ^{1}H NMR (500 MHz; CDCl$_3$): Galactose unit 3.53 (t, 1 H, $J_{5,6}$ = 3 Hz, H-5), 3.75 (dd, 1 H, $J_{5,6a}$ = 3 Hz, $J_{6a,6b}$ = 10 Hz, H-6a), 4.06 (dd, 1 H, $J_{5,6b}$ = 3 Hz, $J_{6a,6b}$ = 10 Hz, H-6b), 4.43 (d, 1 H, $J_{3,4}$ 3 Hz, H-4), 4.67 (d, 1 H, $J_{1,2}$ = 8 Hz, H-1), 5.27 (dd, 1 H, $J_{3,4}$ = 3 Hz, $J_{2,3}$ = 10 Hz, H-3), 5.85 (dd, 1 H, $J_{1,2}$ = 8 Hz, $J_{2,3}$ = 10 Hz, H-2); sphingosine unit δ 0.88 (t, 3 H, CH$_3$), 1.2-1.4 (m, 22 H, 11 CH$_2$), 2.02 (m, 2 H, H-6), 3.69 (dd, 1 H, $J_{1a,2}$ = 3 Hz, $J_{1a,1b}$ = 12.5 Hz, H-1a), 3.86 (dd, 1 H, $J_{1b,2}$ = 4 Hz, $J_{1a,1b}$ = 12.5 Hz, H-1b), 4.43 (m, 1 H, H-2), 5.50 (dd, 1 H, $J_{3,4}$ = 7.5 Hz, $J_{4,5}$ = 15 Hz, H-4), 5.83 (m, 1 H, H-3), 6.02 (td, 1 H, $J_{5,6}$ = 6 Hz, $J_{4,5}$ = 15 Hz, H-5), 6.88 (d, 1 H, NH); protecting group unit δ 5.78 (d, 1 H, CHCl$_2$), 7.37–8.02 (m, 15 H, H-Bz); FABMS: (C$_{47}$H$_{59}$Cl$_2$NO$_{11}$; 884.89 g/mol) MNa$^+$ at m/z 906, 908, 910; anal. calcd. for C$_{47}$H$_{59}$Cl$_2$NO$_{11}$ × 1/2 H$_2$O (893.90): C, 63.15; H, 6.77; N, 1.56. Found: C, 63.01; H, 6.94; N, 1.45.

O-(2,3,6-Tri-O-benzoyl-β-D-galactopyranosyl)-(1 → 1)-(2S,3R,4E)-3-O-benzoyl-2-dichloroacetamido-4-octadecen-1,3-diol (**24**). *O*-(2,3-Di-*O*-benzoyl-β-D-galactopyranosyl)-(1 → 1)-(2S,3R,4E)-3-*O*-benzoyl-2-dichloroacetamido-4-octadecen-1,3-diol (**23**, 155 mg, 0.175 mmol) is dissolved in dry pyridine (0.5 ml) and cooled to −30°. Benzoyl chloride (22.4 μl; 0.193 mmol) is then added and the mixture is first stirred for 2 hr at −30°. The mixture is then raised to room temperature. The product is extracted from the reaction mixture and washed as described for **22**. The crude product is then purified on silica gel Si 60 (200 g) using n-hexane/ethyl acetate (8 : 2, v/v) to yield **24** (111 mg, 0.112 mmol, 64%); R_f 0.35 (n-hexane/ethyl acetate, 7 : 3, v/v). ^{1}H NMR (500 MHz, CDCl$_3$): Galactose unit δ 2.1, 2.3 (2s, 2 OH), 4.00 (t, 1 H, $J_{5,6}$ = 6.5 Hz, H-5), 4.27 (dd, 1 H, $J_{5,6a}$ = 6.5 Hz, $J_{6a,6b}$ = 11.5 Hz, H-6a), 4.29 (d, 1 H, $J_{3,4}$ = 3 Hz, H-4), 4.50 (dd, 1 H, $J_{5,6b}$ = 6.5 Hz, $J_{6a,6b}$ = 11.5 Hz, H-6b), 4.78 (d, 1 H, $J_{1,2}$ = 8 Hz, H-1), 5.35 (dd, 1 H, $J_{3,4}$ = 3 Hz, $J_{2,3}$ = 10 Hz, H-3), 5.77 (dd, 1 H, $J_{1,2}$ = 8 Hz, $J_{2,3}$ = 10 Hz, H-2); sphingosine unit δ 0.88 (t, 3 H, CH$_3$), 1.18-1.38 (m, 22 H, 11 CH$_2$), 2.00 (m, 2 H, H-6), 3.73 (dd, 1 H, $J_{1a,2}$ = 3 Hz, $J_{1a,1b}$ = 10 Hz, H-1a), 4.23 (dd, 1 H, $J_{1b,2}$ = 4 Hz, $J_{1a,1b}$ = 10 Hz, H-1b), 4.41 (m, 1 H, H-2), 5.49 (dd, 1 H, $J_{3,4}$ = 7.5 Hz, $J_{4,5}$ = 15 Hz, H-4), 5.67 (dd, 1 H, $J_{2,3}$ = 7.5 Hz, $J_{3,4}$ = 7.5 Hz, H-3), 5.89 (td, 1 H, $J_{5,6}$ = 6.5 Hz, $J_{4,5}$ = 15 Hz, H-5), 6.80 (d, 1 H, NH); protecting group unit δ 5.70 (s, 1 H, CHCl$_2$), 7.34-8.02 (m, 20 H, H-Bz); FABMS: (C$_{54}$H$_{63}$Cl$_2$NO$_{12}$, 988.99 g/mol) MNa$^+$ at m/z 1010, 1012, 1014.

O-(2,3,6-Tri-O-benzoyl-4-O-trifluoromethanesulfonyl-β-D-galactopy-ranosyl)-(1 → 1)-(2S,3R,4E)-3-O-benzoyl-2-dichloroacetamido-4-octa-decen-1,3-diol (25). A solution of **24** (75 mg, 76 μmol) in dry pyridine (100 μl) and dry dichloromethane (0.40 ml) is cooled under argon atmosphere to −20°. Trifluoromethanesulfonic anhydride (44 μl, 0.189 mmol) is then added slowly. After 2 hr, when TLC in *n*-hexane/ethyl acetate (7:3, v/v) shows that the reaction is completed and a new product (R_f 0.39) has formed at the expense of educt **24**, the mixture is diluted with dichloromethane (2 ml) and washed with cold 1 *M* hydrogen chloride (4 ml). Following phase separation, the aqueous layer is extracted twice with dichloromethane (3 ml). The combined organic phases are concentrated under a nitrogen stream and the residue is purified on silica gel Si 60 (50 g) using a linear gradient from 300 ml *n*-hexane/ethyl acetate (9:1, v/v) to 300 ml *n*-hexane/ethyl acetate (7:3, v/v). Fractions with the pure product (R_f 0.39, *n*-hexane/ethyl acetate, 7:3, v/v) are dried to yield **25** (63.5 mg, 57 μmol, 75%) as a white powder.^{1}H NMR (500 MHz, CDCl₃): Galactose unit δ 4.16 (dd, 1 H, $J_{5,6a}$ = 7 Hz, $J_{6a,ab}$ = 11.5 Hz, H-6a), 4.26 (t, 1 H, $J_{5,6}$ = 7 Hz, H-5), 4.52 (dd, 1 H, $J_{5,6b}$ = 7 Hz, $J_{6a,6b}$ 11.5 Hz, H-6b), 4.81 (d, 1 H, $J_{1,2}$ = 8 Hz, H-1), 5.49 (d, 1 H, $J_{3,4}$ = 3 Hz, H-4), 5.56 (dd, 1 H, $J_{3,4}$ = 3 Hz, $J_{2,3}$ = 10.5 Hz, H-3), 5.72 (dd, 1 H, $J_{1,2}$ = 8 Hz, $J_{2,3}$ = 10.5 Hz, H-2); sphingosine unit δ 0.88 (t, 3 H, CH₃), 1.20–1.39 (m, 22 H, 11 CH₂), 2.00 (m, 2 H, H-6), 3.77 (dd, 1 H, $J_{1a,2}$ = 4 Hz, $J_{1a,1b}$ = 9.5 Hz, H-1a), 4.22 (dd, 1 H, $J_{1b,2}$ = 3.5 Hz, $J_{1a,1b}$ = 9.5 Hz, H-1b), 4.42 (m, 1 H, H-2), 5.48 (dd, 1 H, $J_{3,4}$ = 7.5 Hz, $J_{4,5}$ = 15 Hz, H-4), 5.62 (dd, 1 H, $J_{2,3}$ = 7.5 Hz, $J_{3,4}$ = 7.5 Hz, H-3), 5.90 (td, 1 H, $J_{5,6}$ = 6.5 Hz, $J_{4,5}$ = 15 Hz, H-5), 6.75 (d, 1 H, NH); protecting group unit δ 5.67 (s, 1 H, CHCl₂), 7.34-8.05 (m, 20 H, H-Bz).

S-(2,3,4,6-Tetra-O-acetyl-β-D-galactopyranosyl)-(1 → 4)-O-(2,3,6-tri-O-benzoyl-4-thio-β-D-glucopyranosyl)-(1 → 1)-(2S,3R,4E)-3-O-benzoyl-2-dichloroacetamido-4-octadecen-1,3-diol (26). To the solution of 2-S-(2,3,4,6-tetra-O-acetyl-β-D-galactopyranosyl)-2-thiopseudourea hydro-bromide **(27)**[28] (130 mg, 0.267 mmol) in 0.5 ml of acetone/methanol (1:1, v/v) potassium carbonate (100 mg) and **25** (60 mg, 53 μmol) are added. After stirring, under argon, for 2 hr at room temperature dichloromethane (3 ml) is added, and the salts are removed by washing the mixture three times with water (3 ml). After evaporation of the organic layer, the residue is chromatographed on silica gel Si 60 (50 g) using *n*-hexane/ethyl acetate (7:3, v/v) as eluent. Fractions that yield one single spot on TLC (R_f 0.59, *n*-hexane/ethyl acetate, 1:1, v/v) are dried and freeze-dried from benzene to obtain **26** (48.12 mg, 36 μmol, 68% with respect to **25**). ^{1}H NMR (500 MHz; CDCl₃): Galactose unit δ 3.89 (t, 1 H, $J_{5,6}$ = 6 Hz, H-5), 3.97 (dd, 1 H, $J_{5,6a}$ = 6 Hz, $J_{6a,6b}$ = 11 Hz, H-6a), 4.03 (dd, 1 H, $J_{5,6b}$ = 6 Hz, $J_{6a,6b}$ =

11 Hz, H-6b), 4.85 (d, 1 H, $J_{1,2}$ = 9.5 Hz, H-1), 4.92 (dd, 1 H, $J_{3,4}$ = 3.5 Hz, $J_{2,3}$ = 9.5 Hz, H-3), 5.04 (dd, 1 H, $J_{1,2}$ = 9.5 Hz, $J_{2,3}$ = 9.5 Hz, H-2), 5.37 (d, 1 H, $J_{3,4}$ = 3.5 Hz, H-4); glucose unit δ 3.29 (dd, 1 H, $J_{3,4}$ = 11 Hz, $J_{4,5}$ = 11 Hz, H-4), 4.07 (m, 1 H, H-5), 4.55 (dd, 1 H, $J_{5,6a}$ = 2 Hz, $J_{6a,6b}$ = 12 Hz, H-6a), 4.70 (d, 1 H, $J_{1,2}$ = 8 Hz, H-1), 4.75 (dd, $J_{5,6b}$ = 4 Hz, $J_{6a,6b}$ = 12 Hz, H-6b), 5.42 (dd, 1 H, $J_{1,2}$ = 8 Hz, $J_{2,3}$ = 8.5 Hz, H-2), 5.66 (dd, 1 H, $J_{2,3}$ = 8.5 Hz, $J_{3,4}$ = 11 Hz, H-3); sphingosine unit δ 0.88 (t, 3 H, CH$_3$), 1.17-1.32 (m, 22 H, 11 CH$_2$), 1.95 (m, 2 H, H-6), 3.69 (dd, 1 H, $J_{1a,2}$ = 4 Hz, $J_{1a,1b}$ = 10 Hz, H-1a), 4.16 (dd, 1 H, $J_{1b,2}$ = 3 Hz, $J_{1a,1b}$ = 10 Hz, H-1b), 4.38 (m, 1 H, H-2), 5.43 (dd, 1 H, $J_{3,4}$ = 7.5 Hz, $J_{4,5}$ = 15 Hz, H-4), 5.58 (m, 1 H, H-3), 5.84 (td, 1 H, $J_{5,6}$ = 7 Hz, $J_{4,5}$ = 15 Hz, H-5), 6.70 (d, 1 H, NH); protecting group unit δ 1.53, 1.91, 2.02, 2.10 (4s, 12 H, 4 CH$_3$CO), 5.67 (s, 1 H, CHCl$_2$), 7.31–8.02 (m, 20 H, H-Bz); FABMS: (C$_{68}$H$_{81}$Cl$_2$NO$_{20}$S, 1335.35 g/mol) MNa$^+$ at m/z 1356, 1358, 1360.

S-(β-D-Galactopyranosyl)-(1→4)-O-(4-thio-β-D-glucopyranosyl)-(1 → 1)-(2S,3R,4E)-2-amino-4-octadecen-1,3-diol (**28**). A solution of *S*-(2,3,4, 6-tetra-*O*-acetyl-β-D-galactopyranosyl)-(1 → 4)-*O*-(2,3,6-tri-*O*-benzoyl-4-thio-β-D-glucopyranosyl)-(1 → 1)-(2*S*,3*R*,4*E*)-3-*O*-benzoyl-2-dichloroacetamido-4-octadecen-1,3-diol (**26**, 45 mg, 33.7 μmol) in 10 ml of 0.2 *M* potassium hydroxide in methanol is treated, under argon, for 3 hr at 65°. After cooling to room temperature, the solution is buffered with acetic acid (0.15 ml) and diluted with water (15 ml). The mixture thus obtained is passed over LiChroprep RP 18 to adsorb the saponified product. Salts and other water-soluble materials are washed out with water and the retained lipid is subsequently eluted with methanol. Compound **28** is purified on silica gel Si 60 (50 g), using chloroform/methanol/2.5 *M* ammonia (65:25:4, v/v/v) as eluent, to afford 17.5 mg (27.2 μmol, 81%) of **28** (R$_f$ 0.12, chloroform/methanol/2.5 *M* ammonia, 65:25:4, v/v/v). FABMS: (C$_{30}$H$_{57}$NO$_{11}$S, 639.84 g/mol) MH$^+$ at m/z 640 and MNa$^+$ at m/z 662.

S-(β-D-Galactopyranosyl)-(1→4)-O-(4-thio-β-D-glucopyranosyl)-(1 → 1)-(2S, 3R,4E)-2-(8-N-(7-nitrobenz-1,3-diazol-2-oxa-4-yl)amino)octanamido-4-octadecen-1,3-diol (**29**). A small amount (10 μmol) of *S*-(β-D-galactopyranosyl)-(1 → 4)-*O*-(4-thio-β-D-glucopyranosyl)-(1 → 1)-(2*S*,3*R*,4*E*)-2-amino-4-octadecen-1,3-diol (**28**) is N-acylated with *N*-succinimidyl-8-*N*-(7-nitrobenz-2-oxa-1,3-diazol-4-yl)aminooctanoate essentially as described for the N-acylation of the fluorescent glucosylthioceramide **18**. The crude product is purified by HPLC on ProSep C18 (1 × 25 cm) with methanol/water (85:15, v/v). The eluent is monitored by TLC in chloroform/methanol/15 m*M* calcium chloride (60:35:8, v/v/v), and fractions containing the pure product (R$_f$ 0.58) are collected and freeze-dried to yield **29** (8.5 μmol, 85%). FABMS: (C$_{44}$H$_{73}$N$_5$O$_{15}$S, 944.14 g/mol) MH$^+$ at m/z 944, MH$^+$-water at m/z 926, and MNa$^+$ at m/z 966.

*S-(β-D-Galactopyranosyl)-(1→4)-O-(4-thio-β-D-glucopyranosyl)-(1 →
1)-(2S,3R,4E)-2-[1-^{14}C]-octanamido-4-octadecen-1,3-diol* (**30**). A small
amount (5 μmol) of the lysolipid *S-(β-D-galactopyranosyl)-(1 → 4)-O-(4-
thio-β-D-glucopyranosyl)-(1 → 1)- (2S,3R,4E)-2-amino-4-octadecen-1,3-
diol* (**28**) is acylated with *N*-succinimidyl-[1-^{14}C]octanoate as described for
the radioactive derivative **19**. The product is purified by HPLC on ProSep
C18 (1 × 25 cm) with methanol/water (85 : 15, v/v) to afford **30** (4.4 μmol,
88%). R_f 0.52 (chloroform/methanol/15 mM calcium chloride, 60 : 35 : 8,
v/v/v). FABMS: ($C_{38}H_{71}NO_{12}S$, 766.04 g/mol) MH$^+$ at m/z 766, MH$^+$-water
at m/z 748, and MNa$^+$ at m/z 788.

As an alternative to the synthesis of catabolically stable lactosylcer-
amides described here, a different approach can be chosen. This would
first involve the synthesis of a thiolactose[30] that contains sulfur as the
interglycosidic bond that, after appropriate protection and activation, could
be used for the glycosylation of ceramide or sphingosine derivatives. This
approach would, however, involve the formation of two glycosidic linkages.

Acknowledgment

Support for this work was provided by the Deutsche Forschungsgemeinschaft SFB 284.

[30] L. Reed and L. Goodman, *Carbohydr. Res.* **94**, 91 (1981).

[49] Synthetic Soluble Analogs of Glycolipids for Studies of Virus–Glycolipid Interactions

By JACQUES FANTINI

Introduction

Glycosphingolipids (GSL) are ubiquitous membrane components located almost exclusively at the outer leaflet of the plasma membrane of
mammalian cells. All GSL share a common hydrophobic backbone dipped
in the membrane, i.e., ceramide (Cer), which consists of a fatty acid chain
linked to the sphingosine base.[1] In contrast, the hydrophilic oligosaccharide
residues of GSL protrude into the extracellular space. GSL are classified
into three main series, i.e., ganglio, globo, and lacto, according to their

[1] T. E. Thompson and T. W. Tillack, *Annu. Rev. Biophys. Biophys. Chem.* **14**, 361 (1985).

carbohydrate structure, which may include one of 200 different oligosaccharides.[2]

Because many pathogens use carbohydrate-binding proteins to recognize and adhere to the surface of host cells, a number of GSL have been selected as cellular receptors by various bacteria and viruses.[2] For some viruses, GSL may also play an active role during the fusion between the viral envelope and the plasma membrane.[3-6] Thus, the development of antiviral drugs based on the structure of the oligosaccharidic moiety of GSL receptors is of primary importance. Naturally occurring oligosaccharides, such as those found in human milk, have been shown to inhibit the binding of influenza virus to host target cells.[7] However, a growing number of experimental evidences suggest that the hydrophobic moiety of GSL (particularly the length, saturation, and hydroxylation of the fatty acid chain) influences the conformation of the carbohydrate moiety.[8-10] In this respect, the interaction between HIV-1 surface envelope glycoprotein gp120 and galactosylceramide (GalCer)[11] is not inhibited by free galactose.[12] Moreover, binding of gp120 to liposomes containing GalCer is dependent on the concentration of GalCer, suggesting that the viral glycoprotein interacts preferentially or even exclusively with glycolipid-rich domains in the target membrane.[13] This may indicate that several GalCer molecules are needed for gp120 binding or, alternatively, that the conformation of the GalCer polar group is affected by adjoining GalCer molecules, leading to a higher affinity for gp120. The ability of GSL such as GalCer to form microdomains or patches in biological membranes[14] illustrates the difficulty of synthesizing water-soluble GSL analogs able to prevent virus attachment and separate viruses already attached.

[2] S. I. Hakomori and Y. Igarashi, *J. Biochem.* **118,** 1091 (1995).

[3] Y. Suzuki, Y. Nagao, H. Kato, M. Matsumoto, K. Nerome, K. Nakajima, and E. Nobusawa, *J. Biol. Chem.* **261,** 17057 (1986).

[4] J. L. Nieva, R. Bron, J. Corver, and J. Wilschut, *EMBO J.* **13,** 2797 (1994).

[5] A. Puri, P. Hug, I. Munoz-Barroso, and R. Blumenthal, *Biochem. Biophys. Res. Commun.* **242,** 219 (1998).

[6] D. Hammache, N. Yahi, G. Pieroni, F. Ariasi, C. Tamalet, and J. Fantini, *Biochem. Biophys. Res. Commun.* **246,** 117 (1998).

[7] D. Zopf and S. Roth, *Lancet* **347,** 1017 (1996).

[8] N. Strömberg, P. G. Nyholm, I. Pascher, and S. Normark, *Proc. Natl. Acad. Sci. U.S.A.* **88,** 9340 (1991).

[9] A. Pellizzari, H. Pang, and C. A. Lingwood, *Biochemistry* **31,** 1363 (1992).

[10] D. H. Jones, C. A. Lingwood, K. R. Barber, and C. W. Grant, *Biochemistry* **36,** 8539 (1997).

[11] J. M. Harouse, S. Bhat, S. L. Spitalnik, M. Laughlin, K. Stefano, D. H. Silberberg, and F. Gonzalez-Scarano, *Science* **253,** 320 (1991).

[12] J. Fantini, unpublished observations (1995).

[13] D. Long, J. F. Berson, D. G. Cook, and R. W. Doms, *J. Virol.* **68,** 5890 (1994).

[14] G. van Meer, *Annu. Rev. Cell Biol.* **5,** 247 (1989).

Ideally, the design of synthetic soluble analogs of glycolipids may preserve the biologically active conformation of the carbohydrate part of the GSL while increasing the polarity of the ceramide moiety in order to improve solubility in water. The latter can be done, for instance, by omitting the fatty acid chain or by adding a polar group such as carboxylate at the end of the ceramide-like hydrophobic chains (Fig. 1). This strategy has proved useful for obtaining synthetic soluble analogs of GalCer possessing

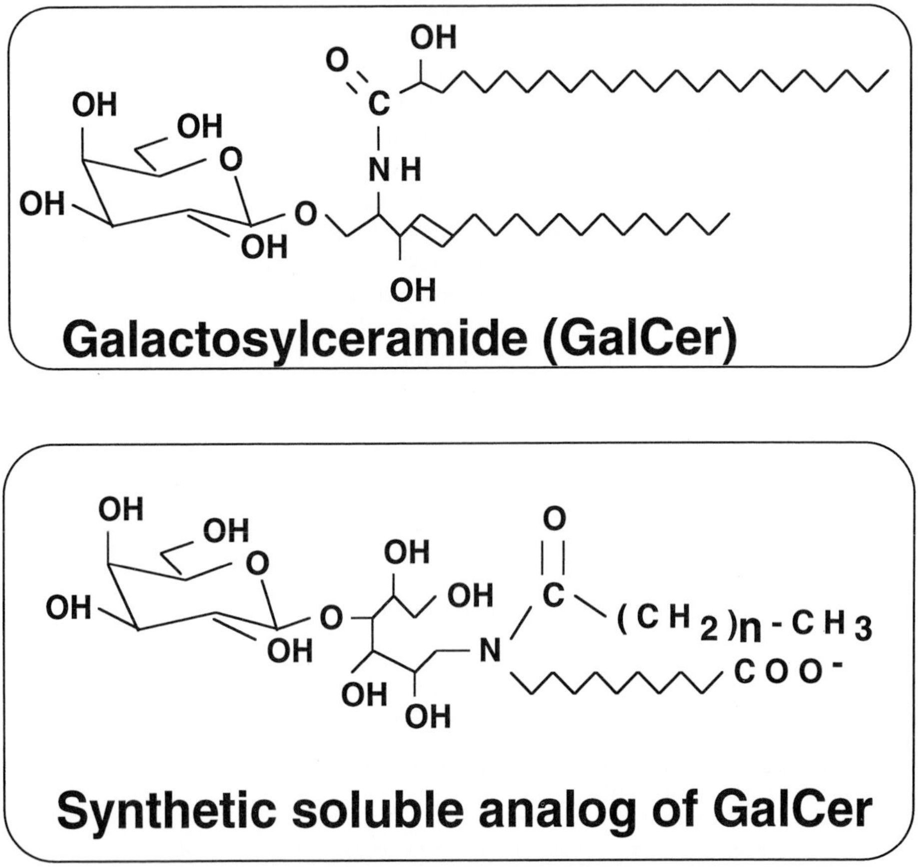

Galactosylceramide (GalCer)

Synthetic soluble analog of GalCer

FIG. 1. Chemical structures of GalCer and a synthetic soluble analog. GalCer is shown with the α-hydroxylated fatty acid C24:0, a form expressed by both neural and intestinal epithelial cells. The synthetic analog of GalCer has been designed and synthesized in the laboratory of I. Rico-Lattes (Toulouse, France). It is the prototype of a series of soluble GalCer-like molecules differing by the number of atoms of carbon ($n = 14$ for the most potent derivative) in the hydrophobic chain. These analogs are obtained from lactose in three stages.[16]

significant anti-HIV activity.[15,16] The physicochemical behavior of GSL analogs in aqueous solution is a major parameter that determines their biological activity.[17] Sulfatide (3'-sulfogalactosylceramide), the natural sulfated derivative of GalCer, aggregates in aqueous solution to form either lamellar or micellar phases, according to its concentration.[17] In contrast, the synthetic soluble analog of GalCer shown in Fig. 1 is expected to form stable vesicles and not micelles in aqueous solution, as one of its hydrophobic chains bears a charged carboxylate group. From a geometrical perspective, the monolayer arrangement of synthetic analogs in the vesicle may mimic a glycolipid patch of the plasma membrane, whereas a micellar phase may not. Synthetic analogs with a very short hydrophobic domain may be present as monomers in aqueous solution. Unfortunately, the gain in solubility may be correlated with a total lack of antiviral activity, as these compounds may not aggregate in water. On the opposite, synthetic analogs with a large hydrophobic part may be almost insoluble in water. The biological activity of such analogs should be tested after incorporation of the molecule in liposomes, which may increase the background values of the solid-phase assays described in this article. For the same reasons, the addition of organic solvents to facilitate the solubilization of synthetic GSL analogs is not indicated.

Immunological Characterization of Synthetic Soluble Analogs of Glycolipids

When available, antibodies against GSL can be used to check the structure of the synthetic analogs. Enzyme-linked immunosorbent assay (ELISA) is a first choice method, although difficulties with coating of GSL to microtiter plates have been reported.[18] In our assay, the protocol for GSL adsorption on ELISA plates is based on the solubilization of the lipid with a mixture of chloroform, ethanol, and n-hexane, which, following evaporation of the solvent, optimizes the orientation of GSL with the hydrophobic moiety bound to the ELISA plate and the polar head accessible to the ligand.[19]

[15] C. R. Bertozzi, D. G. Cook, W. R. Kobertz, F. Gonzalez-Scarano, and M. D. Bernadski, J. Am. Chem. Soc. **114**, 10639 (1992).
[16] J. Fantini, D. Hammache, O. Delézay, N. Yahi, C. André-Barrès, I. Rico-Lattes, and A. Lattes, J. Biol. Chem. **272**, 7245 (1997).
[17] W. Curatolo, Biochim. Biophys. Acta **906**, 111 (1987).
[18] W. H. Frey II, J. W. Schmalz, P. A. Perfetti, T. L. Norris, C. R. Emory, and T. A. Ala, J. Immunol. Methods **164**, 275 (1993).
[19] J. Fantini, D. Hammache, O. Delézay, G. Piéroni, C. Tamalet, and N. Yahi, Virology **246**, 211 (1998).

Assay

A solution of the lipid (natural GSL or synthetic analog) is prepared in hexane:chloroform:ethanol: (11:5:4, v:v:v) at a concentration of 1 mg/ml. This stock solution can be stored at $-20°$ for several days. Immediately before use, the lipid is resuspended in serial dilutions of methanol and coated on ELISA plates (100 μl/well). Several ELISA plates must be tested because marked differences in coating efficiency have been noticed. Polystyrene ELISA plates purchased from Greiner (Osi, Elancourt, France) give reliable results with various GSL, including GalCer, LacCer, and monosialosylgangliosides (GM1, GM3). The plate is left under the flow of a chemical hood until total evaporation of the solvent. The wells are saturated for 2 hr at $37°$ in phosphate-buffered saline (PBS) containing 2% bovine serum albumin (BSA). For detection of GalCer and its synthetic analog, the monoclonal anti-GalCer R-mAb[20] is used at a concentration of 2.5 μg/ml in PBS containing 2% BSA. Anti-GSL antibodies are incubated for 2 hr at $37°$. The wells are then rinsed three times with PBS and incubated further with peroxidase-conjugated rabbit antimouse antibodies (Sigma, St. Louis, MO) at a dilution of 1:1000 in PBS containing 2% BSA. After three washes in PBS, reaction products are developed with o-phenylenediamine and the absorbance is measured at 490 nm. Results of a typical experiment are shown in Fig. 2.

Interaction of Soluble Analogs of Glycolipids with
  Viral Envelope Glycoproteins

The binding of viral envelope glycoproteins to natural and synthetic glycolipid receptors can be measured by two methods.

Direct Assay

In this assay, the glycolipids are coated on ELISA plates and probed with the viral protein. The binding is revealed by an antibody directed to a domain of the viral protein located outside the receptor-binding site. The plate is treated with glycolipids and is saturated as described earlier. To measure the binding of HIV-1 surface envelope glycoprotein gp120, recombinant gp120 (10 μg/ml in PBS containing 2% BSA) is incubated for 2 hr at $37°$. The plates are washed with PBS containing 0.05% Tween 20, incubated with an anti-gp120 mAb (Immunotech, Marseille, France), and revealed with peroxidase-conjugated goat antimouse IgG as described pre-

[20] B. Rantsch, P. A. Clapshaw, J. Price, M. Noble, and W. Seifert, *Proc. Natl. Acad. Sci. U.S.A.* **79**, 2936 (1982).

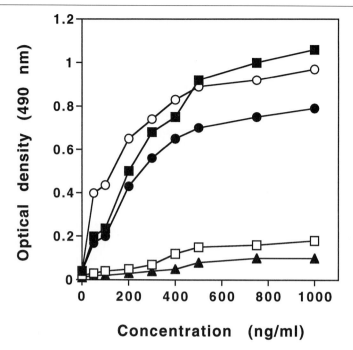

FIG. 2. Immunological recognition of a GalCer analog by an anti-GalCer antibody. The following molecules were coated on polystyrene ELISA plates (Greiner): GalCer (●), sulfatide (■), the synthetic soluble analog of GalCer with $n = 14$ (○), glucosylceramide (□), and ceramide (▲). The binding of the R-mAb monoclonal antibody was revealed by ELISA.

viously. Recombinant gp120 (IIIB isolate) can be purchased from Intracel Corp. (London) or obtained through the Medical Research Council (UK). HIV-1 and HIV-2 surface envelope glycoproteins can also be purified from concentrated viruses by lectin affinity chromatography.[21]

Inhibition Assay

In this assay, the viral protein is coated on ELISA plates and probed with a panel of antibodies recognizing different domains of the protein. The synthetic soluble analog of the glycolipid receptor is used as a competitor. In this case, a stock solution of the analog (2 mg/ml) is prepared in pure water and stored at $-20°$ for several months. This technique may allow delineation of the site of the viral protein involved in the interaction with the glycolipid

[21] D. Hammache, G. Piéroni, N. Yahi, O. Delézay, N. Koch, H. Lafont, C. Tamalet, and Jacques Fantini, *J. Biol. Chem.* **273,** 7967 (1998).

receptor. In the experiment shown in Fig. 3, the synthetic soluble analog of GalCer competes with the binding of anti-gp120 monoclonal antibodies recognizing the V3 domain of the viral glycoprotein. The binding of antibodies directed to other regions of gp120 is not affected by the analog. To ensure accessibility of the V3 domain to anti-V3 antibodies, recombinant gp120 is incubated in Immulon 1 multiwell plates (Nunc) precoated with polyclonal sheep antibodies directed to the C-terminal domain of gp120 (D7326 from Aalto Bioreagents, Dublin, Ireland). For precoating, 100 μl of these sheep antibodies (10 μg/ml in 5 mM Tris–HCl, pH 7.4, 150 mM NaCl) is incubated overnight at 4°. The plates are washed three times with 25 mM Tris–HCl, pH 7.6, 144 mM NaCl and saturated with 5 mM Tris–HCl, pH 7.4, 150 mM NaCl, 2% low-fat dry milk (saturation buffer) for 2 hr at room temperature. Recombinant gp120 (0.2 μg/ml) is then added to the wells and incubated for 2 hr at room temperature. The plates are washed three times and are then incubated with monoclonal anti-gp120 antibodies (5 μg/ml) in either the absence or the presence of various concentrations

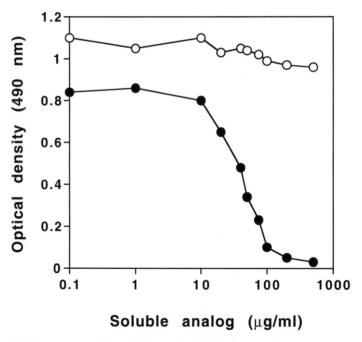

Fig. 3. Binding of anti-gp120 antibodies to recombinant gp120: effect of the soluble analog of GalCer. The soluble analog of GalCer (with $n = 14$) was incubated in competition with monoclonal antibodies recognizing the V3 domain (●) or the N-terminal part of recombinant gp120 (○). The binding of these antibodies to immobilized gp120 was revealed by ELISA.

of the soluble analog of GalCer (2 hr at room temperature). Alternatively, the analog can be incubated *before* adding the anti-gp120 antibodies for at least 1 hr at room temperature. The plates are then revealed with peroxidase-conjugated goat antimouse IgG as described earlier.

If anti-V3 antibodies are not available, an alternative assay can be used to measure the interaction of soluble analogs of GalCer with the V3 domain of HIV-1 gp120. This assay is based on the ability of polysulfonated compounds, such as suramin, to bind to V3 amino acid residues.[22] Recombinant gp120 (2.5 μg/ml) is incubated in polyvinylchloride 96-well plates [number 3911 from Falcon-Becton Dickinson (Le Pont de Claix, France)] overnight at 4°. After three washes in PBS, the plates are saturated with PBS 1% gelatin for 90 min at 37°. Bovine serum albumin is not indicated as a saturating agent because it binds to suramin, although with less afinity than gp120. Also note that gelatin must be used at 37° to avoid gel formation in the wells. [³H]Suramin (49 Ci/mmol, Isotopchim, Ganagobie-Peyruis, France) diluted at 0.1–1 μCi/ml is then added in either the absence or the presence of synthetic GalCer analogs for 2 hr at 37°. After five washes in PBS, each well is individualized and dipped into a vial containing 5 ml of Packard Ultima Gold scintillation liquid. The binding of [³H]suramin to recombinant gp120 is measured by counting the radioactivity in a β scintillation counter (counting time 2 min per sample). Background values of [³H]suramin binding to control wells (i.e., without gp120) are usually below 1000 counts per minute.

Anti-HIV-1 Activity of Synthetic Soluble Analogs of GalCer

HIV-1 can use different cell surface molecules to gain entry into various CD4⁺ and CD4⁻ cell types. The interaction of HIV-1 surface envelope glycoprotein gp120 with the plasma membrane of CD4⁺ cells involves at least two distinct binding sites: (i) CD4 and (ii) a member of the chemokine receptor family, among which the SDF-1 receptor CXCR4 and the RANTES receptor CCR5 have been characterized as major HIV-1 coreceptors. CXCR4 serves a fusion cofactor for T-lymphotropic strains of HIV-1,[23,24] whereas CCR5 mediates the entry of macrophage-tropic iso-

[22] N. Yahi, J. M. Sabatier, P. Nickel, K. Mabrouk, F. Gonzalez-Scarano, and J. Fantini, *J. Biol. Chem.* **269,** 24349 (1994).

[23] Y. Feng, C. C. Broder, P. E. Kennedy, and E. A. Berger, *Science* **272,** 872 (1996).

[24] J. F. Berson, D. Long, B. J. Doranz, J. Rucker, F. R. Jirik, and R. W. Doms, *J. Virol.* **70,** 6288 (1996).

lates.[25,26] In both cases, however, a common mechanism leads to the formation of a trimolecular complex among gp120, CD4, and the selected coreceptor. Following a primary interaction with CD4, a conformational change in gp120 renders the V3 domain (a variable disulfide-linked loop) of the viral glycoprotein available for secondary interactions with either CXCR4[27] or CCR5.[28] In the absence of CD4, the V3 loop is not exposed correctly to allow direct binding of gp120 to chemokine receptors. However, HIV-1 can infect *in vitro* several CD4$^-$ cell types, including neural[11] and epithelial cells[29] that coexpress CXCR4 and high levels of GalCer. Because the V3 domain of gp120 is involved in the fusion of HIV-1 with both CD4$^-$ and CD4$^+$ cells, soluble GalCer analogs may theoretically inhibit HIV-1 fusion in all these cell types.

HIV-1 Infection of Mucosal Epithelial Cells

HT-29 is a CD4$^-$ human colon epithelial cell line expressing high levels of GalCer and CXCR4.[29] These cells are infected by a subclass of T-cell line-adapted and primary HIV-1 isolates possessing a V3 domain able to interact with both GalCer and CXCR4. The cells are grown routinely in 25-cm^2 flasks (Costar, Brumath, France) in DMEM/F12 medium supplemented with 10% fetal bovine serum, penicillin (100 U/ml) and streptomycin (100 μg/ml). Forty-eight hours before infection, the cells are trypsinized and seeded in six-well plates at a mean density of 10^5 cells/cm^2. The virus is preincubated with the indicated concentration of the synthetic soluble analog of GalCer for 30 min at 37°. The mixture is then deposited onto the cells (2 ml per well) for 16 hr at 37°. The multiplicity of infection (m.o.i.) is 0.1 median tissue culture infectious dose (TCID$_{50}$) per cell. After thorough washing to remove excess inoculum, the cells are cultured for 7 days before analysis. The level of infection is determined by measuring the concentration of HIV-1 core antigen p24 in cell-free culture supernatants using an antigen capture assay (Dupont, Les Ulis, France). The supernatants can be stored at $-20°$ for 1–2 weeks without loss of reactive p24. As shown in Fig. 4A, treatment with the synthetic soluble analog of GalCer ($n = 14$)

[25] G. Alkhatib, C. Combadiere, C. C. Broder, Y. Feng, P. E. Kennedy, P. M. Murphy, and E. A. Berger, *Science* **272**, 1955 (1996).
[26] H. Deng, R. Liu, W. Ellmeier, S. Choe, D. Unumatz, M. Burkhart, P. Di Marzio, S. Marmon, R. E. Sutton, C. M. Hill, C. B. davis, S. C. Peiper, T. J. Schall, D. R. Littman, and N. R. Landau, *Nature* **381**, 661 (1996).
[27] C. L. Lapham, J. Ouyang, B. Chandrasekhar, N. Y. Nguyen, D. S. Dimitrov, and H. Golding, *Science* **274**, 602 (1996).
[28] L. Wu, N. P. Gerard, and R. Wyatt, *Nature* **384**, 179 (1996).
[29] O. Delézay, N. Koch, N. Yahi, D. Hammache, C. Tourres, C. Tamalet, and J. Fantini, *AIDS* **11**, 1311.

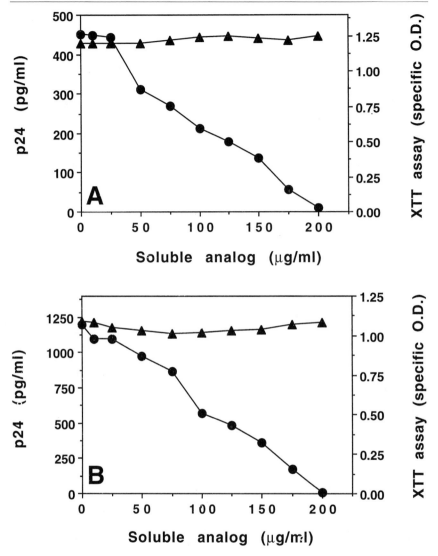

Fig. 4. Anti-HIV-1 activity of the synthetic soluble analog of GalCer. Exponentially growing HT-29 cells in six-well plates (A) or PBMC in 15-ml tubes (B) were exposed to HIV-1 (NDK and LAI isolates, respectively) in the presence of the indicated concentration of the synthetic soluble analog of GalCer with $n = 14$. The level of infection was determined by measurements of p24 HIV-1 core antigen in cell-free supernatants (●). The toxicity of the synthetic analog on both cell types was evaluated by the XTT assay (▲).

results in a dose-dependent inhibition of HIV-1 infection. The concentration of the analog resulting in an inhibition of infection of 50% (EC_{50}) can be extrapolated easily from data.

HIV-1 Infection of Peripheral Blood Mononuclear Cells

Peripheral blood mononuclear cells (PBMC) are obtained from healthy donors by density-gradient centrifugation, activated with phytohemagglutinin, and cultured in RPMI 1640 medium containing 10% fetal calf serum and interleukin 2. The virus is preincubated with the indicated concentration of the synthetic soluble analog of GalCer ($n = 14$) as described earlier. Infections are done in suspension (5×10^6 cells/ml in 15-ml tubes) for 2 hr at 37°. The cells are then rinsed three times with 14 ml of culture medium and cultured for 7 days in 25-cm^2 flasks before p24 measurements (Fig. 4B).

Toxicity Assay

The effects of synthetic analogs of GalCer on the proliferation and viability of HT-29 cells and PBMC have been studied in a colorimetric assay utilizing the tetrazolium salt XTT (sodium 3'-[-1- 1 (phenylamino)carbonyl]-3,4-tetrazolium]-bis(4-methoxy-6-nitro)benzenesulfonic hydrate).[30] This assay is quantitative and far more sensitive than the classical trypan blue exclusion method. For optimal results, the cells are cultured in phenol red-free culture medium to reduce the blank values. Cells are seeded in 96-well plates and cultured in the presence of serial dilutions of the analogs throughout the experiment (in this case, 7 days). On the day of the assay, XTT is dissolved in prewarmed medium at a concentration of 1 mg/ml. Immediately before use, phenazine methosulfate (PMS) is added to the XTT solution (final concentration of PMS, 125 μM). Twenty-five microliters of XTT/PMS mixture is added to 100 μl of culture, giving a final concentration of 0.2 mg/ml XTT and 25 μM PMS. After incubation at 37° for 4–8 hr, the optical density is determined using a test wavelength of 450 nm and a reference wavelength of 650 nm, subtracting blank control values with medium alone (Fig. 4). A decrease of at least 20% of the optical density in the XTT assay indicates cellular toxicity.

Structure–Activity Relationship: Key Role of Hydrophobic Domain in a Series of Synthetic Analogs of GalCer

The biochemical and immunological assays described in the this study are done in 96-well plates, which allows the simultaneous analysis of six

---

[30] N. W. Roehm, G. H. Rodgers, S. M. Hatfield, and A. L. Glasebrook, *J. Immunol. Methods* **142,** 257 (1991).

synthetic molecules per individual plate according to the following scheme: one rank of 12 wells for a negative control (an inactive product such as galactose), one rank for a positive control (an active analog, an antibody, or a polysulfonated compound), and six ranks with the molecules to be tested (six concentrations in duplicate). Considering that 4 plates can reasonably be handled by the same operator, more than 20 molecules can be evaluated per day in a nonautomated way. Similarly, antiviral assays can also be performed at a large scale in 6-well plates. This may allow rapid comparison of the biological activity of series of structurally related analogs. For instance, we have investigated the effect of the length of the hydrophobic moiety in a series of synthetic soluble analogs of GalCer derived from the prototype shown in Fig. 1. We found that the antiviral activity correlates with the level of hydrophobicity of the analog: for analogs with 8, 11, and 15 atoms of carbon in the fatty acid-like motif (i.e., $n = 7$, 10, and 14, respectively), the longer the hydrophobic part, the higher the antiviral activity. Similar data have been reported by Bertozzi et al.[15] on another series of synthetic analogs of GalCer. However, analogs with more than 16 atoms of carbon in the chain are not active, suggesting that a narrow range of hydrophobicity is necessary to achieve antiviral activity. The lack of activity of those analogs with a high degree of hydrophobicity could be tentatively explained by their ability to be incorporated in the plasma membrane, resulting in a significant decrease of concentration. Alternatively, these analogs may form micelles rather than vesicles in aqueous solution. As discussed earlier, the micellar organization of glycolipids in solution may differ significantly from the glycolipid microdomains of the plasma membrane, resulting in a loss of biological activity.

Conclusion and Perspectives

The design of synthetic soluble analogs of GSL receptors is at its beginning. The self-organization of these compounds in aqueous solution, leading to micellar or lamellar phases, is a critical parameter for their antiviral activity. Concerning HIV-1, preliminary data suggest that some synthetic analogs of GalCer have a behavior in solution fully compatible with the recognition of the V3 domain of gp120. Unfortunately, these analogs of first generation may neutralize only a subset of HIV-1 isolates, which use CXCR4 as coreceptor and also bind to GalCer. All other HIV-1 isolates, including those using the major coreceptor CCR5, are not neutralized by these analogs. Interestingly, we have demonstrated that the V3 domain of various HIV-1 isolates binds to GM3,[21] a ganglioside expressed abundantly as patches on the surface of human CD4[+] lymphocytes. Most importantly, GM3 patches can interact sequentially with CD4 and gp120, resulting in the formation of the trimolecular complex CD4–GM3–gp120.[6] The role of GM3 in this multimolecular organization could be to facilitate the migration of the CD4–gp120 complex to an appropriate coreceptor (e.g., CCR5 or CXCR4), as CD4 and these coreceptors are not physically associated in the absence of HIV-1.[27,28] By moving freely in the external leaflet of the plasma membrane, the patch of GM3 may behave as a raft dragging the CD4 receptor and taking aboard the viral particle. The binding of the virion to the raft is stabilized by secondary interactions between the polar head of GM3 and the V3 loop of gp120. The raft may then float on the cell surface until finding an adequate coreceptor that can displace the GM3–V3 loop interactions to its own profit, resulting in the initiation of the fusion process. According to this hypothetical model, which attributes a new role for GSL patches in virus–cell interactions, the affinity of the V3 loop for GM3 must be significantly lower than for the coreceptors. This may explain why the binding site for GM3 is conserved among highly divergent HIV-1 strains (including different HIV-1 subtypes and even an HIV-2 laboratory isolate[21]). Considering that GM3 patches may be used as attachment platforms by a variety of viruses and bacteria (HIV-1,[6] influenza viruses,[31] *Helicobacter pylori*[32]), synthetic soluble derivatives of this ganglioside may behave as molecular decoys that could inhibit the adsorption of these pathogens on mucosal surfaces.

Acknowledgment

This work was supported by the Fondation pour la Recherche Médicale (SIDACTION grant).

[31] T. Sato, T. Serizawa, and Y. Okahata, *Biochim. Biophys. Acta* **1285,** 14 (1996).
[32] T. Saitoh, H. Natomi, W. Zhao, K. Okuzumi, K. Sugano, M. Iwamori, and Y. Nagai, *FEBS Lett.* **282,** 385 (1991).

[50] Preparation of Radioactive Gangliosides, ^{3}H or ^{14}C Isotopically Labeled at Oligosaccharide or Ceramide Moieties

By SANDRO SONNINO, VANNA CHIGORNO, and GUIDO TETTAMANTI

Nomenclature

Ganglioside nomenclature according to Svennerholm[1] and IUPAC-IUB recommendations[2]: GM3, II^{3}Neu5AcLacCer, α-Neu5Ac-(2-3)-β-Gal-(1-4)-β-Glc-(1-1)-Cer; GM2, II^{3}Neu5AcGgOse$_3$Cer, β-GalNAc-(1-4)-[α-Neu5 Ac-(2-3)]-β-Gal-(1-4)-β-Glc-(1-1)-Cer; GM1, II^{3}Neu5AcGgOse$_4$Cer, β-Gal-(1-3)-β-GalNAc-(1-4)-[α-Neu5Ac-(2-3)]-β-Gal-(1-4)-β-Glc-(1-1)-Cer; AGM1, GgOse$_4$Cer, β-Gal-(1-3)-β-GalNAc-(1-4)-β-Gal-(1-4)-β-Glc-(1-1)-Cer; Fuc-GM1, IV^{2}FucII^{3}Neu5AcGgOse$_4$Cer, α-Fuc-(1-2)-β-Gal-(1-3)-β-GalNAc-(1-4)-[α-Neu5Ac-(2-3)]-β-Gal-(1-4)-β-Glc-(1-1)-Cer; GD1a, IV^{3}Neu5AcII^{3}Neu5AcGgOse$_4$Cer, α-Neu5Ac-(2-3)-β-Gal-(1-3)-β-GalNAc-(1-4)-[α-Neu5Ac-(2-3)]-β-Gal-(1-4)-β-Glc-(1-1)-Cer; GD1b, II^3(Neu5Ac)$_2$ GgOse$_4$Cer, β-Gal-(1-3)-β-GalNAc-(1-4)-[α-Neu5Ac-(2-8)-α-Neu5Ac-(2-3)]-β-Gal-(1-4)-β-Glc-(1-1)-Cer; Cer, ceramide; Neu5Ac, N-acetylneuraminic acid; [3-^{3}H($sphingosine$)]Gx, a ganglioside ^{3}H labeled at position 3 of 2S, 3R-sphingosine; [3-^{3}H($sphinganine$)]Gx, a ganglioside ^{3}H labeled at position 3 of 2S, 3R-sphinganine; [6-^{3}H(Gal-IV)]Gx, a ganglioside ^{3}H labeled at position 6 of external galactose; [6-^{3}H($GalNAc$)]Gx, a ganglioside ^{3}H labeled at position 6 of N-acetylgalactosamine; [11-^{3}H($Neu5Ac$)]Gx, a ganglioside ^{3}H labeled at the sialic acid acetyl group; [10-^{14}C($Neu5Ac$)]Gx, a ganglioside ^{14}C labeled at the carbonyl group of the sialic acid acetyl group; and [1-^{14}C($stearoyl$)]Gx, a ganglioside ^{14}C labeled at the carbonyl group of stearoyl chain.

Introduction

The simplest way to produce isotopically radioactive gangliosides is to administer ^{3}H or ^{14}C-labeled biosynthetic precursors to animals or to cells in culture. The resulting ganglioside specific radioactivity is low, however, using precursors such as sugars, sugar derivatives, sphingosine, or serine. Moreover, the radioactive ganglioside mixture must be purified and fractionated to give single homogeneous gangliosides. Thus, chemical and enzy-

---

[1] L. Svennerholm, *Adv. Exp. Biol. Med.* **125**, 11 (1980).
[2] IUPAC-IUB Commission on Biochemical Nomenclature, *Lipids* **12**, 455 (1982); *J. Biol. Chem.* **257**, 3347 (1977).

matic procedures were developed for the radiolabeling of highly purified glycosphingolipids. Radioactive gangliosides are necessary to develop research methods aimed at understanding the physiological role played by gangliosides at the cell surface, therefore, the radiochemical procedures have been developed following a specific design: (1) they must be capable of introducing radioactivity into specific positions of both the oligosaccharide and the ceramide moieties without changing the natural structure in any way, i.e., the labeling must be isotopic; (2) they must be applicable, sometimes with minor modifications, to a large number of ganglioside structures; and (3) they must yield gangliosides with high specific radioactivity.

Table I shows the radiochemical features of isotopically labeled gangliosides and Fig. 1 shows the general scheme of the procedures for introducing radioactivity into ganglioside structures. As an example, Fig. 1 refers to GM1 ganglioside.

1. Remarks and Recommendations on Materials

The preparation of radioactive compounds requires special attention to the purity of the materials, bearing in mind that the final amount of radioactive compound is, at most, in the order of a few milligrams. The commercial chemicals must be the purest available, checked by thin-layer chromatography (TLC) or high-performance liquid chromatography (HPLC) in different solvent systems for homogeneity and, if necessary, purified and crystallized before use. The initial gangliosides must be as pure as possible. Purification of radioactive gangliosides often requires the use of solvents where radiolytic reactions that yield radioactive by-products occur very fast. Thus, it is more useful to purify the cold natural compounds to the best homogeneity. The use of a vacuum oil pump must be avoided and, if absolutely necessary, a liquid nitrogen trap should be used to condense the oil vapors. The common solvents for column chromatography should be distilled before use, and deionized water should be freshly distilled in a glass apparatus. Silica gel chromatography columns for radioactive ganglioside or ganglioside derivative purifications should be washed before use with a few hundred milliliters of the solvent system to elute lipidic components adsorbed on the silica gel.

Some of the reactions described in the following labeling ganglioside procedure sections require anhydrous conditions that must be carried out in a controlled atmosphere chamber (Aldrich). Special attention must be paid to removing water and avoiding water contamination. Glass apparatus should be dried overnight in an oven at 120°, taken out just before use and connected to a tube filled with dryer granules, and heated over a flame. The dehydration of acetone is carried out using a 4-Å molecular sieve, of triethylamine and dimethylformamide by shaking and standing over KOH

TABLE I

RADIOLABELING PROCEDURE AS NUMBERED IN ARTICLE SECTIONS, TYPICAL AMOUNT OF STARTING COLD GANGLIOSIDE SUBJECTED TO THE PROCEDURE, AMOUNT OF PURIFIED GANGLIOSIDE DERIVATIVE SUBJECTED TO RADIOLABELING, RADIOACTIVE COMPOUND USED TO INTRODUCE RADIOACTIVITY AND ITS RADIOACTIVITY FEATURES, AND GANGLIOSIDE RADIOACTIVITY FEATURES AFTER PURIFICATION REPORTED FOR EACH ISOTOPIC RADIOACTIVE GANGLIOSIDE

Radioactive ganglioside	Method	Amount of treated ganglioside (mg)	Amount of purified substrate for radiolabeling (mg)	Radiolabeling compound	Radiolabeling compound radioactivity		Ganglioside radioactivity after purification	
					mCi	Ci/mmol	mCi	Ci/mmol
[6-^{3}H(Gal-IV)]GM1	2	20	13–15	NaB^{3}H$_4$	100	5–10	8–15	1–2
[6-^{3}H(Gal-IV)]AGM1	2	20	13–15	NaB^{3}H$_4$	100	5–10	8–15	1–2
[6-^{3}H(Gal-IV)]GD1b	2	20	11–13	NaB^{3}H$_4$	100	5–10	8–15	1–2
[6-^{3}H($GalNAc$)]GM2	2	20	4–5	NaB^{3}H$_4$	100	5–10	4–8	1–2
[3-^{3}H(C_{18}-$sphingosine$)]GM3	3	50	20–25	NaB^{3}H$_4$	100	5–10	2–4	1–2
[3-^{3}H(C_{20}-$sphingosine$)]GM3	3	50	20–25	NaB^{3}H$_4$	100	5–10	4–8	1–2
[3-^{3}H(C_{18}-$sphingosine$)]GM2	3	50	20–25	NaB^{3}H$_4$	100	5–10	3–6	1–2
[3-^{3}H(C_{20}-$sphingosine$)]GM2	3	50	20–25	NaB^{3}H$_4$	100	5–10	3–6	1–2
[3-^{3}H(C_{18}-$sphingosine$)]GM1	3	50	20–25	NaB^{3}H$_4$	100	5–10	3–6	1–2
[3-^{3}H(C_{20}-$sphingosine$)]GM1	3	50	20–25	NaB^{3}H$_4$	100	5–10	3–6	1–2
[3-^{3}H(C_{18}-$sphingosine$)]Fuc-GM1	3	50	20–25	NaB^{3}H$_4$	100	5–10	3–6	1–2
[3-^{3}H(C_{20}-$sphingosine$)]Fuc-GM1	3	50	20–25	NaB^{3}H$_4$	100	5–10	3–6	1–2
[3-^{3}H(C_{18}-$sphingosine$)]GD1a	3	50	20–25	NaB^{3}H$_4$	100	5–10	3–6	1–2
[3-^{3}H(C_{20}-$sphingosine$)]GD1a	3	50	20–25	NaB^{3}H$_4$	100	5–10	3–6	1–2
[11-^{3}H($Neu5Ac$)]GM3	5	20	7–8	NaB^{3}H$_4$	25	4–10	7–10	2–5
[11-^{3}H($Neu5Ac$)]GM2	5	20	7–8	(C^{3}H$_3$CO)$_2$O	25	4–10	7–10	2–5
[11-^{3}H($Neu5Ac$)]GM1	5	20	7–8	(C^{3}H$_3$CO)$_2$O	25	4–10	7–10	2–5
[10-^{14}C($Neu5Ac$)]GM1	5	20	7–8	(CH$_3$^{14}CO)$_2$O	0.5	0.10–0.12	0.1–0.2	0.05–0.06
[1-^{14}C($stearoyl$)]GM3	6	20	5–6	CH$_3$(CH$_2$)$_{16}$^{14}COOH	0.25	0.05–0.06	0.2–0.3	0.05–0.06
[1-^{14}C($stearoyl$)]GM2	6	20	5–6	CH$_3$(CH$_2$)$_{16}$^{14}COOH	0.25	0.05–0.06	0.2–0.3	0.05–0.06
[1-^{14}C($stearoyl$)]GM1	4	15	5–6	CH$_3$(CH$_2$)$_{16}$^{14}COOH	0.25	0.05–0.06	0.2–0.3	0.05–0.06

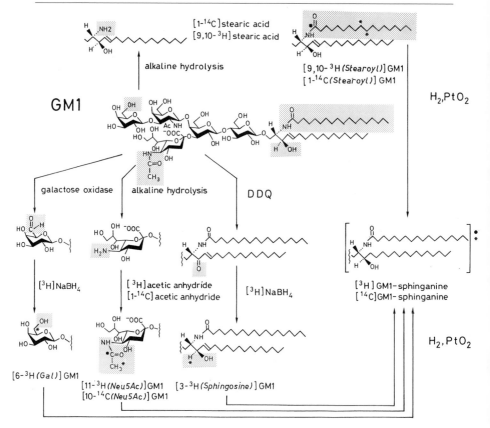

FIG. 1. Scheme for the isotopic labeling of GM1. Each radioactive compound can be prepared as a homogeneous molecular species containing C_{18}- or C_{20}-LCB by HPLC (see Section 3). *, ^{3}H; ●, ^{14}C.

pellets, of methanol, and of dimethyl sulfoxide by refluxing repeatedly and distilling from metallic magnesium and calcium hydride, respectively.

2. Preparation of AGM1, GM2, GM1, and GD1b ^{3}H
Labeled at Position 6 of the Terminal Galactose and
N-Acetylgalactosamine Residues

Galactose oxidase (EC 1.1.3.9.) specifically oxidizes the primary alcohol group of galactose and its N-acetylamino derivative to aldehyde. By utilizing tritium-labeled sodium borohydride, the aldehyde is reduced back to the original alcohol group, causing the tritium to be specifically introduced at sugar carbon 6.

$$R-CH_2-OH \xrightarrow{\text{Galactoseoxidase}} R-CHO \xrightarrow{\text{NaB}^3\text{H}_4} R-CH^3H-O^3H$$

$$R-CH^3H-O^3H \xrightarrow{\text{H}_2\text{O}} R-CH^3H-OH$$

The enzyme shows very low activity on Gal- and GalNAc-containing substrates, which, in solutions such as pure glycolipids, form aggregates of a high molecular mass.[3-5] Catalase (EC 1.11.1.6.) and peroxidase (EC 1.11.1.7) can be added to the reaction mixture to remove the enzyme inhibition exerted by the oxygen peroxide formed during oxidation.[6]

Enzyme reactivity becomes quite good when mixed micelles of glycosphingolipid and Triton X-100 are used as substrates.[7-10] Under this condition, we have the following oxidation reaction yield: 90% on GM1, 80% on GD1b, 30% on GM2, and 88% on AGM1. The enzyme has no, or very low, activity on compounds where the terminal Gal is substituted with Neu5Ac: 2-3% on GD1a.

The same experimental conditions have been applied to the preparation of tritium-labeled AGM1, GM2, GM1, and GD1b and of a GM1 derivative containing a 12-aminododecanoyl chain.

Thirteen micromoles of sphingolipid dissolved in 1 ml of chloroform-methanol (2:1 by volume) is mixed with 180 mg of Triton X-100 and dissolved in 10 ml of the same solvent. The mixture is dried under vacuum and the residue is dissolved in 8 ml of 5 mM EDTA–25 mM sodium phosphate buffer, pH 7.0, containing 450 units of galactose oxidase. The mixture is incubated at 37° under continuous stirring for 24 hr; an additional 450 units of galactose oxidase is added after the first 12 hr. The mixture is dried under vacuum and chromatographed on a 110 × 2-cm silica gel 100 column, equilibrated, and eluted with the solvent system chloroform-methanol–water (60:35:5 by volume). Column elution is monitored by TLC (silica gel HPTLC plates developed in chloroform–methanol–0.2% aqueous CaCl$_2$, 50:42:11 by volume, and visualized with an anisaldehyde spray reagent). Triton X-100 is found in the first eluates. Fractions con-

[3] A. K. Hajra, D. M. Bowen, Y. Kishimoto, and N. S. Radin, *J. Lipid Res.* **7,** 379 (1966).
[4] H. R. Sloan, B. W. Uhlendorf, C. B. Jacobson, and D. S. Fredrickson, *Pediatr. Res.* **3,** 532 (1969).
[5] Y. Suzuki and K. Suzuki, *J. Lipid Res.* **13,** 687 (1972).
[6] A. Novak, J. A. Lowden, Y. I. Gravel, and L. S. Wolfe, *J. Lipid Res.* **20,** 678 (1979).
[7] R. Ghidoni, G. Tettamanti, and V. Zambotti, *Biochem. Exp. Biol.* **13,** 61 (1977).
[8] G. Gazzotti, S. Sonnino, R. Ghidoni, P. Orlando, and G. Tettamanti, *Glycoconj. J.* **1,** 111 (1984).
[9] K. C. Leskawa, S. Dasgupta, J. L. Chien, and E. L. Hogan, *Anal. Biochem.* **140,** 172 (1984).
[10] M. A. Fra, M. Masserini, P. Palestini, S. Sonnino, and K. Simons, *FEBS Lett.* **375,** 11 (1995).

taining the oxidized ganglioside are collected, pooled, and evaporated until completely dry.

The stability of the oxidized ganglioside is quite high, but the purification must be carried out quickly, as otherwise a by-product having a higher TLC mobility is formed (Fig. 2). When stored properly as dry powder, at −20° and under argon, it is stable for at least 2–3 months.

The oxidized sphingolipid is dissolved in propanol–0.1 M NaOH (7:3 by volume) in a screw-capped tube and is treated at room temperature

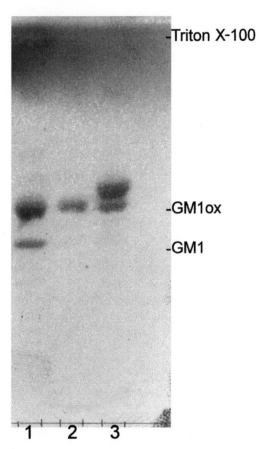

FIG. 2. Silica gel TLC in chloroform : methanol : aqueous 0.2% CaCl$_2$ (50 : 42 : 11 by volume) of the reaction mixture obtained by galactose oxidase treatment of GM1. Lane 1, reaction mixture; lane 2, purified oxidized GM1; and lane 3, purified oxidized GM1 after 1 week at room temperature under air atmosphere.

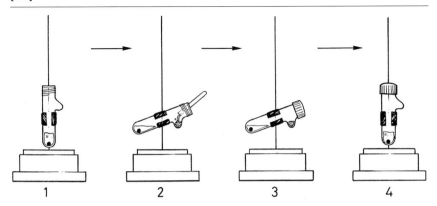

FIG. 3. Scheme for the addition of [^{3}H]NaBH$_4$ to a solution of oxidized ganglioside. (1) The oxidized ganglioside is dried at the bottom of a high-quality glass tube provided with a side arm near the top; the tube is clamped over a magnetic stirrer (off); a small magnet and the appropriate solvent are added; the magnetic stirrer is switched on until complete solubilization of the ganglioside, and the stirrer is turned off. (2) The tube is rotated, and NaB ^{3}H$_4$ is placed carefully in the side arm. (3) The tube is closed with a screw cap provided with a rubber septum. (4) The screw-capped tube is rotated back so that NaB^{3}H$_4$ falls into the solution (sometimes a few taps with a finger are necessary) and the magnetic stirrer is turned on, carefully.

with solid NaB^{3}H$_4$ according to the scheme shown in Fig. 3. After 5 hr, under continuous stirring, 10 mg of cold NaBH$_4$ is added and the reaction is allowed to continue for a further 30 min, still under stirring. The reaction mixture is then treated according to one of the following procedures. (1) The reaction mixture is evaporated under nitrogen pressure to a very small volume and the wet residue is dissolved in 1 ml of water and dialyzed. (2) The reaction mixture is evaporated under nitrogen pressure to a very small volume and the wet residue is solubilized in 1–2 ml of water, passed through a 2-ml ion-exchange resin and evaporated, dissolved in 1–2 ml of methanol, and evaporated to a very small volume and under nitrogen. The water and methanol vapors are condensed carefully in a liquid nitrogen trap (Fig. 4). For procedures 1 and 2, it is always necessary to keep one or two traps of 5 M sulfuric acid to oxidize tritium and keep back tritiated water at the end of the evaporating apparatus (Fig. 4).

The distribution of radioactivity into the ganglioside structure has been determined in a few cases by chemical hydrolysis of the radioactive ganglioside, followed by determination of the specific radioactivity of the formed neutral glycosphingolipids fractionated by TLC (see Table II). The results were 83–85% on GalNAc and 15–17% on Gal for GM2; 97–98% on external Gal; and 2–3% on GalNAc for GM1 and GD1b.

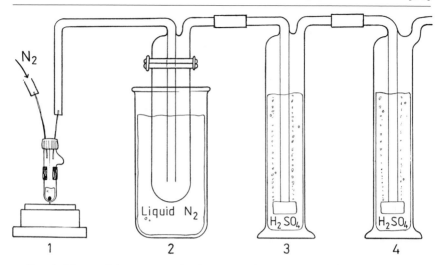

Fig. 4. Scheme of the apparatus for the evaporating procedures. (1) Two stainless-steel needles are inserted into the septum of the screw-cap reaction tube. One needle is connected to a gentle flow of nitrogen gas and the other to the trap apparatus. (2) A liquid nitrogen condenser is used to keep the volume of radioactive solutions small (see Section 2). (3) Primary sulfuric acid bottle to trap residual tritium and tritiated water. (4) Secondary sulfuric acid bottle connected to a vacuum water pump.

TABLE II

DISTRIBUTION OF RADIOACTIVITY INTO SUGAR UNITS OF GM1 TRITIUM LABELED BY GALACTOSEOXIDASE OXIDATION FOLLOWED BY REDUCTION WITH $NaB^3H_4{}^a$

	nmoles	μCi	Ci/mmol	Radioactivity content (%) on GM1 radioactivity
AGM1	3.011	3.676	1.221	100
AGM2	1.801	0.065	0.036	2.95
LacCer	2.220	0.008	0.004	0.33
GlcCer	2.152	—	—	—
Cer	0.816	—	—	—

[a] Twelve microCuries of GM1 (1.221 Ci/mmol) are subjected to partial chemical hydrolysis (0.5 M HCl, 80°, 45 min) to obtain a mixture of AGM1, AGM2, LacCer, GlcCer, and Cer. These compounds were separated by HPTLC [chloroform–methanol–water (110:40:6 by volume)] and analyzed for radioactivity (digital autoradiographic detection) and mass content (densitometry after colorimetric detection with anisaldehyde reagent).

3. Preparation of GM3, GM2, GM1, Fuc-GM1, and GD1a Tritium Labeled at Position 3 of Sphingosine

2,3-Dichloro-5,6-dicyanobenzoquinone (DDQ) specifically oxidizes allylic hydroxyl groups. Through reduction with tritium-labeled sodium borohydride, the formed carbonyl group is reduced back to the original alcohol group so that tritium is introduced at the corresponding carbon position.

$$R-CHOH-CH=CH-R \xrightarrow{DDQ} R-CO-CH=CH-R \xrightarrow{NaB^3H_4} R-C^3HO^3H-CH=CH-R$$

$$R-C^3HO^3H-CH=CH-R \xrightarrow{H_2O} R-C^3HOH-CH=CH-R$$

The oxidation procedure cannot be applied directly to gangliosides for the preparation of compounds containing tritium at carbon 3 of sphingosine due to the formation of many by-products, probably from partial oxidation of sialic acid. This drawback is overcome by preparing mixed micelles of ganglioside and Triton X-100 dispersed in toluene.[11] The yield of the reaction is 60–65% for gangliosides GM3, GM2, GM1, Fuc-GM1, and GD1a. It is unlikely that the method could be extended to gangliosides containing a disialosyl chain, such as GD1b or GT1b. This is probably due to the strong repulsive forces between the highly charged oligosaccharide chains that prevent the formation of inverted micelles.

The natural configuration of the sphingolipid sphingosine is 2S,3R. Reduction of the planar carbonyl group of the 3-ketosphingolipid is not stereospecific and produces a mixture of *erythro* and *threo* diastereoisomers at a ratio of about 2.5:1. After radiolabeling, the ganglioside mixture is purified by reversed-phase HPLC.[8,12,13] The reversed-phase HPLC also fractionates the species containing C_{18}- and C_{20}-sphingosine, thus yielding ganglioside species with homogeneous ceramide moiety.

Sodium borohydride does not act on double bonds, resulting in the radioactive ganglioside maintaining the unsaturation at positions 4 and 5 of long chain bases.

Thirty-three micromoles of ganglioside GM3, GM2, GM1, Fuc-GM1, or GD1a (this amount can be reduced to 13 μmol without compromising the final reaction yield) is dissolved in 25 ml of chloroform–methanol(2:1) by volume and mixed with 25 ml Triton X-100 in the same solvent (60 mg/ml). The solvent is evaporated and the residue is dissolved in 25 mL of

[11] R. Ghidoni, S. Sonnino, M. Masserini, P. Orlando, and G. Tettamanti, *J. Lipid Res.* **22,** 1286 (1981).
[12] S. Sonnino, G. Ghidoni, G. Gazzotti, G. Kirschner, G. Galli, and G. Tettamanti, *J. Lipid Res.* **25,** 620 (1984).
[13] E. Negroni, V. Chigorno, G. Tettamanti, and S. Sonnino, *Glycoconj. J.* **13,** 347 (1996).

DDQ solution in dry toluene (60 mg/ml). The mixture is maintained at 37° for 40 hr under constant vigorous stirring in screw-capped flask. The solvent is then evaporated under vacuum at 37°. The dark brown residue resulting from the oxidation of GM2, GM1, Fuc-GM1, or GD1a is suspended in 25 ml of cold acetone, whereas that from the oxidation of GM3 is suspended in 25 ml of cold acetonitrile and sonicated in an ultrasonic water bath. After centrifugation at 12,000g, the supernatant, which contains Triton X-100 and DDQ, is discarded. The treatment is repeated four times to obtain a white pellet. The oxidized ganglioside is separated from the unreacted compound by chromatography on a 110 × 2-cm silica gel 100 column, equilibrated, and eluted with the solvent system chloroform–methanol–water (60:35:3 by volume) for GM3 and GM2, and 60:35:5 by volume for GM1, Fuc-GM1, and GD1b. Column elution is monitored by TLC (silica gel HPTLC plates developed in chloroform–methanol–0.2% aqueous $CaCl_2$, 50:42:11 by volume and visualized with an anisaldehyde spray reagent). Fractions containing the oxidized ganglioside are collected, pooled, and evaporated until completely dry.

The stability of the oxidized sphingolipid is quite high, but purification must be carried out quickly to avoid β elimination (see Fig. 5). If properly stored as dry powder at −20° and under argon, it is stable for weeks.

The purified 3-ketoganglioside is dissolved in 5 ml of propanol–water (7:3 by volume) in a screw-capped tube and treated at room temperature with solid NaB^3H_4, following the scheme reported in Fig. 4. Ten milligrams of cold $NaBH_4$ is then added and the reaction is continued for 30 min. The solution containing the radioactive ganglioside is treated as reported earlier in Section 2.

Three to 5 milligrams (or up to 4–5 mCi) of radioactive GM3, GM2, GM1, Fuc-GM1, GD1a, or GD1b is dissolved in 0.5 ml of acetonitrile–water (1:1 by volume) and is introduced into an HPLC sample injector equipped with a 1-ml loop. The tube is washed five times with 100 μl of acetonitrile–water (1:1 by volume) and these washings are added to the injection loop to minimize loss of material. The gangliosides are then chromatographed on a reversed-phase LiChrosphere RP8 column (250 × 4 mm i.d, 5 μm average particle diameter; Merck, Darmstadt, FRG), equilibrated, and eluted at a flow rate of 10 ml/min with the solvent system acetonitrile–5 mM phosphate buffer, pH 7.0, 60:40 by volume for GM3, GM2, GM1, and Fuc-GM1 and 50:50 by volume for GD1a and GD1b. The separation of GM3 species is carried out at 45° and that of all the other gangliosides at room temperature. The elution profile is monitored by flow-through detection of both UV absorbance at 195 nm and radioactivity. Fractions containing the same species are collected, pooled, evaporated to a very small volume, and dialyzed. Reversed-phase TLC and reversed-phases HPLC of

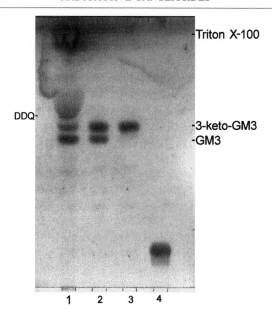

FIG. 5. Silica gel TLC in chloroform:methanol:aqueous 0.2% $CaCl_2$ (50:42:11) of the reaction mixture obtained by DDQ oxidation of GM3. Lane 1, reaction mixture; lane 2, reaction mixture after removal of DDQ and Triton X-100, lane 3, purified 3-keto-GM3; and lane 4, purified 3-keto-GM3 after a long period of time in solution.

GM1 and GM3 gangliosides labeled at position 3 of sphingosine are shown in Figs. 6 and 7.

4. Preparation of GM1 Containing ^{3}H- or ^{14}C-Labeled Fatty Acid Moiety

Gangliosides subjected to strong alkaline treatments in organic solvent systems yield a series of derivatives lacking in the sugar acetyl groups, the fatty acyl chain, or both.[14–20] GM1 hydrolysis in KOH-deoxygenated

[14] S. Sonnino, D. Acquotti, G. Kirschner, A. Uguaglianza, L. Zecca, F. Rubino, and G. Tettamanti, *J. Lipid Res.* **33,** 1221 (1992).
[15] T. Taketomi and N. Kawamura, *J. Biochem.* **68,** 475 (1970).
[16] J. Holmgren, J. E. Mansson, and L. Svennherolm, *Med. Biol.* **52,** 229 (1974).
[17] J. L. Tayot and M. Tardy, *Adv. Exp. Med. Biol.* **125,** 471 (1980).
[18] H. Higashi and S. Basu, *Anal. Biochem.* **120,** 159 (1982).
[19] S. Nevenhofer, G. Schwarzmann, H. Egge, and K. Sandhoff, *Biochemistry* **24,** 525 (1985).
[20] S. Sonnino, G. Kirschner, R. Ghidoni, D. Acquotti, and G. Tettamanti, *J. Lipid Res.* **26,** 248 (1985).

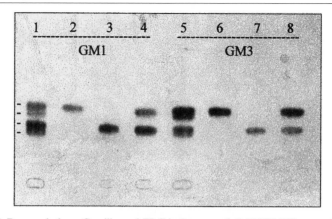

FIG. 6. Reversed-phase C_{18}-silica gel TLC in 2-propanol:5 M NH$_4$OH:water (7:2:1 by volume) of radioactive ganglioside molecular species; autoradiographic detection. Lane 1, DDQ/NaB ^{3}H$_4$-labeled GM1; lane 2, [3-^{3}H(C_{18}-*sphingosine*)]GM1 purified by reversed-phase HPLC; lane 3, [3- ^{3}H(C_{20}-*sphingosine*)-]GM1 purified by reversed-phase HPLC; lane 4, GM1 tritium labeled at the sialic acid residue; lane 5, DDQ/NaB^{3}H$_4$-labeled GM3; lane 6, [3- ^{3}H(C_{18}-*sphingosine*)]GM3 purified by reversed-phase HPLC; lane 7, [3- ^{3}H(C_{20}-*sphingosine*)]GM3 purified by reversed-phase HPLC; and lane 8, GM3 tritium labeled at the sialic acid residue.

anhydrous propanol produces lyso-GM1, the GM1 derivative lacking only the fatty acid chain, in a yield of about 55%.[14] By acylation with radioactive fatty acid the GM1 structure is reconstituted.

$$R-NH-CO-(CH_2)_n-CH_3 \xrightarrow{(OH)^-} R-NH_2$$

$$R-NH_2 \xrightarrow{CH_3-(CH_2)_{16}-^{14}COOR} R-NH-^{14}CO-(CH_2)_{16}-CH_3$$

Propan-1-ol is deoxygenated by a 30-min bubbling with argon (20–30 liters/min) and maintained in a closed bottle under moderate argon pressure. A solution of GM1 (1.4 mM) in deoxygenated propan-1-ol, warmed to 90°, is mixed in a macrovial provided with an open-top screw cap and a natural rubber septum, with a 90° prewarmed deoxygenated propan-1-ol potassium hydroxide solution (1 M) to obtain a final KOH concentration of 0.2 M. Two stainless-steel needles are introduced into the vial septum, one connected to the argon bottle, and the reaction mixture is maintained under an argon flux of 5 ml/min for 30 min. The needles are then removed and the reaction is continued at 90° under continuous stirring for 6 hr. The reaction mixture is dried, and the residue is dissolved in water (5 mg GM1/ 0.1 ml), dialyzed for 3 hr, and freeze-dried. The purification of lyso-GM1 is carried out by silica gel column chromatography (60 × 1.5 cm/50 mg of starting GM1) using chloroform–methanol–water, (60:35:5 by volume) as

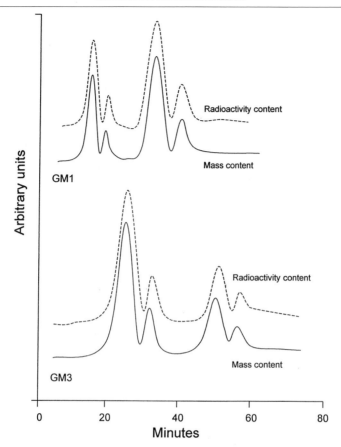

FIG. 7. Reversed-phase HPLC of DDQ/NaB ^{3}H$_4$-labeled GM1 (top) and GM3 (bottom). The eluted molecular species, from left to right, contain *erythro*-C$_{18}$-sphingosine, *threo*-C$_{18}$-sphingosine, *erythro*-C$_{20}$-sphingosine, and *threo*-C$_{20}$-sphingosine. Dotted lines represent radioactivity detection; filled lines represent UV detection. The HPLC C$_{18}$- and C$_{20}$-species distribution reflects that of the starting purified cold ganglioside; it depends on both the natural species distribution and on a partial liquid chromatography enrichment of one species along the ganglioside purification procedures. Ganglioside labeling yields the same specific radioactivity for the different species.

the eluting solvent. The elution profile is monitored by TLC using chloroform–methanol–30 mM aqueous CaCl$_2$–100 mM aqueous KCl (50:50:4:8 by volume) as the developing solvent.

Radioactive stearic acid, pentafluorophenol, chloromethylpyridinium iodine, and tributylamine at a molar ratio of 1:1.2:1.4:2.8 are dissolved in anhydrous dichloromethane and refluxed under nitrogen for 3 hr. The reaction mixture is evaporated and filtered on a 15×1-cm silica gel column, equilibrated, and eluted with hexane. The filtered solution containing the activated radioactive stearic acid is dried, and the residue is solubilized in anhydrous dimethylformamide (0.04 mmol/ml). The solution is added to lyso-GM1 (final stearic acid–ganglioside derivative molar ratio of $1:1$) and the reaction is left at room temperature for 24 hr under continuous stirring. The solution is then lyophilized and the residue is purified by chromatography on a 50×1-cm silica gel column equilibrated and eluted with the solvent system chloroform–methanol–water ($50:42:11$ by volume). Elution is monitored by TLC [silica gel HPTLC plates developed in chloroform–methanol–0.3% aqueous $CaCl_2$ ($50:42:11$ by volume) and revealed through autoradiography]. Fractions containing the radioactive GM1 are collected, pooled, and evaporated until completely dry.

5. Preparation of GM3, GM2, and GM1 with ^{3}H or [14]C Labeling at the Sialic Acid Acetyl Group

Alkaline hydrolysis of monosilogangliosides GM3, GM2, or GM1 in butanol–aqueous tetramethylammonium hydroxide yields a mixture of the derivative lacking the acetyl group of sialic acid and the derivative lacking both the acetyl group of sialic acid and the fatty acyl chain.[20–23] The acetyl group of GalNAc of GM2 and GM1 is not affected by this hydrolysis. The same procedure has been attempted with polysialogangliosides but yielded several derivatives that were not characterized.

$$\xrightarrow{} H_2N-R-NH-CO-(CH_2)_n-CH_3$$

$$CH_3-CO-HN-R-NH-CO-(CH_2)_n-CH_3 \xrightarrow{\ (OH)^-\ }$$

$$\xrightarrow{} H_2N-R-NH_2$$

$$H_2N-R-NH-CO-(CH_2)_n-CH_3 \xrightarrow{(C^3H_3-CO)_2O} C^3H_3-CO-HN-R-NH-CO-(CH_2)_n-CH_3$$

$$H_2N-R-NH-CO-(CH_2)_n-CH_3 \xrightarrow{(CH_3-^{14}CO)_2O} CH_3-^{14}CO-HN-R-NH-CO-(CH_2)_n-CH_3$$

N-Acetylation of the purified neuraminic acid containing ganglioside

[21] V. Chigorno, M. Pitto, G. Cardace, D. Acquotti, G. Kirschner, S. Sonnino, R. Ghidoni, and G. Tettamanti, *Glycoconj. J.* **2,** 279 (1985).

[22] S. Sonnino, L. Cantu, D. Acquotti, M. Corti, and G. Tettamanti, *Chem. Phys. Lipids* **52,** 231 (1990).

[23] R. Navon, R. Khosravi, T. Korczyn, M. Massom, S. Sonnino, M. Fardeau, B. Eymard, M. Lefevre, J. C. Turpin, P. Rondot, and N. Baumann, *Neurology* **45,** 539 (1995).

with acetic anhydride is a simple reaction. Nonetheless, radioactive ace-tic anhydride is a very unstable compound that degrades easily to acetic acid.

The following procedure is applied to the preparation of radioactive GM3, GM2, or GM1. Thirteen micromoles of ganglioside is dissolved in 2 ml of butan-1-ol at 100° and is mixed with 0.2 ml of aqueous 1 M tetramethyl-ammonium hydroxide. The reaction mixture is refluxed at 100° under con-tinuous stirring. After 13 hr, the pH is adjusted to 6–7 by dropwise addition of acetic acid, and the solution is evaporated under vacuum at 50°. The residue is dissolved in a few milliliters of water and is applied to a 5 × 2-cm RP-18 column. The column is washed with 300 ml of water to remove salt and with 500 ml of chloroform–methanol (2:1 by volume) to recover the two ganglioside derivatives, i.e., the deacetyl ganglioside and the deace-tyldeacyl ganglioside. The chloroform–methanol is evaporated and the two compounds are separated by chromatography on a 50 × 1-cm silica gel 100 column, equilibrated, and eluted with chloroform–methanol–water (50:42:11 by volume). Elution is monitored by TLC [silica gel HPTLC plates developed in chloroform–methanol–0.3% aqueous $CaCl_2$ (50:42:11 by volume) and visualized with anisaldehyde spray reagent]. Fractions con-taining the deacetyl ganglioside or the deacetyldeacyl ganglioside are col-lected, pooled, and evaporated until completely dry.

The purified deacetyl ganglioside is dissolved (2 mg/ml) in dry methanol and mixed with radioactive acetic anhydride to a final molar ratio of 1:1. After strring for 30 min at room temperature, 10 μl of cold acetic anhydride is added, the mixture is stirred for an additional 30 min, and the solution is treated as reported earlier except that a trap of NaOH rather than sulfuric acid is used to catch radioactive acetic acid. The addition of cold acetic anhydride can be avoided to increase the ganglioside specific radioactivity. In this case, radioactive ganglioside and the unreacted deacetyl ganglioside are separated by chromatography on a 50 × 1-cm silica gel 100 column, equilibrated, and eluted with chloroform–methanol–water (60:35:8 by volume).

6. Preparation of GM3 and GM2 ^{3}H or [14]C Labeled at the Fatty Acid
   Moiety, Preparation of ^{3}H/[14]C Double-Labeled GM3, GM2, and
   GM1

The preparation of GM3 and GM2 gangliosides containing a radioactive fatty acid chain requires the preparation of the deacetyldeacyl ganglioside derivative as described in the preceding section.

$$H_2N-R-NH_2 \xrightarrow{CH_3-(CH_2)_{16}-{}^{14}COOR} H_2N-R-NH-{}^{14}CO-(CH_2)_{16}-CH_3$$

$$H_2N-R-NH-{}^{14}CO-(CH_2)_{16}-CH_3 \xrightarrow{(C^3H_3CO)_2O} C^3H_3-CO-HN-R-NH-{}^{14}CO-(CH_2)_{16}-CH_3$$

Acylation deacetyldeacyl ganglioside is also carried out as reported earlier. Under these conditions, the acylation is selective for the amino group of sphingosine. The solution is then lyophilized and the residue is purified by chromatography on a 50 × 1-cm silica gel column equilibrated and eluted with the solvent system chloroform–methanol–water (50 : 42 : 11 by volume). Elution is monitored by TLC [silica gel HPTLC plates developed in chloroform–methanol–0.3% aqueous CaCl$_2$ (50 : 42 : 11 by volume), and revealed by autoradiography]. Fractions containing the radioactive deacetyl ganglioside are collected, pooled, and evaporated until completely dry.

The purified radioactive deacetyl ganglioside is dissolved (2 mg/ml) in dry methanol and mixed with 10 μl of acetic anhydride. After stirring for 30 min at room temperature, the solution is concentrated to a very small volume, 1 ml of water is added, residual methanol is eliminated under vacuum, and the solution is dialyzed. The use of radioactive acetic anhydride (see preceding Section) instead of cold compound allows the preparation of double-labeled GM3, GM2, and GM1.

7. Preparation of Radiolabeled Gangliosides Containing Sphinganine

Ganglioside species that contain sphinganine as the long chain base are minor cell components. To prepare radioactive gangliosides containing sphinganine, one performs catalytic hydrogenation of radioactive gangliosides that contain sphingosine.[19] The yield of the reaction is high (100%) (see Fig. 8) for gangliosides containing radioactivity in either the oligosaccharide or lipid moieties.

$$[^3H]R-CH=CH-R \xrightarrow{H_2, PtO_2} [^3H]R-CH_2-CH_2-R$$

Two micromoles of radioactive ganglioside is dissolved in 3–4 ml of 96% ethanol and treated with hydrogen gas at a pressure of 1.05 atm for 36 hr at room temperature in the presence of 1–2 mg of platinum dioxide as the catalyst under constant stirring. At the end of the reaction, the mixture is centrifuged at 9000 rpm for 20 min at 4°, and the supernatant is filtered through paper to remove the insoluble catalyst. The solution is

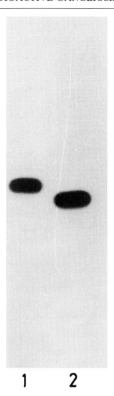

1   2

FIG. 8. Reversed-phase C_{18}-silica gel TLC in acetonitrile–water (20:1 by volume) of radioactive GM3; autoradiographic detection. Lane 1, GM3 containing C_{18}-sphingosine tritium labeled at position 3; and lane 2, GM3 containing C_{18}-sphingosine tritium labeled at position 3 after catalytic hydrogenation. The GM3 molecular species containing C_{18}-sphingosine was prepared by reversed-phase HPLC.

then evaporated until completely dry, solubilized in 1–2 ml, dialyzed at 4°, and mixed with 2–3 volumes of propanol.

8. Stability of Radioactive Gangliosides

Radioactive gangliosides, like all radioactive compounds, require the maximum of care as they degrade very quickly in the solid state, in acid solutions, and in solutions where radicals form easily (such as in the presence of chloroform). The rate of the degradation process increases with time, with increasing ganglioside specific radioactivity and solution specific radioactivity, and when the solutions are frozen. Note that the degradation process is not usually proportional with time sometimes being slow over the first weeks (or days) and then becoming faster. The structural characterization of degraded radioactive gangliosides is difficult and, to our knowledge, has never been carried out.

When radioactivity is introduced into gangliosides, the resulting ganglioside specific radioactivity is related only to the specific radioactivity of the labeling reagent and the reaction stoichiometry. Care is necessary when using reagents with a high specific radioactivity because gangliosides with a specific radioactivity over 10 Ci/mmol are very unstable. Experience has shown that gangliosides with a specific radioactivity of 1–5 Ci/mmol, stored at 4° in propanol–water (7:3 by volume) or in methanol, at a solution specific radioactivity of 50–100 μCi/mL, are stable for several months. In any case, the purity of the ganglioside must be tested frequently and always before any key experiments.

Acknowledgment

This work was supported by the "Coofinanziamento del MURST 1997. Glicobiologia: struttura fine dei glicoconiugati come base del loro coinvolgimento in processi biologici e patologici."

[51] Estimating Sphingolipid Metabolism and Trafficking in Cultured Cells Using Radiolabeled Compounds

By Laura Riboni, Paola Viani, and Guido Tettamanti

I. Introduction*

The use of radioactive metabolites, particularly radiolabeled complex sphingolipids, represents an experimental approach to investigating different aspects of sphingolipid metabolism. The rationale of this approach is based on the fact that when complex sphingolipids (sphingomyelin and

---

* Abbreviations: Cer, ceramide; Cer-1-P, ceramide-1-phosphate; DAG kinase, *sn*-1,2-diacylglycerol kinase; DMSph, *N,N*-dimethylsphingosine; D-PDMP, D-threo-1-phenyl-2-decanoylamino-3-morpholino-1-propanol; Gal, galactose; GalNAc, *N*-acetylgalactosamine; gangliosides are named according to Svennerholm[1]; GD1b lactone, β-Gal-(1 → 3)-; β-GalNAc-(1 → 4)-[α-NeuAc-(2 → 8)-(1 → 9)-α-NeuAc-(2 → 3)]-β-Gal-(1 → 4)-β-Glc-(1 → 1)-Cer; GD1b-ol, α-nonulosamine-(2 → 8)-α-NeuAc-(2 → 3)-containing GD1b; Glc, glucose; Glc-Cer, glucosylceramide; HPTLC, high-performance thin-layer chromatography; Lac-Cer, lactosylceramide; NeuAc, *N*-acetylneuraminic acid; O-Ac, O-acetylated; PBS, phosphate-buffered saline; Sph, D-*erythro*-sphingosine; Sph-1-P, sphingosine-1-phosphate; SM, sphingomyelin.

gangliosides) are added to cultured cells, they are partly taken up and inserted into the outer plasma membrane leaflet (where most cell sphingolipids are located); they are then processed metabolically through the intracellular traffic and metabolizing pathways of the corresponding endogenous components. Assessing the metabolic products formed by cells from an exogenously administered sphingolipid can be instrumental to (a) recognizing the metabolic processing and intracellular trafficking of a specific sphingolipid; (b) ascertaining cell specificity in metabolic processes; (c) establishing the subcellular aspects of metabolic events involved in sphingolipid processing; (d) investigating the role of sphingolipids in signaling processes; and (e) evaluating the effects exerted by different molecules on sphingolipid metabolism.

This approach, compared to administering radiolabeled simple precursors (widely used to investigate sphingolipid metabolism), offers the following major advantages: (a) high sensitivity, even less than 1 pmol/mg cell protein of a metabolite can be revealed and such sensitivity is fundamental to investigating sphingolipid metabolism in cultured cells; (b) specificity, the metabolic processing of single sphingolipids and also the different molecular species of the same molecule can be investigated; (c) selectivity, specific experimental protocols can be planned in order to discriminate between the different metabolic pathways; and (d) the possibility of obtaining information about intracellular sphingolipid trafficking and the subcellular location of the metabolic routes.

All this information can be gained using appropriately radiolabeled complex sphingolipids, adequate pulse–chase conditions, and carefully separating, identifying, and quantifying the radiolabeled metabolites.

This article describes the experimental procedures useful for a global investigation of radiolabeled sphingolipid metabolism in cultured cells and then the specific protocols suitable for studying specific pathways involved in sphingolipid metabolism and intracellular sphingolipid trafficking.

II. Reagents and Radiolabeled Molecules

HPTLC silica gel and cellulose plates, common chemicals, several inhibitors (e.g., conduritol B-epoxide, L-cycloserine, β-chloro-L-alanine, N-butyldeoxynojirimycin, chloroquine, D-PDMP, monensin, fumonisin B_1, nocodazole, tunicamycin), and enzymes (β-galactosidase from jack beans, recombinant β-N-acetylhexosaminidase, and sphingomyelinase from *Staphylococcus aureus*) are from Sigma (St. Louis, MO); ceramide glycanase (from *Macrobdella decora*) and *Vibrio cholerae* sialidase are from Boerhinger Mannheim (Mannheim, Germany); *Escherichia coli* DAG kinase, *Arthrobacter ureafaciens* sialidase, brefeldin A, and (1S,2R)-D-*erythro*-2-

(N-myristoylamino)-1-phenyl-1-propanol are from Calbiochem (La Jolla, CA); and VSWP filters (0.025-μm pore size) and Sep-Pak C_{18} cartridges are from Millipore (Bedford, MA).

Gangliosides, radiolabeled isotopically at the level of (a) C-3 of the long chain base ([^{3}H-Sph]GM1, [^{3}H-Sph]GM2); (b) terminal saccharide [^{3}H-Gal]GM1, [^{3}H-Gal]GD1b, or [^{3}H-GalNac]GM2; (c) sialic acid [^{3}H-NeuAc]GM1 or [^{3}H-NeuAc]GD1a; or (d) fatty acid [^{14}C-stearoyl]GM1, are obtained and stored properly as described by Sonnino et al.[2] Sphingomyelin is radiolabeled isotopically at the level of the sphingosine moiety ([^{3}H-Sph]SM) as described by Iwamori et al.[3] The specific radioactivity of the different radiolabeled sphingolipids varies from 0.05 to 2 Ci/mmol, depending on the type of labeling and the radiolabeled sphingolipid.

Standard radioactive neutral glycolipids ([^{3}H]Glc-Cer, [^{3}H]Lac-Cer, [^{3}H]asialo-GM2, [^{3}H]asialo-GM1) are prepared from [^{3}H-Sph]GM1 by controlled acid hydrolysis. In particular, 0.2 μCi of [^{3}H-Sph]GM1 is dried under nitrogen stream in a screw-capped Pyrex tube and resuspended in 200 μl 0.5 N HCl. After incubation at 80° for 60–150 min, the sample is neutralized with 200 μl 0.5 N NaOH and vortexed. Partitioning is achieved by the addition of 1.6 ml of chloroform/methanol (2:1, by volume). After mixing, samples are centrifuged at 2000g for 5 min. Radiolabeled neutral glycolipids are recovered in the chloroform phase and stored at −20°.

Standard radioactive GD1b-lactone and its reduced form (GB1b-ol) are prepared from [^{3}H-Gal]GD1b as described previously.[4]

Standard [^{3}H-Sph]Cer is prepared from [^{3}H-Sph]GM1 by treatment with ceramide glycanase.[5] The incubation mixture (final volume, 1 ml) containing 0.2–0.5 μCi of [^{3}H-Sph]GM1, 50 mM acetic acid–sodium acetate buffer, pH 5.0, 2 mM sodium cholate, and 2 mU enzyme is incubated at 37° for 16–24 hr. The reaction is stopped by the addition of 4 volumes of chloroform/methanol (2:1, by volume). The sample is then vortexed and centrifuged (3000g for 15 min), and the formed [^{3}H]Cer is recovered in the lower organic phase and stored at −20°.

Standard [^{3}H]Cer-1-P can be prepared from [^{3}H]Cer by treatment with DAG kinase according to Preiss et al.[6] [^{3}H]Cer-1-P can be used for the

[1] L. Svennerholm, Adv. Exp. Med. Biol. 125, 11 (1980).
[2] S. Sonnino, V. Chigorno, and G. Tettamanti, Methods Enzymol. 311 [50] 1999 (this volume).
[3] M. Iwamori, H. W. Moser, and Y Kishimoto, J. Lipid Res. 16, 332 (1975).
[4] S. Sonnino, L. Riboni, D. Acquotti, G. Fronza, G. Kirschner, R. Ghidoni, and G. Tettamanti, in "New Trends in Ganglioside Research" (R. W. Ledeen et al., eds.), Vol. 14, p. 47. Liviana Press, Padova, 1988.
[5] B. Zhou, S. C. Li, R. A. Laine, R. T. C. Huang, and Y. T. Li, J. Biol. Chem. 264, 12272 (1989).
[6] J. E. Preiss, C. R. Loomis, R. M. Bell, and J. E. Niedel, Methods Enzymol. 141, 294 (1987).

preparation of [^{3}H]Sph-1-P after acidic butanol treatment.[7] Standard [3-^{3}H]Sph can be obtained from NEN (Boston, MA).

III. Experimental Procedures

A. Preparation of Radiolabeled Sphingolipid Solutions for Pulse Experiments

Prior to experiments, it is mandatory that the radiochemical purity of the sample be checked by HPTLC. For this purpose, $1-2 \times 10^5$ dpm of the radiolabeled sphingolipid is spotted as a 0.5-cm band on HPTLC plates together with $1-2$ μg of the unlabeled standard. The radiochemical purity of the radioactive sphingolipid [evaluated after HPTLC separation in different solvent systems and quantification (see later for details)] must be over 99%.

At the time of the experiment, an aliquot of the stock radiolabeled sphingolipid, evaporated to dryness under nitrogen stream in a glass tube, is resuspended in a small volume of distilled ethanol and sonicated (10 sec) three times in a bath sonicator. The sample is then concentrated under a nitrogen stream and finally diluted with culture medium without fetal calf serum to the desired final concentration (final ethanol concentration never exceeding 0.1%). The sample is then vortexed and incubated at 37° for 30 min. If pulse at a temperature lower than 37° (16 or 4°) is chosen, the solution is refrigerated at the desired temperature prior to use. Prior to pulse, aliquots of the solution are counted for radioactivity by liquid scintillation counting.

B. Pulse–Chase Protocols

Metabolic studies are performed on cells (generally corresponding to $1-5 \times 10^6$) grown as a monolayer on 60-mm dishes. At the time of the experiment, the medium is removed gently from the plates by a Pasteur pipette connected to a suction pump, and the plates are then washed twice with 3 ml of temperature-conditioned culture medium without fetal calf serum. The cells are then incubated (pulse) for a given period of time in 2 ml of the same medium containing 0.5–2 μM radiolabeled sphingolipid carrying 0.2–2 μCi/ml. At the end of the pulse period the medium is removed and the cells are washed (2 ml/dish, three times) with a medium containing 10% fetal calf serum to eliminate the pericellular, loosely bound pool of radiolabeled sphingolipid. If a pulse–chase approach is chosen, the cells are incubated further in 10% FCS medium for different times (chase).

[7] H. Kaller, *Biochem. Z.* **334,** 451 (1961).

At the end of the pulse or chase period, the medium is collected carefully by a Pasteur pipette and the cells are rinsed twice with cold PBS, harvested by a rubber scraper in Eppendorf tubes, frozen in liquid nitrogen, and lyophilized. The pulse and chase media are centrifuged at 7000g for 10 min to remove dislodged cells and the supernatants can be stored at $-20°$ until processed.

C. Extraction of Radiolabeled Sphingolipids

Radiolabeled sphingolipids are extracted from cultured cells along with other lipids with the use of organic solvents. Different variations in sphingolipid extraction and partitioning from tissues and cells have been reported. The following procedure is a general one and is adequate to analyze, with acceptable reliability and reproducibility and different radioactive metabolites after the administration of complex radioactive sphingolipids to cultured cells.

Cells are extracted in Eppendorf tubes, (all steps are performed at $4°$ unless specified otherwise). Lyophilized samples (generally corresponding to $1–5 \times 10^6$ cells) are resuspended in 50 μl distilled water by sonication and vigorous vortexing, with great care being taken to reach an efficient disaggregation of the lyophilized sample, essential for quantitative lipid extraction. Thus it is useful to have access to an ultrasonic processor for low-volume application, equipped with a stepped titanium probe (e.g., Vibra-cell, Sonics & Materials Inc., Danbury, CT). After resuspension, 350 μl methanol is added to the tubes. After vortexing for 10 min, 700 μl chloroform is added, and the tubes are vortexed again for 10 min and centrifuged at 10,000g for 5 min. The supernatant is collected carefully in Pyrex screw-capped tubes and the precipitate is reextracted with 500 μl of chloroform/methanol (2 : 1, by volume). This is further mixed and centrifuged as described earlier, and the second supernatant is added to the first extract, with the two combined supernatants representing the total lipid extract. An aliquot of the total lipid extract (generally corresponding to 5–10 μl/1.6 ml final volume) is counted for radioactivity and can be analyzed directly by HPTLC. Furthermore, an aliquot of the total lipid extract can be used to determine the phospholipid content.[8,9] The total amount of cellular phospholipids or protein (see analysis of the delipidized pellet) is used for the normalization of radioactivity data.

Because the efficiency of the subsequent partitioning step is dependent on the solvent ratio, care must be taken throughout the extraction procedure to minimize solvent loss and change in solvent composition by evaporation.

---

[8] G. R. Bartlett, *J. Biol. Chem.* **234**, 466 (1959).

[9] J. T. Dodge and G. B. Phillips, *J. Lipid Res.* **8**, 667 (1967).

D. Bulk Separation of Radiolabeled Sphingolipids

The total lipid extract is partitioned by adding 0.25 volumes of 0.1 M KCl. After vigorous shaking, the two phases are separated by centrifugation at 3000g for 5 min. The chloroform (lower) phase is washed once with chloroform/methanol/0.1 M KCl (3 : 48 : 47, by volume). The two upper phase fractions are pooled (aqueous phase), and both upper and lower phases are evaporated to dryness under nitrogen stream.

The dried aqueous phase is then resuspended in 10 μl distilled water and dialyzed through a VSWP filter against 10 ml distilled water for 3 hr at 4°. Alternative to dialysis, the removal of polar contaminants can be accomplished by reversed-phase chromatography using Sep-Pak C_{18} cartridges.[10]

Aliquots of the dried chloroform phase [that may contain, as radioactive metabolites, most of the sphingoid molecules (including Cer, neutral glycolipids, Sph, DMSph, and SM) together with glycerophospholipids] can be processed directly by HPTLC after radioactivity counting. This procedure can be useful in investigating the possible formation of radiolabeled glycerophospholipids, e.g., after administration of [^{14}C-stearoyl]GM1. If this is not the case, the chloroform phase is subjected to a mild alkaline methanolysis to remove glycerophospholipids. To do this, the dried residue is redissolved in 100 μl chloroform. After the addition of 100 μl 0.2 N methanolic KOH, the sample is vortexed and incubated at 37° for 1 hr. The reaction is stopped by neutralization with 100 μl 0.2 N methanolic HCl. In order to reach the final chloroform/methanol ratio (2 : 1, by volume), 200 μl methanol and 700 μl chloroform are added. After mixing, the phases are separated by the addition of 170 μl water and centrifugation (3000g for 5 min). After removing the aqueous phase, the lower phase is washed with the same volume of methanol/water (1 : 1, by volume).

Most gangliosides are recovered quantitatively in the upper, aqueous phase (Figs. 1A–1D), and Sph as well as most sphingolipids are found in the methanolyzed chloroform phase (Fig. 1E). However, a portion of less polar gangliosides (i.e., GM3 and GM2) remains in the chloroform phase, and Sph-1-P partitions almost equally between the two phases. When investigating these molecules, care must be taken to check the recovery of the corresponding radiolabeled standards run through the same procedures. Alternatively, the total lipid extract can be partitioned by the diisopropylether/1-butanol method[11] or in alkaline conditions[12] for the

[10] M. A. Williams and R. H. McCluer, *J. Neurochem.* **35,** 266 (1980).
[11] S. Ladish and B. Gillard, *Anal. Biochem.* **146,** 220 (1985).
[12] Y. Yatomi, F. Ruan, H. Ohta, R. J. Welch, S.-I. Hakomori, and Y. Igarashi, *Anal. Biochem.* **230,** 315 (1995).

A: [³H-Sph]GM1

B: [³H-NeuAc]GM1

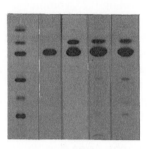
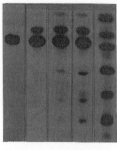

GM3
GM2
GM1

GD1a

GD1b
O-Ac-GT1b
GT1b

St1   0'   30'   2h   4h

0'   30'   2h   4h   St1

C: [³H-Sph]GM2

D: [³H-GalNAc]GM2

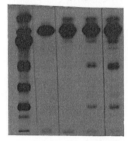
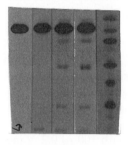

GM3
GM2
GM1

GD1a

GD1b
O-Ac-GT1b
GT1b

St1   0'   30'   2h   4h

0'   30'   2h   4h   St1

E: [³H-Sph]GM1

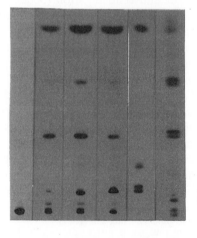

Cer

Glc-Cer

Lac-Cer

Sph

SM
asialo-GM1

0'   30'   2h   4h   St2   St3

quantitative recovery of GM3 and GM2 gangliosides and Sph-1-P, respectively.

The dialyzed aqueous phase and the methanolyzed organic phase are then dried under a nitrogen stream, resuspended in chloroform/methanol (2:1, by volume), and counted for radioactivity by liquid scintillation counting.

E. Thin-Layer Chromatography Separation and Quantification of Radiolabeled Sphingolipids

HPTLC can be used to determine both qualitatively and quantitatively the radiolabeled organic metabolites recovered from the total lipid extract, dialyzed aqueous phase, or the methanolyzed organic phase. To do this, $5-20 \times 10^3$ dpm of the sample is dried under a nitrogen stream, resuspended in 5–10 μl of chloroform/methanol (2:1, by volume) by vortexing, and applied to the HPTLC plate. The sample is applied by Pyrex disposable micropipettes (Corning, NY) or microliter syringes. After air drying (heat should be avoided), the plate is developed in a chromatographic chamber, preincubated with the developing solvent. Examples of the different solvent systems that can be used in mono- or two-dimensional HPTLC, together with indications for their application, are shown in Table I. After HPTLC, the plates are dried by a forced air dryer and analyzed for radioactivity (see later).

FIG. 1. Kinetics of the metabolic processing of different radiolabeled gangliosides by cerebellar granule cells differentiated in culture. GM1 and GM2 gangliosides, radiolabeled at the level of the sphingoid moiety ([³H-Sph]GM1 and [³H-Sph]GM2) or sialic acid ([³H-NeuAc]GM1) or terminal saccharide ([³H-GalNAc]GM2), were administered (1 μM, 1 μCi/ml) to cerebellar granule cells for 30 min, 2 hr, or 4 hr. At the end of the pulse period, cell lipids were extracted, partitioned, and separated by HPTLC. The plates were then sprayed with En³Hance and submitted to fluorography at −80° for 1–2 months. (A–D) HPTLC separation of aqueous phases (10,000–20,000 dpm/lane were spotted; solvent system G, Table I); St1: standard radioactive GM3, GM2, GM1, GD1a, GD1b, GT1b (from top to bottom). (E) HPTLC separation of the methanolyzed organic phase after pulse with [³H-Sph]GM1 (4000–7000 dpm/lane were spotted; solvent system A, Table I). St2: standard radioactive Cer, Sph and SM (from top to bottom), St3: standard radioactive Cer, Glc-Cer, Lac-Cer, asialo GM1 (from top to bottom). Lanes identified as 0′ correspond to controls of the extraction and separation procedures. To do this, 50,000 dpm of the radiolabeled ganglioside was added to control cells immediately before lipid extraction. Note the formation of (a) gangliosides simpler than the administered one (in A–D) derived from degradation (GM2 and GM3 from GM1; GM3 from GM2); (b) neutral glycolipids, Cer and Sph (in E) derived from degradation of [³H-Sph]GM1 (a similar pattern can be obtained with of [³H-Sph]GM2); (c) gangliosides more complex than the administered one (in A–D) and sphingomyelin (in E) that derive from the direct glycosylation or the recycling of the degradation fragments (Sph, NeuAc, GalNAc).

TABLE I
HPTLC SEPARATION OF SPHINGOLIPIDS[a]

Solvent system	Solvent ratio (by volume)	Application to[b]	Separation of
A. Chloroform/methanol/water	55:20:3	TLE, MCP	Cer, neutral glycolipids, Sph, DMSph, SM
B. Chloroform/methanol/water	60:35:8	TLE, MCP	Cer, neutral glycolipids, Sph, SM
C. Chloroform/methanol/32% NH4OH	40:10:1	TLE, MCP	Cer, Sph, DMSph, Glc-Cer
D. Chloroform/methanol	95:5	TLE, MCP	Cer
E. Chloroform/methanol/acetone/acetic acid/water	10:2:4:2:1	TLE, CP, MCP	Cer, glycerophospholipids, Cer-1-P, SM
F. Chloroform/methanol/0.2% CaCl2	55:45:10	TLE, AP	Mono- and disialogangliosides
G. Chloroform/methanol/0.2% CaCl2	50:42:11	TLE, AP	Mono-, di-, and polysialogangliosides
H. n-Butanol/acetic acid/water	3:1:1	TLE, AP	Monosialogangliosides, Sph-1-P
I. Chloroform/methanol/0.2% CaCl2/32% NH4OH	60:50:9:1	AP	Mono-, di-, and polysialogangliosides
J. Chloroform/methanol/acetic acid/water	25:15:4:3.5	TLE, MCP	Cer, Cer-1-P, Sph-1-P, SM

[a] Examples of the solvent systems useful for mono- or two-dimensional HPTLC separation of different radiolabeled lipids that can be formed after the administration of radiolabeled sphingolipids.

[b] AP, aqueous phase; CP, chloroform phase; MCP, methanol-chloroform phase; TLE, total lipid extract.

F. Determination of Radioactivity

Radioactivity is measured by liquid scintillation counting: aliquots of the radioactive sample are counted in a liquid scintillation cocktail (e.g., Optiphase "HighSafe 3," Wallac, Finland) using a liquid scintillation analyzer.

To detect and quantify incorporation of the radiolabel into the different lipids separated by HPTLC, the plates, after migration, are dried and submitted to radioactivity quantification using appropriate position-sensitive counters. Instruments with high sensitivity (e.g., Digital Autoradiograph, Berthold, Germany, or Beta Imager, Biospace Mesures, France) are essential for the rapid and accurate detection of radioactivity, particularly of tritium (Fig. 2; see color insert); such instruments detect very low radioactive levels (50–200 dpm/cm^2) and are great time savers (they detect tritium 50–500 times faster than film). After assessment the plate is sprayed with En^{3}Hance or Amplify and exposed to Kodak XAR-5 films at $-80°$ (Fig. 1).

G. Analysis of the Delipidized Pellet

The pellet obtained after lipid extraction (protein pellet) is dried under nitrogen stream and digested with 100 μl 1 N NaOH at 50° for 1 hr. After dilution with 900 μl of distilled water and vigorous mixing, aliquots are (a) submitted to protein determination,[13] using bovine serum albumin as standard, and (b) counted for radioactivity.

H. Analysis of Culture Media from Pulse and Chase

Analysis of the pulse or chase medium allows the identification and quantification of hydrophilic low molecular weight radioactive metabolites (such as monosaccharides and water) as well as the detection of radioactive organic metabolities that may be released from cells into the culture medium.

Because of the overwhelming preponderance of the administered [^{3}H]sphingolipid in the pulse medium, radioactive metabolites released by cells in the culture medium can be determined more accurately in the chase than in the pulse medium.

Media collected after the pulse or chase period are centrifuged at 7000g for 10 min at 4°. The supernatants are collected accurately and can be stored at $-20°$ until processed.

Depending on the labeling site of the administered sphingolipid and the experimental approach, the medium can be submitted to one or more of the following processes.

[13] O. H. Lowry, N. J. Rosebrough, A. L. Farr, and R. J. Randall, *J. Biol. Chem.* **193,** 265 (1951).

a. Lipid Extraction. This is performed by lyophylizing 0.1–0.5 ml of the medium and submitting it to lipid extraction and analysis as described for cells. Remarkably, when a pulse treatment with a [^{3}H]sphingolipid is followed by a period of chase, a substantial amount of radioactivity is released in the medium. This radioactivity is mainly constituted by the administered [^{3}H]sphingolipid. The amount of [^{3}H]sphingolipid released during chase in serum containing medium is related inversely to pulse time. For example, in cerebellar granule cells submitted to pulse with 1 μM radioactive GM1, radioactive GM1 released in the medium after a 4-hr period of chase accounted for 55 and 30% of total cell-associated radioactivity at 30-min and 4-hr pulses, respectively.

b. Determination of Tritiated Water by Fractional Distillation. In particular, ^{3}H$_2$O produced during [^{3}H-Sph]-, [^{3}H-Gal]-, or [^{3}H-NeuAc]ganglioside degradation, is determined by fractional distillation of the culture medium under carefully controlled conditions. The fraction distilling at 100° is collected and submitted to partitioning with 3 volumes of *n*-hexane. After mixing and centrifugation, the aqueous phase is counted for radioactivity by liquid scintillation.

c. Determination of Nonvolatile Low Molecular Weight ^{3}H Metabolites. ^{3}H neutral sugars and [^{3}H]NeuAc produced by [^{3}H]ganglioside degradation, possibly present in the medium, can be evaluated by lyophilizing 1 ml of medium, resuspending it in 0.2 ml distilled water, and submitting it to dialysis against 2 ml water at 4° (3 hr, repeated twice). The two dialysate fractions are collected, and aliquots (0.5–1 ml) are counted for radioactivity and processed by HPTLC using silica gel plates using propanol/water (7:3, by volume) or propanol/acetic acid/water (85:1:15, by volume) as the solvent systems or cellulose plates (prewashed in 0.1 M HCl) with *n*-butanol/*n*-propanol/0.1 M HCl (1:2:1, by volume) as the solvent system.

I. Isolation and Identification of Radiolabeled Sphingolipids

Single radiolabeled sphingolipid metabolites can be isolated from the final lipid preparation (obtained from cells as described in Sections III,C and III,D, after appropriate pulse/chase times) after separation by mono- or two-dimensional HPTLC (Section III,E). For this purpose, it is always advisable to add an aliquot (about 5–10 μg/10^4 dpm) of the unlabeled standard carrier to the lipid extract. After HPTLC separation and appropriate visualization, the spot comigrating with the authentic standard is scraped off the plate. The radiolabeled metabolite is then eluted from the gel with sequential 1-ml washings of chloroform/methanol (2:1, 1:1, and 1:2, by volume). The pooled extracts of each compound are evaporated

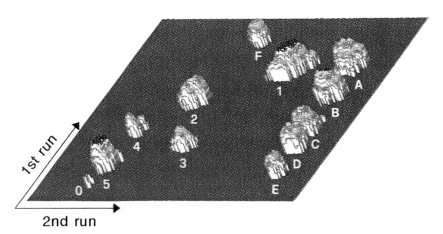

FIG. 2. Two-dimensional HPTLC separation of radiolabeled metabolites recovered in the methanolyzed chloroform phase from cultured cells after administration of [³H-Sph]GM1. Two milliliters of [³H-Sph]GM1 (1 μM, 1.8 Ci/mmol) was administered to preconfluent HeLa cells (grown on a 60-mm dish) for a 2-hr pulse, followed by a 4-hr chase. Cells were scraped off the dish, lipids were extracted and partitioned, and an aliquot (about one-sixth) of the methanolyzed chloroform phase (corresponding to 4000 dpm) was submitted to two-dimensional HPTLC (solvent systems A and C, Table I, for the first and second runs, respectively). Detection by autoradiography (digital autoradiograph, Berthold, Germany, three-dimensional view), acquisition time: 18 hr. O: origin; A–E: standard radioactive Cer (A), Glc-Cer (B), Lac-Cer (C), Sph (D), and SM (E) submitted only to the first run; F standard [³H]Sph submitted only to the second run; 1, 2, 3, 4, 5: [³H]Cer, [³H]Glc-Cer, [³H]Sph, [³H]Lac-Cer, and [³H]SM from HeLa cells, respectively. It is worth noting that the amount of radioactivity in spots 3 and 4 accounts for 200–250 dpm and was measured (standard deviation: 2–3%) by digital autoradiography in 18 hr.

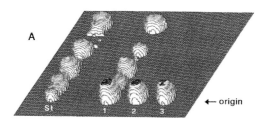

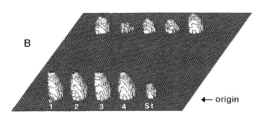

FIG. 6. Effect of low temperature and chloroquine on the metabolic processing of radiolabeled sphingolipids in cultured cells. Radiolabeled molecules recovered in the methanolyzed chloroform phase after a 1-hr pulse with [^{3}H-Sph]GM1 in cerebellar granule cells (A) or [^{3}H-Sph]SM in Neuro2a cells (B) were submitted to HPTLC separation (solvent system A, Table I; detection by digital autoradiography, three-dimensional view). (A) St: standard radiolabeled Cer, Glc-Cer, Lac-Cer, Sph, and SM (from top to bottom). Lanes 1–3: radioactive molecules from granule cells exposed to 1 μM [^{3}H-Sph]GM1 (1.8 Ci/mmol) at 37° in the absence (lane 1) or presence (lane 2) of 50 μM chloroquine or at 4° (lane 3). The major radioactive spot (near the origin) in lane 1 and the only radioactive spot detectable in lanes 2 and 3 are represented by a fraction (about 1–2%) of the incorporated [^{3}H-Sph]GM1 recovered in the organic phase after partitioning. (B) Lanes 1–4: radioactive molecules from neuroblastoma Neuro2a cells exposed to 0.6 μM [^{3}H-Sph]SM (2.9 Ci/mmol) at 37° (lane 1), at 4° (lane 2), or at 37° in the presence of 50 (lane 3) and 100 μM chloroquine (lane 4). St: standard Cer and SM (from top to bottom). Note that low temperature and chloroquine (a) completely abolish the formation of radiolabeled sphingolipids from [^{3}H-Sph]GM1 by granule cells, indicating that the lysosomal compartment is the subcellular site of origin of these molecules and (b) inhibit but not block the formation of [^{3}H]Cer from [^{3}H-Sph]SM in Neuro2a cells; this evidence supports the existence of a plasma-membrane sphingomyelinase operating in these cells.

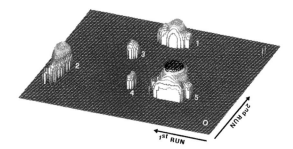

FIG. 7. Formation of a GD1b-carboxyl ester in cerebellar granule cells after the administration of radiolabeled GD1b. Two-dimensional HPTLC of gangliosides extracted from cerebellar granule cells after a 2-hr pulse with [^{3}H-Gal]GD1b, followed by NaBH$_4$ reduction. This treatment allows the reductive cleavage of carboxyl ester linkages, producing GD1b-ol from GD1b lactone (see text for details). (1 and 2) Standard [^{3}H]GD1b-ol subjected only to the first and second runs, respectively. (3–5) [^{3}H]GM1, [^{3}H]GD1b-ol, and [^{3}H]GD1b from cerebellar granule cells. O: origin. First run, solvent F; second run, solvent I (see Table I). Detection by digital autoradiography.

to dryness under nitrogen stream and submitted to appropriate processing (see later). Alternatively, the separation of sphingolipid metabolites can be achieved by silicic acid or silica gel column chromatography according to conventional procedures.

Criteria used for the identification of radiolabeled molecules formed by cells include the following.

a. Comigration on HPTLC with Authentic Standards in Different Solvent Systems. The identity of R_f values in at least two different solvent systems (Fig. 2) represents the first step, a prerequisite for further analysis.

b. Chemical Hydrolysis under Carefully Controlled Conditions. Mild acid hydrolysis is useful in obtaining different lower homologs from a presumed glycolipid. The extract is dried under a nitrogen stream in an Eppendorf tube and resuspended in 50 μl 0.5 N HCl. After incubation at 80° for 120 min, the sample is neutralized with 50 μl 0.5 N NaOH and vortexed. Partitioning is achieved by the addition of 0.4 ml of chloroform/methanol (2 : 1, by volume). After mixing, samples are centrifuged at 3000g for 5 min and processed by HPTLC (using solvent system A or B, Table I). Radiolabeled neutral glycolipids are recovered in the chloroform phase (Fig. 3, lane 5).

Radioactive alkali-labile gangliosides (including *O*-acetylated gangliosides and ganglioside lactones) can be characterized by mild alkaline treatment.[4] For this purpose, a two-dimensional HPTLC (using solvent system G in Table I for both runs) is performed, exposing the plate to ammonia vapors between the first and the second runs (see Section IV,E for further details).

Alkaline hydrolysis in KOH/*n*-butanol at 100° (see Section IV,F for details) can be used to release Sph from radiolabeled Cer.

c. Selective Chemical Derivatizations. For example, radiolabeled GD1b lactone can be derivatized to GD1b-ol after treatment with NaBH$_4$ (see Section IV,E) and sphingosine to its *N*-acetyl or dinitrophenyl derivative by treatment with acetic anhydride or 2,4-dinitrofluorobenzene, respectively (see Section IV,F for experimental details).

d. Enzymatic Treatments. In particular, the radiolabeled metabolite extracted from the silica spots as described earlier is dried completely under a nitrogen stream and submitted to one (or more) of the following enzymatic treatments, depending on the putative metabolite and its site of labeling. All reactions can be performed with aliquots corresponding to 10^4–10^5 dpm of the unknown metabolite. In all cases, it is advisable to process the standard sphingolipid parallely.

β-GALACTOSIDASE. This enzymatic assay is useful for the identification of a sphingolipid carrying a terminal β-Gal residue such as GM1 and asialo-GM1.

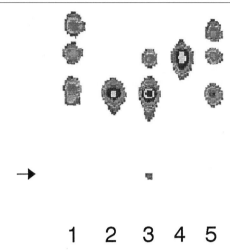

1 2 3 4 5

Fig. 3. Identification of a radiolabeled metabolite produced from cultured cells after administration of a radiolabeled ganglioside. [³H-Sph]GM2 (2 μM) was administered to Neuro2a cells for 4 hr. After lipid extraction and separation, a radiolabeled metabolite, partitioning in the chloroform phase and comigrating with standard asialo-GM2 in different solvent systems, was isolated from the methanolyzed organic phase by scraping off the spot comigrating with authentic standard asialo-GM2 after two-dimensional HPTLC (solvent systems B and C, for the first and second runs, respectively). The radiolabeled metabolite was then eluted form the gel with two subsequent washings (1 ml each) with chloroform/methanol (1:1 and 1:2, by volume, respectively). The pooled extracts were evaporated to dryness under a nitrogen stream and submitted to enzymatic and chemical characterization. Lane 1: standard radioactive Glc-Cer, Lac-Cer, and asialo-GM2 (from top to bottom); lanes 2–5: putative [³H]asialo-GM2 isolated from Neuro2a cells after pulse with [³H-Sph]GM2 prior (lane 2) and after β-N-acetylhexosaminidase treatment for 1 hr (lane 3) or 5 hr (lane 4) or after mild acid hydrolysis (lane 5) (see Section III,I for experimental details). Arrow: origin. The plate is developed in solvent system B (Table I) and spots are visualized by digital autoradiography.

Procedure. Mix the dried sample with 50 μg of sodium taurocholate in chloroform/methanol (2:1, by volume). Dry under a nitrogen stream and redissolve in 80 μl of 84 mM Na_2HPO_4/58 mM citric acid/2 mM EDTA, pH 4.2. After mixing, add 6 μl of β-galactosidase (from jack beans) containing 30 mU. Vortex and incubate the mixture at 37° for 24 hr. The sample is then partitioned with 400 μl chloroform/methanol (2:1, by volume) and processed by HPTLC (solvent system B or F, Table I).

β-N-ACETYLHEXOSAMINIDASE. This enzyme cleaves the terminal N-acetylhexosamine residue from a glycosphingolipid (e.g., GM2 or asialo-GM2).

Procedure. Resuspend the dried sample in 25 μl of 0.6 mM sodium–taurodeoxycholate/100 mM sodium phosphate buffer, pH 5.0, by vigorous shaking. Add 25 μl (2 mU) of β-N-acetylhexosaminidase and incubate at 37° for 5 hr. The sample is then partitioned with 250 μl chloroform/methanol (2:1, by volume) and processed by HPTLC (solvent system B or F, Table I) (Fig. 3, lanes 3 and 4).

VIBRIO CHOLERAE SIALIDASE. This enzyme cleaves sialic acid linkages other than NeuAcα2 → 3Galβ_{int} in the ganglio-series gangliosides (it yields Lac-Cer from GM3 and GD3 or GM1 from GD1a, GD1b, GT1b, and GQ1b).

Procedure. Resuspend the dried residue in 15 μl of 80 mM sodium acetate/10 mM CaCl$_2$, pH 5.5, and leave the tube in a bath sonicator for 30 sec for the complete solubilization of the ganglioside. Start the reaction by adding 10 μl (1 mU) of sialidase from *V. cholerae* and incubate at 37° for 30 min to 1 hr. At the end of the incubation, the mixture is dialyzed through a VSWP filter against distilled water at 4° for 2 hr and processed by HPTLC (solvent system G, Table I).

ARTHROBACTER UREAFACIENS SIALIDASE. This enzyme hydrolyzes all sialic acid residues present in gangliosides, including the NeuAc-α2 → 3Galβ_{int} linked resistant to *V. cholerae* sialidase (e.g., that occurring in GM1 and GM2).

Procedure. Resuspend the sample in 10 μl 0.5% (w/v) sodium deoxycholate by bath sonication. Add 20 μl of 25 mM sodium acetate buffer, pH 4.8, and mix. Start the reaction by adding 20 μl (10 mU) of sialidase from *A. ureafaciens* and vortex. Incubate at 37° for 24 hr. Stop the reaction and partition by the addition of 250 μl chloroform/methanol (2:1, by volume). Desialylated gangliosides, recovered in the organic phase, are processed by HPTLC (solvent system B, Table I).

ENDOGLYCOCERAMIDASE. This enzyme specifically hydrolyzes the linkage between the oligosaccharide chain and the ceramide of various neutral and acidic glycosphingolipids.

Procedure. Dissolve the dried sample in 20 μl of 75 mM acetic acid–sodium acetate buffer, pH 5.0, containing 3 mM sodium cholate. After vortexing, add 10 μl (0.1 mU) of ceramide glycanase from *Macrobdella decora,* mix, and incubate at 37° overnight. The reaction is stopped by the addition of 4 volume of chloroform/methanol (2:1, by volume). The sample is then vortexed and centrifuged (3000g for 10 min). Depending on the site of labeling, process the lower (solvent system B or D in Table I) or the aqueous phase (solvent system: ethanol/1 M ammonium acetate, 7:3, by volume) by HPTLC.

SPHINGOMYELINASE. This enzyme cleaves sphingomyelin to release ceramide and phosphorylcholine.

Procedure. The dried residue is resuspended in 40 μl of 1.25% (w/v) Triton X-100/7.5 mM MgCl$_2$/125 mM Tris–HCl, pH 7.4. After mixing and sonication, add 10 μl (0.05 U) of sphingomyelinase (from *Staphylococcus aureus*) and submit to vortexing. Incubate the reaction mixture at 37° for 3 hr. The reaction is then stopped by the addition of 250 μl chloroform/methanol (2:1, by volume). The sample is then vortexed and centrifuged (3000g for 10 min). The organic phase, containing the released ceramide, is then dried and submitted to HPTLC (solvent system A or D, Table I).

DIACYLGLYCEROL KINASE. A putative radiolabeled ceramide can be phosphorylated to ceramide-1-phosphate after treatment with DAG kinase.

Procedure. The dried sample is resuspended in 10 μl of 7.5% octyl-β-D-glucoside/5 mM cardiolipin/1 mM diethyenetriamine pentaacetic acid (DETAPAC), pH 7.0 (prepared according to Preiss *et al.*[6]). After vortexing, tubes are put in a bath sonicator for 30 sec to achieve consistent solubilization of the lipids. To this mixture add 25 μl of 100 mM imidazole–HCl/100 mM NaCl/25 mM MgCl$_2$/2 mM EGTA, pH 6.6, 5 μl of 20 mM dithiothreitol, and 5 μl of DAG kinase from *E. coli*. After mixing the reaction is started by the addition of 5 μl 10 mM ATP/100 mM imidazole/2 mM DETAPAC, pH 6.6. After mixing, incubate at 30° for 30 min. The reaction is stopped by the addition of 1.5 ml chloroform/methanol (1:2, by volume) and 0.35 ml of 1% (w/v) HClO$_4$. After mixing, the phases are separated by the addition of 0.5 ml chloroform and 0.5 ml 1% (w/v) HClO$_4$ and centrifugation at 5000g for 5 min. The upper phase is discarded, and the lower phase, transferred to a new tube, is evaporated under a nitrogen stream and processed for HPTLC (solvent system E or J, Table I).

IV. Investigation of Specific Pathways and Trafficking Routes in Sphingolipid Metabolism

This section reports on experimental protocols and gives suggestions on how to investigate the different pathways of sphingolipid metabolism and obtain information on the intracellular traffic of sphingolipid molecules. A schematic view of some of these approaches (together with the radiolabeled metabolites recovered in granule cells after different pulse–chase protocols[14–21]) is shown in Table II.

[14] L. Riboni, R. Bassi, S. Sonnino, and G. Tettamanti, *FEBS Lett.* **300,** 188 (1992).
[15] L. Riboni, A. Caminiti, R. Bassi, and G. Tettamanti, *J. Neurochem.* **64,** 451 (1995).
[16] L. Riboni and G. Tettamanti, *J. Neurochem.* **57,** 1931 (1991).
[17] L. Riboni, A. Prinetti, R. Bassi, and G. Tettamanti, *FEBS Lett.* **287,** 42 (1991).
[18] L. Riboni, R. Bassi, M. Conti, and G. Tettamanti, *FEBS Lett.* **322,** 257 (1993).
[19] L. Riboni, R. Bassi, and G. Tettamanti, *J. Biochem.* **116,** 140 (1994).
[20] L. Riboni, A. Prinetti, R. Bassi, and G. Tettamanti, *FEBS Lett.* **352,** 323 (1994).
[21] L. Riboni, R. Bassi, A. Prinetti, and G. Tettamanti, *FEBS Lett.* **391,** 336 (1996).

TABLE II

PROTOCOLS FOR THE INVESTIGATION OF DIFFERENT METABOLIC PATHWAYS OF SPHINGOLIPID METABOLISM IN CULTURED CELLS USING RADIOLABELED SPHINGOLIPIDS[a]

Investigated metabolic pathway	Administered sphingolipid	Pulse–chase conditions	Analysis of[b]	Radiolabeled metabolites of the investigated pathway	Control experiment
Intralysosomal degradation	[3H-Sph]GM1	15- to 120-min pulse	AP	GM2, GM3	Inhibition by chloroquine and conditions impairing vesicle traffic and fusion
	[14C-Stearoyl]GM1		MCP	Lac-Cer, Glc-Cer, Cer, Sph	
	[3H-Sph]sphingomyelin		Me	H_2O	
			AP	GM2, GM3	
			MCP	Lac-Cer, Glc-Cer, Cer	
			MCP	Cer, Sph	
			Me	H_2O	
Extralysosomal degradation	[3H-Sph]GD1a	120-min pulse + chloroquine	AP	GM1	Possible detection at 4°
	[3H-Sph]sphingomyelin		MCP	Cer	
Recycling	[3H-GalNAc]GM1	120-min pulse/0- to 4-hr chase	AP	GD1a, GD1b, O-Ac-GT1b, GT1b	Inhibited by chloroquine and fumonisin B1
	[3H-NeuAc]GM1		AP	GD1a, GD1b, O-Ac-GT1b, GT1b	
	[3H-Sph]GM1		DP	Sialoglycoproteins	
	[14C-Stearoyl]GM1		AP	GD1a, GD1b, O-Ac-GT1b, GT1b	
			MCP	SM	
			CP	Phospholipids	
Direct glycosylation	[3H-Gal]GD1b	2-hr pulse/0- to 4-hr chase	AP	GT1b	Detectable with chloroquine and fumonisin B1; inhibited at 4°
	[3H-NeuAc]GM1		AP	GD1a	

[a] Conditions and formed radiolabeled metabolites refer to experiments performed on rat cerebellar granule cells.
[b] AP, aqueous phase; CP, chloroform phase; MCP, chloroform phase submitted to mild alkaline methanolysis; DP, delipidized pellet; Me, medium.

A prerequisite for these investigations is to know the composition of the major sphingolipids of the cell type(s) used. In fact, because different cells usually show different sphingolipid patterns and are of varying capacity and efficiency in sphingolipid metabolism, it is essential to know the composition in order to choose the appropriate radiolabeled sphingolipid to administer, and to plan adequate pulse–chase protocols.

In the investigation of sphingolipid metabolism, further information can be obtained by using several drugs or experimental conditions known to interfere with this process. One can plan different protocols, enabling specific studies into the role of sphingolipid trafficking and that of different enzymes involved in sphingolipid metabolism. Experimental details for these purposes are reported in Tables III and IV, respectively. Although the information given in Tables III and IV can be used profitably in investigating sphingolipid metabolism, care must be taken before drawing any conclusions on the effect of a drug. In fact, the availability, metabolism, and short-

TABLE III
EXPERIMENTAL CONDITIONS AFFECTING SPHINGOLIPID TRAFFICKING IN CULTURED CELLS[a]

ATP depletion:	Add 2-deoxy-D-glucose (0.2 mM) and sodium azide (2 mM) to the culture medium 30 min prior and during pulse
Microtubule disruption:	Add nocodazole (20 μg/ml from a stock solution in distilled dimethyl sulfoxide) 1 hr prior to pulse and maintain during pulse
Inhibition of intracellular vesicle flow:	Incubate cells at 16° 30 min prior to and during pulse
Inhibition of intracellular vesicle flow and block of intracellular vesicle fusion:	Incubate cells at 4° 30 min prior to and during pulse
Inhibition of vesicle flow out of the cis-Golgi (and vesicular traffic disruption):	Add brefeldin A (1–10 μg/ml from a stock solution in distilled methanol) 1 hr prior to and during pulse
Inhibition of vesicle flow between Golgi and trans-Golgi network:	Add monensin (1–5 μM from a stock solution in distilled dimethyl sulfoxide) 1 hr prior to and during pulse

[a] Because the toxicity of the inhibitors can vary in different cell types, preliminary experiments must be conducted to set up conditions in the used cell type. In case of a molecule insoluble (or poorly soluble) in aqueous solvents, a condition for its complete solubilization is reported. Stock solutions of inhibitors in organic solvents can be prepared 100–1000× its final concentration and stored at −20°; at the time of the experiment, aqueous solutions are prepared by diluting the stock solution with culture medium. Take care to perform control experiments with cells treated with the same final concentration of the used organic solvent.

[b] Different protocols useful in investigating the role of sphingolipid traffic in the metabolic processing of radiolabeled sphingolipids are described.

TABLE IV

EXPERIMENTAL CONDITIONS AFFECTING ENZYMES OF SPHINGOLIPID METABOLISM IN
PULSE–CHASE EXPERIMENTS WITH RADIOLABELED SPHINGOLIPIDS IN CULTURED CELLS[a]

A. Inhibition of enzymes involved in sphingolipid degradation

Inhibition of lysosomal enzymes: Add chloroquine (50–100 μM) to the culture medium 1 hr prior to and during pulse

Inhibition of ceramidase: Add (1S,2R)-D-erythro-2-(N-myristoylamino)-1-phenyl-1-propanol (5–10 μM from a stock solution in ethanol) to the culture medium 30 min prior to pulse and maintain thereafter

Inhibition of lysosomal glucocerebrosidase: Add conduritol B-epoxide (50–100 μM from a stock solution in distilled dimethyl sulfoxide) to the culture medium 30 min to 1 hr prior to pulse and maintain thereafter

B. Inhibition of enzymes involved in sphingolipid biosynthesis

Inhibition of sphinganine biosynthesis (by inhibition of serine-palmitoyl transferase): Add L-cycloserine (1–2.5 mM) or β-chloro-L-alanine (2–5 mM) 2 hr prior to pulse and maintain thereafter

Inhibition of ceramide biosynthesis (by inhibition of sphinganine/sphingosine-N-acyltransferase): Add fumonisin B$_1$ (10–25 μM from a stock solution in methanol) 1 hr prior to pulse and maintain thereafter

Inhibition of Glc-Cer biosynthesis (by inhibition of ceramide : UDP-Glc transferase): Add N-butyldeoxynojirimycin (10–50 μg/ml) 3 hr prior to pulse or D-PDMP (20–50 μM from a stock solution in ethanol added as equimolar fatty acid bovine serum albumin complex) 1 hr prior to pulse and maintain thereafter

Inhibition of the biosynthesis of Glc-NAc-containing gangliosides (by inhibition of sugar nucleotide transport): Add tunicamycin (0.2–1 μg/ml from a stock solution in distilled dimethyl sulfoxide) 1 hr prior to pulse and maintain thereafter

[a] Different protocols useful for the inhibition of specific pathways involved in sphingolipid metabolism are described. For technical details, see Table III.

and long-term effects of a drug can vary from cell to cell, and multiple effects, even from a highly specific inhibitor, can occur. Thus it is always advisable to perform different protocols to independently confirm results with inhibitors.

A. Degradation

Complete catabolism of cell gangliosides and sphingomyelin is known to occur in lysosomes following endocytosis or other internalization mechanisms. The administration of radiolabeled sphingolipids to cultured cells for short periods of time allows an accurate evaluation of the degradation pathway(s) characteristic of the cell type used (Fig. 1).

Administration of a sphingolipid radiolabeled at the level of the lipid moiety (sphingosine or stearic acid) allows demonstration of the catabolic enzymatic equipment of a specific cell type. With this approach, the label

TABLE V

IDENTIFICATION OF DEGRADATION PATHWAY OF GANGLIOSIDE GM1 IN DIFFERENT CELLS[a]

| Cell type | Identified radiolabeled catabolites | | Degradation pathway[b] |
	Aqueous phase	Chloroform phase	
Granule cells			
Fibroblasts	GM2, GM3	Lac-Cer, Glc-Cer, Cer, Sph	A
HeLa cells			
Neuro2a cells	GM2	AsialoGM1, asialoGM2, Lac-Cer, Glc-Cer, Cer, Sph	B or C
C6-glioma	GM2	None	D

[a] Cells were submitted to a 2-hr pulse with 1 μM of [^{3}H-Sph]GM1, and the formed radiolabeled catabolites were analyzed, after extraction and partitioning, by HPTLC in the aqueous phase (solvent system D, Table I) and in the chloroform phase (solvent systems A, B, C, and F in Table I).

[b] Degradation pathways:
A: GM1 → GM2 → GM3 → Lac-Cer → Glc-Cer → Cer → Sph
B: GM1 → GM2 → asialo-GM2 → Lac-Cer → Glc-Cer → Cer → Sph
C: GM1 → asialo-GM1 → asialo-GM2 → Lac-Cer → Glc-Cer → Cer → Sph
D: GM1 → GM2

is maintained as each catabolic product is produced during the step-by-step breakdown of the complex glycosphingolipid. Table V shows how this tool discriminates among the different sphingolipid catabolic pathways occurring in the cell types used.

Experimental Conditions. To evaluate the roles of vesicular traffic, endocytosis, and lysosomal function in the formation of catabolites (Figs. 5 and 6; see color insert), experiments are conducted under conditions known to interfere with these processes (see Tables III and IV). A 30-min to 2-hr pulse with 2 μM ^{3}H-Sph-labeled ganglioside or [^{3}H-Sph]-sphingomyelin is followed by analysis of (a) radioactive organic metabolites recovered in the aqueous phase from cell lipid extracts (Figs. 1A–1D and Fig. 5), (b) radioactive organic metabolites recovered in the methanol–chloroform phase (Figs. 1E, 2, and 6); and (c) radioactive water in the culture medium. Parallel experiments are carried out with the same amount of exogenous [^{3}H-Sph]- and [^{14}C-stearoyl]sphingolipid to reveal if there is a different pathway of catabolism (i.e., if the same concentration of a sphingolipid is administered, the molecule formed exclusively from degradation will be produced in equal amounts, regardless of the labeling site of the sphingolipid) (Fig. 4).

B. Extralysosomal Degradation Processes

Some complex sphingolipids may be partially degraded by extralysosomal enzymes. For example, such reactions include the removal of phos-

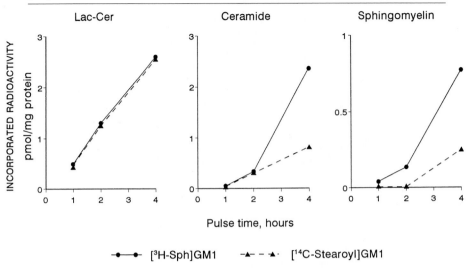

FIG. 4. Formation of radiolabeled metabolites derived from different metabolic pathways in cultured cells after administration of differently labeled GM1. Two milliliters of 1 μM [^{3}H-Sph]GM1 (1.2 μCi/ml) or [^{14}C-Stearoyl]GM1 (0.055 μCi/ml) was administered to cerebellar granule cells for 1, 2, or 4 hr. Radioactive Lac-Cer, Cer, and SM were extracted, separated, and quantified as described in the text (Section III). In these experimental conditions, (a) identical amounts of radioactive Lac-Cer are produced with both labels; hence, this metabolite originates from the catabolism of the administered ganglioside; (b) radioactive Cer originates from the catabolism of the administered ganglioside until the 2-hr pulse. Thereafter, recycling of [^{3}H]Sph (and possibly also of [^{14}C]stearic acid) is also active. (c) Salvage pathways of catabolic fragments ([^{3}H]Sph and [^{14}C]Stearic acid) represent the metabolic pathway leading to radiolabeled sphingomyelin.

phocholine from sphingomyelin and sialic acid from GD1a ganglioside, which may occur at the level of the plasma membrane in some cells. The investigation of these degradation pathways could be useful in studies focusing on the response of cells to extracellular modulators.

Experimental Conditions. To clarify whether internalization of the administered sphingolipids is necessary for catabolism, parallel experiments (Fig. 6B) are carried out with inhibitors of vesicular trafficking (see Table III). Other useful experiments are to utilize inhibitors of enzymes that play a role in sphingolipid metabolism (Table IV).

For example, to determine the participation of intracellular acidic compartments, cells are given a 30-min to 2-hr pulse with 2 μM ^{3}H-Sph-labeled GD1a ganglioside or [^{3}H-Sph]sphingomyelin in the presence or absence of 50 to 100 μM chloroquine, followed by analysis of (a) the aqueous phase from cell lipid extracts—in the case of ^{3}H-Sph-labeled GD1a, evaluate the formation of [^{3}H]GM1; or (b) the methanol–chloroform phase after a pulse

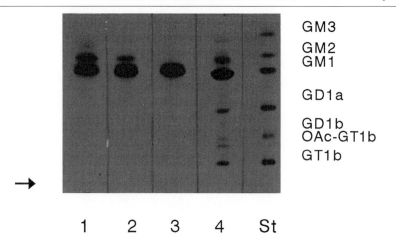

GM3
GM2
GM1

GD1a

GD1b
OAc-GT1b
GT1b

1   2   3   4   St

FIG. 5. Influence of experimental conditions affecting intracellular vesicle traffic, or the activity of lysosomal enzymes on the metabolism of [³H-Sph]GM1 in cerebellar granule cells. The metabolic processing of [³H-Sph]GM1 in a pulse–chase protocol is also shown. Radiolabeled gangliosides recovered in the dialyzed aqueous phase from granule cells after pulse with 1 μM [³H-Sph]GM1 were separated by HPTLC (solvent system G, Table I; detection by fluorography after En³Hance spraying). Lanes 1–4: gangliosides extracted from granule cells after a 1-hr pulse at 37° (lane 1), at at 16° (lane 2), or at 37° in the presence of 50 μM chloroquine (lane 3) or after a 1-hr pulse followed by a 6-hr chase at 37° (lane 4). St: standard radioactive GM3, GM2, GM1, GD1a, GD1b, and GT1b (from top to bottom). Arrow: origin. Note that (a) the formation of gangliosides produced from GM1 degradation after a 1-hr pulse (lane 1) is reduced or absent in conditions inhibitory of intracellular vesicle traffic or lysosomal enzymes (lanes 2 and 3); and (b) in the pulse–chase protocol (lane 4) di- and trisialogangliosides are actively produced from recycling of [³H]Sph.

with [³H-Sph]sphingomyelin to evaluate the formation of [³H-Sph]ceramide (Fig. 6B). If catabolism has been blocked, it requires an acidic compartment.

C. Recycling Processes

After the administration of radiolabeled sphingolipids, some products of lysosomal degradation may leave the lysosomes and be recycled for biosynthetic purposes. Among such fragments are sialic acid, N-acetylgalactosamine, sphingosine, and fatty acid. These recycling pathways can be investigated by administering compounds specifically radiolabeled at the level of these residues and investigating those radiolabeled metabolites different from catabolic ones. One way of making an assessment is by comparing the amount of the same metabolite produced by cells after administration of the same but differently radiolabeled sphingolipid (Fig. 4). In addition, parallel experiments in the presence of inhibitors of enzymes involved in sphingolipid biosynthesis (see Table IV,B) can be useful.

Experimental Conditions. To evaluate the extent of recycling of catabolic intermediates, cells are treated with a 4-hr pulse with 1 μM sphingolipid radiolabeled at the level of the headgroup {e.g., the terminal saccharide, such as [^{3}H-NeuAc]GM1 or [^{3}H-GalNeuAc]GM2 (Figs. 1B and D)} versus the level of the label in the lipid moiety, e.g., ^{14}C-stearoyl- or ^{3}H-Sph-labeled ganglioside GM1 (Figs. 1A and 1E; Fig. 4, right), with analysis of (a) radioactive organic metabolites recovered in the aqueous phase from cell lipid extracts (Figs. 1A–1D and Fig. 5, lane 4); (b) radioactive organic metabolites recovered in the methanol–chloroform phase (Fig. 1E); and (c) volatile and nonvolatile low molecular weight radioactive metabolites in the medium. When the experiments are performed in the presence of specific inhibitors (Table IV), such as fumonisin B1 (which inhibits acylation of sphingosine and sphinganine) or D-PDMP (which inhibits GlcCer synthase and other activities), this will reveal if the radiolabeled portion of the sphingolipid has been hydrolyzed to that downstream metabolite (in these examples, sphingosine and ceramide, respectively) and before being converted back to a complex sphingolipid.

D. Direct Glycosylation of [^{3}H]Gangliosides

The administration of gangliosides radiolabeled at the terminal sugar residue ([e.g., [^{3}H-Gal]) is an experimental approach suitable for investigating the possible direct utilization of the intact molecule for the biosynthesis of higher homologs. In fact, in [^{3}H-Gal]gangliosides such as [^{3}H-Gal]GM1, the radiolabel is lost during the first step of the major GM1 degradation pathway (GM1 → GM2), during which the original radioactive Gal is removed. Although the main fate of free ^{3}H-Gal in cells is represented by degradation, the biosynthesis of a ganglioside that is more complex than the administered one can also occur through the recycling of ^{3}H-Gal. To clarify this point, the radiolabeled higher homolog of the administered ganglioside, e.g., [^{3}H]GD1a from [^{3}H-Gal]GM1, is purified after HPTLC by plate scraping and elution from the gel. Purified [^{3}H]GD1a is then subjected to mild acid hydrolysis in 0.5 M HCl at 80° for times varying from 10 to 60 min to produce asialo-GM1, asialo-GM2, Lac-Cer, and Glc-Cer. In order to calculate the specific radioactivity of each glycolipid, the reaction products are subjected, in parallel, at different acid hydrolysis times to quantification by densitometry, and to radioactivity counting by radiochromatoscanning. Any difference in specific radioactivity between the terminal and the internal galactose (i.e., of asialo-GM1 and Lac-Cer) accounts for the direct glycosylation process. In the event of there being no difference, it is concluded that the [^{3}H]GD1a produced by the cells is mainly due to the reutilization of the released [^{3}H]Gal.

Experimental Conditions. Cells are pulsed for 4–6 hr with 2 μCi/ml of [³H-Gal]GM1 or [³H-Gal]GD1b to investigate their direct glycosylation to [³H]GD1a and [³H]GT1b, respectively, with analysis of the aqueous phase by mono- or two-dimensional HPTLC (solvent system F and I, Table I) for the identification of radioactive gangliosides.

E. Direct Utilization of [³H]Gangliosides for Biosynthesis of Ganglioside Lactones

Radiolabeled sphingolipids can also be used to estimate less common metabolic pathways such as the biosynthesis of carboxyl esters, particularly inner esters (lactones) of gangliosides. This process leads to a variation of the negative charge of a ganglioside without modifying its degree of sialylation and could be relevant in modulating ganglioside-binding properties and surface behavior at the membrane level. Particular attention must be paid to the detection and quantification of the lactonization process of a ganglioside as the inner ester linkages of ganglioside lactones can be destroyed or even produced artificially during the procedures used to extract and quantify gangliosides from cultured cells. In fact, under mild alkaline conditions, ganglioside lactones are labile, whereas under acidic or drastic dehydrating conditions, they are produced spontaneously. Thus, particular care must be taken when evaluating radiolabeled metabolites. The following protocol is suitable for investigating the formation of GD1b-lactone from radiolabeled GD1b in cultured cells.

Experimental Conditions. Cells are pulsed for 1–4 hr with 2 μCi/ml of [³H-Gal]GD1b to investigate its possible direct lactonization to [³H]GD1b lactone.[22] At the end of the pulse periods, the cells are washed twice in PBS and, after scraping, are centrifuged at 5000g for 5 min at 4°. Immediately after centrifugation, parallel duplicate experiments are performed by resuspending the cell pellets in (a) 50 μl of PBS at 4° or (b) 50 μl of a freshly prepared solution of NaBH$_4$ (2 mg/ml in PBS) and then incubated at 37° for 30 min. This causes the reductive cleavage of the inner ester linkage, the sialic acid carboxyl group, being transformed into an alcoholic group.[23] This avoids creating any possible artifacts during the detection of ganglioside lactones.

Both tubes (a and b) are then extracted and partitioned as described in (Sections III,C and III,D).

For the control it is advisable to perform these procedures using a pellet obtained from untreated cells (not submitted to pulse) and adding 5 × 10⁵

[22] R. Bassi, L. Riboni, and G. Tettamanti, *Biochem. J.* **302,** 937 (1994).
[23] S. K. Gross, M. A. Williams, and R. H. McCluer, *J. Neurochem.* **34,** 1351 (1980).

dpm of [³H-Gal]GD1b and [³H-Gal]GD1b lactone after the addition of chloroform/methanol in the first lipid extraction (Section III,C), with analysis of the dialyzed aqueous phase.

For condition (a), check for the formation of [³H]GD1b lactone. This is performed by two-dimensional HPTLC [using solvent system G (Table I)] for both runs; between the two runs, expose the plate to ammonia vapors for 5 hr and leave it in a sealed chamber saturated with ammonia vapor but avoid contact with the NH₄OH solution. Before the second run, remove any ammonia by leaving the plate under a flow of warm air. During the ammonia treatment, lactones undergo two processes: a major one, ammoniolysis, with formation of the amide derivative, and a minor one, hydrolysis, with formation of the parent alkali-stable gangliosides. Thus lactones can be recognized by the presence of two spots located under the diagonal of the alkali-stable gangliosides and on the same vertical line.[4] The faster moving spot corresponds to the amide derivative, whereas the slower moving one corresponds to the hydrolyzed lactone. The ratio between the two spots is about 9:1.

For conditions (b), check for the formation of [³H]GD1b-ol. Perform two-dimensional HPTLC using solvent systems F and I (Table I) for the first and second runs, respectively (Fig. 7; see color insert). After isolating the presumed [³H]GD1b-ol, check its resistance to V. cholerae sialidase and that it produces an oligosaccharide after endoglycoceramidase treatment that migrates on HPTLC in the same position as that from standard GD1b-ol submitted to the same treatment.

Note that the formation of the "ol" derivative indicates the presence of a carboxyl ester, but does not exclude the possible occurrence of external ester linkages (although these have never been shown to occur in nature). The parallel detection and identification of GD1b lactone in condition (a) and GD1b-ol in condition (b) is evidence of a natural lactonization of GD1b ganglioside in the studied cell type.

F. Formation of Bioactive Sphingoid Molecules

A further and particularly intriguing application of the use of radiolabeled sphingolipid in cultured cells lies in their potential role in biosignaling processes. Complex sphingolipids are metabolic precursors of different sphingoid bioactive molecules, which include Cer, Sph, Cer-1-P, Sph-1-P, and DMSph. The administration of a complex sphingolipid radiolabeled at the level of the long chain base and the estimation of the production of these bioactive sphingoids in cells stimulated by various regulators (hormones, growth factors, cytokines, etc.) can provide evidence of the direct involvement of a specific metabolic pathway in signaling events.

Experimental Conditions. Two different pulse protocols, both appropriate for this research, can be followed.

(a) A 10-min to 2-hr pulse with 2 μM of [^{3}H-Sph]sphingomyelin or [^{3}H-Sph]ganglioside in the absence (control) or presence of the selected stimulus. For some cells it has been reported that replacing cell-conditioned medium with fresh medium causes a rapid, transient stimulation of sphingolipid metabolism; therefore, it is advisable to perform the pulse by adding a 10–20× stock solution of the radiolabeled sphingolipid to cells preincubated in the pulse medium. By counting the radioactivity in the the total lipid extract from the control and treated cells, check that the stimulus does not interfere with the uptake of the administered sphingolipid.

(b) A 2- to 4-hr pulse with 2 μM of [^{3}H-Sph]sphingomyelin or [^{3}H-Sph]ganglioside at 16 or 4°, followed by a 10-min to 4-hr chase in the absence (control) or presence of the selected stimulus at 37°. This protocol, involving cell stimulation independently of pulse, eliminates any possible interference by the stimulus on the uptake of the exogenous sphingolipid. However, incubation at low temperature decreases the uptake of exogenous sphingolipid and may interfere with the kinetics and the magnitude of the cell response to stimuli. Thus, prior to performing this protocol, preliminary experiments must be conducted to set up conditions that allow sufficient labeling and efficient response to stimulus in the cell type used.

In both cases, proof of the subcellular location of the enzyme involved in the detected metabolic step can be achieved in parallel experiments with inhibitors of vesicle traffic (see Table III) or of specific (see Table IV).

Analysis of (a) the total lipid extract for the formation of radioactive Sph-1-P and of (b) the methanolized organic phase for the formation of radioactive Cer, Cer-1-P, Sph, and DMSph. Perform appropriate HPTLC separations, using different solvent systems (Table I). Experimental protocols for the identification of the different bioactive sphingoids are described here.

Identification of Cer. Comigration with an authentic standard after two-dimensional HPTLC in solvent systems A and C (Table I) for the first and second runs, respectively (Fig. 2). After isolation of the presumed [^{3}H]Cer, proceed to its characterization (Fig. 8) by detecting (a) the release of radiolabeled Sph after alkaline hydrolysis. For this purpose, resuspend an aliquot of the presumed [^{3}H]Cer (corresponding to about 5000 dpm), diluted with 5 μg standard Cer, in 100 μl of 10 N aqueous KOH/n-butanol (1:9, by volume). Incubate the sample at 100° for 3 hr. At the end of the incubation, stop the reaction by rapid out dropping the temperature to 4° and add 200 μl of water and 200 μl of diethylether. [^{3}H]Sph, recovered in the ether phase, can be identified by HPTLC (Fig. 8, lane B); (b) the

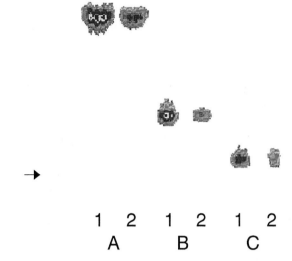

FIG. 8. Identification of Cer as a radiolabeled metabolite produced from cultured fibroblasts after a pulse with [³H-Sph]GM1. [³HSph]GM1 (2 μM) was administered to preconfluent human fibroblasts for 4 hr. Putative [³H]Cer was isolated from the methanolyzed chloroform phase after two-dimensional HPTLC (solvent systems B and C—see Table I—for the first and second runs, respectively) by scraping off the spot comigrating with standard Cer and eluting the gel twice with 1 ml chloroform/methanol (2 : 1, by volume). Lanes 1 and 2: standard [³H]Cer (lane 1) and [³H]Cer from human fibroblasts (lane 2) prior (lanes A) and after alkaline hydrolysis (formation of Sph) (lanes B) or treatment with DAG kinase (formation of Cer-1-P) (lanes C). Arrow: origin. See Sections III,I and IV,F for experimental details. Monodimensional HPTLC, solvent system A (Table I), detection by digital autoradiography.

formation of [³H]Cer-1-P after treatment with DAG kinase (see Section III,I,*d* for experimental details) by HPTLC (Fig. 8, lane C).

Identification of Sph. Comigration with an authentic standard after two-dimensional HPTLC in solvent systems A and E (Table I) for the first and second runs, respectively (Fig. 1). After isolation of the presumed [³H]Sph and addition of a standard Sph (10 μg/10,000 dpm), proceed to its characterization by detecting (a) its *N*-acetyl derivative after treatment with acetic anhydride. For this purpose, resuspend the dried [³H]Sph sample in 80 μl of anhydrous methanol. After vortexing, add 5 μl of acetic anhydride, mix, and incubate at room temperature for 1 hr. Process the sample prior and after treatment by HPTLC in solvent system A (Table I); (b) its 2,4-dinitrophenyl derivative after treatment with dinitrobenzene. In particular, resuspend the dried [³H]Sph sample in 200 μl methanol containing 1 μl 1-fluoro-2,4-dinitrobenzene. Add dropwise 800 μl potassium borate buffer, pH 10.5, and incubate the mixture at 60° for 30 min. After cooling the mixture to room temperature, add 0.75 ml of methanol and 2 ml chloroform. Process the sample prior and after treatment by HPTLC using chloroform/ hexane/methanol (5:5:2, by volume) as solvent system.

Identification of Cer-1-P. Comigration with an authentic standard after two-dimensional HPTLC in solvent systems B and E (Table I) for the first and second runs, respectively. After isolation, the presumed [³H]Cer-1-P can be identified by detecting the release of radiolabeled Sph-1-P after deacylation in acidic butanol. For this purpose, after dilution of the radioactive metabolite with standard Cer-1-P and drying under N_2 in a screw-capped Pyrex tube, resuspend the sample in 500 μl of 6 N HCl/1-butanol (1:1, by volume). Mix and incubate at 100° for 1 hr. Dry under a nitrogen stream and resolve [³H]Sph-1-P by HPTLC using solvent system J (Table I). About 60–70% of Cer-1-P is recovered as Sph-1-P by this procedure.

Identification of Sph-1-P. Comigration with an authentic standard after two-dimensional HPTLC in solvent systems B and C (Table I) for the first and second runs, respectively. After isolation, the presumed [³H]Sph 1-P can be identified by detecting its *N*-acetyl derivative after treatment with acetic anhydride (see earlier discussion). Resolve the *N*-acetyl derivative by HPTLC in solvent system H (Table I).

[52] Enzymatic Synthesis of [¹⁴C]Ceramide, [¹⁴C]Glycosphingolipids, and ω-Aminoceramide

By MAKOTO ITO, SUSUMU MITSUTAKE, MOTOHIRO TANI, and KATSUHIRO KITA

Introduction

We have isolated a novel enzyme that hydrolyzes the *N*-acyl linkage between fatty acids and sphingosine bases in ceramides of various sphingolipids and have tentatively named it sphingolipid ceramide *N*-deacylase (SCDase).[1] SCDase was found to catalyze the hydrolysis, and its reverse reaction as well, under different conditions.[2] Under acidic conditions including a high concentration of Triton X-100, SCDase catalyzes the hydrolysis reaction efficiently. In contrast, at neutral pH with a low concentration of the detergent, the reverse reaction tends to be favored (Fig. 1). We

[1] M. Ito, T. Kurita, and K. Kita, *J. Biol. Chem.* **270,** 24370 (1995).

[2] K. Kita, T. Kurita, and M. Ito, *in* "Proceedings of the XVIIIth International Carbohydrate Symposium," p. 549. Milan, Italy, 1996.

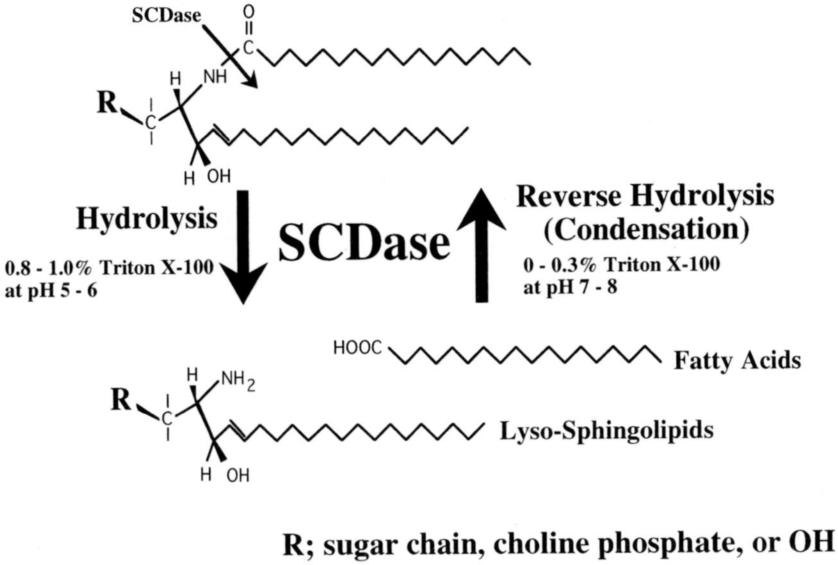

R; sugar chain, choline phosphate, or OH

Fig. 1. Mode of action of sphingolipid ceramide *N*-deacylase (SCDase). SCDase catalyzes the reversible reactions in which the *N*-acyl linkage of ceramide in various sphingolipids is hydrolyzed or synthesized. The direction of the reaction depends mainly on the concentration of detergents. The condensation reaction tends to be favored with a decrease in detergents, and in the absence of detergent substantially no hydrolysis reaction occurs.

can therefore utilize the desired reaction of SCDase if the conditions are changed. In Chapter 32, the enzymatic *N*-deacylation of various sphingolipids was described. This article describes the synthesis of [^{14}C]ceramide,[3] [^{14}C]glycosphingolipids,[4] and ω-aminoceramide[5] by the reverse hydrolysis (condensation) reaction of SCDase.

Materials

SCDase is purified from the culture fluid of *Pseudomonas* sp. TK4 as described previously[1] or purchased from Takara Shuzo Co., Japan. Various lysosphingolipids are prepared by SCDase, as shown in Chapter 32, and some are obtained from Takara Shuzo Co., Japan. 1-^{14}C-labeled fatty acids (50–55 mCi/mmol) are purchased from American Radiolabeled Chemicals. D-Sphingosine is obtained from Sigma and ω-aminododecanoic acid from Wako Chemical Co., Japan.

[3] S. Mitsutake, K. Kita, N. Okino, and M. Ito, *Anal. Biochem.* **247,** 52 (1997).
[4] S. Mitsutake, K. Kita, T. Nakagawa, and M. Ito, *J. Biochem.* (*Tokyo*) **123,** 859 (1998).
[5] M. Tani, K. Kita, H. Komori, T. Nakagawa, and M. Ito, *Anal. Biochem.* **263,** 183 (1998).

Definition of Enzyme Unit

One unit (U) of SCDase is defined as the amount capable of hydrolyzing 1 μmol GM1a per minute under the conditions described in the preceding chapter and 10^{-3} and 10^{-6} U of enzyme are expressed as 1 mU and 1 μU, respectively, in this study.

General Properties of Condensation Reaction by SCDase

The optimum pH for the condensation reaction by SCDase is found to be pH 7–8,[2] whereas the optimum pH for the hydrolysis reaction is determined to be pH 5–6.[1] The concentration of detergents affects the direction of the reaction strongly, i.e., the optimum concentration of Triton X-100 for the synthesis of glycosphingolipids by SCDase is around 0.1%, and a higher concentration inhibits the condensation reaction, whereas 0.8% Triton X-100 is found to be optimal for the hydrolysis of glycosphingolipids by the enzyme. It should be noted that the synthesis of ceramide by SCDase requires a higher concentration of Triton X-100 (0.3%) due possibly to the insolubility of the substrate.[3] SCDase can also catalyze the exchange (transfer) reaction in which sphingolipids are used instead of their lyso forms as an acceptor for fatty acids. However, the exchange reaction is found to proceed much slower than the condensation reaction (unpublished data).

Specificity of Condensation Reaction by SCDase

The specificity of SCDase for polar head groups is quite broad, i.e., not only neutral and acidic lysoglycosphingolipids, but also lysosphingomyelin, are found to be good acceptors for the condensation reaction. This tendency is also found in the specificity for the hydrolysis reaction by the enzyme.[1] Interestingly, SCDase efficiently catalyzes the condensation of stearic acid to sphingosine to produce ceramide,[3] although ceramide is somewhat resistant to hydrolysis by the enzyme.[1] For saturated fatty acids, lauric acid (C12:0) to lignoceric acid (C24:0) are efficiently condensed to galactosylsphingosine (GalSph), but caproic acid (C6:0) is found to be a very poor substrate.

Preparation of [^{14}C]Ceramide by SCDase[3]

Enzymatic Synthesis of [^{14}C]Ceramide

A ^{14}C-labeled fatty acid (100 nmol) and 200 nmol of sphingosine dissolved in ethanol are dried completely in an Eppendorf tube with nitrogen gas. Five hundred microliters of 50 mM phosphate buffer, pH 7.0, containing 0.6% Triton X-100 is added, mixed well, and sonicated in a sonic bath for 5 min. Five hundred microliters of SCDase (1 mU/ml) is added to the mixture and incubated at 37° for 20 hr (final concentration of Triton X-100 is 0.3% in 25 mM phosphate buffer, pH 7.0). After incubation, the reaction mixture is dried by a Speed-Vac concentrator. Under these conditions, about 60 (for lauric acid)–80% (for stearic acid) of fatty acid is condensed to sphingosine (d18:1) to produce [14C]ceramide. [14C]Palmitic acid is also found to be condensed to sphinganine (d18:0, dihydrosphingosine), but the reaction efficiency is reduced by half in comparison with the reaction using sphingosine.

Purification of [14C]Ceramide

The dried material described earlier is dissolved in 1 ml of hexane/diethyl ether/acetic acid (50/50/1, v/v/v) and applied to a Sep-Pak Plus Silica cartridge that has been equilibrated previously with the same solvent. The fatty acid is then washed out from the cartridge with 10 ml of the same solvent. Bound [14C]ceramide is then eluted from the cartridge with 10 ml of chloroform/methanol (2/1, v/v). The eluate is dried under N$_2$ gas and suspended in distilled water, sonicated, and applied to a Sep-Pak C$_{18}$ cartridge for desalting that has been equilibrated with distilled water. After washing the cartridge with 20 ml of distilled water, [14C]ceramide is eluted with 3 ml of methanol and 10 ml of chloroform/methanol (2/1, v/v). The eluate is dried under N$_2$ gas, dissolved in 1 ml of chloroform/methanol/water (90/10/1, v/v/v), and applied to a Sep-Pak CM cartridge that has been equilibrated with the same solvent. [14C]Ceramide is passed through the cartridge by the same solvent, whereas cold sphingosine is trapped. Sphingosine is eluted with chloroform/methanol/0.5 N HCl (90/10/1, v/v/v). In a typical experiment, 66 nmol of [14C]ceramide (C16:0, d18:1) is obtained from 100 nmol of [14C]palmitic acid and 200 nmol sphingosine. The contamination of 14C-labeled fatty acids and sphingosine in the final preparation is less than 1%.

Preparation of [14C]Glycosphingolipids by SCDase[4]

Enzymatic Synthesis of [14C]Glycosphingolipids

To determine the optimum conditions for glycosphingolipid synthesis by SCDase, various factors (pH, concentration of lysoglycosphingolipids and fatty acids, concentration of detergents, enzyme amount, and incubation time) are examined.[4] The following are optimum conditions for the synthesis of each glycosphingolipid. For synthesis of [^{14}C]galactosylceramide (GalCer), 100 nmol psychosine (GalSph) and 100 nmol of [^{14}C]stearic acid are incubated in 1 ml of 25 mM phosphate buffer, pH 6.5, containing 0.1% Triton X-100. For [^{14}C]globotetoraosylceramide (Gb4Cer), 200 nmol of lyso-Gb4Cer and 100 nmol of [^{14}C]stearic acid are incubated in 1 ml of 25 mM phosphate buffer, pH 6.0, containing 0.1% Triton X-100. For [^{14}C]sulfatide, 400 nmol of lysosulfatide and 100 nmol of [^{14}C]stearic acid are incubated in 1 ml of 25 mM phosphate buffer, pH 7.0, containing 0.1% Triton X-100. For [^{14}C]GM1a, 400 nmol of lyso-GM1a and 100 nmol of [^{14}C]stearic acid are incubated in 1 ml of 25 mM phosphate buffer, pH 7.0, containing 0.05% Triton X-100. All incubations are carried out with 1 mU SCDase at 37° for 20 hr. In a typical experiment, the reaction yield of GalCer, Gb4Cer, sulfatide, and GM1a by SCDase is 67.1, 71.6, 70.3, and 54.4%, respectively.

Purification of [^{14}C]Glycosphingolipids

The purification of [^{14}C]glycosphingolipids consists of three steps. First, [^{14}C]stearic acid is removed from the reaction mixture by a normal-phase Sep-Pak Plus Silica. Second, samples are desalted by using a reversed-phase Sep-Pak C$_{18}$ cartridge. Third, free lysoglycosphingolipids are removed by a cation-exchange Sep-Pak CM (for GalCer and Gb4Cer) or an anion-exchange cartridge Sep-Pak QMA (for sulfatide). The method using Sep-Pak cartridges is simple, time-saving, and reproducible. However, it is not suitable for GM1a; HPLC using a silica column is required for purification of GM1a.

A reaction mixture containing ^{14}C-labeled GalCer, Gb4Cer, or sulfatide is dried with a Speed-Vac concentrator, suspended in 1 ml of hexane/diethyl ether/acetic acid (50/50/1, v/v/v, solvent A), and applied to a Sep-Pak Plus Silica cartridge previously equilibrated with solvent A. Synthesized [^{14}C]glycosphingolipids and free lysoglycosphingolipids are adsorbed on the cartridge, whereas free [^{14}C]stearic acids are eluted with 10 ml of solvent A. [^{14}C]Glycosphingolipids and free lysoglycosphingolipids are then eluted from the cartridge with 10 ml of chloroform/methanol (2/1, v/v). The eluate is dried, suspended in 1 ml of distilled water, and applied to a Sep-Pak C$_{18}$ cartridge to remove salts. After washing with 10 ml of distilled water, [^{14}C]glycosphingolipids and lysoglycosphingolipids are eluted from the cartridge with 3 ml of methanol and 10 ml of chloroform/methanol (2/1, v/v). The eluate containing [^{14}C]GalCer is dried, dissolved in 1 ml of chloroform/methanol/distilled water (90/10/1, v/v/v), and applied to a Sep-Pak CM cartridge equilibrated with the same solvent. [^{14}C]GalCer passes through the cartridge, whereas psychosine is trapped. Gb4Cer, dissolved in chloroform/methanol/distilled water (65/25/4, v/v/v, solvent B), is applied to a Sep-Pak CM cartridge and eluted with solvent B. Lyso-Gb4Cer is trapped and eluted with solvent B, but distilled water is replaced by HCl. The sample containing [^{14}C]sulfatide is dissolved in 1 ml of solvent B and is applied to a Sep-Pak QMA cartridge. Lysosulfatide passes through the cartridge, whereas [^{14}C]sulfatide is trapped and eluted with 5 ml of chloroform/methanol/0.5 *N* NaOH (65/25/4, v/v/v). [^{14}C]Sulfatide is then desalted using a Sep-Pak C$_{18}$ cartridge. The sample containing [^{14}C]GM1a is desalted by a Sep-Pak C$_{18}$ cartridge as described earlier, dissolved in solvent B, and applied to HPLC using a normal-phase silica column (Inertsil SIL100-5, 4.6 × 250 mm, GL Science Inc.) to remove the ^{14}C-labeled fatty acid and lyso-GM1a. The column is equilibrated and eluted with solvent B. In a typical experiment, the yield after purification is 45.4% for GalCer, 43.1% for Gb4Cer, 41.1% for sulfatide, and 39.4% for GM1a. The contamination of lysoglycosphingolipids or ^{14}C-labeled fatty acids in final preparations is less than 1%.

Preparation of ω-Aminoceramide by SCDase[5]

ω-Aminoceramide, the fatty acyl moiety in ceramide replaced by ω-amino fatty acid, is useful not only as a precursor of various fluorescence-labeled ceramides, but also for immobilization of ceramide for affinity chromatography.

Enzymatic Synthesis of ω-Aminoceramide

Free fatty acids are efficiently condensed by SCDase to sphingosine to produce ceramide containing different fatty acids.[3] However, this reaction hardly progresses when an ω-aminododecanoic acid is used instead of normal fatty acids. Interestingly, once the ω-amino group is blocked with trifluoroacetate (TFAc), the reaction proceeds efficiently. The block of TFAc is removed easily from the *N*-TFAc-aminoceramide with mild alkaline treatment to produce ω-aminoceramide. The most striking difference between a *N*-TFAc-blocked ω-amino fatty acid and normal fatty acids for the condensation reaction by SCDase is the optimum pH for the reaction. The optimum pH for the synthesis of ceramide by SCDase is found to be around pH 7 when free dodecanoic acid (lauric acid, C12:0) is used. This result is consistent with that obtained previously using stearic acid and palmitic acid.[3] The optimum pH shifts to pH 10–11 when *N*-TFAc-blocked ω-aminododecanoic acid is used for the reaction.

The reaction mixture containing 40 μmol of *N*-TFAc-aminododecanoic acid, 20 μmol of sphingosine, and 100 mU of SCDase in 20 ml of 25 m*M* glycine–NaOH buffer, pH 10, containing 0.3% Triton X-100 is incubated at 37° for 24 hr. The reaction efficiency is not less than 90%. The extent of reaction for sphinganine (d18:0) is about 60–80% of that for sphingenine (d18:1) or phytosphingosine (t18:0). SCDase can be used for synthesis not only of ω-aminoceramide but also various ω-aminoglycosphingolipids and sphingomyelin.[5]

Purification of ω-Aminoceramide

After incubation, the reaction mixture is applied directly to a C_{18} silica column (2.4×3.7 cm, 16.7 ml) to remove salts that had been equilibrated previously with distilled water. After washing the column with 200 ml of distilled water, N-TFAc-aminoceramide is eluted with 20 ml of methanol and 200 ml of chloroform/methanol (2/1, v/v). Both eluates are combined, dried under N_2 gas, dissolved in 10 ml of chloroform/methanol (95/5, v/v), and applied to a Sep-Pak Plus Silica cartridge to remove unreacted sphingosine. N-TFAc-aminoceramide is eluted with 20 ml of the same solvent, whereas sphingosine is trapped and eluted with chloroform/methanol (2/1, v/v). The fraction containing N-TFAc-aminoceramide is dried under N_2 gas, dissolved in 10 ml of chloroform/methanol/water (60/30/5, v/v/v), and applied to a Sep-Pak QMA (OH^- form) cartridge to remove contaminating N-TFAc-amino fatty acid. A small amount of ω-amino fatty acid, which is generated during incubation at pH 10, is also removed at this step. N-TFAc-aminoceramide is eluted from the cartridge with 20 ml of the same solvent, whereas N-TFAc-aminododecanoic acid and aminododecanoic acid are trapped and eluted with chloroform/methanol/0.5 N NaOH (60/30/5, v/v/v). The fraction containing N-TFAc-aminoceramide is dried under N_2 gas and is dissolved in 5 ml of chloroform/methanol/water (60/30/5, v/v/v). To remove the N-TFAc group, 50 µl of sodium methoxide (28% in methanol) is added. After stirring at room temperature for 18 hr, the mixture is neutralized with glacial acetic acid and dried under N_2 gas, dissolved in 10 ml of distilled water, and applied to a Sep-Pak C_{18} cartridge for desalting. After washing the column with 100 ml of distilled water, ω-aminoceramide is eluted successively with 5 ml of methanol and 20 ml of chloroform/methanol (2/1, v/v). Both eluates are combined, dried under N_2 gas, dissolved in 5 ml of chloroform/methanol (95/5, v/v), and applied to a Sep-Pak Plus Silica cartridge to remove Triton X-100 and N-TFAc-aminoceramide. After washing the cartridge with 20 ml of the same solvent, ω-aminoceramide is eluted with 20 ml of chloroform/methanol (2/1, v/v). The yield of ω-aminoceramide is 70–80%. The final preparation of ω-aminoceramide shows no contamination of sphingosine, ω-amino fatty acid, and N-TFAc derivatives.

Comments

To date, literature on the preparation of [^{14}C]sphingolipids[6,7] and of ω-aminoglycosphingolipids[8] seems to be restricted to those using purely chemical procedures. In comparison with the methods reported previously, the present method using SCDase is novel, much easier, and gives a high yield. Because the specificity of the enzyme is very specific to the amino group of sphingosine bases, no by-products are generated during the reaction. Furthermore, the fact that a wide range of sphingolipid species and sample amounts (from the nanomole to millimole level) can be used is significantly advantageous.

Acknowledgment

This work was supported in part by a Grants-in Aid for Scientific Research Priority Areas (09240101) and Scientific Research (B) (09460051) from the Ministry of Education, Science, and Culture of Japan.

[6] J. K. Anand, K. K. Sadozai, and S. Hakomori, *Lipids* **31**, 995 (1996).
[7] S. Sonnino, M. Nicolini, and V. Chigorno, *Glycobiology* **6**, 479 (1996).
[8] K. Kamio, S. Gasa, and A. Makita, *J. Lipid Res.* **33**, 1227 (1992).

Author Index

Numbers in parentheses are footnote reference numbers and indicate that an author's work is referred to although the name is not cited in the text.

A

Abbas, H. K., 131, 140(7)
Abbas, S., 549, 551, 552(24)
Abe, A., 42, 48, 50, 105, 106, 373, 375, 376, 377(5), 378, 379, 384, 385
Abe, H., 460
Abe, T., 537
Abousalham, A., 233, 235, 237(17), 241(3)
Abouzied, M. M., 372
Acevedo, E., 82
Achenbach, H., 346
Ackermann, T., 372
Acquotti, D., 583, 649, 650(14), 652, 652(20), 658, 667(4), 679(4)
Adachi, H., 122
Adam, D., 165, 167(6), 168
Adam, M. A., 421, 423(63)
Adam-Klages, S., 165, 167(6), 168
Adams, J., 286
Adler, D. A., 548
Aebersold, R., 573, 580, 596, 597, 599(133)
Aerts, J. M., 202
Aggarwal, B. B., 235
Agranoff, A. B., 384
Agranoff, B. W., 123
Ahmed, S. N., 184
Ahn, K. A., 464(19d), 465
Ahn, N. T., 396
Aignesberger, A., 259
Aitchison, S., 104
Aizawa, S., 82
Akahori, Y., 26
Akamatsu, Y., 4, 8, 307
Akanuma, H., 138
Akino, T., 23, 29(10), 30(10), 63, 248
Akiyama, H., 371
Akiyama, T., 433, 460
Ala, T. A., 629

Alberts, J. F., 364, 367(19)
Albrecht, B., 602, 609, 610(24)
Alcami, J., 589
Alcami, J., 589
Alekseeva, S. G., 538, 538(28)
Alemany, R., 216
Alexander, C., 15, 19(2), 20, 38, 195, 199(16), 200(16), 480(11), 481, 488(11), 497(11)
Alkhatib, B., 634
Allan, M. L., 558
Allen, J. M., 216
Allen, K. Y., 265
Almenoff, J. S., 298
Alon, R., 549
Alvarez, L., 241
Alving, C. A., 73(9), 74
Ameen, M., 95
Amemiya, M., 433, 460
American Radiolabeled Chemicals, 493, 497(44)
Ames, B. N., 35, 178, 502
Amidon, B., 275
Anand, J. K., 504, 506(30), 689
Anand, K. J., 443, 444(27)
Anderer, F. A., 31, 32, 32(2)
Anderson, J. M., 204
Anderson, R. G. W., 185, 186
Andersson, K., 65
Ando, T., 549, 550
André-Barrès, C., 628(16), 629
Andrew, S. W., 558
Andries, T., 503
Andrieu, N., 169
Anhand, J. K., 591
Aoki, J., 122, 307
Apostolski, S., 65
ApSimon, J. W., 363
Aquotti, D., 504
Aragozzini, F., 349, 350, 350(5), 357(20)
Arai, H., 307

C

Subject Index

A